The IACUC HANDBOOK

Third Edition

The IACUC HANDBOOK

Third Edition

Edited by
Jerald Silverman
Mark A. Suckow
Sreekant Murthy

CRC Press is an imprint of the
Taylor & Francis Group, an **informa** business

CRC Press
Taylor & Francis Group
6000 Broken Sound Parkway NW, Suite 300
Boca Raton, FL 33487-2742

Printed on acid-free paper
Version Date: 20131217

International Standard Book Number-13: 978-1-4665-5564-8 (Hardback)

Library of Congress Cataloging-in-Publication Data

The IACUC handbook / edited by Jerald Silverman, Mark A. Suckow, and Sreekant Murthy. -- Third edition.
p. ; cm.
Includes bibliographical references and index.
ISBN 978-1-4665-5564-8 (hardcover : alk. paper)
I. Silverman, Jerald, editor of compilation. II. Suckow, Mark A., editor of compilation. III. Murthy, Sreekant, editor of compilation.
[DNLM: 1. National Institutes of Health (U.S.). Institutional Animal Care and Use Committee. 2. Animal Care Committees--organization & administration. 3. Animal Welfare. 4. Animal Experimentation--legislation & jurisprudence. 5. Animals, Laboratory. QY 54]

HV4708
179'.3--dc23 2013049106

Visit the Taylor & Francis Web site at
http://www.taylorandfrancis.com

and the CRC Press Web site at
http://www.crcpress.com

Preface to the Third Edition

Since the publication of the first (2000) and second (2006) editions of this book there have been a number of revisions to those federal laws, regulations, and policies that affect IACUCs. In some cases, these changes have simply reflected common, best practices of IACUCs, while others have directed IACUCs to implement new practices. Of note, the 8th edition of the *Guide for the Care and Use of Laboratory Animals*, published in 2011, added significant context for IACUC policies and practices. Examples of new expectations for IACUCs outlined in the 8th edition of the *Guide* include: application of experimental and humane endpoints; establishing mechanisms for communication of unexpected outcomes to the IACUC; enhanced review of studies involving physical restraint, multiple survival surgery, and food and fluid regulation, and the use of non-pharmaceutical grade compounds in animals; and post-approval monitoring.

The 8th edition of the *Guide* offers substantial definition with respect to animal care program management responsibility. In this regard, a collaborative relationship between the IACUC, the attending veterinarian, and the institutional official is expected, with the institutional official bearing ultimate responsibility for the program. As a result, the role of many IACUCs may evolve from being the presumed agent with sole responsibility, to that of a key team member with shared responsibility. Consequently, the IACUC retains an essential and central role that is supported by the attending veterinarian and institutional official.

Earlier editions of *The IACUC Handbook* exclusively focused on the IACUC within the context of the United States. While still U.S.-centric, this edition includes a chapter that offers the reader an international perspective on how an IACUC or other style of review can function beyond the borders of the United States. Though many general principles of animal care and use are the same, operational aspects may vary between countries or regions; thus, it is hoped that this approach will provide a basis for consideration of interpretation of relevant issues in a way that can be applied broadly.

In 2013 the American Veterinary Medical Association published updated guidelines for the euthanasia of animals. Those guidelines have been incorporated into this edition as needed. As with many guidelines, those related to euthanasia have evolved in response to new scientific information and the cumulative professional experience of those in the field.

This third edition of *The IACUC Handbook* builds upon the foundation laid in the first two editions and provides not only new information on existing laws, regulations, policies and "how to" advice, but also features an updated survey of IACUC practices from institutions around the nation. In this way, readers benefit from the combined experience of numerous IACUCs and some sense of 'best practices' can be ascertained. In some chapters we have permitted the repetition of items that are similar to those found in other chapters. These have been cross-referenced to provide a broader perspective of expert opinions. We hope all of this will be of value to you.

To ensure that the contributions of our authors are precise, outside review was conducted beyond the editorial review we provided. In this regard we thank Carol Clarke of USDA/APHIS/AC and Patricia Brown, Susan Silk, and Lori Hampton of NIH/OLAW who provided reviews of all chapters for concordance with their respective regulations and policies. APHIS/AC policies, which are in the online *Animal Care Resource Guide*, are interpretive rules that for practical purposes have the legal standing of the regulations they help clarify. NIH/OLAW offers guidance on the interpretation of the PHS Policy and the recommendations of the *Guide* by providing online responses to Frequently Asked Questions. In some cases NIH/OLAW will not approve an animal welfare Assurance until the Assurance is consistent with the information in the Frequently Asked Questions.

Though intended to be a central part of the animal care responsibility triad, along with the attending veterinarian and the Institutional Official, the IACUC remains a key operational element. The IACUC has a unique place in the oversight of animal care and use activities for most organizations in the U.S. It is a self-regulatory committee, representing its institution to the federal government. Equally importantly, it helps represent to the public the basic standards of animal care and concerns for animal welfare to which researchers, institutions, and laboratory animal science professionals subscribe. Yet, if self-regulation is to be continued, then we must go the extra mile and strive to reach even higher standards of animal care and use. For that to occur, institutions and their IACUCs must be leaders, not followers.

Jerald Silverman, DVM
Mark A. Suckow, DVM
Sreekant Murthy, PhD

Introduction to the Third Edition

The Animal Welfare Act was amended in 1985 to require research facilities to establish a Committee to assess animal care treatment and practices in experimental research and represent society's concerns regarding the welfare of the animals used in those facilities. This Committee was later termed the Institutional Animal Care and Use Committee (IACUC) with its duties more specifically outlined in the 1989 revisions to the Animal Welfare Act Regulations under 9 CFR Part 2. The Public Health Service (PHS) Policy on Humane Care and Use of Laboratory Animals which was implemented in 1985 also required the Institutional Animal Care and Use Committees (IACUCs) to oversee animal care and use at research facilities. As a result of these regulations, the IACUC has become an empowered premier administrative instrument of animal welfare oversight in biomedical research today.

Over nearly three decades, IACUCs and the larger biomedical research community have amassed a wealth of experience in the operation of animal care and use programs that support quality biomedical and behavioral research within the context of humane animal care and use. This extraordinary progress is due in large part to the tradition within the community of addressing issues and sharing experiences. As a result, IACUC oversight of humane animal care and use in research has been able to keep pace with the exceptional growth of biomedical research and technology in recent decades.

This handbook provides a comprehensive collection of best practices; all of which are compliant with the federal guidelines and requirements. The sections labeled "regulatory" have been reviewed by staff at NIH Office of Laboratory Animal Welfare (OLAW) and USDA APHIS/Animal Care (AC) for consistency and compliance with the PHS Policy and the USDA Animal Welfare Act Regulations (AWAR).

The best practices presented were developed by conscientious IACUCs and research teams from academia, industry, the private and public sectors, and institutions of various sizes. Answers to a series of operational, policy, and management questions are first addressed from the perspective of the applicable regulatory language, followed by opinions from knowledgeable and experienced professionals in the field, and in some cases ultimately followed by responses to informal surveys on selected institutional policies and practices.

Every effort has been made to apply correct interpretations within the context of the specific issues being discussed. However, in this highly nuanced field, readers are cautioned not to apply these interpretations out of context or to extend them beyond their intended meaning. Accordingly, such interpretations should not be seen as formulating new federal regulations or policies. Readers are therefore advised to refer to source documents and engage in direct consultations with NIH/OLAW and APHIS/AC when in doubt. In addition, professional standards and policies change and evolve over time as a result of new discoveries and new regulations. In light of this, the reader is also advised to keep abreast of developments while using this handbook for guidance.

The information provided in the sections labeled "survey" gives an interesting view of how the surveyed institutions address certain issues, but a note of caution is advised here as well. Some responses may appear to indicate deviations from federal policy and regulation. This is a result of several variables (e.g., type of institution, species involved, applicable oversight) that affect applicability or lead to differences in interpretation of the survey question. In light of this, survey responses should not be construed as a noncompliance with the regulations.

For the experienced reader and novice alike, this book provides a wealth of useful information and insight into the collective experience and wisdom of the numerous expert editors and authors. Both NIH/OLAW and APHIS/AC commend the authors and editors for their outstanding efforts and for moving the biomedical research community forward in its formulation of best practices and commonly accepted professional guidance in this complex arena.

Dr. Patricia A. Brown
Director of the Office of Laboratory Animal Welfare

Dr. Chester A. Gipson
Deputy Administrator USDA/APHIS/Animal Care

Some Comments about the Surveys

The surveys in this text are informal reports of how different IACUCs around the nation approach some of the questions presented in the chapters. The intent of the surveys is to provide the reader with a general sense of how various IACUCs approach day-to-day concerns, in particular when issues arise that are not clearly defined by federal laws, regulations or policies.

Using a database of NIH/OLAW Accredited institutions, a completely anonymous questionnaire with 281 questions was e-mailed to approximately 1100 IACUC chairpersons throughout the United States. Once a week, for three weeks, reminder e-mails were sent to the entire database. No identifying information was requested of respondents. 297 questionnaires were returned and all provided usable data. Each chairperson responded to almost all of the questions. Data were collected by one of the editors (J.S.) and sent to the appropriate chapter authors.

All questions were multiple choice and usually there was only one allowable answer. For those types of questions (e.g., the ones shown below) the total number of individual respondents is represented by the denominator and the numerator represents the number of those respondents who chose the answer shown. The sum of the percentages shown will always add up to 100%. However, some questions allowed for multiple answers and we have indicated that on the survey responses found in this book. For multiple answer questions, the total number of individuals responding to the question is represented by the denominator and the numerator represents the total number of respondents who chose the particular response shown. Because one survey participant may have chosen more than one answer, the sum of the percentages shown after each answer can easily total more than 100%.

Five demographic questions were asked of all respondents.

1. *What is the nature of your organization?*

- Primarily academic. 184/297 (61.9%)
- Primarily pharmaceutical (with drug discovery). 18/297 (6.1%)
- Primarily a private non-pharmaceutical research or testing company (including CROs). 40/297 (13.5%)
- Primarily zoological or other exhibitor, with associated research. 0/297 (0%)
- Primarily laboratory animal breeding. 5/297 (1.7%)
- Primarily governmental research or testing. 30/297 (10.1%)
- Other. 20/297 (6.7%)

2. *Does your organization currently have any level of AAALAC accreditation?*

- Yes. 201/296 (67.9%)
- No. 95/296 (32.1%)

3. *What is the approximate size of your organization? If your organization has multiple sites consider only the site(s) for which your IACUC has regulatory oversight responsibility.*

- About 1–100 full-time-equivalent (FTE) employees. 63/296 (21.3%)
- About 101–500 FTE employees. 63/296 (21.3%)
- About 501–1,000 FTE employees. 48/296 (16.2%)
- About 1,001–5,000 FTE employees. 57/296 (19.3%)
- Over 5,000. 65/296 (22.0%)

4. *Is your organization legally required to comply with the Animal Welfare Act for all or some of its animal related work?*

- Yes. 282/296 (95.3%)
- No. 14/296 (4.7%)

5. *Is your organization legally required to comply with the PHS Policy for all or some of its animal related work?*

- Yes. 278/294 (94.6%)
- No. 16/294 (5.4%)

Editors

Jerald Silverman, DVM, is director of the department of Animal Medicine and professor of pathology at the University of Massachusetts Medical School. He is also an adjunct professor at the Cummings School of Veterinary Medicine at Tufts University. He received his degrees in vertebrate zoology and veterinary medicine from Cornell University and has a master's degree in nonprofit organization management from the New School for Social Research. Dr. Silverman is a Diplomate of the American College of Laboratory Animal Medicine, past president of the American Society of Laboratory Animal Practitioners, and a full member of the American Association for Cancer Research. In 2011 he was honored by the Massachusetts Society for Medical Research for his work as a teacher, mentor, author, and strong advocate of laboratory animal science. He has served on numerous professional and community committees and editorial boards and continues to be an active teacher, writer, and researcher. He has over 50 peer-reviewed journal publications and 17 books and chapters on subjects as diverse as cancer research to public relations tactics in the debate over animal experimentation.

Mark Suckow, DVM, is assistant vice president for research at the University of Notre Dame, Notre Dame, Indiana. He received his DVM from the University of Wisconsin in 1987 and completed a post-doctoral residency program in laboratory animal medicine at the University of Michigan in 1990. He is a diplomate of the American College of Laboratory Animal Medicine.

Dr. Suckow has published more than 100 scientific papers, patents, books, and book chapters. He has created two companies to commercialize technology based on intellectual property he developed. Dr. Suckow is active in professional organizations and has served as president of both the American Association for Laboratory Animal Science (AALAS) and the American Society of Laboratory Animal Practitioners (ASLAP). He currently serves as a member of the AAALAC Council on Accreditation.

Sreekant Murthy, PhD, is a professor of medicine, an adjunct professor of biomedical engineering, and senior associate vice provost for research compliance (retired) at Drexel University. He recently joined Rowan University, Glassboro, New Jersey, as their Chief Research Compliance Officer. In 1973, he earned his doctoral degree from Philadelphia College of Pharmacy, currently known as University of Sciences, Philadelphia. He completed his postdoctoral research assistantship at Temple University in 1974 and joined Hahnemann University (Drexel University College of Medicine) in the same year. He has been at the same institution since then. As a faculty member he simultaneously chaired two IACUCs in the same institution for five years. As the senior associate vice provost, he managed research compliance in human subject research, animal welfare, institutional biosafety committee, radiation safety and research integrity. He has been the institutional official for the IACUC since 2006. At Rowan University he continues to be responsible for all aspects of research compliance, research integrity, and export control.

Dr. Murthy is a well-recognized researcher in the area of inflammatory bowel diseases. He was instrumental in characterizing and popularizing an animal model of experimental colitis that has been extensively used in various laboratories around the world. He has published 70 papers in peer-reviewed journals, over 150 abstracts in meetings and symposia, and has made nearly 200 invited presentations nationally and internationally in the area of gastroenterology. He has edited two books, written several chapters in books, and served as a peer reviewer for several journals and grant applications.

Contributors

Kenneth P. Allen, DVM, DACLAM
Assistant Professor and Staff Veterinarian
Biomedical Resource Center
Medical College of Wisconsin
Milwaukee, Wisconsin

David R. Archer, PhD
Associate Professor of Pediatrics
Aflac Cancer and Blood Disorders Center
Emory University School of Medicine
Atlanta, Georgia

Ron E. Banks, DVM, DACAW, DACLAM, DACVPM, Fellow NAP, CPIA
Director, Office of Animal Welfare Assurance
Duke University
Durham, North Carolina

Kathryn Bayne, MS, PhD, DVM, DACLAM, DACAW, CAAB
Global Director
AAALAC International
Frederick, Maryland

Robert M. Bigsby, PhD
Professor Emeritus of Obstetrics and Gynecology and of Cellular and Integrative Physiology and of Pharmacology and Toxicology
Indiana University School of Medicine
Indianapolis, Indiana

Christine A. Boehm, DVM, DACLAM
Assistant Director, Laboratory Animal Research Center
Indiana University School of Medicine
Indianapolis, Indiana

Marcy Brown, BS, MA, CPIA
Regulatory Compliance Lead, Comparative Medicine
Pfizer Worldwide Research and Development
La Jolla, California

Jennifer N. Camacho, LVT, RLATg, CMAR
Animal Enrichment Program Manager
Massachusetts General Hospital
Center for Comparative Medicine
Charlestown, Massachusetts

Larry Carbone, DVM, PhD, DACLAM, DACAW
Senior Veterinarian and Associate Director, Laboratory Animal Resource Center
University of California, San Francisco
San Francisco, California

Marilyn J. Chimes, DVM, DACLAM, JD
Counsel
Scharf Banks Marmor LLC
Chicago, Illinois

Susan Stein Cook, DVM, MS, DACLAM
Consultant in Laboratory Animal Medicine
Williamston, Michigan

and

Attending Veterinarian
Central Michigan University
Mount Pleasant, Michigan

Robin Crisler, DVM, MS, DACLAM
Associate Director, Laboratory Animal Research Center
Indiana University School of Medicine
Indianapolis, Indiana

Peggy J. Danneman, VMD, MS, DACLAM
Senior Staff Veterinarian, Courses and Conferences
The Jackson Laboratory
Bar Harbor, Maine

Bernard J. Doerning, DVM, MBA, DACLAM
Director, Safety Assessment and Laboratory Animal Resources
Merck Research Laboratories
Kenilworth, New Jersey

Melissa C. Dyson, DVM, MS, DACLAM
Associate Professor and Assistant Director, Unit for Laboratory Animal Medicine
University of Michigan Medical School
Ann Arbor, Michigan

Diane J. Gaertner, DVM, DACLAM
Director, University Laboratory Animal Resources
University of Pennsylvania
and
Professor, C-E, Department of Pathobiology
School of Veterinary Medicine
University of Pennsylvania
Philadelphia, Pennsylvania

Ed J. Gracely, PhD
Associate Professor, Department of Family, Community, and Preventive Medicine
Drexel University College of Medicine
and
Drexel University School of Public Health
Philadelphia, Pennsylvania

William G. Greer, BS, CPIA, LAT
Associate Director, Office for Research Protections
Pennsylvania State University
University Park, Pennsylvania

Javier Guillén, DVM
Senior Director and Director of European Activities
AAALAC International
Pamplona, Spain

Debra L. Hickman, DVM, MS, DACLAM
Director of Laboratory Animal Resources
School of Medicine
Indiana University
Indianapolis, Indiana

Michael J. Huerkamp, DVM, DACLAM
Director, Division of Animal Resources and Attending Veterinarian
Professor, Pathology and Laboratory Medicine
Emory University
Atlanta, Georgia

Larry Iten, DVM
Director, IACUC Office
Office of Research Administration
Emory University
Atlanta, Georgia

Alicia Z. Karas, MS, DVM, DACVAA
Assistant Professor of Clinical Sciences
Tufts Cummings School of Veterinary Medicine
North Grafton, Massachusetts

Kathy Laber,* DVM, MS, DACLAM
Professor
Medical University of South Carolina
and
Animal Program Director
Ralph H. Johnson VA Medical Center
Charleston, South Carolina

* Currently, Dr. Laber is the chief of the Comparative Medicine Branch at the National Institute of Environmental Health Sciences (NIEHS), Research Triangle Park, North Carolina.

Neil S. Lipman, VMD
Director and Professor, Center of Comparative Medicine and Pathology
Memorial Sloan-Kettering Cancer Center
and
Weill Medical College of Cornell University
New York, New York

Michael D. Mann, PhD
Professor Emeritus of Cellular and Integrative Physiology
University of Nebraska Medical Center
Omaha, Nebraska

Angela M. Mexas, DVM, PhD, DACVIM
Staff Veterinarian, University Laboratory Animal Resources
University of Pennsylvania
Philadelphia, Pennsylvania

Sreekant Murthy, PhD
Chief Research Compliance Officer
Rowan University
Glassboro, New Jersey

Christian E. Newcomer, VMD, MS, DACLAM
Executive Director
Association for Assessment and Accreditation of Laboratory Animal Care International
Frederick, Maryland

Gwenn S. F. Oki, MPH, CIP
Compliance Director
Van Andel Institute
Grand Rapids, Michigan

Scott E. Perkins, VMD, MPH, DACLAM
Director, Division of Laboratory Animal Medicine
Tufts University and Tufts Medical Center
Boston, Massachusetts

and

Associate Professor, Department of Environmental and Population Health
Tufts Cummings School of Veterinary Medicine
North Grafton, Massachusetts

Ernest D. Prentice, PhD
Associate Vice Chancellor for Academic Affairs
University of Nebraska Medical Center
Omaha, Nebraska

Lester L. Rolf Jr., PhD, DVM
Attending Veterinarian
University of the Sciences
Philadelphia, Pennsylvania

and

Attending Veterinarian
Cooper Medical School of Rowan University
Camden, New Jersey

Harry Rozmiarek,* DVM, PhD, DACLAM
Director, Laboratory Animal Medicine
Fox Chase Cancer Center
and
Professor Emeritus, Laboratory Animal Medicine
University of Pennsylvania
Philadelphia, Pennsylvania

* Deceased.

Howard G. Rush, DVM, MS, DACLAM
Associate Professor
Unit for Laboratory Animal Medicine
Medical School
University of Michigan
Ann Arbor, Michigan

Priya Sankar
Associate General Counsel
Drexel University College of Medicine
Philadelphia, Pennsylvania

Mary Jo Shepherd, DVM, CPIA
Director of the Office of the IACUC
Columbia University
New York, New York

Jerald Silverman, DVM, MPS, DACLAM
Director, Department of Animal Medicine
Professor, Department of Pathology
University of Massachusetts Medical School
Worcester, Massachusetts

Alison C. Smith, DVM, DACLAM
Professor, Comparative Medicine
Medical University of South Carolina
Charleston, South Carolina

Harold F. Stills Jr., DVM, DACLAM
Director, Division of Laboratory Animal Resources
Professor of Microbiology, Immunology and Molecular Genetics
College of Medicine
University of Kentucky
Lexington, Kentucky

Mark A. Suckow, DVM, DACLAM
Assistant Vice President for Research
University of Notre Dame
Notre Dame, Indiana

Joseph D. Thulin, DVM, MS, DACLAM
Director, Biomedical Resource Center
Medical College of Wisconsin
Milwaukee, Wisconsin

Robin Lyn Trundy, MS, CBSP
Assistant Director and Biosafety Officer
Vanderbilt Environmental Health and Safety
Vanderbilt University
Nashville, Tennessee

Contents

1

Origins of the IACUC

Harry Rozmiarek

Introduction

The need to provide regulatory oversight to assure animal welfare in the research laboratory was recognized in Victorian England over 120 years ago and led to the 1876 Cruelty to Animals Act. Modifications of this law that exist today include "Home Office" oversight and registration of individuals who conduct research procedures using animals. In the U.S., the earliest national law addressing animal welfare was the 28-hour law, enacted in 1887. This law primarily governs animals being transported for market and does not specifically address animals in the research laboratory.

The New York Anticruelty Bill of 1866 and the formation of Humane Societies in New York (1866), Pennsylvania (1867), Massachusetts (1868), and Washington, D.C. (1870), addressed the use of animals in research, as did the formation of the American Antivivisection Society in Jenkintown, PA, in 1883. Nevertheless, none of these actions had national authority or scope. The first official law addressing the care and use of laboratory animals in the U.S. was the Laboratory Animal Welfare Act of 1966. An amendment in 1970 changed the name to the Animal Welfare Act (AWA), by which it is known today.

Background

Prior to the 20th century the responsibility for animals used in research in the U.S. was placed directly in the hands of the research investigator. The quality of animal care and animal welfare varied tremendously among research laboratories. Laboratories even within the same school or institution had different animal care policies and standards of care. Animal care was frequently the responsibility of a diener, who provided food, basic care, and much of the animal manipulation. Basic nutrition and sanitation were often inadequate and no environmental or housing standards were available. In many instances, the animal handling staff meant well but was not adequately trained or qualified. In other instances, animal welfare and even adequate care were of low priority. Rodents were commonly fed cereal grains and dogs and cats received table scraps. Backyard and part-time breeders provided rodents of variable genetic background, and the animals harbored a variety of parasites, pathogenic bacterial agents, and viruses. Long-term studies with rodents were impossible, as animals suffered from malnutrition and clinical disease and accurate research data were difficult to obtain. Dogs and cats were obtained individually, often provided in crates or gunnysacks and obtained from variable sources. Breeders of good-quality, standardized animals for research did not exist.

Central Management and Organization of Laboratory Animal Care

One of the first major research institutions in the U.S. that provided central management and care for research animals directed by a highly trained and qualified individual was the University of Chicago, where in 1945 Dr. Nathan Brewer was hired to direct a centrally managed program of laboratory animal care. Dr. Brewer had provided animal facility management at the University of Chicago as early as 1930 but had since received formal training as a veterinarian and obtained a Ph.D. in physiology. A number of other institutions in the country were following similar paths as the need for better and more standardized animal care and welfare was recognized. However, it was not until the 1940s that a number of professionals with a major interest in laboratory animal care began meeting in the Chicago area; this was undoubtedly stimulated and encouraged by Dr. Brewer. Among these was Dr. W.T.S. Thorpe, director of the newly organized Laboratory Aids Branch at the National Institutes of Health (NIH), which eventually developed into the National Center for Research Resources. In 1950 these meetings led to the formal establishment of the Animal Care Panel, which in 1967 became the American Association for Laboratory Animal Science (AALAS). Dr. Brewer served as the first national president of this organization and with other colleagues stimulated many additional activities and organizations instrumental in shaping animal care and use policies and laws in the U.S.[1]

Officially established and conducting annual conferences on animal care, the Animal Care Panel appointed a committee in 1961 that was headed by Dr. Bennett Cohen and charged with providing animal care and use guidelines for research facilities. Their product was the publication in 1963 of the first edition of the *Guide for Laboratory Animal Facilities and Care* (later titled *Guide for the Care and Use of Laboratory Animals*). Future editions of this publication were supported by the NIH and guided by the Institute for Laboratory Animal Research (ILAR). The National Academy Press, under the auspices of the National Research Council, published the eighth and current edition of the *Guide* in 2011.[2] This single document serves as the "bible" for laboratory animal care and use policies and guidelines in the U.S. It is excerpted from and referenced in all major guidelines and regulations and has been translated into, and is being published in, at least 11 other languages. Over 550,000 copies have been distributed throughout the world. The *Guide* expects every institution that has an animal care and use program to have an IACUC. It draws upon federal law which creates a statutory basis for the IO, AV, and the IACUC to have shared responsibility and authority for the animal care and use program[3] with the IACUC being responsible for the assessment and oversight of the program.

AAALAC

In 1963, the Animal Facilities Accreditation Board (AFAB) previously known as American Facilities Certification Board (AFCB) saw a need to evaluate the standard of animal care and use that was developed in its new *Guide*. This was accomplished in 1964 when two-person teams from the AFAB visited 26 major institutions in the U.S. that conducted research with animals and evaluated their policies and programs using the standards in the new *Guide*. This board soon saw the need to function independently and in 1965 changed its name and incorporated in the state of Illinois as the American Association for Accreditation of Laboratory Animal Care (AAALAC). The first two institutions accredited by AAALAC were the University of Louisville in Kentucky and Howard University in Washington, D.C. This independent accrediting agency changed its name in 1996 to the Association for Assessment and Accreditation of Laboratory Animal Care International. Still using the most recent edition of the *Guide* as one of its primary reference documents, AAALAC now has nearly 700 accredited institutions in the U.S. and 188 in 36 other countries under its umbrella of accreditation. The AAALAC program has always expected a mechanism of regular review and assurance (an IACUC or similar committee) as part of a good animal care and use program.

Animal Welfare Act

Prior to 1966 the U.S. did not have a national law addressing laboratory animals. Local humane societies were actively promoting protection for pets and sometimes for farm animals as well. Concurrently, the scientific community was improving the quality of animal care and well-being in the research laboratory. The increasing need for research dogs and cats was partially fulfilled by animal dealers who obtained these animals in various ways and sold them to research laboratories. A series of articles and news reports on animal neglect and abuse at the hands of animal dealers culminated in two major articles in *Life* (November 27, 1965) and *Sports Illustrated* (February 4, 1966) magazines. This was accompanied by pictures showing animal neglect. It suggested a need for regulation and a system of enforcement. Catalyzed in part by these articles, the Laboratory Animal Welfare Act was passed by Congress in 1966 (Public Law 89–544) and for the first time in the U.S. there were legal standards for laboratory animal care and use. The USDA was named the responsible agency for implementing and enforcing this new law and it promptly began providing regulations.[4] Although the act covered all laboratory animals, it was enacted primarily to protect dogs and cats and to counteract the business of pets being stolen by dealers and then sold for research purposes. Research laboratories and dealers were now required to register their facilities, be licensed, and undergo inspection by USDA personnel, who prepared reports and issued citations for noncompliance. These early inspections did not extend into the research laboratory, where animal care and use remained under the direction of the research investigator. Public groups concerned about the acquisition and welfare of animals destined for research use continued their activities, which ranged from local and national humane societies concerned about animal welfare and well-being to radical animal rights groups opposed to the use of animals for any reason. The activity of such groups seemed to escalate and become more vocal in the early 1980s. This activity peaked in a series of illegal break-ins and vandalism by a terrorist group and moved to the forefront of public opinion soon after two incidents involving "animal cruelty" and insensitivity in two well-known research institutions.[5,6] This climate raised public concern and visibility of animals in research and served as a catalyst for amendments and clarifications of guidelines and regulations providing for animal welfare.

A series of legislative bills and meetings in the early to mid-1980s, spearheaded by Senators Robert Dole, George Brown, John Melcher, and others, stimulated interest in including a requirement for an IACUC as part of the USDA regulations. Public Law 99-198, the Food Security Act of 1985, Subtitle F—Animal Welfare, also called the Improved Standards for Laboratory Animals Act, was enacted December 23, 1985. This clarified the meaning of *humane care* by mentioning specifics such as sanitation, housing, and ventilation. It directed the Secretary of Agriculture to establish regulations to provide exercise for dogs and an adequate physical environment to promote the psychological well-being of nonhuman primates. It also specified that pain and distress must be minimized in experimental procedures and that alternatives to such procedures be considered by the PI. The act also defined practices that were considered to be painful and stated that no animal could be used in more than one major operative procedure from which it recovered (with listed exceptions). Significantly, it established the IACUC, with a description of its roles, composition, and responsibilities to APHIS/AC. It also included the formation of an information service at the National Agricultural Library to assist those regulated by the act in training employees, searching for ways to reduce or replace animal use, preventing the unintended duplication of research; it also provided information on how to decrease pain or distress. The final section of the amended act explained the penalties for release of trade secrets by regulators and the regulated community. In a *Federal Register* publication of August 31, 1989, the USDA published final rules for Parts 1 and 2 of the Animal Welfare Act Regulations, as required by the 1985 Animal Welfare Act amendment, and, effective October 30, 1989, the USDA for the first time required that each registered research institution appoint an IACUC of no fewer than three members, which "serves as the agent of the research facility that ensures that the facility is in full compliance with the Act." The final rules for 9 CFR Part 3 (Standards) were published on February 15, 1991, and became effective on March 18, 1991. The USDA still regularly inspects registered institutions but relies heavily on the IACUC to ensure that institutional practices are in regulatory compliance.

A number of amendments to the AWA have led to regulations that now include animal transportation, marine mammals, and animals in the research laboratory. However, the AWA, via the Farm Security and Rural Investment Act of 2002, still excludes common laboratory rats (*Rattus*) and mice (*Mus*), birds bred specifically for research, and farm animals used in production agriculture. The most probable reasons for these exclusions are insufficient funds and staffing. Because nearly 90% of laboratory animals are rats and mice, their inclusion would have a significant impact on the workload of the USDA inspection staff.

U.S. Public Health Service Policy

The first *Public Health Service Policy on Humane Care and Use of Laboratory Animals* went into effect in 1973, was revised in 1979, and was revised again in 1986.[7] All PHS policies on this subject evolved from an NIH policy published in 1971.[8] That policy referenced several NIH and PHS statements on appropriate care and humane treatment of laboratory animals, among them the then existing version of the *Guide*. It introduced the animal care committee as a means of local assurance and included all live vertebrate animals. Each revision placed more specific responsibilities on that committee.

The 1971 NIH policy stated that institutions or organizations using warm-blooded animals in research or teaching supported by NIH grants, awards, or contracts would "assure the NIH that they will evaluate their animal facilities in regard to the maintenance of acceptable standards for the care, use, and treatment of such animals." The institution could either show that it was accredited by a recognized professional laboratory animal accrediting body or show that it had established its own committee to carry out that assurance function. The minimal number of committee members was not stated, but at least one member had to be a doctor of veterinary medicine. Guidelines to be followed included the *NIH Guide*, all applicable portions of the AWA, and an appended set of guidelines known as the *Principles for the Use of Laboratory Animals*. The committee was required to inspect the animal facilities of the institution at least once a year and report its findings and recommendations to responsible officials of the institution. Records of activities and recommendations were required and were to be available for inspection by NIH representatives. Under this policy, institutions accredited by AAALAC were not required by NIH to have an evaluation committee.

The first PHS Policy replaced the NIH policy on July 1, 1973, and continued to accept AAALAC accreditation in lieu of an institutional committee. The minimal number of committee members was now set at three, and unaccredited or partially accredited institutions were required to have a committee. Only when institutions used a "significant number of animals" was a veterinarian required on the committee. If the number of animals used in activities supported by the Department of Health, Education, and Welfare (DHEW) was not "significant," a veterinarian was not required, but one of the committee members must then be "a scientist with demonstrated expertise in the care and use of laboratory animals." If such a person was not available, a veterinarian employed on a consultant basis was an acceptable alternative. A "significant number of animals" was not defined and the final determination was made by the DHEW from animal inventory information provided as part of the institution's Assurance statement. The Policy for institutional review of applications and proposals stated, "Grantee and contractor institutions are encouraged to review their applications and proposals in the light of the pertinent provisions of the Animal Welfare Act, the standards set by the Institute of Laboratory Animal Resources (the NIH *Guide*), and the DHEW Principles for the Use of Laboratory Animals, and to familiarize their staff with these provisions, standards, and principles." However, there was no requirement under this Policy that institutional committees perform review of individual proposals or regularly provide to the DHEW summaries or certifications of such committee actions. The Policy did not specify who should perform the reviews. Under a different action, there was a requirement to keep records of committee activities for at least 3 years after the budget period. Assurance statements had to list the facilities and components of the program and committee. Activities included periodic facility inspections and reports to responsible officials at least once a year. The PHS Policy was revised on January 1, 1979, and now required all animal-using grantee institutions to have a "committee to maintain oversight of its animal care program." The Policy also required an institution to submit an Assurance statement to the NIH/Office for Protection from Research Risks, now the NIH/Office of Laboratory Animal Welfare (OLAW), and have it found to be acceptable before receiving a PHS grant for studies in which animals or animal facilities were used. In addition, to assure compliance with the edition of the *Guide* then in use, the AWA, and other applicable laws and regulations, it also required that "such assurance must also indicate that the institution has appointed and will maintain a committee to maintain oversight of its animal care program."

AAALAC accreditation was again recommended as the best means of demonstrating conformance with good animal care and use provisions. No explicit name was given to this committee, but the key words and all references to it were *animal care*. Grantee institutions, as in the past, were obliged to keep records of committee activities, including recommendations and determinations, with those records being available for inspection by authorized PHS officials. Review of individual proposals or projects by the institution's committee was encouraged, but not required. Even though the review remained merely a recommendation, the suggestion that the committee act as a reviewing body further strengthened its role. A list of committee members was required on the Assurance form and information on each member was to include degrees, position, title, and a short description of the member's relevant background. A sample statement on the form was preceded by the following:

> The samples given below for committee members are not intended to dictate numbers of or qualifications for committee members. However, except in unusual circumstances, the committee should be of at least five members and include at least one veterinarian. Any such unusual circumstances should be explained in a statement accompanying the assurance.

The following was contained in the sample statement:

> We have appointed and will maintain a committee of at least five members to maintain oversight of our animal care program. The members have appropriate education and experience to perform their duties with respect to the types of animals and species used and the kinds of projects to be undertaken. If the conduct of a specific project is to be reviewed, the quorum will not include any member having an active role in the project. Changes in membership will be reported annually to the Office for Protection from Research Risks (OPRR), National Institutes of Health (NIH), and PHS.

Three options were provided for reporting an institution's degree of compliance: AAALAC accreditation, certification by its own institutional committee, or committee recommendations of improvements needed for compliance.

If the committee recommended any immediate or future improvements, the NIH/OPRR would expect to receive an annual report of the progress of these improvements toward compliance. The Policy now required that the committee would review its facilities and procedures at least once a year and that responsible officials would receive and consider all reports from the committee. They also would make an annual review of committee activities for compliance and would keep records of committee meetings and related administrative actions. A new Assurance statement was required every five years. After allowing five years to study the effectiveness of the Assurance system, it was concluded in 1984 that the committees frequently seemed "less than fully assertive" in carrying out their responsibility.[7] A recommendation after this study suggested that "the PHS Policy should be further modified to define more precisely the responsibilities of the awardee institutions, particularly the role of the animal care committee. It is imperative that the experience and expertise of the members of such committees be used to conduct full and effective reviews of proposals involving research with animals. The appointment of a nonscientist and an individual unaffiliated with the institution should be given serious consideration."[9] This study resulted in the latest revision of the PHS Policy, which further defines and outlines requirements of an animal care committee. This Policy now includes provisions of the Health Research Extension Act of 1985, which was enacted on November 20, 1985, as Public Law 99-158. The most significant changes required by this action are that the Policy now applies to intramural research conducted by the PHS and that the IACUC is appointed by the chief executive officer of the institution.[7] The PHS Policy, revised in 2002, requires that the program description include an explanation of the training or instruction available to scientists, animal technicians, and other personnel involved in animal care, treatment, or use. This training or instruction must include information on the humane practice of animal care and use and the concept, availability, and use of research or testing methods that minimize the number of animals required to obtain valid results and minimize animal distress. The IACUC must now evaluate and prepare twice-yearly reports on all of the institution's programs and facilities (including satellite facilities) for activities involving animals instead of once each year. The IACUC, through the IO, is responsible for reporting requirements and minority views filed by members of the IACUC must be included in reports filed under this Policy. The revised policy permits institutions with PHS Animal Welfare Assurances to submit verification of IACUC approval for competing applications or proposals subsequent to peer review but prior to an award. Also, new footnotes 6 and 12 of the PHS Policy provide institutions the option of coding the names of the IACUC members in materials routinely submitted to NIH/OLAW.

Department of Defense

While the scientific community and the public sector were gradually evolving guidelines and policies to assure animal welfare and good animal care[10–19] research conducted by the Department of Defense (DOD) had similar concerns. As early as 1961, the DOD issued the *Policy on Experimental Animals*,[20–22] which directed that "all aspects of investigative programs involving the use of laboratory animals and sponsored by Department of Defense agencies will be conducted according to the Principles of Laboratory Animal Care as promulgated by the National Society for Medical Research." While this early Policy provided few specifics, a number of revisions followed and included all animals used in DOD laboratories, both in the United States and abroad. A joint regulation issued by the Army, Navy, Air Force, Defense Nuclear Agency, and Uniformed Services University on June 1, 1984, *Use of Animals in DOD Programs*, required that "all DOD organizations having animals (other than military working, recreational, and ceremonial) will seek accreditation by AAALAC."[23] It further required that "the local commanders of each DOD organization conducting or sponsoring activities involving animals in Research, Development, Teaching and Education (RDTE), clinical investigations, diagnostic procedures, or instructional programs will form a committee to oversee the care and use of animals." Such committees were to be appointed by the local commander, include a doctor of veterinary medicine, be made up of at least three members, review protocols and assure compliance with policies, standards, and regulations. The concept and practice of such committees to review and assure appropriate animal care and use, while not known by the acronym *IACUC*, were in place at many military installations prior to their being regularly formed at most academic institutions.

Principles for Animal Care in Research

The *International Guiding Principles for Biomedical Research Involving Animals*[24] were developed by the Council for International Organizations of Medical Sciences (CIOMS) as a result of extensive international and interdisciplinary consultations spanning the three-year period 1982–1984 and were published in 1985. The International Guiding Principles have gained acceptance and recognition internationally. The European Medical Research Councils (EMRC), an international association that includes all the West European medical research councils, fully endorsed the Guiding Principles in 1984 and in the same year they were endorsed by the World Health Organization's (WHO) Advisory Committee on Medical Research at its 26th session. *The U.S. Government Principles for the Utilization and Care of Vertebrate Animals Used in Testing, Research, and Training* were formulated in 1984 by the U.S. Interagency Research Animal Committee and were to a large extent based on the CIOMS Guiding Principles. These U.S. Interagency Principles are published in both the PHS Policy and the *Guide*. The need for an IACUC-like body was stated in the International Principles in the following Basic Principle.

> VII. Where waivers are required in relation to the provisions of article VII, the decisions should not rest solely with the investigators directly concerned but should be made, with due regard to the provisions of articles IV, V, and VI, by a suitable constituted review body. Such waivers should not be made solely for the purposes of teaching or demonstration.

This need for an IACUC was repeated more specifically in the U.S. Government Principles by the following.

> IX. Where exceptions are required in relation to the provisions of these Principles, the decisions should not rest with the investigators directly concerned but should be made, with due regard to Principle II, by an appropriate review group such as an institutional animal care and use committee. Such exceptions should not be made solely for the purposes of teaching or demonstration.

A joint committee of CIOMS and the International Council for Laboratory Animal Science (ICLAS) has been addressing the updating and revision of the International Principles and this revision was accepted in final form in 2012 by both ICLAS and CIOMS. The need for an IACUC-type review body with such responsibilities again appears in this new document and will continue to encourage all countries to require such review and assurance.

Evolution of the IACUC

The 1986 PHS Policy first described the IACUC in its present form and membership and only provided general guidance on how such a committee should be formed and operated. With an amendment to the AWA effective October 30, 1989, USDA regulations required a similar IACUC. Most institutions now have a single committee that satisfies both PHS and USDA requirements. The Scientists Center for Animal Welfare (SCAW) was instrumental in providing early guidance to institutions on IACUC functions and organization through regional conferences and workshops, culminating in a special AALAS publication, *Effective Animal Care and Use Committees.*[25] Since 1983, training and regular conferences of this type are provided by annual animal care and use conferences sponsored by the Public Responsibility in Medicine and Research (PRIM&R), regional workshops supported by the NIH/OLAW, and numerous similar activities. As described in the AWAR (§2.31,b,2) this committee "shall be composed of a chairman and at least two additional members" and is to "prepare reports of its evaluations" ... "and submit the reports to the Institutional Official of the research facility."[4] While originally borrowed from the human Institutional Review Board structure, the establishment of local animal care committees (i.e., IACUCs) to review and assure animal welfare and well-being is now common practice in the animal research community. The goal of each committee is compliance with guidelines and regulations while allowing for the flexibility of individual institutional tailoring to effectively meet the unique needs of the institution. Active participation by research scientists allows for the unique needs of research investigators to be voiced, participation by members unaffiliated with the institution protects the public conscience, and veterinarians assure appropriate medical care and provisions. This committee has become the primary oversight mechanism for animal care and use, responsible through an IO and reporting directly to regulatory and granting agencies on animal care and use matters. The PHS, AAALAC, and the USDA continue to provide additional guidance and clarification of the expectations and responsibilities of the IACUC through respective on-line guidance and "frequently asked questions." There is a continuing need for education to assure that this concept works as well as possible.

References

1. Rozmiarek, H. 1986. Current and future policies regarding animal welfare. *Invest. Radiol.*, 22: 175, 1986.
2. Committee to Revise the Guide for the Care and Use of Laboratory Animals. 2011. *Guide for the Care and Use of Laboratory Animals.* Washington, D.C.: National Academy Press.
3. Van Sluyters, R.C. 2008. A guide to risk assessment to animal care and use programs. The metaphor of the 3-legged stool. *ILAR J.* 49:372–378.
4. Office of the Federal Register, Code of Federal Regulations, Title 9, Animals and Animal Products, subchapter A, parts 1, 2, and 3, Animal Welfare, Washington, D.C., 2002.
5. Fraser, C. The Raid at Silver Spring. *The New Yorker* 66: April 19, 1993.
6. McCabe, K. Who will live, who will die. *The Washingtonian* 112: August 1986.
7. National Institutes of Health. Office of Laboratory Animal Welfare. 2002. *Institutional Animal Care and Use Committee Guidebook*, 2nd edition Bethesda: The National Academies Press.
8. Whitney, R.A., Jr. 1987. Animal care and use committees: History and current national policies in the United States. *Lab. Anim. Sci.*, special issue, 18–21.
9. National Institutes of Health. 1971. Care and Treatment of Animals (NIH 4206). NIH Guides for Grants and Contracts, No. 7.
10. Federation of Animal Science Societies. 2010. *Guide for the Care and Use of Agricultural Animals in Agricultural Research and Teaching*, 3rd edition. Savoy, IL: Federation of Animal Science Societies.
11. Committee on Animal Research and Experimentation of the Board of Scientific Affairs. 1991. *Guidelines for Ethical Conduct in the Care and Use of Animals.* Washington: American Psychological Association.
12. U.S. Department of Health and Human Services. 1988. *National Institutes of Health, Institutional Administrators' Manual for Laboratory Animal Care and Use.* NIH Publication No. 88-2959.
13. ILAR Committee on the Use of Laboratory Animals in Neuroscience and Behavioral Research. 2003. *Guidelines for the Use of Mammals in Neuroscience and Behavioral Research.* Washington, D.C.: National Academy Press.
14. ILAR Committee on the Use of Animals in Precollege Education. 2006. *Principles and Guidelines for the Use of Animals in Precollege Education.* Washington, D.C.: Institute for Laboratory Animal Resources.

15. Joint AAMC-AAU Committee. 1985. *Recommendations for Governance and Management of Institutional Animal Resources.* Washington, D.C.: Association of American Medical Colleges and the Association of American Universities.
16. U.S. Department of Health and Human Services. 1991. *Preparation and Maintenance of Higher Mammals during Neuroscience Experiments.* NIH Publication No. 91-3207, Washington, D.C.: National Institutes of Health.
17. Universities Federation for Animal Welfare. 1999. *The UFAW Handbook on the Care and Management of Laboratory Animals,* 7th edition, South Mimms: Potters Bar, U.K.
18. International Air Transportation Association. 2012. *Live Animal Regulations.* http://www.iata.org/publications/Pages/live-animals.aspx.
19. National Association for Biomedical Research. 2004. *State Laws Concerning the Use of Animals in Research.* Washington, D.C.: National Association for Biomedical Research.
20. U.S. Department of Defense. 1959. *Directive 5129.1.* Washington, D.C.: Department of Defense Instruction.
21. U.S. Department of Defense. 1961. *Directive 51225.5.* Washington, D.C.: Department of Defense Instruction.
22. U.S. Department of Defense. 1961. *Policy on Experimental Animals in Department of Defense Research.* Washington, D.C.: Department of Defense Instruction.
23. U.S. Department of Defense, United States Army. 1984. *Army Regulation 70-18, SECNAVINST 3900.388, AFR 100-2, DARPAINST 18, DNAINST 3216.1B, USUHSINST 3203. The Use of Animals in DOD Programs.* Washington, D.C.
24. International Guiding Principles for Biomedical Research Involving Animals. 1985. http://cioms.ch/publications/guidelines/1985_texts_of_guidelines.htm.
25. Orlans, F.B., R.C. Simmonds, and W.J. Dodds. 1987. Effective animal care and use committees. *Lab. Anim. Sci.* special issue, January 1987 (published in collaboration with the Scientists Center for Animal Welfare).

2

Circumstances Requiring an IACUC

Marcy Brown and Mary Jo Shepherd

Introduction

APHIS/AC and NIH/OLAW have overlapping regulatory functions codified in the AWA, the AWAR, the HREA, and the PHS Policy. These documents specify the various situations in which an IACUC is required. Although all components of an animal care and use program—the IO, the AV, the animal care staff, the research community and the IACUC—have important roles to play, unquestionably, federal law, regulation, and policy place the IACUC in a pivotal position for ensuring animal welfare. In addition, the *Guide* and the *Institutional Animal Care and Use Committee Guidebook*[1] describe the key role IACUCs have in the assessment and oversight of the animal care and use program. This chapter explains the various circumstances in which institutions are required to have an IACUC and summarizes the regulatory requirements of the AWA, AWAR, HREA, and PHS Policy for an IACUC. The fact that the AWA and the HREA have separate mandates concerning the creation of an IACUC can be a source of confusion since the circumstances under which these two laws require IACUCs differ. In an effort to alleviate this confusion, the reader is directed to the specific sections of the AWA, AWAR, and PHS Policy that describe the regulatory requirements for the IACUC. The remainder of the chapter provides the answers to a series of questions designed to reveal whether IACUCs are always required at research institutions, and whether IACUCs are ever required at grade schools, secondary schools, science fairs, zoos, aquaria, animal shelters, or local humane organizations.

2:1 What regulatory agencies require the appointment of an IACUC?

Reg. There are two: the USDA and the PHS (see 2:2). APHIS/AC, through the AWAR, requires an IACUC at any institution that uses animals in research, testing or teaching, provided the species is among those listed under the definition of "animal" in §1.1 of the AWAR (see 12:1). The PHS, through NIH/OLAW, requires an IACUC at any institution that conducts PHS-supported activities involving any live vertebrate animal (PHS Policy IV,A,1).

2:2 What specific documents describe the regulatory requirements for the IACUC?

Reg. The specific documents that describe the regulatory requirements for the IACUC are the AWA, Title 7 of the U.S. Code (7 U.S.C.) §2131–2159 (Public Law 89-544), and the HREA (Public Law 99-185). Detailed regulations and standards for implementing the AWA are set forth by the USDA in Title 9 of the Code of Federal Regulations

(9 CFR), Chapter 1, Subchapter A, Parts 1, 2, and 3. Detailed regulations for implementing the HREA are set forth in PHS Policy IV,A,3; IV,B; IV,C; and IV,E-F.

Although most of the AWAR that specifically govern the IACUC are found in §2.31, references to the IACUC are scattered throughout the AWAR as follows:

PART 2 — REGULATIONS

Subpart C — Research Facilities

- §2.31 — Institutional Animal Care and Use Committee
- §2.33,a,3 — Attending Veterinarian and Adequate Veterinary Care
- §2.35,a–§2.35,f — Recordkeeping Requirements
- §2.36,b,3 — Annual Report
- §2.37 — Federal Research Facilities
- §2.38,f,2,11; §2.38,k,1 — Miscellaneous

PART 3 — STANDARDS

Subpart A — Specifications for the Humane Handling, Care, Treatment, and Transportation of Dogs and Cats

- §3;6,d — Primary Enclosures
- §3.8,b,1; §3.8,c; §3.8,d,2 — Exercise for Dogs

Subpart B — Specifications for the Humane Handling, Care, Treatment, and Transportation of Guinea Pigs and Hamsters

- §3.28,c,3 — Primary Enclosures

Subpart C — Specifications for the Humane Handling, Care, Treatment, and Transportation of Rabbits

- §3.53,c,3 — Primary Enclosures

Subpart D — Specifications for the Humane Handling, Care, Treatment, and Transportation of Nonhuman Primates

- §3.80,b,2,iii; §3.80,c — Primary Enclosures
- §3.81,c,3; §3.81,d; §3.81,e,ii — Environment Enhancement to Promote Psychological Well-Being
- §3.83 — Watering

The PHS Policy also sets forth detailed requirements governing IACUCs. References to IACUCs are found primarily in the following sections:

- IV,A,3 — Institutional Animal Care and Use Committee
- IV,B — Functions of the IACUC
- IV,C — Review of PHS-Conducted or PHS-Supported Research Projects

- IV,E Recordkeeping Requirements
- IV,F Reporting Requirements

The PHS Policy (IV,A,1) requires institutions to use the *Guide* as a basis for developing and implementing an institutional program for activities involving animals.

2:3 What is the difference between the Animal Welfare Act and the Animal Welfare Act regulations in regard to requirements for an IACUC?

Reg. The AWA (Public Law 89-544, U.S. Code 7 §2131–2159) is a statute enacted by Congress in 1966 and governs the care and use of animals in research for both government and nongovernment facilities. The AWAR is 9 (CFR Chapter 1, Subchapter A). Every regulation listed in the CFR must have an enabling statute (in this case the AWA). The purpose of the AWAR is to describe in greater detail how an agency should interpret the law.

Opin. Both the AWA and the AWAR require research facilities using regulated species to establish an IACUC (see 2:1; 2:2). However, the AWAR set forth a more extensive list of duties for the IACUC.

2:4 What is the difference between the HREA and the PHS Policy in regard to the requirements for an IACUC?

Opin. The HREA is the statutory authority for the PHS Policy. Both the HREA and the PHS Policy require institutions using animals in research, research training, and biological testing activities conducted or supported by the PHS to establish an IACUC (see 2:1; 2:2). The PHS Policy sets forth an extensive list of functions of the IACUC.[2]

2:5 What are the regulatory charges of the IACUC?

Reg. The regulatory charges of the IACUC are described in the AWA §2143,b and §2143,c; the AWAR §2.31,c; HREA §495,b,3; and PHS Policy IV.

Opin. In summary, the above require the IACUC to:

- Review, at least once every six months, the institution's program for animal care and use, using the AWAR and the *Guide* as a basis for evaluation.
- Inspect, at least once every six months, all of the institution's animal facilities, including animal study areas and satellite facilities, using the AWAR and the *Guide* as a basis for evaluation.
- Submit to the IO reports of the above evaluations which distinguish significant deficiencies from minor deficiencies, contain a reasonable and specific plan and schedule for correcting each deficiency, describe and justify any departures from the *Guide* and PHS Policy, and which are signed by a majority of the committee and include any minority views.[3]
- Review and investigate concerns involving the care and use of animals at the institution resulting from public complaints or from reports of noncompliance received from personnel at the institution.

- Make recommendations to the IO regarding any aspect of the institution's animal program, facilities, or personnel training.
- Review and approve, require modifications in (to secure approval), or withhold approval of those components of proposed activities related to the care and use of animals.
- Review and approve, require modifications in (to secure approval), or withhold approval of proposed significant changes regarding the care and use of animals in ongoing activities.
- Suspend an activity involving animals if, according to PHS Policy, it is not being conducted in accordance with the applicable provisions of the AWA, the *Guide*, PHS Policy, or the institution's Animal Welfare Assurance as approved by NIH/OLAW, or, according to the AWAR, is not being conducted in accordance with the description of the activity provided by the PI and approved by the IACUC.

2:6 How might records for IACUCs be effectively maintained?

Reg. Institutions are required to maintain various records related to the animal care and use program. The AWAR (§2.35) and PHS Policy (IV,E) require each institution to maintain:

- Minutes of IACUC meetings, including records of attendance, activities of the committee, and committee deliberations (see 6:25–6:27)
- Records of applications, proposals, and proposed significant changes in the care and use of animals, and whether IACUC approval was given or withheld
- Records of semiannual IACUC reports and recommendations, including minority views, that have been forwarded to the IO

In addition, PHS Policy (IV,E) requires each institution to maintain:

- A copy of its Animal Welfare Assurance
- Records of accrediting body determinations

The AWAR (§2.35,f) and PHS Policy (IV,E) stipulate that all records must be maintained for at least 3 years and that records relating to applications, proposals, and proposed significant changes must be maintained for 3 years after completion of the activity. All records must be accessible for inspection and copying by authorized representatives of APHIS/AC, NIH/OLAW, other PHS representatives, and (for accredited institutions) AAALAC site visitors.

Opin. The IACUC or IACUC staff should either maintain the records listed above or ensure that they are maintained in a secure and accessible location at the institution. Additional records that should be kept include:

- IACUC correspondence, including e-mail records
- Institutional policies, guidelines, and SOPs for animal care and use
- Reports of IACUC investigations of concerns involving animals

Methods of maintaining records will vary, depending upon the organization of the institution and the size and complexity of its animal use program. However, any method that is used to maintain records must be efficient, reliable, and secure. While some institutions choose to use traditional paper records, many institutions use electronic protocol management systems to maintain records of review and approval of animal use protocols. Such systems conduct the entire protocol workflow including authorship, submittal, review, comments, revision and approval, as well as amendments, ongoing annual and continuation reviews. Systems may also include IACUC meeting agenda and minutes automation, e-mail notifications, information alerts, and historical reporting, thereby facilitating regulatory compliance. Electronic systems have many advantages, including the ability to sort and retrieve information quickly by various parameters such as species, type of procedure, protocol expiration date, and location of the procedure or animal housing. Many systems also manage the purchase of animals to ensure congruence between number of animals ordered and number of animals approved on a protocol. Information stored in an electronic protocol system may also facilitate preparation of the institution's annual report to APHIS/AC and preparation of the Program Description that is required as part of the application for AAALAC accreditation or reaccreditation.

2:7 Is an IACUC *always* required where animals are used for research, teaching, or product safety evaluation?

Reg. No. An IACUC is required only if one of the following applies (see 8:4–8:6):

- The species used for research, teaching, or testing is covered by the definition of animal given in §1.1 of the AWAR. The AWAR definition of animal specifically excludes birds, rats of the genus *Rattus*, and mice of the genus *Mus*, bred for use in research, horses not used for research purposes and other farm animals used in agricultural research. (See 12:1.)
- The research, research training, or biological testing is supported by the PHS, or the activity will be performed at an institution with an Animal Welfare Assurance that commits the institution to comply with the PHS Policy, regardless of the source of funding.
- The institution is receiving support for animal research, research training, or biological testing from an agency that requires compliance with PHS Policy (e.g., the National Science Foundation and the Veteran's Administration).[4]

Opin. If an institution is or desires to be accredited by AAALAC, it must have an IACUC.

2:8 A pet dog is used in a research or teaching hospital as nothing more than a model in an art therapy class. Is it necessary for the hospital to have (or establish) an IACUC to approve the use of this animal?

Reg. (See 2:7.)

Opin. No. An IACUC is only required when APHIS/AC or NIH/OLAW regulated species are being used in experimentation, teaching, and testing. A hospital that allows dog owners to bring their pet animals into the hospital for use in an art therapy class does not own the animals. Therefore, the AWAR do not apply and an IACUC review is not required. Assuming this activity is not taking place as part of a PHS-funded research project, an IACUC review also is not required under PHS Policy, since it applies only to institutions that receive support for vertebrate animal activities from an agency of the PHS or from an agency that requires compliance with PHS Policy. (See 2:7 and 2:12.)

2:9 Do elementary or secondary schools need to have an IACUC to keep pet animals?

Reg. No, the AWA does not regulate the keeping of pets, whether in a private home or by a school. In addition, both the AWAR (§1.1, Research facility) and the AWA (§2132,e) specifically exclude elementary and secondary schools from being designated as research facilities. Therefore, the regulations do not apply and an IACUC is not required.

Opin. An elementary or secondary school that keeps pet animals is also not considered an exhibitor under the definitions in §1.1 of the AWAR because it does not purchase animals and exhibit them to the public for compensation. An IACUC also is not required under PHS Policy, since it applies only to institutions that receive support for vertebrate animal activities from an agency of the PHS or from an agency that requires compliance with PHS Policy. (See 2:8; 2:10–2:12.)

2:10 Is it necessary to have IACUC oversight of elementary or secondary school science fair projects or other educational activities?

Reg. No. Elementary and secondary schools are specifically excluded from the definition of research facility in §1.1 of the AWAR and §2132,e of the AWA. Accordingly, the regulations do not apply and an IACUC is not required. An IACUC also would not be required under PHS Policy, since it applies only to institutions that receive funds for vertebrate animal activities from an agency of the PHS or from an agency that requires compliance with PHS Policy.

2:11 What guidelines exist for use of animals in elementary or secondary school educational activities?

Opin. Although elementary or secondary schools are not covered by the AWA, AWAR, HREA or the PHS Policy, they may choose to follow the guidance contained in the *Guide* and AALAS Position Paper *Use of Animals in Precollege Education.*[5] The latter document presents guidelines and resources for the humane care and responsible use of animals in precollege education and offers recommendations on classroom dissection and on the use of animals in science fair projects. As part of its broader educational mission to ensure that all animal use is performed responsibly and humanely, AALAS has also developed a series of species-specific informational pamphlets about species commonly found in classrooms settings such as mice, rats, hamsters, guinea pigs, rabbits, reptiles, and amphibians.[6] The National Science Teachers Association has also developed a Position Statement on Responsible Use of Live Animals and Dissection in the Science Classroom[7] and supports including live animals as part of instruction in the K-12 science classroom.

Other guidelines that may be useful in elementary or secondary school educational activities are:

- The *Use of Animals in Biology Education* developed by the National Association of Biology Teachers[8]
- *Guidelines for Ethical Conduct in the Care and Use of Animals* from the American Psychological Association[9]
- *Principles and Guidelines for the Use of Animals in Precollege Education* from the Institute of Laboratory Animal Resources[10]

In addition to these guidelines, individual state governments or local school districts may develop standards that are either guidelines or requirements. Also, in many states there are nonprofit biomedical research societies that schools can consult for information on guidelines.[11] In general, these societies promote public education and understanding in support of biomedical research.

2:12 Do colleges or universities with live mascots need to have the care and use of these animals approved by the IACUC?

Reg. Under the AWA (§2143,a,1) the care and treatment of a regulated-species mascot would come under APHIS/AC jurisdiction if the animal is used in research or exhibited to the public during football games, etc. In the event the animal is used in research, an IACUC would be required per AWAR §2.31. The animal would not fall under APHIS/AC jurisdiction if it is merely a pet.

Opin. Institutions receiving PHS funding that use live animals as mascots are technically not required to have the care and use of these animals approved by the IACUC since the PHS Policy applies only to activities supported by PHS funds. However, to the extent the mascot is housed in university facilities, it should be cared for and housed in a manner consistent with the *Guide* and AWAR, and the housing and care of a regulated-species would be subject to APHIS/AC inspection.

Therefore, the question of whether the care and use of a mascot must be approved by the IACUC depends upon an institution's policies since it is not required under the AWAR or PHS Policy. For example, it is the policy at some institutions that all uses of live vertebrate animals must be approved by an IACUC, regardless of whether the animals are an APHIS/AC regulated species or are used in PHS-supported activities. In these cases, it is important for the institution to have clear written policies or SOPs regarding animals used as mascots, exhibits or for purposes other than research, research training, and biological testing. Such policies or SOPs should include a provision for IACUC review and approval if the institution requires it, as well as stipulations for animal care (e.g., stating who provides housing and husbandry) and veterinary services. For institutions participating in AAALAC's accreditation program, if institutional policies require some degree of IACUC oversight of animals used for a purpose other than research, testing or teaching, then AAALAC would expect adherence to those policies.[12]

2:13 Do zoological gardens and aquariums fall under the jurisdiction of the AWAR or PHS Policy?

Reg. Yes. Zoos and aquariums that maintain AWAR-regulated species (e.g., marine mammals, exotic rodents, etc.) are defined as exhibitors under §1.1 of the AWAR and must comply with all regulations governing exhibitors, as described in §2.1–2.5 of the AWAR.

Opin. Strictly speaking, only zoos and aquariums that receive support for research using animals from an agency of the PHS, such as the NIH, are required by NIH/OLAW to comply with its policy. However, other federal and private funding agencies (e.g., the Veterans Administration and the National Science Foundation) may require adherence to the PHS Policy as a condition of receiving their support. (See 2:7.)

2:14 Under the AWAR, are zoological gardens and aquariums expected to adhere to the same research criteria as biomedical research institutions? (See 2:15.)

Reg. APHIS/AC Policy 10,[13] which provides clarification on the AWA (§2132) and the AWAR (§1.1; §2.6,c), states "Licensed exhibitors [such as zoos and aquariums] occasionally collect information on their animals with the intent to improve the nutrition, breeding, management, or care of such animals. APHIS has determined these programs may be exempted from the registration requirements of the regulations as long as the collection methods:

- Are performed as an adjunct to normal husbandry or veterinary procedures for the benefit of the animal or species (e.g., routine veterinary care, embryo transfer, artificial insemination, electroejaculation); or
- Are not invasive (e.g., feed studies); or
- Do not cause pain or distress to the animal (e.g., behavioral observations)

However, if the licensed exhibitor is conducting biomedical research and/or testing (using the animals as models for human applications), conducting invasive or painful/distressful procedures for nonhusbandry purposes or if the research involves domestic dogs or cats, then the licensee is *not* exempt from the need for registration." (For further information on this topic, see reference[14].)

2:15 Is research conducted at zoological gardens and aquariums required to come under the purview of an IACUC?

Reg. (See 2:14.) If a zoological garden or aquarium is required to be registered as a research facility under the AWA, that research is required to be reviewed by an IACUC, just as any other AWA-registered research facility.

Opin. If the activity involves an APHIS/AC-regulated species and is considered regulated research by APHIS/AC (see 2:13), it comes under the purview of an IACUC. Similarly, if the activity is supported by the PHS or other agencies that require compliance with PHS Policy, it must be reviewed and overseen by an IACUC (see 2:7).

2:16 Do private and municipal animal shelters and humane societies fall under the jurisdiction of the AWA and its regulations or the PHS Policy and therefore require an IACUC?

Reg. The legal requirements for businesses regulated by the AWA are described in the AWA (§2133; §2136) and the AWAR (§1.1, Dealer; §2.1,a,3,i–viii).

Opin. Generally, animal shelters and humane societies do not fall under the jurisdiction of the AWAR unless they sell dogs or cats to a research facility, licensed dealer, exhibitor, or sell animals wholesale as pets (AWAR §1.1, Dealer). If a private shelter provides animals to a research facility, it must be licensed with APHIS/AC as a dealer and must comply with all AWA regulations governing dealers, which do not include a requirement for an IACUC. A municipal animal shelter or humane society that provides animals to a research facility is not defined as "a person" by APHIS/AC and, consequently, does not meet the definition of dealer under §1.1 of the AWAR. Accordingly, municipal shelters and humane societies are exempt from the regulations governing dealers.

Animal shelters and local humane organizations fall under the jurisdiction of PHS Policy only if they receive support from the PHS or other agency that requires compliance with PHS Policy, which would include the requirement for an IACUC.

2:17 The AWAR (§2.31,d,1) exempts studies defined as field studies from the requirement for IACUC review. However, the AWAR definition of a field study *excludes* any study that involves an *invasive procedure, harms* or *materially alters the behavior of the animals under study*. In determining whether a field study is exempted from the requirement for IACUC review, what is meant by an "invasive procedure" and "materially alters the behavior of the animals"?

Reg. The AWAR (§1.1) state, "Field study means a study conducted on free-living wild animals in their natural habitat. However, this term excludes any study that involves an invasive procedure, harms, or materially alters the behavior of an animal under study." The AWAR do not further define the phrases "invasive procedure," "harms" or "materially alters the behavior." (See 2:18.)

Opin. Although the phrases are not further defined, the context in which they appear in the AWAR gives guidance as to their intended meaning. Clearly, the intent of this regulation is to require IACUC review of any projects that have the potential to disrupt animals or to alter their environments. Included in these would be field studies that require animals to be trapped or confined in any way, handled (see AWAR §1.1 for a definition of handling), anesthetized, tagged or marked in any other way, or subjected to blood or tissue sampling. This leaves strictly observational studies—ones in which neither the animals nor their environments are disturbed or permanently altered—as being exempt from the requirement for IACUC review.

2:18 The AWAR (§2.31,d,1) exempt field studies from IACUC review. Does this mean that an institution only conducting field studies is *not* required to have an IACUC?

Reg. (See 2:17.)

Opin. It depends; if the field studies are PHS-supported and involve vertebrate animals, an IACUC is required for oversight in accord with PHS Policy. According to NIH/OLAW FAQ A.6, "IACUCs must know where field studies will be located, what procedures will be involved, and be sufficiently familiar with the nature of the habitat to assess the potential impact on the animal subjects. If the activity alters or influences the activities of the animal(s) that are being studied, the activity must be reviewed and approved by the IACUC (e.g., capture and release, banding). If the activity does not alter or influence the activity of the animal(s), IACUC review and approval is not required (observational, photographs, collection of feces)."[15]

2:19 The AWAR (§2.31,d,1) exempt studies defined as field studies from the requirement for IACUC review. However, the AWA definition of a field study (§1.1) includes the phrase "free-living wild animals in their natural habitat." In determining whether a field study is exempted from the requirement for IACUC review, can an animal be kept for any length of time outside of its natural habitat before it is no longer considered free living?

Reg. (See 2:17.) The AWAR definition of a field study makes it clear that any study that involves an invasive procedure or has the potential to harm or materially alter an animal's behavior is not exempt from IACUC review.

Opin. Removing an animal from its natural environment for any length of time would appear always to have the potential to materially alter its behavior. Such removal would typically involve capture, trapping or herding, handling, restraint and/or physical confinement. These activities, no matter how carefully they are performed, would carry a high likelihood of altering an animal's behavior in a significant fashion, at least in the short term.

References

1. Applied Research Ethics National Association/Office of Laboratory Animal Welfare. 2002. *Institutional Animal Care and Use Committee Guidebook*. Bethesda: National Institutes of Health. http://grants.nih.gov/grants/olaw/GuideBook.pdf.
2. National Institutes of Health, Office of Laboratory Animal Welfare. 2002. *Public Health Service Policy on Humane Care and Use of Laboratory Animals.* Section IV. Implementation by Institutions. National Institutes of Health. http://grants.nih.gov/grants/olaw/references/phspol.htm.
3. National Institutes of Health, Office of Extramural Research. 2011. *Semiannual Report to the Institutional Official.* http://grants.nih.gov/grants/olaw/sampledoc/ioreport.htm.
4. National Science Foundation. NSF Grant Proposal Guide. 2011. Chapter II.D.6. http://www.nsf.gov/pubs/policydocs/pappguide/nsf11001/gpg_2.jsp#IID6.
5. American Association for Laboratory Animal Science. *Use of Animals in Precollege Education.* http://www.aalas.org/association/position_statements.aspx#PrecollegeU. Accessed January 2, 2013.
6. American Association for Laboratory Animal Science Caring for Animals in the Classroom. http://www.aalas.org/resources/classroom_animals.aspx. Accessed Jan. 2, 2013.
7. National Science Teachers Association. 2008. NSTA Position Statement: Responsible use of live animals and dissection in the science classroom. http://www.nsta.org/about/positions/animals.aspx.
8. National Association of Biology Teachers. *The Use of Animals in Biology Education.* http://www.nabt.org/websites/institution/index.php?p=97. Accessed January 2, 2013.
9. American Psychological Association. 2012. *Guidelines for Ethical Conduct in the Care and Use of Animals.* http://www.apa.org/science/leadership/care/guidelines.aspx.
10. National Academy of Sciences. 2004. Institute of Laboratory Animal Resources. *Principles and Guidelines for the Use of Animals in Precollege Education.* http://www.nabt.org/websites/institution/File/Principles and Guidelines for the Use of Animals in Precollege Education.pdf.
11. States United for Biomedical Research. 2012. http://www.statesforbiomed.org/content/national-network-0.
12. Bayne, K. Global Director, AAALAC International, personal communication, 2013.
13. U.S. Department of Agriculture. 2011. *Animal Care Policy Manual.* Policy 10, Specific activities requiring a license or registration. http://www.aphis.usda.gov/animal_welfare/policy.php.

14. Kohn, B. and S.L. Monfort. 1997. Research at zoos and aquariums: Regulations and reality. *J. Zoo Wildlife Med*. 28: 241–250.
15. National Institutes of Health, Office of Laboratory Animal Welfare. 2012. Frequently Asked Questions A.6, Does the PHS Policy apply to animal research that is conducted in the field? http://grants.nih.gov/grants/olaw/faqs.htm#App_6.

3

Creation of an IACUC

Marcy Brown and Mary Jo Shepherd

Introduction

This chapter provides guidance on how institutions may form an IACUC. It describes whose interests the IACUC should serve and whose responsibility it is to appoint the members. Since most IACUCs conduct their meetings in private, the nonaffiliated member of the committee can be the general public's only link to the federally mandated oversight process for animal welfare. Accordingly, information is provided on the critical role of the nonaffiliated member and how institutions can go about finding an effective one. For an IACUC to fulfill its regulatory requirements properly, its members must be informed of their responsibilities under federal laws, regulations, and policies. In addition to these federally mandated obligations, IACUCs may be charged with additional responsibilities by their state, city, or institution. Given the complexity of this environment, institutions have a responsibility to provide their IACUC members with specialized training. This chapter also describes some methods for IACUC member training. The latter part of the chapter suggests means by which larger institutions may utilize multiple IACUCs to reduce the workload associated with reviewing hundreds of animal use protocols and inspecting dozens of animal facilities. Finally, there is a discussion regarding whether a small institution may choose to use the services of the IACUC at a larger institution with a more extensive animal care and use program.

3:1 In general, what is expected of the IACUC, and whose interests does it represent?

Reg. The AWA (§2143,b,1) charges the members of the IACUC with representing "society's concerns regarding the welfare of animal subjects." In addition, the nonaffiliated member is specifically charged with representing general community interests in the proper care and treatment of animals (AWA §2143,b,1,B,iii; AWAR §2.31,b,3,ii; *Guide* p. 24). The PHS Policy (IV,B) references the IACUC as "an agent of the institution" that will "oversee the institution's animal program, facilities, and procedures" (PHS Policy IV,A,3,a).

Opin. The IACUC is expected to oversee and evaluate the institution's animal care and use program to ensure consistency with the recommendations of the *Guide,* the requirements of the AWAR, and PHS Policy. The authority, composition, and functions of the IACUC are also described in the *Institutional Animal Care and Use Committee Guidebook*[1] (pp. 11–18). The Committee represents multiple interests, including those of the institution, the research community, the animals, and the public. It also serves as the self-monitoring agent of federal agencies and accrediting bodies, such as APHIS/AC, NIH/OLAW, and AAALAC.

3:2 Who is responsible for appointing the IACUC?

Reg. The Chief Executive Officer (CEO) is charged with appointing an IACUC (AWA §2143,b,1; AWAR §2.31,a; PHS Policy IV,A,3,a). The CEO is defined in PHS Policy (IV,A,3,a, footnote 5) as the highest operating official of the organization, such as the President of a university. PHS Policy allows the CEO to delegate the authority to appoint the IACUC but requires that such delegation be specific and in writing. The *Guide* (p. 12) requires institutions to comply with the AWAR and PHS Policy. (See 5:13.)

Opin. In larger institutions it is not unusual for the CEO to delegate authority for appointment of the IACUC to a senior administrator who is more directly responsible for the institution's research program, such as a Vice Chancellor, Dean, or Vice President for Research. This individual frequently serves as the organization's IO (PHS Policy III,G) and signs and submits documents such as the PHS Animal Welfare Assurance. It is important to note that the CEO and the IO do not have to be the same individual.

Surv. For practical purposes, who really appoints the IACUC at your institution (i.e., the person actually makes the choice, not the person who rubber stamps that choice)?

• The CEO	38/296 (12.8%)
• The Institutional Official (if other than the CEO)	97/296 (32.8%)
• The IACUC chairperson	63/296 (21.3%)
• The IACUC or an IACUC subcommittee makes the real decision	61/296 (20.6%)
• A dean or other administrative persons(s) who is neither the CEO nor the Institutional Official	15/296 (5.1%)
• An independent animal users committee	0 (0%)
• I don't know	1/296 (0.3%)
• Other*	21/296 (7.1%)

*"Other" included some of the following comments:

- Partially depends on individuals' willingness to serve on the IACUC; Committee and Chair may pass information on volunteers to the IO, who gets official approval from CEO
- Volunteers are requested, then approved by Provost
- The Chair seeks qualified candidates and offers a choice; candidates may also come from the Chair of a department, the dean of a college, or volunteers
- The IACUC Chair and the Associate VP for Research
- The IACUC
- The Faculty Steering Committee

- The President
- The IO in consultation with IACUC Administrator, Attending Veterinarian, and occasionally the IACUC Chair
- The Institute Director
- The IACUC, Veterinarian, and Director of Compliance
- The IACUC Chair in consultation with an IACUC subcommittee and the dean(s)
- The IACUC Chair in consultation with the Attending Veterinarian
- The Head of Research
- Divisional management (each division has 2 representatives)
- The corporate veterinarian makes recommendations to IO
- A "Committee of Management," including the IACUC Chair provides a list to the IO for consideration
- The center director
- The Board of Governors
- The Associate VP for Research
- The *ad hoc* Committee on Committees

3:3 For practical purposes, should the IACUC Chair be given the option of accepting or rejecting a proposed appointment to the committee?

Reg. Both the AWAR (§2.31,a) and the PHS Policy (IV,A,3,a) mandate that the IACUC be appointed by the CEO.

Opin. Allowing the IACUC Chair the option to accept or reject appointment of the committee members would be a violation of the AWAR and PHS Policy. However, input from the Chair, as well as the IACUC staff, the AV, and even members of upper management or the research community may be a useful method of informing the CEO of specific needs of the IACUC in order to ensure it possesses the expertise necessary to fulfill its responsibilities. In addition, many readers may be familiar with the tactic of appointing to the IACUC PIs who have been identified as being somewhat refractory with respect to animal welfare regulations, policies, and guidelines as a method for rendering them more tractable.

3:4 Are there specific membership requirements for an IACUC?

Reg. Yes, the AWAR (§2.31,b,2) state that the committee shall be composed of a Chair and at least two additional members. The AWAR (§2.31,b,3) also state that at least one member shall be a Doctor of Veterinary Medicine, with training or experience in laboratory animal science and medicine, and at least one member not affiliated in any way with the facility other than as a member of the Committee.

PHS Policy (IV,A,3,b), on the other hand, states that the committee shall include at least one veterinarian, one practicing scientist experienced in research involving animals, one member whose primary concerns are in a nonscientific area, and one individual who is not affiliated with the institution in any way and is not a member of the immediate family of a person who is affiliated with the institution. An individual may fulfill more than one requirement; however, no IACUC constituted under the PHS Policy may consist of fewer than five members (PHS Policy IV,A,3,b).

The *Guide* (pp. 24–25) and APHIS/AC Policy 15 echo these requirements and add that nonaffiliated members "should not be laboratory animal users…"

Opin. The final composition and number of members on an IACUC will depend on a number of factors, including the nature of the research and funding, the size of the institution, and the species involved.

3:5 What is the definition of the nonaffiliated member of the IACUC?

Reg. The nonaffiliated member is defined as an individual who represents general community interests in the proper care and use of animals, is not a laboratory animal user, is not affiliated with the institution, and is not an immediate family member of a person who is affiliated with the institution (AWA §2143,b,1,B; AWAR §2.31,b,3,ii; PHS Policy IV,A,3,b,4; *Guide* pp. 24–25; APHIS/AC Policy 15).[2] (See 5:31–5:38.)

3:6 How might IACUCs find individuals to serve as nonaffiliated members?

Opin. There are no specific regulations that prescribe how nonaffiliated members have to be selected. Nonaffiliated members may be found in a number of ways. Perhaps the most common source is through contacts of the CEO, members of the IACUC, AV, or other personnel in departments at the institution, such as the Public Relations office. Other useful sources for recruiting nonaffiliated members may include professional societies (e.g., lawyers, ethicists, clergy, teachers, librarians, health care professionals, police, firefighters), local humane organizations, state biomedical advocacy organizations, local veterinarians and nonprofit service organizations.

3:7 What is the suggested duration of IACUC membership for the Chair and committee members?

Opin. There are no specific regulations that require how long a member has to serve on the IACUC. Serving as an effective member of an IACUC is a difficult task. To perform their duties competently, the members of an IACUC must become familiar with a large body of regulations, policies, and guidelines. They also must learn about the history, special problems, research strengths, goals, objectives, and idiosyncrasies of their institution and its investigators. Experience has shown that a new IACUC member requires 6 months to a year to gain the experience required to function as a fully effective member of an IACUC. In light of this, most institutions ask members to serve for at least 2 years, and many for 3 years. Some institutions also ask capable and willing members to serve a second 2- or 3-year term, or longer. While NIH/OLAW does not address length of service for IACUC members, they do point out that if a required member (i.e., scientist, nonscientist, veterinarian, or unaffiliated) leaves the committee so that position is no longer filled, the IACUC is not properly constituted and may not conduct official business until a member who fulfills the required position is appointed.[3] To avoid vacancies in these critical membership positions, many institutions have more than one scientist, nonscientist, veterinarian, or nonaffiliated member serving on their IACUC.

Serving as an effective IACUC Chair requires all of the knowledge and expertise needed to become an effective committee member, plus leadership skills to manage the committee effectively and to interact with the institution's administration. For these reasons, the person selected to Chair the IACUC has usually served at least one term as a committee member. Many institutions ask the IACUC Chair to serve multiple terms in order to retain someone with a high level of regulatory and institutional knowledge in this position. (See Chapter 5 for additional information.)

Surv. What is the typical length of service of an IACUC member (non-chairperson) at your institution?

- We have not yet had any turnover 7/296 (2.4%)
- 1–2 years 9/296 (3.0%)
- 3–4 years 117/296 (39.5%)
- >4 years 160/296 (54.1%)
- I don't know 3/296 (1%)

3:8 What is the typical length of service for an IACUC Chairperson at your institution?

Opin. There are no regulations that require a specific term of service for the IACUC Chair (see 3:7). However, in addition to serving as an IACUC member, the Chair has the added responsibility of keeping abreast of regulatory trends and interpretations, ensuring that the IACUC carries out its required mandates, educating and supporting the IACUC members, serving as the IACUC spokesperson and leader, and communicating regularly with the IO, AV, IACUC Administrator and researchers. For these reasons, many institutions ask the IACUC Chair to serve several years in order to retain someone with a high level of regulatory and institutional knowledge in this position. This is evident in the survey results where the greatest percentage of responses (63%) indicated that the typical length of service for the IACUC Chair is more than 4 years.

Surv. What is the typical length of service for an IACUC chairperson at your institution?

- We have had only one chairperson 21/297 (7.1%)
- 1–2 years 17/297 (5.7%)
- 3–4 years 68/297 (22.9%)
- Over 4 years 187/297 (63.0%)
- I don't know 4/297 (1.3%)

3:9 What is the faculty rank of your chairperson? Include persons with annotated titles such as "Clinical Assistant Professor."

Opin. There are no specific regulations requiring a particular faculty rank for the IACUC chairperson. The majority (57.2%) of survey respondents indicated that the Chair was at the Associate or Full Professor level. This suggests that most academic institutions believe the Chair should have sufficient stature to effectively lead the IACUC. In addition, many academic institutions do not allow non-tenured faculty members to serve on the IACUC, in order to avoid overburdening them in their early careers or negatively impacting their progress to achieving tenure.

Surv. What is the faculty rank of your chairperson? Include persons with annotated titles such as "Clinical Assistant Professor."

- Not applicable as we are not an academic institution 92/297 (31.0%)
- Instructor or assistant professor 19/297 (6.4%)
- Associate professor 61/297 (20.5%)
- Full professor 109/297 (36.7%)
- I don't know 0/297 (0%)
- We are academic but our chair does not have a faculty title 16/297 (5.4%)

3:10 Does your institution provide compensation of any kind (other than reimbursed parking fees, lunch during a meeting, or similar minor compensation) to either the IACUC chairperson or IACUC members? (See 5:9.)

Opin. There are no specific regulations regarding compensation for IACUC members. However, APHIS/AC has stated that "compensation of the nonaffiliated member is permissible only when it does not jeopardize the member's status as a nonaffiliated member… The dollar amount of compensation, if any, should not be so substantial as to be considered an important source of income or to influence voting on the IACUC."[2] NIH/OLAW provides similar guidance in its FAQ B.11, as follows: "Nominal compensation for service on the IACUC, or reimbursement for expenses such as parking and travel costs, is generally not viewed as jeopardizing the nonaffiliated status of a member. Any compensation for participation should not be so substantial as to influence voting or reflect an important source of income. It is acceptable for the institution to pay for IACUC training of nonaffiliated members (e.g., attendance at IACUC 101)."[4] Lastly, the *Guide* (pp. 24–25) echoes this sentiment: "The public member may receive compensation for participation and ancillary expenses (e.g., meals, parking, travel), but the amount should be sufficiently modest that it does not become a substantial source of income and thus risk compromising the member's association with the community and public at large."

The majority of survey respondents (65.4%) indicated that their IACUC members, including the Chair are *not* compensated. Only 12% indicated that the Chair received direct compensation, while 6% indicated their IACUC members received direct compensation. This direct compensation varied and was described as a "nominal" or "small" fee/stipend/honorarium, or items such as free parking, meals, travel (mileage) reimbursement and educational support, none of which appear to be substantial enough to influence voting or be an important source of income.

Surv. Does your institution provide compensation of any kind (other than reimbursed parking fees, lunch during a meeting, or similar minor compensation) to either the IACUC chairperson or IACUC members? Check as many responses as appropriate.

• I don't know about the compensation to the chairperson	0/297 (0%)
• I don't know about the compensation to the IACUC members	2/297 (0.4%)
• No, the chairperson is not compensated	155/297 (30.1%)
• No, IACUC members are not compensated	182/297 (35.3%)
• Yes, the chairperson receives direct compensation	62/297 (12.0%)
• Yes, the chairperson receives indirect compensation to his/her home department	23/297 (4.5%)
• Yes, IACUC members receive direct compensation	31/297 (6.0%)
• Yes, IACUC members receive indirect compensation via a credit to their home department	5/297 (1.0%)
• The method of compensation is variable*	55/297 (10.7%)

*Variable methods of compensation include the following:

- The IACUC Chair gets salary support.
- The IACUC Chair is employed for the position.
- The previous Chair received salary compensation but the current Chair does not.
- The Chair and Vice Chair receive salary support; other members receive no compensation.
- Prior Chairs have been compensated but the current chair is not. Faculty members are not compensated.
- IACUC Chair is a salaried, FTE position. The Chair performs few other duties, and has no staff.
- 50% of Chair's salary is covered (the home department pays the other 50%).
- The Chair is relieved of some teaching assignments during the time serving as the Chair.
- The chairperson receives a salary supplement.
- Compensation is restricted to approved administrative costs for the Chair's research.
- Small percent of salary is covered.
- Administrative salary supplement.
- Members are provided support for attendence at outside meetings on IACUC functions (e.g., educational support for conferences).
- Compensation takes the form of an award.
- Only the veterinarian is compensated.
- The veterinarian and nonaffiliated member receive nominal compensation.
- Nonaffiliated members get a small stipend for their time. Breakfast is provided for IACUC meetings. We have been trying to secure release time for other IACUC members but budgets are tight.
- The nonaffiliated member receives a small honoraria/stipend.
- The nonaffiliated member receives parking and a small stipend.

- Nonaffiliated members are reimbursed for mileage to attend meetings and receive free parking.
- Nonaffiliated members are compensated for time of service, meals and travel expenses.
- Only nonaffiliated members are compensated to account for the fact that they are not eligible to participate in the occupational health and safety program.
- Compensation to nonaffiliated member only, American Express gift cards (typically $25–100/visit).

3:11 Are institutions required to train its IACUC members?

Reg. Both the AWAR and PHS Policy are silent on a specific method or type of training that should be provided to IACUC members to ensure they are qualified. However, the AWAR (§2.31,a) and the PHS Policy (IV,A,3,a) state that the IACUC must be qualified through the experience and expertise of its members to assess the research facility's animal program, facilities, and procedures, while the AWA (§2143,b) states that members shall possess sufficient ability to assess animal care, treatment and practices in experimental research. In addition, the AWAR (§2.32,a) states "It shall be the responsibility of the research facility to ensure that all scientists, research technicians, animal technicians, and other personnel involved in animal care, treatment, and use are qualified to perform their duties." IACUC members are "involved in animal care, treatment, and use" because they oversee it, so they are included in the training requirements. This is clarified in APHIS/AC Policy 15, which states "IACUC members must be qualified to assess the research facility's animal program, facilities and procedures. The research facility is responsible for ensuring their qualification, and this responsibility is filled in part through the provision of training and instruction. For example, IACUC members should be trained in understanding the Animal Welfare Act, protocol review, and facility inspections."

Recognizing the complexity of the responsibilities that IACUC members are asked to fulfill, the *Guide* (p. 17) states, "It is the institution's responsibility to ensure that IACUC members are provided with training opportunities to understand their work and role. Such training should include formal orientation to introduce new members to the institution's Program; relevant legislation, regulations, guidelines and policies; animal facilities and laboratories where animal use occurs; and the processes of animal protocol and program review. Ongoing opportunities to enhance their understanding of animal care and use in science should also be provided." The *Guide* (p. 15) also states "All Program personnel training should be documented." Accordingly, PHS-Assured and AAALAC-accredited institutions are expected to provide adequate training for IACUC members and to document that such training has been provided.

Opin. There are many ways that institutions can train IACUC members. A common method is to prepare an IACUC member's handbook that contains copies of relevant regulations, policies, and guidelines, as well as copies of the institution's forms, guidelines, and SOPs. New members typically have one or more orientation sessions under the guidance of the IACUC staff, the Chair, or the AV. New committee members may also be assigned a mentor who is a more experienced member of the IACUC and they may be invited to attend one or two IACUC meetings as observers before their membership term starts. In addition, members may attend an ever-growing number of meetings, workshops and webinars that address animal care and use issues of importance to IACUCs. Other useful information for training new IACUC members includes a tutorial on the PHS Policy (http://grants.nih.gov/grants/olaw/tutorial/index.htm.)

AALAS sponsors a variety of online animal care and use courses in the AALAS Learning Library[5] which is an excellent source of training materials for IACUC members. The IACUC 101™ series is a nonprofit educational organization that presents programs designed to provide IACUC members, administrators, veterinarians, animal care staff, researchers, regulatory personnel, compliance officers, and others with information on the role and responsibilities of IACUCs and an understanding of federal policies and regulations governing the care and use of animals in research and education.[6] Public Responsibility in Medicine and Research (PRIM&R) offers webinars throughout the year and other educational opportunities in conjunction with its annual IACUC conference. The Scientists Center for Animal Welfare (SCAW)[7] offers an Educational CD that includes a variety of resources useful for IACUC members.* SCAW[8] also sponsors an educational Winter Conference and other IACUC Training Workshops each year. The Massachusetts Society for Medical Research (MSMR)[9] offers on-line training for IACUC community members that is also good adjunct training for IACUC members and administrators. The Collaborative Institutional Training Initiative (CITI) offers member institutions an on-line "Essentials for IACUC Members" course.[10]

Several Institute for Laboratory Animal Research (ILAR) publications also contain information of value to IACUC members.[11–16] *Lab Animal*, a peer-reviewed journal for professionals in animal research, publishes an IACUC Resources link,[17] as well a monthly *Protocol Review* column useful for training IACUC members.

3:12 May an institution have more than one IACUC?

Reg. The AWA (§2143,b,1) states, "The Secretary shall require that each research facility establish at least one committee." The AWAR (§2.31,a), which implement the AWA, read, "The Chief Executive Officer of the research facility shall appoint an Institutional Animal Care and Use Committee (IACUC)," language that is echoed in PHS Policy (IV,A,3,a).

* Includes: *Public Health Service Policy on Humane Care and Use of Laboratory Animals; 8th edition of the Guide for the Care and Use of Laboratory Animals; Institutional Animal Care and Use Committee Guidebook*; USDA/APHIS/ Animal Welfare Regulations; USDA/APHIS/Animal Welfare Act; USDA/APHIS/Animal Care Policy Manual; AVMA *Guidelines on the Euthanasia of Animals*; ACLAM Education and Training Guidelines Website; ACLAM Adequate Veterinary Care Guidelines, Veterinary Medical Program (published by ILAR); Agriculture Guidelines (FASS); *Guidelines for the Care and Use of Mammals in Neuroscience and Behavioral Research* (ILAR Publication); *Guidelines for the Capture, Handling, and Care of Mammals (2007); Guidelines for Use of Live Amphibians and Reptiles in Field and Laboratory Research* (Second Edition, 2004); *Guidelines for the Use of Fishes in Research* (2004 American Fisheries Society); *Guidelines for the Use of Wild Birds in Research* (The Ornithological Council); Federal Insecticide Fungicide, Rodenticide Act (EPA); VSTA USDA Biologics Act (FDA); Information Resources for IACUC's 1985-2000; NICEATM-ICCVAM Five-Year Plan; Working at Animal Biosafety Levels 1, 2 and 3; List of Acronyms; Canadian Council on Animal Care: Guidelines on *The Care and Use of Fish in Research, Teaching and Testing; Guidelines on: The Care and Use of Wildlife*; Manual for Community Representatives.

Opin. There is no regulatory language that precludes establishment of multiple IACUCs at one institution. The statements above leave the door open for appointing more than one committee at an institution. In practice, a number of institutions have opted to appoint more than one IACUC, and their systems have been approved by both NIH/OLAW and APHIS/AC. These institutions have separate physical areas of jurisdiction in their animal care and use program. Separate jurisdiction prevents overlap of oversight among the committees, thereby preventing conflict. Institutions with multiple IACUCs should contact their regional APHIS/AC office for guidance and those with PHS funding should contact NIH/OLAW for advice on how to describe their IACUC structure and function in their Assurance. Multiple IACUCs are found most commonly at large institutions that have multiple campuses or locations, diverse scientific missions, many animal facilities, and or large numbers of investigators and protocols.

3:13 Under what circumstances might multiple IACUCs be advantageous?

Opin. Multiple IACUCs may be advantageous at large institutions where the sheer volume of research activity exceeds the capacity of a single IACUC to review the associated protocols and program in an effective and timely manner. Dividing the work among multiple IACUCs, each appointed to oversee a discrete portion of the research program, may be an optional way to handle the workload. An added advantage of dividing the workload is that it allows committees to gain expertise in reviewing protocols in particular research areas, since they usually are appointed within academic units or for coverage of specific research disciplines. Existence of more localized IACUCs also makes it easier for IACUC members to maintain a personal relationship with investigators and a sense of ownership and pride in the portion of the program they oversee.

Another situation in which the existence of more than one IACUC may be advantageous is in institutions that have disparate research programs, such as those with Colleges of Medicine, Veterinary Medicine, and Agriculture. In others, it occurs when the research program is physically divided between two locations. In both situations, the organization may find it effective to appoint individual IACUCs to review protocols from and oversee the separate components of its overall program. In these cases, it is important that the responsibility for the ongoing review (e.g., amendment, annual or triennial review) of an individual protocol remain with the original IACUC that approved it.

3:14 What might be the disadvantages of multiple IACUCs?

Opin. A major challenge for a multiple IACUC system is to develop mechanisms for maintaining consistency and quality in program review, facility inspection, and protocol review across the institution. If there is not an overarching coordinating group, the institution may become fragmented and inconsistent, or conflicting practices may arise. However, one IACUC may not have authority over another IACUC. Each IACUC must report directly to the IO, as per the HREA and AWAR (§2.31,c,3). Therefore the overarching group must carefuly confine its activities to policy guidance and may not assume the functions of the IACUC as described in the PHS Policy (IV,B,1- 8) and AWAR (§2.31,c). Another disadvantage is the need for increased IACUC staffing and funding to sustain the necessarily redundant components of multiple committees. The cost of maintaining an IACUC may be considerable and, in the absence of centralized support, can place a large burden on smaller administrative units such as colleges or departments.

Another potential problem with the multiple IACUC system is that each committee must meet all the membership requirements for an individual IACUC (e.g., veterinarian, nonaffiliated member, nonscientist), including the requirement that no more than three members from the same administrative unit may serve on the IACUC (AWAR §2.31,b,4; *Guide* p. 25). It can be difficult to appoint a properly constituted committee from within a small administrative unit. Moreover, difficulties can arise with locally constituted IACUCs because no member may participate in the IACUC review or approval of a research project in which the member has a conflicting interest, nor can any member who has a conflicting interest contribute to the constitution of a quorum during the review of the activity in which the person has a conflict (PHS Policy IV,C,2; AWAR §2.31,d,2). Even when there is no obvious conflict of interest, it is important to recognize that there can appear to be a conflict when IACUC members drawn from a small administrative unit review each other's protocols. Finally, when IACUC members are drawn from a limited pool, individuals may find that they have an excessive time commitment to the IACUC. A number of the difficulties cited, in particular issues related to consistency of actions, can be mitigated by having one or more of the same individuals serve on each IACUC and/or by having an experienced IACUC staff that supports all the IACUCs.

3:15 Should multiple IACUCs all report to a central IACUC, advisory committee, or other authority?

Opin. No. All IACUCs must have a direct reporting channel to the IO according to PHS Policy (IV,B,3; IV,B,5) and the AWAR (§2.31,c,3). The IO must be the same individual for all IACUCs under one NIH/OLAW Assurance. Parts of the organization that have different Assurances could have different IOs.

3:16 What are some methods for multiple IACUCs to effectively interact?

Opin. In institutions where multiple IACUCs report to a single IO, the IO may appoint an IACUC advisory committee to address general issues that affect the entire animal care and use program. The committee could consist of the IO or the IO's designee, the Chairs of the individual IACUCs, the AV, and the IACUC Administrator. Some simple strategies for multiple IACUCs to effectively interact can include:

- Having some of the same members serve on more than one committee;
- Using a standardized animal use protocol form;
- Sharing the development/modification of common guidelines, SOPs, procedures and policies;
- Sharing an IACUC administrator and/or IACUC staff;
- Developing a common IACUC training program;
- Coordination of an AAALAC site visit, including sharing responsibility for the Program Description;
- Developing and maintaining centralized information resources for all the IACUCs.

3:17 What strategies can be used to ensure that protocol review is consistent throughout an institution with multiple IACUCs?

Opin. The most important step in ensuring consistent protocol review is to develop a single animal use protocol form that can be used throughout the institution. The form should be designed to allow sufficient flexibility in describing proposed procedures, so as to encompass the full range of activities at the institution (e.g., antibody production, survival surgery, product safety testing, field research, breeding) while requiring all investigators to provide mandated information for proposed uses of animals (e.g., rationale for animal use, justification for animal species and numbers, alternatives to painful procedures, humane end points). NIH/OLAW has posted a comprehensive animal study proposal sample document on their website at http://grants.nih.gov/grants/olaw/sampledoc/animal_study_prop.htm. AALAS's IACUC.org has posted a variety of sample protocol forms on its website at http://www.iacuc.org/protocol.htm. The sample forms from both of these resources may be downloaded and customized by institutions for use in their program.

Another important strategy to ensure consistent protocol review is for each IACUC to share the same guidelines, SOPs, and policies, especially those that pertain to the protocol review process and special considerations for IACUC review, such as those described in the *Guide* (pp. 27–31, experimental and humane endpoints, physical restraint, multiple survival surgical procedures, food and fluid regulation, and use of non-pharmaceutical-grade chemicals).

It is also important for the IO to ensure that the membership of each IACUC not only meets the minimal regulatory requirements, but includes individuals who have sufficient expertise to competently evaluate the protocols they review. Each committee needs to have scientists familiar with the kinds of animal use they will be asked to review and veterinarians knowledgeable in the health and husbandry of the proposed animal species. If a committee is asked to review a protocol in an area outside of the expertise of its members, it may use non-voting ad hoc consultants to assist in its evaluations. Depending on the nature of the activities to be reviewed, sufficient technical expertise must be available for each IACUC to evaluate potential risks associated with protocols (e.g., biosafety, radiation safety, occupational health) and to ensure that appropriate safeguards are proposed to mitigate these risks. This can be accomplished by including a health and safety specialist on every committee or by providing a centralized health and safety resource for use by all IACUCs.

Lastly, a formal, coordinated IACUC member training program for the multiple IACUCs, using methods described in 3:11 will enhance consistency of protocol review.

3:18 What mechanism might be used to assure consistent IACUC semiannual inspection and review of animal care facilities and programs at institutions that have multiple IACUCs?

Opin. One way for institutions that have multiple IACUCs to ensure that semiannual inspections and reviews are consistent is to utilize an institution-wide committee composed of at least two members from each individual IACUC to perform these functions. However, a majority of the members of each individual IACUC would have to review and sign the report for it to be a valid semiannual report under the AWAR. Also, any individual IACUC member wanting to participate in the inspections or initial review could not be excluded. The institution could also have an institution-wide committee to review the reports of the individual IACUCs that perform semiannual reviews and to make recommendations to improve consistency. However, the content of the semiannual report is the responsibility of the IACUC with jurisdiction over that area. An institution-wide committee may not conduct oversight editing or influence the content of the reports. The AWAR (§2.31,c,3) require the IACUC to perform this function. The development of institution-wide forms to be used in conducting inspections and preparing reviews also can help assure consistency. One tool to assist IACUCs in conducting semiannual facility and program reviews is the NIH/OLAW *Semiannual Program Review and Facility Inspection Checklist.*[18] Another strategy is for personnel from a central unit (e.g., AV, institutional compliance officer) to participate as ad hoc consultants in the facility inspections and program reviews conducted by each IACUC and participate in all IACUCs to ensure consistency. The PHS Policy (IV,B,3, footnote 8) states that the IACUC "may, at its discretion… invite *ad hoc* consultants to assist in conduction the evaluation. However the IACUC remains responsible for the evaluation and report." The AWAR (§2.31,c,3) has similar wording. Again, a formal, coordinated IACUC member training program will ensure all members understand the semiannual inspection and program review process.

Lastly, institutions that require all of their components to be accredited by AAALAC should have an independent method for ensuring uniform compliance with applicable regulations, policies, and guidelines.

3:19 What is the least common denominator (e.g., school, college, department, individual investigator) around which an IACUC can form?

Opin. Institutions with multiple IACUCs could form them around:

- Discrete units within the institution's administrative structure (e.g., medical school, veterinary school, graduate school, college of agriculture)
- The nature of the proposed research (e.g., agricultural, biomedical, field research, product testing)
- The animal species being used (e.g., farm animals, laboratory animals, nonhuman primates, exotic species)
- Geographically separate facilities (i.e., if an institution has more than one IACUC, each IACUC must oversee different physical areas. See 3:12.)

3:20 What needs to be considered when selecting the size of a unit upon which to base an IACUC?

Opin. When selecting the size of a unit upon which to base an IACUC, consideration needs to be given to:

- The potential workload of the IACUC (e.g., number of protocols per year; number, size, and location of animal facilities)

- The distance of the proposed IACUC from the IO in the reporting lines of the institution
- Whether the unit has the resources to support an IACUC
- Whether the unit has sufficient breadth of animal use to prevent frequent potential conflicts of interest within the IACUC
- Whether the pool of animal users from which IACUC members can be drawn is sufficiently large
- The AWAR requirement (§2.31,b,4) that no more than three members of the IACUC can be from the same administrative unit. The *Guide* (p. 25) echoes this language.

Surv. How many regular voting members are on your IACUC? (See 5:1.)

- 3–5 37/297 (12.4%)
- 6–10 158/297 (53.2%)
- 11–15 76/297 (25.6%)
- 16–20 18/297 (6.1%)
- >20 8/297 (2.7%)
- I don't know 0/297 (0%)

3:21 Can a small institution without an IACUC use the services of a nearby institution with an IACUC to review and approve research supported by a PHS grant?

Reg. PHS Policy (IV,A,3) stipulates that the IACUC must be appointed by the CEO (or the CEO's designee) and mandates the number and qualifications of the IACUC members (see Chapter 5). The AWAR (§2.31,a; §2.31,b) essentially echoes PHS Policy, although the minimal number of IACUC members required differs between the two. Neither the PHS Policy nor the AWAR stipulate that any of the *non*-veterinary IACUC members must also be employees of, or in any other way affiliated with, the institution (see 5:19). However, PHS Policy (IV,A,3,b) and the AWAR (§2.31,b,3,i) state that the IACUC veterinarian must have "direct or delegated program authority and responsibility for activities involving animals at the institution (PHS Policy) [or] research facility (AWAR)." The AWAR (§2.33,a,1) further stipulate, "Each research facility shall employ an attending veterinarian under formal arrangements. In the case of a part-time attending veterinarian or consultant arrangements, the formal arrangements shall include a written program of veterinary care and regularly scheduled visits to the research facility." The AWAR note (§2.33,a,3), "The attending veterinarian shall be a voting member of the IACUC; Provided however, that a research facility with more than one Doctor of Veterinary Medicine (DVM) may appoint to the IACUC another DVM with delegated program responsibility for activities involving animals at the research facility." Per PHS Policy (IV,A,3,B) the veterinarian with direct or delegated program authority, named in the Assurance as such, is a member of the IACUC.

Opin. Although there are some potential pitfalls to this approach (e.g., see 5:19), it might be worth considering in certain situations, such as when a small institution with only a few animal users has as its neighbor a larger institution with a more extensive and experienced animal care and use program. In such a case, the larger institution could agree to include the smaller institution under its NIH/OLAW Assurance if the smaller institution is receiving PHS funding. The CEO (or the IO, if the authority to appoint the IACUC has been delegated) of the smaller institution could agree to appoint the larger institution's AV to serve as the smaller institution's AV and the larger institution's IACUC to serve as the smaller institution's IACUC, in each case with the proper program responsibilities. Other variations of this approach are also possible, but in each situation care must be taken by both institutions to ensure that the arrangements they make do not compromise their overall compliance with the requirements of the PHS Policy and the AWAR regarding the responsibilities of the IACUC, AV, CEO, and IO (see also Chapter 5). Institutions considering any such relationship are well advised to consult representatives of NIH/OLAW and APHIS/AC about the appropriateness of their proposed arrangements.

3:22 How might an IACUC be effectively constituted at a very small institution where all or most personnel are likely to be directly or indirectly involved with the research to be reviewed?

Opin. As pointed out in 5:31, the use of the use of persons not employed by the institution as members on the IACUC is compatible with AWA and PHS Policy. In this situation, supplementing the affiliated members with a number of nonaffiliated members whose expertise is well suited to the needs of the institution would prevent the problems inherent in trying to populate an IACUC with the institution's entire staff of animal users.

References

1. Office of Laboratory Animal Welfare, Applied Research Ethics National Association. 2002. Institutional Animal Care and Use Committee Guidebook. Bethesda: National Institutes of Health. http://grants.nih.gov/grants/olaw/GuideBook.pdf.
2. U.S. Department of Agriculture. 2011. Animal Care Policy Manual. Policy 15. Institutional Official and IACUC Membership. http://www.aphis.usda.gov/animal_welfare/policy.php.
3. National Institutes of Health, Office of Laboratory Animal Welfare. 2013. Frequently Asked Questions B.1, What are the IACUC membership criteria? http://grants.nih.gov/grants/olaw/faqs.htm#IACUC_1.
4. National Institutes of Health, Office of Laboratory Animal Welfare. 2013. Frequently Asked Questions B.11, May the institution pay or reimburse expenses incurred by nonaffiliated members? http://grants.nih.gov/grants/olaw/faqs.htm#IACUC_11.
5. American Association for Laboratory Animal Science. AALAS Learning Library. https://www.aalaslearninglibrary.org. Accessed Jan. 2, 2013.
6. IACUC 101™ Series. http://iacuc101.org/. Accessed Jan. 2, 2013.
7. Scientists Center for Animal Welfare. Publications & Training Materials. http://www.scaw.com/publications--training-materials/. Accessed July 9, 2013.
8. Scientists Center for Animal Welfare, Conferences & Workshops. http://www.scaw.com/conferences--workshops/. Accessed July 9, 2013.
9. Massachusetts Society for Medical Research. *Introduction to Institutional Animal Care and Use* Committees on-line training. http://www.msmr.org/index.html. Accessed Jan. 2, 2013.

10. Collaborative Institutional Training Initiative. https://www.citiprogram.org/. Accessed July 9, 2013.
11. van Zutphen, B. 2007. Education and Training for the Care and Use of Laboratory Animals: An Overview of Current Practices. *ILAR J.* 48(2):72–74.
12. Dobrovolny, J., J. Stevens, and L.V. Medina. 2007. Training in the laboratory animal science community: Strategies to support adult learning. *ILAR J.* 48(2):75–89.
13. Anderson, L.C. 2007. Institutional and IACUC responsibilities for animal care and use education and training programs. *ILAR J.* 48(2):90–95.
14. Medina, L.V., K. Hrapkiewicz, M. Tear et al. 2007. Fundamental training for individuals involved in the care and use of laboratory animals: A review and update of the 1991 NRC Core Training Module. *ILAR J.* 48(2):96–108.
15. Greene, M.E., M.E. Pitts, and M.L. James. 2007. Training Strategies for Institutional Animal Care and Use Committee (IACUC) Members and the Institutional Official (IO). *ILAR J.* 48(2):131–142.
16. Foshay, W.R. and P.T. Tinkey. 2007. Evaluating the Effectiveness of Training Strategies: Performance Goals and Testing. *ILAR J.* 48(2):156–162.
17. *Lab Animal (NY).* IACUC Resources. http://www.labanimal.com/laban/links/RL13.html. Accessed July 9, 2013.
18. National Institutes of Health, Office of Laboratory Animal Welfare. Semiannual Program Review and Facility Inspection Checklist. http://grants.nih.gov/grants/olaw/sampledoc/cheklist.htm. Accessed July 9, 2013.

4

Reporting Lines of the IACUC

Marcy Brown and Mary Jo Shepherd

Introduction

Federal law, regulation, and policy stipulate that the IACUC must report to the individual who has been designated to serve as the IO. This chapter clarifies what it means to be the IO by reviewing this individual's responsibilities under the PHS Policy, the AWAR and the *Guide*. Examples of common job titles for IOs in various types of institutions are used to provide the reader with a sense of the level of authority that this person should have within an institution. The chapter concludes with brief discussions of useful methods for educating the IO about the functions of its IACUC. The chapter also provides suggestions on which individuals may contact NIH/OLAW, APHIS/AC, or AAALAC for guidance on IACUC related issues.

4:1 To what person is the IACUC required to report?

Reg. The IACUC must report to the institution's IO. According to PHS Policy (IV,B,3; IV,B,5) and the AWAR (§2.31,c,3–5), the IACUC is required to submit reports of its semiannual program evaluations and facility inspections to the IO and to make recommendations to the IO regarding any aspect of the institution's program, facilities, or personnel training. PHS Policy (IV,F,1; IV,F,2) requires that the IACUC, through the IO, submit an annual report to NIH/OLAW. This report must describe any changes in the program, facilities, or IACUC membership and list the dates when the IACUC conducted its semiannual evaluations of the institution's program and facilities and submitted the evaluations to the IO. In addition, PHS Policy (IV,C,7; IV,F,3) requires the IACUC, through the IO, promptly to provide NIH/OLAW with a full explanation of the circumstances and actions taken with respect to any serious or continuing noncompliance with PHS Policy, any serious deviation from the *Guide*, or any IACUC suspension of an activity. In its *Guidance on Prompt Reporting to OLAW under the PHS Policy on Humane Care and Use of Laboratory Animals*[1] NIH/OLAW provides examples of situations that must be reported by the IACUC, through the IO, including the "failure to correct deficiencies identified during the semiannual evaluation in a timely manner."

The AWAR (§2.31,c,3) require the IACUC to report through the IO to APHIS/AC and any federal funding agency, in writing, within 15 days whenever the institution fails to adhere to a reasonable plan and schedule for correcting a significant deficiency in the institution's program or facilities (see 25:12; 26:16). In addition, if the IACUC suspends an activity involving animals, PHS Policy (IV,C,7) and the

AWAR (§2.31,d,7) require that the IO, in consultation with the IACUC, review the reasons for the suspension, take appropriate corrective action, and report that action with a full explanation to NIH/OLAW and/or APHIS/AC and any federal agency funding the project if the animal activity is regulated under the AWAR.

4:2 How is the Institutional Official defined?

Reg. The AWAR (§1.1, Institutional Official) state that the IO is the individual at a research facility who is authorized legally to commit on behalf of the research facility that it will meet the requirements of the AWAR. Similarly, PHS Policy (III,G) defines the IO as the individual who signs and has the authority to sign the institution's Assurance, which commits the institution to meet the requirements of PHS Policy. If not done by the Chief Executive Officer (CEO), the IO signs and submits the annual report to APHIS/AC (§2.36,a). The CEO may or may not be the IO as defined by PHS Policy (IV,A,3,a, footnote 5). The *Guide* (pp. 13–14) notes that the IO is "The individual who, as a representative of senior administration, bears ultimate responsibility for the Program and is responsible for resource planning and ensuring alignment of Program goals with the institution's mission."

Opin. NIH/OLAW FAQ G.5[2] also describes the differences between the CEO and IO and states that "In some institutions, the IO and the CEO may be one and the same..." If the CEO does not designate an IO, it is the authors' opinion that the CEO becomes the IO by default and must be identified as such on the NIH/OLAW Assurance and any other pertinent documents. NIH/OLAW will not approve an Assurance without the signature of an individual identified as the IO; the CEO or IO also must sign and submit the annual report to APHIS/AC (AWAR §2.36,a).

4:3 What are the responsibilities of the IO?

Reg. The IO is the individual who is responsible for ensuring that an institution complies with all applicable animal welfare laws, regulations, and policies. The IO signs forms, reports, and letters on behalf of the institution and interacts with the IACUC in overseeing the institution's animal care and use program.

According to PHS Policy, the IO makes commitments on behalf of the institution that the requirements of the Policy will be met:

- Signs and has the authority to sign the Assurance (III,G)
- Receives semiannual evaluation reports and recommendations from the IACUC (IV,B,3; IV,B,5), including minority views (IV,E,1,d)
- In consultation with the IACUC, determines whether deficiencies are significant or minor (IV,B,3)
- Receives notification of the IACUC's decision to approve or withhold its approval of animal activities (IV,C,4)
- Receives and submits annual reports to NIH/OLAW (IV,F)
- Consults the IACUC regarding suspensions and corrective actions; reports to regulatory and funding agencies (IV,C,7)

- May subject protocols that have been approved by the IACUC to further review and approval but may not approve an activity that has not been approved by the IACUC (IV,C,8)
- Ensures that the institution maintains required records for the specified period (IV,E)

According to the AWAR (§1.1, Institutional Official), the IO legally commits the institution to meet the requirements of the AWAR as follows:

- Signs and submits the registration form (§2.30,a)
- If not done by the CEO, signs and submits the annual report to APHIS/AC (§2.36,a)
- Receives semiannual evaluation reports along with recommendations and minority views, from the IACUC (§2.31,c,3; §2.35,a,3)
- In consultation with the IACUC, determines whether deficiencies are significant or minor (§2.31,c,3)
- Reports to APHIS/AC and any federal agency funding the research, in writing, within 15 business days, any significant deficiency the IACUC found uncorrected within the prescribed time that was indicated on the semiannual inspection report (§2.31,c,3)
- Consults with the IACUC regarding suspensions and corrective actions; reports to regulatory and funding agencies (§2.31,d,7)
- May subject protocols that have been approved by the IACUC to further review and approval but may not approve an activity that has not been approved by the IACUC (§2.31,d,8)

4:4 Who can serve as the IO if that role is delegated by the Chief Executive Officer?

Reg. (See 4:2.)

Opin. The AWAR and PHS Policy describe no restrictions nor do they stipulate any minimum qualifications for individuals who serve as IOs. However, to be effective, an IO must have the level of authority described in 4:6. (See 3:2; 4:5.)

4:5 Who typically fills the role of IO in academia and in industry?

Opin. Some common titles of individuals who serve as IOs in academia and industry include Chancellor; President; Provost; Vice Provost; Vice Chancellor for Health Affairs; Vice Chancellor, Administration; Vice Chancellor for Academic Affairs; Vice Chancellor for Research; Executive Director, Research Resources; Director, Research Administration; Director of Sponsored Research; Dean, School of Medicine; Associate Dean, Research and Sponsored Programs; Chief Executive Officer; Executive Vice President for Health Affairs; Senior Vice President and Vice Provost for Health Affairs; Vice President for Health Affairs; Vice President for Research; Vice President for Research and Graduate Studies; Vice President for Clinical Affairs; Vice President for Worldwide Toxicology.

4:6 What institutional power or authority should the IO have?

Reg. (See 4:2.)

Opin. The IO must be authorized to legally commit on behalf of the institution that the requirements of the AWAR and/or PHS Policy will be met. To be effective, the IO must have sufficient administrative authority to promulgate, implement, and enforce policies across departmental lines. In addition, the IO must have sufficient fiscal authority to approve and fund a level of staffing and resources that are adequate to meet "the goals of quality animal care and use" (*Guide* p. 14) and any needed program improvements, facility repairs, and renovations. (See 4:3.)

4:7 How involved does the IO need to be in the general activities of the IACUC?

Opin. The IO's level of involvement in the activities of the IACUC varies widely from institution to institution. Some factors that can influence the IO's involvement include the size and complexity of the institution's research program, whether the IO has a background in biomedical research, and the management styles of the organization and the IO. In some institutions, the IO or a representative from the IO's office actually serves as a member of the IACUC and is directly involved in the general activities of the committee. However, NIH/OLAW recommends that the IO not serve on the IACUC, since the IACUC reports to the IO.[3] At a minimum, the IO must understand the functions of the IACUC as they are defined by the AWAR and PHS Policy. Furthermore, IOs that are organizationally distant from the IACUC must ensure that there is a mechanism by which they are promptly informed of any potential threats to animal welfare, or any violations of the AWAR, PHS Policy, the *Guide*, or the institution's PHS Assurance.

4:8 What means might be useful for educating the IO with respect to the functions of the IACUC and related policies of the institution?

Opin. The IO should be aware of the laws and regulations governing the use of animals in research, teaching or testing, understand and support the key components of the animal care and use program, and know where to find information when needed. Training of the IO may be provided by the IACUC Chair, the IACUC Administrator and/or the AV and may be formal, informal, or a combination of both. Training should include information on the specific IO responsibilities as well as IACUC and AV responsibilities, a broad overview of the program and key reference documents. Useful information regarding training for both the IACUC and the IO may be found in the article *Training Strategies for Institutional Animal Care and Use Committee (IACUC) Members and the Institutional Official (IO.)*[4] The NIH/OLAW online tutorial on the PHS Policy is also a useful training resource.[5] Finally, NIH/OLAW conducts a free online webinar series to help both IACUCs and IOs "explore their responsibilities in the oversight of PHS-funded research that involves the use of live vertebrate animals."[6] Webinar recordings are available to the public through the NIH/OLAW website. (See also 3:11). Numerous workshops and meetings that provide IACUC training to IOs, members and administrators are also listed on the NIH/OLAW website.

4:9 Is it appropriate for a research facility to establish guidelines on who may contact either NIH/OLAW or APHIS/AC for guidance on IACUC- related issues, or must this always be the IO?

Opin. It is wise for institutions to have designated spokespersons for communicating with external entities such as federal regulating agencies and the media. Such designations help to prevent inadvertent errors and miscommunications. In situations in which an institution is making an official report of an animal welfare incident to APHIS/AC, NIH/OLAW, or AAALAC, the AWAR (§2.31,d,7), PHS Policy (IV,F), and AAALAC's Rules of Accreditation (Sect. 2)[7] stipulate the mechanisms for submitting a formal report. Adherence to these requirements is, of course, not optional. Efforts to designate official spokespersons must not, however, infringe upon the rights granted to individuals to independently report alleged animal welfare incidents under institutional, state, and federal whistleblower and whistleblower protection policies.

References

1. National Institutes of Health, Office of Laboratory Animal Welfare. 2005. Notice NOT-OD-05-034, Guidance on Prompt Reporting to OLAW under the PHS Policy on Humane Care and Use of Laboratory Animals. http://grants.nih.gov/grants/guide/notice-files/not-od-05-034.html.
2. National Institutes of Health, Office of Laboratory Animal Welfare. 2012. Frequently Asked Questions, G.5. What is the difference between the Institutional Official and the Chief Executive Officer? http://grants.nih.gov/grants/olaw/faqs.htm#instresp_5.
3. National Institutes of Health, Office of Laboratory Animal Welfare/Applied Research Ethics National Association. 2002. Institutional Animal Care and Use Committee Guidebook. Bethesda: National Institutes of Health. http://grants.nih.gov/grants/olaw/GuideBook.pdf.
4. Greene, M.E., M.E. Pitts and M.L. James. 2007. Training Strategies for Institutional Animal Care and Use Committee (IACUC) Members and the Institutional Official (IO). *ILAR J*. 48(2):131–142.
5. National Institutes of Health, Office of Laboratory Animal Welfare. 2011. Tutorial on PHS Policy. http://grants.nih.gov/grants/olaw/tutorial/index.htm.
6. National Institutes of Health, Office of Laboratory Animal Welfare. 2013. OLAW online seminars. http://grants2.nih.gov/grants/olaw/e-seminars.htm.
7. Association for Assessment and Accreditation of Laboratory Animal Care International. Rules of accreditation. http://www.aaalac.org/accreditation/rules.cfm. Accessed April 8, 2013.

5

General Composition of the IACUC and Specific Roles of the IACUC Members

Christian E. Newcomer and William G. Greer

Introduction

The general composition requirements for IACUCs, established by federal regulations since 1985, give institutions considerable latitude in fashioning their IACUCs to reflect the scientific expertise and meet the needs of their animal care and use programs. Although the specific IACUC composition requirements vary according to the regulatory oversight agency involved, both the AWAR and the PHS Policy require the IACUC to have a diversified membership, including a veterinarian and a member who is unaffiliated with the institution. The PHS Policy further extends the diversification by requiring the institution to include a nonscientist on the IACUC. Membership diversification on the IACUC is intended to broaden the perspective and add depth to the important IACUC review processes. For example, in institutions conducting agricultural animal research programs, IACUCs should endeavor to incorporate members with appropriate agricultural expertise as recommended by the *Guide for the Care and Use of Agricultural Animals in Research and Teaching*;[1] other specialty areas of research animal use should be similarly addressed.

Most institutions recognize that developing and sustaining an effective IACUC requires planning and an ongoing commitment. The selection of individuals to meet specific membership requirements, new IACUC member orientation or education activities, and successful integration of new IACUC members as contributing members demand considerable effort. Even in a fully functional and effective IACUC, self-assessment and periodic adjustments are necessary for continued high-quality service to the institution and for maintenance of an appreciation for and understanding of the emerging trends in animal care and use programs. The *Guide*[2] (pp. 12–13) emphasizes this point stating, "The body of literature related to animal science and use of animals is constantly evolving requiring programs to remain current with the information and best practices."

To assist in IACUC development and self-review efforts, this chapter is intended to examine the overall composition of the IACUC and provide information on the roles, responsibilities, and issues involving the IACUC and its various member categories. Some of the trend information provided was acquired through the author's (Newcomer) review of numerous institutions during the conduct of AAALAC site visits and the association with AAALAC Council site visit deliberations for more than two decades, from the authors' personal experiences as IACUC members and/or administrators at more than a dozen institutions, and from over 300 IACUC Administrators that attended past IACUC Administrators Best Practice Meetings.

5:1 How many members must an IACUC have?

Reg. The number of IACUC members required depends upon whether the institution is a recipient of funding from the PHS or other cooperating federal agencies, such as the Department of Veterans Affairs. In these instances, the institution must have at least five IACUC members (PHS Policy IV,A,3,b). Institutions that use animal species regulated under the AWAR and funded through internal or private sources are required to have at least three IACUC members (AWAR §2.31,b,2). The *Guide* (p. 5) states, "The size of the institution and the nature and extent of the Program will determine the number of members of the committee and their terms of appointment." However, the minimum number of members allowable must also have the appropriate roles on the IACUC as described in 5:2.

Opin. In the authors' experience organizations usually exceed the minimal membership requirements set by the federal regulations and this impression is affirmed by the survey data provided by 297 institutions queried in 2012 about the characteristics of their IACUC membership in this survey. When responders were asked "How many regular voting members are on your IACUC?" the majority of institutions (53.2%) reported that the committee size was between six and ten members. Several institutions (25.6%) reported larger committees of 11–15 members. They comprised the second most prevalent size category followed by smaller committees of three to five members in 12.4% of institutions. Six and three percent of institutions reported committee sizes of 16–20 and greater than 20, respectively. Thus, it remains uncommon (as reported in the previous edition of this book)[3] for institutions to operate with a committee size hovering at the regulatory minimums. Although a minimum of five IACUC members may be adequate for a small organization with a focused research program and a limited spectrum of animal use, generally a minimum of seven or eight members is required in larger, more complex programs and to meet the growing demands of responsible oversight. (See 3:10.)

5:2 What specific members are required for the IACUC?

Reg. The AWAR (§2.31,b) state that the committee shall be composed of a Chair and at least two additional members. The AWAR do not describe the qualification of the Chair, but it states that the other two members of the committee will include at least one veterinarian and at least one shall not be affiliated in any way with the facility and shall not be a member of the immediate family of a person who is affiliated with the facility.

The AWAR (§2.31,b,4) state that if the committee consists of more than three members, not more than three members shall be from the same administrative unit of the facility. The administrative unit is defined as the organizational or management unit at the departmental level of a research facility (AWAR §1.1, Administrative Unit), which, for example, may include the Office of Research Administration and the university (institutional) laboratory animal resources.

The *Guide* (p. 24) states that the IACUC should have a veterinarian who is "either certified (e.g., by ACLAM, ECLAM, JCLAM or KCLAM) or with training and experience in laboratory animal science and medicine or in the use of the species at the institution." In addition, there should be "at least one practicing scientist experienced in research involving animals." Finally, there should be "at least one member from a nonscientific background, drawn from inside or outside the institution" and "at least one public member to represent general community interests in the proper care and use of animals. Public members should not be laboratory animal users, affiliated in any way with the institution, or members of the immediate family of a person who is affiliated with the institution."

PHS Policy (IV,A,3,b), on the other hand, states that membership must consist of not fewer than five members. The PHS Policy (IV,A,3,b,1–IV,A,3,b,4; IV,A,3,c) states that the committee shall include at least one veterinarian, one practicing scientist experienced in research involving animals, one member whose primary concerns are in a nonscientific area (for example, ethicist, lawyer, member of the clergy), and one individual other than a member of the IACUC who is not affiliated with the institution in any way and is not a member of the immediate family of a person who is affiliated with the institution. One individual may fulfill more than one requirement. However, no IACUC constituted under the PHS Policy may consist of fewer than five members. Also note that PHS Policy II requires institutions to comply with the AWAR, as applicable, and to follow the *Guide*. Readers should refer to the NIH/OLAW FAQ B.1 for a succinct discussion of IACUC membership criteria.[4]

Opin. Organizations involved in agricultural research and teaching should be aware that there are guidelines for the composition of animal care and use committees overseeing agricultural research activities.[1] The Agricultural Guide[1] recommendations are used as a standard by AAALAC but have no regulatory empowerment. They closely parallel the PHS Policy, differing only by specifying two types of scientific individuals on the committee. One should be a scientist from the institution who has experience in agricultural research or teaching involving agricultural animals, and the other should be an animal, dairy, or poultry scientist who has training and experience in the management of agricultural animals. This document further recommends that a separate committee not be established for this purpose, but that the composition of the IACUC should be modified according to the recommendations just noted to provide for the centralized and uniform oversight of the institution's animal care and use program. In an institution that conducts both agricultural research and PHS-funded research in nonagricultural species, the requirement for two committee members to have the agricultural expertise specified would likely necessitate exceeding the five-person minimal membership requirement of the IACUC under PHS Policy. However, in the authors' experience, the IACUCs serving these dual arenas have taken this approach.

5:3 Must an IACUC report to NIH/OLAW and APHIS/AC the names and titles of all members of the IACUC?

Reg. The general requirements to NIH/OLAW in an institution's Animal Welfare Assurance include the provision for the reporting of IACUC members' names, position titles, and credentials; however, a footnote clarifies that institutions may, at their discretion, represent the names of members other than the IACUC Chair and veterinarian with numbers or symbols (PHS Policy IV,A,3,b, footnote 6).[5,6] The ability to withhold the names of IACUC members may be beneficial for the individuals and institutions in the event that Federal Freedom of Information Act requests are later pursued through NIH/OLAW. The AWAR do not contain a comparable provision and therefore there is no reporting requirement to APHIS/AC.

5:4 Can one person fill more than one position on the IACUC?

Reg. The PHS Policy (IV,A,3,c) provides a written assent to dual representation as long as the IACUC totals at least five members.[3,7] The AWAR offer no direct discussion of this matter and do not preclude this practice.

Opin. It is permissible for an individual to fill more than one of the membership categories on an IACUC under both the AWAR and the PHS Policy. While it is permissible, APHIS/AC strongly discourages the practice of one person's filling more than one role, citing the "potential for conflicts of interest and/or undue influence by one person over the facility's program."[8] The authors concur with this assessment and identify the significant IACUC responsibilities, workload and benefits of diversification as other reasons to not consolidate roles in an individual. Very few institutions have exploited dual category representation as a long-term strategy for their IACUC, and, when it occurs, it frequently involves the unaffiliated member serving in the dual capacity as a nonscientist. It is also conceivable, but exceedingly rare in the authors' experience, that an entrepreneurial veterinarian in a small contract laboratory setting could serve as the IO, scientist, and AV within the AWAR. In each of the foregoing circumstances, the objectivity and independence of the IACUC's decisions could be compromised, particularly if the committee is small. In the authors' experience, dual capacity appointments as the nonscientist and unaffiliated member are most commonly used only transiently to bridge a gap in the mandated membership after the departure or resignation of a member.

5:5 What problems can occur if the number of IACUC members is too large or too small?

Opin. There can be significant problems related to the size of the IACUC. IACUCs that have a small membership often have a narrower base of expertise and, depending upon the size of program they service, may encounter an onerous workload. Also, small IACUCs may often have a more difficult time making a quorum, and, even in the presence of a legitimate quorum, member absences can have a marked effect on decision making. The challenges for IACUCs that have large memberships are very different. In large IACUCs, the principal problem is that members may become disengaged from the activities of the IACUC. This might be due to impediments to open communication in the IACUC or to the daunting scope and size of the program and its challenges. Thus, members have no sense that their efforts are having any impact on program quality or progress. In the depths of ennui, these members contribute to the quorum in number only. The challenges for the IACUC Chair are to ensure that all members have the opportunity for real contribution and an opportunity to be heard on issues, to help members feel a sense of accomplishment, and, when necessary, to inform the IO that the IACUC may be too large for the tasks at hand. Training opportunities are important for the maturation and effective operation of all IACUCs and can help keep committee members engaged.

5:6 What institutional constituencies should be represented on the IACUC?

Opin. The majority of organizations go beyond the federally mandated composition of the IACUC (see 5:2) to diversify the IACUC membership. The first priority in this effort generally is to enlarge the complement of scientific members on the IACUC to match the predominant areas of scientific expertise in the program. In addition to the potential political benefits this has for the IACUC within the scientific community it serves, it confers tangible benefits for the IACUC's deliberations on protocol matters. It also may improve the quality of correspondence exchanged between the IACUC and PIs.

Other institutional constituencies that the authors find to be favored are the following:

- IACUC administrators
- Health and safety personnel (including occupational health and safety)
- Senior animal management or supervisory personnel
- Statisticians
- Research laboratory technicians (with extensive animal involvement)
- Librarians
- Legal counsel
- Public relations personnel
- Grants and financial personnel
- Student body representatives

The merits of having these people on the IACUC should be evaluated in the context of the program for which they are intended. In the authors' opinion, IACUC administrators, health and safety personnel, statisticians, librarians, legal counsel, and public relations personnel should be given priority. IACUC administrators can provide consistency, serve as a source for regulatory knowledge, enhance institutional memory and action on important issues/decisions, improve IACUC processes and efficiency, offer principal investigator training and assistance and solidify the IACUC's reputation in "customer service." Health and safety personnel are comfortable working at the interface of science and society and can make a significant contribution to the IACUC by keeping it apprised of the status of health and safety regulations, and recommendations relating to the animal care and use environment.[9] Statisticians and librarians can help the IACUC and investigators address the appropriateness of animal numbers and availability of alternatives. Legal counsel and public relations personnel also can be of great assistance in the consideration of issues relating to science, policy, perception, and the public.

5:7 Should the selection of IACUC members depend upon the institution's research goals and expertise?

Opin. Two of the most important functions of the IACUC, protocol review and programmatic review, should be major driving forces for the IACUC to select members with a share in the institution's research goals and expertise. Protocol review should be insightful, thorough, and objective. Members who have scientific expertise in the areas under review can help focus the discussion of relevant issues, improving the quality of the review. IACUC programmatic review is a mandated responsibility of the IACUC (PHS Policy IV,B,1–IV,B,5; AWAR §2.31,c,1–§2.31,c,5) but it really is best performed as a bidirectional process involving the community served. This implies that the IACUC should foster conduits for the flow of information from all participants in the animal care and use program. The appointment of IACUC members who represent specialty interests or areas of expertise can prove advantageous to this process. Most institutions make a concerted effort to staff their IACUCs with members who reflect the research expertise of the program, at the same time making sure to prevent possible conflicts of interest.

5:8 What is the typical IACUC composition in a large versus a small institution, and are the members' roles different in these settings?

Opin. There are different trends in the composition of the IACUC and in the ways the IACUC uses its members in large versus small institutions. In small commercial institutions or independent research organizations based around a particular area of scientific inquiry, the IACUCs tend to be smaller (five to seven members) because the number and diversity of issues requiring consideration by the IACUC are limited. Data reported in the previous edition of this text[3] indicated that 81% of nonacademic institutions had fewer than ten members, whereas only 58% of academic institutions had this profile. Small academic institutions tend to have slightly larger IACUCs (seven to ten members). Although the research portfolio may be limited compared to that of a large academic institution, the relative diversity of the research activities can be quite high even though each department may only have a few active faculty researchers. The intimacy of the small institutional environment sometimes imposes a higher level of expectation that the IACUC members will be able to speak authoritatively about the research and programmatic needs of their colleagues. IACUC members in small institutions also are more likely to be active participants in developing and implementing programmatic components (e.g., biosafety review and IACUC educational efforts) that may fall under the IACUC's purview.

Large institutions sometimes have 15 to 20 or more IACUC members simply to provide ample representation of the animal user groups, scientific areas under study and number of investigators (e.g., 150 to 200) in the program. Data from the 2012 survey of institutions indicated that almost 9% operated with IACUCs of 16 or more members compared to 8% in the previous edition.[3] (See 3:20.) In the authors' experience, these IACUCs tend to be more formal, impersonal, and insular. Frequently, institutionally provided administrative support obviates the need for the personal involvement of the IACUC members in ushering new initiatives forward. As noted in 5:5, a significant challenge for IACUC members in this position is to combat these tendencies and convince the faculty that the IACUC is a receptive, helpful, and responsive body that works for the institution in support of the scientific mission by encouraging improvement of the animal care and use program. Another issue germane to extremely large IACUCs is the number of veterinarians appointed to that committee. Although federal law has stipulated that an IACUC need only have one veterinary member, most institutions that have large IACUCs (greater than 15 persons) have increased the number of veterinarians appointed. This action helps ensure that the veterinary perspective will be represented consistently and affords the veterinarians an opportunity to divide the workload and concentrate on specialty areas of interest and needs.

5:9 Should the IACUC Chair, individual IACUC members, or their departments receive compensation for their efforts?

Reg. Although PHS Policy has not commented on the general topic of IACUC member compensation, the *Guide* (pp. 24–25) does suggest that compensation of the public member should, "be sufficiently modest that it does not become a substantial source of income and thus risk compromising the member's association with the community and the public at large." NIH/OLAW FAQ B.11 is also instructive on this matter. The AWAR are silent on this issue.

Opin. In contrast to earlier comments on this topic in the previous edition of this book which were based upon anecdotal information, compensation for the IACUC Chair has not been established as a common practice and occurs relatively infrequently. For those that have received this acknowledgment of their efforts, salary support, funding to support professional development, research program development, research technical support, and a variety of other approaches have emerged from faculty–administrator negotiations. With less success than the Chair to date, other IACUC members with faculty appointments are also beginning to pursue and gain some compensation for their efforts. Examples include travel support to national meetings, reduced instructional commitments, or release from ancillary academic obligations such as service on other committees. Members may find that documentation of their time and effort on IACUC activities may help advance the argument for reciprocity of benefits.

Surv. According to the 2012 survey of 296 respondents, approximately 16% acknowledged providing direct (12%) or indirect (4.5%) support to the IACUC chairperson. In contrast, 7% of institutions reported compensating IACUC members through direct (6%) or indirect (1%) mechanisms. The 2012 survey clarified that compensation such as parking fees, local travel, lunch during the meeting and other minor compensations should be excluded in responding. (See 3:10.)

5:10 How can the performance of the IACUC be evaluated?

Opin. Several mechanisms are available and should be utilized to evaluate the performance of the IACUC on an ongoing basis. The IO should meet periodically with the Chair of the IACUC to discuss the content, style, and timeliness of IACUC reports and correspondence with members of the institution's research community. This is the most common method reported by the institutions known to the authors, although few institutions couch such meetings in terms of an evaluation. Also, without the inducement of a particular incident, the IO should consider soliciting the comments of the faculty or staff who use laboratory animals as well as of the members of the IACUC. The assessment of the IACUC as a working group seems particularly important because of the profound potential impact of group dynamics on IACUC morale, attention to detail, and, ultimately, decision making.

Programmatic review contains the essential element of self-assessment that should be practiced by every IACUC. The *Guide* (p. 13) sets the expectation that responsible resource planning by the IO will provide the IACUC with the authority and resources to fulfill its responsibilities including, "specific training to assist IACUC members in understanding their roles and responsibilities and evaluating issues brought before the committee." Such training ordinarily entails member attendance at national conferences followed by the communication and discussion of pertinent topics by the entire committee. Simple measures such as the quality of meeting minutes and the attendance records for meetings (i.e., just making or failing to make a quorum repeatedly portends disaster) can be useful. Also, the level of member participation in post-approval monitoring activities and the overall record and severity of indiscretions found during post-approval monitoring indirectly reflects on the performance of the committee. Lack of programmatic oversight, inadequate training programs for personnel, lack of rigor in protocol review and/or failure to provide guidance to PIs through post-approval monitoring activities can increase the number of reportable events, a metric that reflects negatively on the IACUC.

Participation in the voluntary peer review process of the AAALAC entails a substantial review of IACUC function through interviews and the review of written materials. Generally, AAALAC site visits are conducted at triennial intervals and should be supplemented by the institution's internal evaluation mechanisms. In some instances, expert consultants in the area of IACUC function also have been used by institutions.

5:11 Who can remove the IACUC chairperson or other IACUC members?

Reg. The PHS Policy (IV,A,3,a) requires the Chief Executive Officer (CEO) to appoint the IACUC, and the AWAR (§2.31,a) specifically directs the CEO to appoint the Chair. Thus, although once established the IACUC possesses autonomous review and reporting functions, service on the IACUC is at the discretion of the CEO. Under the PHS Policy (IV,A,3,a, footnote 5), the CEO may delegate the authority to appoint the IACUC if the delegation is specific and in writing.

Opin. It is very fortunate that institutions are rarely, if ever, confronted with this quandary because very few institutions have clearly written bylaws for their IACUCs that stipulate who has jurisdiction or what the process is in this matter, and no clarification is offered in the PHS Policy or the AWAR. For the most part, informal discussions originating from the Chair, in the case of other IACUC members, or from the CEO, in the case of the Chair, apparently have proved adequate for the silent and amicable departure of IACUC members in the vast majority of cases. In the absence of bylaws clearly delineating who has authority in this matter, the CEO retains that authority. In a related matter, without bylaws the extent to which, and the situations in which parliamentary procedure should be applied in IACUC meetings and deliberations and the ability of the IACUC to self-regulate through the use of censure or expulsion procedures are also questionable.[10] (See 6:12.)

It is reasonable and advisable for institutions to establish performance standards, i.e., participation expectations, for the Chair or other IACUC members to ensure that the discussions with individuals about termination of service are not regarded as arbitrary and capricious. Attendance at meetings, participation in other IACUC activities and ability to perform work assignments adequately in a timely manner are fair performance measures. In recognition of workload and the demanding schedules of IACUC members, the appointment of alternative members on the IACUC, as well as the formal designation of a Vice Chair may have value as an aid to IACUC function and in succession planning.

5:12 What criteria can be applied for removing the IACUC chairperson or other IACUC members?

Opin. Most organizations agree that inappropriate personal conduct, poor attendance or lack of participation in IACUC activities, or repeated inadequate preparation for assigned IACUC duties might constitute sufficient reason to seek the removal of a member. Ideally, these broad areas would be included in the bylaws developed for the IACUC, and the IACUC members would be informed during their initial IACUC training of the general performance criteria for committee members. The IACUC Chair or IO should inquire whether there are any mitigating circumstances for those members who cannot regularly attend a high percentage (e.g., 75%) of the IACUC meetings/activities held annually and consider the replacement of these individuals if better attendance is not forthcoming. Committee membership agreements can also be used effectively for this purpose. These agreements define expectations of service in each of the areas of IACUC responsibility and serve as a tangible measure of member engagement and contribution and can be used by some programs for the periodic review of IACUC member contribution. (See 6:9–6:11.)

5:13 What is the procedure for selecting and appointing the IACUC Chair?

Opin. Most IACUCs attempt to prepare for a transition in the position of Chair through the appointment of a Vice Chair or an ad hoc Chair-in-Training (e.g., Vice-chair position, see 5:11). In this regard, the IACUC usually has the opportunity to serve in an advisory capacity to the CEO, who ultimately has the responsibility of appointing the Chair (AWAR §2.31,a; §2.31,b). (See 3:2; 3:3.) Other groups that can prove useful in identifying and supporting candidates for the position of IACUC Chair in academic institutions include the faculty senate and faculty research advisory committees. The CEO can also solicit the input of IACUC administrator who may have valuable insights on the members who have been especially conscientious and made valuable contributions to the IACUC.

5:14 What traits or qualifications are desirable for the IACUC Chair?

Opin. In most institutions, a concerted effort is made to recruit a Chair who is a well-respected scientist and who has had significant experience using laboratory animals in research. Regardless of whether or not the Chair has a scientific background, he or she should be regarded as a good colleague in the context of the institution's scientific mission and have exhibited good leadership qualities. The Chair should be patient, tolerant, diplomatic, tactful, and efficient in the handling of the IACUC's business and sensitive issues. He or she should be able to dedicate the time necessary with sufficient flexibility to plan, oversee, and/or participate in all critical IACUC functions and activities. The notion of adequate participation in critical IACUC functions presupposes that the Chair will be knowledgeable and conversant in current regulatory interpretations, contemporary practices, and trends pertaining to animal use and welfare and committed to leading a successful institutional animal care and use program. In addition to providing a continuing and stabilizing presence, the Chair is likely to encounter occasional urgent matters that require immediate attention.

Some institutions have been slow to recognize that they have appointed a Chair who has not been performing adequately or may be constitutionally unfit for this important position. A partial list of problems indicative of such an appointment includes disorganization of IACUC meetings and activities; inattention to issues raised by the faculty, IACUC members, or veterinary staff in IACUC meetings; a lack of understanding of the regulations and standards and reluctance to consult with other knowledgeable IACUC members or staff or members who do; resistance to the adoption of needed programmatic changes as new standards emerge; proclivity for unbalanced communications (written or verbal) with investigators without the collaborative input or endorsement of the IACUC; or inappropriate and unilateral supplication, or alternatively, rigidity in dealing with investigators.

5:15 What is the Chair's role in the oversight of IACUC activities?

Opin. The Chair should play an active role in the oversight of all IACUC activities regardless of his or her role as an actual participant. The Chair serves six important constituent groups:

- The senior administration (embodied in the CEO and IO)
- The scientific community
- Other members of the IACUC
- The IACUC office administration and support
- The federal government
- The public

The immense commitment within the institution of time, resources, and interest to the IACUC's activities should compel the Chair's enthusiastic involvement in IACUC oversight. (See 3:1.)

5:16 Can the AV serve as the IACUC Chair?

Opin. While there is no prohibition in the PHS Policy or AWAR of the AV serving as the IACUC Chair, this practice is strongly discouraged.[5] Only a few institutions among the hundreds known to the authors are aware of have taken this approach. NIH/OLAW had issued an explanation that strongly suggested that having either the AV or IO serve as the IACUC Chair may be inappropriate because of real or perceived conflicts of interest, the consolidation of important leadership roles and the disruption of the necessary checks and balances intended in the institutional reporting structure and IACUC review processes.[11] The programmatic review activities of the IACUC encompass laboratory animal management and veterinary care, both of which are often under the direct purview of the AV. Thus, the appointment of the AV as the Chair affords the AV an opportunity to influence the IACUC openly or subtly. Efforts to stimulate institutions to reflect upon the advisability of continuing to use this approach have remained common during AAALAC site visits discussions. AAALAC has often observed that the responsibilities of the AV were already sufficiently demanding and considerably compounded by adding the complex job of the IACUC Chair. Moreover, AAALAC has recognized some instances in which the AV did not carry the title of IACUC Chair but was functioning *de facto* as the Chair and carrying the workload of the IACUC.[12] Institutions should be wary of this devolutionary development.

5:17 Should the manager of the institution's laboratory animal resource unit serve as the IACUC Chair?

Opin. The appointment of the animal facility manager as the IACUC Chair is much more common than the appointment of the AV in this capacity but still quite rare. Facility managers have a broad understanding of the various aspects of the animal care and use program but generally have a narrower scope of responsibilities than the AV. This lessens the conflict of interest issue discussed in 5:16; the issue is not entirely eliminated, however, since the review of facilities and the diverse provisions for laboratory animal management are critical elements of the IACUC's programmatic review that generally fall within the domain of the facility manager. The involvement of the facility manager in the financial management of the resource may be another deterrent to the appointment of this individual as the Chair in some institutions.

5:18 In an academic setting, what should be the faculty rank of the IACUC Chair? Is a person who has academic tenure preferred?

Surv. We queried the faculty appointment status of the Chair in 297 institutions (See 3:9 for complete survey responses). Resulting data are summarized in Table 5.1.

Academic institutions demonstrated an overwhelming preference for the appointment of a faculty member to Chair the IACUC and a majority of the faculty appointed had achieved a rank of Full Professor; tenure status was not solicited. In the previous edition,[3] we established that the percentage of tenured Full Professors serving as the Chair was higher than non-tenured Professors and that regardless of faculty rank, 62% of Chairs had attained tenure. Also, more Associate Professors occupy the position of Chair than Assistant Professors (20.5% versus 6.4%) according to the 2012 survey.

Opin. The data support the interpretation that most academic institutions believe that high-ranking faculty will best command the respect and wield the authority to lead the IACUC through its federally mandated review activities impacting institutional resources and faculty satisfaction. In many large academic institutions the appointment of a Full Professor (and one who has an active research program or prior history of laboratory animal use) is deemed desirable. Academic rank, per

TABLE 5.1

Faculty Status of the Chair in Reporting Institutions

Academic Title	Total (n = 297)
Non-faculty title or unreported[a]	108
Instructor/Assistant Professor	19
Associate Professor	61
Full Professor	109

[a] Non-academic institutions and Chairs without faculty title are combined.

se, is a less critical factor than an individual's scientific stature, collegiality, and reputation for fairness within the institution's scientific community. Institutions should recognize that the faculty often directs its antagonism toward adverse decisions made by the IACUC to the Chair. Hence, the appointment of a nontenured faculty member (in a tenure track position) as the IACUC Chair conceivably might enhance that individual's vulnerability during tenure review decisions. Most institutions (or faculty members) appear to recognize this and avoid placing lower-ranking, untenured academic faculty in this precarious position.

5:19 Can an IACUC be composed entirely of persons not employed by the university (or other institutions)?

Reg. Both the AWAR and the PHS Policy (see 5:2) require that at least one member of the IACUC not be affiliated with the institution in any way other than as a member of the IACUC. Other members are not required to be affiliated except that the veterinarian must have direct or delegated program authority (PHS Policy IV,A,3,b,1) and responsibility (AWAR §2.31,b,3,i) for activities involving animals at the institution. Also, pursuant to PHS Policy (IV,B), the IACUC functions as an "agent of the institution."

Opin. The AWAR (§2.33,a,1) state that the research facility must employ an AV under formal arrangements. Under the PHS policy, the AV must have programmatic authority, so it is unlikely that the research facility will have an animal care and use program without some financial arrangements with the AV. Even if the AV is not employed by the organization, he/she would be affiliated by acceptance of critical responsibilities within the program. However, neither the PHS Policy nor the AWAR stipulate that some (or any) of the remaining members appointed must be institutional employees. Nevertheless, the authors have never encountered an IACUC that did not contain at least one member who was an employee. (See 3:21.) However, the authors are aware of companies that have contracted with contract research organizations or other animal program service providers to conduct their animal research studies without representation on the IACUC. A more common occurrence is that an organization participating in a cooperating research partnership with another organization may elect to co-appoint a knowledgeable IACUC that serves the research missions of both organizations, irrespective of the employment associations of the membership. The pivotal issue in the matter of IACUC appointment is not whether the CEO holds any financial sway over the committee members through employment, but rather, whether the CEO is capable of disbanding the IACUC if it proves to be incompetent, inattentive to animal welfare issues or fails to discharge its federally mandated responsibilities with due diligence in a competitive scientific environment. The growth of inter-institutional collaborations has prompted a *Guide* recommendation (p. 15) that organizations develop a formal written understanding of the responsibilities of each party in the partnership including IACUC oversight.

5:20 Can persons who do not have prior experience or expertise with laboratory animals be appointed as members of an IACUC?

Reg. The regulatory requirements for personnel with experience with laboratory animals are minimal (PHS Policy IV,A,3,a; IV,A,3,b; AWAR §2.31,a; §2.31,b). Both sets of regulations require that the veterinarian have experience with laboratory animals and the PHS Policy stipulates that, in addition, one practicing scientist has experience in research involving animals. Members who do not have laboratory animal experience can be called into IACUC service.

Opin. Most organizations easily exceed the regulated number of those with laboratory animal experience, but IACUCs often have many members who have laboratory animal experience related to one or a few species involving applications in a particular area of research. Much is to be said for the vicarious experience these individuals garner from talking with colleagues, participating in active research programs, reading the literature, and gaining experience on the IACUC. However, introspection by the IACUC is also desirable to allow the self-identification of areas of IACUC weakness and foster the recruitment of internal or external consultants in areas of need.

5:21 Can or should an outside consultant serve as the IACUC Chair?

Reg. The regulatory references for an IACUC Chair appear in PHS Policy (IV,A,3,b) and in the AWAR (§2.31,c,3; §2.31,b,2). The regulatory references for consultant(s) also appear in the PHS Policy (IV,C,3) and the AWAR (§2.31,c,3; §2.31,d,3).

Opin. This practice is not prohibited by the PHS Policy or the AWAR and the authors are aware of organizations that have used outside expert consultants in this leadership position. Although an outside consultant may initially lack an appreciation for the scientific landscape and key players of the institution, this is not an insurmountable deficit and might be easily offset by the consultant's superior knowledge of the regulations and best practices. This may also afford some institutions a fiscal advantage in the final analysis and help preserve valuable time in the laboratory for a scientist who would otherwise be assigned this important duty.

5:22 What are the specific duties of the IACUC Chair?

Opin. The IACUC Chair has the responsibility for overseeing the coordination and implementation of effective, efficient systems for protocol review by the IACUC in compliance with the PHS Policy and the AWAR. These review activities can only be performed at a properly convened meeting of the IACUC unless the review to approve or require modifications (to secure approval) or request Full Committee Review (FCR) is done using Designated Member Review (DMR).[13] Thus, the Chair should:

- Ensure that a quorum of the IACUC is present
- Declare the loss of a quorum, resulting in the end of official business if a sufficient number of members depart
- Prepare or oversee the preparation of meeting minutes and reports and submit these documents to the IO in accordance with PHS Policy (IV, E; IV,F) and the AWAR (§2.31,c,3; §2.31,d,2; §2.31,d,4; §2.35,a)
- Report to the IO any activities that have been suspended by the IACUC for noncompliance with PHS Policy or the AWAR (see Chapter 29)
- Establish a sound system of written communication for the IACUC with investigators concerning the approval status of protocols and the steps necessary to secure approval

- Designate, as specified in the PHS Policy (IV,C,2), at least one qualified member of the IACUC to conduct designated member review[13] if full committee review is not requested.

Beyond these relegated duties that the Chair performs on behalf of the committee, most institutions expect the Chair to keep abreast of new regulatory trends and interpretations and evaluate and champion policy and practice initiatives (e.g., new training and educational programs) to improve the animal care and use program. This process involves the Chair's regular interaction with other areas of expertise within the organization, ranging from other institutional committees to occupational health and safety, human resources, and the physical plant. Moreover, because these onerous diverse duties intensively consume time, energy, and require consistent relationship-building within and outside the organization, organizations are increasingly relying on the recruitment of professional IACUC administrators to partner with the Chair to ensure the detailed, competent and timely completion of many tasks related to IACUC function.

5:23 When a quorum is lost because of the mandated nonparticipation of a member that results from a conflict of interest, may that person's alternate reconstitute the quorum and vote?

Opin. Situations such as this should provide a strong impetus for institutions to articulate the role, responsibility, and authority of alternate members clearly in a charter document. The NIH has issued guidance on this issue (NOT-OD-11-053) indicating that a qualified alternative can be appointed to represent one or more IACUC members and/or more than one alternative can be appointed to represent a particular IACUC member.[14] However, alternative members cannot serve simultaneously with the member they represent in any committee activity.

Notice NOT-OD-11-O53 accommodates the instance where an IACUC loses a quorum due to a member's unavailability (i.e., disqualification) by virtue of conflict of interest; a qualified alternative member may be used to reconstitute the quorum to proceed with a vote. The authors agree that such a substitution to make a quorum would be appropriate because the alternative member represents a redundancy in position and not of perspective. The underlying assumption should be that an alternative member will be knowledgeable about the facts under review and evaluate them independently and autonomously, understand all relevant IACUC issues, and vote his or her conscience. The AWAR do not address alternate members, but APHIS/AC does not preclude this policy.

5:24 Can alternate members of the IACUC perform semiannual inspections that involve AWA-regulated species?

Reg. APHIS/AC and NIH/OLAW have agreed in a guidance document (NIH/OLAW FAQ B.2) that alternate members can participate by sharing their expertise and perspectives in IACUC meetings and review activities even when their primary IACUC member is present. However, they may only contribute to a quorum in place of, and not as a supplement to, the primary member they are intended to represent.

Opin. Appropriately appointed alternate members would be eligible to participate in semiannual inspections involving APHIS/AC-regulated species provided that the member and his or her alternate are not participating simultaneously. Notice NOT-OD-11-053[14] (which provides guidance from both NIH/OLAW and APHIS/AC) encourages alternate member attendance at IACUC meetings and other activities even when the regular member is in attendance, however, this does not expand the opportunity for simultaneous participation during semiannual inspections, even if the member and his or her alternate are inspecting different facilities. Similarly the member and his or her alternate cannot constitute the two-person team required under the AWAR (§2.31,c,3) during the facility inspection process, because they represent only a single vote on the IACUC. As suggested in 5:23, creation of a charter document governing these activities would be advisable to clarify the institution's expectations of its IACUC members. (See 23:17.)

5:25 An IO appoints a scientist as an alternate member of the IACUC, to fill in when any scientific member is "unavailable." Under what conditions can this individual participate in IACUC functions? Should institutions place any conditions on what constitutes the "unavailability" of a member?

Opin. According to NIH Notice NOT-OD-11-053[14] unavailability encompasses instances in which the member is unable to attend the meeting, needs to leave the meeting early or arrive late, or is recused from participating due to a real or potential conflict of interest. The Notice represents a change in position on this issue since the earlier edition of this volume was published and confirms the growing recognition that obligations in regulatory oversight compete with scientific productivity and scholarship in many instances. Institutions may find it useful to communicate a performance standard indicating that members who are regularly "unavailable" at the last moment or otherwise appear to be avoiding participation are not meeting their institutional and IACUC commitments. Alternate IACUC members should resist the capricious or compromised decision making that might be associated with short notice circumstances.

5:26 Must the institutional AV also serve as the veterinary member of the IACUC?

Reg. The PHS Policy (IV,A,3,b,1) and the AWAR (§2.31,b,3,i) state that the veterinarian on the IACUC shall have training or experience in laboratory animal science and medicine and have direct or delegated program authority and responsibility for activities involving animals at the institution. The AWAR (§2.33,a,3) state, "The attending veterinarian shall be a voting member of the IACUC; Provided, however, That a research facility with more than one Doctor of Veterinary Medicine (DVM) may appoint to the IACUC another DVM with delegated program responsibility for activities involving animals at the research facility."

Opin. Written guidance jointly provided by NIH/OLAW and APHIS/AC[15] recognizes that institutions with large and complex programs may wish to appoint more than one AV to address the needs of the research program and the specialty care of animal species involved. Not all of the AVs need to serve on the IACUC and service on the IACUC implies nothing about the authority hierarchy in veterinary care in the institution as a whole. Therefore, programs may appoint several AVs but only one has direct or delegated program authority and responsibility and *must* be a voting member of the IACUC (see below).

The wording of AWAR §2.31,b,3,i, as indicated in the Regulatory paragraph, describes the IACUC veterinarian with the wording found in AWAR §1.1, Attending Veterinarian. Although this suggests that the AV must be an IACUC member, AWAR §2.33,a,3 states that if the research facility has more than one veterinarian, then a veterinarian other than the AV can serve on the IACUC in place of the AV as long as that other veterinarian has delegated program responsibility for animal activities. PHS Policy IV,A,3,b,1 appears to offer concurrence with the AWAR by stating that the IACUC veterinarian is one who has "direct or delegated program authority and responsibility." Nevertheless, NIH/OLAW, in reviewing this chapter, provided guidance by stating that "For PHS Assured institutions, the veterinarian with program authority must serve on the IACUC. Other veterinarians with delegated clinical responsibilities may also serve, but the senior veterinarian responsible for the program must serve on the IACUC."

NIH/OLAW and APHIS/AC have jointly issued a very helpful clarification on the appointment and expected role of the AV on the IACUC and the opportunities for how to approach the matter of AV appointment in complex institutions.[15] Clear lines of authority and responsibility should be sought in the appointment of the AV to best effectuate programmatic oversight, essential programmatic changes if needed and quality protocol procedures. Institutions with complex programs may wish to discuss their options with NIH/OLAW and APHIS/AC.

5:27 What training or experience is useful for the veterinary member of the IACUC?

Reg. The PHS Policy (IV,A,3,b,1) and AWAR (§2.31,b,3,i) stipulate that the veterinary member of the IACUC should have training or experience in laboratory animal medicine and science. (See 27:2.) The AWAR (§1.1) further clarify, under the definition of the attending veterinarian, that this individual should have either graduated from a veterinary school accredited by the American Veterinary Medical Association Council on Education, acquired a certificate issued by the American Veterinary Medical Association's Education Commission for Foreign Veterinary Graduates, or received equivalent formal education as deemed appropriate by the APHIS administrator.

Opin. The intent of the PHS Policy and AWAR is to help ensure that the IACUC can rely upon the veterinary member not only for competent clinical insights, but also for information on ancillary areas such as unusual animal models, zoonoses, other occupational health and safety concerns, hazard containment, genetics, and unique nutritional and husbandry requirements of laboratory animal species. Proficiency in these diverse subject areas can be demonstrated formally by board certification in the American College of Laboratory Animal Medicine (ACLAM), and it is advisable for one or more of the veterinarians responsible for a large and complex program of research animal use to be board certified by ACLAM or comparable national laboratory animal medicine specialty (e.g., ECLAM, JCLAM or KCLAM). However, the needs of research animal care and use programs that are smaller or of a more limited scope may be met by veterinarians who are not ACLAM board certified. Regardless of their ACLAM certification status, all veterinary personnel involved in the oversight of research animal use should be aware and stay abreast of the central research, clinical, and regulatory concerns pertaining to the species under their care. This can be accomplished by reviewing the literature and attending the national meetings sponsored by various organizations, including AALAS, ACLAM, American Society of Laboratory Animal Practitioners, Institute for Laboratory Animal Research, Association of Primate Veterinarians, American Veterinary Medical Association, AAALAC and other international laboratory animal science/medicine organizations.

5:28 What is the role of the veterinarian on the IACUC?

Reg. The role of the veterinarian within the context of the IACUC is defined under the AWAR (§2.31,d,iv,B; §2.31,d,vi; §2.31,d,vii; §2.31d,ix; §2.31,d,x,B) and by the *Guide* (p. 14) which serves as an extension and amplification of the PHS Policy. The *Guide* (p. 14) and PHS Policy (IV,A,3,b,1) indicate that the AV is responsible for the health and well-being of all the laboratory animals used in the institution.[15] In summary, the AWAR include provisions for

- Veterinary consultation on the recognition and palliation of pain
- Direction of animal care and use (a specific role of the AV)
- Medical care
- Aseptic surgery and postoperative care
- Oversight of multiple major survival surgery resulting from a veterinary condition in an animal that also had experimental surgery

Specific provisions within the *Guide* (pp. 27–28, 30) include

- Advising the IACUC on new procedures or procedures with the potential to cause pain and distress that cannot be reliably controlled
- Ensuring that veterinary care is available to mitigate the illnesses, lesions, or behavioral abnormalities associated with animal restraint

These sources also expand on the responsibilities of the AV to the institutional animal care and use program that are unrelated to IACUC membership per se. (See 27:7.)

Opin. Veterinarians play a unique role on IACUCs, as a result of their expertise in laboratory animal medicine and science and broader understanding of the physiological, behavioral, nutritional, and husbandry needs of laboratory animal species. Although by no means a lone voice on the IACUC on matters of animal health and welfare, the veterinarian should make the concerted effort to address these areas with precision and authority. Special areas of emphasis for the IACUC veterinarian should include the discussion of the use of proper anesthesia, analgesia and euthanasia in laboratory animals in the relief of pain and distress; assurance of veterinary involvement in pre-surgical planning and surgical skills/program assessment; the analysis of possible iatrogenic complications to the procedures used or the disease model proposed; the selection of humane endpoints and response to unexpected outcomes; and a review of the plans for appropriate and timely medical intervention.[15] (See 27:7.)

5:29 Under what conditions can the veterinarian delegate some of his or her responsibilities to other IACUC members?

Reg. According to the AWAR (§2.31,d,1,iv,B) the AV may delegate his or her responsibility for playing a role in the planning and consultation on procedures that may cause more than momentary or slight pain or distress.

Opin. Depending on the background, training, and experience of the other IACUC members, and the size, complexity, and maturity of the animal care and use program, many of the veterinarian's responsibilities can be delegated to another veterinarian or other members for defined periods (e.g., a particular IACUC meeting or semiannual review). For the most part the appointment of an alternate for the veterinarian is used to address this issue in most institutions. Very few of the institutions known to the authors would proceed with an IACUC meeting in the absence of the veterinarian, particularly if this involved neglecting significant issues because of insufficient veterinary input. The strategy most often used when the IACUC veterinarian is absent is to invite another veterinarian with relevant skills and knowledgeable, to participate as a guest in an advisory capacity or require evidence of veterinary approval before proceeding. Secondarily, IACUCs can table issues requiring the veterinarian's special analysis and comment. (See 6:7; 27:7.)

5:30 Does the attending veterinarian have any mandated training responsibilities under the AWAR or PHS Policy?

Reg. Training is not assigned as a specific requirement to the AV in either the AWAR or the PHS Policy. The AWAR (§2.32,a; §2.32,b) indicate that training "shall be the responsibility of the research facility." The PHS Policy (IV,C,1,f) places the responsibility of assuring that personnel are appropriately qualified and trained to conduct the procedures proposed on research animals on the IACUC, through its protocol review process. AVs, or their designee, however, must provide "guidance to principal investigators or other personnel involved in the care and use of animals regarding handling, immobilization, anesthesia, analgesia, tranquilization and euthanasia; and … adequate pre-procedural care and post-procedural care in accordance with current established veterinary medical and nursing procedures," as stipulated in the AWAR (§2.33,b,4; §2.33,b,5). IACUCs should work collaboratively in support of the program of veterinary care to ensure that ample resources are available to the veterinarians to meet this critical objective. The *Guide* (p. 106) has a requirement for the AV to provide guidance "to investigators and all personnel involved in the care and use of animals to ensure appropriate husbandry, handling, medical treatment, immobilization, sedation, analgesia, anesthesia and euthanasia. In addition, the AV should provide guidance... to surgery programs and perioperative care involving animals." With regard to surgical training, the *Guide* (p. 116) states the "IACUC, together with the AV, is responsible for determining that personnel performing surgical procedures are appropriately qualified and trained in the procedures."

Opin: In most institutions, species specific training initiatives are run by the veterinarian(s) or under the direct management of the veterinary team operating either through the animal resource or the IACUC administrative office. Collaborative ventures with the veterinary staff augment, or in cases supplant, this effort in some institutions. In these cases, most use a multidisciplinary approach involving institutional scientists, capable technicians with advanced skills, dedicated training personnel within the IACUC, or veterinary resource group as well as the veterinarian(s). In the authors' view, verification of training and documentation of proficiency continue to be areas of vulnerability in many research animal programs.

5:31 What is meant by the nonaffiliated (outside) member of the committee, and what is the intent of including such an individual on the committee?

Reg. The intent of including a nonaffiliated member on the IACUC is to ensure that someone who is not affiliated with (or beholden to) the institution in any manner is involved in the review of the institution's animal care and use activities. This individual also may not be a member of the immediate family of a person affiliated with the institution (AWAR §2.31,b,3,ii; PHS Policy IV,A,3,b,4; *Guide* p. 24). According to the AWAR (§2.31,b,3,ii), this individual is intended to represent the "general community interests in the proper care and treatment of animals." (See 3:5; 3:6.) The language in the PHS Policy is the most restrictive, providing for an individual "who is not affiliated with the institution in any way other than as a member of the IACUC, and is not a member of the immediate family of a person who is affiliated with the institution."[7]

Opin. According to the language in PHS Policy, an individual would be disqualified as an unaffiliated member if they served voluntarily in any other capacity for the institution (e.g., service on an IBC). NIH/OLAW does not allow individuals who have used laboratory animals in the past to be committee members. Ostensibly, because the nonaffiliated member is not expected to benefit personally from any of the activities proposed and is not dependent on the institution for his or her livelihood, she or he enhances the public's confidence in the unfettered objectivity of the IACUC review processes. Institutions should consider other situations that might be perceived conflicts of interest in the appointment process, however tenuous, such as a retiree receiving a pension from the institution and should be prepared to explain their decision process in these cases. The unaffiliated member should be regarded as a full IACUC member and should be afforded opportunities to participate equivalent to those of all other IACUC members. However, in the authors' view, the unaffiliated member should not be asked to take the lead in any areas in which he or she may not be comfortable (e.g., as the primary reviewer in the Designated Member Review process for a protocol). (See 3:22.)

5:32 As an extension and example of 5:31, if two colleges, A and B, reciprocate in course admissions for students and exchange faculty for course instruction, would it be appropriate for the unaffiliated IACUC member at College A to be a spouse of a faculty member at College B?

Opin. In the authors' view, institutions should exercise common sense in the application of appointments when the "affiliation" is extremely tenuous as it is in this case. If the faculty member at College B is not personally involved in some aspect of the animal care and use program at College A or is not dependent upon the success of College A for salary and livelihood, there is no compelling reason to believe that the spouse would not be able to perform this important public service admirably.

5:33 What are the typical backgrounds of individuals who serve as nonaffiliated members of IACUCs?

Opin. There is no typical background of these individuals. Institutions have chosen a wide variety of individuals to serve in this capacity. Without implying that the following categories are mutually exclusive, the range of individuals includes broadly educated, erudite humanists; businesspersons; public servants; educators; persons involved in other types of animal care and use; physicians, veterinarians, scientists, and technicians who do not use laboratory animals; and concerned citizens who seek to make a public contribution.[7] (See 3:6.)

5:34 Can the nonaffiliated member be a scientist, veterinarian, ethicist, or biostatistician who is not affiliated with the institution?

Reg. Yes.

Opin. Any of these individuals would qualify as long as he or she is not involved in the care and use of laboratory animals, either directly or through consulting and similar tangential service. (See 5:31.)

5:35 Can a retiree from an institution serve as the unaffiliated member of the same institution's IACUC?

Opin. In the authors' experience, very few institutions exercise the opportunity to appoint either retirees or alumni as unaffiliated members perhaps because they are disinclined to create the perception of a conflict of interest (see 5:31). Even in the case of pensioners, the retirement income was earned previously, and the unaffiliated member's livelihood is not in immediate jeopardy if they offer opinions at odds with institutional interests. Appointments of this type are not prohibited by federal regulations nor discouraged in guidance documents. In the few cases of this type known to one of the authors, fidelity to the institution appeared to engender robust participation, yielding very honest and critical questions, review, and commentary.

5:36 Can or should the nonaffiliated member be compensated for service on the IACUC?

Opin. Neither the PHS Policy nor the AWAR prohibit the compensation of the outside member. NIH/OLAW has stated that nominal compensation is permissible without jeopardizing a member's nonaffiliated status if compensation is only in conjunction with service on the IACUC and is not so substantial as to be considered an important source of income or to influence voting on the IACUC.[16] (See 5:9). The *Guide* (pp. 24–25) also endorses this approach.

In accordance with the findings of the previous survey on this subject, the 25 organizations offered narrative comments in the 2012 survey indicating that the nonaffiliated members were compensated minimally. Support for the travel and parking for the nonaffiliated member was frequently mentioned and several organizations indicated that the nonaffiliated members were compensated for their time or were given honoraria using the descriptor "small." Only one institution offered information on the level of compensation, noting that the member received a payment per meeting and per protocol. Other types of compensation mentioned included support for attendance at meetings on IACUC function and provision of gift cards periodically. In the previous edition,[3] one institution mentioned making a charitable contribution on behalf of the nonaffiliated members. These types of innovative approaches illustrate the wide latitude institutions can exercise to acknowledge the valuable contributions and important public service provided by the unaffiliated member. Institutions should be cautioned that compensation of their nonaffiliated members at or exceeding the rate of the member's primary income source is likely to be perceived by the lay public and by regulatory agencies as a conflict of interest and cannot be recommended.

5:37 Can the nonaffiliated member serve on more than one IACUC, in the same or in different institutions?

Opin. There is no proscription against nonaffiliated members' serving on more than one IACUC if they have an interest in that level of involvement, and the authors are aware of individuals who have done this with laudable interest and commitment. PHS-Assured institutions with two or more IACUCs should contact NIH/OLAW for advice if they intend to utilize one nonaffiliated member for service on more than one of these. However, NIH/OLAW, in reviewing this chapter, has indicated that it is not allowing a single nonaffiliated member to serve on multiple IACUCs at the same institution. Skeptics may speculate that the excessive involvement of an individual as a nonaffiliated member on several IACUCs is an indication of a shift in the individual's neutrality and objectivity, which might be a reason for institutions to avoid the "overzealous" contributor. Also, scientific staff may be concerned that the individual could be a potential source for the leakage of scientific ideas to other institutions. This matter could be readily addressed by assuring that the nonaffiliated member has received training in issues concerning confidentiality and nondisclosure of proprietary information. Institutions may also want to consider asking their nonaffiliated members to sign formal agreements to this effect.

5:38 What is the role of the nonaffiliated member on the IACUC?

Opin. The role of the nonaffiliated member is not really defined other than as described in 5:31. Most institutions are hopeful that this individual will play an active role in all IACUC activities and will be comfortable with and become adept at making persistent, straightforward, and disarming inquiries about matters that are undetected by the institutional members on the IACUC. The nonaffiliated member also projects the institution's commitment to transparency by affording the member a full understanding of the organization's scientific programs involving animals and ensuring that control measures for research posing a risk to the public are adequate.

5:39 What are the background and qualifications of individuals who typically fill the role of the nonscientist?

Reg. The PHS Policy (IV,A,3,b,3) indicates that the "primary concerns" of individuals serving in this capacity should be "in a nonscientific area," and the specific examples cited (e.g., ethicist, lawyer, member of the clergy) have no obvious connections to any area of science. The AWAR do not specify the need to appoint an IACUC member in this particular category.[7]

Opin. In many institutions, an individual who has been chosen has some scientific training and perhaps even some responsibilities in a scientific area but clearly does not qualify as a "practicing scientist with experience in research involving animals" as noted in PHS Policy (IV,A,3,b,2). The types of individuals seen in this role among the institutions surveyed for the previous edition[3] included lawyers, clergy, health and safety personnel, business and human resources personnel, public relations personnel, quality assurance/control personnel, and technicians not involved with animal care or use. In recent years, the authors have noted more institutions relying on IACUC administrators serving in this role.

5:40 Is a biostatistician considered a nonscientist?

Opin. A FAQ issued by NIH/OLAW describes a nonscientist as an individual who is "not a practicing scientist experienced in research involving animals."[7] Clearly, statisticians are scientists because statistics is a branch of mathematics, a pure science. However, as noted, most institutions would classify a biostatistician as a "nonscientist" by the NIH/OLAW definition, even if his or her work may entail the analysis of data from animal studies. Although the biostatistician might be involved in the application of a mathematical science to the analysis of animal studies, few people would embrace the proposition that a biostatistician would be tempted to encourage animal studies simply to perpetuate the data needed for the practice of his or her science. (See 5:46.)

5:41 What interests does the nonscientist represent and what role does he/she play on the IACUC?

Opin. This individual serves to diversify the IACUC membership further, adding balance for the scientific members who may be regarded as having a vested interest in the promotion of animal studies. As with the nonaffiliated member, the nonscientist should participate in and contribute to all of the IACUC's mandated activities. With regard to protocol review, this member can be especially valuable by working to ensure that the IACUC meets its new obligation under the *Guide* (p. 25) to provide a "clear and concise sequential description of the procedures the use of animals that is easily understood by all members of the committee."

5:42 Should the scientist on the IACUC have animal research experience?

Reg. PHS Policy (IV,A,3,b,2) and the *Guide* (p. 9) but not the AWAR, indicate that an IACUC should include in its membership at least one practicing scientist who has laboratory animal experience.

Opin. This is a universal practice among the institutions known to the authors. In large institutions, typically a number of scientists are appointed to the IACUC to reflect the interests of different user groups in the organization. The appointment of scientists who have laboratory animal experience aids the IACUC's discussion of relevant issues during protocol and program review. It helps the IACUC better understand the selection, use, and limitations of animal models and certain aspects of experimental design. Institutions conducting research and teaching in agricultural animals should also be aware of and follow recommendations of the *Ag Guide* concerning the experience of the scientific members. The *Ag Guide* recommends two scientists should be included in a basic IACUC: one is a scientist with experience in conducting research or teaching with agricultural animals and the other is an animal, dairy or poultry scientist with experience in the management of agricultural animals.

5:43 Should the scientific member of the IACUC be a senior-level scientist, or is a junior-level scientist acceptable?

Opin. Scientists at any level are appropriate, but many institutions feel that senior scientists confer more authority and credibility on the IACUC. However, most academic organizations have found the recruitment of senior scientists to the IACUC to be difficult because these individuals usually have already made significant committee contributions to the institution in other areas and have other significant obligations within the institution. In academic institutions, junior faculty appointed to the IACUC may find the extensive time commitment an impediment to their efforts to develop a viable portfolio in research and scholarship for promotion and tenure consideration. Also regarding promotion and tenure, institutions should be aware of the conflicts of interest that may arise when a junior faculty member reviews the protocols of senior faculty member involved in salary, promotion and tenure decisions.

5:44 What is the role of the scientist on the IACUC?

Opin. The principal roles of the scientist on the IACUC are to ensure that the interests of scientific colleagues are being fairly represented in the review process and to aid in the IACUC's assessment of the relevance, validity, and technical aspects of the studies proposed. Of course, the scientist recognizes the confluence of animal health, welfare, and scientific interests in research animal studies and plays an important role as a proponent for the prudent, ethical, and humane use of animals. In broader issues involving program development and implementation, the scientist can give to the IACUC perspectives on how best to launch new initiatives to engender the support of the scientific community and others involved in the care and use of laboratory animals.

5:45 Should the IACUC include an ethicist as a member?

Opin. Some institutions have been successful in recruiting one or more individuals knowledgeable in ethics with an education and work focus in the arts and humanities or in business, but few have been able to identify a bona fide ethicist. Clearly, issues in ethics may be important in the IACUC's deliberations and there may be occasions when an ethicist can be helpful in leading this process.

5:46 Should the IACUC include a biostatistician as a member?

Opin. Most IACUCs have not pursued the option of including a biostatistician to ensure that investigators are using "the minimum number (of animals) to obtain valid results" (PHS Policy IV,D,1,a–IV,D,1,b; U.S. Government Principle III; AWAR §2.31,e,1; §2.31,e,2) because they expect the scientist(s) and veterinarian on the IACUC to evaluate this area. The *Guide* (p. 25) emphasizes that animal care and use protocols should include a discussion of statistical justification for the numbers of animals and the experimental group sizes. Scientific, veterinary, or other IACUC members should be capable of performing this function if they can commit the effort and are given sufficient information by the investigator about the assumptions on the data anticipated in the experimental study. However, confirming that the investigator is proposing to use the minimal number of animals necessary can involve an extensive and potentially redundant effort of the IACUC. If an IACUC routinely encounters weak or dubious explanations for the numbers of animals requested in protocols, the addition of a biostatistician to the IACUC may be helpful. (See 5:40.)

5:47 Should animal care technicians be considered for IACUC membership?

Opin. In the authors' experience, several institutions have appointed animal care technicians to their IACUCs in an effort to develop a better sense of how the basic animal care program functions. An added benefit is that it may promote a sense of empowerment and inclusion in the IACUC review processes for the entire animal care staff.

5:48 If an institution is performing Good Laboratory Practices (GLP) studies, can a quality assurance (QA) person also serve on the IACUC?

Opin. In the authors' opinion, it would be permissible to have a QA person serve on the IACUC. QA personnel have responsibility for all processes influencing the integrity, quality, validity, and documentation of data. IACUC functions are not included in the QA bailiwick, but some IACUCs might benefit from a heightened scrutiny of their processes as well as the general contributions an individual who has a QA background would be expected to add as his or her experience accumulates.

5:49 Should individuals who have specialty expertise be included as voting, nonvoting, or ad hoc IACUC members?

Reg. PHS Policy (IV,C,2; IV,C,3) and the AWAR (§2.31,d,2; §2.31,d,3) clearly indicate that voting privileges are reserved for full IACUC members who are appointed by the CEO. Nevertheless, the PHS Policy and AWAR do not preclude the use of nonvoting consultants.

Opin. Having the individuals available who have specific expertise to bring the best decision-making to the IACUC in different areas is always in the institution's interest. Individuals who have specialty expertise might include properly appointed ad hoc members who are willing to participate in the full range of IACUC activities. Individuals who have specialty expertise and are not interested in assuming all of the responsibilities of full membership but who would be helpful periodically to the IACUC can be invited to participate in an advisory capacity as consultants without voting privileges. The types of ad hoc appointees vary by function and are diverse. Institutions might consider including librarians for alternative searches; environmental or occupational health and safety personnel for hazardous studies; and ichthyologists, herpetologists, or ornithologists for studies involving species in their respective disciplines. Some institutions maintain a list of consultants willing to participate at any time as experts in different areas of importance to rigorous and thoughtful protocol review. These contingencies should be addressed in the charter developed by the institution for the IACUC.

5:50 Should the IACUC seek the advice of consultants for issues that may require special expertise, such as pain and distress concerns?

Reg. PHS Policy (IV,C,3) and the AWAR (§2.31,d,3) describe the use of consultants and how they may not "approve or withhold approval of an activity or vote with the IACUC."

Opin. IACUCs have the freedom to exercise their ability to use consultants whenever necessary. However, in most institutions, their use is reserved for truly extraordinary issues as opposed to any areas in which the IACUC feels it might improve its decision making. In the authors' opinion, many IACUCs would potentially benefit from the more liberal use of consultants.

5:51 Should consultants be from inside or outside the institution, and should they be anonymous?

Opin. Consultants can be selected from either inside or outside the institution, depending upon the availability of relevant expertise and the resolution of concerns about objectivity. Most organizations have not found it practical, necessary, or desirable to maintain the anonymity of the consultant. However, if the need for a consultant is likely to involve a highly contentious and disputed high-stakes issue, anonymity may be elected by the IACUC or imposed by the consultant as a condition of participation. Regardless of the circumstances, the IACUC has the obligation to foster an environment conducive for the investigator to respond to the consultant's critique.

5:52 Should consultants to the IACUC be compensated?

Opin. External consultants frequently are compensated for their efforts, and it seems reasonable that these individuals should be compensated according to the institutional policies established for other types of external academic review.

5:53 Should there be a confidentiality agreement between the consultant and the institution?

Opin. It is advisable for institutions to have a confidentiality agreement with consultants to prevent the disclosure of important, and potentially patentable, scientific or technical information. In addition, confidentiality agreements are important because they establish or reaffirm a precedent for the manner in which all materials relevant to the IACUC's deliberations are handled. This issue becomes particularly important from the legal standpoint if the institution is faced with the prospect of retrieving documents that may have been obtained illegally by adversarial parties. In a broader sense, this issue is also important for all members on the IACUC. It is highly advisable that all institutions assure that IACUC members are aware of and adhere to the requirements for confidentiality pertaining to protocol review and other committee business. Training of the IACUC should encompass this important area.

5:54 What support staff is useful for the IACUC?

Opin. Many institutions have developed an animal care and use compliance office to provide the administrative, monitoring, and training activities of the IACUC. The composition of this office depends upon the size and scope of the animal care and use program and the extent to which the IACUC members are willing and able to become personally involved in the fine details of committee function. In general, the growth of IACUC administrative offices nationally affirms that scientific (and other) members of IACUCs have increasingly found significant support from their IOs for divesting the responsibility for program implementation in many areas of IACUC function. In the opinion of the authors, an effective IACUC office can greatly benefit an institution ensuring timely efficient IACUC administrative processes, creating a partnership with the scientific community/programs and building a reputation for customer service by the IACUC. A survey from within the IACUC administrator community indicated a 17% growth in IACUC offices from 2005 (44%) to October 2012 (61%).[17] The types of positions that have proved to be very useful on the IACUC office staff in different settings include a director/administrator or senior assistant to the IACUC Chair; training, monitoring (e.g., post-approval monitoring) and compliance personnel; computer/database specialists; biostatisticians; and clerical staff. In small institutions with limited programs, a small staff of 0.5 to 1 full-time equivalent (FTE) or less may be sufficient, but in large, complex programs it is not uncommon for the IACUC office staff to include six to ten FTEs.

5:55 Who typically provides the administrative "support" staff to the IACUC?

Opin. The support staff is generally recognized and provided as an administrative cost that is provided under the authority of the CEO or IO of the organization.

5:56 What are some of the typical responsibilities of the administrative office?

Opin. Administrative staff is often involved in conducting facility inspections and preparing the correspondence related to semiannual reviews, protocol matters, post-approval monitoring activities and policies and procedures of the IACUC, and in maintaining the records thereof. The staff also may be responsible for making the arrangements for all IACUC activities. If staff members have appropriate training and technical expertise, they may be involved in conducting institutional training seminars and visiting laboratories and animal use areas to provide instruction and monitor ongoing activities. This staff may maintain databases of IACUC protocol approval and of the appropriate ancillary approvals in other areas such as use of biohazards, and the enrollment in occupational health and safety activities. In many cases they may be a key liaison with other important institutional functions that support the animal care and use program such as occupational health and safety and physical plant personnel. In a few institutions, support staff has been available to conduct literature searches for alternatives to animal use and to provide a statistical analysis to confirm that the investigator has properly justified the number of animals requested.

Although many institutions draw their IACUC support staff from technical backgrounds involving experience with laboratory animals, others have had success recruiting from diverse disciplines without this experience. Institutions should promote the development activities of this staff by supporting their attendance at workshops and conferences dedicated to IACUC-related topics, animal care and use program improvements, and/or the training of technical staff. These types of programs are offered by AAALAS, AAALAC, Public Responsibility in Medicine and Research (PRIM&R), IACUC Administrators Association, the Laboratory Animal Welfare Training Exchange (LAWTE), IACUC 101, Scientists Center for Animal Welfare (SCAW) and NIH/OLAW in partnership with regional academic sponsors and also by commercial enterprises. This investment in staff training will make the staff more effective in general. Training by these programs can efficiently give staff that does not have prior experience in the laboratory animal arena a quick and balanced understanding of key issues in this new environment as well as enhance the effectiveness of IACUC professional that already have experience.

References

1. Committees to Revise the Guide for the Care and Use of Agricultural Animals in Agricultural Research and Teaching. 2010. *Guide for the Care and Use of Agricultural Animals in Agricultural Research and Teaching*, 3rd edition. Champaign: Federation of Animal Science Societies. http://www.fass.org/docs/agguide3rd/Ag_Guide_3rd_ed.pdf.
2. Committee for the Update of the Guide for the Care and Use of Laboratory Animals. 2011. *Guide for the Care and Use of Laboratory Animals*, 8th edition. Washington, D.C.: National Academies Press.
3. Newcomer, C.E. 2007. General composition of the IACUC and specific roles of the IACUC members. In *The IACUC Handbook*, 2nd edition, eds. J. Silverman, M. Suckow and S. Murthy, 37–60. Boca Raton: CRC Press.
4. National Institutes of Health, Office of Laboratory Animal Welfare. 2012. Frequently Asked Questions B.1, What are the IACUC membership criteria? http://grants.nih.gov/grants/olaw/faqs.htm#IACUC_1.
5. National Institutes of Health, Office of Laboratory Animal Welfare. 2002. *Public Health Service Policy on Humane Care and Use of Laboratory Animals*. Bethesda: National Institutes of Health. http://grants.nih.gov/grants/OLAW%20/references/phspol.htm#ReportingRequirements.
6. National Institutes of Health, Office of Laboratory Animal Welfare. 2002. *Public Health Service Policy on Humane Care and Use of Laboratory Animals*. Bethesda: National Institutes of Health. http://grants.nih.gov/grants/olaw/references/phspol.htm#FOOTNOTES.
7. National Institutes of Health, Office of Laboratory Animal Welfare. 2012. Frequently Asked Questions B12, How does OLAW define nonscientific and nonaffiliated IACUC members? http://grants.nih.gov/grants/olaw/faqs.htm#IACUC_12).
8. U.S. Department of Agriculture, Animal and Plant Health Inspection Service. 2011. Policy 15, Institutional Official and IACUC Membership. http://www.aphis.usda.gov/animal_welfare/policy.php?policy=15.
9. Committee on Occupational Safety and Health in Research Animal Facilities. 1997. *Occupational Health and Safety in the Care and Use of Research Animals*. Washington, D.C.: National Academy Press.
10. Robert, H.M. and S.C. Robert. 1990. *The Scott, Foresman Robert's Rules of Order Newly Revised*, eds. H.M. Robert III and W.J. Evans. Glenview: Scott, Foresman.
11. National Institutes of Health, Office for Protection from Research Risks. 1993. Frequently asked questions about the Public Health Service Policy on Humane Care and Use of Laboratory Animals. *ILAR News* 35(3–4):47.

12. Newcomer, C.E. 1997. Behold! The animal care and use magician! *Connection* [newsletter of AAALAC]: 10–11. http://www.aaalac.org/publications/Connection/Where_is_the_institution.pdf.
13. National Institutes of Health, Office of Laboratory Animal Welfare. 2009. Notice NOT-OD-09-035, Guidance to IACUCs Regarding Use of Designated Member Review (DMR) for Animal Study Proposal Review Subsequent to Full Committee Review (FCR). http://grants.nih.gov/grants/guide/notice-files/NOT-OD-09-035.html.
14. National Institutes of Health. 2011. Notice, NOT-OD-11-053, Guidance to Reduce Regulatory Burden for IACUC Administration Regarding Alternative Members and Approval Dates. http://grants.nih.gov/grants/guide/notice-files/NOT-OD-11-053.html.
15. Brown, P. and C. Gibson. 2009. Multiple campuses, one IACUC; how many AVs? A word from OLAW and USDA. *Lab Anim.* (N.Y.) 38:114.
16. National Institutes of Health, Office of Laboratory Animal Welfare. 2012. Frequently Asked Questions B.11, May the institution pay or reimburse expenses incurred by nonaffiliated members? http://grants.nih.gov/grants/olaw/faqs.htm#IACUC_11.
17. Greer, W.G. 2012. Personal communication. Summary of data obtained from a survey conducted by the IACUC Administrator Association.

6

Frequency and Conduct of Regular IACUC Meetings

Sreekant Murthy

Introduction

The goal of the IACUC is to ensure the humane care and use of animals in research, testing and teaching. Additional goals are for the IACUC to remain in compliance with regulations while maintaining flexibility to meet the unique needs of the institution's academic and research mission. There are four fundamental components for the effective and efficient functioning of the IACUC. They are: 1) active participation of the IACUC members to meet investigator's needs in conducting research; 2) participation of community members to bring transparency and public conscience; 3) participation of AVs to provide appropriate medical care and monitor the well-being of animals and 4) firm commitment from the institution to comply with regulations and their Assurance (if applicable) by establishing a committee to oversee the institution's animal program, facilities, and research and educational projects involving the use of animals.

The IACUC derives its authority from laws mandated by the 1985 HREA and the AWA. These laws mandate the proper constitution and functions of the IACUC. The institution and researchers play a significant role in the composition and functioning of the IACUC through self- governance and monitoring. These laws require an adequate veterinary care program and provision of an appropriate environment and housing for animals by maintaining facilities that are for the management of institution's animal care and use program. A critical part of this program is a requirement by the *Guide* for establishing an occupational health and safety program[1] to protect the health and safety of employees involved in animal-related activities.

A comparison of the general issues for IACUC composition and functions shows that the PHS endorses the *U.S. Government Principles for the Utilization and Care of Vertebrate Animals Used in Testing, Research, and Training.* Principle I states: "The transportation, care, and use of animals should be in accordance with the Animal Welfare Act (7 U.S.C. 2131 *et. seq.*) and other applicable Federal laws, guidelines, and policies." The guidelines include those in the *Guide*. The AWA requires that proposed activities using animals be conducted in accordance with the AWAR. Based on these acts, regulations, policies, and principles, IACUCs are constituted to protect the health and well-being of the animals used in testing, research, research training, and biological testing. The AWAR (§2.31,a) state that "nothing in this part shall be deemed to permit the Committee or IACUC to prescribe methods or set standards for the design, performance, or conduct of actual research or experimentation by a research facility." This statement, and Principle II of the U.S. Government Principles noted above, ensures animal care and use without compromising research for the good of society. Therefore, a carefully composed IACUC that oversees humane use and well-being of the animals and facilitates animal research is fundamental to a research facility's teaching and research objectives and ultimately to the good of society. This chapter addresses critical issues that are pertinent to the functioning of the IACUC.

6:1 How is the frequency of meetings of the IACUC decided?

Reg. IACUCs are required by the AWAR (§2.31,c,1; §2.31,c,2) and PHS Policy (IV,B,1-IV,B,3) to conduct a review of its animal care and use program every 6 months. APHIS/AC requires that a majority of the IACUC review and sign the ensuing semi-annual inspection report (AWAR §2.31,c,3) but does not require that this be done at a convened meeting of the IACUC. Likewise, NIH/OLAW requests that the semiannual report has the signatures of a majority of the IACUC members but there is no requirement to do this at a convened meeting of a quorum of the IACUC.[2] The *Guide* (p. 25) states that "the IACUC must meet as often as necessary to fulfill institutional responsibilities and records of committee meetings and results of deliberations should be maintained." There is a requirement that if full committee review of an agenda item is required, then approval of such items must be made at a convened meeting of a quorum of the IACUC with the approval vote of a majority of the quorum present.[3]

The IACUC is empowered by PHS Policy IV,B,8 and AWAR §2.31,c,8 to suspend an activity if it finds noncompliance with the PHS Policy, *Guide*, Assurance, or violations of the AWAR. Suspension may occur only after review of the matter at a convened meeting of a quorum of the IACUC, and with the suspension vote of a majority of the quorum present.

The AWAR (§2.31,d,5) requires that all IACUC approved protocols be reviewed no less frequently than annually. The PHS Policy (IV,C,5) requires that the IACUC conduct a complete review at least once every 3 years. However, it is up to the IACUC to determine the best way to conduct this review. Although not required, some institutions prefer to conduct semiannual program reviews as a full committee in a convened meeting with a quorum of the IACUC present. Thus, the number of convened meetings may vary from none to as many as needed to meet regulatory and institutional requirements.

Opin. There is a fundamental responsibility for the IACUC to meet as often as necessary to oversee the coordination and implementation of effective, efficient systems for protocol review and program review to remain in compliance with PHS Policy and the AWAR. Each institution must evaluate its responsibilities to the investigator and funding organizations. Based on the number of IACUC applications, the institution should conduct as many meetings as necessary to review protocols. Many institutions decide on the frequency based on NIH or other sponsor deadlines that may include industry-related and internal applications. Full committee reviews and suspension of an activity are required to be conducted in a convened meeting. It is this author's opinion that animal welfare is best served by having the semiannual review of the program of animal care and use (including facility inspections) at a convened meeting of the IACUC with a quorum present.

6:2 Who determines the frequency of IACUC meetings?

Opin. There is no regulation or policy that authorizes specific individuals to determine the frequency of meetings. In a survey published in *The IACUC Handbook* (second edition),[4] in 61% of the institutions the IACUC Chair or the Chair after consulting members of the IACUC, determined the frequency of meetings. In some instances (11%), the institution's internal policy or standard operating procedures or the IACUC administrator determined the frequency. Approximately 6% of the institutions either did not have a policy or the IO determined the frequency. It is recommended that the scheduling of a meeting be done by the IACUC after consultation with the Chair. In this author's opinion, it is a good practice to write a policy on the frequency of meetings and publish the schedule of meetings and the deadlines for the submission of protocols. This will allow the investigators to be informed about the schedules and prepare for submissions on time. If a meeting is going to be cancelled either due to lack of protocol submissions or for any other reasons, the IACUC Chair or administrator should inform the members of the IACUC and the investigators as early as practicable that the regularly scheduled meeting will not be held and provide a date for when the next meeting is going to be held.

6:3 What is the frequency and conduct of Full Committee Review (FCR) IACUC meetings?

Opin. There is no regulation or policy that requires how often the IACUC should meet as a full committee. In circumstances where the IACUC is conducting FCR or the IACUC is contemplating suspending an activity, it can only be done with the use of a full committee. FCR must also be used when any member of the IACUC requests FCR during the process of preparing to conduct DMR. PHS Policy IV,C,2 states that each IACUC member shall be provided with a list of proposed research projects to be reviewed, with written descriptions available. Any member may obtain, upon request, full committee review of those research projects. It is important to reiterate here that IACUCs must strive to conduct as many convened meetings as necessary to comply with federal regulations in the situations stated above and to ensure that the institution's animal welfare program is fulfilling its responsibility to provide humane care and that the procedures used are supported by practical, ethical and scientific principles. The following survey response illustrates the practices in various institutions on frequency and conduct of IACUC meetings.

Surv. How frequently does your IACUC have a full committee meeting to review protocols?

- Never 1/297 (0.3%)
- Only when requested by the chairperson or an IACUC member 24/297 (8.1%)
- Approximately monthly or more often 176/297 (59.3%)
- Approximately six times a year 13/297 (4.4%)
- Approximately four times a year 24/297 (8.1%)
- Approximately two times a year 50/297 (16.8%)
- Approximately one time a year 4/297 (1.4%)
- Other* 5/297 (1.7%)

*Not specified

6:4 What parliamentary rules are used in conducting IACUC meetings?

Reg. The PHS Policy (e.g., IV,C,2; IV,C,6) and AWAR (e.g., §2.31,d,2; §2.31,d,6) address some aspects of parliamentary procedure, such as conducting FCR of protocols at a convened meeting of a quorum of the IACUC and voting in case of a suspension. NIH/OLAW has noted in a publication that the validity of IACUC activities is always predicated on the existence of a properly constituted IACUC, although there is no requirement that all of the members be present at all meetings.[3,5]

Opin. None of the regulatory agencies or the *Guide* prescribes a regulation for conducting the IACUC parliamentary procedures. Nevertheless, survey results published in the previous edition of this book[4] indicated that 69% of institutions use *Robert's Rules of Order.* Approximately 24% of institutions use internal institutional rules combined with Robert's Rules and a few institutions follow their own internal rules. It is imperative that institutions not allow *Robert's Rules of Order* or other parliamentary or internal rules to supplant those procedures which are specified in PHS Policy or the AWAR.

6:5 What information should be provided to IACUC members prior to a meeting?

Reg. The AWAR (§2.31,d,2) and PHS Policy (IV,C,2) state: prior to IACUC review, each member of the committee shall be provided with a list of proposed activities to be reviewed. Written descriptions of all proposed activities that involve the care and use of animals shall be available to all IACUC members, and any member of the IACUC may obtain, upon request, full committee review of those activities.

Opin. Some IACUCs provide their members with all correspondence concerning the review of a protocol. This includes amendments, correspondence concerning allegations of protocol violations or animal mistreatment, and results of subcommittee investigations (particularly those that may lead to suspensions and therefore require review and approval by the full committee). Also, reports of any task forces or subcommittees that the IACUC has requested to review or modify a policy or procedure, updates on institutional policies and procedures that affect the conduct of an IACUC meeting, membership changes, updates on APHIS/AC regulations and NIH/OLAW policies and guidance on the PHS Policy and relevant changes to the NIH Grants Policy Statement, copies of all pertinent publications, and meeting notices that may enhance the review process.

Some institutions are moving the IACUC submission and review process from paper to electronic methods. There is no regulation to prevent or allow electronic reviews so long as reviewers have the information about all of the proposed activities. The entire electronic submission and review process depends upon the robustness of the system to permit the reviewer to view all pertinent documents specific to an existing continuing review, initial review submission, or a third year *de novo* review of a protocol renewal. The system should permit viewing specific information about the investigator and study personnel to ensure that there is no reason for concern whether they have an outstanding non-compliance issue. The system should also be sufficiently robust to document the process of approval especially when conducting designated member review (DMR), amendments and training updates.

6:6 What additional verbal or written information might be given to IACUC members at the time of a regular meeting, but before research protocols are discussed?

Opin. Verbal information that can be presented at a convened meeting may include reports from the Chief Executive Officer, IO, Director of the Office of Research, Chair of the IACUC, AV, supervisor of the animal care facilities, reports from other committees and presentations from invited guests. Other written materials may include opinions by external consultants and any other reports that the Chair or the institution deems necessary for the proper conduct of IACUC functions. Written information may include minutes of a previous meeting for approval, the agenda for the convened meeting, summaries of protocol reviews, external consultant's reports, post-approval monitoring reports, subcommittee reports, updates on policies and procedures, educational materials, regulatory updates, and any other handouts that are pertinent for the conduct of IACUC functions. Very few IACUCs conduct separate quarterly, semiannual, or annual business meetings to discuss various issues that have been the general concern to the IACUC (see 6:7 below). When held, these meetings are for the purpose of discussing IACUCs administrative procedures. They may include

- Evaluation of surveys conducted by the IACUC
- Recruitment of new members
- Review of attendance records
- Replacing a member
- Training new members
- Creating and updating standard operating procedures
- Creating/reviewing institution's emergency/disaster plans
- Policy on whistle blowers
- Updating IACUC forms
- Educational matters
- Methods to train and certify animal users
- Creation and dissemination of newsletters for animal users
- Security issues
- Legal opinions
- Ethical issues for the appropriate use of animals in research and education
- Post-approval monitoring reports
- Review of a disaster plan
- Review of concerns about the animal care and use program

An additional purpose of business meetings is to give the IACUCs undivided attention at regular meetings for the review of the institution's animal care and use program, protocols for animal use, and flexibility to meet the unique needs of the institution.

6:7 What specific business should be conducted at regular IACUC meetings? Is a formal vote required for any business conducted?

Reg. The AWAR (§2.31,c,1-8; §2.31,d,1), PHS Policy (IV,B,1-IV,B,8; IV,C,1-IV,C,8) and the *Guide* (p. 25) indicate that the following things must be done by the IACUC:

- Review and approval of reports of an institution's program for the humane care and use of animals and inspections of animal facilities, including animal study areas (laboratories), before submitting them to the IO.
- Discuss and distinguish significant deficiencies from minor deficiencies included in the above report.
- Discuss and provide a reasonable and specific plan and schedule (with dates) for correcting each deficiency.
- Review, and if review warrants, investigate concerns involving animal care and use complaints from all sources.
- Make recommendations to the IO on all aspects of institutional animal program and training of personnel who handle animals.
- Review and approve, or require modifications or clarifications, or withhold approval of proposed and ongoing activities with significant changes related to animal use.
- Suspend an activity involving animals that compromise the health and well-being of an animal used in research (PHS Policy IV,C,6) in a convened meeting.
- Any business that the Committee considers threatening to, or has already affected, the health and well-being of the animals used in their research facility.
- Continuing oversight of animal after the IACUC's initial protocol review to ensure the well-being of the animals and to refine research procedures.

Opin. Many of the above items are pertinent for the semiannual review of an institution's animal care and use program. Any institutional program that is reviewed at a convened meeting, whether the program does or does not have deficiencies (minor or significant) and with a plan and schedule for corrections recommended by the committee, should be formally approved by a majority of the quorum present. All minority opinions also must be recorded. This is a recommended procedure to follow but federal regulations do not require a formal vote at a convened meeting of the IACUC to approve these actions. These items are distinctly different from those that are discussed in a regular meeting. Regular meetings are usually reserved for reviewing research protocols. However, there is no regulatory reason for not conducting both aspects of IACUC business at a single convened meeting.

Many IACUCs have agendas that include discussions of the institutional program of animal care and use, review of protocols, and other general business of the IACUC. All business matters that pertain to institutional program evaluations, protocol reviews, sanctions (if the institution has authorized the IACUC to do so), and any other matters that are part of the AWAR and PHS Policy (or institutional policy) can be approved in a convened meeting. However, to suspend an animal activity or to conduct a full committee review of a protocol with a subsequent vote, the IACUC must meet in a convened meeting of a quorum of the IACUC with a majority of the quorum present.[3] A majority of the quorum present must vote to approve or suspend a project. All other business matters should be discussed and, if the committee is developing a policy, such policies should be approved by a majority of members. The survey results published in the previous edition of this book[4] showed that in most institutions the IACUC discussed all business matters and protocol reviews in the same meeting. However, it is important that the IACUC reserve the committee's time to conduct protocol reviews to aid the investigator and possibly meet at a separate meeting to discuss business matters, unless the business matter is related to an activity being discussed.

6:8 At a full committee meeting, does the IACUC require a formal vote for items other than the approval of a protocol or the suspension of an animal activity?

Reg. The formal votes required for the approval of items other than the protocol itself are described in question 6:1. Compliance with the PHS Policy IV,B,5 requires written recommendations to the IO regarding any aspect of the institution's animal program, facilities, or personnel training. This policy does not state that the written recommendations to the IO require a formal vote and approval. Federal regulations do not require a formal vote to approve meeting minutes. The regulations only speak of maintaining records (AWAR §2.35,a,1-3; PHS Policy IV,E). NIH/OLAW guidance regarding use of designated member review subsequent to full committee review states that "All IACUC members agree in advance in writing that the quorum of members present at a convened meeting may decide by unanimous vote to use DMR subsequent to FCR when modification is required to secure approval."[6]

Opin. Each PHS Assured or APHIS/AC registered institution, acting through its IACUC, is responsible for all animal-related activities at the institution. All of these activities are not always protocol specific and some agenda items may require discussion and a majority vote of a quorum of the members present at a convened meeting of a properly constituted IACUC for the suspension of an animal activity. Examples of these include allegations of mistreatment or noncompliance, suspension of researchers' other animal activities unrelated to the protocol in question, substantiation or rejection of a whistle blower's allegations, changes in policies and standard operating procedures. Such discussions may have to be formally approved by the IACUC as they may impact animal welfare. Many institutions take a formal vote on business items discussed at a convened meeting. Some institutions, in their internal SOPs, require approval of the minutes at a subsequent meeting. In order to use DMR subsequent to FCR, the quorum of members must approve the use of DMR[6] (see 6:20 for details).

The following survey response illustrates the practices in various institutions when IACUCs require a formal vote for items other the approval of a protocol or the suspension of an animal activity.

Surv. At a full committee meeting, does your IACUC require a formal vote for items other than the approval of a protocol or the suspension of an animal activity? Example: A formal vote to approve meeting minutes.

- Yes, for some but not all administrative items 102/297 (34.3%)
- Yes, for just about all administrative items 157/297 (52.9%)
- No 36/297 (12.1%)

- I don't know 0/297 (0%)
- Not applicable 1/297 (0.3%)
- Other* 1/297 (0.3%)

*Not specified

6:9 For a full committee meeting to occur, what should be your IACUC's policy with regard to the attendance of specific members?

Reg. The AWAR (§2.31,b,1-§2.31,b,4) require that the IACUC consist of at least three members, a Chair, a veterinarian and a person not affiliated with the institution. PHS Policy (IV,A,3,b; IV,A,3,c) and the *Guide* (p. 24) require a minimum of five members that includes a veterinarian, a scientist, a nonscientist, and a nonaffiliated member. However, neither the PHS Policy nor the *Guide* specifies the role of the fifth member. *The Institutional Animal Care and Use Committee Guidebook*[7] under "Committee composition" (pp. 12–13) does include a Chair as one of the IACUC's members and provides a clear description of the Chair's role as an effective leader who is crucial to an effective IACUC and this individual needs full support of the IO. At the same time, the *Guidebook* also states that one member can fulfill more than one of the roles in the IACUC membership. Be that as it may, for a quorum, the minimum attendance is two under the AWAR and three under the PHS Policy. (See 6:14 for a definition of quorum.)

Opin. Ideally, the AV (or designee), nonaffiliated and nonscientist (under PHS Policy) members (or their respective alternates) should be present at meetings to address ethical issues and appropriate use of animals from the non-institutional and non-scientific perspectives. Also, it is ideal to have the IACUC membership at more than the minimum number so that the quorum can be maintained if one or two members are not available for the meeting. This will also help provide the expertise necessary for thorough and efficient review of protocols. The presence of a statistician will add additional strength in evaluating one of the three "R"s (Reduction). However, it is not legally required for any specific member to be present at all meetings, as long as the committee membership is properly constituted and a quorum is present.[5,8]

Surv. What is your IACUC's policy with regard to attendance of specific IACUC members?

- Not applicable as we do not have full committee meetings 1/297 (0.3%)
- All voting members (or their alternate) must be present 6/297 (2%)
- The attending veterinarian (or designee) must be there 34/297 (12%)
- At least some of the members (or alternates) required by regulation or policy must be there 22/297 (7%)
- As long as a quorum is present we will have the meeting 229/297 (77%)
- I don't know 0/297 (0%)
- Other* 5/297 (2%)

*Not specified

6:10 What should be your IACUC's expectation with regard to the frequency of attendance of an IACUC member at meetings?

Opin. Attendance of a committee member at the scheduled meetings is an important consideration in selecting a member. In a previous survey conducted on this question, the majority of institutions did not have a formal policy, but observed that one-half to two-thirds of regular members generally attended the meeting.[4] Attendance of certain members such as the AV (or designee), a member with subject matter expertise (e.g., a scientist) and a non-scientist is critical for the proper review of a protocol or a policy. The absence of specific members may result in unnecessary delays in the approval of a study or an inadequate or poor review due to a lack of reviewer expertise. Determination of the validity of the absence should be at the discretion of institution's officials and particularly the committee Chair. The IO or the Chair may set guidelines for members missing consecutive or too many meetings. Likewise, the institution in its bylaws and standard operating procedures may set up policies when and how far in advance of a committee meeting a member should notify the committee of his/her absence, a policy for unacceptable absences and reporting absences to the institution's officials. A point to consider at the time of appointing the committee is to select appropriate alternates for each of the members so that there will be availability of expertise to review protocols. This will also help to prepare alternates to become full members when an opportunity arises for changing or replacing a member. The IACUC must always strive for 100% attendance so that the quality of discussions that occurs in the meeting is enhanced to an extent that the committee can speak with one voice or differing opinions are heard in order to take appropriate actions. While 100% attendance may not be achievable or possible, what is possible is that absences from meetings are kept at a minimum and when they do occur they are clearly unavoidable.

6:11 What actions are taken if an IACUC member does not attend an adequate number of meetings?

Opin. There is no regulation that requires actions to be taken when members do not attend an adequate number of meetings. NIH/OLAW FAQ B.4 states that "Attention should be paid to attendance at IACUC meetings to ensure that an appropriate mix of members attends meetings. Chronic nonattendance by IACUC members, especially those explicitly required by PHS Policy or USDA regulations, implies a lack of participation in the oversight responsibility of the IACUC."[9] Thus, the institution and the IACUC should pay considerable attention to attendance of PHS- and USDA-specified members so when federal auditors come for inspection, they find that the institution is in good standings with respect to members' attendance.

A previous survey[4] conducted on this question showed that majority of institutions do not have a formal policy on actions to be taken for lack of attendance by a member. It is important that the institution or the IACUC to establish a policy on actions to be taken toward a member who fails to attend two consecutive meetings or groups of meetings without an explanation acceptable to the committee Chair. Institutions should establish policies on what constitutes and grounds for removal following a request by the Chair or a recommendation by the Chair or the IACUC to the institution's appointing official to remove such members. It is not unusual to recognize certain members typically attending only those meetings in which their own or their colleague's (departmental) protocol is being reviewed. In such cases, careful attention should be paid to make sure that there is no direct conflict in that member attending and voting on those protocols. In those cases, it may be better for the member to voluntarily abstain or for the Chair to request the member to abstain from voting. An abstention does not affect the outcome of a vote for approval made by a properly constituted quorum unless the abstention is due to a conflict of interest. A conflict of interest abstention (more properly called a recusal), decreases the size of the quorum present.[10] PHS Policy (IV,C,2) states that "no member of the IACUC may participate in the IACUC review or approval of a research project in which the member has a conflicting interest (e.g., is personally involved in the project) except to provide information requested by the IACUC; nor may a member who has a conflicting interest contribute to the constitution of a quorum." In the same survey[4] some institutions responded that the institution should send an informal notification or a memo requesting that the member attend more meetings before any formal action is taken.

6:12 If member is to be asked to leave the committee, who would formally take this action?

Opin. Removing a committee member is also a relatively uncharted and controversial territory unless the evidence is clear that the member has violated certain IACUC policies including attendance requirements or breach of confidential committee information. Generally the Chair or the IACUC may make a recommendation to the appointing official (e.g., the Chief Executive Officer, CEO) to remove or replace a member. In general, the institutional officer appointing members honors such requests if the requests are well substantiated. A previous survey[4] conducted on this topic showed that in approximately 50% of the institution, the IO formally took the action. The overall response in the current survey is similar to the previous survey. Having the IO take the formal action may be acceptable as long as the institutional policy to remove a member from the committee formally gives that responsibility to an administrative official such as the IO. (See 5:11.)

Surv. If IACUC members (or alternates) were to be asked to resign from the committee, who would formally take this action?

• We have no policy	49/296 (16.6%)
• The Institutional Official	164/296 (55.4%)
• The IACUC chairperson	51/296 (17.2%)
• I don't know	17/296 (5.7%)
• Other*	15/296 (5.1%)

*Other ranged from university president or Chief Executive Officer (CEO) taking the action, IO and CEO, the IO, Chair and IO, Chair and Dean, to it has not happened.

6:13 When does your IACUC veterinarian first have the opportunity to evaluate pain and distress on a protocol and provide any input back to the investigator?

Reg. The AWAR (§2.33,b,4; §2.33,b,5) state that the AV should provide "Guidance to principal investigators and other personnel involved in the care and use of animals regarding handling, immobilization, anesthesia, analgesia, tranquilization, and euthanasia;" and "Adequate pre-procedural and post-procedural care in accordance with current established veterinary medical and nursing procedures."

The AWAR (§2.31,d,1,iv,B) require that for procedures involving more than momentary or slight pain or distress, either the veterinarian or his or her designee must be consulted in the planning of the research project. The *Guide* (p. 106) states that the AV "should provide guidance to investigators and all personnel involved in the care and use of animals to ensure appropriate husbandry, handling, medical treatment, immobilization, sedation, analgesia, and euthanasia." It also states that the AV should provide guidance and oversight of surgery programs and perioperative care.

Opin. When research projects involve pain or distress, the veterinary consultation on analgesia or anesthesia is normally provided prior to the IACUC review. It is generally better if the institution establishes a policy when this consultation should occur. It is preferable to have this consultation before the meeting and before a protocol is submitted for review as it may prevent unnecessary delays in approval. If for extraordinary reasons a veterinarian is unavailable another approach to this requirement is to have the AV identify one or more designees who can be involved in planning and consultation. Another veterinarian with knowledge in this area is an appropriate choice for such designation. Non-veterinarians also may be involved in the consultation at the discretion of the AV. (See Chapters 5 and 27 for additional information.)

The author and the *Guide* endorse the recommendations of the American College of Laboratory Animal Medicine (ACLAM).[11] ACLAM states that "the institution bears responsibility and must assure, through authority explicitly delegated to the veterinarian or to the IACUC that only facilities with programs appropriate for the intended surgical procedures are utilized and that personnel are adequately trained and competent to perform the procedures. The veterinarian's inherent responsibility includes monitoring and providing recommendations concerning preoperative procedures, surgical techniques, the qualifications of institutional staff to perform surgery and the provision of postoperative care."[11]

ACLAM also recommends that "The veterinarian must be involved in the review and approval of all animal care and use in the institutional program. This includes advising on the design and performance of experiments using animals as related to model selection, collection and analysis of samples and data from animals, and methods and techniques proposed or in use. This responsibility is usually shared with investigators, the IACUC, and external peer reviewers."[11]

The author also endorses the APHIS/AC policy that the AV retains the authority to alter post-operative care if unexpected pain and/or distress occur in an animal. However, the IACUC must approve a significant change to the protocol if the AV requests to alter post-operative care for the remaining animals.[12]

According to a previous survey,[4] in many institutions the AV evaluated pain and distress in sufficient time before the protocol was reviewed by the IACUC. In some institutions both the AV and IACUC members had the opportunity to evaluate the pain at the same time. In this author's opinion, it is best for the PI to involve the AV or the AV's designee in planning the type of care and appropriate use of drugs to relieve pain or distress in order to avoid possible delays in securing approval.

The NIH Grants Policy Statement (http://grants.nih.gov/grants/policy/nihgps_2012/nihgps_ch2.htm#additional_review_criteria) and the Worksheet for Review of the Vertebrate Animal Section (VAS) provided to NIH applicants[13] require that the investigator describe the procedures for ensuring that the discomfort, distress, pain and injury will be limited to that which is unavoidable in the conduct of scientifically sound research. They require investigators to describe the use of analgesic, anesthetic, and tranquilizing drugs and/or comfortable restraining devices, where appropriate, minimize discomfort, distress, pain, and injury. In many institutions investigators involve the AV prior to submitting an NIH grant. By doing this pain or distress and the appropriate use of drugs are properly discussed prior to an IACUC review. This allows the PI to secure timely approval for "just-in-time" reviews. (See 8:22–8:24.)

6:14 What constitutes the quorum needed to conduct an IACUC meeting?

Reg. A quorum is an assembly of a majority (more than 50%) of the voting members of the IACUC. For the AWAR (§2.31,b), the minimum number is two and for PHS Policy (IV,A,3,b), the minimum number is three (provided the IACUC has only three and five members, respectively). Otherwise, it requires more than 50% of the members to be present in a convened meeting to constitute a quorum. Both the AWAR and the PHS Policy state that a member who has a conflict of interest may not contribute to the constitution of a quorum. No member may participate in the IACUC review of an activity in which that member has a conflicting interest (e.g., is personally involved in the activity), except to provide information requested by the IACUC; nor may a member who has conflicting interest contribute to the constitution of a quorum at which a proposed activity relevant to the conflict is being considered for approval (AWAR §2.31,d,2; PHS Policy IV,C,2). The IACUC may invite consultants to assist in the review, but they may not approve or withhold approval of an IACUC activity; therefore, they cannot vote and be counted towards meeting quorum.[14]

NIH/OLAW and APHIS/AC[15] have accepted the practice of using designated alternates for IACUC members, if the alternates are formally appointed by the Chief Executive Officer and identified in the NIH/OLAW Animal Welfare Assurance. In these instances, the individual(s) serving as alternate(s) do not add to the total number of committee members unless the primary member is not at the meeting.

Opin. Often, many institutions face the problem attaining a quorum of the IACUC. Some institutions apply many methods to attain or maintain quorum at a convened meeting. These methods can include calling members of the committee ahead of the meeting date to alert them about an upcoming meeting or arranging a meeting on a day that is most suitable for all or a majority of the members. Other ideas include designating alternative members or providing food at the meetings so that the members can devote their lunch, breakfast, or dinner time for the meeting. All of these methods are workable and there are no AWAR or PHS Policy limitations for applying these methods.

Generally, it is advisable to have the IACUC membership at more than the minimum number mandated by either the AWAR or PHS Policy. For example, it is possible that one or more members may leave the meeting at the most inopportune time due to prior commitments or unavoidable circumstances. This might result in the loss of a quorum if only the minimum number of persons are on the IACUC and present at the meeting. Persons with conflicting interests may affect the quorum composition if IACUC membership is small as they cannot be used to constitute a quorum (AWAR §2.31,d,2; PHS Policy IV,C,2). The problem of a member leaving a convened meeting can often be avoided if protocols reviewed by that individual are placed on the top of the agenda.

The simple polling of IACUC members for their votes does not, however, satisfy the definition of a meeting of a convened quorum and is not to be used for conducting IACUC business that requires the vote of a convened quorum of the committee.[15] For example, polling should not be considered a valid method of voting under the "full committee review" method of protocol review and is not an acceptable substitute for having a vote of a convened quorum to consider the suspension of a previously approved activity involving animals. E-mail, polling, conference calls, etc., are not normally acceptable methods for full committee review (see 6:18; 6:19 for exceptions). A vote on a possible suspension of an animal activity requires a face-to-face meeting of a quorum of the IACUC (see 6:11). It is acceptable to poll members to see if any want a full committee review before relegating a protocol to Designated Member Review.

IACUCs often encounter problems when members abstain from voting due to conflict of interest or other valid reasons. In such cases quorum will be lost if the abstention is due to a conflict of interest.[10]

Approximately 15% of institutions previously surveyed[4] indicated that their IACUCs required all five members mandated by PHS Policy to be present at each meeting. The AWAR require only three members to constitute an IACUC. However, the majority of institutions that submit federal grants must adhere to the PHS Policy, which requires the institution to constitute an IACUC consisting of at least five members.

6:15 If a quorum is in place at the beginning of an IACUC meeting, but the quorum is lost during the meeting, is there any way that the meeting can continue?

Opin. There are no regulations that stipulate whether a meeting should be suspended if the quorum is lost in the middle of a full committee meeting. However, the AWAR and PHS Policy (see 6:14) state that a quorum of members must be present to approve, withhold approval, or perform other IACUC actions such as suspending an animal activity. Therefore, all IACUC activities that require voting by a majority of the quorum will become suspended. In such exceptional circumstances, according to PHS Policy[17] the IACUC with a quorum of members present may unanimously vote to require modifications to secure approval and have the revised protocol reviewed and approved by DMR or returned for FCR at a convened meeting.

If all members of the IACUC are not present at a meeting, the committee may use DMR subsequent to FCR by following the following stipulations.[17]

a. All IACUC members agree *in advance in writing* that the quorum of members present at a convened meeting may decide by unanimous vote to use DMR subsequent to FCR when modification is needed to secure approval. However, any member of the IACUC may, at any time, request to see the revised protocol and/or request FCR of the protocol.

b. In order to conduct reviews by DMR subsequent to FCR, the institution should specify its intention to conduct reviews in this manner in its Assurance with OLAW. (IACUCs that newly elect to utilize a standard operating procedure for DMR subsequent to FCR should provide information about this program change to OLAW in the next Annual Report.)

If all members are not present and the IACUC lacks written standard procedures as described above, the committee has the option to vote to return the protocol for FCR at a convened meeting or to employ DMR. If electing to use DMR, all members, including the members not present at the meeting, must have the revised research protocol available to them and must have the opportunity to call for FCR. A DMR may be conducted *only* if all members of the committee have had the opportunity to request FCR and none have done so (PHS Policy IV,C,2).[17]

Alternatively, to avoid issues related to the maintenance of a quorum, the institution may amend its NIH/OLAW Assurance to conduct a review by electronic means; however, even such electronic reviews require stringent application of policies that are described in 6:17 and 6:18. APHIS/AC does not provide any direction for audiovisual conferencing as an alternate to face-to-face meeting. However, APHIS/AC concurs with NIH/OLAW that "... the functional equivalent of a convened quorum in the full committee mode might be emulated through real-time, interactive, video-conferencing added to the end of the described electronic review process."[18]

6:16 Should any or selected investigators be invited to the IACUC meeting?

Opin. It is customary for some IACUCs to invite investigators to respond to questions in a face-to-face meeting. Since the committee has the dual responsibilities of thoroughly reviewing protocols and executing decisions, the PI can answer specific questions raised by the IACUC. This often saves time for the IACUC and the PI. The PI should leave the meeting before a vote is taken. If the protocol involves a committee member that has a conflicting interest, the member is usually asked to leave the meeting while the committee is discussing that protocol (although the person may give factual information to the committee if requested to do so). The member with a conflicting interest should leave the meeting when the committee votes.

6:17 What methods do IACUCs use to formally review a protocol?

Reg. To approve an animal activity the criteria to be met are covered under PHS Policy (IV,c,1,a-IV,c,1,g; IV,d,1,a-IV,d,1,e) and the AWAR (§2.31,d; §2.31,e). The method to accomplish this is left up to each individual institution. According to these regulations, there are only two lawful methods of IACUC review allowed by PHS Policy and AWAR: (1) full committee review (FCR) by a convened quorum of the IACUC and (2) designated member review (DMR) by one or more members chosen by the chairperson, employed only after all voting members have been provided an opportunity to call for full committee review. The NIH Guide for Grants and Contracts NOT-OD-06-052[18] provides guidance regarding the use of tele-video-conferencing when a convened meeting is required. The guidance is also provided in an article published by NIH/OLAW concerning IACUC functions and the use of technological advances in communication technology.[17]

Opin. Many institution use FCR to review protocols. FCR protocols require a convened meeting of a quorum of members and protocols can only be approved by an affirmative vote of a majority of quorum present in a meeting. Some institutions also use an informal pre-review process. During this pre-review process a primary member or a team of members may generate some questions and the PI can respond to those questions ahead of a meeting and the modified protocol is reviewed by the IACUC. Pre-review in any institution is a voluntary process that aids the investigator and the IACUC in resolving some of the major issues related to the protocol, cuts down the committee time in reviewing the protocol and improves the turnaround time. However, even though a protocol has gone through a pre-review process, the IACUC must still grant approval prior to the research team initiating the activities described in the protocol. (See 9:4–9:10.)

According to the NIH Guide for Grants and Contracts NOT-OD-06-052,[18] methods of telecommunication (e.g., telephone or video conferencing) that are acceptable for the conduct of official IACUC business requiring a quorum are as follows:

- All members are given notice of the meeting.
- Documents normally provided to members during a physically convened meeting are provided to all members in advance of the meeting.
- All members have access to the documents and the technology necessary to fully participate.
- A quorum of voting members is convened when required by PHS Policy.
- The forum allows for real time verbal interaction equivalent to that occurring in a physically convened meeting (i.e., members can actively and equally participate and there is simultaneous communication).
- If a vote is called for, the vote occurs during the meeting and is taken in a manner that ensures an accurate count of the vote. A mail ballot or individual telephone polling cannot substitute for a convened meeting.
- Opinions of absent members that are transmitted by mail, telephone, fax or e-mail may be considered by the convened IACUC members but may not be counted as votes or considered as part of the quorum.
- Written minutes of the meeting are maintained in accord with the PHS Policy (IV,E,1,b).

It should be noted that APHIS/AC concurs with Notice NOT-OD-06-052.

Surv. 1 What methods do your IACUC use to formally review a protocol?

- Entirely full committee review — 69/296 (23%)
- Mostly full committee review but some designated member review — 94/296 (32%)
- Entirely designated member review — 9/296 (3%)
- Mostly designated member review but some full committee review — 95/296 (32%)
- A fairly even mix of full committee review and designated member review — 29/296 (10%)

Surv. 2 How does your IACUC perform full committee review (FCR)?

• We do not use FCR	4/297 (1%)
• We have 1 or 2 members (non-chairperson) present the protocol and then the remainder of the committee can join the discussion before we vote	141/297 (48%)
• The Chair presents the protocol and then remainder of the committee can join the discussion before vote	113/297 (38%)
• The full committee rubber stamps a prior designated member review decision	2/297 (0.7%)
• Other*	37/297 (13%)

*Not specified

6:18 What are the requirements for conducting other than face-to-face full committee review IACUC meetings ("exceptional circumstances" meetings)?

Reg. Guidance from NIH/OLAW states that "the traditional meeting is still seen as the optimum environment for fulfilling the intent of the PHS Policy, OPRR (now OLAW) recognizes the new communications tools available and the need for flexibility in the ways that institutions may comply with the PHS Policy in the many diverse settings encountered."[17] APHIS/AC and NIH/OLAW concur that a traditional full committee meeting with a convened quorum of the IACUC leads to a review in real-time; however, the functional equivalent of a convened quorum may be emulated through real-time, interactive video conferencing.[16] For these reasons, several criteria have been provided for establishing alternate methods that may be considered functionally equivalent to meetings of a convened quorum under exceptional circumstances. Two of those criteria are that the alternate approach must include a high degree of interactivity and allow for careful deliberation of sensitive issues. Another is that a quorum of IACUC members must be in direct communication with each other and be given full opportunity to participate for the duration of the meeting. Institutions are reminded that details of their IACUC procedures, especially those that may vary from those outlined in the PHS Policy should be thoroughly described in the institutional Assurance and submitted for OLAW review. Thus, the requirements are essentially the same as face to face meetings.

Opin. In exceptional circumstances, telephone and audio-visual conferencing may be appropriate alternatives to face-to-face meetings for full committee review of protocols and other IACUC matters The requirements for exceptional meetings are[17,18]

- All members must be given ample prior notice to participate.
- At least a quorum of the voting members (>50%) must be convened on the same conferencing line whether it is telephone or an audio-visual line.
- The quorum of IACUC members must be in direct communication with each other and be given full opportunity to participate for the duration of the meeting.
- The minutes of the meetings must be compiled and maintained on file as required by federal regulatory agencies.

6:19 How can a designated member review (DMR) of protocols be accomplished using the electronic media meeting format?

Reg. The PHS Policy (IV,C,2) states "If full committee review is not requested, at least one member of the IACUC, designated by the chairperson and qualified to conduct the review, shall review those research projects and have the authority to approve, require modifications in (to secure approval) or request full committee review of those research projects." The AWAR (§2.31,d,2) has identical language. In both cases, prior to IACUC review, each member of the committee must be provided with a list of proposed activities to be reviewed. Written descriptions of all proposed activities that involve the care and use of animals shall be available to all members, and any member of the IACUC may obtain, upon request, full committee review of those activities.[5] The regulations do not say how the proposed activities list needs to be disseminated so long as the list of activities are provided or made accessible to the IACUC members.

Opin. A DMR means that the IACUC Chair appoints a reviewer(s) to review an IACUC protocol. DMR also can be conducted using electronic media to facilitate the review process as long as the requirements in the PHS Policy (IV,C,2) are met. (See chapter 8 for additional information.) Many institutions have a submission webpage interface that allows researchers to submit new protocols, continuing review, amendments, protocol closures and three year protocol renewals electronically to the IACUC. Investigators can expect faster service with electronic submission. Some institutions use the electronic submission methods to automatically streamline protocols destined for FCRs whenever protocols involve unalleviated pain and distress, surgery, food or water restriction, paralytics, or prolonged restraint. All other protocols are sent for DMR. Some recent IACUC software is equipped for primary and secondary reviewers to electronically enter their review comments at the time of the review. Such review comments are fully visible to all members of the committee prior to IACUC meetings. The process allows designated reviewers to make either unanimous decisions or call for a full committee review. Certainly electronic DMR serves the intended purpose of speeding up the protocol review process. It may also decrease the overall workload relevant to full committee review.

NIH/OLAW and APHIS/AC have endorsed the use of electronic communications for certain IACUC functions based on 2006 guidance.[18] According to this publication "The conveyance, by fax or email, of information such as the institutional Assurance; animal study proposals; agendas and minutes of meetings; institutional policies and standard operating procedures; reports, announcements or correspondence from oversight or regulatory agencies; and other matters related to the institutional animal care and use program for consideration and review by IACUC members would be regarded as appropriate." Institutions should exercise caution when electronic methods are used to prevent security breaches and should assure that privileged and confidential information is communicated through a secure site that has controlled access and is encrypted.

6:20 What procedures are appropriate for using designated member review (DMR) after a full committee review (FCR)?

Reg. NIH/OLAW has provided guidance to PHS awardee institutions and IACUCs concerning the acceptable actions of an IACUC to meet the requirements of PHS Policy.[6] This guidance relates to IACUC required modifications to secure approval, and actions the IACUC may take for the use of DMR subsequent to FCR. APHIS/AC concurs with NIH/OLAW on this guidance.[6] The guidance indicates that the IACUC may take one of the following three actions:

1. If all members of the IACUC *are present* at a meeting, the committee may vote to require modifications to secure approval and have the revised research protocol reviewed and approved by DMR, or returned for FCR at a future convened meeting.
2. If all members of the IACUC *are not present* at a meeting, the committee may use DMR subsequent to FCR according to the following stipulations:
    a. All IACUC members agree in advance, in writing, that the quorum of members present at a convened meeting may decide by unanimous vote to use DMR subsequent to FCR when a modification is needed to secure approval. However, any member of the IACUC may, at any time, request to see the revised protocol and/or request FCR of the protocol.
    b. In order to conduct reviews by DMR subsequent to FCR, the institution should specify its intention to conduct reviews in this manner in its Assurance with OLAW. (IACUCs that newly elect to utilize a standard operating procedure for DMR subsequent to FCR should provide information about this program change to OLAW in the next Annual Report.)
3. If all members *are not* present and the IACUC lacks written standard procedures as described above, the committee has the option to vote to return the protocol for FCR at a convened meeting or to employ DMR. If electing to use DMR, all members, including the members not present at the meeting, must have the revised research protocol available to them and must have the opportunity to call for FCR.[6] A DMR may be conducted only if all members of the committee have had the opportunity to request FCR and none have done so (PHS Policy IV,C,2).

Opin. Federal regulations allow only two methods of protocol review, FCR and DMR. It is entirely up to the institution to decide whether DMR can be used subsequent to FCR. The institution, in its NIH/OLAW Assurance, must include that they will be using DMR subsequent to FCR. Despite this Assurance, the IACUC, under specific circumstances where animal welfare issues are of significant concern, should exercise the option of using FCR. Even though the Assurance allows DMR subsequent to FCR, under no circumstances does the guidance or the Assurance prevent the use of FCR. While DMR after FCR speeds up the review time and saves committee time, a thorough FCR is essential when animal welfare is paramount. The survey below indicates that while 38% of the institutions have prior agreement, approximately 21% of institutions use only FCR to secure approval.

Surv. With reference to using Designated Member Review (DMR) after a Full Committee Review (FCR) please check all appropriate boxes.

- We do not use DMR after FCR 74/294 (21%)
- We occasionally or often use DMR after FCR 78/294 (22%)
- When we use DMR after FCR the quorum present votes to send protocols for DMR and subsequently the full committee is given the opportunity to agree to use DMR 66/294 (19%)

- When we use DMR after FCR the quorum present votes to send the protocol to DMR but all members have previously agree to allow the DMR process to proceed — 135/294 (38%)
- Other* — 2/294 (0.6%)

*Not specified

6:21 Can IACUC reports be endorsed by electronic methods and can signatures of members be obtained by using electronic methods?

Opin. Since most reports, including semiannual program review and inspection reports, require IACUC approval (see 6:1), alternative methods that exactly follow those used at a legally convened face-to-face meeting of the IACUC may be appropriate; however, semiannual reports can be randomly signed by the members. Electronic FCR meetings may not be conducted via e-mail, as this does not provide the opportunity for real-time communication. IACUC activities generally require documenting votes or verifying committee approval. They also may require signatures to be legally binding. Electronic methods of voting are conducted quite rapidly, but signatures are difficult to obtain by this method. Therefore, in a meeting using telecommunication the Chair calls for a vote, the IACUC coordinator records the vote, and then prepares and distributes the final document by mail to solicit original signatures for the permanent record. Some electronic methods have the ability to validate and authenticate credentials of investigators and committee members. Such methods have been used in a few institutions. In this author's opinion, if electronic signatures are going to be used for IACUC administrative activities, the institution's NIH/OLAW Assurance should specify such use.

6:22 Must the minutes of IACUC meetings be recorded? If so, what precautions should be taken to maintain confidentiality and what is the impact on the federal Freedom of Information Act (FOIA)?

Reg. PHS Policy (IV,E,1,b) and the AWAR (§2.35,a,1) require records of IACUC meetings to be kept, whether the meeting is face-to-face or held alternatively using electronic communications (see 6:18; 6:19). They must be compiled and maintained on file for three years after the study ends (AWAR §2.35,f; PHS Policy IV,E,2). Minutes must include records of attendance, activities of the committee, and committee deliberations (PHS Policy IV,E,1,b). The federal FOIA, at the present time, only applies to documents in the possession of a federal agency.

Opin. The general advice is that meeting minutes should be written carefully with appropriate usage of words so that the lay public and any outside auditors would not be alarmed by reports obtained through the FOIA or pertinent state laws (see Chapter 22). The *Institutional Animal Care and Use Committee Guidebook*[19] cautions that the person assigned to prepare reports should be aware of the FOIA and their state's open record laws. The FOIA does not require a private organization or business to release any information directly to the public, whether it has been submitted to government or not. However, information submitted to the government by such organizations are made available through a FOIA request. There are some exemptions to FOIA requests such as the one covering trade secrets and confidential business information.[20] Many of such reports are accessible under such laws; therefore, particular care must be taken to prevent use of unnecessary language that may be detrimental to the institution or the investigator.

While recording meeting minutes satisfies regulatory needs, special care must be taken to ensure proper usage of words since meeting minutes can, under certain circumstances, be obtained through a state's FOIA or any state open record laws (the federal FOIA is at present only applicable to documents in the possession of a federal agency). (See Chapter 22.) The best practice is not to include extraneous information in the minutes so that the information is not taken out of context by any readers. The minutes should contain sufficient detail for an outside person to determine the discussions and conclusions on routine and controverted issues discussed by the committee. It is not required to specifically identify a member involved in the deliberations. Follow your state and federal laws and regulations on recording minutes to remain in compliance. Institutions should seek advice from their legal counsel regarding what information needs to be included in the minutes and reports.

It is important that each IACUC take into account potential security problems of any privileged and trade-secret confidential information which may be protected by law (AWA §2157). Since minutes are often generated using computers, special care should be taken to prevent unauthorized access to computer records. The mandatory enforcement of passwords, controlled access and encryption should be practiced to prevent unauthorized access. IACUC members should be advised to return copies of the minutes to the IACUC secretary once the minutes are ratified by the committee. Unwanted and used copies of the minutes should be destroyed using appropriate methods.

Many IACUCs record meetings on audiotape. This raises a question as to whether these tapes should also be saved as part of recordkeeping. This is internal policy unique to each IACUC—whether to save or erase the tape. In general, these tapes also contain irrelevant discussions; therefore, it is customary for the IACUC to transcribe portions of the discussion that are relevant and then erase the tape. There are no regulations that explicitly deal with audio taping and archiving of such tapes. (See 8:19; 8:22; and Chapter 22 for additional information.)

Surv. Does the Freedom of Information Act (FOIA) or a similar state law impact on what is placed in your IACUC minutes?

- We do not record minutes — 2/297 (0.7%)
- FOIA has no impact on our minutes — 145/297 (48.8%)
- FOIA may have an impact but it depends on the topic being discussed — 98/297 (33%)
- FOIA seems to have a major impact on our IACUC's minutes — 26/297 (8.8%)
- I don't know — 26/297 (8.8%)

6:23 What voting information should be recorded by your IACUC?

Reg. The AWAR (§2.35,a,1) and PHS Policy (IV,E,1,b) state that the minutes of IACUC meetings include records of attendance and activities of the committee and committee deliberations. The AWAR (§2.31,d,2) and PHS Policy (IV,C,2) do not allow an IACUC member who has a conflicting interest to review, approve, or contribute to the constitution of quorum. The AWAR and the PHS Policy do not state whether the names of individuals who voted, approved, opposed, or abstained need to be recorded.

Opin. It is a common practice in many institutions to record the names of individuals attending a meeting. It is customary to record the actual number of individuals approving, withholding approval, or abstaining in a vote, but not the names of members.

6:24 Should each IACUC protocol be voted on individually?

Opin. In general, each protocol is unique and it is best to approve them individually as they are discussed. This method inherently avoids problems that arise from maintaining a quorum. An exception for collective approval arises if two similar (or dissimilar protocols) from the investigator are discussed, and the investigator is called to answer specific questions pertaining to one or more of the protocols.

6:25 Who records IACUC meeting minutes?

Opin. The institution should assume responsibility for selecting a person to record meeting minutes, based on input from the IO, the Office of Research, and the Chair of the IACUC. In some institutions the IACUC administrator or his/her designee records the minutes. Other examples are the Chair, a recording secretary, or a member of the committee recording the minutes. It is important that persons assigned to this task be knowledgeable of federal and other requirements and the institution's animal care and use program. If not, someone who is knowledgeable of IACUC regulations, policies and procedures should read the minutes before they are presented to the committee or administrative officials.

6:26 What should be included in the meeting minutes?

Reg. (See 6:21)

Opin. Because the AWAR (§2.35,a) and PHS Policy (IV,E,b) state that the minutes require the records of attendance, activities of the committee, and committee deliberations, it is reasonable that the following items be included in IACUC meeting minutes:

- Inspection reports distinguishing significant and minor deficiencies with plans and schedules for correcting each deficiency.
- Investigations of complaints and issues of noncompliance received from the public or research facility personnel.
- Approval, conditions such as modifications and clarifications, withheld approvals, and suspensions of protocols.
- Approval or withheld approval of proposed significant changes to protocols.
- Training and certification guidelines approved by the committee.
- Consultants' reports on protocols.
- New policies or guidance voted on by the committee.

The minutes should include sufficient detail of the discussions and language that is generally understood by the nonscientific and lay members of the committee and persons from outside agencies.

6:27 How does your IACUC use the minutes of its meetings?

Reg. The IACUC minutes must be maintained for the duration of the project plus at least 3 years and be accessible for inspection and copying by authorized NIH/OLAW or other PHS representatives at reasonable times and in a reasonable manner (PHS Policy IV,E,2). The AWAR (§2.35,a,1) lists the IACUC minutes as a record and the AWAR (§2.35,f) requires all records and reports to be maintained for a minimum of 3 years.

Opin. There are no guidelines for the way the recorded minutes of the meetings are to be used. In general, minutes are maintained as the official and primary record for NIH/OLAW, APHIS/AC and AAALAC inspections and site visits to document animal welfare issue, pain or distress issues, and programs and facilities issues discussed at the meeting. Minutes can be used for bringing consistency to program management, review procedures, confirming or modifying policies, and general procedures of the committee.

6:28 IACUCs often review many protocols at a meeting. The concern is that those protocols considered at the beginning of the meeting may be reviewed more thoroughly than those at the end of the meeting. How can IACUCs avoid this problem?

Opin. The AWAR (§2.31,c,4-§2.31,c,7) and PHS Policy (IV,C) provide appropriate guidelines for reviewing research projects, but the guidelines do not provide a time schedule for discussing reports and research projects in a meeting. The *Guide* (p. 25) suggests that each IACUC should meet as often as necessary to fulfill its responsibility and provides topics that should be considered in the review of animal care and use protocols (*Guide* pp. 24–34). Various practices used to ensure a thorough review of protocols, irrespective of the number of protocols and the agenda sequence, include lengthening the meeting time to accommodate review of all protocols in a single convened meeting without losing a quorum and postponing the review of some protocols without delaying proposed research. This generally happens in large institutions with many protocols to review. However, for small institutions with few protocols and those institutions that effectively and efficiently use DMR, this is not an issue.

6:29 Should an IACUC have a written policy on what constitutes a conflict of interest which may lead to a member being excused from voting?

Reg. The AWAR (§2.31,d,2) and the PHS Policy (IV,C,2) state that no IACUC member may participate in the IACUC review or approval of an activity in which that member has a conflicting interest except to provide information as requested by the IACUC.

Opin. Conflict of interest may arise in many ways. A common potential or apparent conflict of interest is when a PI comes from the same department as the IACUC member. It is not considered acceptable to serve as a designated reviewer in those circumstances as the member may have sole say on approving the protocol. In full committee consideration a member must make up her/his own mind whether to vote or abstain. However, if there is any apparent, perceived, or potential conflict of interest, the member should declare so before joining any review, discussion, or vote. If assigned as one of a team of reviewers, the member can either participate in the review or ask to be relieved from that duty. It is recognized that a member from the same discipline can often provide the best review of a protocol but may then have to abstain from any vote.

The responsibility of determining if a member has a conflict with the protocol to be reviewed may fall on the Chair. Therefore, it is a best practice for the Chair to start the meeting asking members whether they have a conflict with any of the protocols or animal issues that are going to be discussed at the meeting. This will help determine how the agenda needs to be navigated while paying attention to the quorum.

The overall guiding principle is that members should act professionally and impartially in considering IACUC matters. Likewise, resolution of a conflict of interest may arise in many ways. An SOP to resolve the conflict is critical for the growth of a research enterprise. The SOP should describe the procedures for resolving conflicts arising from the investigator as well as a committee member. Investigator-initiated conflicts generally arise at the time of submitting a protocol. Here the investigator may request that a certain member be excluded from the review of the protocol, providing evidence that substantiates the claim that a conflict exists with that member. If that member's expertise is essential for the review, the IACUC can exercise the option of using a consultant for the review. If a member thinks that he/she has a conflict with a protocol, the member should notify the IACUC Chair and not participate in the IACUC review or approval except to provide information when requested by the committee. Potential conflict is implicit if the member has competing research, access to funding information which may give an unfair advantage to the member, and occasionally personal bias against an investigator's research.

References

1. Committee for the Update of the Guide for the Care and Use of Laboratory Animals. 2011. *Guide for the Care and Use of Laboratory Animals*, 8th ed., 17–23. Washington, D.C.: National Academies Press.
2. Office of Laboratory Animal Welfare. 2011. Semiannual report to the Institutional Official. http://grants.nih.gov/grants/olaw/sampledoc/ioreport.htm.
3. Wolff, A. 2002. Correct conduct of full-committee and designated-member review. *Lab. Animal (NY)*. 31:28–31.
4. Murthy, S. 2002. Frequency and Conduct of Regular IACUC Meeting. In *The IACUC Handbook*, 2nd ed. Silverman J, M. Suckow, and S. Murthy. Boca Raton: Taylor & Francis.
5. Office of Laboratory Animal Welfare. 2012. Frequently Asked Questions B.3. Must certain members be present in order to conduct official business? http://grants.nih.gov/grants/olaw/faqs.htm#IACUC_3.
6. Office of Laboratory Animal Welfare. 2009. Guidance to IACUCs Regarding Use of Designated Member Review (DMR) for Animal Study Proposal Review Subsequent to Full Committee Review (FCR). Notice Number: NOT-OD-09-035. http://grants.nih.gov/grants/guide/notice-files/NOT-OD-09-035.html.
7. Office of Laboratory Animal Welfare. 2002. *Institutional Animal Care and Use Committee Guidebook*, 2nd ed., 12–13. Bethesda: National Institutes of Health.

8. Ellis, G. and N.L. Garnett. 1997. Maintenance of properly constituted IACUCs. OPRR Reports, No. 97–03. http://grants.nih.gov/grants/olaw/references/dc98-01.htm.
9. Office of Laboratory Animal Welfare. 2012. Frequently Asked Questions B.4. Is a certain level of meeting attendance required of IACUC members? http://grants.nih.gov/grants/olaw/faqs.htm#IACUC_4.
10. Office of Laboratory Animal Welfare. 2012. Frequently Asked Questions B.5. What is a quorum and when is a quorum required? http://grants.nih.gov/grants/olaw/faqs.htm#IACUC_5.
11. American College of Laboratory Animal Medicine, Public Statements, Adequate Veterinary Care. Accessed Sept. 25, 2012. http://www.aclam.org/Content/files/files/Public/Active/position_adeqvetcare.pdf.
12. Animal and Plant Health Inspection Service. 2011. *Animal Care Resource Guide.* Policy 3. Pre- and Post-Procedural Care. http://www.aphis.usda.gov/animal_welfare/downloads/policy/Policy%203%20Final.pdf.
13. Office of Laboratory Animal Welfare. Worksheet for Review of the Vertebrate Animal Section (VAS). Accessed Sept. 25, 2012. http://grants.nih.gov/grants/olaw/vaschecklist.pdf.
14. Office of Laboratory Animal Welfare. 2002. *Institutional Animal Care and Use Committee Guidebook,* 2nd ed., 85. Bethesda: National Institutes of Health.
15. Office of Extramural Research. 2001. Guidance Regarding Administrative IACUC Issues and Efforts to Reduce Regulatory Burden. 2001. Notice NOT-OD-01-017. http://grants.nih.gov/grants/guide/notice-files/NOT-OD-01-017.html
16. Garnett, N. and S. Potkay. 1995. Use of electronic communications for IACUC functions. *ILAR J.* 37: 190–192.
17. Office of Laboratory Animal Welfare. 2012. Frequently Asked Questions D.19. May an IACUC use designated member review (DMR) to review an animal study protocol subsequent to full committee review (FCR) when modifications are needed to secure approval? http://grants2.nih.gov/grants/olaw/faqs.htm#proto_19.
18. Office of Laboratory Animal Welfare. 2006. Guidance on Use of Telecommunications for IACUC Meetings under the PHS Policy on Humane Care and Use of Laboratory animals. http://grants.nih.gov/grants/guide/notice-files/not-od-06-052.html.
19. Office of Laboratory Animal Welfare. 2002. *Institutional Animal Care and Use Committee Guidebook,* 2nd ed., 32–33. Bethesda: National Institutes of Health.
20. Freedom of Information Act. 5 U.S.C. § 552, As Amended By Public Law No. 104-231, 110 Stat. 3048. Accessed Oct. 11, 2012. www.justice.gov/oip/foia_updates/Vol_XVII_4/page2.htm.

7

General Format of IACUC Protocol Forms

Christian E. Newcomer and William G. Greer

Introduction

Protocol review was mandated by federal law for most institutions using animals in research, teaching, and testing by the passage in 1985 of an amendment to the AWA known as the Improved Standards for Laboratory Animals Act (Public Law 99-198) and by a component of the legislative reauthorization for the NIH, known as the Health Research Extension Act of 1985 (Public Law 99-158). In advance of this requirement, and as early as 1980, several academic institutions already had begun to implement the basic concepts of protocol review. The veterinarians in these institutions had made a compelling case for needing some information about the nature of the ongoing animal use to be able to sign their institutional APHIS/AC annual report in good conscience. In the context of the contemporary standards for protocol review in most institutions, the forms used were rudimentary. The information retrieved in these pioneering efforts was extremely scant and generally would now be insufficient for the issuance of an IACUC approval. Nevertheless, these initial efforts were important because they introduced the setting, players, and ethos of an evolving process that eventually would be adapted and embellished at diverse institutions across the country as the federal mandates for protocol review took effect.

Many of the factors that influenced the early approaches to protocol review continue to be important today. These have shaped the character and content of the forms used in the protocol review process according to institutional preference and stimulated organizations to meet regulatory and ethical considerations with clarity, consistency, and completeness of all documents. For example, most institutions and investigators want the protocol review process to be objective, consistent, and efficient. This orientation fosters the support of research investigators and IACUC members who must commit precious time to fulfill this institutional requirement. However, the actual protocol review forms used to facilitate this process are markedly different from institution to institution. Institutional philosophy, the types of research conducted, pragmatic considerations related to administrative and regulatory detail, and many other ancillary factors can influence the format of the IACUC protocol review form.

The purposes of this chapter are to examine the various strategies used by institutions in the information collection phase of the protocol review process and to highlight the strategies that appear to be the most successful and widely adaptable. The chapter also discusses alternative approaches that have proved to be well suited to particular institutional environments. Information for the preparation of this chapter was acquired through insights into the practices of numerous institutions from conducting of AAALAC site visits or reviewing other materials related to AAALAC activities, from over 300 institutions during the conduct of IACUC Administrators Best Practice Meetings (see Chapter 5) and from the authors' personal experiences as IACUC members and/or administrators at more than a dozen institutions.

A well-designed protocol form can assist investigators in their efforts to provide the clear, concise, and comprehensive information necessary for the review process. It is very important to the smooth functioning of the IACUC. Most IACUCs have revised their protocol forms several times since the inception of the protocol review process in response to new or augmented regulatory requirements, the need to improve the quality and detail of the information provided, or the necessity to enhance the retrieval and sharing of information contained in these documents. While the protocol review form should be considered a living document and an IACUC should not be reluctant to change the protocol review format for due cause, format changes are likely to be disruptive and burdensome during a transition period. For this reason, format changes should be carefully planned and coupled with investigator outreach and education efforts.

7:1 Should there be a standard protocol form for submission to the IACUC?

Opin. The large majority of institutions require investigators to submit their animal care and use protocols to the IACUC for review on an institutionally developed, standardized protocol review form. This approach is usually regarded as an essential first step in the fulfillment of the IACUC's mission to establish a review process that is perceived as unbiased, consistent, efficient, and, under optimal conditions, user-friendly. Most institutions have PIs who file multiple studies with the IACUC and the use of a standardized form allows these individuals to develop a familiarity with the type of information required by the IACUC for successful completion of that committee's review process. Most institutions have found that the quality of information provided by PIs using standardized protocol forms improves over time. Nevertheless, when changes are incorporated into existing forms to meet newly defined needs of the IACUC, a period of coaching and re-acclimatization for investigators is necessary. Additionally, standardization is generally regarded as helpful and critical to the IACUC members in meeting their review obligations. The compartmentalization of information into predictable areas enables the IACUC members to focus their review and discussion on sensitive areas with greater alacrity and facilitates the retrieval of information pertinent to any subsequent IACUC monitoring and compliance activities.

It was quite unusual for institutions to conduct IACUC protocol review without the use of a standardized form, but this practice was used by some institutions at the time of the first edition of this book. The authors are not aware of any institutions that have continued to use this method. It was used mostly by academic institutions with smaller animal care and use programs and a limited research profile, usually serving ten or fewer PIs. In those instances, the IACUC reviewed the PI's grant application and a supplemental narrative concerning specific areas deemed necessary for IACUC review. The narratives were responsive to a list of queries provided by the IACUC to ensure that all of the federally mandated aspects of protocol review were covered. While it is conceivable that this approach could occasionally produce a reasonably integrated, relevant, and coherent document for the IACUC review process, in practice this rarely happened. In addition to requiring IACUC members to demonstrate a unique flexibility and commitment, this approach to protocol review challenged IACUC members to wade through much extraneous information and maintain a running inventory of the quality and location of the investigator's responses (or nonresponses) to important issues. In order for the information gleaned from the grant application and protocol at the time of the review to remain conveniently accessible to the IACUC for later use, a synopsis of the review becomes necessary. For these reasons, a free-form protocol IACUC review process will likely remain limited, if indeed it continues, to animal care and use programs in small institutions imbued with a sense of collegial courtesy, reciprocity, and the good faith principle.

7:2 What general information is necessary for the IACUC to review a protocol adequately?

Opin. Several authoritative documents reiterate the basic components of an animal care and use activity that should be assessed by the IACUC during the protocol review process. These documents include the AWAR, the PHS Policy, U.S. Government Principles (VI to VIII) and the *Guide* (pp. 25–35) which added new elements for the IACUC to consider. The AWAR impose specific review requirements on regulated animal species and many institutions have chosen as a matter of policy to extend these requirements to all vertebrate species. Although the practice appears to becoming rare, other institutions have maintained a dichotomy in the review criteria based upon exclusions provided in the definition of the term *animal* in the AWAR (§1.1). (See 12:1.) The authors endorse the following list of elements delineated in the *Guide* as a complete, comprehensive, and convenient resource for achieving a sound protocol review process serving the researcher and the IACUC:

- Rationale and purpose of the proposed use of animals
- A clear and concise sequential description of the procedures involving the use of animals that is easily understood by all members of the committee
- Availability or appropriateness of the use of less-invasive procedures, other species, isolated organ preparation, cell or tissue culture, or computer simulation
- Justification of the species and number of animals proposed
- Whenever possible, the number of animals and experimental group sizes should be statistically justified (e.g., provision of a power analysis)
- Unnecessary duplication of experiments
- Non-standard housing and husbandry requirements
- Impact of the procedures performed on the animals' well-being
- Appropriate sedation, analgesia, and anesthesia. (indices of pain or invasiveness might aid in the preparation and review of protocols)
- Conduct of surgical procedures including multiple operative procedures
- Post-procedural care and observation (for example, inclusion of post-treatment or post-surgical animal assessment forms)
- Description and rationale for anticipated or selected endpoints
- Criteria and process for timely intervention, removal of animals from a study, or euthanasia if painful or stressful outcomes are anticipated

- Method of euthanasia or disposition of animal, including planning for care of long-lived species following study completion
- Adequacy of training and experience of personnel in the procedures used, and roles and responsibilities of the personnel involved
- Use of hazardous materials and provision of a safe working environment
- Weighing study objectives against potential animal welfare concerns
- Response to unexpected outcomes
- Physical restraint
- Food and fluid regulation
- Use of non-pharmaceutical-grade chemicals and other substances

The reader should refer to other chapters of this book where many of these topics are explored in detail. NIH/OLAW has a sample animal study protocol document.[1] This document was written to address the needs of numerous diverse animal programs. NIH/OLAW intends for the institution to customize the sample document, adding or removing content to convert it into a form that is appropriate for the IACUC's needs and the institution's animal program.

7:3 Should the standard protocol form be used to gather other information needed for the institution to ensure program compliance with federal guidance documents on other topics?

Opin. Documents governing the development and operation of animal care and use programs typically identify two responsible parties; the IACUC and the institution. By law and policy the IACUC is typically responsible for those activities directly related to the proposed care and use of vertebrate animals for research, teaching or testing. Although the IACUC must review and assume responsibility for the entire animal program, the institution carries the ultimate responsibilities for establishing a functional occupational health and safety program, ensuring federal funding to support animal research is administered according to the terms and conditions of the grant or contract, establishing an effective training program, and ensuring complete and accurate records are maintained.

The protocol review form and review process are convenient and adaptable for capturing the information needed by the institution in fulfillment of these other responsibilities. A submission to the IACUC may not be considered complete unless the PI's submission includes a copy of his/her NIH grant application, for example, to allow the administrative office to complete the protocol and grant congruency process.[2] Institutions also can use the submission to ensure everyone listed on the protocol has completed federally required training in humane animal care and use. The submission form may include a question that asks the PI to verify the current training status of laboratory personnel. Final protocol approval may then be contingent on successful completion of the training. In addition, organizations often facilitate the distribution and completion of personal risk assessment forms by personnel with animal contact in conjunction with the protocol review process to aid the risk assessment review conducted by the program's occupational medicine service.

7:4 Is there an optimal format for an IACUC protocol form?

Opin. A central question in the protocol review process is how to optimize information retrieval from the areas identified in 7:2 and 7:3, in order for the IACUC protocol review process to be sufficiently rigorous to ensure high-quality animal care, safety, and other considerations and still improve a PI's prospect for receiving protocol approval on a first attempt. There is no apparent consensus on this matter as evidenced by the many different approaches taken to protocol reviews by IACUCs, which in turn are mirrored by the variety of forms used by different IACUCs. The forms used by institutions generally fall into the following categories:

- Those that state a requirement or use a question (regarding the areas defined in 7:2) to elicit a free-form narrative response and consolidate all relevant information onto a single form
- Those that use a free-form narrative approach but are divided into modules allowing the form to be tailored to the particulars of the submission (e.g., those not involving surgery would delete the sections aligned with this activity)
- Those that emphasize a checklist and fill-in-the-blank approach whenever possible, either in the modular or consolidated format: forms generally intended to expedite completion and review of straightforward protocols. The use of specific protocol templates developed to accommodate specific research activities common in the institution are also aligned with this approach.

Institutions using a free-form narrative for protocol review simply ask questions related to each of the items in 7:2, eliciting the investigator's response. The protocol form includes questions covering all areas and answers are typically provided only for the segments of the protocol that apply. Thus, in studies involving minimal animal procedures, the majority of the form may not be used. However, in protocols involving complicated studies with extensive animal use and entailing specialized expertise (e.g., surgery, post-procedural care, or the use of biohazardous agents), all information deemed necessary for the IACUC review is provided. Most institutions realize that investigators cannot be relied upon to provide the kind of qualitative and quantitative information necessary for protocol approval without assistance, even though this subject may have been extensively covered in institutional training and education activities. Therefore, many institutions using this format generally make an effort to enhance the PI's appreciation for the "hot button" issues or nuances particular to each area of review. This is accomplished through the distribution of guidelines for the completion of a protocol review form, example (mock) protocols, IACUC position statements, regulatory announcements on issues, or articles emphasizing the IACUC's interest or perspective on a topic. In a variant of this approach, some IACUCs incorporate this type of information into sidebar or parenthetical comments in the protocol form to channel the investigator in the right direction.

Institutions using the modular version of the free-form narrative protocol work toward paperwork reduction by capitalizing on the fact that many studies only involve a limited, and somewhat predictable, spectrum of activities around which the "core" protocol form is designed. This core form is linked to appendices that are designed for specific topics such as surgery and postoperative care, prolonged restraint, multiple major survival surgery, other procedures potentially involving significant pain and distress for laboratory animals, and the use of biohazards. The psychological benefits attendant to the reduction in paperwork for investigators and IACUC members are additional reasons given by institutions that use this approach.

Checklists and "fill-in-the-blanks" formats are used by some IACUCs as the predominant modes of information collection in a form in an effort to expedite completion of the forms through a series of discrete, informative inquiries into aspects of the protocol. Some areas of inquiry and some research settings are better suited to this approach. For example, organizations having a narrow research mission in which many of the studies follow a predictable format and involve techniques that are well described and repetitive often use this format to allow investigators to respond quickly. In some institutions specific protocol templates are developed for specific activities. For example, some institutions create templates that focus more on issues unique to the use of animals in educational exercises if they have large teaching programs, or on breeding colony husbandry and management if they maintain substantial populations of genetically modified animals. Pharmaceutical companies and commercial laboratories involved in testing compounds in conformance with federal regulations often have extensive standard operating procedures for animal care and use. Many aspects of their IACUC protocol form are addressed simply by citing these procedures or other documents used in project development if these primary sources are relevant and have been reviewed and approved by the IACUC. Some larger animal care and use programs, with diversified research endeavors, have incorporated this approach into their protocol form to service research activities that can benefit. A significant pitfall in this approach, when it is applied to complicated studies, is that it tends to disperse the information about the different types of procedures being performed on the animals and can limit the reviewer's appreciation for the temporal sequencing and impact of procedures on the animals. Flowcharts illustrating the various uses of animals in an application often can remedy this problem. A second problem is that these forms tend to be lengthy because they are structured to capture numerous small bits of information. If the forms are poorly worded, the information gathered may be voluminous but disconnected and confusing. Also, when SOPs are changed in the course of a study, the PI must assure that research personnel are following the SOP as currently written to achieve compliance with the IACUC approved protocol.

Many institutions have blended the narrative, checklist, and fill-in-the-blank strategies in their protocol review efforts to optimize these efforts. This is the approach to protocol review that the authors deem optimal for most animal care and use programs. The experience of the IACUC often leads to identifying portions of the protocol form that investigators repeatedly do not answer correctly or adequately. These areas of the form can be transformed into a checklist format or narrative blocks with supplemental instruction on the appropriate content of the blocks. For example, in the segment on the search for alternatives to painful procedures, the form might include a list of commonly used databases, along with prompts for the keywords used and the date of the search. As another example, in the discussion on the provisions for postoperative care, the questions might stimulate responses for each important aspect of care (e.g., physiological monitoring, analgesic therapy, fluid and antibiotic therapy, record keeping). Similarly, in polyclonal antibody production protocols, the questions could channel the investigator into methods that conform to IACUC guidelines.

An approved IACUC protocol form should be regarded as the investigator's contract with the institution to conduct animal care and use activities in conformance with applicable laws, regulations, and sound clinical and scientific practices. In summary, any approach, or mix of approaches, has merit if it satisfies the investigators and IACUC members and produces a well-defined contract through a thorough and appropriate IACUC review and approval process.

7:5 How specific should the information on the protocol form be to secure an approval? Is it acceptable for an IACUC to approve a generic protocol?

Reg. For PHS Policy purposes, IACUC approvals must be project specific in order to address the required review criteria at PHS Policy IV,C,1. The use of a more generic protocol, with project-specific amendments to cover significant changes, is one method that meets this requirement.

Opin. There are many investigators who mistakenly view the IACUC protocol review and approval function only as an approval of the animal procedures. Thus, if they use the same animal procedures over an extended period, they prefer to work under a generic approval to reduce their "unnecessary" interactions with the IACUC "bureaucracy." Nevertheless, most IACUCs have not acquiesced to this preference because protocol approval is not simply an approval of procedures. In general, IACUCs require PIs to submit very specific information on their protocol forms, to allow the IACUC to link the scientific objectives with the justification for a particular animal model, the number of animals requested to complete the studies, and the particular procedures that are involved. Accordingly, in the few IACUCs that do use this approach for certain types of protocols, the PIs are required to submit supplementary information for each study conducted under the procedure-based protocol to define its objectives, assure the validity of the animal model, assure the lack of alternatives, and explain the need for the number of animals requested. Information pertaining to any new animal welfare or personnel safety concerns that may have been introduced by new variables in the studies proposed also must be provided and reviewed.

There are times when the approval of a generic protocol, abiding by the contingencies stated, may be appropriate and efficient. An example is a protocol with a standard set of procedures that is used repetitively to test compounds and materials in a particular animal model with recognized validity of the study of a class of compounds. The only variable in these protocols is the material or compound under investigation. Also, there is an opportunity for scientists to reduce some of the specificity in their protocol details if this is acceptable to the IACUC based upon their assessment that satisfactory animal welfare outcomes will not be compromised. Since PIs are expected to conduct the procedures approved by the IACUC as written, they may afford themselves some latitude when preparing their submissions for committee review if their IACUC is receptive to this approach. For example if an IACUC approved blood collection from a dog at 2 milliliters every three hours for a 24-hour period, a fixed and discrete expectation has been established in the protocol. In this example, if the PI later determined that 3 milliliters of blood was better suited to thoroughly complete their laboratory assay and provide for sample back-up storage, a protocol modification would need to be submitted to the IACUC for review and approval before additional blood could be collected. Had the PI prepared his submission and been approved by the IACUC to allow for some minor adjustment of study parameters (i.e., bleeding intervals and blood volumes) to optimize the data and assay by requesting permission to collect up to 4 milliliters of blood at 12 or less times in a 24 hour period and by providing the proper scientific rationale for the request, he/she would have had the flexibility to proceed without registering an amendment to the protocol.

7:6 Should the protocol form be used as a vehicle for the continued instruction of investigators in aspects of the PHS Policy, AWAR, and ethical considerations in animal research?

Opin. Most IACUCs use protocol forms that refer to the PHS Policy, AWAR, and ethical issues in one or more areas. Some include a statement signed by the investigators assuring that all of the activities described will be conducted in conformance with the aforementioned regulations. These references to PHS Policy and the AWAR serve to remind PIs of the regulatory underpinnings of the protocol review process. They also emphasize the contractual nature of an approved protocol. Similarly, many IACUC protocol review forms attempt to evoke the discussion of ethical considerations in response to a variety of experimental practices such as prolonged restraint, food and water restriction, use of aversive stimuli, surgery, multiple major survival surgery, disease induction, post-procedural care, and the selection of study end points. While using the protocol form in this fashion to give these issues continued visibility to the investigators is well intended, in the authors' opinion the use of the protocol form as a significant tool for this instruction is a superficial approach with little merit. Most IACUCs would likely agree that investigators tend to focus their attention on filling out protocol forms for the purpose of securing of approval and not to be educated about potentially complicated matters.

7:7 Could inclusion of a questionnaire on the form on animal welfare considerations be useful to the IACUC?

Opin. Fewer than half of the IACUCs the authors are familiar with have attempted to evaluate animal welfare considerations in a separate section or questionnaire. When this is done, it usually is limited to querying the investigator about the allocation of animals to the various APHIS/AC pain categories (if used) to aid the PI's understanding of the AWAR requirement for the search for alternatives and scientific justification for conducting APHIS/AC Category E animal studies. (See 16:5.) Even at this basic level, most institutions encounter occasions in which the IACUC does not agree with the PI's assessment or the IACUC is not able to reach a full consensus on an animal welfare matter. On most IACUC protocol forms, this type of information is integrated into the sections involving the discussion of procedures performed and the overall management of the animal model. If the information given in these areas is insufficient for the IACUC to identify and assure the

PI's attention to animal welfare issues, changes in the protocol form or the training of investigators is warranted. It seems very likely with increased attention given to animal welfare issues in the *Guide* (see 7:2), the statement in the *Guide* (p. 27) that the, "IACUC is obliged to weigh the objectives of the study against potential animal welfare concerns," and AAALAC's discussion of the "harm-benefit" analysis[3] that institutions will be prompted to use the protocol form to capture animal welfare information.

7:8 Should IACUC protocol forms require the PI to include a statement written in language understandable to the lay public about the purpose and relevance of the proposed studies?

Reg. This inquiry has become standard on most IACUC protocol forms. The basis for making this request is U.S. Government Principle II, which states, "Procedures involving animals should be designed and performed with due consideration for relevance to human or animal health, the advancement of knowledge, or the good of society."

Opin. In the age of highly specialized research, most IACUCs regard this type of statement as essential for the lay member of the IACUC, institutional public relations or development personnel, and scientists on the IACUC who work in areas unrelated to the protocol under consideration. In the authors' experience, weak responses to this inquiry on the IACUC protocol form often reflect the general inexperience of scientists in addressing a scientifically unsophisticated public. Some institutions do not provide this type of inquiry on the form because all animal protocols are subsumed under an organizational or corporate research mission that assures this matter.

7:9 What information is appropriate to be included in the nontechnical description of the project(s) and the statement on the significance of the animal use to the scientific objectives of the project(s)?

Opin. The *Guide* (p. 25) states that the PI should provide "a clear and concise sequential description of the procedures involving the use of animals that is easily understood by all members of the committee" to help ensure the participation of all IACUC members in the review process. Writing a cogent lay summary for the proposed research can be a challenging task and the effort required is well invested in the authors' view. The function of the lay summary is to provide the nonscientific members of the IACUC, and particularly the nonaffiliated member, with an understanding of the scientific questions being asked and how the animals are allocated and used to answer those questions. The lay summary should indicate how the animal model contributes to the hypothesis being tested, the exploration of a new and potentially important area of science, or the resolution of an important scientific dispute. Obscure scientific jargon and the lack of effort to connect the studies to conditions or phenomena that are widely known or easily appreciated by the lay audience are common weaknesses in these statements that have been observed by the authors. To alleviate problems related to the use of scientific terms that are unfamiliar to lay members of the committee, some institutions have elected to provide a glossary of lay terms as a reference (e.g., http://research.

uthscsa.edu/iacuc/Glossary%20of%20IACUC%20terms.pdf).[4] Investigators and their institutions should keep in mind that ultimately science competes for the attention, recognition, and continuous stream of financial support from a public that has other avid interests and recognizes the competing needs of other worthy endeavors.

7:10 Should IACUC protocols submitted for educational activities have the same basic structure as a research protocol?

Opin. To provide for uniformity in oversight, some institutions use the same protocol form for all activities involving animal care and use. Alternatively, in institutions using animals in teaching activities that do not necessarily involve painful or distressful procedures, a unique and streamlined protocol template is often used. For example, life sciences and agricultural animal husbandry courses are frequently taught and may entail only approved methods of euthanasia or common farm animal husbandry practices. In situations such as these, institutions often create a protocol template to capture what routine, "non-invasive" techniques will be conducted, the number of animals needed per student, and the number of students participating in the activity. With this information and the understanding that the course content encompasses this educational practicum, IACUCs can still approve and oversee the activity as they would any other project. This practice often optimizes the IACUC's administrative efforts and efficiency in the protocol review process and aids PIs who may be laden with the preparation of protocols from many other active research endeavors. Institutions also have found the development of special attachments to the protocol form to address areas regarded as unique to the use of animals in educational exercises necessary to fulfill adequate IACUC review requirements. The areas of special emphasis include a review of

- Course syllabus to understand how and why animal studies are integrated into the course
- All relevant handouts provided to students pertaining to animal use
- Student profile information (e.g., preparatory coursework and degree major)
- Summary of animal care and use educational discussion or materials provided to the students
- Backup provisions for students not fulfilling their obligations to laboratory animals
- Occupational health and safety provisions
- Summary of students' written critiques of the educational activity from the previous year
- Opportunities or plans for the course director to develop alternatives to the use of animals for subsequent sessions of the course

A similar approach allowing some modification of the protocol review template might also be applicable to studies for medical professionals to provide instruction or develop proficiency on particular procedures or the management of clinical emergencies such as laparoscopic surgery, endoscopy, bronchoscopy, endotracheal intubation, and Advance Trauma Life Support Training. The use of animals in these procedures is invasive and requires highly competent procedural and post-procedural clinical management of the animal model, and the assessment of the team providing the clinical care is clearly crucial. Also, the incorporation of alternative methods and the use of simulacra to prepare trainees for in vivo components would be important to ascertain. Further, the measures taken to ensure that these trainees, who typically have only transient animal contact, are afforded knowledge of the hazards and risks they may be exposed to and the practices used to achieve risk mitigation are essential.

7:11 Is a PI required to provide information on the protocol review form that may be proprietary in nature?

Reg. Under the AWA (§2143,a,6,B), "No rule, regulation, order, or part of this chapter shall be construed to require a research facility to disclose publicly, or to the Institutional Animal Committee during its inspection, trade secrets or commercial or financial information which is privileged or confidential." §495,e of the HREA imparts a similar position to the PHS Policy.

Opin. PHS Policy and the AWA do not require IACUCs to review proprietary information. Most IACUCs have designed their protocol forms and implemented procedures to prevent the unwitting and unnecessary disclosure of proprietary information by investigators. On the other hand, IACUCs must have sufficient information about all aspects of the study to make an informed decision about the assurances for animal welfare during all phases of animal care and use. Where PHS Policy is concerned, the IACUC must be assured that the occupational health and safety program has identified compounds and other materials, equipment, practices, and procedures that may have an impact on occupational health and safety considerations for personnel involved in the project (*Guide* pp. 18–23). Hence, most IACUCs do not require the disclosure of the name of a proprietary compound but do require information about the chemical class of the compound, toxicity data in the target animal species or other animals, special handling requirements, and the basis for dosage selection. The development of new techniques and the use of new instrumentation in animals for application in humans pose more of a quandary. For example, if a new surgical technique is an integral component of a proprietary package involving a new device, an IACUC would be remiss to waive its responsibility to assess the impact of the procedure on experimental animals. Also, most IACUCs require sufficient information about the purported advantages of the new equipment to support the claim of potential benefit. (See 22:3–22:6.) The training of IACUC members should include the expectation and importance of nondisclosure of proprietary information available to them through the review process.

Protections for nondisclosure of confidential information are also built into the AWAR (§2.35,f) which states: "APHIS inspectors will maintain the confidentiality of the information and will not remove the materials from the research facilities' premises unless there has been an alleged violation, they are needed to investigate a possible violation, or for other enforcement purposes. Release of any such materials, including reports, summaries, and photographs that contain trade secrets or commercial or financial information that is privileged or confidential will be governed by applicable sections of the Freedom of Information Act." See Chapter 22 for additional information on propriety and confidentiality issues.

7:12 Should information such as the PI's name, title of the application, and research sponsor be included on the IACUC protocol form, or could that information bias the IACUC's decisions?

Opin. While there is some variation in the content of the protocol signalment among the institutions known to the authors, all identified the PI by name and included a title for the proposed activities. PHS Policy (IV,C,1,f) and the AWAR (§2.31,d,1,viii) require the IACUC to verify that personnel conducting procedures are appropriately qualified and trained. This would be difficult, if not impossible, without the identification of the PI. Most protocol forms also asked the PI to provide information about the funding source, particularly when the IACUC was responsible for sending the letters of protocol approval to funding agencies. Nevertheless, in some institutions a response to this inquiry is at the discretion of the PI. In institutions that fund their research programs internally, this inquiry is not relevant. Virtually none of the institutions believe that this type of information is likely to compromise their objectivity, notwithstanding the admission by several institutions that internally funded projects have to provide evidence of peer scientific review via an internal mechanism or consultants, whereas externally funded projects do not. Other information about the PI and associated laboratory personnel often collected in the protocol form includes position or academic title; office, laboratory, and home (emergency) phone numbers; and office and laboratory location. In the authors' view, emergency contact information is of paramount importance for critical correspondence on animal welfare matters and should always be provided. In many institutions the ability to serve as a PI is reserved for individuals with particular positions or academic titles. (See 22:6; 22:7.)

7:13 What are some methods used by IACUCs to identify protocols for the purpose of correspondence, timely periodic reviews, and aiding of the oversight of ongoing animal-based research activities?

Opin. Most institutions use a numerical or alphanumerical identification system for the protocols that are submitted to and approved by the IACUC. This type of system offers several advantages and there are numerous variations in this approach. It provides a brief, content-neutral reference that links both the animals and when necessary the grant application or award to a set of approved animal care and use procedures. Many institutions indicate the year of protocol submission in their log or accession system. This method provides rapid identification of those protocols that are due for annual or triennial renewal. Other numbers are added to reflect the sequence of protocol submission within the year, and sometimes letter or decimal extensions are used to indicate the renewal year within the triennial period. As an example, 13-38.16 might indicate the 38th protocol submitted to the IACUC in 2013, with a full review of the protocol scheduled in the year 2016.

It is not uncommon for the approval of problematic protocols submitted at or near the end of the calendar year to be delayed until the following year. Computer databases are used routinely to track the actual dates of submission and approval to ensure that the requirements for periodic review or renewal are met in a timely fashion. Also, organizations vary in their requirements for the number of animal species that are permitted in a particular protocol application; some allow multiple species, whereas others allow only one. Most institutions are now following the species-dedicated protocol approach to allow ease of activity tracking and post-approval monitoring efforts. Institutions that allow only one species per protocol application sometimes retain the same number but follow it with a letter designator for species (e.g., *R* = rat, *Rb* = rabbit, *M* = mouse) to associate the procedures described with a particular grant application or award. The same tracking number may be retained, along with a modifying extension, to indicate that animal care and use procedures are identical to those described in the parent protocol, but an alternate project title or funding source is associated with the protocol. (See 7:12.)

7:14 When protocols are due to expire, should the IACUC retain the original number assigned or assign a new number to the protocol?

Reg. The AWAR (§2.31,d,5) require review of IACUC protocols at least annually, while the PHS Policy (IV,C,5) requires the same at least every 3 years.

Opin. Many institutions apply a modifying extension to the same protocol number when an annual review of animal care and use activities is mandated for a species, such as occurs under the AWAR. However, most institutions that receive PHS support for their research programs prefer to retire protocol numbers when the protocols have reached the limit of their 3-year PHS approval period.[5] It should be noted, though, that at least a general review of each protocol involving AWAR-regulated species must be performed on an annual basis (AWAR §2.31,d,5). This approach is convenient in that it ensures individuals involved in animal care and use (and research oversight) that the review process is current. It is becoming common in the authors' experience, for institutions to retain the original protocol number through the life of the project for efficiency as well as to indicate the continuity and longevity of a research program. This approach is also satisfactory as long as the IACUC has performed a rigorous review at the appropriate review cycle. (See 7:13.) PIs and animal facility managers have found that changing their protocol numbers in the course of a project necessitates, for example, updating hundreds of cage cards on mouse cages or developing methods to ensure procedure and training details are consistently and adequately documented under the new protocol number. Retaining the protocol number might reduce the need to reaffirm and update details with unnecessary repetition.

7:15 Are there advantages to using an "electronic" protocol e-submission format?

Opin. The use of an electronic protocol format requires institutions to either develop a homegrown system or purchase a commercially available product. In both cases, the process can be found to be cost prohibitive. Institutions using e-protocol management systems usually have large animal care and use programs and budgets that can support the associated expenses.

Institutions using e-protocol templates are afforded many benefits. Typically the data base used to manage the system is accessible to IACUC members thereby giving committee members unrestricted access to all protocols, including those under review. In addition, the e-submission template is typically programed to gather only the relevant information the IACUC needs to conduct the project review. For example, a PI may be asked to check a series of boxes identifying the procedures they will conduct. For conducting surgeries, the survival surgery box would be checked and PIs performing survival surgery would need to answer all questions relating to the topic. As e-submission programs become more affordable they likely will help maximize the efficiency of the PIs' protocol completion process, the IACUC's review process, and document management.

References

1. National Institutes of Health, Office of Laboratory Animal Welfare. Sample Animal Study Proposal. http://grants2.nih.gov/grants/olaw/sampledoc/oacu3040-2.htm#admindata. Accessed April 3, 2013.
2. National Institutes of Health, Office of Laboratory Animal Welfare. 2012. Frequently Asked Questions D.10, Is the IACUC required to review the grant application? http://grants.nih.gov/grants/olaw/faqs.htm#proto_10.
3. Association for Assessment and Accreditation of Laboratory Animal Care International. 2013. Frequently Asked Questions C.3, Harm-benefit analysis. http://www.aaalac.org/accreditation/faq_landing.cfm#B3.
4. University of Texas Health Science Center at San Antonio. UTHSCSA IACUC Glossary. http://research.uthscsa.edu/iacuc/Glossary%20of%20IACUC%20terms.pdf. Accessed March 9, 2013.
5. National Institutes of Health, Office of Laboratory Animal Welfare. 2012. Frequently Asked Questions D.1, How frequently should the IACUC review research proposals? http://grants.nih.gov/grants/olaw/faqs.htm#proto_1.

8

Submission and Maintenance of IACUC Protocols

Kathy Laber and Alison C. Smith*

Introduction

The authors would like to preface this chapter with the observation that there are no laws requiring that an animal use protocol be written and approved by the institution's IACUC; rather, specific components related to the care and use of animals in biomedical research, biological testing, and research training must meet specific requirements detailed in applicable regulations and laws. Exceptions to such standards may be made only when specified by research protocol and any exception shall be filed with the Institutional Animal Committee (AWA §2143,a,3,E). Nevertheless, the most common approach to obtaining, documenting, and reviewing the required information is, as alluded to in the *Guide* (p. 25), through the completion of a "protocol" that provides the legally required details of the research involving animals that is then reviewed by the IACUC. Questions often arise about when and how best to utilize such a system to ensure that the legal requirement of review of animal use in research, teaching and testing is met. This chapter helps to address some of the more commonly asked questions on the submission and maintenance of the commonly used "animal care and use protocol."

8:1 When is an IACUC-approved protocol needed?

Reg. The laws from which the animal care and use protocol arose are the AWA and the HREA. As defined by these laws and their associated explanatory regulations and policies, IACUC-approved protocols are required for the following:

- All live or dead warm-blooded animals used in research, teaching, or testing in registered research facilities as defined by the AWAR (§1.1). Specific animals excluded from the AWAR are birds, rats of the genus *Rattus*, and mice of the genus *Mus*, bred for use in research, and farm animals used as a source of food or fiber or for research designed to improve the quality of food or fiber. Specific research studies excluded from the requirement for a protocol are field studies. Field study (AWAR §1.1) means a study conducted on free-living wild animals in their natural habitat. However, the term excludes any study that involves an invasive procedure harms, or materially alters the behavior of an animal under study vertebrate animals used or intended for

* The authors thank Farol N. Tomson for his contribution to this chapter in the first edition of The IACUC Handbook.

use in research, research training, experimentation or and biological testing activities conducted or supported by any PHS agency (PHS Policy II; III,A). Note that the PHS Policy expands the definition of animals to include all vertebrates, not just warm-blooded vertebrates.

The *Guide* (p. 2) expands the definition to state "any vertebrate animal (i.e., traditional laboratory animals, agricultural animals, wildlife, and aquatic species) produced for or used in research, teaching, or testing."

Opin. An IACUC-approved protocol is also needed when projects supported by other public or private agencies have a requirement for IACUC approval (see 8:4–8:6). Alternately, the individual institution may require IACUC approval even when the animal is not an APHIS/AC-regulated species, and the study is not supported by a PHS agency. Many institutions with Assurances accepted by NIH/OLAW do not differentiate between vertebrate animal activities supported by PHS funds and those that have other funding sources regarding the need for an IACUC-approved protocol. It is common practice for institutions with Assurances to require IACUC-approved protocols for all non-PHS-supported activities involving vertebrate animals. For example, a breeding colony of mice may not be funded through the PHS and common laboratory mice are not currently a regulated species covered by the AWAR. Therefore, the colony is not required by any of the regulations to have an IACUC approval; however, if an institution wishes to be AAALAC accredited, the breeding colony would need to have an IACUC approved protocol. Regardless of the nuances of which vertebrate animals may or may not be covered by the laws and guidance documents, it is recommended that institutions extend IACUC review to projects that fall in this category to ensure that an appropriate standard of animal care and use is uniformly applied.

8:2 What activities are specifically exempt from the requirement for an IACUC-approved protocol?

Reg. AWAR (§2.31,d,1) state that field studies, as defined in 8:1 above, are exempt from required IACUC review of an animal study related activity. Although the *Guide* (p. 2) does not address invertebrate animals (e.g., cephalopods) in detail, it does establish general principles and ethical consideration that are also applicable to these species and situations. The following statement is from the Rules of Accreditation as found on the AAALAC website.[1]

> All animals used or to be used in research, teaching or testing at accreditable units are to be included and evaluated in accordance with the standards set forth in Section 2 of these Rules. This includes traditional laboratory animals, farm animals, wildlife, and aquatic animals. Nontraditional animals, inclusive of invertebrate species, are also included where they are relevant to the unit's mission.

AAALAC acknowledges that a formal protocol for invertebrate work may not be needed and that a SOP, or policy statement/guideline reviewed by the IACUC, would suffice. It further states that intensity of review for lower level invertebrates such as nematodes would differ from higher level invertebrates, such as squid.[2]

Opin. IACUC-approved protocols are not required if the animals to be used are

- Not under the auspices of the AWAR or PHS Policy (see 8:1), that is, not a regulated species (e.g., a horseshoe crab, not supported by PHS funding, or not used for research, biological testing, or research training)
- Not supported by any additional private or public agency requiring IACUC approval
- Not required by the home institution to be IACUC reviewed

Institutions may own and use animals for a variety of reasons that do not include research, biological testing, or training in research. Animals can play a role as institutional mascots, as key participants in rodeos, as featured stars in a theatrical play, or the focus of a student's art project. People often keep companion animals on site as pets or in cages or aquariums in their offices or dormitory rooms. Institutions may house animals for specific purposes such as guard dogs, pet therapy animals, mounted police horses, or classroom pets. Institutions also deal with pest animals, such as wild rodents and certain birds. If these animals are not involved in biomedical research or intended to be used in a research project, they are exempt from IACUC approval. (See Chapter 2.) However, to help ensure appropriate care of all living creatures under the institution's purview (refer to the *Guide* p. 2) it may be wise to extend the concept of IACUC review to some of these situations (the format of the information being requested could be tailored to fit the species and the role the animal is filling).

8:3 What information is required on an IACUC protocol?

Reg. The AWA and the HREA do not detail protocol content requirements; however, the AWAR and the PHS Policy are very specific.

1. PHS Policy (IV,C,1,a–IV,C,1,g):

> In order to approve proposed research projects or proposed significant changes in ongoing research projects, the IACUC shall conduct a review of those components related to the care and use of animals and determine that the proposed research projects are in accordance with this Policy. In making this determination, the IACUC shall confirm that the research project will be conducted in accordance with the Animal Welfare Act insofar as it applies to the research project, and that the research project is consistent with the *Guide* unless acceptable justification for a departure is presented. Further, the IACUC shall determine that the research project conforms to the institution's Assurance and meets the following requirements:
>
> - Procedures with animals will avoid or minimize discomfort, distress, and pain to the animals, consistent with sound research design.
> - Procedures that may cause more than momentary or slight pain or distress to the animals will be performed with appropriate sedation, analgesia, or anesthesia, unless the procedure is justified for scientific reasons in writing by the investigator.
> - Animals that would otherwise experience severe or chronic pain or distress that cannot be relieved will be painlessly killed at the end of the procedure or, if appropriate, during the procedure.

- The living conditions of animals will be appropriate for their species and contribute to their health and comfort. The housing, feeding, and non-medical care of the animals will be directed by a veterinarian or other scientist trained and experienced in the proper care, handling, and use of the species being maintained or studied.
- Medical care for animals will be available and provided as necessary by a qualified veterinarian.
- Personnel conducting procedures on the species being maintained or studied will be appropriately qualified and trained in those procedures.
- Methods of euthanasia used will be consistent with the recommendations of the American Veterinary Medical Association (AVMA) Panel on Euthanasia,[3] unless a deviation is justified for scientific reasons in writing by the investigator."

2. PHS Policy (IV,A,1) requires institutions to use the *Guide* as a basis for developing and implementing an institutional program for activities involving animals. The *Guide* was authored by the National Research Council of the National Academies and provides guidance on caring and using animals in keeping with the highest scientific, humane, and ethical principles. This document is a reference resource used by AAALAC, the primary agency that accredits programs that utilize animals in biomedical research and biological testing. To be in compliance with PHS Policy when AWA-regulated species are used, both the AWA and the recommendations in the *Guide* must be followed (PHS Policy II). Additional requirements in the *Guide* (pp. 25–33) that are not specifically detailed in PHS Policy or in the AWA include the following:

NOTE: New concepts introduced in the 2011 *Guide* are italicized.

- Rationale and purpose of the proposed use of animals (and see PHS Policy IV,D,1,c; AWAR §2.31,e,2)
- A clear and concise sequential description of the procedures involving the use of animals that is easily understood by all members of the committee
- Availability/appropriateness of less invasive procedures/species/non-animal use. NOTE: this applies to ALL animal procedures, not just the painful/distressful ones required by the AWA
- Justification of the species and number of animals requested. Statistical justification of animal numbers and experimental group size whenever possible (and see PHS Policy IV,D,a-b; AWAR §2.31,e,2)
- Unusual housing and husbandry requirements
- Impact of the proposed procedures on the animals' well-being
- Post-procedural care and observation
- Description and rationale for anticipated or selected endpoints
- Criteria and process for timely intervention, removal of animals from a study, or euthanasia if painful or stressful outcomes are anticipated
- Method of euthanasia or disposition of animal *including planning for care of long-lived species after study completion* (and see PHS Policy IV,D,1,e; AWAR§2.31,e,5)

- Adequacy of training and experience of personnel in the procedures used, *and roles and responsibilities of the personnel involved* (and see PHS Policy IV,C,1,f; AWAR §2.32,a-c)
- Safety of working environment for personnel
- Special Considerations:
  - If there is a potential for unrelieved pain or distress, or other animal welfare concerns, weigh the objectives of the study against potential animal welfare concerns
  - When novel studies are proposed and information is lacking about endpoints (recommend pilot studies), a system of communication with the IACUC should be in place during and after the study
  - IACUC should consider the potential for unexpected outcomes if using novel experimental variables
  - Physical restraint
  - Food and fluid regulation
  - Use of non-pharmaceutical-grade chemicals

3. AWAR (§2.31)

   Additional requirements that need to be addressed within the AWAR include the following:

- *(§2.31,d,1,ii) The PI has considered alternatives to procedures that may cause more than momentary or slight pain or distress to the animals, and has provided a written narrative description of the methods and sources, e.g., Animal Welfare Information Center used to determine that alternatives were not available.
- (§2.31,d,1,iii) The PI has provided written assurance that the activities do not unnecessarily duplicate previous experiments.
- (§2.31,d,1,iv,B) Procedures that may cause more than momentary or slight pain or distress to the animals will involve in their planning consultation with the attending veterinarian or his or her designee.
- *(§2.31,d,1,iv,C) Procedures that may cause more than momentary or slight pain or distress to the animals will not include the use of paralytics without anesthesia.
- (§2.31,d,1,ix) Activities that involve surgery include appropriate provision for the pre-operative and post-operative care of the animals in accordance with established veterinary medical and nursing practices. All survival surgery will be performed using aseptic procedures, including surgical gloves, masks, sterile instruments, and aseptic techniques. Major operative procedures on non-rodents will be conducted only in facilities intended for that purpose which shall be operated and maintained under septic conditions. Non-major operative procedures and all surgery on rodents do not require a dedicated facility, but must be performed using aseptic procedures. Operative procedures conducted at field sites need not be performed in dedicated facilities, but must be performed using aseptic procedures.

- (§2.31,d,1,x,A-C) No animal will be used in more than one major operative procedure from which it is allowed to recover unless: justified for scientific reasons by the PI or required as routine veterinary procedure, or to protect the health or well-being of the animal as determined by the AV, or- in special circumstances as determined by the Administrator on an individual basis. (Written request for this caveat must be submitted to the APHIS Administrator.)

* These principles are supported in the U.S. Government Principles for the Utilization and Care of Vertebrate Animals Used in Testing, Research and Training.

Opin. The *Institutional Animal Care and Use Committee Guidebook*[4] summarizes the information and items that are legally required to be included on IACUC forms. However, the mechanisms that institutions use to solicit the appropriate information in a concise manner from the investigative community are as varied as the institutional cultures. Each operating IACUC should determine for itself the most effective way of obtaining and then evaluating the required information. In addition, institutions may opt to require information about animal-related activities that extends beyond the legal requirements mentioned, such as Drug Enforcement Administration registration numbers, location of controlled substances, and the individual responsible for controlled substances. Many IACUCs have found that the use of a protocol form helps research investigators to delineate the information that the IACUC requires for reviewing a proposal and helps the IACUC achieve greater consistency in its reviews. Many of these forms are available from institutional web sites. Given the additional concepts that the IACUC needs to review per the *Guide*, it is likely that institutions will need to revamp their current protocol format to mine the information needed. A sample animal study protocol is provided on the NIH/OLAW website.[5] Institutions may customize the sample document for their animal care and use program.

8:4 Do government agencies conducting or funding animal research require IACUC-approved protocols?

Reg. (See 2:6; 8:1.)

Opin. In 1985 the Interagency Research Animal Committee, with representatives from various federal agencies, developed a set of principles (U.S. Government Principles for the Utilization and Care of Vertebrate Animals Used in Testing, Research, and Training) that applies to institutions and persons using animals in any type of federally funded program.[6] The HREA is congruent with those principles. The PHS Policy, which implements and supplements the aforementioned principles and law requires that all PHS-supported activities involving animals, be it at a PHS agency or an awardee U.S. institution, follow the PHS Policy guidance which states "Institutions in foreign countries receiving PHS support for activities involving animals shall comply with this Policy, or provide evidence to the PHS that acceptable standards for the humane care and use of the animals in PHS-conducted or supported activities will be met." This does not, however, state that IACUC approval must be obtained. In addition, many federal agencies, such as the National Science Foundation and the Veterans Administration, embrace the PHS Policy requirements and all federal agencies that use animals adhere to the U.S. Government Principles. Regardless of the funding source, when AWAR species are used, the requirement for IACUC-approved protocols as detailed in the AWAR applies.[7]

8:5 Do private agencies funding biomedical research require IACUC-approved protocols?

Reg. (See 8:1 for animals and animal activities requiring IACUC approval.)

Opin. If the animal is used for research, biological testing, or research training and is an AWAR-regulated species, then an IACUC approval is required by law. However, there are no federal or state laws that require private funding agencies to obtain IACUC approval of animal use projects if the animals are not AWAR-regulated covered species. Some, but not all, private agencies require IACUC approval before releasing funds for research; for example, the American Heart Association, American Cancer Society, and American Lung Association require IACUC approval for all vertebrate animals.[8–10]

8:6 How can institutions require IACUC approval of projects funded from agencies that do not require it, or that involve animals that are not covered by either the AWAR or the PHS Policy?

Reg. There are many valid reasons for institutions to perform program oversight institution-wide, using uniform and consistent standards for animal care and use. Likewise, it is generally impractical to separate activities based on the source of funding. Institutions must implement the PHS Policy for all supported activities involving animals and must ensure that any standards that might not be consistent with PHS Policy do not affect or pose risks to PHS-supported animal activities.[11,12]

Opin. Savvy institutions recognize—and others have been reminded by the public, which holds them accountable—that the ethical responsibility for the use of animals extends well past the legally defined types of research and animal species boundaries assigned by regulatory agencies. It is the option of each institution to determine how best to shoulder that responsibility. If an institution chooses to have its IACUC review and approve non-biomedical (e.g., agricultural) research using animals, or research using animals not regulated under the limitations of the AWAR or PHS Policy (e.g., invertebrates), they can simply make an IACUC review of those species an institutional policy. In addition, they may choose to announce this policy to NIH/OLAW, the oversight office for the PHS Policy, as part of their Assurance statement. NIH/OLAW advises institutions that the maintenance of uniform and consistent standards is an essential component of the development and implementation of a quality animal care and use program.[11] In addition, if an institution wants to signal to the public that it has a high quality peer reviewed program through achieving AAALAC accreditation, it must follow the *Guide* which requires IACUC approval of projects (see 8:1).

8:7 Must the IACUC review and approve protocols using animal carcasses and parts of carcasses?

Reg. As stated in 8:1, AWAR (§1.1, Animal) define regulated animals as "live or dead." A research facility (AWAR §1.1, Research Facility) is defined as one using or intending to use only live animals. There is no reference to using dead animals in research facilities. It is only in the definition of a dealer (§1.1, Dealer) that the AWAR refer to live or dead animals. (See 12:21; 14:28.)

Only live animals are mentioned in PHS Policy III,A. A proposal involving animals to be killed for the purpose of using their tissues or one that involves project-specific antemortem manipulation is not exempt from protocol review. This is explained further in NIH/OLAW FAQ A.3[13] and as shown below:

> PHS grant applicants using shared animal tissues or slaughterhouse materials are advised to specify the origins of the tissues when describing their proposed use in an application [that is available online[14]], especially if the "no" box is checked on the vertebrate animal block of the face page. Any reference to the use of animal tissues in the application is likely to trigger questions about IACUC approval. Therefore, an explanation in the [narrative part of the grant] application that the tissues will come from dead animals as a byproduct of other IACUC-approved studies, or from a slaughterhouse, will help avoid delays in the peer-review process.[13,15]

The *Guide* does not specify 'live or dead' animal which leaves room for AAALAC Council interpretation on its applicability to animal carcasses and parts. However, Council will bring issues of non-USDA compliance to the attention of the institution.

Opin. Ordinarily, neither the AWAR nor the PHS Policy protocol review requirements apply to dead animals in the research facility setting; however, IACUCs often choose to review the use of carcasses and carcass parts. This practice may best serve the interests of the institutions for a variety of non-regulatory reasons such as improved public relations, occupational safety, and decreased institutional liability. The review of carcasses and carcass parts can also help to protect the investigators from a public that is critical about PIs obtaining certain samples. For example, researchers who receive biopsy material (or carcasses) from endangered species. The IACUC review is an effective way of informing the institution about these projects. IACUC members can discuss and advise the institution on projects that may become controversial and warrant additional consideration.

8:8 Must the IACUC review and approve protocols using animal tissues, cells, or biological fluids derived from living animals?

Reg. (See 8:1.)

Opin. The IACUC must address issues related to animal care, animal handling, and distress that can be associated with obtaining the specimens, and Occupational Health and Safety (OHS) issues for the personnel collecting the specimens. If the specimen collection technique is not invasive and does not require the animal to be restrained (e.g., feces on the floor, remnants of placentas) or use of preserved tissues, OHS issues may still require consideration. A greatly simplified protocol format addressing only the potential issues related to specimen collection could be easily processed via a designated reviewer on the IACUC. Often, these types of collections are "add-ons" from an investigative group that does not own the animal or is not involved with the primary reason the animals are in the facility. As an alternative, the "specimen collection" could simply be added on as an amendment to the PI's protocol.

8:9 Must the IACUC review and approve protocols using tissues, cells, or biological fluids acquired from sources other than the IACUC's institution?

Reg. (See 8:1.)

Opin. This issue is not directly addressed in the AWAR, but a similar issue is addressed by the NIH/OLAW as it relates to purchasing antibodies from a commercial source (see 8:10). The outside source (vendor) providing the tissues, cells, or fluids must comply with the conditions of 8:1 and must have a NIH/OLAW Assurance if live animals are being used to produce these products specifically for the institution, versus a standard "off-the-shelf" product the company sells commercially. (See 8:10 and NIH/OLAW FAQ A.2.)

8:10 Does the purchase of antibodies from commercial sources require IACUC approval?

Reg. NIH/OLAW has made the following interpretation of PHS Policy:

> The generation of custom antibodies is an activity involving vertebrate animals and covered by PHS Policy. Antibodies are considered customized if produced using antigen(s) provided by or at the request of the investigator (i.e., not purchased off-the-shelf). An organization producing custom antibodies for an awardee must have or obtain an Assurance, or be included as a component of the awardee's Assurance. In addition, the awardee must provide verification of project-specific IACUC approval for the production of the antibodies.[16]

Opin: When standard reagent antibodies (e.g., mouse-antihuman) are produced by a commercial supplier using their own resources and offering them for general sale, for example, through a catalog, the institution may consider the antibodies to be "off-the-shelf" reagents and the supplier is not required to file an Assurance with NIH/OLAW. If, on the other hand, a supplier or contractor produces custom antibodies using antigen(s) provided by or at the request of a PI, the antibodies are considered "customized" and the vendor or subcontractor must file an Assurance with NIH/OLAW. Alternately, the receiving institution can include the vendor in its own institutional Assurance.[16]

Usually, it is known in advance that someone needs custom antibodies when writing a PHS grant. In such cases, the applicant must mark the PHS grant application (PHS Form 398)[13] "yes" for vertebrate animal involvement and include the appropriate Animal Welfare Assurance number(s), verification of project-specific date of IACUC protocol review, and the identification of all project performance sites. All animal-related activities supported by the PHS must be conducted at Assured institutions and must be reviewed and approved by an IACUC. When both the PHS grantee and its contractor hold NIH/OLAW-approved Assurances, some latitude is allowed in determining which IACUC (if not both) will review the proposal. However, the institution which subcontracts or subgrants (also known as the prime grantee) any animal activity is accountable for providing effective oversight mechanisms to ensure compliance with the PHS Policy.[17] Part of that responsibility includes ensuring that subgranted/subcontracted animal-related activities are conducted only at an Assured institution.

There are no AWAR requirements for a research facility to review protocols of researchers using commercial companies producing their antibodies. However, as noted in APHIS/AC Policy 10,[18] if that commercial company uses rabbits (or any other regulated species) to produce antibodies, it must be registered as a research facility and have its own IACUC review of relevant procedures. It is advisable, although not necessary, for the research facility to have a record of the commercial company's NIH/OLAW Assurance and/or APHIS/AC registration number. (See 16:22; 18:13.)

8:11 Does the use of fertilized eggs (birds, reptiles, and amphibians) or fetuses (mammalian) require an approved protocol?

Reg. If the eggs and fetuses remain as part of the female (dam), it is the dam that is the object of IACUC interest. If the dam is under the auspices of the AWAR or PHS Policy (see 8:1), then IACUC approval is required. If they are chicken eggs, NIH/OLAW has made the following interpretation of the PHS Policy:

> Although avian and other egg-laying vertebrate species develop backbones prior to hatching, OLAW interprets the PHS Policy as applicable to their offspring only after hatching. The egg-laying adult animal is covered by the PHS Policy. OLAW expects Assured institutions to have policies and procedures in place that address the care or euthanasia of animals that hatch unexpectedly.[19]

Opin. If the eggs are allowed to hatch (or if the eggs hatch inadvertently) and if the animals are under the considerations described in 8:1 (i.e., birds, reptiles or amphibians in addition to mammals), then IACUC approval is required. The AWAR does not regulate most birds or cold-blooded (poikilothermic) animals. Therefore, at this time there are no AWAR requirements for the IACUC to review the use of eggs (unhatched or hatched) from these species.

The use of mammalian fetuses or fertilized eggs cannot be done without use of the dam or egg-producing female. An IACUC approval is needed if the female is included in the conditions described in 8:1, 8:4, or 8:5. (See 13:11; 14:19.)

8:12 Is an IACUC protocol required for breeding colonies, independently of any experimental procedures performed on the animals?

Reg. If the breeding animals are included in the considerations described in 8:1, then IACUC approval is required. This would be considered an element of research, testing, or teaching that involves the care and use of animals (as defined in the AWAR §1.1, Activity) and a PHS-supported activity involving animals that must be reviewed and approved by an IACUC (PHS Policy II, III,A). (See 8:14; 14:16.)

Opin. In some instances, such as the production of genetically unique rodents, some IACUCs might consider the breeding itself as an animal "research" project and require IACUC approval. In other instances, breeding colonies (rodents, fish, birds, domestic livestock, etc.) that are not involved with any activities described in the regulatory section of 8:1 (i.e., are species not regulated by the AWAR and not involved in PHS-supported activities) are not required to have IACUC approval unless the institution requires it (see 8:6; 8:14; 14:22), or the animal activity is included as part of an institution's AAALAC accreditation. In the previous edition of this *Handbook*, the survey results showed that 54% of respondents did not require a separate breeding protocol, and only 20% of respondents required a breeding protocol.[20]

8:13 What information should be included for a protocol describing the maintenance of a breeding colony?

Reg. If IACUC approval is required of a breeding colony (see 8:12), then colony managers or responsible investigators must submit certain information to the IACUC (see 8:3). The *Guide* (p. 28) makes specific reference to the potential for unexpected outcomes to develop when breeding genetically modified animals (particularly mice and fish), and the need to monitor these outcomes. It would be beneficial for the protocol to contain information on the anticipated impact of the genetic alteration on animal well-being. The *Guide* (pp. 75–77) also notes that it is important to maintain the genetic integrity of the colony whether the animals are inbred or outbred and that the use of standard nomenclature is important.

Opin. In addition, IACUC members may want to know more about the colony husbandry practices. This information may be requested in special forms or formats, or can be submitted as a separate document for review. Conventional IACUC research protocol forms can also be used, but because research is usually not being performed in a breeding colony, the forms include many unrelated questions.

Breeding animal programs require special attention to certain animal welfare concerns, such as male-to female-ratios, artificial insemination, chemicals synchronizing estrus, length of time the male is left with the female and the type of caging to be used. Additional variables influence breeding results and the IACUC may want very detailed information on file (e.g., light cycles, nutrition, sanitation, bedding material, cage size and type). Other information often requested involves the age of the breeders, when to cull old breeders, number of litters permitted per dam, and ways birthing difficulties are managed. The information requirement will vary from IACUC to IACUC and from species to species.

Some institutions have determined that it is a best practice for them to develop forms to be used for animal breeding programs. Additional information related to breeding colonies may include some or all of the following (in addition to items mentioned previously):

- The name of the veterinarian to be contacted if and when needed
- The source of the breeding animals
- The number of offspring per year produced or number of breeding pairs per year maintained

- The disposition of animals when no longer useful or if there is a failure to have genetic trait required
- The method of euthanasia when needed
- The justification for maintaining animals
- A description of painful or stressful husbandry procedures (if any)
- A list of personnel who will handle animals and their training and experience

The *Institutional Animal Care and Use Committee Guidebook*[4] is a useful resource for IACUC review of breeding colony operations.

8:14 Do sentinel animals or blood donors need IACUC approval? Since these animals are not used in specific research projects, is it necessary to develop protocols for the review and approval of these projects by the IACUC?

Reg. If the animals in question fall under the considerations described in 8:1, then IACUC approval is required. NIH/OLAW has made the following interpretation of the PHS Policy:

> PHS Policy applicability is not limited to research. It also includes all activities involving animals including testing and teaching. [OLAW] has determined that although animals used as sentinels, breeding stock, chronic donors of blood and blood products, or for other similar objectives may not be part of specific research protocols, their use for these purposes contributes significantly to the institutional research program and constitutes activities involving animals. Consequently, the IACUC must receive and approve of protocols and appropriate systems to monitor the use of animals prior to the commencement of such activities, and should then perform reviews at the appropriate intervals (IV.C of the PHS Policy).[21] See NIH/OLAW FAQ D.16.

The AWAR (§1.1) broadly define activities involving animals as "those elements of research, testing, or teaching procedures that involve the care and use of animals."

Opin. The regulations apply to those animals involved in biomedical or other types of research projects. By inference, the AWAR cover animals that contribute to an institution's research program. There are, however, many animal uses that are not considered to be research. For example, blood donor animals used for veterinary clinical procedures (dogs, cats, birds, ferrets, etc.) in veterinary clinics (within registered research facilities) may not be part of any research or teaching project and are not required by regulations to be reviewed by the IACUC. Many management-related decisions are based on sample trials of products on animals to determine whether they serve the intended purpose; for example, different bedding materials, caging, foods, or enrichment devices are often tested on resident animals. Data collection consists of observations of the animals with which they are used, and the results aid management to make purchase decisions. Small animal projects that impact on housing and management of the animals are typically not required by regulation to be reviewed by the IACUC. If these projects, however, have a true experimental design that has potential physiological impact on the animals and the results are submitted for publication, such projects require IACUC review. (See 8:31.)

8:15 If an investigator wishes to conduct research using animals at another institution, must the IACUC at the investigator's home institution approve the protocol? Must the other institution's IACUC approve it?

Reg. There are many circumstances that involve partnerships between collaborating institutions or relationships between institutional animal care programs. It is imperative that institutions define their respective responsibilities. The PHS Policy requires that all awardee institutions and performance sites hold an approved Animal Welfare Assurance. NIH/OLAW negotiates an Inter-institutional Assurance[22] when an institution receives PHS funds through a grant or contract award and the institution has neither its own animal care and use program, facilities to house animals, nor an IACUC and does not conduct animal research on site, but will conduct the animal activity at an Assured institution (named as a performance site). Assured institutions also have the option to amend their Assurance to cover non-Assured performance sites, which effectively subjugates the performance site to the Assured institution and makes the Assured institution responsible for the performance site. If both institutions have Domestic Assurances, they may exercise discretion in determining which IACUC reviews research protocols and under which institutional program the research will be performed. It is recommended that if an IACUC defers protocol review to another IACUC, then documentation of the review should be maintained by both committees. Similarly, an IACUC conducting a semiannual evaluation should notify the other IACUC if significant issues are raised during a program inspection of a facility housing a research activity for which that IACUC bears some oversight responsibility as clarified in NIH/OLAW FAQ D.8.[23]

Furthermore, the *Guide* (p. 15) stresses that collaborations between institutions have the potential to allude to vague responsibilities regarding animal care and use. To prevent any ambiguities, institutions should have a formal written understanding (e.g., memorandum of understanding) that addresses responsibilities for offsite animal care and use, ownership, and IACUC review and oversight. The institutions should be aware that their respective IACUCs might also review the protocols in reference to the collaborations at hand.[24] (See 9:53.)

8:16 If an investigator wishes to conduct research using animals at a *foreign* institution, must the IACUC at the investigator's home institution approve the protocol? Must the other institution's IACUC approve it? What about samples that are obtained by citizens of the foreign country and sent to a domestic institution?

Reg: All animal activities supported by the PHS must be reviewed by the IACUC of the domestic-Assured awardee institution that receives such support. Foreign institutions that serve as performance sites must also have Foreign Assurances on file with NIH/OLAW. NIH/OLAW considers institutions whose scientists are engaged in such collaborative work accountable for the animal-related activities from which they receive animals or animal parts. When a foreign institution holds a PHS Assurance, it also is expected that the institution will conduct the study in accordance with the applicable host-nation's laws, policies and regulations governing animal care and use and the *International Guiding Principles for Biomedical Research Involving Animals.*[25–27]

In the specific case of sample collection, the review should take into account the species involved, nature of the specimen, and the degree of invasiveness of the procedure, giving appropriate consideration to the use of anesthetics and analgesics. In instances in which samples are obtained directly by citizens of a foreign country for subsequent shipment, recipient PHS-supported investigators should determine the proposed methods of collection and present that information to their IACUC for review. Prior to sample collection and regardless of whether specimens are obtained by an awardee institution's investigator directly or by persons in a foreign country, NIH/OLAW strongly recommends that each awardee institution consult with other agencies of the U.S. government concerning importation requirements. Depending on the species involved and the nature of the specimen, the following may be of assistance:

- U.S. Fish and Wildlife Service (USFWS)[28]
- Department of the Interior (for compliance with the International Convention on Trade in Endangered Species of Fauna and Flora (CITES)[29]
- U.S. Department of Agriculture (regarding potential animal pathogens)[30]
- The Centers for Disease Control and Prevention (concerning importation of nonhuman primates and potential pathogens of human beings)[31]

Opin. In collaboration with a domestic PHS-assured institution, duplicative IACUC review is not required; however, as indicated, a domestic IACUC must review protocols conducted in foreign locations.

8:17 What minimal qualifications does a person need in order to submit a protocol to the IACUC? For example, can students, postdoctoral fellows, medical residents, or visiting scientists submit a protocol?

Reg. Personnel conducting procedures on the species being maintained or studied will be appropriately qualified and trained in those procedures (AWAR §2.31,d,1,viii; PHS Policy IV,C,1,f). NIH/OLAW provides further guidance on what kind of training is necessary and how frequently it should be provided in their FAQs.[32] The AWA also states that it is the responsibility of the institution to ensure that all personnel involved in animal care and use are qualified to perform their duties (AWAR §2.32,a).

The *Guide* (p. 15) places increased emphasis on the need for adequate education and training for all personnel involved with the care and use of animals.

Opin. There is a difference between the person submitting the protocol and the person actually performing the animal work. There are no federal or state minimal requirements or qualifications for anyone to "submit" a protocol to the IACUC. Such requirements are left to the discretion of each institution. Some institutions allow only their own scientists or faculty to submit protocols. Others may allow outside scientists to submit protocols. Some find a middle ground and have local faculty "sponsor" non-faculty members. However, there are definite requirements detailed below on who can work with animals on a protocol.

AWAR §1.1 (Principal Investigator) defines the PI's responsibility. According to PHS Policy II the awardee institution is held responsible for the conduct of the study: "No PHS support for an activity involving animals will be provided to an individual unless that individual is affiliated with or sponsored by an institution which can and does assume responsibility for compliance with this Policy." NIH/OLAW and APHIS/AC expect the PI to assume responsibility for conducting his or her study according to an approved protocol. It is important for the IACUC, as an agent of the institution, to know who is working with the animals and ensure that they are adequately trained and which individual has ultimate responsibility for the conduct of the study.[33]

Review of research personnel qualifications is an institutional responsibility which typically falls under the purview of the IACUC.[32] The IACUC is best equipped to evaluate individuals' qualifications when very specific information is submitted regarding their duties and training pertinent to their roles in the protocol. Per the *Guide* (p. 15) institutions are responsible for providing appropriate resources to support adequate training, but the IACUC is responsible for providing oversight and evaluating the effectiveness of the training.

8:18 Who typically maintains protocols and for how long should they be kept?

Reg. For AWA-regulated species, animal projects and protocol records must be held for 3 years after completion of a study, as are the IACUC minutes and records of deliberations. The PHS Policy (IV,E,1-2) requires that awardee institutions maintain records for 3 years after completion of the animal activity, including: a copy of the approved Assurance, minutes of IACUC meetings, records of applications, proposals and proposed significant changes, records of semiannual reports and recommendations to the IO, and records of accrediting body determinations. These records must be made available to APHIS/AC or NIH/OLAW when requested (AWAR §2.35,a; §2.35,f; PHS Policy IV,E,1; IV,E,2). (See Chapter 22.)

Opin. A separate IACUC office or a research administration office usually maintains the protocol files. Investigators should also keep a copy. These records are also considered institutional documents and subject to whatever institutional requirements are established (above and beyond the federal regulations).

8:19 Who should have access to IACUC protocols?

Reg. For AWA-regulated species and PHS-funded animal projects, protocol records must be made available to APHIS/AC or NIH/OLAW when requested (AWAR §2.35,f; PHS Policy IV,E,2).

Opin. Individuals have the right to access records in possession of the federal government under the Freedom of Information Act (FOIA). Unless information is specifically exempted from release as described in Section 522,b of the FOIA, NIH/OLAW is required to release information in its possession.[34,35] There is no requirement for institutions outside the federal government to comply with the federal FOIA. However, all 50 states as well as the District of Columbia have some form of freedom of information, sometimes referred to as "Sunshine Laws" that may govern release of information by institutions.[36] Activities conducted at public institutions inclusive of IACUC operations may be covered by "sunshine" laws (open meetings or open records laws). (See Chapter 22.)

IACUC documents can be subject to the state's public records law, and, except when exempted under the law, IACUC documents are available to the public. These vary from state to state. For example, Florida's Statute Section 240.241(2) exempts the following information from being released:[37] "[M]aterials that relate to methods of manufacture or production, potential trade secrets, potentially patentable material, actual trade secrets, business transactions, or proprietary information received, generated, ascertained, or discovered during the course of research conducted within the state universities shall be confidential and exempt from the provisions of s.119.07(1), except that a division of sponsored research shall make available upon request the title and description of a research project, the name of the researcher, and the amount and source of funding provided for such project."

In addition, IACUC meetings may also be subject to the state's open meetings law; therefore, students, faculty, staff, media, and the public at large can have access to them. Individual states' freedom of information acts govern who has access to IACUC records and meetings. Questions that arise regarding the legal access to this information should be addressed by legal counsel from the institution. (See Chapter 22.)

In privately funded institutions not subject to sunshine laws, access to meetings and access to records become policy matters for the institution. Researchers, certain administrators, as well as the IACUC, veterinary, and husbandry staffs are most involved and affected by this information and, therefore, should be afforded access. Conducting open meetings and distributing nonexempt information to the public can be in the best interest of animal research. This is an effective means of public education and helps to negate the veil of secrecy that often hinders the public image of animal research.

Surv. Who has ready access to IACUC protocol forms other than the IO, IACUC office, the IACUC Chairperson, the Principal Investigator, and the Attending Veterinarian? Check as many boxes as appropriate.

- Nobody else 28/297 (9.4%)
- Certain high level institutional officials 41/297 (13.8%)
- Certain animal facility administrators 116/297 (39.0%)
- Certain animal care or animal health technicians 107/297 (36.0%)
- The Principal Investigator's research team 157/297 (52.9%)
- USDA/Animal Care inspector (for Animal Welfare Act covered species) 129/297 (43.4%)
- Other IACUC members or alternates 156/297 (52.5%)
- Our legal counsel 30/297 (10.1%)
- There is open access to anybody 50/297 (16.8%)
- I don't know 10/297 (3.4%)

8:20 May protocols be submitted electronically to the IACUC and can protocols be electronically distributed to IACUC members for review?

Reg. Neither the AWAR nor PHS Policy prohibits protocols from being electronically submitted to the IACUC and electronically posted for members to review.

Opin. Technology is available for protocols to be completed and submitted to secure web sites. Members can then gain access to this site to read or download the information and return e-mail messages with their results. It is emphasized that the final approval of protocols submitted in this fashion must not vary from the regulations and requirements set forth by the AWAR and PHS Policy (IV,C,2-3) regarding full committee review and designated member review. The use of electronic communication simply speeds up the review process by replacing the slower mail delivery systems. (See 6:17–6:19; 6:21; 8:34.)

The use of electronic signatures of the investigator and administrators (if they are required) have become commonplace. During the interagency cooperative NIH/OLAW Online seminar entitled *"Emerging Issues, USDA Perspective"*, broadcast on September 17, 2009, Dr. Betty Goldentyer, Eastern Region Director, APHIS/AC, responded to a series of questions regarding electronic systems for conducting IACUC business. When asked if electronic signatures were acceptable, she replied, "Yes. Electronic signatures are just fine."[38]

Surv. Does your IACUC currently use a reasonably sophisticated electronic protocol submission and/or review system? This refers to commercial or in-house developed systems other than basic word processing.

- Yes, for submission only. 13/296 (4.4%)
- Yes, for review only. 0
- Yes, for submission and review. 69/296 (23.3%)
- Yes, but it's hard to describe. It's a "hybrid" system. 19/296 (6.4%)
- No, we do not use a sophisticated submission or review system. 195/296 (65.9%)

8:21 What are some recommendations for maintaining confidentiality if electronic media are used for submission and review?

Reg. The unauthorized release of confidential IACUC information by members is already prohibited by law (AWA §2157,a; §2157,b):

> It shall be unlawful for any member of an Institutional Animal Committee to release any confidential information of the research facility including information that concerns or relates to (1) the trade secrets, processes, operations, style of work, or apparatus; or (2) the identity, confidential statistical data, amount or source of any income, profits, losses, or expenditures, of the research facility. It shall be unlawful for any member of such Committee (1) to use or attempt to use to his advantages, or (2) to reveal to any other person any information which is entitled to protection as confidential information under subsection (a) of this section.

Opin. Security is always a concern when dealing with animal research records, especially those protocols under discussion and not officially approved. NIH/OLAW provides the following perspective: "Institutions utilizing innovative modes of communication must be aware of the potential security problems inherent in the method chosen. Some material considered by IACUCs should be treated as privileged or confidential, especially prior to final committee action. In the case of trade secrets, such information may be protected by law (7 U.S.C. 2157, Section 27). Because of widespread reports of unauthorized access to computer records, the use of available computer security measures such as passwords, controlled access, and encryption should be considered."[39] The use of the electronic mail at public institutions is not private. Files of public employees may be considered public documents and may be subject to inspection under a state's public records law (see Chapter 22).

8:22 What is meant by "just-in-time" review of an IACUC protocol?

Reg. Effective September 1, 2002, the NIH changed the NIH Grants Policy Statement to permit institutions with PHS Animal Welfare Assurances to submit verification of IACUC approval for an application subsequent to peer review but prior to award, as described in NOT-OD-10-120.[40] This process is referred to as just-in-time. The just-in-time policy reduces regulatory burden on investigators and IACUCs, allowing resources to be focused only on those projects likely to be funded. Key to this change are the following expectations:[41]

- PHS still requires that the institution hold an approved Assurance and the animal use component must receive IACUC approval prior to the release of grant funds.
- Institutions may still choose to review the animal protocol at any time during the grant submission process.
- Under no circumstances may an IACUC be pressured to approve a protocol or be overruled on its decision to withhold approval.
- NIH peer review of the animal component of the grant will continue and will supplement, but not replace, the IACUC review and approval.
- It is incumbent upon investigators to convey to their IACUCs all modifications in animal usage that may result from NIH peer review feedback.
- It is incumbent upon institutions to communicate to the NIH any IACUC-imposed changes that impact on the scope of the funded study.
- It is incumbent upon the NIH to ensure that institutions are given adequate notice to allow for IACUC review prior to award.

To provide clarification of the just-in-time review process, NIH/OLAW has provided specific guidance regarding the roles of NIH's Scientific Review Groups (SRG) with respect to review of the Vertebrate Animal Section (VAS) of applications and the oversight role of the IACUC. The Notice not only describes how the VAS section is evaluated as part of the peer review process and contributes to the overall scoring, but also defines responsibilities of the SRG, the IACUC, and the investigator.[42] NIH/OLAW has posted a VAS fact sheet at http://grants.nih.gov/grants/olaw/VASfactsheet_v12.pdf.

Opin. Many funding agencies (private, nonprofit organizations, etc.) still require animal use approval (or pending approval) prior to accepting a grant proposal. Some, however, apply the "just-in-time" process, akin to the NIH funding procedure. As the approach required by the individual funding agency may change, the authors advise investigators to check with the funding agency routinely prior to submitting a request for funds. (See 9:13.)

8:23 Is it necessary to submit an IACUC application at the time a grant is submitted to PHS?

Reg. (See 8:22.)

Opin. No, under the just-in-time policy (see 8:22) it is not necessary to submit evidence of IACUC approval at the time of grant submission. However, confirmation that the proposed experiments have been reviewed and approved by the institution's

IACUC must be submitted to the PHS funding component before grant funds to conduct animal activities are released. Nevertheless, the majority of grant applications that propose to use animals contain preliminary data based on animal use. Therefore, an approved IACUC protocol reflecting the preliminary data collection must be on file at the institution.

8:24 With the just-in-time process in place, after an NIH grant is approved for funding, how much time does an investigator have to submit a protocol to the IACUC, and how much time does the IACUC have to review the protocol?

Reg. (See 8:22.) The regulations do not specify a time frame; they only state that an IACUC approval must be obtained prior to release of funds.[41]

Opin. NIH will send a request to both the institution and the PI for just-in-time information after the grant has been reviewed and received an impact score of 40 or less.[43] If the grant is subsequently funded, the release of funds usually occurs within 2 months of NIH's request for the information. Additionally, the investigator receives a Summary Statement sheet after the peer review process that signals whether or not the application is in the fundable range. Given that both investigators and their supporting institutions are eager to receive the grant monies, investigators are advised to submit their animal use protocol for IACUC approval at this time. On rare occasions, a grant with a score that initially appears not to be within the fundable range is funded. When this happens the institution and PI receive notification of the need for just-in-time review of the animal use protocol congruent with the release of funds notification. In this situation, a restriction is placed on the Notice of Grant Award indicating that no grant-related work involving the use of animals can be initiated until verification of IACUC approval has been provided to the NIH Grants Management Officer. (See 9:51; 9:52.) The range of time lines allowed by funding agencies to provide verification of IACUC approval of protocols has been surveyed.[44]

8:25 Investigator Jones has a PHS grant and permanently moves from Institution A to Institution B. The PHS allows him to transfer his grant to Institution B. He has an IACUC-approved protocol at institution A but will complete the animal-related research at Institution B. Must Institution B's IACUC review and approve the proposed research before it can begin?

Reg. The transfer of a grant requires prior approval of the PHS funding agency. Once the transfer has been approved an updated Vertebrate Animal Section (VAS) is submitted describing any changes in the animal activity and how the veterinary medical care will be provided at the new performance site (Institution B). Institution B must have an Assurance with NIH/OLAW and must provide verification of IACUC approval of the proposed animal activity.[45] The institution relinquishing the grant must do so prior to the transfer of the award.

Opin. Because many of the policies and regulations that govern the use of animals in research are performance based, individual institutions vary in their approaches to approving animal care and use. Because accountability transfers to the receiving institution, the outcome of protocol review at Institution B may require the investigator to modify his or her protocol to meet the institution's own animal care policies.

8:26 Which people should be listed on an IACUC protocol?

Reg. (See 8:3.)

Opin. The IACUC must fulfill its charge that all personnel conducting procedures on animals are adequately trained and qualified, and that they are enrolled in applicable OHS programs. Therefore, the funded PI as well as all individuals who will conduct animal related procedures on behalf of the investigator should be identified on the information provided to the IACUC.

Surv. Which people are typically listed on your IACUC protocol form?

In a previous survey,[20] 71% of the respondents said that the PI and other persons handling or performing procedures on animals were listed on the protocol form.

8:27 What are some ways for the IACUC to minimize the potentially overwhelming amount of information within the protocols that investigators and IACUC members need to complete or review?

Opin. The process of protocol review can be streamlined for IACUCs by utilizing a pre-review and/or designated reviewer system. An administrative review by IACUC support staff can help to pick up glaring errors or information omissions. Pre-review can also be conducted by a qualified veterinarian and an investigator who are familiar with the protocol format, further helping to reduce errors and make the reviewing process less onerous for the rest of the committee members. (See 9:4–9:12.) Finally, utilization of a designated reviewer for a specified classification of protocols (e.g., those that do not require invasive procedures, those that involve sacrifice for tissues only) helps to minimize the number of reviews for which each committee member is responsible. (See 9:21–9:27.)

The process of protocol completion can be streamlined for investigators by utilizing a protocol format that is limited to the necessary information required to review the process; and that is clear and concise, thereby supporting accurate completion of the required information. It is an option to have the IACUC and veterinary unit provide online SOPs,[46] policies, and veterinary information for the more common, universally applied animal use techniques (see 8:29). Investigators can then, if applicable, "cut and paste" information such as anesthetic and analgesic regimes, blood withdrawal frequencies and volumes, and DNA sampling procedures. However using SOPs provides other challenges to the review system, in that when changes are made to an SOP, the research team must alter their procedures to accommodate the change. If they continue to use a procedure that is not described in the modified SOP referred to in their approved protocol, they will be working outside of an approved protocol, a noncompliance that must be reported to NIH/OLAW.[46]

Surv. Does your IACUC have a general administrative review to assure that the protocol form is filled in correctly before the protocol is either pre-reviewed by a veterinarian or sent for formal IACUC review?

- Yes 203/297 (68.4%)
- No 89/297 (30.0%)
- Other 5/297 (1.7%)

8:28 What is a reasonable sequence of time and events that begins when a PI submits a protocol and ends when the protocol is ready for a vote?

Opin. The time frame for the process of approving protocols hinges largely on the number of times per year the institution's IACUC meets. Although an average seems to be monthly, (see 6:3) IACUCs at larger institutions may meet more frequently, and IACUCs at smaller institutions may meet less frequently. Typically the completed protocol is required to be submitted by the investigator to the IACUC administrative staff at some set date prior to the convened meeting. The interval set between protocol submission and committee review is in part dictated by the number of protocols the committee has to process, the infrastructure support provided to the committee, and the availability of individuals serving on the committee (e.g., institutions can opt to provide dedicated time and support for faculty members to complete committee assignments).

The protocol is distributed through the channels that the institution has established (e.g., some variation of pre-review; see 8:27) or it is distributed to the entire committee for their determination if the proposal can be reviewed by DMR before it is referred to the Chair for assignment of designated reviewers (see NIH/OLAW FAQ D.19). With a pre-review process, comments can be returned to the investigator for correction or discussion prior to deliberation by the IACUC. Use of the designated reviewer system can shorten the approval process if modifications to obtain approval are agreed to by the PI. If not, the protocol must be reviewed at a convened meeting of the full committee (AWAR §2.31,d,2; PHS Policy IV,C,2). The length of time required to resolve protocol issues is somewhat dependent on the nature of the committee's concerns, but in the authors' opinion, it is more dependent on the communication skills of the PI and the working relationship between the committee members and the investigative community.

Surv. 1 If your IACUC uses an informal veterinary pre-review process, how long does that process typically take from the time the veterinarian receives the protocol until s/he returns it to either the IACUC or the investigator?

- We do not use a veterinary pre-review 116/295 (39.3%)
- Usually less than one calendar week 129/295 (43.7%)
- Usually one to two calendar weeks 41/295 (13.9%)
- Usually more than two but less than three calendar weeks 2/295 (0.7%)
- Usually more than three but less than four calendar weeks 2/295 (0.7%)
- Usually more than four calendar weeks 0
- I don't know 5/295 (1.7%)

Surv. 2 If your IACUC ever uses a Designated Member Review (DMR) for protocol review, how long does it take from the time the designated reviewer receives the protocol until the protocol is first returned to either the investigator or the IACUC office?

- We never use DMR 44/294 (15.0%)
- Usually less than one calendar week 137/294 (46.6%)
- Usually one to two calendar weeks 91/294 (31.0%)

- Usually more than two but less than three calendar weeks — 17/294 (5.8%)
- Usually more than three but less than four calendar weeks — 4/294 (1.4%)
- Usually more than four calendar weeks — 0
- I don't know — 1/294 (0.3%)

Surv. 3 If your organization ever uses Full Committee Review (FCR), how long does it take from the time committee members first receive a protocol until it is discussed at a full committee meeting? This includes the time it may require to have back and forth discussions with the investigator.

- We do not use FCR — 13/266 (4.9%)
- Usually less than one calendar week — 33/266 (12.4%)
- Usually one calendar week but less than two — 102/266 (38.4%)
- Usually more than two calendar weeks but less than three — 60/266 (22.6%)
- Usually more than three calendar weeks but less than four — 40/266 (15.0%)
- Usually more than four calendar weeks but less than six — 11/266 (4.1%)
- Usually more than six calendar weeks — 1/266 (0.4%)
- I don't know — 6/266 (2.3%)

8:29 If the exact same procedure is repeatedly used (such as a surgical procedure), can the IACUC or PI have an approved "procedure bank" so a PI can reference the approved procedure without having to write it out in detail every time? (See 8:30)

Opin. Yes. SOPs describing a surgical procedure can be inserted within a protocol or referenced in a protocol rather than providing a written narrative of a common animal use procedure. It is recommended that such SOPs be reviewed and approved for changes or concerns by the IACUC at least on an annual basis (AWAR §2.31,d,1,xi,5), with a 3-year *de novo* review for institutions following the PHS Policy (IV,C,5) or semiannually, if they involve AIPHIS/AC-regulated species.[47]

Surv. Does your IACUC have a collection of IACUC approved research techniques or policies which an investigator can reference in his or her application without having to rewrite the entire technique/policy?

- No — 126/297(42.4%)
- Yes — 171/297(57.6%)

8:30 If a procedure is repeatedly performed, but with a single variable (e.g., a new drug to be tested), can the IACUC have an approved "procedure bank" so a PI can reference the approved procedure without having to write it out in detail every time? If yes, how should the IACUC address the issue of the single variable?

Opin. As stated in 8:29, SOPs can be used within a protocol to streamline the process, such as the approach to drug delivery. Nevertheless, a new protocol or protocol amendment must be reviewed and approved with each new drug that is proposed for use in the animal. An alternative approach to attaching or inserting SOPs within a newly formatted protocol would be to continuously amend a standard protocol with the necessary information about the new drug. The committee can then provide appropriate review for the amendments. Some have argued that if the compound administered is in the same class (e.g., an antigen), the PI and staff are the same, and the type of animal used is identical, there should be the ability to have one "global approval" for work done. Others may argue that objectives of the study should also be considered even though the compound approved is of the same class, same PI and same species. This approach does not address animal usage or the potential for variability in drug reactions even within a similar class of agents.

8:31 A PI wishes to use a very small amount of blood (for research) that will be taken for clinical purposes by a veterinarian from a pet dog. Is IACUC approval needed for the veterinarian to give the blood to the researcher?

Opin. The purpose for which the blood is drawn is a critical component in the response to this question. Because a portion of the blood is drawn to obtain data for research purposes, an IACUC-approved protocol is necessary. It is recommended for the protection of the institution that an informed consent be obtained from the client as the representative of the animal (the subject).[48,49] (See 14:42; 29:65.)

If the blood is extra material, not specifically obtained for the research, NIH/OLAW would not require the investigator to follow the PHS Policy. Similarly, NIH/OLAW would not require IACUC approval for the activity and would not require the veterinary clinic be covered by an Assurance because this is tissue harvest rather than work with live vertebrate animals.

8:32 A veterinarian draws blood from a cow that is to be slaughtered the next day. The blood is specifically destined for use in an in vitro biomedical research project. Is IACUC approval needed to draw and use the blood?

Reg. IACUC approval is required for all regulated species used for teaching, testing and experimentation (AWAR §1.1, Research Facility; AWAR §2.31,d,1; PHS Policy IV,C, 1-2).

Opin. Yes. This is considered antemortem specimen collection and the time of death after the blood collection is irrelevant. If, however, the blood was drawn after the animal was slaughtered, then the discussion found in 8:7 is relevant.[13,15]

8:33 Live mice are fed to snakes as a source of food, and the snakes are the research subject supported by PHS funds. The mice arrive in the morning and are fed by the end of the day. Does the use of the mice as food require IACUC approval?

Opin. Some oversight by the IACUC is very necessary. The source of the incoming mice could have a significant impact on the health of other rodents housed within the facilities. The source of food for the snakes should be reviewed and described in the IACUC protocol that covers the snakes (i.e., an unusual husbandry issue) or may be covered in a husbandry SOP that the IACUC reviews as part of the overall animal care and use programmatic review.[50] (See 12:29.)

8:34 Is an electronic signature acceptable for approving semiannual reviews?

Opin. The regulations (AWA and PHS Policy) do not prohibit the use of electronic signatures. NIH/OLAW allows the use of electronic signatures if they can be validated. Both the AWAR and the PHS Policy require research facilities to maintain minutes of IACUC meetings, including records of attendance, activities of the committee and committee deliberations. (AWAR §2.31, 2.35; PHS Policy IV,E,1). The sole legal requirement for signatures is stated in the AWA in reference to the semiannual reports of the IACUC which "shall be reviewed and signed by a majority of the IACUC members and must include any minority views." (AWAR §2.31,c,3). The use of electronic signatures in the conduct of IACUC business has become commonplace. The APHIS/AC has stated that the use of electronic signatures is acceptable during a 2009 online seminar in response to a series of questions regarding electronic systems for conducting IACUC business (See 8.20).[38] Both APHIS/AC and NIH/OLAW have stated that latitude is afforded IACUCs in terms of conduct of IACUC business where the regulations are silent.[51] It may be assumed that electronic signatures are acceptable for approving the semiannual reviews as well.

References

1. Association for Assessment and Accreditation of Laboratory Animal Care International. 2013. Rules of Accreditation. http://www.aaalac.org/accreditation/rules.cfm#definitions.
2. Association for Assessment and Accreditation of Laboratory Animal Care International. Frequently asked questions B.1, Animal Ownership. 2013. http://www.aaalac.org/accreditation/faq_landing.cfm#A1.
3. American Veterinary Medical Association. 2013. *AVMA Guidelines for the Euthanasia* of *Animals*: 2013 edition. https://www.avma.org/KB/Policies/Documents/euthanasia.pdf.
4. Applied Research Ethics National Association/Office of Laboratory Animal Welfare. 2002. *Institutional Animal Care and Use Committee Guidebook,* 2nd ed. Bethesda: National Institutes of Health. http://grants1.nih.gov/grants/olaw/GuideBook.pdf.
5. National Institutes of Health, Office of Extramural Research. 2012. Animal Study Proposal. http://grants.nih.gov/grants/olaw/sampledoc/animal_study_prop.htm.
6. Office of Science and Technology Policy, Interagency Research Animal Committee. May 20, 1985. *U.S. Government Principles for Utilization and Care of Vertebrate Animals Used in Testing, Research, and Training*. Federal Register, Washington, D.C. Agencies represented include Veterans Administration; Department of Energy; National Aeronautics and Space Administration; Environmental Protection Agency; Department of the Interior; Department of State; Department of Defense; National Science Foundation; U.S. Department of Agriculture, and Consumer Product Safety Commission; Department of Health and Human Services, including the National Institutes of Health, Fogarty International Center, Centers for Disease Control and Prevention, Office of International Health, Health Research Services Administration, and Food and Drug Administration). http://www.grants.nih.gov/grants/olaw/references/phspol.htm#USGovPrinciples.
7. DeHaven, W.R. 1998. Personal communication to Farol Tomson.
8. American Heart Association. Policies governing all research awards. Ethical aspects of research with human subjects and animals. http://my.americanheart.org/professional/Research/FundingOpportunities/ForScientists/Policies-Governing-All-Research-Awards_UCM_320256_Article.jsp#Ethical. Accessed Jan. 28, 2013.

9. American Cancer Society. Research programs and funding. http://www.cancer.org/research/researchprogramsfunding/index. Accessed Jan. 28, 2013.
10. American Lung Association. 2012. Biomedical research grant (RG) program description. http://www.lung.org/assets/documents/rg-program-description.pdf.
11. Potkay, S., N.L. Garnett, J.G. Miller et al. 1995. Frequently asked questions about the Public Health Service Policy on Humane Care and Use of Laboratory Animals. *Lab Anim. (NY)* 24(9): 24–26. http://grants.nih.gov/grants/olaw/references/laba95.htm.
12. National Institutes of Health, Office of Laboratory Animal Welfare. 2012. Frequently Asked Questions A.1, Should institutions apply the PHS Policy for all animal activities regardless of the source of funding? http://www.grants.nih.gov/grants/olaw/faqs.htm#App_1.
13. National Institutes of Health, Office of Laboratory Animal Welfare. 2013. Frequently Asked Questions A.3, Does the PHS Policy apply to use of animal tissue or materials obtained from dead animals? http://grants.nih.gov/grants/olaw/faqs.htm#App_3.
14. U.S. Department of Health and Human Services, Public Health Service. Grant Application. http://grants.nih.gov/grants/funding/phs398/phs398.html. Accessed Jan. 28, 2013.
15. Garnett, N.L. and W.R. DeHaven. 1997. OPRR and USDA Animal Care response on applicability of the animal welfare regulations and the PHS Policy to dead animals and shared tissues. *Lab Anim. (NY)*, 26(3): 21.
16. National Institutes of Health, Office of Laboratory Animal Welfare. 2012. Frequently Asked Questions A.2, Does the PHS Policy apply to the production of custom antibodies or to the purchase of surgically modified animals? http://www.grants.nih.gov/grants/olaw/faqs.htm#App_2.
17. National Institutes of Health, Office of Laboratory Animal Welfare. 2013. Frequently Asked Questions D.8, When institutions collaborate, or when the performance site is not the awardee institution, which IACUC is responsible for review of the research activity? http://grants.nih.gov/grants/olaw/faqs.htm#d8.
18. U.S. Department of Agriculture. 2011. Animal Care Policy Manual. Policy 10, Specific activities requiring a license or registration. http://www.aphis.usda.gov/animal_welfare/policy.php?policy=10.
19. National Institutes of Health, Office of Laboratory Animal Welfare. 2012. Frequently Asked Questions A.4, Does the PHS Policy apply to live embryonated eggs? http://grants.nih.gov/grants/olaw/faqs.htm#App_4.
20. Laber, K. and A. Smith. 2007. Submission and Maintenance of IACUC Protocols in *The IACUC Handbook*, 2nd ed., eds. J. Silverman, M. Suckow and S. Murthy S, 93-112. Boca Raton: Taylor & Francis.
21. National Institutes of Health, Office of Laboratory Animal Welfare. 2013. Frequently Asked Questions D.16, Is IACUC approval required for the use of animals in breeding programs, as blood donors, as sentinels in disease surveillance programs, or for other non-research purposes? Available at: http://grants.nih.gov/grants/olaw/faqs.htm#proto_16.
22. National Institutes of Health, Office of Laboratory Animal Welfare. 2012. Interinstitutional Assurance. http://grants.nih.gov/grants/olaw/sampledoc/interinstitutional_assurance.htm.
23. National Institutes of Health, Office of Laboratory Animal Welfare. 2013. Frequently Asked Questions D.8, When institutions collaborate, or when the performance site is not the awardee institution, which IACUC is responsible for review of the research activity? http://grants.nih.gov/grants/olaw/faqs.htm#proto_8.
24. National Institutes of Health, Office of Extramural Research. 2001. Guidance regarding administrative IACUC issues and efforts to reduce regulatory burden. Notice NOT-OD-01-017, http://grants.nih.gov/grants/guide/notice-files/NOT-OD-01-017.html.
25. National Institutes of Health, Office of Laboratory Animal Welfare. 2013. Frequently Asked Questions D.13, If an animal activity will be performed outside of the U.S. (either by a foreign awardee or by a foreign institution as a subproject for a domestic awardee), is the awardee's IACUC required to review and approve that activity. http://www.grants.nih.gov/grants/olaw/faqs.htm#proto_13.

26. Potkay, S., N.L. Garnett, J.G. Miller et al. 1997. Frequently asked questions about the Public Health Service Policy on Humane Care and Use of Laboratory Animals. *Contemp. Topics Lab. Anim. Sci.* 36(2): 47–50.
27. Council of International Organizations of Medical Sciences. 2012. Text of Guidelines and Other Normative Documents http://www.cioms.ch/index.php/publications/texts-of-guidelines.
28. U.S. Fish and Wildlife Service. 2012. How to obtain a permit. http://www.fws.gov/permits/instructions/ObtainPermit.html.
29. Convention on International Trade in Endangered Species of Wild Fauna and Flora. www.cites.org. Accessed Jan. 28, 2013.
30. U.S. Department of Agriculture. 2006. Guidelines for Importation of human and non-human primate material. http://www.aphis.usda.gov/import_export/animals/animal_import/downloads/ihnhum.html.
31. Centers for Disease Control and Prevention. 2012. Request to import biological agents or vectors of human disease. http://www.cdc.gov/od/eaipp/importApplication/agents.htm.
32. National Institutes of Health, Office of Laboratory Animal Welfare. 2012. Frequently Asked Questions G.1, What kind of training is necessary to comply with PHS Policy, and how frequently should it be provided? http://grants.nih.gov/grants/olaw/faqs.htm#instresp_1.
33. National Institutes of Health. What investigators need to know about the use of animals. NIH Publication 06-6009. http://grants.nih.gov/grants/olaw/InvestigatorsNeed2Know.doc. Accessed Jan. 28, 2013.
34. U.S. Department of Justice. 2011. What is FOIA? http://www.foia.gov/about.html.
35. National Institutes of Health, Office of Laboratory Animal Welfare. 2013. Frequently Asked Questions. C.4, Are all documents submitted to OLAW subject to the Freedom of Information Act? http://grants.nih.gov/grants/olaw/faqs.htm#report_4.
36. Wikipedia. 2012. Freedom of Information in the United States. http://en.wikipedia.org/wiki/Freedom_of_information_in_the_United_States.
37. Florida Statutes, Sect. 240.241(2).
38. National Institutes of Health, Office of Laboratory Animal Welfare. 2013. Educational Resources. http://grants.nih.gov/grants/olaw/educational_resources.htm.
39. Garnett, N. and S. Potkay. 1995. Use of electronic communications for IACUC functions. *ILAR J.* 37:190–192. http://grants.nih.gov/grants/olaw/references/ilar95.htm.
40. National Institutes of Health. 2010. Notice NOT-OD-10-120, Revised policy on applicant institution responsibilities for ensuring just-in-time submissions are accurate and current up to the time of award. http://grants.nih.gov/grants/guide/notice-files/NOT-OD-10-120.html.
41. Federal Register, Vol. 67, No. 152, 51289. August 7, 2002. Laboratory Animal Welfare: Change in PHS Policy on Humane Care and Use of Laboratory Animals http://grants.nih.gov/grants/olaw/fed_reg_v67n152.pdf.
42. National Institutes of Health., Office of Laboratory Animal Welfare. 2010. Notice NOT-OD-10-128, Clarification on the roles of NIH scientific review groups (SRG) and Institutional Animal Care and Use Committees (IACUC) in review of vertebrate animal research. http://grants.nih.gov/grants/guide/notice-files/NOT-OD-10-128.html.
43. National Institutes of Health. 2012. Request for just-in-time information. http://grants2.nih.gov/grants/peer/jit.pdf.
44. Mann, M.D. and E.D. Prentice. 2007. Verification of IACUC approval and the just in time PHS grant process. *ILAR J.* 48(1):12–28.
45. National Institutes of Health, Office of Laboratory Animal Welfare. 2013. Frequently Asked Questions F.3, May an investigator transfer animals and research to an institution different than the grantee institution http://grants.nih.gov/grants/olaw/faqs.htm#useandmgmt_3.
46. National Institutes of Health, Office of Laboratory Animal Welfare. 2012. Frequently Asked Questions D.14, May standard operating procedures (SOPs) or blanket protocols that cover a number of procedures be utilized in lieu of repeating descriptions of identical procedures in multiple protocols? http://grants.nih.gov/grants/olaw/faqs.htm#proto_14.
47. Brown, P. and C.A. Gipson. 2012. A word from OLAW and USDA. *Lab Anim. (NY)* 41:41–43.

48. National Institutes of Health, Office of Laboratory Animal Welfare. 2012. Frequently Asked Questions A.7, Does the IACUC need to approve research studies that use privately owned animals, such as pets? http://grants.nih.gov/grants/olaw/faqs.htm#App_7.
49. National Institutes of Health, Office of Laboratory Animal Welfare. 2012. Frequently Asked Questions A.8, How can the IACUC determine if activities involving privately owned animals constitute veterinary clinical care or research activities? http://grants.nih.gov/grants/olaw/faqs.htm#App_8.
50. Garnett, N. 2003. A Word from OLAW. *Lab Anim. (NY)* 32(10):19. http://grants.nih.gov/grants/olaw/references/laban32_10_1103.pdf.
51. Brown, P. and C. Gipson. 2010. A word from OLAW and USDA. *Lab Anim. (N.Y.)* 30:299. http://grants.nih.gov/grants/olaw/references/39_10_1010.pdf.

9

General Concepts of Protocol Review

Ernest D. Prentice, Gwenn S. F. Oki, and Michael D. Mann

Introduction

Reviewing protocols involving the use of animals is one of the most important responsibilities of the IACUC. Although the format employed for this review varies across institutions, the criteria used in reviewing and approving protocols should be as consistent as possible. A thorough and up-to-date review is a particularly important consideration with regard to animal welfare issues, such as the use of non-animal model alternatives and the application of refinement techniques to reduce animal pain, discomfort, and distress.

The purpose of this chapter is to present general concepts of protocol review that will help IACUC administrators, IACUC members, AVs, and other interested individuals gain an understanding for how IACUCs conduct protocol review. Particular attention is given to issues such as scientific merit review, prereview, use of consultants, designated member review versus full committee review, and IACUC actions. Because the answers to the questions posed in this chapter are designed to be as succinct as possible, the reader is encouraged to consult the PHS Policy, the AWAR, and the *Guide*.

9:1 What guidelines do the AWAR, the PHS Policy, and the *Guide* provide with respect to IACUC review of protocols?

Reg. PHS Policy (IV,C,1,a–IV,C,1,g; IV,D,1,a–IV,D,1,e) and the AWAR (§2.31,d; §2.31,e) specify criteria that must be met before an IACUC can approve a proposed activity involving the use of animals (e.g., research) as described in a protocol. These requirements address issues such as

- Avoidance or minimization of discomfort, distress, and pain to the animals
- Appropriate living conditions for species used in the project
- Availability of medical care to be provided by a qualified veterinarian
- Euthanasia consistent with the recommendations of the American Veterinary Medical Association (AVMA) *Guidelines on Euthanasia*[1]
- Qualifications of personnel conducting procedures on the species being studied

PHS Policy (IV,A,1) requires that Assured institutions also base their IACUC review of protocols on the *Guide* and that they comply with the AWAR which apply to APHIS/AC regulated species. The *Guide* (pp. 25–26) lists 15 topics that IACUCs should review.

The AWAR (§2.31,d; §2.31,e) list a total of 16 requirements that must be met before the IACUC can approve a protocol. Some of the requirements mirror the PHS Policy, whereas others are not found in the PHS Policy. For example, the AWAR require an investigator to consider alternatives to procedures that may cause more than momentary pain or distress and has provided a written narrative description of the methods and sources used to determine that alternatives are not available. The AWAR also require the investigator to provide written assurance that the animal-related activities do not unnecessarily duplicate previous experiments. The PHS Policy requires that the Investigator consider alternatives, but does not require a description of the methods and sources used in that consideration. However, it should be noted that if the activity is both PHS funded and APHIS/AC regulated, compliance with the AWAR is an absolute requirement (PHS Policy II).

While the AWAR are silent regarding the U.S. Government Principles, the PHS Policy includes the nine *U.S. Government Principles for the Utilization and Care of Vertebrate Animals Used in Testing, Research, and Training* that provide additional guidance concerning the use of alternatives that embrace the concepts of Replacement, Reduction, and Refinement (3 Rs). The *Guide* (p. 4) also endorses the nine Principles and further emphasizes the 3 Rs. Of particular note is Principle II which states that "procedures involving animals should be designed and performed with due consideration of their relevance to human or animal health, the advancement of knowledge, or the good of society." Principle II represents the scientific and societal justification for using animals in research which is also addressed as "Scientific Review" in the *Guide* (p. 26). (See 9:13.)

Opin. It should be noted that neither the PHS Policy nor the AWAR provide specific guidance in all areas of protocol review. It is, therefore, left to the institution to develop specific guidelines for implementation of federal requirements. Many institutions have, therefore, developed investigator handbooks that include information concerning local policies for anesthetics, analgesics, blood sampling, tail sampling, antibody production, pre- and post-operative care for non-rodent animals, and other aspects of animal research.

Finally, it should be mentioned that NIH periodically publishes guidance on selected topics, as notices in the NIH *Guide for Grants and Contracts* that provide information concerning NIH/OLAW's interpretation of the requirements of the PHS Policy.[2] It is particularly useful to access NIH/OLAW's FAQs which provide guidance on many topics, including protocol review.[3] In the interest of harmonization, APHIS/AC concurs with NIH/OLAW guidance provided in these FAQs, where applicable. In addition, in *Lab Animal's* Protocol Review column, NIH/OLAW and APHIS/AC often respond to questions posed in various scenarios.[4] APHIS/AC also issues its Animal Care Policy Manual that provides further guidance for IACUCs.[5] The manual, however, primarily emphasizes animal care as opposed to the elements of protocol review. Finally, the *Institutional Animal Care and Use Committee Guidebook,* 2nd edition[6] is a helpful but somewhat outdated resource. Publication of a 3rd edition is anticipated.

9:2 What are some useful pathways to disseminate protocols for review once they have been submitted to the IACUC?

Reg. PHS Policy (IV,C,2) requires that each IACUC member be provided with a list of proposed projects to be reviewed, and that written descriptions of projects be available to all IACUC members. The AWAR (§2.31,d,2) have the same requirement.

Opin. There are many methods used to disseminate protocols for review including hard copies that are mailed, or distributed by fax, standard electronic mail, phone, and smart phones (see 6:5). When a protocol is assigned to full committee review (FCR), it is common practice for IACUCs to assign each protocol to either one reviewer or to a primary and secondary reviewer (chosen based upon their expertise in the subject matter of the protocol) who are given principal responsibility for in-depth review. With this method, the other members of the IACUC receive either a complete copy of the protocol or a protocol summary for review. Some IACUCs, particularly at institutions that have a large amount of research using animals, utilize a pre-review system where protocols are clarified and modified as necessary prior to full committee review (see 9:4–9:12).

Many IACUCs also utilize a designated member review (DMR) system, sometimes incorrectly referred to as "expedited review." The use of the latter term is strongly discouraged by APHIS/AC, NIH/OLAW and AAALAC because it has been incorrectly interpreted as "light review." DMR helps to decrease the workload of the IACUC during full committee meetings. In general, the method used to disseminate protocols for review is dictated by the size, diversity of research, electronic capability, and specific needs of the institution. (See 9:21–9:32.)

9:3 What are the IACUC and institutional obligations for reviewing a protocol for an internally funded project?

Reg. If an institution has elected, in its NIH/OLAW Assurance, to apply the requirements of the PHS Policy to all activities involving animals (regardless of the source of funding), then the IACUC obligations for reviewing internally funded projects are the same as those obligations under the PHS Policy.

Opin. Internally funded projects may pose special problems for the review process, particularly with respect to scientific merit review. Even IACUC members who maintain steadfastly that the committee has no responsibility or right to review scientific merit agree that the IACUC has some obligation to do so when the proposal will not be subjected to external peer review. A minimal level of merit that justifies animal use must be assured before protocols are approved. Needless to say, animal welfare considerations should be the same, and the same rigorous review process with regard to regulatory compliance should be uniformly applied to both internally and externally funded projects.

9:4 What is meant by prereview of a protocol?

Opin. Prereview refers to the informal review of a protocol by one or more individuals prior to formal review by the IACUC, which can be conducted only by FCR and DMR. The use and scope of prereview varies at institutions. In some institutions, prereview is conducted by an IACUC administrator or staff who checks for typographical errors and to verify that the form is complete. In other institutions, veterinarians or subject matter experts conduct a rigorous review of animal welfare concerns, raising questions that they anticipate will be asked by the IACUC. Prereview is not a substitute for full committee review or designated member review. Prereview is optional; a review, not a prereview, must be conducted before an action can be taken (approval, seek modifications to secure approval, or withholding approval). A good prereview can certainly help reduce the workload of the IACUC. It should be noted that scientific merit review by an external body prior to IACUC review can considered a specialized form of prereview. (See 9:12.)

9:5 What are the reasons for a prereview of protocols?

Opin. Protocols are sometimes incomplete when they are submitted for IACUC consideration. Information that the IACUC needs in order to effectively review the protocol might be absent or unclear. For example, the investigator may be proposing to use a method of anesthesia involving injectable agents without specifying the dose. Clearly, this information must be provided before the IACUC can approve the protocol. Prereview saves the IACUC time, because IACUC members do not have to read protocols that are obviously not ready to be reviewed. Prereview is also helpful to the investigator in terms of speeding up the review process, because the protocol is less likely to be referred back for review by the full committee pending further information.

9:6 Is prereview of protocols required by AWAR or PHS Policy?

Opin. Prereview of protocols is not a PHS Policy requirement, however, the AWAR (§2.31,d,1,iv,B) require PIs to consult with a veterinarian when procedures are planned that may cause more than momentary pain or distress. Prereview of the protocol by a veterinarian is one way that such a consultation can be provided.

9:7 Can the prereview team be an IACUC subcommittee composed of IACUC members, non-IACUC members, or a combination of these?

Opin. Because prereview is not a mandated function of an IACUC, the prereview process can be structured in a way that is most helpful to the IACUC and investigators. Prereview can be carried out by designated IACUC members, qualified IACUC office staff, campus veterinarians, or any other experienced individuals who are able to adequately review the protocol and do not have a conflict of interest.

9:8 What is a workable mechanism for the prereview of protocols?

Opin. The best mechanism depends on the institutional structure and the composition of the prereview team. For example, an institution with multiple vivaria may choose to have incoming protocols immediately sent to a prereviewer. Each prereviewer handles all of the submitted protocols from the particular vivarium on the campus, so the prereviewer becomes very familiar with the investigators, facilities, and types of research conducted in that unit. The prereviewer reviews the proposal, notes any points of inadequacy in the document, and communicates with the PI. The prereviewer may be able to suggest to the PI any needed corrections or suggest adding clarifications to the protocol (for example, changing the route, dose or type of anesthetic agent) or suggest adding a sentence to the protocol (for example, indicating the volume of blood to be withdrawn). In the case of more substantial corrections, the prereviewer interactively discusses the problems with the PI and may recommend that the investigator rewrite the protocol and submit a corrected version. There is the potential problem that the PI may interpret prereview as actual IACUC review, and changes made may be interpreted as the only action that will be necessary to obtain IACUC approval. The PI should clearly understand that prereview is not full committee or designated member review and that additional changes may be required (requested) by the IACUC.

9:9 What should the PI do following prereview?

Opin. Assuming that the prereviewer suggests changes that are in accordance with IACUC policies and concerns, it is in the PI's best interests to make the necessary corrections before the IACUC meeting or DMR so that the protocol is suitable for review.

9:10 What happens to a protocol after it is prereviewed?

Opin. After a protocol is prereviewed, the investigator submits a modified or unmodified protocol that must be processed for formal review by the IACUC. The formal review may be accomplished by DMR (see 9:21–9:27) or by FCR, which is review at a convened meeting of a quorum of the IACUC. (See 9:28–9:32.)

9:11 Are the results of the prereview presented to the full IACUC? If so, by whom?

Opin. In institutions that still use paper submissions, the IACUC may be presented with a completely revised application form including all of the changes made by the PI as a result of the prereview process. Alternatively, the changes may be presented as a separate document along with the original application form. The former method is preferred because it yields a "clean," complete protocol application, which is a benefit to both the IACUC and the investigator and his or her staff. In institutions that use electronic submissions, the PI can simply alter the online form, especially if the submission software is able to track all the changes. Unfortunately, not all of the commercial software packages allow such tracking. From the point of view of minimizing the workload of the IACUC it is preferable to have a single updated document for review. The PI may find that a single accurate copy of the approved protocol facilitates compliance by his or her staff when they are conducting the experiment. Conducting work that has not been approved by the IACUC is a frequently occurring form of noncompliance reported to NIH/OLAW. NIH/OLAW has a webinar on reporting non-compliance which provides helpful guidance.[7]

9:12 Are prereview comments binding on investigators?

Opin. No. Prereview is simply a method for giving investigators advice on their protocols. A possible exception may arise when the AV determines during prereview that a particular anesthetic or surgical procedure is not appropriate (AWAR §2.31,d,iv,B; §2.33,b,4). Investigators may occasionally disagree with the prereviewer's suggestions and choose to provide an explanation and send an unmodified or a partially modified protocol to the IACUC for formal review. However, an investigator who insists on presenting an inadequate document to the IACUC is likely to find that the IACUC will not approve the protocol and the protocol will be returned for modifications or clarification for the same reasons identified by the prereviewer. This will result in unnecessary delays in obtaining IACUC approval of the project.

9:13 Should the IACUC review protocols for scientific merit? Do the AWAR or the PHS Policy specifically require (or prohibit) review of scientific merit?

Opin. In our experience, many IACUCs review protocols for some level of "scientific merit" or "scientific relevance." It is not clear, however, that these terms have the same or similar meaning to all IACUCs. Whether IACUCs should perform such review is open to debate. Prentice et al.[8] and, more recently, Mann and Prentice[9] reviewed the PHS Policy, in which the term "scientific merit" is not used. Rather, it refers to "relevance" in U.S. Government Principle II, in a manner that strongly suggests the terms are synonymous. The PHS Policy also uses the terms "sound research design" (IV,C,1,a) and "scientifically valuable research" (IV,D,1,d). When all of these terms are considered together, they support the position of NIH/OLAW for PHS-funded research activity that an IACUC should consider the scientific relevance of a proposal. Indeed, NIH/OLAW has written in a FAQ, the following guidance which makes it clear that it is seldom possible to separate considerations of scientific value from animal welfare:[10]

> Peer review of the scientific and technical merit of an application is considered the purview of the NIH Scientific Review Groups (SRGs), which are composed of scientific experts from the extramural research community in a particular area of expertise. However, SRGs also have authority to raise specific animal welfare concerns that can require resolution prior to a grant award.
>
> Although not intended to conduct peer review of research proposals, the IACUC is expected to include consideration of the U.S. Government Principles in its review of protocols. Principle II calls for an evaluation of the relevance of a procedure to human or animal health, the advancement of knowledge, or the good of society. Other PHS Policy review criteria refer to sound research design, rationale for involving animals, and scientifically valuable research. Presumably a study that could not meet these basic criteria is inherently unnecessary and wasteful and, therefore, not justifiable.

The AWAR appear inconsistent in their reference to "scientific merit review." In the Public Comment Section of the regulations, APHIS/AC stated, "we added the term 'animal care and use procedure'... to avoid any misunderstanding or implication that APHIS intends to become involved in the evaluation of the design, outlines, guidelines, and scientific merit of proposed research."[11] Also, the AWAR (§2.31,a) and the AWA itself (§2143,a,6,A,i–ii) state that "except as specifically authorized by law or these regulations, nothing in this part shall be deemed to permit the Committee or IACUC to prescribe methods or set standards for the design, performance, or conduct of actual research or experimentation by a research facility." One way to interpret this apparent inconsistency is to accept there may be many scientific aspects of a study that do not relate to the animal activities, and are, therefore, outside the boundaries of IACUC review. On the other hand, the AWAR (§2.31,e,4) like the PHS Policy, make reference to "scientifically valuable research."

It appears that neither NIH/OLAW nor APHIS/AC explicitly require IACUC review of scientific merit. Nevertheless, according to NIH/OLAW,[12] an institution cannot defer scientific relevance review to the funding agency. IACUC approval, using criteria stated in the PHS Policy, must (in the original policy statement[13] not under the just-in-time criteria noted below) precede NIH peer review. Therefore, approval by an IACUC of a proposed activity that is conditional upon successful peer review by the funding agency would not be in keeping with the PHS Policy requirements. Such conditional action does not constitute IACUC approval required by the PHS Policy either prior to review by the NIH Scientific Review Group (SRG) or under just-in-time[14] requirements. The NIH review of merit should, therefore, be viewed as additional assurance rather than the only assurance that the research has value.[8,12] Based on this, Prentice et al.[8] concluded that the IACUC does have a responsibility to review the scientific relevance of animal projects that are subject to the requirements of the PHS Policy. There is not, however, general agreement with this conclusion. Black[15] argues that merit and relevance are not the same thing; that IACUC review does not constitute peer review, the consequences of IACUC review are different from those of external review, and IACUC review may constitute a violation of researchers' academic freedom. Prentice et al.[16] and Mann and Prentice[9] offer counter arguments. Clearly there is not general agreement.

The advent of just-in-time certification of IACUC approval of research proposals (see 8:22) seems to add another level to the arguments above. It would seem that the NIH tacitly allows the IACUC to refrain from reviewing scientific merit or relevance. On the other hand, referring to the just-in-time process, the NIH wrote, "The fundamental PHS Policy requirement that no award may be made without an approved Assurance and without verification of IACUC approval remains in effect. This change only affects the timing of the submission of the verification of that review."[17] Given this, and the information in the *Guide* (p. 26), it seems that the PHS holds the IACUC responsible for at least some review of scientific relevance.

Surv. 1. Does your IACUC consider the scientific merit of a protocol application that has had external scientific review by a peer group?

• Not applicable	62/296 (21.0%)
• No, we believe it is the external peer reviewers who have that responsibility	61/296 (20.6%)
• Yes, we ask our members to provide at least a cursory review of scientific merit	151/296 (51.0%)
• Yes, but only for species covered by the Animal Welfare Act	6/296 (2.0%)
• I don't know	7/296 (2.4%)
• Other	9/296 (3.0%)

Surv. 2. Please respond to this question only if your IACUC is a research organization that receives federal or private grant monies as the primary source of its research income. Does your IACUC provide scientific review for internally funded research proposals (i.e., research proposals that have not undergone external peer review)?

• Yes	135/226 (59.7%)
• No, but they are reviewed by another group within our organization such as a departmental or other committee	51/226 (22.8%)
• No, and they are not reviewed by any other group within our institution	22/226 (9.8%)
• I don't know	7/226 (3.1%)
• Other	11/226 (4.9%)

9:14 Can the IACUC approve judicious use of animals without consideration of scientific merit?

Opin. Prentice *et al.*[8] note that there are two levels of review for scientific merit. They refer to a "fundamental level" of review in which scientists form "basic judgments about the adequacy and appropriateness of experimental design in terms of the ability to test the hypothesis, use controls, sample size, statistical analysis, and the training and experience of investigators." This first-level merit review is necessary for the IACUC to approve the judicious use of animals. The other level of review, "knowledge-based level," requires "an assessment be made of the scientific importance of the study." An assessment of merit at the fundamental level could be made by any appropriately constituted IACUC, even in the absence of special expertise in the topic of the protocol. It would seem that a judgment of the merit at the knowledge-based level would be necessary to determine the appropriateness of the proposed animal use. Of course, the latter judgment requires that an IACUC, involved in reviewing proposals over a range of subjects, have suitable expertise in that range of subject matters, or employ ad hoc consultants. (See 9:13.)

9:15 The AWAR (§2.31,a) state that the IACUC shall not prescribe methods or set standards for the design, performance, or conduct of actual research or experimentation by a research facility. What does this actually mean since the IACUC has the authority to approve or withhold approval of animal research activities?

Reg. AWAR(§2.31,a) state "Except as specifically authorized by law or these regulations, nothing in this part shall be deemed to permit the Committee or IACUC to prescribe methods or set standards for the design, performance, or conduct of actual research or experimentation by a research facility."

Opin. This apparent contradiction in the AWAR was discussed in detail by Prentice *et al.*[8] and Mann and Prentice.[9] Consider an animal use protocol that did not contain suitable controls. One could argue that if such a protocol contained no other animal welfare issues, it should be approved by the IACUC because the quoted text above seems to say that scientific design should not be considered. On the other hand, the AWAR (§2.31,e,2) says that a protocol should contain "A rationale for involving animals and for the appropriateness of species and numbers to be used." In this case, the numbers would be inappropriately low because of the missing controls. Lack of controls is an animal welfare issue. This is a poorly designed experiment that will not produce usable data and therefore leads to the unnecessary loss of life of the experimental animals with no scientific benefit. Therefore, the IACUC should not approve the protocol.

It seems that the phrase in AWAR (§2.31,a) "Except as specifically authorized by law or these regulations" is critical. The IACUC should not "prescribe methods or set standards for the design, performance, or conduct of actual research or experimentation" except as specifically stated in the AWAR itself or relevant laws.

9:16 At what point in the review of a protocol are consultants best brought in?

Reg. PHS Policy (IV,C,3) and the AWAR (§2.31,d,3) allow the use of consultants for the review of a protocol. Consultants may not vote with the IACUC, and the IACUC remains responsible for its actions and decisions when consultants are used.

Opin. The use of consultants is highly variable among IACUCs. In some institutions, it is common practice for IACUC reviewers to use colleagues within the institution as consultants for clarification of issues raised during prereview of a protocol (see 9:4; 9:5). This prereview clarification can greatly speed the final review process. Outside consultants may be engaged when the animal proposal involves a highly novel and controversial, cutting edge research or for research using wildlife when the institution has no other expert in the field. Outside consultants may be used when there is disagreement among the IACUC members as to the scientific relevance of a given proposal or if unresolved concerns exist with regard to animal welfare. Consultants also are useful if an investigator asks the committee to reconsider its decision (see 9:56; 29:37).

9:17 Is it appropriate for the IACUC to use consultants to review scientific merit?

Reg. (See 9:4.)

Opin. If the IACUC assesses the scientific merit of a study (see 9:13; 9:14), then it is appropriate for the IACUC to use consultants, particularly if the IACUC does not have the requisite expertise to assess scientific merit. Also, outside reviewers would likely be most useful when there is disagreement among the IACUC members as to the scientific merit of a protocol or if an investigator appeals the decision of the IACUC (see 9:14; 9:56; 29:37). Indeed, the IACUC could seek advice from consultants with regard to any aspect of the project that is problematic.

9:18 What types of consultants might be useful for review of scientific merit? Who picks them?

Opin. Consultants can be

1. Experts on the topic of the protocol under review
2. Experts in the use of animals, with respect to a particular procedure proposed in the protocol
3. Both

An example of the first use specified above is an expert who is asked to review a protocol because the IACUC is uncertain whether the use of the ascites method for making monoclonal antibodies is required in a particular project, as the PI claims. Or, the consultant can be asked to judge whether any useful information will be forthcoming once the antibody is made. An example of the second use is an expert in primate behavior and care who is asked to review a protocol because the IACUC is uncertain whether a rhesus monkey may reasonably be restrained continuously for a specified period of time. Or, the expert can be asked whether any useful information could be derived from the experiments that require the monkey to be restrained for that period of time.

Consultants should be selected by the IACUC, but it may also be reasonable to ask the PI to suggest possible experts. This allows the committee to maintain objectivity in the review process while considering the PI's point of view. It should also be noted that the IO could appoint a consultant after IACUC review in order to assist him/her in any decision whether or not to overturn IACUC approval.

However, it must be remembered that no one, including the IO, can overturn an IACUC disapproval (AWAR §2.31,d,8; PHS Policy IV,C,8). It should be noted that the IO should not overturn an IACUC's decision to approve a protocol based on the criteria that the IACUC uses to judge the protocol. Although the IO has the power to do this, he or she will soon have a dysfunctional IACUC if their decisions are overturned based on animal welfare issues. It is more common for the IO to disallow IACUC-approved research based on personnel, funding, or resource issues. Finally, some institutions require the consultant to sign a Confidentiality Agreement to protect the proprietary nature of the proposed study. This agreement will be useful in terms of protecting anonymity of the consultant and the confidential nature of the study.

9:19 Should the identity of consultants be kept anonymous to the investigator?

Opin. The same arguments apply to anonymity of consultants and to the anonymity of grant proposal reviewers. It has been argued that reviewers refrain from making negative comments about a protocol if their identity is known to the investigator and may fear reprisals. It also is possible to argue that reviewers may make unfair statements about a protocol because their identity is withheld. Whether this will happen likely depends on the individual reviewers and consultants and on the culture of animal care and use at the institution. In any case, if the reviewers of a protocol remain anonymous, then certainly consultants should as well. On the other hand, anonymity may be difficult if the investigator is asked to suggest possible experts. Regardless, the IACUC must determine whether the consultant is conflicted before appointment.

9:20 An IACUC retains the services of a consultant suggested by the PI to help evaluate a complex protocol. Based in part on the consultant's recommendation, the protocol is approved. The IACUC subsequently learns that the consultant was a former student of the PI who submitted the protocol. Should the IACUC revisit the protocol's approval?

Opin. The PI should have informed the IACUC of this consultant's conflict of interest. But, even if consultants are used in the review process, it is still the IACUC that must make the decision for approval. In this case, the IACUC determined that the protocol should be approved. The occurrence of the conflict of interest would not necessarily change the fact that the IACUC had found the protocol acceptable. However, if the IACUC's decision to approve the protocol was largely based on the consultant's review then the IACUC might well want to review the protocol again and even retain the services of another consultant.

9:21 What is "designated member review" and what is its intent?

Reg. Both the PHS Policy (IV,C,2) and the AWAR (§2.31,d,2) recognize a method of "designated member review" (DMR) that is widely implemented by research institutions.[18] Some institutions inaccurately and inappropriately refer to this review process as expedited review, but designated member review is the correct term. DMR must conform to the following process: written descriptions of research projects that involve the care and use of animals must be made available to all

IACUC members, and any member of the IACUC must have the opportunity to request full committee review of those research projects. If full committee review (FCR) is not requested, at least one member of the IACUC, designated by the chairman and qualified to conduct the review, shall review those activities, and shall have the authority to approve, require modifications in (to secure approval), or request full committee review of any of those activities. (PHS Policy IV,C,2; AWAR §2.31,d,2).

Opin. Although the exact procedures frequently vary between institutions, the DMR process may enable IACUCs to review and approve protocols faster than those presented for FCR. It is important to mention, however, that DMR in no way implies that the quality of review is less stringent than a protocol reviewed by the full committee. A successful DMR process allows institutions, particularly larger institutions that process a large number of protocols, the opportunity to reduce the workload of the IACUC at convened meetings, thereby, allowing members to focus on protocols that may warrant more time and attention. (See 11:10.) It should also be noted that the DMR process does not reduce the IACUC office staff workload that may include follow-up reminders to the designated reviewer, correspondence with the PI, etc.

There is no federal requirement that limits the number of protocols that may be reviewed by the DMR process or specifies the number of full committee meetings. The *Guide* (p. 25), however, states "the committee must meet as often as necessary to fulfill its responsibilities." Additionally, in accordance with best practices the expectation is that an IACUC will meet in a convened meeting at least twice per year to approve the semiannual program review and facility inspections, but this is not an NIH/OLAW or APHIS/AC requirement. Clearly, progressive IACUCs meet on a regular basis to conduct whatever business is appropriate to their programs. This would, of course, be described in the PHS Assurance.

It is possible that all protocols could be reviewed, at least initially, by DMR. This is made more likely by use of an effective prereview process. FCR would then only be applied to protocols for which FCR was requested by an IACUC member. However, such a review system would not be considered best practice. On the other hand, some IACUCs review all protocols by the full committee at convened meetings. This also does not reflect best practice unless the number of protocols is very small.

In the DMR process, the members of the IACUC do not vote. If a FCR is not called for, the designated reviewer has the authority to approve a protocol, require modifications to secure approval or refer it to the full committee. If there is more than one designated reviewer, the reviewers must agree on a course of action or refer the protocol to the full committee.

Surv. For the initial formal review of IACUC protocols (designated member review or full committee review), which review method does your IACUC use?

- We use designated member review 100% of the time 33/296 (11.3%)
- We use designated member review 75–99% of the time 59/296 (19.9%)
- We use designated member review 50–74% of the time 36/296 (12.2%)
- We use designated member review 25–49% of the time 22/296 (7.4%)
- We use designated member review 1–24% of the time 55/296 (18.6%)
- We only use full committee review 91/296 (30.7%)

9:22 Under what circumstances might a protocol be assigned to a designated member review process?

Reg. A protocol may be assigned to one or more designated reviewers only after all IACUC members have been provided the opportunity to call for full committee review (PHS Policy IV,C,2; AWAR §2.31,d,2). (See 9:21; 9:23.)

Opin. Because the AWAR and the PHS Policy do not define the criteria for assigning protocols to the DMR process, each institution must develop an internal policy, tailored to the individual institution's needs and idiosyncrasies. Even if an institution develops criteria to determine the kinds of activities or protocols that it will permit to conduct review by DMR, all IACUC members must still be provided with the opportunity to call for full committee review of each individual project.

An institution should first decide whether it is advantageous to establish a DMR process. The IACUC should create a list of well-defined criteria to determine the types of protocols qualifying for DMR versus FCR. For example, some institutions restrict DMR to protocols involving noninvasive or acute procedures that do not cause more than momentary pain or distress to the animals (e.g., the procedures only involve euthanizing the animals in order to harvest tissues), or to protocol amendments that involve minor changes to an approved study. (See 9:23.)

If an institution elects to adopt a DMR process, the IACUC should document the assignment criteria and the procedures for processing the protocols. The IACUC should develop a set of SOPs in order to ensure that the DMR process is not misused. The description of the DMR process and its use should be included in the institution's Animal Welfare Assurance, approved by NIH/OLAW and the SOP should be included in the institution's IACUC policy and procedures. It should be noted that APHIS/AC has no such Assurance requirement.

9:23 Under what circumstances is the designated member review not appropriate?

Opin. Although each institution must determine the criteria appropriate for its own program, DMR may be unacceptable under many circumstances, including the following:

- IACUC members are not provided with sufficient information concerning the proposed research activities.
- IACUC members are not given an opportunity to call for FCR prior to the approval of the proposed protocol via the DMR process (PHS Policy IV,C,2; AWAR §2.31,d,2).
- If any IACUC member requests FCR (PHS Policy IV,C,2; AWAR §2.31,d,2).
- One or more of the designated reviewers has a conflict of interest (PHS Policy IV,C,2; AWAR §2.31,d,2).
- The proposed protocol does not meet the requirements for DMR delineated in the institution's NIH/OLAW-approved Animal Welfare Assurance or the institution's IACUC policy and procedures. For example, the invasive nature of the research may not permit the protocol to be reviewed by the DMR method according to the institution's policies and its Animal Welfare Assurance. Other examples include procedures which involve unalleviated pain or distress in APHIS/AC regulated species or major changes in on-going research with APHIS/AC regulated species. (See 9:30.)

9:24 Should the entire IACUC be involved in the designated member review process or can a subcommittee conduct designated member review?

Reg. (See 9:21.)

Opin. The entire committee must have the opportunity to review the information provided on each DMR protocol review list and to request FCR. If there is no call for FCR, one or more members of the IACUC who are qualified to review the protocol can be assigned by the chairperson to review the protocol in its entirety. Questions and concerns raised by any member of the IACUC should always be addressed prior to approval by DMR. Even though an IACUC member is not assigned as a designated reviewer if he/she raises an issue without calling for FCR, that issue should be brought to the attention of the designated reviewer(s) or cause the protocol to be brought to FCR, if that is the desire of the member who raised the issue.

9:25 What is required for approval of a protocol via the designated member review process?

Reg. The designated reviewer(s) must use the same criteria that are applicable to protocols undergoing FCR. Neither the PHS Policy nor the AWAR make any distinction regarding approval criteria for DMR vs. FCR.[19] The designated reviewer(s) must ensure that the review method and criteria for approval are in full compliance with all of the applicable requirements of the PHS Policy (IV,C,1,a–g; IV,D,1,a–e) and the AWAR (§2.31,d,1,i–xi; §2.31,e,1–5).

Opin. One of the intents of the DMR process is to conduct rapid review of routine protocols, but by no means does this process require less information than the FCR process. Another intent could be to redistribute IACUC workload or to enable IACUCs to focus on thorny issues without the distraction of routine protocols. Before review via a DMR process, each member of the IACUC must have access to written descriptions of research projects and a list of proposed research and the opportunity to call for FCR. If a subcommittee is being utilized for this review process, all designated reviewers *must* agree on the decision. It is not acceptable to use a majority vote.[18] If one of the reviewers disagrees, the protocol must be referred to the full committee for review.

9:26 What is an effective means for administering a designated member review?

Opin. One significant difference between DMR and FCR is that the DMR process does not require a convened meeting of a quorum of the IACUC. Despite this fact, after a protocol is assigned to designated review, at least one or more IACUC members should review the entire protocol with the same degree of thoroughness that is given to a protocol reviewed by the full committee.

Another difference between the two processes involves the way in which protocol information can be disseminated to the IACUC members. Unlike the FCR protocols, the information pertinent to protocols presented for DMR can be more easily disseminated through the use of electronic means (e.g., fax or e-mail) in order to facilitate the most expeditious type of review. (See 6:5; 6:15; 6:19.)

The following is an example of a DMR process. This particular example goes beyond the requirements of the AWAR and PHS Policy in that it requires more protocol specific information to be sent to all members and documentation of each designated reviewer's decision with regard to DMR:

- The IACUC administrator reviews the protocol submitted to the committee and decides whether it qualifies for DMR based upon the IACUC's predetermined criteria.
- A summary of the protocol qualifying for DMR (including title of project; species, number of animals requested; type of experimental procedures) is forwarded to members of the IACUC via postal mail, fax, or email. In some instances, additional information may be included (e.g., supporting grant materials or parts of the original protocol).
- Each IACUC member reviews the summary and, if necessary, requests a copy of the complete protocol to determine whether clarification or changes are needed, and has the opportunity to call for FCR.
- If no member calls for FCR in a reasonable, specified time period, the Chair appoints one or more designated reviewers. For example, a subcommittee, composed of the IACUC Chair and the AV, is authorized to review and approve the protocol. In a case where the Chair or the AV has a conflict of interest (e.g., is personally involved in the project), he/she must be replaced by another member qualified to conduct the review (PHS Policy VI,C,2; AWAR §2.31,d,2). (See 6:11; 6:14; 6:29.) In our experience, 3–5 days is a sufficient time for the members to decide whether the protocol should be referred to FCR. The protocol can be reviewed and either approved, require that modifications be made to secure approval, or the protocol can be referred to the full committee.
- Documentation, including review sheets, is maintained as evidence of each designated reviewer's decision with regard to these protocols. The designated reviewer sheets are kept in the protocol file in the event the DMR is ever questioned. *No matter what process an institution adopts, the DMR procedure must be documented from assignment to approval each time it is used.*

9:27 An IACUC chairperson appoints designated member reviewers. Can an IACUC member who was not chosen as a designated reviewer demand to be made a designated reviewer?

Reg. Both AWAR (§2.31,d,2) and PHS Policy (IV,C,2) specify that at least one member of the IACUC be designated *by the chairperson* and be qualified to conduct the review.

Opin. There is no requirement for the chairperson to honor such a request. Although this may be an unusual request, an IACUC Chair would be wise to find out why an IACUC member has made this demand and perhaps refer the protocol to full committee review. In fact, the IACUC member who made the demand can call for full committee review. (See 9:24.)

9:28 What is "full committee" review?

Reg. Full committee review (FCR) of an IACUC protocol is one that is conducted by a quorum of the IACUC at a regularly scheduled or specially convened meeting (PHS Policy IV,C,2; AWAR §2.31,d,2).

Opin. Due to the volume of protocols reviewed at many institutions, a good portion of the actual work involved in protocol review may be done before the meeting via a mechanism such as prereview (see 9:4–9:12). Thus, at the meeting, the committee members are able to concentrate on the proposed protocol, its "scientific merit," experimental design, and the humane care and use of laboratory animals without the necessity of obtaining clarifications related to incomplete information and/or confusing points.

9:29 What is the purpose of a full committee review?

Opin. The purpose of FCR is to have all IACUC members involved in reviewing and making decisions regarding the disposition of protocols during an interactive meeting. This in turn allows the IACUC to utilize the expertise of its members in a discussion-based format, thus facilitating the resolution of protocol related issues.

9:30 When is full committee review of a protocol appropriate?

Reg. Full committee review is appropriate at any time and required when requested by any member of the IACUC (PHS Policy IV,C,2; AWAR §2.31,d,2). Each member must have the opportunity to review any protocol or significant change before approval may be granted.[18] There are no other federal requirements specifying when a protocol should receive FCR or the type of protocol that should receive FCR.

Opin. Many IACUCs have developed criteria to determine the types of protocols or changes which require FCR. Items that may be considered in developing such criteria include

- Invasiveness of procedures
- Level of pain or distress
- Species and number of animals requested
- Experimental design
- The nature of the animal use, e.g., research project versus educational use
- "Controversial" procedures, e.g., death as an endpoint
- Procedures that request exceptions to regulations, e.g., multiple survival surgeries that require appropriate justification and which the IACUC may want to review only at the time of a fully convened meeting
- Whether the protocol will receive peer review by the funding agency or other group prior to funding
- Major protocol amendments involving procedures that would require FCR at initial review

9:31 What is an effective full committee review process?

Reg: PHS Policy (IV,C,1,a–g; IV,D,1,a–e) and AWAR (§2.31,d; §2.31,e). While the PHS Policy and AWAR only require a quorum (simple majority) to officially convene a full committee meeting, it is obviously advantageous to have a greater number of IACUC members present, particularly key individuals such as the AV and, of course, the non-affiliated member. It should also be noted that consultation with the AV is required during design of the study, prior to IACUC review, for APHIS/

AC regulated research that involves procedures that may cause more than momentary or slight pain or distress to the animals (AWAR §2.31,d,1,iv,B). NIH/OLAW provides excellent guidance on the correct conduct of full committee review as well as designated member review.[18]

Opin. Each IACUC should establish SOPs regarding the conduct of FCR of protocols. One effective method of full committee review includes the following process:

- Prereview (see 9:4–9:12).
- Review by IACUC staff or designated member of the committee for completeness of the application and compliance with information requirements, as indicated in PHS Policy (IV,C,1,a–g; IV,D,1,a–e; AWAR §2.31,d,1,i–xi; §2.31,e,1–5).
- Review by the AV in addition to the consultation required during design of the study (AWAR §2.31,d,1,iv,B).
- Review by outside consultants if the IACUC members do not possess relevant expertise.
- Review by committee members assigned as primary and secondary reviewers.
- Primary and secondary reviewers communicate with the investigator prior to the meeting, during which the reviewers may recommend modifications in the protocol (PHS Policy IV,C,2; AWAR §2.31,d,2).
- Every member of the committee should be provided with a copy of the protocol.
- Presentation of the protocol involving all the IACUC members in attendance by the primary and secondary reviewers followed by a general discussion of the protocol.
- Following discussion at a convened quorum of the committee, a vote is taken to determine final disposition of the protocol.

9:32 What are the PHS Policy and AWAR requirements for approval of a protocol by full committee review?

Reg. The PHS Policy and AWAR require

- Each member of the IACUC must be provided with, at the minimum, a list of protocols to be reviewed (PHS Policy IV,C,2; AWAR §2.31,d,2) and written descriptions of research projects are available to all IACUC members.
- Any member upon request may obtain a full committee review of proposed activities (AWAR §2.31,d,2).
- Both the PHS Policy (IV,C,1,a–g; IV,D,1,a–e) and the AWAR (§2.31,d; §2.31,e) have specific requirements about information that must be included in the protocol as well as the items the IACUC must consider. These are described below.
- No member of the IACUC may participate in the review or approval of a research project in which the member has a conflict of interest (e.g., is

personally involved in the project) except to provide information requested by the IACUC (PHS Policy IV,C,2; AWAR §2.31,d,2). Some IACUCs require this member to leave the room during final discussion and voting on the protocol in question.

- Approval of a protocol considered by the full committee "may be granted only after review at a convened meeting of a quorum of the IACUC and with the approval vote of a majority of the quorum present" (PHS Policy IV,C,2). The AWAR (§2.31,d,2) have the same requirement.
- A member who has a conflict of interest in a protocol under review may not contribute to the constitution of a quorum (PHS Policy IV,C,2; AWAR §2.31,d,2). This person may not vote on the protocol in question.
- Institutions must maintain written documentation of committee deliberations (PHS Policy IV,E,1,2; AWAR §2.35,a,1–2).
- Both the investigator and the institution must be notified in writing of the committee's decision (PHS Policy IV,C,4; AWAR §2.31,d,4).

Opin. While most IACUCs include documentation of committee deliberations in the minutes, such documentation can also be maintained on file with the protocol. Filing records of particularly sensitive deliberations along with the protocol is of importance to institutions subject to state open records laws (see Chapter 22). It is up to the IO, IACUC Chair, or IACUC staff to explain to APHIS/AC, NIH/OLAW, or AAALAC, as appropriate, the rationale for the institution's recordkeeping methods. This should be documented in the PHS Assurance and in the institutional policy and procedures.

9:33 What constitutes "administrative review," and when may it be used?

Reg. The PHS Policy (IV,C,1) states that "In order to approve proposed research projects or proposed significant changes in ongoing research projects, the IACUC shall conduct a review of those components related to the care and use of animals and determine that the proposed research projects are in accordance with this Policy." It goes on to say that such review must use either the DMR or the FCR method. However, the Policy does not say what should be done if there are non-significant changes. The AWAR (§2.31,c,6–7) contain essentially the same requirements as the PHS Policy. Neither the PHS Policy nor the AWAR uses the term "administrative review" or discusses anything to which the term could reasonably be applied.

In NIH/OLAW Notice NOT-OD-03-046,[20] it is stated that "IACUCs may, by institutional policy, classify certain proposed additions or changes in personnel, other than Principal Investigator, as 'minor' provided an appropriate administrative review mechanism is in place to ensure that all such personnel are appropriately identified, adequately trained and qualified, enrolled in applicable occupational health and safety programs, and meet other criteria as required by the IACUC."

Opin. According to NIH/OLAW, administrative review is possible in some cases provided the process and its application are documented.[20] It is, however, inappropriate for initial and continuing review of animal projects, with the possible exception of some wildlife or other observational studies where no interventions are applied to animals.[21] Such administrative review may be applied to requests for changes that are not significant. This would include, for example, changes in personnel other than the PI. NIH/OLAW has no written endorsement of the following examples, but it seems that replacement of a small number of animals due to a technical, vivarium or vendor-related problem or a change in a procedure that occurs after euthanasia could be approved by administrative review. (See 10:4; 10:5.) However, according to APHIS/AC during their review of this chapter, "any changes to an approved protocol require Committee member involvement either through FCR or DMR." Certainly, such a rigid interpretation of the term "any changes" is problematic.

NIH/OLAW has indicated in a FAQ[22] that changes in the objectives of a study; proposals to switch from non-survival to survival surgery; changes in the degree of invasiveness of a procedure or discomfort to an animal; changes in species or in approximate number of animals used; changes in anesthetic agents, use or withholding of analgesics or methods of euthanasia; and changes in the duration, frequency or number of procedures performed on an animal would not qualify for administrative review. Wolff et al.,[23] speaking for NIH/OLAW, have clearly stated that administrative review may not be used to grant continuance to a project that has expired because the investigator failed to seek continuing review in a timely fashion.

Finally, administrative review under the conditions described above, may be performed by the IACUC Chair or administrator, by the AV or by an IACUC member designated by the IACUC Chair.

9:34 An IACUC protocol was properly approved. Can an investigator begin research with animals based on an oral approval from the IACUC office or must there be written documentation of the approval?

Opin. Once approval of the protocol has been documented in the IACUC records, which can mean written in the IACUC minutes by the recording secretary, recorded on a summary check sheet or some other process approved by the IACUC, the research may begin but not until the date of an official approval letter that has been issued to the investigator. Once all requirements are satisfied and documented accordingly, IACUC approval letters should be issued as soon as possible. Oral conveyance of IACUC approval is not sufficient.

9:35 Is it important for the IACUC to know whether a protocol is new or a resubmission with a change in title?

Opin. If a protocol is submitted and approval is withheld, it should be treated as a new protocol upon resubmission regardless of whether the title or funding source is changed or not and reviewed by either DMR or FCR as specified in the PHS Assurance or institutional policy and procedures. If reviewed by DMR, the requirements for DMR must be met for this version of the proposal. If a protocol has been previously reviewed and approved, then the title or funding source may be changed without re-review if it is the policy of the institution to handle changes in title administratively.

9:36 What procedures can the IACUC use to determine if a protocol is new or a resubmission with a change in title?

Opin. If a protocol is determined to be a resubmission with a change in the title only, the IACUC may decide to verify this by comparing the resubmission with the previously approved protocol on file. Investigators sometimes modify protocols during the course of resubmission without recognizing that the IACUC must approve all proposed significant changes (PHS Policy IV,B,7; AWAR §2.31,c,7). Indeed most IACUCs require approval of any change prior to implementation. It is perhaps a better procedure to make changing the title of a protocol a separate process with its own form, or treat it as just one of several types of amendments to a protocol. That way the changes in the protocol would be obvious.

For IACUCs that use electronic submissions of amendments, it is possible to detect changes in the protocol electronically through use of software with document comparison features.

9:37 Should IACUC decisions be influenced by the source or size of a research grant?

Reg. The PHS Policy and the NIH Grants Policy Statement (Part II, Terms and Conditions) require the institution to verify, before award, that the IACUC has reviewed and approved those components of grant applications and contract proposals related to the care and use of animals.

Opin. On reflex, most IACUC members would probably answer "no" to this question. In general, all protocols should be reviewed with the same rigor regardless of how much money is involved or the source of funding. The size of the grant should not impact IACUC decisions.

Whereas political pressures for IACUC approval of large or prestigious grants are real, studies should be approved based on appropriateness of the proposed research. Certainly, animal welfare should never be compromised in the interest of increasing grant funds. IACUCs must be constantly vigilant for inconsistencies in their deliberations and decisions, especially those affected by such pressures or perceptions.

Institutions may specify in their Animal Welfare Assurance that research that is not funded by the PHS will be conducted according to the PHS Policy and NIH/OLAW guidance. Institutions choose to do this for several reasons:

- Having a single set of standards for the animal program simplifies operations oversight,
- They wish to provide state of the art care to all their research animals—a choice that appeals to students, donors, the public, and strengthens recruitment efforts,
- An outside authority regulates the behavior of difficult animal users.

9:38 Should the decisions of the IACUC be influenced by the potential scientific importance of a project proposed in a protocol?

Reg. The IACUC has the authority to approve a proposed project. NIH/OLAW and APHIS/AC have stated that under no circumstances is an IACUC required to approve a project against its will.[24] (See 16:8.)

Opin. IACUC decisions should not be influenced by the investigator, species, funding source, or "hotness" of the research topic. Review should be based on animal welfare issues in consideration of the regulatory requirements. Nevertheless, both the dollar value of the grant and a species-specific view of an animal's societal worth potentially could, but should not, affect IACUC deliberations.[25] (See 9:68.)

9:39 If a protocol is completely novel and contains untested surgical or experimental procedures, how could such a protocol be reviewed and approved when the procedures cannot be referenced?

Opin. Few protocols actually contain untested surgical procedures. In some cases, the IACUC could recommend that a pilot study be conducted involving only the part of the protocol using the untested procedures. With these pilot data derived from successful use of the previously untested procedures, the investigator could then obtain IACUC approval for the complete protocol. Pilot studies, however, must also be reviewed and approved by the IACUC (see NIH/OLAW FAQ D.11).

Alternatively, the IACUC could approve the protocol as submitted but with a reduced number of animals, the remainder being approved when the investigator has tested the new procedure. Both of these actions allow the researcher to proceed with the study while the animal subjects are being protected. The protocol could also be reviewed based on knowledge of the technique most closely related to the proposed procedure—the ones from which this was extended. Also the IACUC may factor in the effect this procedure might have on human pain and distress and extrapolate to animals. (See 13:6; 13:8; 13:9.)

9:40 What are the possible decisions an IACUC can make after protocol review?

Reg. Both the PHS Policy (IV,B,6) and the AWAR (§2.31,c,6; §2.31,d,4) allow the IACUC to approve, require modifications (in order to secure approval), or withhold approval of proposed activities related to the care and use of animals. This includes significant changes to ongoing animal activities (AWAR §2.31,c,6; §2.31,d,4; PHS Policy IV,B,7). The IACUC can also vote to suspend a protocol (AWAR §2.31,c,8; §2.31,d,4; PHS Policy IV,C,6).

Opin. In determining the disposition of a protocol submitted for review by the full committee review method, an IACUC has multiple options including

- Approval
- Require modifications to secure approval (see 9:42)
- Referring back to the full committee for re-review
- Withhold approval
- No decision

When the DMR method is used, the only options available are

- Approval
- Require modifications to secure approval
- Refer to the full committee for review

It should be noted that during re-review a PI's protocol and/or response to the IACUC's review the committee or designated reviewer may identify additional concerns or even change the requirements in consideration of new information.

9:41 What constitutes "approval" of a protocol? Does this action necessarily mean that no further changes or information are needed?

Reg. Per PHS Policy and the AWAR, IACUCs either approve, require modifications in (to secure approval), or withhold approval of protocols (AWAR §2.31,c,6; PHS Policy IV,B,7). The same holds true for designated reviewers with an additional option of referring the protocol for FCR but without the option of withholding approval. Anything short of final approval is not adequate for initiation of animal activities or submission of an IACUC approval date to the NIH as part of a grant application.

Opin. Generally, an approved protocol is one that contains all the required information, has been judged by the IACUC to be acceptable, has satisfied all approval criteria, and whose approval has been recorded in the IACUC minutes or other records.

9:42 What constitutes "requires modifications to secure approval"?

Reg. (See 9:41.)

Opin. The IACUC, during a full committee meeting, may require modifications of a protocol in order for the investigator to secure approval. This action may be inappropriately referred to as "approval with conditions" which will be addressed shortly. In any event, often, the Chair, the AV or another member is assigned as a designated reviewer to review the investigators response or the revised protocol. He or she is empowered by the IACUC to approve the protocol without further review by the full committee. Or, a subcommittee may be formed and empowered to review the response and approve the protocol via the DMR process, however, depending upon the conditions imposed, the entire committee may wish to see the response.

In 2009 NIH/OLAW issued guidance on DMR subsequent to FCR.[26,27] This guidance states that if *all* members of the IACUC are present at a meeting they may vote to require modifications to secure approval, and the protocol will then be reviewed by DMR. However, if not all members are present at a meeting, a quorum of the members may decide by unanimous vote to use DMR subsequent to FCR providing *all* IACUC members agree in advance, in writing, to a policy which allows this review procedure. (See 9:47.)

Approval with "conditions" does not mean that the study can be initiated. The PI must first comply fully with all conditions arising from the IACUC's review, and then final approval can be granted. The reader is cautioned that terms such as "conditional approval," "provisional approval," or "approved pending clarification" frequently cause confusion. NIH/OLAW and APHIS/AC strongly advise IACUCs to avoid these terms, as they may lead to noncompliance. (See 9:47.)

IACUCs might determine that a protocol is approvable, contingent on receipt of a specific modification or confirmation (e.g., receipt of assurance that the PI will conduct the procedure in a fume hood). The IACUC can handle this modification (or clarification) as an administrative detail that is documented in the protocol record. On the other hand, protocols that are missing substantive information necessary for the IACUC to make a judgment (e.g., justification for withholding analgesics in a painful procedure) are incomplete. If the protocol is incomplete, it is not possible to satisfy the protocol review criteria. IACUCs should devise effective ways of differentiating between substantive omissions and administrative issues.

9:43 What constitutes "tabling a protocol" for reconsideration by the full committee?

Opin. "Tabling" of a protocol is a term often used by IACUCs, but it is not an action described in the PHS Policy, AWAR, or in any related guidance. The IACUC may "table" or "postpone" further review of a protocol pending receipt of additional substantive information or a significant revision of the protocol. The action of tabling a protocol is generally used during the process of FCR when the IACUC decides it is necessary for the whole committee to re-review the protocol before further action can be taken. Tabling a protocol is usually reserved for proposals that do not contain sufficient information or those in which the IACUC has identified a serious animal welfare concern. The term "tabling", however, may have a negative impact on investigators. A more diplomatic term is "refer back to the full committee" for re-review by the IACUC.

9:44 What might cause an IACUC to withhold approval of a protocol?

Reg. Approval may be withheld if any of the PHS Policy (IV,C,1) or AWAR (§2.31,d,1; §2.31,e) criteria are not met.

Opin. Withholding approval of protocols is infrequent. Generally, approval is withheld when the PI and the IACUC are not able to come to an agreement on issues in which the IACUC requires modifications to secure approval.

Both the PI and the IO must be notified in writing of the committee's decision. This written notification should include a statement of the reasons for the decision and provide the PI with an opportunity to respond in person or in writing. Certainly, the IACUC may reconsider its decision with documentation in the committee minutes, in light of new information provided by the PI (PHS Policy IV,C,4; AWAR §2.31,d,4). In some cases, the IACUC may want to consult an outside expert in order to obtain a supplemental review.

9:45 How should an investigator respond to questions or conditions from the IACUC?

Opin. Responses from the PI to questions or concerns raised by the IACUC should be in such a format as to result in a protocol that contains a complete, easy to discern description of the proposed activities. The response from the PI should clearly answer each issue raised. This can be achieved via a point-by-point letter that serves as a means for the PI to respond to the committee's questions and concerns accompanied by a revised protocol which incorporates all required changes and clarifications. Requiring the PI to provide a revised protocol has the advantage of ensuring that a complete and up-to-date master protocol is embodied in one document. Copies of the final approved IACUC protocol should be maintained in the IACUC administrative office, in the animal facility, and in the investigator's laboratory. This master protocol also has the advantage of facilitating the management of the protocol by the animal care staff and for compliance review purposes. In the case of electronic protocol submissions, such responses are probably best made directly to the online protocol document. Finally, if PIs submit clear protocols and IACUCs perform equally clear reviews, then the overall efficiency of the review process improves accordingly.

Surv. Approximately what percentages of your IACUC protocols are approved the first time they undergo either full committee review or designated member review?

• About 0–2%	34/296 (11.5%)
• About 3–10%	38/296 (12.8%)
• About 11–25%	34/296 (11.5%)
• About 26–50%	52/296 (17.6%)
• Greater than 50%	134/296 (45.3%)
• I don't know	4/296 (1.4%)

9:46 Should there be a time limit to receive responses to queries raised by the IACUC?

Opin. Establishment of time limits and other constraints should be considered when the IACUC develops an SOP. Many IACUCs have found it useful to set specific time constraints and deadlines which can be of benefit to both the investigator and the committee when the expectations are clear. Ironclad deadlines and inflexible staff, however, can present a major point of contention for faculty already juggling multiple projects and attempting to comply with much paperwork and many deadlines.

Surv. What does your IACUC do if an investigator does not respond to IACUC queries in a reasonable length of time?

• We have never encountered this problem	88/297 (29.6%)
• The protocol remains open for a response, forever	19/297 (6.4%)
• Our IACUC has a time limit, after which the protocol is considered invalid	119/297 (40.1%)
• We have no firm policy	68/297 (22.9%)
• I don't know	1/297 (0.3%)
• Other	2/297 (0.7%)

9:47 Once a response from an investigator is received, following full committee review, can the IACUC Chair or his/her designee approve protocols if all required modifications to secure approval are met?

Opin. This should be determined at the meeting when the decision is made to grant further review and approval authority via the DMR process (see 9:42). That is, approval may be granted on condition that the PI meet certain requirements, and the committee unanimously agrees that the assigned reviewer(s), which may include the Chair, can determine if the conditions have been satisfactorily met (PHS Policy IV,C,2; AWAR §2.31,d,2). Again, it should be noted that the term "conditional approval" is not mentioned in either the AWAR or PHS Policy and its use is strongly discouraged.

9:48 Is a majority vote required for any IACUC actions related to protocol approval?

Reg. Approval of a protocol being reviewed by FCR "may be granted only after review at a convened meeting of a quorum of the IACUC and with the approval vote of a majority of the quorum present" (PHS Policy IV,C,2). The AWAR (§2.31,d,2) has the same requirement. This is also true for votes to suspend an activity. (See 9:21 for designated-member review requirements.)

Opin. The PHS Policy and AWAR represent the minimum requirements and an institution may implement more stringent requirements. These should be clearly stated in the institution's policy and procedures and its PHS Assurance (if applicable). It is important to consider the decision to exceed the requirements carefully, as the procedures stated in the institution's PHS Assurance become the legal requirements for PHS funding upon NIH/OLAW approval of the Assurance. It is also important to remember that all approval criteria, including institutional specific requirements, must be satisfied in order for the protocol to be approved and authorized for implementation.

9:49 At a full committee meeting of a 20-member IACUC, 15 members are present. Six vote to approve a protocol, six abstain, and three vote against approval. Is the protocol considered approved according to the AWAR and PHS Policy?

Opin. No. In this example, the 15 members present constitute the quorum. In order to approve a protocol, a simple majority of those members present, i.e., eight, would have to vote for approval of the protocol. Because only six members voted to approve, the vote count fell short by two votes, and the protocol cannot be approved according to the AWAR and PHS Policy.[28] (See 9:48.)

9:50 How should the IACUC handle minority opinions to IACUC actions on the review of protocols?

Reg. PHS Policy (IV,E,1,d) and AWAR (§2.31,c,3) refer to semiannual facility inspections and program evaluations and for those purposes minority views must be included. For minority views expressed during the review of an animal use protocol, NIH/OLAW has provided guidance on a dissenting vote versus a minority view: "Both protocol approval and suspension of animal study protocols by the IACUC require a majority vote of a quorum of the IACUC. Although an IACUC member's dissenting vote on these issues must be recorded in the minutes, this does not constitute a minority view for reporting purposes. Any IACUC member may submit a minority view to OLAW addressing any aspect of the institution's animal program, facilities, or personnel training. Whether OLAW receives a minority view as part of an annual report, renewal Assurance document materials, or directly from the dissenting IACUC member, it carefully reviews the information provided in accordance with requirements of the PHS Policy and provisions of the Guide."[29]

Opin. Because neither the PHS Policy nor the AWAR address this issue relative to specific IACUC protocols, it should be considered when the IACUC develops an SOP. A common practice is to record dissenting opinions in the minutes. However, another option is to allow the dissenter to write a letter expressing his/her opinion. This is then filed along with the IACUC protocol. When deciding on the method of dissent documentation, the IACUC should review any applicable state open records laws (see Chapter 22).

9:51 A protocol has been approved with changes requested by the IACUC. The protocol is associated with a grant application to the NIH. Does the approval letter sent by the Research Administration Office to the NIH have to detail the changes made in the animal use portions of the grant?

Reg. The PHS Policy (IV,D,2) requires "verification of approval (including the date of the most recent approval) by the IACUC of those components related to the care and use of animals. ... If verification of IACUC approval is submitted subsequent to the submission of the application or proposal, the verification shall state the modifications, if any, required by the IACUC."

Since 2002, NIH Grants Policy has allowed the verification of IACUC approval to be submitted later in the application process to reduce the time to award and ensure the accuracy and timeliness of the information.[14] Such a request for information later in the review cycle is referred to as "just-in-time." (See 8:22–8:24; 9:13.) NIH/OLAW has described the oversight role of the IACUC by stating:

> An institution that elects to proceed according to "just-in-time" procedures for IACUC approval bears the responsibility for supporting the decisions of the IACUC. Under no circumstances may an IACUC be pressured to approve a protocol or be overruled on its decision to withhold approval. The PHS Policy requires that modifications required by the IACUC be submitted to the NIH with the verification of IACUC approval, and it is the responsibility of institutions to communicate any IACUC-imposed changes to NIH staff. It is incumbent upon investigators to be totally forthcoming and timely in conveying to the IACUC. any modifications related to project scope and animal usage that may result from the NIH review and award processes. Should an institution find that one of its investigators disregards his/her responsibilities, the institution may, for example, determine that all animal protocols from that investigator be subject to IACUC approval prior to allowing that investigator to submit an application.[30]

Surv. Please respond to this question only if your institution receives grants from the U.S. Public Health Service. Does your IACUC inform the granting institute or center when a procedure or species that is not on a funded grant is substituted for one that is on the funded grant?

- Yes we do so routinely 27/232 (11.6%)
- Yes, we do so on occasion for any such substitution 21/232 (9.1%)
- Yes, we do so but only when species covered by the Animal Welfare Act are involved 11/232 (4.7%)
- No, we do not 87/232 (37.5%)
- I don't know 45/232 (19.4%)
- Other 41/232 (17.7%)

9:52 Should the IACUC review just the IACUC protocol or should the IACUC also review the animal care and use sections of an associated grant proposal?

Reg. Verification of IACUC approval submitted to the NIH means that the institution certifies to the NIH that the IACUC has reviewed and approved those components of the grant application related to the care and use of animals. "Applications or proposals (competing and non-completing) covered by this Policy from institutions which have an approved Assurance on file with OLAW shall include verification of approval... by the IACUC of those components related to the care and use of animals." (PHS Policy IV,D,2). The signature of the institutional representative on the PHS 398 form is a legally binding statement that the IACUC has approved all animal activities covered in the grant application. The submission of just-in-time verification also constitutes a binding statement. It should be noted that the signature on the PHS 398 is an attestation of assurance provided by the institution and not the IACUC.

NIH/OLAW in its webinar on "Grants Policy and Congruence" states: "The institution must document the IACUC's approval of the animal activities proposed in the grant. The institution, when asked by NIH, must be able to associate each grant or grants with a relevant protocol or protocols. If the institution uses a protocol numbering system, it must be able to link protocol numbers to grant numbers. But a 1:1 ratio is not required."[31]

Opin. If the IACUC protocol accurately reflects the information contained in the grant application, then the verification is valid. The IACUC, however, cannot be assured of such validity unless it also reviews the vertebrate animal sections of the associated grant application, or the committee relies on another mechanism such as a valid comparison of the protocol(s) with the grant performed by a qualified administrator such as a grant specialist, compliance officer or personnel within the IACUC office. With the advent of just-in-time (see 8:22; 9:13: 9:51), review of the grant proposal need not take place until it appears that the proposal will be funded.

Some institutions do not review the grant proposals at all. They simply rely upon a statement by the PI or department chair that the protocol accurately reflects what is in the grant proposal. However, if the PI's or chair's certification is inaccurate, the resulting institutional verification to the PHS will be invalid. If such a discrepancy were to be discovered, there are potential civil and criminal penalties for submitting false statements.[31]

Finally, it should be noted that all animal use on a specific grant must be approved by the IACUC but does not have to be contained in only one IACUC protocol. It is an institutional decision as to whether the procedures will be covered in one or several (many) IACUC protocols.

Surv. Please respond to this question only if your organization receives grants from the U.S. Public Health Service. How does your institution assure that what is written on a grant application is covered by one or more IACUC protocols?

- We make no such determinations 14/236 (5.9%)
- The Principal Investigator attests to congruency between the grant and protocol(s) 69/236 (29.2%)
- The IACUC office compares the grant to one or more IACUC protocols 80/236 (33.9%)
- Our institution, but not the IACUC office, makes the determination 35/236 (14.8%)
- Members of our IACUC are asked to make the determination 27/236 (11.4%)
- I don't know 6/236 (2.5%)
- Other 5/236 (2.1%)

9:53 Should the IACUC accept an approval statement from an IACUC at another institution?

Reg. An IACUC may accept the approval of an IACUC at another institution if that institution has an approved NIH/OLAW Animal Welfare Assurance. This practice is normally limited to collaborations or subawards involving performance sites outside the awardee institution. In most instances, the IACUC of the performance site assumes responsibility for animal activities in its facilities. Both institutions should have a clear understanding and a written agreement of what their respective responsibilities are in this situation, particularly if one IACUC agrees to abide by the determinations of another IACUC.[32] (See 8:9; 8:15.)

Opin. This is an area for which the IACUC should establish an SOP. The IACUC should have some "comfort level" about the other institution, its committee, and the quality of its research. To avoid problems, a collaborative protocol approved by an IACUC at another institution should probably receive at least the equivalent of DMR by the local IACUC. In some instances, FCR is warranted. Issues to consider in accepting an approval statement from an IACUC at another institution include

- Does the institution have an approved NIH/OLAW Animal Welfare Assurance?
- Is the institution a USDA-registered research facility?
- Is the institution accredited by AAALAC?

For example, this situation may arise when an institution's faculty conducts research at a nearby Veterans Administration (VA) facility. Because the VA facility may not apply for or receive funding from other federal agencies, the grants must be made to the investigator's home institution. The problem is created because VA-IACUC-approval must be granted before research is initiated at the VA facility. The home institution has jurisdiction by virtue of receiving the grant. Thus, either redundant application or this kind of "external" approval is necessary.

Acceptance of an external IACUC review and approval also includes the responsibility for continuing review. The external institution should agree to supply all continuing review documents relative to the protocol as they are reviewed and approved. Failure to do this can result in withholding a notice of approval to the granting agency and possible loss of grant funds. In some cases the home institution's sponsored programs office may require an assurance of continuing IACUC approval for the grant funds to be released. This can be another source of problems for the PI's research.

9:54 Can an invvestigator rely on approval of a protocol reviewed and approved by an IACUC at another institution in order to initiate his or her research? What should be the conditions and limitations for this reciprocity of IACUC approval?

Reg. (See 8:15; 9:53.)

Opin. An IACUC may choose to accept approval for a protocol approved by an IACUC at a different institution. However, any conditions regarding acceptance of an approval statement from an IACUC at another institution should be described in the IACUC's written policies and in the written agreement between the two institutions. For example, in order to facilitate research, an IACUC may choose to allow animals to be purchased or transferred to its institution, but procedures that involve animals cannot be performed until the local IACUC has reviewed and approved the protocol. (See 9:53.) Because local institutional requirements vary, automatic approval is not advisable.

9:55 An investigator subcontracts part of a research project to another institution where animals will be used. What oversight and paperwork responsibilities does the primary institution have relative to the PHS Policy and the AWAR?

Reg. NIH/OLAW requires that any subcontracted work involving animals that is supported by PHS funds be conducted only at other NIH/OLAW-Assured institutions (PHS Policy V,B). If the performance site (collaborating institution or subcontractor) is not NIH/OLAW-Assured, then NIH/OLAW requires that an Assurance for the performance site be negotiated in order for the work to go forward (PHS Policy V,B). Alternatively, an institution can choose to accept responsibility for the work under its own Assurance and include the performance site as a covered component with a written agreement between the institutions. It should be recognized, however, that the latter arrangement means that the NIH/OLAW-Assured institution assumes full responsibility for ensuring that all the animal work conducted at the performance site is in full compliance with the PHS Policy and the AWAR. This arrangement necessitates that the NIH/OLAW-Assured institution's IACUC conduct semiannual program reviews and facility inspections of the performance site. (See 8:15.)

In addition, NIH Grants Policy Statement on written agreements requires that awardees have a formal written agreement with consortium participants that includes agreement for meeting the PHS Policy requirement for review and approval of proposed animals activities and significant changes to animal activities, and semiannual facilities review by an IACUC.

The AWAR does not address subcontracting. The institution which conducts the animal activity will be held accountable for their care, hence a protocol approved by the IACUC of the institution conducting the animal activity needs to be in place. The approval process is outlined in AWAR §2.31,c,6 and §2.31,c,7.

Opin. When an institution subcontracts part of a research project to another institution, there should be a clearly defined written agreement (*Guide* p. 15) concerning the responsibilities of each institution to comply with the PHS Policy and AWAR regardless of whether both institutions have approved Assurances on file with NIH/OLAW. If a serious compliance problem arises at a subcontract site, it will likely impact the primary institution. However, if the subcontract site is registered with APHIS/AC and is the facility whose IACUC approved the protocol, that site would be held primarily responsible. Written agreements can help minimize and resolve potential problems.

Surv. If your IACUC subcontracts or otherwise collaborates for animal care or use with another institution, do you require a Memorandum of Understanding to delineate areas of responsibility for animal care and use?

• Not applicable	118/296 (39.9%)
• No, we do not	20/296 (6.8%)
• Rarely	18/296 (6.1%)
• Yes, but only if we are the primary grant or contract holder	20/296 (3.8%)

• Yes, but only if we are dealing with species covered by the Animal Welfare Act	0/296 (0%)
• Yes, under most circumstances	107/296 (36.2%)
• I don't know	5/296 (1.7%)
• Other	8/296 (2.7%)

9:56 Can an investigator appeal the decision of the IACUC relative to a protocol review decision?

Reg. The PHS Policy (IV,C,4) and the AWAR (§2.31,d,4) require that written notification of withholding approval include a statement of the reasons for the decision, "to give the investigator an opportunity to respond in person or in writing." Nevertheless, IACUC decisions to withhold approval may not be overturned by a higher institutional authority (PHS Policy IV,C,8; AWAR §2.31,d,8).

Opin. The above regulatory information suggests that a reconsideration by the IACUC should be an option. Furthermore, the "Supplementary Information" that accompanies the August 31, 1989 Federal Register containing 9 CFR Parts 1, 2, and 3 Final Rules (page 36132), states that "on the basis of the response, the committee may reconsider its decision." This is another area for which the IACUC is advised to establish a written SOP for exactly how an investigator appeal will be handled with attention to best detail and due process. Certainly, investigators should be invited to meet with the committee during the appeal process. (See 9:57.)

9:57 What is an appropriate and effective mechanism for appeal of IACUC decisions. (See 9:56; 29:37; 29:38.)

Opin. Some suggestions to consider in developing an SOP for appeals include

- Assignment of an IACUC member to work with the investigator to facilitate the process
- Involvement of an expert consultant to advise the IACUC

It should be remembered that according to both the AWAR and PHS Policy an IO may not overturn a decision of the IACUC to withhold approval (see 9:56). Therefore, the IACUC must alter its own decision, which it can do by a vote of a majority of a quorum at a convened meeting or via the DMR review process, although the latter method may not do justice to an appeal that really should be considered by the full committee.

9:58 Can a protocol for which approval has been withheld be resubmitted with modifications?

Reg. Yes, neither PHS Policy nor the AWAR precludes resubmission.

Opin. Resubmission of a modified protocol should be encouraged. A resubmission provides the investigator with an opportunity to respond to the IACUC's review as allowed by PHS Policy (IV,C,4) and the AWAR (§2.31,d,4). The IACUC also may consider assigning a member to work with the investigator to develop a protocol that can be approved by the committee and allow the investigator's research program to progress. (See 9:57.)

9:59 Is there any institutional authority that can reverse the decision of the IACUC to withhold approval?

Reg. No. The AWAR (§2.31,d,8) state that "officials of the research facility ... may not approve an activity involving the care and use of animals if it has not been approved by the IACUC." The PHS Policy (IV,C,8) is nearly identical. (See 9:56; 29:39.)

9:60 At times, PIs who do drug testing may receive a test drug, but the manufacturer does not want to reveal what the drug is. Can the IACUC approve the use of the drug in animals?

Opin. It is not uncommon that a pharmaceutical company contracts testing of a new drug to an investigator in a research institution. Because such drug development is highly competitive, the company may not want to divulge the exact identity of the drug. This poses a problem for the IACUC charged with evaluating the effect of the drug on the welfare of animals. In order to approve the use of the drug, the IACUC must know at least the general class of the drug as well as its dose and route of administration, and any known risks to the animals. With this much information it is usually possible for the IACUC to approve use of the drug. It is often advantageous to have all IACUC members sign a binding confidentiality agreement. (See 9:61, 27:21; 27:24.)

9:61 Sometimes PIs do not know which of a class of drugs they will use, either because there are many or because they don't know which will be available. Can the IACUC approve the use of a class of drugs in animals?

Opin. Sometimes an investigator wants to use a particular class of drug in a study, but he or she is unsure if or when a given drug will be used. The IACUC can approve the use of a general class of drugs if a list of drugs that might be used along with dose range, frequency and route of administration for each one is provided. It would also be helpful for the committee to know the conditions under which each would be used, the mechanism of action, potential side effects, and toxicity. The IACUC may choose to require that the PI notify the committee of which drugs are actually used either at the time of annual or continuing review or in a special interim or periodic report, as deemed appropriate by the IACUC. (See 9:60.)

9:62 The IACUC has a responsibility for initial review and continuing oversight of animal research projects. Sometimes projects involve adverse events or unexpected problems. What actions should the IACUC take, if any, when notified of these adverse events?

Opin. If the unexpected problem results in morbidity or the death of animals, the IACUC may require the PI to submit interim reports, i.e., to report more often than annually or triennially as required by AWAR and PHS Policy. The nature and duration of such reporting would be determined by the severity of the problem and its frequency. The IACUC may also ask the PI to alter the procedures of the project and to submit a revised protocol for review. (See 27:42.)

9:63 An IACUC member requests that the IACUC reopen the discussion of a previously approved protocol. He was away when the protocol was approved, and he has concerns that were not considered during the review process. Can an IACUC re-examine the approval of a protocol under these circumstances?

Opin. The IACUC is charged with continuous oversight of projects and to review concerns involving the care and use of animals at the institution. Therefore, the IACUC *can* re-examine a protocol at any time it thinks it is necessary. If an absent IACUC member raises valid concerns that were not addressed at the time of initial or continuing review, these should be brought before the full IACUC. It should be noted, however, that the project may have already been started and therefore the IACUC should include this fact in their reconsideration of a project. Requiring major changes in the project may invalidate research that has already been conducted.

Finally, any IACUC member at any time should express any concerns they have regarding animal welfare or compliance. NIH/OLAW and USDA/APHIS have advised: "Review of concerns is not a matter of choice; concerns must be brought to the IACUC's attention… Concerns about animal activities at an institution must be reviewed at a convened meeting of the IACUC."[33] All concerns warrant referral to the full committee and consideration for re-review of the protocol as necessary.

9:64 Can an IACUC develop and/or accept SOPs in lieu of the investigator describing procedures in detail within a protocol? What special concerns might be generated by doing so?

Opin. The AWAR and the PHS Policy do not specifically preclude the use of SOPs. Nevertheless, there are some issues that should be considered. First, if the SOP is to be part of an animal use protocol, then it should be specifically reviewed together with the protocol that references its use. This is an important facet of the review process and clarifies what will happen to an animal from the beginning to the end of the study. If not, then all aspects of the protocol have not been reviewed by the IACUC. Secondly, when the SOP is modified, this would constitute a change in every protocol that references its use, and all associated protocols should be amended each time the SOP is modified. In addition, every investigator who uses the SOP should be aware of the change and adopt it to their research protocol. Changing an SOP without adoption by the investigator might well have the effect of putting him/her into non-compliance.

It might be possible for the institution to maintain different versions of the same SOP, but it appears that keeping track of which version goes with which protocol would pose a difficult process. The investigator of each study must take responsibility for use of any SOP in their research. NIH/OLAW and APHIS/AC issued an NIH/OLAW FAQ D14[34] on SOPs or blanket protocols that provides guidance on this issue. Additional guidance has been published in *Lab Animal*.[35]

9:65 For commonly used, well established clinical procedures, such as taking a blood sample from a dog's cephalic vein, should the IACUC request details of how the procedure itself will be performed?

Reg. The PHS Policy (IV,C, 1,f) and AWAR (§2.32,a) require the personnel conducting procedures on animals to be appropriately qualified and trained. It is the responsibility of the institution to provide necessary resources for training and it is the responsibility of the IACUC to ensure that personnel are, in fact, qualified. Even if a procedure is well established, the protocol should contain sufficient detail about the procedure(s) which, in turn, helps the IACUC in their evaluation of personnel qualifications.

Opin. The IACUC should ask for information about parameters of preparation of the skin for the sample, a range of the volume of blood to be drawn, the interval between draws and the range of frequency of draws. It would not be necessary to ask for the angle of needle insertion and such items unless the IACUC was not convinced that the person doing the blood draw knew how to do it. Training should be required if personnel lack sufficient expertise prior to initiation of the procedure(s).

9:66 Rather than providing details of an experimental procedure on an IACUC form, a PI provides literature references that clearly describe the details of what he will do. Should the IACUC consider approving this study with just the literature reference or should the actual procedure be part of the IACUC protocol?

Opin. The actual procedure should be part of the IACUC protocol so that it is clear what will be done as part of the experimental procedures. As necessary, the IACUC reviewer may choose to read the literature references.

9:67 Can the IACUC use teleconferences or other electronic communication in its review of animal use protocols? (See 6:17–6:19.)

Reg. According to PHS Policy, "If full committee review is requested, approval of those research projects may be granted only after review at a convened meeting of a quorum of the IACUC and with the approval vote of a majority of the quorum present." The AWAR use essentially the same language, and neither document addresses how the convened meeting may occur.

Opin. Traditionally, a quorum of the IACUC members meets together in a room to discuss and vote on a protocol. According to Garnett and Potkay, "The traditional meeting provides the kind of environment which is most conducive to thoughtful deliberation and interaction and is still regarded as the optimum forum for many of the IACUC functions."[36] But, this is not the only possibility. With modern technology, virtually identical face-to-face meetings are possible without the participants being in the same room.

What should be the criteria for such electronic meetings? NIH/OLAW has listed, and USDA/APHIS has concurred, with what are appropriate criteria:[37]

1. Methods of telecommunications (e.g., telephone or video conferencing) are acceptable for the conduct of official IACUC business requiring a quorum, provided the following criteria are met:
2. All members are given notice of the meeting.
3. Documents normally provided to members during a physically-convened meeting are provided to all members in advance of the meeting.
4. All members have access to the documents and the technology necessary to fully participate.
5. A quorum of voting members is convened as required by PHS Policy and AWAR.

6. The forum allows for real time verbal interaction equivalent to that occurring in a physically-convened meeting (i.e., members can actively and equally participate and there is simultaneous communication).
7. If a vote is called for, the vote occurs during the meeting and is taken in a manner that ensures an accurate count of the vote. A mail ballot or individual telephone polling cannot substitute for a convened meeting.
8. Opinions of absent members that are transmitted by mail, telephone, fax or e-mail may be considered by the convened IACUC members but may not be counted as votes or considered as part of the quorum.
9. Written minutes of the meeting are maintained in accord with the PHS Policy, IV,E,1,b and the AWAR §2.35,a,1.

Clearly, it is possible to accomplish full committee review in electronic meetings. There is no real need to use such techniques for designated reviewer methods. The conveyance of information required for the latter method can clearly be handled electronically. There is no voting involved in the process.

9:68 Should the decisions of the IACUC be influenced by the species proposed for use in the project?

Opin. "Speciesism," the treatment of animals differently solely because they are members of a different species, can drift into the IACUC deliberation process. Some species-specific care and use criteria must be part of the protocol, but it is assumed by some individuals that some species have a greater moral standing than others. This can come about because they are more like human beings than other species or because they often serve as pets. Some people justify distinctions between species on the basis of supposed sentience. However, there is little solid evidence for a difference in sentience between a number of species, such as a rat and a mouse or a cat and a dog. According to one survey, only a minority of IACUCs let the species to be used influence their decisions.[38]

There are no reports of tracking time spent in deliberation of protocols by species but a guess is that primate protocols often take much longer for the IACUC than those involving other species for the same level of invasiveness. There is nothing in the AWA, the AWAR, the PHS Policy or the *Guide* to suggest that regulatory agencies expect this. Species selection and IACUC deliberations should be made on the basis of scientific appropriateness for the proposed project. (See 9:37.)

9:69 Is it possible to set up a commercial IACUC that hires-out to review and approve protocols for an institution that doesn't want to set up its own IACUC?

Reg. The AWAR (§2.33) require the AV to be an employee of the institution. The PHS Policy (IV,A,1) requires compliance with the AWA and, like the AWAR, states that the IACUC veterinarian must have programmatic authority and is responsible for animals at the institution. As such, it seems nearly impossible to have an IACUC-for-hire.

NIH/OLAW states: "Domestic Assurances are negotiated with institutions that: control their own animal facilities, conduct animal research on-site, have an animal care and use program with an IO, an IACUC and a veterinarian with program authority and access to all animals."[39]

Opin. There are commercial/independent Institutional Review Boards (IRBs, independent of research institutions) that perform the jobs of institutional IRBs on a fee-for-service basis. There is no mention in the PHS Policy or AWAR of a commercial IACUC. An institution that receives PHS funding, must submit an Assurance outlining its program of animal care and use. It must also name the IO, IACUC Chair and AV. Both the PHS Policy and the AWAR specify a line of authority that would be very difficult to establish with such an external IACUC. In addition, the AV must be a member of the IACUC. That would mean that the AV of every institution that used such a commercial IACUC would have to be a member of that IACUC. This appears to be logistically problematic even using electronic technologies. In addition, according to the PHS Policy and the AWAR, the IACUC must conduct semiannual evaluations of the institution's animal care and use program along with inspections of animal facilities. This would also be difficult to achieve with a commercial IACUC. Finally, IACUCs have responsibilities that can be very institution specific and this includes more than protocol review. IACUCs are expected to address immediate animal concerns and noncompliance and formulate case specific corrective actions and having an outside entity to do this would not be advisable.

9:70 Can polling be used to obtain approval for a protocol in lieu of having a full committee meeting?

Reg. The word 'polling' or the word 'electronic' (associated with protocol review) do not exist in the PHS Policy.

Opin. Whereas it is not referred to directly in the PHS Policy, polling has been addressed by the following NIH/OLAW guidance:[28]

> Polling is defined as sequential, one-on-one communication, either in person or via telephone, e-mail, fax, U.S. mail, or by other similar means. Polling is an appropriate mechanism for providing all committee members with the opportunity to call for full review of a protocol prior to initiating the "designated reviewer" method of protocol review described below. It may also be appropriate as a mechanism for distributing and reviewing drafts of meeting minutes or reports.
>
> The simple polling of IACUC members does not, however, satisfy the definition of a meeting of a convened quorum and should not be used for conducting IACUC business that requires the vote of a convened quorum of the committee. For example, polling should not be considered a valid method of voting under the "full committee" review method of protocol review and is not an acceptable substitute for having a vote of a convened quorum on the suspension of a previously approved activity involving animals.

The bottom line is that polling can be used for setting up meetings and for collecting responses from members as to whether a protocol should be referred for full committee review as a part of the designated reviewer process. It is not acceptable to employ polling to establish a quorum to conduct business or to review or vote on protocols as a part of a full committee review process. Polling is not conducive to the dynamic exchange of ideas that is important in the full committee deliberation process. As a result, polling precludes due consideration of all the aspects for which the full committee review process intends.

Acknowledgment

The authors wish to thank Rebecca Bogatz for her unfailing assistance in typing the chapter, checking references, and providing valuable editorial comments.

References

1. American Veterinary Medical Association. 2013. AVMA guidelines for the euthanasia of animals: 2013 edition. https://www.avma.org/KB/Policies/Documents/euthanasia.pdf.
2. National Institutes of Health, Office of Laboratory Animal Welfare. 2013. Notices. http://grants.nih.gov/grants/olaw/references/notices.htm.
3. National Institutes of Health, Office of Laboratory Animal Welfare. 2013. Frequently Asked Questions (FAQs). http://grants.nih.gov/grants/olaw/faqs.htm.
4. National Institutes of Health, Office of Laboratory Animal Welfare. 2013. Commentary. http://www.grants.nih.gov/grants/olaw/references/commentary.htm.
5. U.S. Department of Agriculture, Animal and Plant Health Inspection Service. 2011. Animal care policy manual. http://www.aphis.usda.gov/animal_welfare/policy.php.
6. Applied Research Ethics National Association/Office of Laboratory Animal Welfare. 2002. Institutional Animal Care and Use Committee Guidebook, 2nd edition. Bethesda: National Institutes of Health. http://grants.nih.gov/grants/olaw/GuideBook.pdf.
7. National Institutes of Health, Office of Laboratory Animal Welfare. 2009. Reporting Noncompliant Events to OLAW. http://grants.nih.gov/grants/olaw/educational_resources.htm.
8. Prentice, E.D., D.A. Crouse and M.D. Mann. 1992. Scientific merit review: The role of the IACUC. *ILAR News* 34:15–19.
9. Mann, M.D. and E.D. Prentice. 2004. Should IACUCs review scientific merit of animal research projects? *Lab Anim. (N.Y.)* 33:26.
10. National Institutes of Health, Office of Laboratory Animal Welfare. 2013. FAQ D.12, Is the IACUC Responsible for Judging the Scientific Merit of Proposals? http://grants.nih.gov/grants/olaw/faqs.htm.
11. Final Rules: Animal Welfare. 9 CFR Parts 1 and 2, Federal Register, Vol. 54. No. 168, August 31, 1989, P. 36114. http://awic.nal.usda.gov/final-rules-animal-welfare-9-cfr-parts-1-and-2 Accessed Jan. 6, 2013.
12. Miller, J.G.,(former director of OLAW) personal communication, 1991.
13. National Institutes of Health, Office for Protection from Research Risks. 1991. The Public Health Service responds to commonly asked questions. *ILAR News* 33:68.
14. National Institutes of Health. 2012. Notice NOT-OD-12-101, Notice of Requirement for Electronic Submission of Just-in-Time Information and Related Business Process Changes Beginning April 20, 2012 http://grants.nih.gov/grants/guide/notice-files/NOT-OD-12-101.html.
15. Black, J. 1993. Letter to the editor. *ILAR News* 35:1.
16. Prentice, E.D., D.A. Crouse, D.A. and M.D. Mann. 1993. Letter to the editor. *ILAR News*, 35:2.
17. Federal Register, August 7, 2002, 67 FR 51289.
18. Wolff, A. 2002. Correct conduct of full-committee and designated-member protocol review. *Lab Anim. (N.Y.)* 31(9):28–31.
19. National Institutes of Health, Office of Laboratory Animal Welfare. 2013. FAQ D.6, What criteria should the IACUC consider when reviewing protocols. FAQ D.7, Should the IACUC consider the three "Rs" of alternatives when reviewing protocols? http://grants.nih.gov/grants/olaw/faqs.htm.

20. National Institutes of Health. 2003. Office of Extramural Research revised guidance regarding IACUC approval of changes in personnel involved in animal activities. http://grants.nih.gov/grants/guide/notice-files/NOT-OD-03-046.html.
21. National Institutes of Health, Office of Laboratory Animal Welfare. 2013. FAQ D.17, What guidelines should IACUCs follow for fishes, amphibians, reptiles, birds, and other nontraditional species used in research? http://grants.nih.gov/grants/olaw/faqs.htm.
22. National Institutes of Health, Office of Laboratory Animal Welfare. 2013. FAQ D.9, What is considered a significant change to a project that would require IACUC review? http://grants.nih.gov/grants/olaw/faqs.htm.
23. Wolff, A., N. Garnett, S. Potkay, et al. 2003. Frequently asked questions about the Public Health Service Policy on Humane Care and Use of Laboratory Animals. *Lab Anim. (N.Y.)* 32(9):33–36.
24. Garnett, N.L. and W.R. DeHaven. 1998. A word from the government. *Lab Anim. (N.Y.)* 27(3):19.
25. Silverman, J. 1997. Do pressure and prejudice influence the IACUC? *Lab Anim. (N.Y.)* 26(5):23–25.
26. National Institutes of Health, Office of Laboratory Animal Welfare. 2009. Notice NOT-OD-09-035, Guidance to IACUCs regarding use of Designated Member Review (DMR) for animal study proposal review subsequent to Full Committee Review (FCR). http://grants.nih.gov/grants/guide/notice-files/NOT-OD-09-035.html.
27. National Institutes of Health, Office of Laboratory Animal Welfare. FAQ D.19, May an IACUC use Designated Member Review (DMR) to review an animal study protocol subsequent to Full Committee Review (FCR) when modifications are needed to secure approval? http://grants.nih.gov/grants/olaw/faqs.htm.
28. Silverman, J. 1996. Majority rules? *Lab Anim. (N.Y.)*, 25(4):22. http://grants.nih.gov/grants/olaw/references/laba96v25n5.htm.
29. National Institutes of Health, Office of Laboratory Animal Welfare. FAQ C.6, What are PHS requirements for recording and reporting minority views? http://grants.nih.gov/grants/olaw/faqs.htm#report_6.
30. National Institutes of Health. 2010. Notice NOT-OD-10-128, Clarification on the roles of NIH Scientific Review Groups (SRG) and Institutional Animal Care and Use Committees (IACUC) in review of vertebrate animal research. http://grants.nih.gov/grants/guide/notice-files/NOT-OD-10-128.html.
31. National Institutes of Health, Office of Laboratory Animal Welfare. June 7, 2012. Transcript of OLAW webinar. Grants Policy and Congruence. http://grants.nih.gov/grants/olaw/120607_seminar_transcript.pdf Accessed 06/19/2013.
32. National Institutes of Health, Office of Laboratory Animal Welfare. 2013. FAQ D.8, When institutes collaborate, or when the performance site is not the awardee institution, which IACUC is responsible for review of the research activity? http://grants.nih.gov/grants/olaw/faqs.htm.
33. Brown, P. and C.A. Gipson. 2010. A Word from OLAW and USDA. *Lab Anim. (N.Y.)* 39:167. http://grants.nih.gov/grants/olaw/references/39_6_0610.pdf.
34. National Institutes of Health, Office of Laboratory Animal Welfare. FAQ D.14, May Standard Operating Procedures (SOPs) or blanket protocols that cover a number of procedures be utilized in lieu of repeating descriptions of identical procedures in multiple protocols? http://grants.nih.gov/grants/olaw/faqs.htm.
35. Gipson, C. and P. Brown. 2012. A word from OLAW and USDA. *Lab Anim. (N.Y.)* 41:43. http://grants2.nih.gov/grants/olaw/references/41_02_0212.pdf.
36. Garnett, N. and S. Potkay. 1995. Use of electronic communications for IACUC functions. *ILAR J.* 37:190–192.
37. National Institutes of Health, Office of Laboratory Animal Welfare. 2006. Notice NOT-OD-06-052, Guidance on use of telecommunications for IACUC meetings under the PHS Policy on Humane Care and Use of Laboratory Animals. http://grants.nih.gov/grants/guide/notice-files/NOT-OD-06-052.html.

38. Silverman, J., S.A. Baker, and C.W. Lidz. 2012. A self-assessment survey of the Institutional Animal Care and Use Committee, Part 1: Animal welfare and protocol compliance. *Lab Anim. (N.Y.)* 41:230–235.
39. National Institutes of Health, Office of Laboratory Animal Welfare. 2012. Obtaining an Assurance. http://grants.nih.gov/grants/olaw/obtain_assurance.htm.

10

Amending IACUC Protocols

Angela M. Mexas and Diane J. Gaertner

Introduction

During the course of any given research project it is incumbent upon all parties involved (the investigative team, the veterinary team, and the IACUC) to continually review the project's progress as data are collected and analyzed. At any point during the experimental process there may be a need to amend or revise a protocol to reflect changes in the intended research plan. Any *significant change* from the work outlined in the approved protocol should not only be communicated to the IACUC, in the form of an amendment, but must also be reviewed by the IACUC *prior* to implementation. In addition, any relevant, secondary parties (e.g., veterinarians, diagnostic services/quality assurance, occupational health and safety) should review such changes prior to implementation. According to the *Guide* (p. 33), "Continuing IACUC oversight of animal activities is required by federal laws, regulations, and policies"... and ... "a variety of mechanisms can be used to facilitate ongoing protocol assessment." The AWAR (§2.31,c,7; §2.31,d,1) require that investigators receive IACUC approval for all *significant changes* to approved animal research protocols. The AWAR (§2.31,d,1,i-xi) further delineate the criteria that should be used to evaluate all proposed changes. In accordance with AWAR (§2.31, d, 8), only the IACUC can approve protocols and protocol changes, although proposed activities and proposed changes that have already been approved by the IACUC may be subject to appropriate review by officials of the institution.

Changes to the protocol may need to be presented for consideration (for example):

- When experimental data suggests a better way to address the experimental question
- When unanticipated findings change the course of the experiment
- When new findings suggest there are improvements to be made
- When clinical complications arise during the course of an experiment

Federal regulations provide little guidance with respect to the actual criteria used in determining what constitutes a significant change to an approved research protocol. NIH/OLAW recently addressed this issue, providing guidance on what is to be considered a significant change that would require IACUC review.[1] In addition, the initial protocol should have carefully delineated the potential and expected complications along with a rate of failure to be expected in any given protocol, and this information should be taken into account by the investigators and the IACUC when considering if a revision is necessary or appropriate in any given circumstance. If experimental procedures vary from the described plan, the protocol should be amended to reflect these findings.

Individual institutions have devised various methods to accommodate for the submission and review of protocol amendments. The guidelines, policies and regulations should always be consulted to determine if any particular amendment constitutes a significant change which *can* be addressed by protocol amendment or if the change is a significant change that requires a *de novo* protocol submission. This chapter will describe the mechanisms being used to amend protocols and suggest times when amending a protocol may be useful, appropriate, or required.

10:1 What is the purpose of a protocol amendment?

Reg. The PHS Policy (IV,B,7) and the AWAR (§2.31,c,7) require PIs to seek IACUC approval for protocol modifications. The IACUC may approve an amendment, require modifications in order to secure approval, or withhold approval of significant changes regarding the care and use of animals in ongoing activities after review of the proposed changes. According to the *Guide* (p. 25), "The committee is responsible for oversight and evaluation of the entire Program... including... review and approval of proposed animal use (protocol review) and of proposed significant changes (protocol amendment) to animal use."

Opin. The purpose of a protocol amendment is to modify a previously approved animal use protocol. By submitting an amendment the investigator notifies the IACUC of a proposed change in the protocol and seeks approval in order to implement the proposed change. All animal procedures, manipulations, and actions *must* have documented IACUC approval *prior* to beginning the proposed animal work (NIH/OLAW FAQ D.5).[2] A protocol amendment provides an opportunity for the investigator to refine the experiment as needed using newly acquired experimental results, clinical expertise, or a proposed change in the experiment. These refinements may eventually lead to a reduction in the total number of animals used, and may be aimed at reducing or eliminating animal pain and distress. The investigator must submit an amendment to his/her protocol in order to keep the IACUC informed of any changes to the original plan for animal use. The IACUC, in turn, is charged with evaluating and approving, disapproving, or requiring modifications in order to secure approval of the proposed changes.

During continuing review of an active protocol any party involved may suggest the need to submit an amendment to the protocol in order to document and communicate protocol revisions for any change proposed. There are no regulations or guidance that prevents amending a protocol at the time of continuing review.

Surv. Members of the laboratory animal community were asked to answer a series of questions regarding a) what protocol changes could be made by the amendment processes and b) the process of amending approved protocols at their respective institutions. As can be seen in Table 10.1 below, the majority of survey respondents are from institutions where less than 100 approved protocols are managed at any given time (61%). Therefore, it is not surprising that most responders manage less than 100 *de novo* protocol submissions and amendments per year. Nevertheless, institutions with greater numbers must rely heavily on a well organized and clearly documented mechanism for processing of protocol amendments in order to satisfy the needs of both the investigators and the regulatory agencies.

TABLE 10.1

Annual Number of Protocols and Protocol Amendments Reviewed

	0–100	101–500	>500	I Don't Know	Not Applicable
How many approved protocols do you currently have?	172/282 (61%)	67/282 (24%)	23/282 (8%)	20/282 (7%)	0/282
How many *de novo* (complete) protocols do you review each year?	222/280 (79%)	45/280 (16%)	2/280 (1%)	11/280 (4%)	0/280
How many total amendments do you review each year?	190/281 (68%)	60/281 (21%)	15/281 (5%)	15/281 (5%)	1/281 (0.4%)

10:2 What are the regulatory requirements for protocol amendment?

Reg. PHS Policy (IV,B,7) and the AWAR (§2.31,c,7) require that the IACUC review and approve, require modifications in (to secure approval) or withhold approval of proposed significant changes regarding the care and use of animals in ongoing activities. PHS Policy (IV,C,1,a–g) states that the IACUC should review animal-related components of any proposed protocol and determine if the proposed research is in accordance with PHS Policy, the AWA, the *Guide*, and the institution's PHS Assurance. The AWAR (§2.31,d,1) explain how to conduct the review and state that the IACUC shall determine that the proposed significant changes meet requirements which are detailed in AWAR §2.31,d,1,i-xi. These requirements include minimizing pain and distress, considering alternatives to painful procedures, animal housing and veterinary care, personnel training and qualifications, surgical standards, and appropriate euthanasia techniques.

Opin. Initiating significant changes to IACUC-approved protocols without prior IACUC review is considered to be noncompliance with PHS Policy, and should be promptly reported to NIH/OLAW.[3,4] It is also a noncompliance with APHIS/AC regulatory requirements in cases pertaining to APHIS/AC regulated species. It is not, however, reportable to APHIS/AC unless the IACUC suspends the protocol in question. Due to the serious implications of noncompliance, establishing a streamlined mechanism for the prompt review of protocol amendments should be a priority in any animal program. Having an effective and time efficient way to document and review these changes is important to minimize the number or incidence of unapproved animal activities within any program. PIs and all members of their scientific teams must be made maximally aware that procedures on animals must follow approved protocols and that any activity that is not outlined in the approved protocol should not be performed prior to submission *and* approval of a protocol amendment.

10:3 What changes constitute a significant modification to an approved animal use protocol?

Reg. In 2003, according to a published notice, NIH/OLAW[5] in consultation with APHIS/AC revised the guidance on review mechanisms for personnel changes on a protocol. According to this guidance, institutions may classify certain proposed additions and changes in personnel (other than the PI) as "minor" provided that an appropriate administrative review mechanism is in place to ensure that all research personnel are appropriately identified, adequately trained, qualified in applicable occupational health and safety programs, and meet other criteria as required by the IACUC. The IACUC remains responsible for confirming that all IACUC review criteria are maintained and documented (PHS Policy IV,C,1; AWAR §2.31,d; §2.31,e; §2.35,a,2). Personnel training and qualifications are the responsibility of the research facility under section §2.32 of the AWAR.

Significant changes requiring IACUC review, as described by NIH/OLAW[1] guidelines include but are not limited to, changes

- In the objectives of a study
- From non-survival to survival surgeries/procedures
- Resulting in greater discomfort or in a greater degree of invasiveness
- In the species or in approximate number of animals used (see 10:8 below)
- In the principal investigator
- In anesthetic agent(s) used or the use or withholding of analgesia
- In the methods of euthanasia
- In the duration, frequency or number of procedures performed on one animal

Opin. Any and all proposed changes to an animal related protocol should be submitted for review and approval by the IACUC prior to their implementation. The IACUC or the IACUC administrator (under the direction of the IACUC) must distinguish between protocol changes in order to determine the course of action required for review and implementation of such changes in the timeliest manner. There are, generally speaking, three categories of changes to be considered:

1. Changes that significantly alter the scope of the project which would best be achieved by a *de novo* protocol submission rather than a protocol amendment
2. Changes that are significant enough to require both submission of a protocol amendment and full IACUC review prior to their implementation (herein referred to as major and minor, significant changes)
3. Changes that are minor enough and have no implications on animal welfare, such as administrative changes, which can be processed by submission of a protocol amendment without requiring IACUC review (not significant changes).

Each IACUC should utilize the list of guidelines and examples provided to further clarify within their own institution as to what types of significant changes can be administered via protocol amendment and which changes require that a new protocol be submitted. As an example, in Table 10.2 below, we categorize significant protocol changes into those we think would require a new protocol and those we think require only an amendment. While this table is intended to help provide examples, it represents only our opinion and not a regulatory requirement. In our opinion, changes limited to those in personnel (other than the PI) and/or changes that add minimally invasive procedures that do not have the potential to cause any pain or distress should be considered minor significant changes requiring an amendment, but not a new protocol. Furthermore, only changes with no implications to animal welfare, such as room designations, removal of qualified personnel, or minor corrections to

TABLE 10.2

Protocol Changes Requiring a New Protocol versus Those Requiring an Amendment

Protocol Changes Requiring a New Protocol	Protocol Changes Requiring a Protocol Amendment
Changes in the purpose or aim of a study	Substitution in personnel (other than the PI)
Change of PI	Small increase in the number of animals (subject to individual IACUC guidelines)
Change from peer-reviewed to non-peer-reviewed funding source	Additional sample collection (non-surgical procedure)
Change in species used	Addition of non-invasive (non-surgical procedures)
Large increase in the number of animals used or any change in the number of cats, dogs, or non-human primates used	Addition of drugs or treatments use to ameliorate pain or suffering from complications associated with an approved surgery/procedure
Need to repeat an experiment utilizing more animals	Addition of a survival surgery or a painful procedure

the protocol should be considered insignificant enough to be processed by the IACUC administration, without review by an IACUC member. In all cases, the IACUC is responsible for carefully delineating which changes are considered significant and for training IACUC staff accordingly. Thus, the IACUC approves all methods of review and only IACUC members can review and approve proposed significant changes.

Examples of significant changes (as defined by APHIS/AC) can also be found in the *Animal Welfare Inspection Guide* (Section 7–41 on Procedure for Protocol Review).[6] Additional significant changes that meet NIH/OLAW criteria for requiring IACUC review, but are not in the NIH/OLAW guidance[1] include the following:

- Unconventional housing or husbandry requests, especially if they do not meet minimum housing requirements
- Requests to house animals outside of the vivarium (e.g., in a lab) for more than 12 consecutive hours
- Addition of any procedures with the potential to cause pain or distress
- Addition of any procedures that may result in unexpected death or other complications not described in the original protocol
- Changes that would render immune competent animals immunocompromised

While there is no limit to the number of amendments that can be submitted for a specific protocol, complex protocol changes or submission of multiple subsequent changes to one protocol are not always suited for the amendment mechanism. The protocol amendment process should be limited to relatively simple changes that do not extensively affect the existing documentation and potentially lead to confusion. For example, if one complex procedure is approved in the original protocol, a substantial or complicated change to the performance of said procedure may be too difficult to delineate in an amendment without contradicting the procedure approved in the original document. If the amended protocol is unclear, it may confuse rather than clarify what is actually approved. If complex changes or multiple subsequent changes will result in confusing protocol documentation, a completely rewritten protocol is advised. Therefore, the reviewer should consider not only the amendment proposed, but also the original submission when evaluating a protocol change. The use of an electronic IACUC protocol management system, which can produce an integrated version of the protocol after each amendment is approved, may allow more complex protocol changes to be clearly documented and may avoid some occasions where complex amendments will require the writing of a new protocol to achieve clarity.

The addition of any surgical procedure should be considered a major significant change and requires a protocol amendment to be submitted and reviewed by the IACUC prior to approval and implementation. A veterinarian should be consulted[7,8] to determine if the procedure constitutes a major or minor surgery in order to advise the IACUC in the process of reviewing the amendment. In terms of multiple survival surgeries on one animal, "Multiple procedures that may induce substantial post-procedural pain or impairment may be conducted on a single animal only if justified by the PI, and reviewed and approved by the IACUC."[8] For animals regulated by the AWAR, multiple major surgical procedures on a single animal are acceptable only if they are: included in and essential components of a single research project or proposal; scientifically justified in writing by the investigator; or necessary for clinical reasons (AWAR §2.31,d,1,x,A-C).[8] Procedures performed for medical reasons must be approved by the Attending Veterinarian or his/her designee. APHIS/AC Policy 14[9] clarifies special circumstances which require APHIS Administrator approval, as mentioned under AWAR §2.31 d,1,x,C. Generally, the circumstance is a desire to conduct a major operative procedure on an animal that already underwent a major operative procedure for research purposes under another study. Should questions arise regarding the addition of a major survival surgery on a single animal, investigators should seek approval by the APHIS Animal Care Administrator. Finally, if a protocol amendment involves the use of laparoscopic surgery, the IACUC should determine if this procedure is considered a major or minor procedure[10] as described in the *Guide* (p. 117). In our opinion most laparoscopic procedures constitute major surgery. The addition of such a procedure under any circumstance should be considered a significant change due to the potential for post-surgical complications and the invasive nature of the procedure. The addition of such a procedure in an animal that is undergoing other surgical procedures should follow the rules for multiple major surgical procedures as described above.

We consider a change in PI to significantly affect the change in scope of a particular study, and therefore we would recommend that institutions should require a new protocol submission in such cases. However, other factors may be considered that could justify a change in PI as a significant change requiring an amendment and IACUC review rather than a new protocol submission. For example, if the proposed new PI was formerly included as a collaborator in the protocol and is a well-known member of the investigative team at this particular institution, perhaps an amendment would be acceptable. In contrast, a new PI with no previous history at the institution, or a PI who proposes to do work in a species in which he/she has no previous experience constitute, in our opinion, a need for submission of a new protocol. Furthermore, a change in PI may have significant implications in regards to changes in the funding source which will require careful review by members of the IACUC regardless of whether the change is submitted as a new protocol or an amendment to an existing protocol.

Surv. Members of the laboratory animal community were asked to categorize individual types of changes to a protocol into those that would require no amendment, an amendment, or a new protocol at their respective institutions. The survey results (Table 10.3) are mostly consistent with recommendations made by regulatory agencies and *The IACUC Handbook* (2nd edition).[11] Of note, only 35.7% of

TABLE 10.3

Indicate Which of the Following Changes Are Considered Appropriate by Your IACUC for Revision by the Protocol Amendment Mechanism

	No Amendment Needed	New Amendment Needed	New Protocol Needed	Not Applicable	Number of Responses
Changes to the objectives of a study	19 (6%)	104 (36%)	157 (54%)	12 (4%)	292
Addition of minor surgery	0	256 (87%)	24 (8%)	14 (5%)	294
Addition of major surgery	0	192 (65%)	82 (28%)	19 (7%)	293
Changing from non-survival to survival surgery	2 (1%)	177 (60%)	98 (33%)	18 (6%)	295
Changing the degree of invasiveness of a procedure	0	229 (78%)	57 (19%)	8 (3%)	294
Changing the expected pain or distress of an experimental animal	1	223 (76%)	64 (22%)	7 (2%)	295
Performing a new procedure or changing a procedure being used on mice or rats	1	224 (76%)	48 (16%)	22 (8%)	295
Performing a new procedure or changing a procedure being used on an USDA species	3 (1%)	195 (66%)	55 (19%)	41 (14%)	294
Changing personnel involved in animal procedures	53 (18%)	226 (77%)	7 (2%)	9 (3%)	295
Changing anesthetic agent or agents	7 (2%)	262 (89%)	17 (6%)	7 (2%)	293
Changing to the use of or withholding of analgesics	2 (1%)	228 (78%)	53 (18%)	11 (4%)	294
Changing the analgesic agent	8 (3%)	263 (90%)	14 (5%)	9 (3%)	294
Changing the methods of euthanasia	5 (2%)	259 (88%)	23 (8%)	7 (2%)	294
Changing the duration, frequency or number of procedures performed on an animal	2 (1%)	233 (79%)	53 (18%)	6 (2%)	294
Addition of a new strain or stock of the same species	30 (10%)	232 (79%)	21 (7%)	11 (4%)	294
A change that increases the numbers of mice or rats	9 (3%)	239 (82%)	18 (6%)	25 (9%)	291
A change that increases the numbers of a USDA-regulated animal	2 (1%)	225 (77%)	21 (7%)	46 (16%)	294
A change that increases the number of animals by 5% or less	52 (18%)	228 (78%)	5 (2%)	9 (3%)	294
A change that increases the number of animals by 5%–10%	30 (10%)	249 (84%)	6 (2%)	10 (3%)	295
A change that increases the number of animals by 10% or more	4 (1%)	231 (79%)	49 (17%)	9 (3%)	293
A change of Principal Investigator	7 (2%)	174 (59%)	105 (36%)	8 (3%)	294
A change or addition of a new species	3 (1%)	169 (59%)	104 (36%)	11 (4%)	287
A change from a peer-reviewed funding source to a non-peer reviewed funding source	109 (37%)	106 (36%)	13 (4%)	65 (22%)	293
Change to a vendor, such as a producer of polyclonal antibody	138 (48%)	110 (38%)	7 (2%)	34 (12%)	289
Change to a location where all or part of the study will be done	24 (8%)	219 (75%)	22 (8%)	29 (10%)	294

(continued)

TABLE 10.3 (Continued)

Indicate Which of the Following Changes Are Considered Appropriate by Your IACUC for Revision by the Protocol Amendment Mechanism

	No Amendment Needed	New Amendment Needed	New Protocol Needed	Not Applicable	Number of Responses
A request for housing conditions that are not typical of vivarial housing or husbandry	4 (1%)	210 (72%)	51 (17%)	28 (10%)	293
A change to reflect more clinical signs or animal deaths than previously anticipated	13 (4%)	216 (74%)	50 (17%)	14 (5%)	293
Addition of a prolonged restraint procedure	2 (1%)	203 (69%)	71 (24%)	18 (6%)	294
The addition of or a change to the use of hazardous agents	3 (1%)	215 (74%)	61 (21%)	12 (4%)	291

responders list a change in the PI as requiring a new protocol submission while 59.2% consider an amendment appropriate for this activity. There was a wide distribution of responses, with no consensus as to whether a protocol amendment is warranted for a change in the funding source or vendors, but a large number of responders replied that the question was not applicable at their institution.

10:4 What format should be used to submit an amendment?

Reg. The format for protocol amendments is determined by the IACUC. According to the *Guide* (p. 34) "Some institutions use the annual review as an opportunity for the investigator to submit proposed amendments for future procedures, to provide a description of any adverse or unanticipated events, and to provide updates on work progress," suggesting that the annual review cycle may be utilized to update protocol contents. The required content for a significant amendment is outlined in AWAR §2.31,e and includes

- Identification of the species and the approximate number of animals to be used
- A rationale for involving animals, and for the appropriateness of the species and numbers of animals to be used
- A complete description of the proposed use of the animals
- A description of procedures designed to assure that discomfort and pain to animals will be limited to that which is unavoidable for the conduct of scientifically valuable research, including provision for the use of analgesic, anesthetic, and tranquilizing drugs where indicated and appropriate to minimize discomfort and pain to animals
- A description of any euthanasia method to be used

Opin. Amendments may be submitted through formal (amendment form) or informal (e-mail or faxed communication) mechanisms, provided that all the necessary information is supplied and all of the members of the IACUC are provided the opportunity to call for review at a convened meeting of the IACUC. If the amendment is a significant change, it may be reviewed by either the full committee or by designated member review. The decision to require a formal submission process or accept an informal (e-mail) communication is likely dependent on the number of submissions the IACUC has to review and the number of IACUC members and staff available to process each review. Thus, institutions that review large numbers of protocol amendments often require a specific form or document to expedite review of each submission and to ensure that all the required information has been provided. For clarity, it may be necessary to require that only a single amendment to an IACUC protocol be under current review at any given time. Questions to be addressed in a request for protocol amendment should include the following:

- What is the purpose or rationale for requesting this protocol amendment?
- Is there a proposed change in personnel associated with the protocol?
  - If adding personnel to the protocol, or changing the roles of personnel on the approved protocol, it is necessary to submit documentation detailing how or why the person is qualified to perform the tasks described.
- Is there a proposed change to the species, sex, or strain of animal used in the protocol? Why?
- Will there be specialized housing requirements associated with the proposed change in the protocol?
  - Is there housing available for the new proposed species, number, and type of animal proposed?
  - Will the experiment require housing that is not standard for the species or facility?
  - Will the housing provided meet experimental requirements and comply with the *Guide's* and AWAR recommendations?
- Are you requesting a change in the number of animals used?
  - If additional animals are needed, justification for the increased number of animals must be provided.
  - If additional animals are *not* needed for new experiments, then will any existing experiments have to be deleted or modified?
- Are you proposing a new procedure be added to the protocol?
  - The description of the new procedure must be sufficiently detailed and justified to enable the IACUC to conduct its review.
  - Is this a minor (e.g., blood draw, data collection) or major (e.g., surgical) procedure?
  - How often and for what duration will the procedure be performed?
  - Will the animal experience pain or distress before, during or after the procedure?
    - How will this be mediated and minimized?
    - If pain will not be minimized, how will this affect the pain and distress category under which your protocol is classified?
    - Justification for the use of this animal model and the lack of alternatives to their use in painful procedures must be provided.

TABLE 10.4

Mechanisms and Forms Used for Amending Protocols

The Following List Includes Minor or Major (Significant) Amendments	Yes	No	Not Applicable	Other
Can protocols be amended by e-mail without the use of an amendment form?	52/294 (18%)	230/294(78%)	9/294 (3%)	3/294 (1%)
Can protocols be amended by a memo without the use of an amendment form?	64/293 (22%)	217/293 (74%)	9/293 (3%)	3/293 (1%)
Do you have a special form for amendments?	206/295 (70%)	81/295 (27%)	6/295 (2%)	2/295 (1%)
Is there a special form simpler/shorter than your "regular" protocol form?	195/290 (67%)	18/290 (6%)	73/290 (25%)	4/290 (2%)
Is a new or additional literature (or related) search required for minor amendments?	39/294 (13%)	223/294 (76%)	14/294 (5%)	18/294 (6%)
Is a new or additional literature (or related) search required for significant amendments?	187/294 (64%)	53/294 (18%)	16/294 (5%)	38/294 (13%)
At your institution, can there be more than one amendment under review at the same time to amend a single protocol?	177/297 (60%)	83/297 (28%)	28/297 (9%)	9/297 (3%)

- Are you proposing a change in the procedures already described in the original protocol (any change in drugs, methods, anesthesia, analgesia, or euthanasia)?
  - Does the change result in re-classification of the original procedure (e.g., minor to major surgery, multiple survival surgeries)?
- Will prolonged restraint of conscious animals be required?
- Will new hazardous materials be used?

Surv. Members of the laboratory animal community were asked to describe the mechanisms and forms by which protocol amendments are processed at their institutions. Table 10.4 summarizes these results. The majority of institutions require a special, albeit shorter, form for submission of a protocol amendment and only major, significant changes require investigators to perform additional literature searches prior to submission at most institutions. A majority of institutions (60%) require that only one amendment is under review any given time for a single protocol.

10:5 Do the designated member review and full committee review processes also apply to amendments?

Reg. PHS policy (IV,C,2) and the AWAR (§2.31,d,2) require the same review procedure for proposed significant changes in ongoing activities (protocol amendments) as they do for new activities (new protocols). (See 9:21; 9:28.)

Opin. All IACUC members must have the opportunity to evaluate proposed significant changes and request full committee evaluation at a convened meeting of the IACUC, as in full committee review. If no member requests full committee review, the IACUC may assign the amendment for designated member review. Amendments should always be reviewed with access to and in the context of the complete original protocol. Whether a full committee review mechanism or a designated reviewer mechanism is used,[12,13] all reviewers must have access to both the amendment documents and the original protocol. IACUC approval of protocol amendments must be documented in the protocol file with a dated signature. Documentation of unapproved amendments associated with the protocol may be kept on record for the reviewers' reference, but should not affect objective review of a proposed amendment in question.

10:6 Can the IACUC Chair or an IACUC subcommittee review and approve significant changes or must the full committee review and approve such submissions?

Reg. PHS Policy (IV,C,2) and the AWAR (§2.31,d,2) identify full committee review and designated member review as the only approved methods to evaluate animal protocols and proposed significant changes to ongoing protocols. IACUC subcommittees are not authorized to approve proposed changes to an ongoing protocol. Only the full committee (or in the case of designated member review, a qualified IACUC member) has the authority to approve proposed significant changes in ongoing activities (AWAR §2.31,d,2; §2.31,d,8). Furthermore, in instances when full committee review is conducted, NIH/OLAW has determined that serial one-on-one meetings, telephone, fax, or electronic mail to poll or obtain members' votes cannot preempt the voting mechanism of a quorum of a convened IACUC.[14] APHIS/AC concurs with PHS on this matter, which is addressed in the *Inspection Guide* under Chapter 7–51.[6] Real-time interactive electronic methods of protocol review are acceptable only as described by NIH/OLAW in FAQ B.8.[15] In general, the methods typically used for protocol review should also be used to determine how amendments are reviewed.

Opin. Institutional policy will determine which types of protocol amendments are eligible for designated member review and which would require a full committee review. Any member(s) of the IACUC, including the Chair, may serve as a designated reviewer. IACUC members will determine which protocol amendments will need to be reviewed by the full committee, since they will always have the option of requesting full committee review for a given protocol amendment. If an institution establishes a policy to have all protocol amendments reviewed by full committee, this decision would obviate the choice between designated member review and full committee review. On the other hand, the IACUC may allow personnel changes (other than the PI) or other not significant administrative changes, such as typographical errors, to be approved by the Chair or an IACUC administrator without review by a formal delegation mechanism. Institutional policies should clearly delineate which changes, if any, to a protocol may be approved by the IACUC Chair without committee involvement. This policy should be made available to all interested parties.

Surv. Members of the laboratory animal community were asked to describe the mechanisms by which protocol amendments are processed at their institutions. Interestingly, there was no consensus as to the specific mechanism by which major and minor amendments would be reviewed (Table 10.5).

TABLE 10.5

Mechanisms of Review for Significant (Major) versus Minor Amendments

	Select the Mechanism(s) That Best Summarizes Your Institution's Review of SIGNIFICANT (Major) Amendments	Select the Mechanism(s) That Best Summarizes Your Institution's Review of MINOR Amendments?
Review and approval by IACUC staff member(s)	30/295 (10.1%)	48/296 (16.3%)
Review and approval by a single member of the IACUC (Designated Member Review)	71/295 (24.1%)	94/296 (31.9%)
Review and approval by two (2) members of the IACUC, with or without review by IACUC staff	67/295 (22.7%)	69/296 (23.4%)
Review and approval by the Full Committee at a convened IACUC meeting	186/295 (63.1%)	75/296 (25.4%)
Not applicable	13/295 (4%)	13/296 (4.4%)
Other	25/295 (8.5%)	19/296 (6.4%)

Note: IACUC staff can certainly review a major amendment but only IACUC members can approve the amendment.

10:7 How should the IACUC review a protocol amendment?

Opin. Each institution must develop its own guidelines regarding significant protocol modifications and ensure the availability of an IACUC for the evaluation and approval of protocol amendments. To determine the significance of a proposed change, the investigator and the IACUC should consider the effect of any proposed change on animal welfare and the overall ethical cost-benefit ratio of the research proposed, as they would when conducting a *de novo* protocol review. Careful consideration to the basic areas of concern in primary protocol review should be revisited. These areas may include (but are not limited to)

- The potential for pain or distress
- The level of invasiveness of procedures described in the amendment
- Their effect on environmental conditions (e.g., diet, water, space, lighting, humidity, temperature)
- The potential for unexpected outcomes
- Humane experimental endpoints
- Use of non-pharmaceutical grade chemicals in an animal protocol
- Physical restraint
- Methods or indications for euthanasia or removal from the protocol

Surv. 1 and 2 Members of the laboratory animal community were asked to categorize each of the proposed changes (from Table 10.3) as a major or minor protocol amendment if their institution made a distinction in the way major and minor protocol changes are handled. The results (Table 10.6) suggest that only

TABLE 10.6

Indicate to Which Category Each of the Following Changes Would Belong, if These Are Considered by Your IACUC

	Minor Amendment	Significant (Major) Amendment	Not Applicable	I Don't Know	Total
Changes to the objectives of a study	17 (8%)	145 (64%)	63 (28%)	0	225
Addition of minor surgery	51 (23%)	132 (59%)	41 (18%)	1 (0%)	225
Addition of major surgery	0	179 (80%)	44 (20%)	1 (0%)	224
Changing from non-survival to survival surgery	2 (1%)	176 (79%)	44 (20%)	1 (0%)	223
Changing the degree of invasiveness of a procedure	6 (3%)	179 (80%)	38 (17%)	1 (0%)	224
Changing the expected pain or distress of an experimental animal	4 (2%)	180 (81%)	39 (17%)	0	223
Performing a new procedure or changing a procedure being used on mice or rats	26 (12%)	143 (65%)	44 (20%)	7 (3%)	220
Performing a new procedure or changing a procedure being used on an USDA species	12 (6%)	151 (69%)	47 (21%)	9 (4%)	219
Changing personnel involved in animal procedures	158 (71%)	17 (8%)	48 (21%)	0	223
Changing anesthetic agent or agents	73 (33%)	110 (49%)	37 (17%)	3 (1%)	223
Changing to the use of or withholding of analgesics	16 (7%)	168 (75%)	38 (17%)	1 (0%)	223
Changing the analgesic agent	80 (36%)	101 (45%)	39 (18%)	2 (1%)	222
Changing the methods of euthanasia	52 (23%)	133 (60%)	37 (17%)	1 (0%)	223
Changing the duration, frequency or number of procedures performed on an animal	17 (8%)	165 (74%)	39 (17%)	2 (1%)	223
Addition of a new strain or stock of the same species	108 (49%)	61 (28%)	49 (22%)	2 (1%)	220
A change that increases the numbers of mice or rats	72 (32%)	96 (43%)	48 (22%)	6 (3%)	222
A change that increases the numbers of a USDA-regulated animal	42 (19%)	119 (54%)	48 (22%)	13 (6%)	222
A change that increases the number of animals by 5% or less	125 (56%)	52 (23%)	44 (20%)	1 (1%)	222
A change that increases the number of animals by 5%–10%	110 (50%)	68 (30%)	43 (19%)	1 (1%)	222
A change that increases the number of animals by 10% or more	27 (12%)	155 (70%)	41 (18%)	0	223
A change of Principal Investigator	38 (17%)	129 (59%)	49 (22%)	4 (2%)	220
A change or addition of a new species	19 (9%)	142 (65%)	56 (25%)	2 (1%)	219
A change from a peer-reviewed funding source to a non-peer reviewed funding source	89 (40%)	15 (7%)	104 (47%)	13 (6%)	221
Change to a vendor, such as a producer of polyclonal antibody	107 (48%)	20 (10%)	83 (37%)	12 (5%)	222
Change to a location where all or part of the study will be done	84 (38%)	78 (35%)	49 (22%)	10 (5%)	221
A request for housing conditions that are not typical of vivarial housing or husbandry	29 (13%)	139 (63%)	47 (21%)	5 (2%)	220

(continued)

TABLE 10.6 (Continued)

Indicate to Which Category Each of the Following Changes Would Belong, if These Are Considered by Your IACUC

	Minor Amendment	Significant (Major) Amendment	Not Applicable	I Don't Know	Total
A change to reflect more clinical signs or animal deaths than previously anticipated	19 (9%)	153 (70%)	43 (20%)	5 (2%)	220
Addition of a prolonged restraint procedure	13 (6%)	160 (72%)	46 (21%)	2 (1%)	221
The addition of or a change to the use of hazardous agents	20 (9%)	159 (72%)	39 (18%)	2 (1%)	220

changes in personnel or the number of animals used (if 5% or less of estimated number) would be generally considered minor amendments. Interestingly, there was no consensus as to the category under which changes in anesthetic or analgesic agents would fall under, even though these items are listed specifically by NIH/OLAW as needing special attention and approval by IACUC review.[1] It is the authors' opinion that such changes should always be considered significant changes and should undergo review by the IACUC because special care must be taken to ensure proper dosing and adequate control of consciousness and pain. In addition, adding a new strain or stock of the same species of animals and changes that increase the number of mice or rats in a protocol were similarly classified as either minor or major amendments by the survey respondents. It is our opinion that there should be a cut-off measure applied to when a change becomes significant. For example, an IACUC may allow an increase of less than 10% of the original approved number of animals but an increase in animal numbers beyond the specified percentage would constitute a major (significant) change and this will be included as part of the institution's Assurance to NIH/OLAW. In the authors' opinion, a change in animal numbers greater than 10% should be considered a major change regardless of whether the strain or stock used is similar or different than what was previously indicated on the protocol. A change in species used, or addition of any animals of an APHIS/AC regulated species should always be considered a significant change. Other changes that did not reach a consensus included changes in funding sources, vendors, or the location where the study would be conducted. While a change in the location where work is performed may sometimes be minor, special care must be taken to ensure such a change will not affect important safety measures, such as animal traffic patterns set to prevent the spread of infectious diseases or occupational health and safety parameters put in place to protect people at their work place.

Survey respondents also were asked about the nature of their protocol review. Responses varied as to the proportion of institutions that separate minor and significant amendments for the method of protocol review (see Table 10.6 for review procedures for major versus minor amendments). The majority of respondents (79 respondents, 60%) indicated that they review less than 100 minor and major (significant) amendments, respectively, which correlates with the overall numbers of protocols reviewed at each institution.

Surv. 3 If your institution separates minor from major (significant) amendments, do they undergo different courses of protocol review by IACUC members or IACUC staff?

- Minor and significant amendments have the same course of protocol review — 73/292 (25%)
- Minor and significant amendments have a different course of protocol review — 131/292 (45%)
- Not applicable, we do not separate minor from significant amendments — 83/292 (28%)
- I don't know — 3/292 (1%)
- Other — 2/292 (1%)

Minor amendments were generally reviewed within one week (82%), while major (significant) amendments took longer (as would be expected) with 30% and 40% of respondents claiming a 1–2 and 3–4 week response time, respectively (see below for numbers and time course for major and minor amendments). A significant number of respondents also stated that amendments were reviewed at the next IACUC meeting, and therefore could take anywhere from 3–30 days, depending on the date of submission. A few of the respondents (12) gave percent estimates rather than numerical values for the relative number of minor vs. significant protocol amendments.

Surv. 4 If your reviews separate minor from significant (major) amendments, of the total number of amendments reviewed per your institution, how many are minor amendments?

- 0–100 amendments — 126/228 (55%)
- 101–500 amendments — 25/228 (10%)
- More than 500 amendments — 8/228 (4%)
- Don't know — 21/228 (9%)
- Do not distinguish — 4/228 (2%)
- Not applicable — 43/228 (19%)

Surv. 5 If your reviews separate minor from significant (major) amendments, of the total number of amendments reviewed per year how many are significant amendments?

- 0–100 amendments — 146/231 (63%)
- 101–500 amendments — 13/231 (6%)
- More than 500 amendments — 6/231 (3%)
- Don't know — 21/231 (9%)
- Not applicable — 45/231 (19%)

Surv. 6 If your reviews separate minor from significant (major) amendments, of the total number of amendments reviewed, how many days does it take to review a minor amendment?

- Less than one week 96/117 (82%)
- One to two weeks 15/117 (13%)
- Up to one month 2/117 (2%)
- Other 4/117 (3%)

Surv. 7 If your reviews separate minor from significant (major) amendments, of the total number of amendments reviewed, how many days does it take to review a significant amendment?

- Less than one week 25/110 (23%)
- One to two weeks 34/110 (31%)
- Up to one month 44/110 (40%)
- Other 7/110 (6%)

10:8 How many animals may be added to an ongoing study using an amendment mechanism?

Reg. PHS Policy (IV,D,1,a) and the AWAR (§2.31,e,1; §2.31,e,2) require that proposals to the IACUC specify and include a rationale for the approximate number of animals proposed for use. Individual institutions, therefore, must establish mechanisms to monitor and document the number of animals acquired and used in approved activities. PHS Policy (via U.S. Government Principle III) also requires that the minimum number of animals needed to obtain valid scientific results be approved for use.

Opin. The cutoff for animals requested above the approved number of animals should be no more than the exact number of animals approved by the IACUC or a small percentage (5%–10%) in excess of the approved number of rodents. In concordance with our opinion, some institutions allow an amendment mechanism for ordering up to 10% more animals (rodents), with any additional animal requests requiring IACUC review.[1] In the authors' opinion, requests for any additional non-rodent mammals, irrespective of the number, should require review and approval by the IACUC.

Surv. Survey responses were somewhat ambiguous in regards to when a change in animal numbers should be considered a minor or major (significant) change (Table 10.6). When asked if a change that increases the number of mice or rats should be considered a minor or major (significant) change, 43% said it should be major and 32% said it should be minor. When asked if a change that increases the number of animals by 5%–10% should be considered a minor or major (significant) change, 30.6% said it should be major and 49.6% said it should be minor. On the other hand, there was a slight majority who considered an increase in the number of APHIS/AC-regulated animals to be a major (significant) change (53.6%) and an increase in the number of animals of less than 5% to be a minor significant change (56.3%). Changes that increase the number of animals by 10% or more are considered by most (69.5%) to constitute a major (significant) change.

10:9 The PI on an IACUC protocol does not perform any hands-on animal work. Hands-on work is done by a research associate. If there is a request for a change in PI to one who also will not do any hands-on animal work, is this considered a significant change? Does this change require a protocol amendment or a new protocol submission?

Reg. Both the PHS Policy (IV,B,7) and AWAR (§2.31,c,7) require the IACUC to review proposed significant changes in ongoing activities using animals. According to NIH/OLAW, a change in PI is always considered significant and therefore requires review and approval by the IACUC.[1] Whereas changes in other personnel may be considered minor changes,[5] changes in PIs are considered a significant change that may require submission of a new protocol. Changes in other personnel may be addressed by a protocol amendment.

Opin. IACUCs are responsible for assuring that any personnel on an animal use protocol have been adequately trained and are qualified to perform the specified procedures on the selected species. In the opinion of the authors, PIs must complete at least the minimal institutional training requirements associated with their animal work regardless of whether or not they handle the animals. Furthermore, a change in PI on an NIH grant would be considered to be a significant "change in scope" such that the grantee must obtain prior NIH approval. Similarly, we suggest that a change in PI should necessitate submission of a new protocol for approval, especially if the new PI was not already working with the species utilized in the protocol or is not a member of the original PI's collaborative group.

10:10 A dog is anesthetized and prepared for surgery. The PI then recognizes that the dosage of an experimental drug that she believed to be safe might actually cause a problem. The potential problem is easily avoided by lowering the dosage. Should she make the correction and then notify the IACUC, or stop the research and obtain IACUC approval before proceeding?

Reg. Both the PHS Policy (IV,B,7) and AWAR (§2.31,c,7) require the IACUC to review proposed significant changes in ongoing activities using animals prior to implementation of such changes. AWAR §2.31,d,1,iv,B requires that the AV be involved in planning and consultation regarding procedures which cause more than momentary or light pain and distress. According to the *Guide* (p. 5) "Veterinary consultation must occur when pain or distress is beyond the level anticipated in the protocol description or when interventional control is not possible," suggesting anticipated complications that may arise from performing procedures as described in the protocol should be discussed with the veterinary team before the procedure is performed, if possible. However, the PHS Policy (IV,C,1,a) requires that procedures conducted with animals avoid or minimize discomfort, distress, and pain to the animals. U.S. Government Principles IV and V also stress the avoidance or minimization of discomfort, distress, and pain. Appropriate changes in a procedure currently underway may be made for the animal currently undergoing surgery, but additional planned procedures (if performed), would constitute noncompliance that must be promptly reported to NIH/OLAW.

Opin. Regulatory agencies emphasize the need to protect animals from unnecessary pain and distress and provide guidelines to minimize unwanted consequences in the use of animals in research. "The IACUC is responsible for assessment and oversight of the institution's Program components and facilities" (*Guide* p. 14) and as such would be charged with establishing guidelines regarding whether specific changes in drug doses are considered major or minor changes for protocol review. These decisions should consider the type of drug in question as much as the difference in dosage or route of administration. Changes in drugs related to anesthesia, analgesia, and euthanasia should be considered significant changes and require attention by the IACUC and AV (or designee) prior to implementation. This is especially true if the change results in a reduced dose to the animal.

In an emergency, or if the investigator detects a potential serious risk during the performance of an experimental procedure, the investigator must change the procedure to benefit the animal. In addition, there should be, if possible, an immediate consultation with a veterinarian in order to determine if the change in dosage is warranted or permissible. The AV has authority to intervene and change dosages for individual animals for medical reasons and "an integral component of veterinary medical care is the prevention or alleviation of pain associated with procedural and surgical protocols," (*Guide* p. 120). This could be considered such a case, since the animal is already anesthetized, and intervention by the AV to adjust the dose would avoid the need for an additional anesthetic event. In such a case, documentation of both the change and consultation, along with the veterinarian's advice and/or consent should be followed by submission of a protocol amendment. The changed procedure must not be conducted on additional animals. Proceeding with the experiment using the change in drugs or dosages on other animals would be considered significant deviations from the approved protocol and would require submission of a protocol amendment and IACUC approval prior to their implementation.

10:11 During the course of research it becomes evident that no pain or distress is occurring in a study that was initially assumed to cause mild pain. The PI wishes to amend his IACUC protocol to indicate no pain or distress. Is this considered a major (significant) or minor amendment?

Reg. The AWAR (§2.36) stipulate the required information to be submitted in an institution's annual report, which includes the number of animals used in each pain category. For this reason, changes in pain categories (whether increasing or decreasing in degree) should be reported to the IACUC through the amendment process. Individual IACUC's have some discretion to define what is considered a major (significant) or minor amendment and whether these categories are handled differently according to their own animal care and use program needs.[1] The AWAR (§2.31,e) provides some guidelines regarding what information should be considered by the IACUC when a significant change in an ongoing activity involving animals is proposed.

Opin. The preceding scenario does not identify the laboratory animal as an AWA-regulated species and annual reports to APHIS/AC do not require the inclusion of data on species that the AWA does not define as animals in their purview. Nevertheless, many IACUCs follow the same procedures for all species to preclude the perception of a double standard regarding animal welfare. It is our opinion that the IACUC would need to evaluate any change in the expected pain or distress as a significant change even if the opinion of the PI was that pain or distress was reduced by the change.

10:12 At 2 years into a 3-year swine study, a PI submits a protocol amendment. The amended portion of the protocol does not involve any animal pain or distress although other parts of the original study included more than momentary pain. Does the PI have to review the most current literature to see whether any alternatives to the unchanged painful procedures have been found during the previous 2 years?

Reg. Farm animals used as models of human disease (in biomedical research) or in development of biologicals for nonproduction are regulated under the AWA. APHIS/AC Policy 12 states that "Significant changes are subject to prior review by the IACUC. If those changes include a painful or distressful procedure, a consideration of alternatives or a revision of the prior search may be required." APHIS/AC Policy 12[16] also states that "Although additional attempts to identify alternatives or alternative methods are not required by Animal Care at the time of each annual review of an animal protocol, Animal Care would normally expect the principal investigator to reconsider alternatives at least once every 3 years, consistent with the triennial de novo review requirements of the Public Health Service Policy on Humane Care and Use of Laboratory Animals (IV,C,5)."

Opin. When submitting an amendment to a protocol that will induce additional pain, the PI must submit an alternatives search and justify the need for the addition of the new pain-causing procedure. If submitting an amendment for a procedure that will not cause pain or distress, then additional alternative searches are not required.

In the scenario presented, the PI would need to submit a new animal use protocol to the IACUC within the next year if he or she plans to continue the experiments beyond the three year mark from the first protocol approval. Given that timetable, the PI should consider, instead, submitting a new protocol that incorporates the planned amendment, along with a documented new search for alternatives for the potentially painful procedures included in the entire protocol submitted. Depending upon the complexity and number of amendments to this protocol, a *de novo* protocol submission may be the most advantageous option for both the PI and the IACUC.

10:13 A group of neurosurgeons is investigating the effects of ovarian hormones on sleep, metabolism, emotion, and cognition. Their experiments require placement of a cerebral intra-ventricular cannula for injection of metabolic inhibitors in rats. The rats usually recover from surgery within 24 hours without complications. However, during their last experiment, the investigators noticed that 10% of the rats became lethargic and obtunded 24–48 hours after surgery. The rats were euthanized and a necropsy evaluation revealed suppurative inflammation within the ventricles of the brain and intra-lesional bacteria, suggestive of contamination through the intra-ventricular catheter. Must the investigator submit an amendment to request additional animals to be added to their protocol in order to replace the animals lost to this unexpected complication?

Reg. NIH/OLAW lists a change in the estimated number of animals used in a protocol as one of the examples of changes that should be considered significant enough to require protocol amendment (FAQ D.9).[1] According to FAQ F.2,[17] "although the PHS Policy does not explicitly require a mechanism to track animal usage by investigators, it does require that proposals specify a rationale for the approximate number of animals to be used and be limited to the appropriate number necessary to obtain valid results. This implicitly requires that institutions establish mechanisms to document and monitor numbers of animals acquired and used, including any animals that are euthanatized because they are not needed. Monitoring should not exclude the disposition of animals inadvertently or necessarily produced in excess of the number needed or which do not meet criteria (e.g., genetic) established for the specific study proposal. Institutions have adopted a variety of administrative, electronic, and manual mechanisms to meet institutional needs and PHS Policy requirements."

Opin. In this scenario the amendment should include not only the additional number of animals used but also the measures that will be taken in order to prevent cannula infections in the future. To justify the need for an increase in the number of animals used, the investigator would describe the unexpected complications that arose in previous experiments, prompting an explanation of how these unanticipated complications would be prevented heretofore (i.e., systemic antibiotics may be added to the protocol or specific procedures for handling and protection of the cannulas should be described). Individual IACUCs can determine the cutoff for what is considered enough of an increase in the number of animals to require an amendment. While this protocol uses rats, not APHIS/AC regulated animals, it is our opinion that an increase of more than 10% would constitute a significant change in the number of animals used and therefore, should require a protocol amendment. NIH/OLAW reviewers of this chapter noted that this incident must be reported to the IACUC even if the number of animals is less than 5% because of the contamination of the cannula.

References

1. National Institutes of Health, Office of Laboratory Animal Welfare. 2012. Frequently Asked Questions D.9, What is considered a significant change to a project that would require IACUC review? http://grants.nih.gov/grants/olaw/faqs.htm#proto_9.
2. National Institutes of Health, Office of Laboratory Animal Welfare. 2012. Frequently Asked Questions D.5, May the investigator begin animal work before receiving IACUC approval? http://grants.nih.gov/grants/olaw/faqs.htm#proto_5
3. National Institutes of Health, Office of Extramural Research. 2005. Notice NOT-OD-05-034, Guidance on Prompt Reporting to OLAW under the PHS Policy on Humane Care and Use of Laboratory Animals. http://grants.nih.gov/grants/guide/notice-files/NOT-OD-05-034.html.
4. Garnett, N.L. and C.A. Gipson. 2003. A mouse isn't a rat, but what's the big deal? A word from OLAW and USDA. *Lab. Anim. (N.Y.)* 32:19. http://grants.nih.gov/grants/olaw/references/lab_animal2003v32n9_Silverman.htm.
5. National Institutes of Health, Office of Extramural Research. 2003. Notice NOT-OD-03-046, Guidance Regarding IACUC Approval of Changes in Personnel Involved in Animal Activities. http://grants.nih.gov/grants/guide/notice-files/NOT-OD-03-046.html.
6. United States Department of Agriculture, Animal and Plant Health Inspection Service. 2013. Animal Welfare Inspection Guide. http://www.aphis.usda.gov/animal_welfare/downloads/Inspection%20Guide%20-%20November202013.pdf.

7. Applied Research Ethics National Association/Office of Laboratory Animal Welfare. 2002. Other protocol review considerations/surgery. In *Institutional Animal Care and Use Committee Guidebook*, 2nd edition, 145, Bethesda: National Institutes of Health.
8. National Institutes of Health, Office of Laboratory Animal Welfare. 2012. Frequently Asked Questions F.9, Are multiple major survival surgical procedures permitted on a single animal? http://grants.nih.gov/grants/olaw/faqs.htm#f9.
9. United States Department of Agriculture, Animal and Plant Health Inspection Service. 2011. Animal Care Policy Manual. Policy #14. http://www.aphis.usda.gov/animal_welfare/policy.php?policy=14.
10. National Institutes of Health, Office of Laboratory Animal Welfare. 2012. Frequently Asked Questions F.13, Are laparoscopic procedures considered major surgery? http://grants.nih.gov/grants/olaw/faqs.htm#useandmgmt_13.
11. Gaertner, D.J. and K.D. Moody. 2002. Amending IACUC Protocols. In *The IACUC Handbook*, ed. J. Silverman, M. Suckow, and S. Murthy S., 2nd edition, 139–148. Boca Raton: Taylor & Francis.
12. Wolff, A. 2002. Correct conduct of full-committee and designated-member protocol reviews. *Lab. Anim. (N.Y.)* 31(9):28–31.
13. National Institutes of Health, Office of laboratory Animal Welfare. 2012. Frequently Asked Questions D.3, What are the possible methods of IACUC approval? http://grants.nih.gov/grants/olaw/faqs.htm#proto_3.
14. Garnett, N. and S. Potkay. 1995. Use of electronic communications for IACUC functions. *ILAR J.*, 37:190. http://grants.nih.gov/grants/olaw/references/ilar95.htm.
15. National Institutes of Health, Office of Laboratory Animal Welfare. 2012. Frequently Asked Questions B.8, May an IACUC conduct business on a teleconference call? http://grants.nih.gov/grants/olaw/faqs.htm#IACUC_8.
16. United States Department of Agriculture, Animal and Plant Health Inspection Service. 2011. Animal Care Policy Manual Policy 12, Consideration of alternatives to painful/distressful procedures. http://www.aphis.usda.gov/animal_welfare/policy.php?policy=12.
17. National Institutes of Health, Office of Laboratory Animal Welfare. 2012. Frequently Asked Questions F.2, Is the IACUC responsible for tracking animal usage? http://grants.nih.gov/grants/olaw/faqs.htm#useandmgmt_2.

11

Continuing Review of Protocols

Gwenn S. F. Oki and Ernest D. Prentice

Introduction

Initial review and approval of a project by an IACUC represent an informed judgment by the committee, that when the protocol is initiated as approved, it will be conducted in full compliance with all applicable federal requirements contained in the AWA, AWAR, HREA, and PHS Policy. The IACUC's review responsibilities do not, however, end with initial approval of the protocol. The IACUC is obligated by both the AWAR and the PHS Policy to conduct ongoing reviews of protocols.

The purpose of this chapter is to present general concepts of continuing review that will be useful to IACUCs. It should be noted that continuing review is just as important as the initial review because it serves to ratify the decisions of the IACUC on the current status of the protocol and helps the investigator maintain compliance.

11:1 What is meant by continuing review?

Opin. APHIS/AC interprets continuing review, performed no less often than annually (AWAR §2.31,d,xi,5) as a monitoring process in an effort to determine that the study remains in compliance, that the activities have been "conducted in accordance with the approved protocol," that significant and (at many institutions) minor modifications to the protocol have received prior IACUC approval,[1] and to ensure that the investigator is informed of new institutional, state or federal requirements. It should be noted that the *Guide* (p. 34) endorses APHIS/AC and NIH/OLAW requirements and further states that "some institutions use the annual review as an opportunity for the investigator to submit proposed amendments for future procedures, to provide a description of any adverse or unanticipated events, and to provide updates on work progress."[2] NIH/OLAW requires triennial (every three years) continuing review to be a *de novo* process,[3,4] meeting all the new proposal review criteria as set forth in PHS Policy IV,C,1–4.

11:2 What is the intent of continuing review?

Opin. In general, the intent of a regulation or policy can be found in the preamble. For example, the preamble to the AWAR makes a single statement indicating that the intent of continuing review is "to provide current information to the research facility regarding all ongoing activities so that it can remain in compliance."[5] Although the intent of continuing review was not articulated in the form of a preamble to the PHS Policy, NIH/OLAW via PHS Policy IV,A,1 has interpreted that continuing review, subject to the AWAR, be performed no less than annually as a monitoring function.[5,6]

11:3 What is the difference between "periodic review" and "continuing review?"

Opin. The regulations do not use the term "periodic review." The *Institutional Animal Care and Use Committee Guidebook*[7] references "periodic review" and "continuing review" as part of a training syllabus for IACUC members.[7] Some institutions have chosen to categorize continuing review as distinct from periodic review, with periodic review signifying that a specific animal protocol requires more frequent review and oversight than routine continuing review which is, at minimum, an annual event under AWAR and a triennial event under the PHS Policy (see 11:1). The need for periodic review or more frequent continuing review could be due to the novelty of the research, high risk potential for unrelieved pain and distress, or use of a sensitive animal model. These decisions are usually made at the time of initial review or at the time of a significant modification. Since the term "periodic review" is a recently coined IACUC term that appears to be interchangeable with "continuing review," for purposes of this chapter, "continuing review" will be used to synonymously include "periodic review."

11:4 Is a protocol initially considered approved on the date when the research funding begins or on the date when the IACUC has taken final action on the protocol?

Reg. PHS Policy (IV,D,2) considers a protocol approved on the date of the IACUC approval, not when research funding begins.

Opin. The PHS Policy and AWAR are silent regarding the date of approval of an animal activity. NIH/OLAW and APHIS/AC offer the following guidance on this topic. In its guidance document (NOT-OD-11-053)[8] NIH/OLAW (with APHIS/AC concurrence) stated: "Date of [IACUC] approval is the date that appears in the written notification of approval to the principal investigator. IACUCs have flexibility to develop a system that functions well for their institution, provided that the approval date occurs within a reasonable period of time after review and approval (see AWAR §2.31,d; PHS Policy IV,C). For example, the IACUC may designate an approval date as the last working day of the month in which the proposal was reviewed, allowing time to make minor corrections or confirm other activities which may affect the approval (e.g., safety committee review, research staff training)."

While it appears to make sense to synchronize the IACUC initial protocol approval date with that of the funding start date, this practice is not permissible in many institutions. If the IACUC approves a protocol and submits a letter to the funding agency certifying that the protocol was approved, the actual date of the IACUC's approval must be indicated as required by the PHS Policy (IV,D,2). This is not a future date with final approval contingent on actual funding. In addition, the just-in-time process for verification of IACUC approval for the animal use portions of the grant application and any IACUC imposed changes must be noted on the IACUC's verification of approval prior to the grant award. (See 8:22.) The institution should have a mechanism and procedures established to verify concordance of the grant application with the approved IACUC protocol.[9,10] Despite the availability of the just-in-time process, it should be noted that institutions have the prerogative to require investigators to obtain IACUC approval of the grant application prior to NIH peer review, thereby not utilizing the just-in-time process.

11:5 What is the maximal life of an approved protocol?

Opin. The AWAR and PHS Policy do not limit the life of a protocol; however, both require that the IACUC conduct continuing reviews of approved studies at specified intervals. (See 11:7; 11:17.) Based on PHS Policy IV,C,5 and according to the *Institutional Animal Care and Use Committee Guidebook*,[3] "The PHS Policy requires that a complete IACUC review of PHS supported protocols be conducted at least once every three years. This triennial review is interpreted by OLAW as a requirement for *de novo* review, meaning that the criteria and procedures for review specified in IV.C. of the PHS Policy must be applied not less than once every three years." The 3-year review begins from the actual date of approval and cannot be extended.[8] Therefore, many institutions require a new submission at least triennially since it is very likely that the science has evolved, that the current protocol may include multiple amendments that have been approved over the 3-year period, and the IACUC may have new policies or a revised protocol application form that asks new questions that need to be specifically addressed. Since the triennial review constitutes a *de novo* review it may be more efficient to require submission of a new protocol.

11:6 What do the AWAR and PHS Policy indicate with regard to continuing review of protocols?

Reg. The frequency of IACUC consideration of approved, ongoing activities is one of the few areas in which the NIH/OLAW and APHIS/AC have differing requirements. The AWAR (§2.31,d,5) state that "the IACUC shall conduct continuing reviews of activities covered by this subchapter at appropriate intervals as determined by the IACUC, but not less than annually." PHS Policy (IV,C,5) states that "the IACUC shall conduct continuing review of each previously approved, ongoing activity covered by this Policy at appropriate intervals as determined by the IACUC, including complete review in accordance with IV,C,1–IV,C,4 at least once every three years."

Opin. Other than the maximal time interval between continuing reviews, the AWAR make no further statements regarding the form or substance of continuing review. The PHS Policy (IV,A,1) indicates that "compliance with the USDA regulations is an absolute requirement of the Policy" for Assured institutions, therefore, requiring annual continuing reviews. The PHS Policy (IV,C,5), however, also indicates that review criteria set forth at PHS Policy IV,C,1–4 must be satisfied at least triennially. This is the same criterion that IACUCs must use to conduct an initial review of a proposed protocol.[11] Hence, triennial review of a protocol constitutes a *de novo* assessment of a currently approved study. Because triennial review of a protocol is considered review of a new study, once approved, this triennial review establishes a new protocol approval period. (See 11:5.)

11:7 How often should continuing review of protocols be performed?

Reg. As mentioned in 11:6, the AWAR requires continuing review be performed at intervals determined appropriate for the study being conducted but no less often than once annually. The PHS Policy requirement is the same, except that the interval is no less than triennially. (See 11.5.)

Opin. Each IACUC should establish the frequency of continuing review at the time of initial review of the protocol. Many institutions have divided continuing reviews, based on review intervals, to meet the AWAR on an annual basis and to require a new protocol submission at the time of triennial review. Because the PHS Policy necessitates *de novo* review at the end of 3 years utilizing the same assessment criteria established for new protocol submissions and because there will more than likely be changes in the experimental methods and procedures, new institutional policies, and IACUC membership, it may be more efficient to require the investigator to submit a completely new protocol at the time of triennial review. It should, however, be noted that there is no AWAR or PHS Policy requirement for resubmission of a new IACUC application for an ongoing study.

The frequency of continuing review should also be evaluated at the time of a significant modification. For example, assume that a nonhuman primate species is added to a protocol and multiple major survival surgeries will be performed. An IACUC may then decide that continuing review should occur more often than annually. Indeed, the IACUC may also decide that periodic review (at more frequent intervals), independent of continuing review, is also warranted.

11:8 What is an effective and appropriate way to conduct continuing review of a protocol?

Opin. Most IACUCs require investigators to use an institution-specific continuing review form. This standardizes the information required for all continuing studies and streamlines the review process. The form should be designed in consideration of the intent of continuing review described previously. One model developed in collaboration with NIH/OLAW and APHIS/AC has been described.[12] In addition, in 1993, NIH/OLAW published the following advice on annual review which is still relevant today.[13]

> A relatively simple monitoring mechanism, which meets USDA requirements and serves to monitor animal activities covered by the PHS Policy, can be implemented by the use of a standard form containing basic protocol information (including title, approval number, date, and species). This form is then sent to the PI to (1) verify active status, (2) verify that completed activities were conducted in accordance with the approved protocol, (3) describe any proposed departures from the approved protocols, and (4) solicit information about activities projected for the upcoming year. Information pertaining to the future would indicate either that no changes were proposed or describe changes that the PI would like to implement. This document would then be brought before the IACUC for its consideration and documented as an official IACUC action. All protocols must be reviewed by the IACUC and those protocols with significant changes must be reviewed and approved by the IACUC. Such a monitoring system, however, does not preclude the requirement for

> triennial review or for PIs to seek IACUC approval when they want to make significant changes in approved protocols at other than the regularly scheduled monitoring periods. Such approval must be obtained *prior* to implementing the changes. Both the USDA and PHS requirements, of course, may also be satisfied by conducting complete *de novo* reviews of all animal study proposals on an annual basis.[13]

It should be noted that significant changes to a protocol must be reviewed and approved by either the full committee at a convened meeting or through designated member review (DMR) prior to implementation of the protocol modification. Minor changes may generally be approved administratively by the IACUC administrator, the IACUC Chair, or AV.[14] (See Chapter 10.)

11:9 Is full committee review by the IACUC necessary in order to reapprove a protocol?

Reg. Review procedures under PHS Policy (IV,C,2) apply to the initial and triennial reviews and IACUC review of proposed significant changes to the use of animals in ongoing activities. Review procedures under the AWAR (§2.31,d,2) apply to initial and continuing reviews and IACUC review of proposed significant changes to the use of animals in ongoing activities.

Opin. Full committee review is not necessary for continuing reviews. However certain IACUCs may decide that full committee review is necessary for continuing studies that involve sensitive species or high-profile controversial studies. Some IACUCs use the designated reviewer process (see 9:21) as indicated in the AWAR (§2.31,d,2), the PHS Policy (IV,C,2) and NIH/OLAW FAQ D.3.[15]

- Each IACUC member is provided with a list of continuing protocols to be reviewed. Any IACUC member shall have available, upon request, the written descriptions of the research protocols.
- Any IACUC member can request full committee review. If full review is not requested, then one or more IACUC members (the designated reviewers), selected by the IACUC Chair, is provided with the specific continuing review form(s), the corresponding approved protocol files and is responsible for conducting the continuing review.
- The designated reviewer is authorized to approve, required modifications to secure approval, or to request full IACUC review.

In most situations, continuing studies involving no changes or minor proposed changes are most efficiently handled by DMR. Alternatively, some IACUCs conduct full committee review of all continuing protocols by utilizing a primary reviewer. Typically, in this scenario, all IACUC members receive a copy of the continuing review form. The primary reviewer, IACUC Chair, AV, and IACUC administrator also are provided a copy of the currently approved protocol. At the time of the convened meeting, the primary reviewer provides a summary of his or her findings with a recommendation for action. After discussion, a vote is taken.

Depending on the size of the institution's animal program, either the DMR or full committee review process may be appropriate. For institutions with smaller animal programs, the latter may be the chosen method; for larger institutions, the former may be more efficient in consideration of time constraints and the volume of continuing reviews.

11:10 Can a designated reviewer be used for continuing review? If so, under what circumstances can the designated reviewer process be used for continuing review?

Opin. As indicated in 11:9, the DMR process (AWAR §2.31,d,2; PHS Policy IV,C,2; NIH/OLAW FAQ D3[15]) can be utilized for both annual and triennial continuing reviews. IACUCs with a lower volume of protocols may elect full committee review for all aspects of animal research conducted at their institution. Some IACUCs may choose the DMR process for continuations that involve no changes or minor proposed changes, thus requiring full committee review for all other continuing protocols. Other IACUCs may decide that full committee review is necessary for continuing studies that involve sensitive species or high-profile controversial studies, thus utilizing the DMR process for all other continuing protocols. Depending upon the institutional policies, protocol volume, and their research portfolio, IACUCs must determine the most appropriate continuing review process. (See 6:19; 9:28; 11:9.)

11:11 Under what circumstances can an alternate member conduct continuing review?

Reg. APHIS/AC and NIH/OLAW have a joint statement that alternates can serve in the capacity of their full member counterpart when the full member is unavailable or absent.[8,16] (See 5:23–5:25.)

Opin. At the time of continuing review, and at a convened meeting of the IACUC, the alternate member would attend the meeting in place of their full member counterpart. The alternate member is designated as an alternate for a specific full member(s) and may not represent more than one member at any one time.[8] The alternate member would be responsible for review of new protocols, modifications, continuing reviews, etc. and would behave as any other voting member at the meeting. Alternate members can also conduct continuing reviews as a DMR as long as the method and processes for DMR are followed and their full member counterpart is not a DMR for the same continuing review.[15,17]

11:12 If a designated reviewer is used (see 6:19; 11:9; 9:21–9:27), must the full IACUC still give final approval to the decision made by that person?

Opin. The DMR process utilized by IACUCs is meant to save time and to promote efficiency, especially for high-volume IACUCs. As indicated in the AWAR (§2.31,d,2) and PHS Policy (IV,C,2), the designated reviewer is authorized to approve, to require modifications, or to request full IACUC review. If the latter is not requested by the designated reviewer or by another IACUC member, then the designated reviewer has the authority to approve the continuation report or protocol. Requiring the full IACUC to give final approval of the designated reviewer's approval is an unnecessary duplication of effort that defeats the purpose of the designated reviewer process. If the full IACUC approves what a DMR has already decided, then the protocol is considered to be approved by the full committee. Thus, either the DMR or full committee approves the protocol, not both.[15] (See 9:21; 9:28; NIH/OLAW FAQ D.3.)

11:13 How do investigators change procedures or personnel on their IACUC-approved protocol?

Opin. IACUCs at most institutions require that investigators submit a protocol amendment form when a protocol alteration is needed. This may include modifications in experimental procedures, an increase in the number of animals required, or a change in personnel involved in the study (i.e., direct handling and use of animals, etc.). As discussed in 11:7, significant changes require IACUC approval prior to implementation and can be submitted either before or at the time of continuing review.[14] (See Chapter 10.)

Utilizing the DMR process (see 11:9), a copy of the original protocol is given to the designated reviewer who will review the protocol and the continuation review form. Depending on the number of amendments, it may be advisable for the PI to submit a completely rewritten protocol. When a protocol has undergone a number of amendments, even those that are minor, the reviewer may have difficulty discerning what constitutes the most recent version of the approved protocol. Indeed, many institutions have instituted the requirement that any amendment be incorporated into a master protocol. This master protocol would constitute an up-to-date protocol that could be used by the IACUC in their review and approval processes. This method can greatly facilitate the IACUC's understanding of the continuation report, any proposed changes, and the way the changes relate to the last approved version of the protocol. In addition, it serves to facilitate the animal care and compliance staffs' management and oversight of the protocol, assist the research staff in the appropriate conduct of the study, and facilitate APHIS/AC inspectors and accreditation site visitors in their review of the study.

11:14 What action should the IACUC take if an investigator does not respond to a request for information concerning the continuing review of his protocol?

Reg. If an investigator permits his or her IACUC approval to expire because of failure to respond to a request for updated protocol information, and PHS-funded animal work is ongoing, then the research is no longer in compliance with PHS Policy. IACUCs are required to report this and other incidents of serious or continuing noncompliance to NIH/OLAW (PHS Policy IV,F,3,a). However, the expiration of a protocol does not require a report to NIH/OLAW or APHIS/AC. If research continues on an expired protocol, that is reportable to NIH/OLAW and the PHS funding component.[18]

Opin. The AWAR (§2.31,d,5) and PHS Policy (IV,C,5) require that reviews of animal protocols be conducted at least annually and triennially, respectively. If the IACUC's request for information is necessary in order for the continuation report to be approved, the IACUC has no choice but to halt further use of any animals until the issue is resolved and the protocol is reinstated. (See 29:50; 29:51.) If the continuation re-approval date has passed with no response from the investigator, appropriate individuals in the institution should be notified (e.g., the animal facility management, the institution's AV, the PI's department head) with a copy to the PI. Operationally, IACUCs should have established policies and procedures for handling these types of situations. It should be noted that both the PHS Policy and the AWAR are silent on the disposition of animals already in the process of experimentation when the protocol lapses and a continuation report has not been submitted. There are a wide variety of ways that institutions are handling this issue. The authors of this chapter think that the animals already in the experimental queue should remain in the study until protocol procedures for their use have been completed. Furthermore, for other animals, no procedures other than routine husbandry care can be carried out until the protocol is reapproved. One possible exception is the maintenance of an animal disease model with IACUC approval. (See 11:15.)

11:15 If a protocol is near completion and the continuation report due date is imminent, can the IACUC vote to extend the protocol for a specified short period without submitting a formal report for continuing review?

Reg. See 11:14.

Opin. No. Both APHIS/AC and NIH/OLAW are clear that the IACUC may not extend the protocol approval dates without the appropriate annual or triennial reviews within the specified period.[19]

11:16 If a protocol expires can the investigator continue use of animals that have already undergone some of the research procedures and are in the queue for completing study-related interventions?

Reg. See 11:14.

Opin. The AWAR and PHS Policy prohibit continuation of a study if the investigator has not secured approval to continue the study. Many IACUCs have established procedures for the transfer of the animals into a holding protocol managed by the animal care facility or AV. The IACUC approved holding protocol does not permit any experiments to be conducted but will allow the animals to be maintained until the investigator secures continuation approval. This is an example of a procedure that will temporarily maintain the animals so that once the investigator secures approval, the research interventions can be completed and wasting of animals is minimized. It should be noted that financial charges cannot be made to a PHS grant during lapses in IACUC approval.[18]

Surv. If a protocol expires, which of the following statements are applicable to your IACUC? Check as many boxes as appropriate.

Numbers represent number of responses from 294 unique respondents.

• No additional experimental work is allowed under any circumstances	233/294 (50.3%)
• We allow an investigator to conclude any ongoing work	29/294 (6.3%)
• We allow the animal facility personnel to conclude any ongoing work on the protocol	7/294 (3.7%)
• We only allow the investigator to continue to breed animals	10/294 (2.2%)

- We only allow animal facility personnel to continue to breed animals for the investigator 31/294 (6.7%)
- We only allow the investigator to euthanize animals for the investigator 21/294 (4.5%)
- We only allow animal facility personnel to euthanize animals 74/294 (16%)
- Other 48/294 (10.4%)

"Other" included

- Each case is treated individually.
- Work may continue if a renewal of the protocol is in process.
- When protocol expires, all animals transferred to a "holding" protocol under the supervision of veterinarian.
- We would only allow any special care or handling necessary for the health of the animals.
- We only allow animal facility personnel to care for the animals.
- We ask the investigator to submit a new protocol if necessary, but to suspend all work on the protocol until approval.
- We amend the protocol for additional time.
- The animal facility takes possession of the animals and no work is allowed until a protocol exists for the animals. If animals are euthanized it is done by the animal facility and the investigator is charged for our time. The above continues only if work is non-invasive in nature. Otherwise an amendment or new protocol is required to continue.
- An assessment is conducted to determine whether some ongoing studies should be allowed to be completed.

11:17 When should an entirely new protocol be submitted? Why?

Opin. Many institutions require investigators to submit a new protocol every 3 years. Although this procedure fits PHS Policy (IV,C,5) for complete *de novo* review at least triennially, it should be emphasized that the AWAR and PHS Policy do not limit the life of a protocol. Therefore, it is possible to allow a protocol to continue indefinitely without a new protocol submission as long as both the AWAR and PHS Policy requirements for continuing review on an annual and *de novo* review on a triennial basis (respectively) are met. For PHS purposes, the triennial review must include a review of all current information relevant to the protocol (e.g., master protocol), although submission of an entirely new protocol is not required. (See 11:5; 11:6.)

11:18 If electronic submission and review processes are used for triennial (*de novo*) review, can the review be done without a new protocol submission?

Opin. The AWAR and PHS Policy do not specify the need for a new protocol submission at any time. At the time of triennial review, however, it makes sense that a new protocol submission be provided by the investigator to facilitate the IACUC's review. Electronic submission and review does not preclude the need for a complete and logically organized protocol. While electronic submission allows for flipping between sections and screens in search of information, in deference to the reviewer, an up-to-date, organized protocol would reduce review time, make the review process less burdensome, and enhance the approval timeline.

11:19 How much information concerning results of a study at the time of continuing review should the IACUC require?

Opin. Reporting the results of an ongoing study is not an AWAR or PHS Policy requirement. However, at the time of triennial review, the IACUC is required to perform a *de novo* review of the continuing study (see 11:5). This necessitates that the protocol meet all review criteria and study conduct requirements as stipulated in PHS Policy (IV,C,5) and AWAR (§2.31,d,i–xi; §2.31,e,1–5). It, therefore, makes sense for the IACUC to require that the investigator provide an informative summary of the results of the study to date, in order to facilitate IACUC review. Indeed, the only way the IACUC can justify use of the animals already included (to date) in the research project is to obtain information on the progress of the study. Admittedly, investigators encounter "blind alleys" and other problems that may affect study results. In some cases, animal experiments must be repeated. The IACUC should be advised of such issues that arise during the course of the study and at time of continuing review in order to fulfill its responsibilities.

11:20 Is a literature search required at the time of continuing review?

Reg. The AWAR (§2.31,d,1,ii) require documentation that alternatives were considered. A literature search is recommended in APHIS/AC Policy 12 as the most effective and efficient way to meet the alternatives search requirement of the AWAR. Although the need for a similar literature search is not specifically indicated, PHS Policy (IV,A) requires that applicable AWAR requirements also be met as part of the institution's Assurance. In addition, APHIS/AC Policy 12 provides guidance indicating that "Animal Care would normally expect the principal investigator to consider alternatives at least once every 3 years, consistent with the triennial review requirements of the PHS Policy (IV,C,5)."[20,21] (See 10:11.)

Opin. If the research is APHIS/AC regulated, a literature search should, therefore, be conducted at the time of initial review, when there are proposed significant changes to an ongoing protocol, and at the time of triennial review. Since the PHS Policy has no literature search requirement for research not subject to AWAR, it is up to the institution's IACUC to decide whether to require a literature search for non-USDA covered species.

References

1. Office for Protection from Research Risks, Division of Animal Welfare. 1993. Issues for Institutional Animal Care and Use Committees (IACUCs): Frequently Asked Questions about the Public Health Service Policy on Humane Care and Use of Laboratory Animals. *ILAR News* 35(3–4):47–49.

2. Committee for the Update of the Guide for the Care and Use of Laboratory Animals. 2011. *Guide for the Care and Use of Laboratory Animals*, 8th edition, 34. Washington: The National Academies Press.
3. Applied Research Ethics National Association/Office of Laboratory Animal Welfare. 2002. *Institutional Animal Care and Use Committee Guidebook*, 2nd edition, 96. Bethesda: National Institutes of Health.
4. Office of Laboratory Animal Welfare. 2013. Frequently Asked Questions D.1, How frequently should the IACUC review research protocols? http://grants.nih.gov/grants/olaw/faqs.htm#IACUC_2.
5. Office of the Federal Register, 9 CFR Parts 1, 2, and 3, Fed. Register. 54(168), 36133.
6. National Institutes of Health, Office of Laboratory Animal Welfare. 2013. Frequently Asked Questions G.6, Is post-approval monitoring required? http://grants.nih.gov/grants/olaw/faqs.htm#IACUC_2.
7. Applied Research Ethics National Association/Office of Laboratory Animal Welfare. 2002. *Institutional Animal Care and Use Committee Guidebook*, 2nd edition, 286. Bethesda: National Institutes of Health.
8. Office of Extramural Research. 2011. Notice NOT-OD-11-053, Guidance to Reduce Regulatory Burden for IACUC Administration Regarding Alternate Members and Approval Dates. http://grants.nih.gov/grants/guide/notice-files/NOT-OD-11-053.html.
9. Applied Research Ethics National Association/Office of Laboratory Animal Welfare. 2002. *Institutional Animal Care and Use Committee Guidebook*, 2nd edition, 22. Bethesda: National Institutes of Health.
10. National Institutes of Health, Office of Laboratory Animal Welfare. 2013. Frequently Asked Questions D.6, What criteria should the IACUC consider when reviewing protocols? http://grants.nih.gov/grants/olaw/faqs.htm#.
11. National Institutes of Health, Office of Laboratory Animal Welfare. 2013. Frequently Asked Questions D.10, Is the IACUC required to review the grant application? http://grants.nih.gov/grants/olaw/faqs.htm#IACUC_2.
12. Oki, G.S.F., E.D. Prentice, N.L. Garnett et al. 1996. Model for performing Institutional Animal Care and Use Committee continuing review for animal research. *Contemp. Topics Lab. Anim. Sci.* 35(4):53.
13. National Institutes of Health, Office of Laboratory Animal Welfare. 1993. Frequently Asked Questions about the Public Health Service Policy on Humane Care and Use of Laboratory Animals. http://grants.nih.gov/grants/olaw/references/ilar93.htm.
14. National Institutes of Health, Office of Laboratory Animal Welfare. 2013. Frequently Asked Questions D.9, What is considered a significant change to a project that would require IACUC review? http://grants.nih.gov/grants/olaw/faqs.htm#IACUC_2.
15. National Institutes of Health, Office of Laboratory Animal Welfare. 2013. Frequently Asked Questions D.3, What are the possible methods of IACUC review? http://grants.nih.gov/grants/olaw/faqs.htm#IACUC_2.
16. Wigglesworth, C. and C. Gipson. 2005. Alternate IACUC Members: What Are the Rules? *Lab Anim. (NY)* 34:12.
17. National Institutes of Health, Office of Laboratory Animal Welfare. 2013. Frequently Asked Questions B.2, May the IACUC have alternate members? http://grants.nih.gov/grants/olaw/faqs.htm#IACUC_2.
18. National Institutes of Health, Office of Extramural Research. 2010. Notice NOT-OD-10-081, Guidance on confirming appropriate charges to NIH awards during period of noncompliance for activities involving animals. http://grants.nih.gov/grants/guide/notice-files/NOT-OD-10-081.html.
19. National Institutes of Health, Office of Laboratory Animal Welfare. 2013. Frequently Asked Questions D.2, May the IACUC administratively extend approval of a project that has expired? http://grants.nih.gov/grants/olaw/faqs.htm#.

20. U.S. Department of Agriculture, Animal and Plant Health Inspection Service. 2011. Animal Care Policy Manual. Policy 12, Consideration of alternatives to painful/distressful procedures. http://www.aphis.usda.gov/animal_welfare/policy.php?policy=12.
21. National Institutes of Health, Office of Laboratory Animal Welfare. 2013. Frequently Asked Questions D.7, Should the IACUC consider the three "Rs" of alternatives when reviewing protocols? (Refinements to research, Reduction of animal numbers, and Replacement with non-animal models). http://grants.nih.gov/grants/olaw/faqs.htm#proto_7.

12

Justification for the Use of Animals

Larry Carbone

Introduction

PHS Policy and the AWA and its regulations call on the IACUC to review the investigator's proposal to use laboratory animals. Justification of the type and number of animals must be reviewed, as must exceptions to certain specific guidelines (including multiple major survival surgery, use or nonuse of painkilling drugs, and choice of any nonstandard euthanasia method).

The word *justification* may imply moral or ethical analysis to some, or at least some sort of cost/benefit analysis, but it is used more broadly in the regulations. For the most part, scientific justification (or the scientific rationale for choosing a particular experimental design) is what the regulations specify; there is little regulatory guidance in the type of justification expected. The case of multiple major survival surgery stands out in that other types of justification are considered by the *Guide* (p. 30): it is acceptable for the IACUC and investigator to consider "conservation of scarce animal resources" but not "cost-savings alone" as justification for multiple surgeries.

While the regulatory language could be read to imply that justification is all-or-nothing, matters are rarely so simple. IACUCs must deliberate how strong a justification is required to allow the use of animals, especially for animals at risk of significant unalleviated pain or distress.

12:1 What is the regulatory definition of an *animal*?

Reg. PHS Policy (III,A) defines an *animal* as "any live, vertebrate animal used or intended for use in research, research training, experimentation, or biological testing or for related purposes." NIH/OLAW explicitly states that unhatched embryos of egg-laying species may have vertebrae, but are not covered by PHS Policy, and that presumably goes for fetal mammals *in utero* as well. Larval forms of amphibians and fish are considered vertebrates.[1] Zebrafish hatch from their chorion at approximately 72 hours post-fertilization.[2]

The AWAR (§1.1, Animal), and AWA (§2132,g) define an *animal* as any live or dead dog, cat, nonhuman primate, guinea pig, hamster, rabbit, or any other warm-blooded animal, which is being used or is intended for use for research, teaching, testing, experimentation, or exhibition purposes, or as a pet. This term excludes birds, rats of the genus *Rattus*, and mice of the genus *Mus*, bred for use in research, and horses not used for research purposes and other farm animals, such as, but not limited to, livestock or poultry used or intended for use as food or fiber, or livestock or poultry used or intended for use for improving animal nutrition, breeding, management, or production efficiency, or for improving the quality of food or fiber. With respect to a dog, the term means all dogs, including those used for hunting, security, or breeding purposes. Thus, under the AWA, the overwhelming majority of animals used in laboratories (common laboratory rats, mice, and increasingly, zebrafish) are not regulated under the AWA. Fetuses and unhatched birds are not regulated under the AWA.

Opin. Institutions are free to expand upon the regulatory definition of animal and charge their IACUC to cover more animals in more settings than the minimum required by regulation. They may review protocols for non-AWAR-covered vertebrates, whether the animals are on publicly funded research or not. They may include embryos or non-vertebrates at their discretion.

Surv. Does your IACUC review protocols for species not covered by the Animal Welfare Act or the PHS Policy (e.g., cephalopods)?

- Not applicable 70/295 (23.7%)
- Yes, routinely 80/295 (27.1%)
- Yes, but only upon request 14/295 (4.8%)
- No 128/295 (43.4%)
- Other 3/295 (1.0%)

12:2 What is the regulatory definition of *justification*?

Reg. Justification is not defined in the PHS Policy, the AWA, the AWAR or the *Guide*. In the AWAR, for example, using animals in more than one major survival operative procedure (AWAR §2.31d,1,x,A) must be "justified for scientific reasons." The *Guide* (pp. 30, 64) also refers to the justification of certain practices for scientific reasons, such as singly housing social species or performing multiple surgeries on the same animal. Cost-savings *alone* [emphasis added] is cited in the *Guide* (p. 30) as an inadequate reason (i.e., justification) for performing multiple major survival surgical procedures on animals but cost savings is not mentioned as adequate or inadequate for other issues. The *Guide* (p. 30) states that "conservation of scarce animal resources may justify the conduct of multiple major surgeries on a single animal." For rodent caging, NIH/OLAW in FAQ F.10 writes: "Blanket, program-wide departures from the *Guide* for reasons of convenience, cost, or other non-animal welfare considerations are not acceptable" and more generally, in FAQ G.11 "Animal welfare and the integrity of research findings, rather than cost alone, should be the primary factors in decisions related to assuring compliance with the recommendations in the *Guide* in PHS-funded research."[1]

Opin. *Justification* can range from strong moral defense of an action to a much simpler rationale or explanation. As an example, an investigator's findings that anti-inflammatory medications may skew data may be the proposed justification (rationale) for withholding this form of pain management. That information alone does not morally justify conducting a particular painful experiment without full pain management. The moral justification requires both more facts (e.g., How painful is the experiment? To what extent will analgesics skew the data? What will the experiment accomplish if successful?) and clear articulation of values (How great must the potential benefit be to allow some or much pain?).[3]

12:3 In what ways is it required that the use of animals be justified on an IACUC protocol?

Reg. Institutions using animals regulated by the AWAR (see 12:1) must assure the following:

- That all proposals to the IACUC contain a rationale for involving animals, and for the appropriateness of the species and numbers of animals to be used (AWAR §2.31,e,2).
- A description of procedures designed to assure that discomfort and pain to animals will be limited to that which is unavoidable for the conduct of scientifically valuable research, including provision for the use of analgesic, anesthetic, and tranquilizing drugs where indicated and appropriate to minimize discomfort and pain to animals; (AWAR §2.31,e,4).
- That the PI has considered alternatives to procedures that may cause more than momentary or slight pain or distress to the animals and has provided a written narrative description of the methods and sources (e.g., the Animal Welfare Information Center) used to determine that alternatives to animal use are not available (AWAR §2.31,d,1,ii).
- That the PI has provided written assurance that the research activities do not unnecessarily duplicate previous experiments (AWAR §2.31,d,1,iii) (more detailed discussion of alternatives in 12:6–12:13).

The HREA (§495,c,2) requires applicants for grants, contracts, or cooperative agreements involving research on animals to submit "a statement of the reasons for the use of animals in the research to be conducted with funds provided under such grant or contract." Applications and proposals (competing and noncompeting) for awards submitted to PHS that involve the use of animals are required to specify the species and approximate number of animals to be used, the rationale for involving animals, and the appropriateness of the species and numbers to be used (PHS Policy IV,D,1,a; IV,D,1,b).

The *Guide* (p. 25) states, "The following topics should be considered in the preparation and review of animal care and use protocols:

- Rationale and purpose of the proposed use of animals.
- Justification of the species and number of animals requested. Whenever possible, the number of animals requested should be justified statistically.
- Availability or appropriateness of the use of less-invasive procedures, other species, isolated organ preparation, cell or tissue culture, or computer simulation."

12:4 Should the IACUC perform an ethical review of animal-use protocols?

Reg. The AWA, AWAR, and PHS Policy do not use the words *ethical* or *moral*. The *Guide* (pp. 25–33) lists several functions of an IACUC; ethical review is not one of those. For certain procedures, the *Guide* (p. 27) states, "the IACUC is obliged to weigh the objectives of the study against potential animal welfare concerns."

Opin. A societal consensus ethic underlies both the use of animals and constraints placed on that use. The *NASA Principles for the Ethical Care and Use of Animals* state that "The advancement of biological knowledge and improvements in the protection of the health and well-being of both humans and other animals provide strong justification for biomedical and behavioral research."[4] That permissive stance is balanced against constraints: "It is wrong to inflict harm on individuals without strong justification."[5] The governmental/societal belief that animal use can be ethically justified leads not just to permitting of, but to active funding of animal studies. The belief that such use *must be* justified leads to the establishment of IACUCs and other safeguards to animal welfare.

Animals have limited rights under this system in that their interests must be weighed in any cost–benefit evaluation that occurs. Potential benefits (primarily in the form of human and animal health, welfare, and basic knowledge) are weighed not just against financial costs (the limited funds disbursed by funding agencies or the research dollars in an industrial setting) but also against costs to animal welfare. Funding agencies and corporate administrators may weigh financial costs against potential benefit in distributing funds, but there is only limited weighing of benefit against harm to animals in any setting.

The 2011 *Guide* (p. 27) introduced new language that the IACUC is "obliged to weigh the objectives of the study against potential animal welfare concerns." AAALAC's posted guidance on this new requirement—that the IACUC "will weigh the potential adverse effects of the study against the potential benefits that are likely to accrue as a result of the research"[6]—transposes weighing objectives against welfare to weighing welfare against likely benefits. It remains unclear whether the IACUC is now to somehow score objectives (perhaps, some scientific questions justify less animal use than others), or evaluate the project's likely success (scientific merit) in meeting those objectives, or simply to continue the IACUC's long-standing commitment to assessing whether the procedures are indeed those necessary for the stated objectives.

Despite the new language of the *Guide*, for the most part the IACUC does not and cannot conduct this explicit ethical review. The IACUC is charged with reviewing the rationale (preferably statistical) for the animal numbers chosen, for instance, but not whether a particular line of research warrants that number. Similarly, the IACUC evaluates a technical claim that nonhuman primates alone are likely to provide the sort of data sought, not whether a particular project ethically merits the use of primates. Because the IACUC does not have the tools (or the regulatory mandate) to conduct a thorough assessment of the scientific merit (i.e., the potential benefits) of a proposed project, it cannot make a thorough cost–benefit ethical analysis. (See 9:14–9:18.)

The IACUC and researchers operate within a societal ethical context in which several principles are evident. These principles include the ideal that animals should only be used, and especially should only be harmed, for the highest-quality and most important research. Thus, U.S. Government Principle II states, "Procedures involving animals should be designed and performed with due consideration of their relevance to human or animal health, the advancement of knowledge, or the good of society."[20]

Beyond this, harm to animals must be minimized through consideration and implementation of replacement, reduction, and refinement alternatives. An ethical hierarchy of species is built into the regulations and guidelines: human good, even the pursuit of basic knowledge with no evident application, is considered sufficient to justify at least some use of some animals, while a burden is placed on investigators to attempt to replace more sentient species with less sentient species.

Though IACUCs generally avoid putting themselves in the position of being ethical arbiters, most probably venture into that territory when a proposal appears to push ethical boundaries. In studies conducted by Franz Stafleu, Rebecca Dresser, and others, IACUCs are most likely to reject or return protocols when at least two of these three conditions apply: there appears to be a great deal of pain or distress anticipated; the scientific project's worth seems particularly suspect; and especially, when the animal species is one (such as primates) that is of very high concern.[7–9]

Surv. Does your IACUC consider the balance between animal welfare and scientific (or educational) importance while reviewing a proposed animal activity?

- Yes, this is routinely incorporated into the protocol review 254/296 (85.81%)
- Yes, but only for species covered by the AWA 5/296 (1.7%)
- No 9/296 (3.0%)
- We have no specific policy on this 25/296 (8.5%)
- Other 3/296 (1.0%)

12:5 How can the IACUC review scientific merit as a component of the justification of use of animals?

Reg. Scientific merit includes both the importance of the scientific question being asked and the likelihood that the proposed project will answer it. U.S. Government Principle II states, "Procedures involving animals should be designed and performed with due consideration of their relevance to human or animal health, the advancement of knowledge, or the good of society." It does not state that an IACUC is the body to make this assessment, but rather that "the responsible Institutional Official shall ensure that these principles are adhered to." Principle IX, however, in allowing exceptions, does in fact place some responsibility back on the IACUC, to review exceptions to the other standards in accord with Principle II: "Where exceptions are required in relation to the provisions of these Principles, the decisions should not rest with the investigators directly concerned but should be made, with due regard to Principle II, by an appropriate review group such as an institutional animal care and use committee." Note that the Principles were written before the 1985 laws that mandated IACUCs for APHIS/AC-registered facilities and for institutions with Assurances on file with NIH/OLAW.

The proscription against unnecessarily duplicating existing work, as found in AWAR §2.31,d,1,iii, relates directly to assessing the importance of the question being asked, and this is a consideration the IACUC is charged with reviewing. PHS policy and AWAR charge the IACUC with reviewing some ancillary components of the project's likelihood of answering the question, such as the qualifications of the personnel involved. Neither requires that the IACUC review whether the basic methodology proposed is likely to answer the research question. (See 9:15.)

The AWA does not call for scientific merit review by the IO, the IACUC or other bodies. It requires that pain be limited to that "which is unavoidable for the conduct scientifically valuable research" (AWAR §2.31,e,4) but does not charge the IACUC with determining what research is scientifically valuable.

NIH/OLAW has provided some guidance on this topic in FAQ D.12[1], at least for PHS-funded research:

- Peer review of the scientific and technical merit of an application is considered the purview of the NIH Scientific Review Groups (SRGs), which are composed of scientific experts from the extramural research community in a particular area of expertise. However, SRGs also have authority to raise specific animal welfare concerns that can require resolution prior to a grant award…
- Although not intended to conduct peer review of research proposals, the IACUC is expected to include consideration of the U.S. Government Principles in its review of protocols. Principle II calls for an evaluation of the relevance of a procedure to human or animal health, the advancement of knowledge, or the good of society. Other PHS Policy review criteria refer to sound research design, rationale for involving animals, and scientifically valuable research. Presumably a study that could not meet these basic criteria is inherently unnecessary and wasteful and, therefore, not justifiable.
- The primary focus of the SRG is scientific merit and the primary focus of the IACUC is animal welfare. The two bodies have differing constitutions, mandates and functions. However, since it is not entirely possibly to separate scientific value from animal welfare some overlap is inevitable. SRGs may raise concerns about animal welfare and IACUCs may question the scientific rationale or necessity for a procedure.

Opin. Any review of the ethical justification of animal use includes some cost–benefit reckoning: is the potential pain, death, or distress of animals worth the benefits that will accrue to the potential beneficiaries of the knowledge gained? Thus, scientific merit certainly is an inherent component of the justification of the use of animals, as non-meritorious research can hardly justify harming animals. However, the IACUC is not necessarily the body to conduct this review. Depending on the field of research under consideration and the makeup of the IACUC, there may not be sufficient expertise to adjudicate at other than a superficial level the value of the research question or the methodology proposed. Even when one or two IACUC members have relevant expertise, this expertise may be inferior to that of a panel of experts reviewing a project. For these reasons, IACUCs may defer in many cases to peer review panels at funding agencies.

The IACUC may also review and approve studies that have not had a peer review process at a funding agency. For example, internal funds may be available to help investigators perform preliminary work necessary for writing a successful grant application. In such circumstances, some in-house departmental review committee or ad hoc expert consultation may be sought. The IACUC may decide to require and accept such reviews rather than try to assess the scientific merit on its own. Roughly 80% of survey respondents report that either the IACUC or another in-house body assesses the merit of projects that have not had external merit review. (See survey 9:14.)

While the IACUC may not be the body that assesses the scientific merit of a project, it should know that peer review has been performed. The IACUC needs to know that this review was favorable and done by a body the IACUC finds credible.

The IACUC may be reviewing a project that has not had scientific merit review, that was judged meritorious but not strong enough to beat out the competition for funding, or that was externally reviewed and found to be non-meritorious.

Although the IACUC may defer to funding agencies' favorable reviews without requiring a look at the peer reviewer's feedback, for projects that were not awarded funding, the IACUC may find it useful to look at the reviewers' positive and negative comments. These could be especially useful to a departmental committee reviewing one of their colleagues' proposals. Unlike limited grant dollars that are intended only for the very best proposals, there is no theoretical limit to the number of studies an IACUC or a departmental review committee can approve. With detailed funding agency comments in hand, the departmental committee and the IACUC would know whether they are being asked to approve a project that has not been deemed good enough for external funding and could know whether the protocol at hand has been revised to address the funding agency's concerns.

"Merit," like harm to animals, is not all-or-nothing but exists on a continuum. IACUCs, even if they have no formally explicit policy, may develop strategies for dealing with protocols of middling merit, especially if the project entails great harm to highly sensitive species. They may insist on pilot studies, or only approve the least harmful components of the project. They may withhold approval altogether until the project has been reconceived and undergone another round of merit review.

12:6 What topics does the *Guide* highlight for the IACUC's special consideration and weighing of objectives against potential animal welfare concerns?

Reg. The *Guide* (p. 27) states that "the topics below are some of the most common requiring special IACUC consideration" but does not rule out special consideration of other topics:

- Experimental and humane endpoints
- Unexpected outcomes
- Physical restraint
- Multiple survival surgical procedures
- Food and fluid regulation
- Use of non-pharmaceutical-grade chemicals and other substances
- Field investigations
- Agricultural animals

12:7 What justifications for using vertebrate animals are considered appropriate?

Reg. The AWAR (§2.31,e) require that a proposal to conduct an activity involving animals, or to make a significant change in an ongoing activity involving animals, must contain the following:

- Identification of the species and the approximate number of animals to be used (§2.31,e,1)
- A rationale for involving animals, and for the appropriateness of the species and numbers of animals to be used (§2.31,e,2)

In addition, the IACUC is specifically tasked with reviewing the investigator's consideration of alternatives and assuring that the work is not unnecessarily duplicative (AWAR §2.31,d,1,ii; §2.31,d,1,iii), that animal pain is being prevented or treated to limit it to what is unavoidable for scientifically valuable research (AWAR §2.31,e,4), that living conditions and veterinary care are appropriate (AWAR §2.31,d,1,vi; §2.31,d,1,vii), that standards for sterile surgery are met (AWAR §2.31,d,1,ix), and that humane euthanasia techniques are employed (AWAR §2.31,d,1,xi).

PHS Policy (IV,D,1,a; IV,D,1,b) requires most of the same information. U.S. Government Principle III states, "The animals selected for a procedure should be of an appropriate species and quality and the minimum number required to obtain valid results. Methods such as mathematical models, computer simulation, and in vitro biological systems should be considered."

The *Guide* (pp. 25–26) expands on this by recommending that "the following topics should be considered in the preparation of the protocol by the researcher and its review by the IACUC:

- Rationale and purpose of the proposed use of animals.
- Availability or appropriateness of the use of less-invasive procedures, other species, isolated organ preparation, cell or tissue culture, or computer simulation.
- Justification of the species and number of animals proposed; whenever possible, the number of animals and experimental group sizes should be statistically justified (e.g., provision of a power analysis;...)."

Opin. The IACUC must gain a good sense that the use of animals is justified by the research question being posed and that the methods chosen to answer it are likely to yield success. Peer review for scientific merit addresses much of this set of concerns (i.e., "Is this project worth this use of animals?"). This is generally not performed by the IACUC itself, but the IACUC needs to know whether this expert review has occurred.

It remains that the brunt of the IACUC's work is evaluating the scientific or technical reasons for using animals, especially in ways that may cause pain or distress, and in particular, the investigator's efforts to minimize pain and distress, through his or her search for alternatives to animals and to painful procedures.

12:8 How can the IACUC evaluate the justification for use of a particular species of animal?

Opin. After the decision is made that non-animal models will not suffice, the IACUC must review the rationale for the particular species (and possibly, strain) to be used. Choice of species is one of the aspects of a project that expert peer reviewers should evaluate, and for the most part, the IACUC places some trust in this review. Legitimate factors include the following:

- Is the species an established model in the literature?
- Is the species an appropriate size for the samples required or the procedures performed?
- Are there particular introduced or natural genes in the species of interest?

- Can the species be maintained easily, safely, and humanely in the animal facility?
- Is the species choice mandated by a regulatory or funding agency?
- Is it an endangered species?

Regardless of the peer review at the funding agency, the PI should explain the choice of species to the IACUC, and the IACUC may have concerns beyond those of the peer review committee. Is the facility appropriate for maintaining a particular species of animal? Do the investigator and support staff have sufficient expertise in working with this species? Though the peer review process may be better suited for evaluating the search for replacement with less sentient or nonsentient subjects, the *Guide* (p. 25) directly charges the IACUC with reviewing this consideration as well.

The author does not consider species differences in regulatory status a sufficient justification for species choice. For example, if hamsters were the better species choice on all of the above criteria, compromising these standards and selecting mice simply to avoid AWAR coverage would not be justified.

12:9 How much detail should the IACUC require relative to animal identification in the IACUC protocol and in records? For example, should the IACUC request that Fischer 344 rats be identified as "rats" or *Rattus norvegicus* or "F344/Crl rat"?

Reg. The *Guide* (p. 25) lists choice of species, not strain, stock, genotype or phenotype, as an object of IACUC review. Other sections of the *Guide* (pp. 27, 31, 121) note times when the strain or stock of animals should be considered and note (p. 75) that this strain, stock and genotype information should be on cage cards.

Opin. While nothing specific in the AWAR, the PHS Policy, or the *Guide* requires identification of animals by strain on IACUC protocols, there are circumstances where consideration of phenotype is essential. Often, the reason a given species is used is the special characteristics of a particular strain or construct. Likewise, many of the challenges of maintaining animal welfare relate to genetically-based developmental problems or susceptibility to illness. When such strains are to be used, if one is to justify animal use, those special characteristics must be described and plans to manage a phenotype that may cause pain or distress must be outlined.

IACUCs have broad authority and if they want to require that investigators use official nomenclature of strains on IACUC applications, or scientific instead of common species names, they are free to do so. The strain or construct should be named by using appropriate nomenclature (see *Guide* p. 75), though this is far less important to the goal of animal welfare than careful characterization of the phenotype.

12:10 How can the IACUC evaluate the justification for causing animals significant unalleviated pain or distress?

Reg. The AWAR (§2.31,d), based on the AWA (§2143,a,3,A), require the IACUC to determine that the proposed activity meet the requirement that "Procedures that may cause more than momentary or slight pain or distress to the animals will… be performed with appropriate sedatives, analgesics or anesthetics, unless withholding such agents is justified for scientific reasons, in writing, by the principal investigator and will continue for only the necessary period of time." This information must be provided to the IACUC (AWAR §2.31,e,4). APHIS/AC Policy 11 details that agency's policy on procedures that may cause more than slight or momentary pain or distress to animals. For pain, it is "pain in excess of that caused by injections or other minor procedures." For nonpainful distress, it lists examples of distressful procedures that *may* meet APHIS/AC criteria:[10]

- Food and/or water deprivation or restriction beyond that necessary for normal presurgical preparation.
- Noxious electrical shock or thermal stress that is not immediately escapable.
- Paralysis or immobility in a conscious animal.
- Forced exercise (e.g., swimming or treadmill protocols).
- Infectious and inflammatory disease models.

The AWAR (§2.36,b,7) require research institutions to file an annual report, including in Column E: "the common names and the numbers of animals upon which teaching, experiments, research, surgery, or tests were conducted involving accompanying pain or distress to the animals and for which the use of appropriate anesthetic, analgesic, or tranquilizing *drugs would have adversely affected the procedures, results, or interpretation of the teaching, research, experiments, surgery, or tests* [emphasis added]. An explanation of the procedures producing pain or distress in these animals and the reasons such drugs were not used shall be attached to the annual report." APHIS/AC Policy 11 includes examples of painful and distressful procedures and Appendix A of the *Animal Welfare Inspection Guide* illustrates when animals should be reported in Column C, D, or E.[10,11,24]

PHS Policy (IV,C,1,b) similarly calls on the IACUC to determine that "Procedures that may cause more than momentary or slight pain or distress to the animals will be performed with appropriate sedation, analgesia, or anesthesia, unless the procedure is justified for scientific reasons in writing by the investigator." It does not set thresholds for "momentary or slight" pain or distress. PHS Policy (IV,C,1,a; IV,C,1,c–d) states additional compliance requirements for research projects.

The *Guide* (p. 5) states that studies that will cause severe or chronic pain or distress should include descriptions of appropriate humane endpoints or provide science-based justification for not using a particular, commonly accepted humane endpoint. The *Guide* does not specifically address situations in which the IACUC might review an investigator's plans to withhold perioperative analgesics or other such sources of transient (but not momentary) pain.

Opin. Roughly 7.6% of AWAR-covered animals were reported in Column E in 2007, and there is no reason to assume rats and mice are used less in painful studies than are the AWAR-regulated species.[3,12] Policy 11 sets a fairly low threshold of what should be considered "more than momentary or slight pain or distress."[10] When procedures cross this threshold and analgesic, anesthetic, or tranquilizing drugs are not used due to research requirements, those procedures are reported in Column E. APHIS/AC does not subdivide Column E studies that have crossed its threshold, so they may range from animals undergoing an 18-hour fast through animals receiving no post-operative analgesics for a minor procedure to animals on chronic studies of painful metastatic cancers. Certainly, even without regulatory guidance or mandate, IACUCs should scrutinize the justification for some Column E studies more closely than others and require a higher justification standard for some.

IACUCs should ask what it means that analgesic or other drugs might "adversely affect" data. Experiments performed with analgesics administered or withheld may yield different results, but "in studies where the use of certain analgesics appears to be contraindicated, investigators should be mindful that unwanted variables from pain-induced perturbation of homeostatic mechanisms can affect the animal model."[13]

A literature search or expert consultation examining the potential adverse effects of pain treatments on the data should likewise explore the potential effects of untreated pain or distress before deciding that withholding pain management is justified. Specific data may not be available for every model, and the IACUC must judge whether to approve such studies only when interference by analgesics has been demonstrated, or when such interference seems plausible but uncertain.

Surv. When an investigator proposes a study in which pain-relieving medications will be withheld during potentially painful procedures, what does your IACUC typically require as scientific justification that analgesic or anesthetic drugs would adversely affect the scientific project?

- Not applicable — 37/296 (12.5%)
- No references are required but the investigator must make a statement that analgesic or anesthetic drugs may interfere with the scientific project — 24/296 (8.1%)
- No references are required but the investigator must make a statement describing how all specific classes of analgesic or anesthetic drugs that could alleviate pain or distress will interfere with the scientific project — 39/296 (13.9%)
- The investigator must make a statement describing how all specific classes of analgesic or anesthetic drugs that could alleviate pain or distress will interfere with the scientific project, and provide references — 140/296 (47.3%)
- Any of the above justifications, plus the investigator must include parallel information on his/her review of how untreated pain or distress may affect the scientific project — 52/296 (17.6%)
- None of the above — 3/296 (1.0%)
- Other — 1/296 (.3%)

12:11 What outside data must an IACUC consider when evaluating an investigator's proposed justification to use animals?

Reg. Under the AWA (§2143,a,3,B) and the AWAR (§2.31,d,1,ii), the IACUC is required to assure that "the principal investigator has considered alternatives to procedures that may cause more than momentary or slight pain or distress" and that "the principal investigator has provided written assurance that the activities do not unnecessarily duplicate previous experiments" (AWAR, §2.31,d,1,iii). Both imply at least some evaluation beyond the investigator's laboratory for comparative information.

The *Guide* (pp. 25–26) and the HREA (§495,c,1,B) have similar requirements that an investigator consider alternatives and prevent unnecessary duplication of experiments. On the topic of alternatives, NIH/OLAW has written in their FAQ D.7 that "The federal mandate in U.S. Government Principle IV to avoid or minimize discomfort, pain, and distress in experimental animals, consistent with sound scientific practices, is synonymous with a requirement to implement refinements (e.g., less invasive procedures or the use of analgesia). Similarly, the mandate in U.S. Government Principle III to use the minimum number of animals necessary to obtain valid results is synonymous with a requirement to reduce animal numbers. U.S. GovernmentPrinciple III further states that mathematical models, computer simulation, and in vitro biological systems should be considered, and is synonymous with a requirement to replace non-animal models wherever possible. Thus, consideration of the three "Rs" should be incorporated into IACUC review, as well as other aspects of the institution's program (e.g., investigator training)."[1] Of particular importance, the *Guide* (p. 25) recommends that animal care and use protocols should consider the "availability or appropriateness of the use of less-invasive procedures, other species, isolated organ preparation, cell or tissue culture, or computer simulation." PHS Policy (IV,C,1,a) states: "In order to approve proposed research ... the IACUC shall determine that... procedures with animals will avoid or minimize discomfort, distress, and pain to the animals, consistent with sound research design."

The regulations do not require the IACUC to review what granting agencies or other peer-review bodies have determined the merit of the project to be. They do not require the IACUC to read the citations found in the investigator's search for alternatives, or to read the investigator's publications from prior approved protocols.

Opin. The investigator must describe literature searches, consultations, and other sources to assure the IACUC that she or he has done sufficient research to justify the following assertions:

- That the work proposed does not unnecessarily duplicate previous work, whether published or not.
- That replacements to the use of live sentient animals, particularly in studies that may cause animal pain or distress, have been sought and considered.
- That alternative methods to reduce the numbers of animals used, particularly in studies that may cause animal pain or distress, have been sought and considered.
- That refinements to experimental procedures that may cause animal pain or distress have been sought and considered.

It is essential to recognize the limitations to published literature using the proposed models, as Materials and Methods (M&M) sections are rarely detailed enough to seriously address all potential refinements. At times, M&M may not even mention the anesthesia used for a surgical model, creating the false impression that anesthesia or analgesia cannot be used.[14]

IACUCs might learn useful information in reading reviewer comments on grant applications. The investigator can be asked to provide these for IACUC review.

12:12 What is meant by the search for alternatives?

Reg. The *Guide* (p. 25) calls on investigators and IACUCs to consider "less invasive procedures, other species, isolated organ preparation, cell or tissue culture, or computer simulation," but does not provide guidance or set standards for how to make or evaluate that consideration. The AWA (§2143,a,3,B; §2143,d,3; §2143,e,3) and the AWAR (§2.31,d,1,ii) go one step further by mandating that the investigator not just consider, but actively search for such alternatives, that the IACUC determine that the investigator has considered such alternatives, and if identified alternatives will not be used, how s/he determined they could not be used. The AWAR focus is on alternatives to "procedures that may cause more than momentary or slight pain or distress to the animals."

Opin. Russell and Burch provided a useful framework for considering alternatives to the use of live sentient animals and to the use of painful or distressful experiments.[15] They elaborated the sometimes-overlapping, sometimes-contradictory "3Rs" framework of replacement, reduction, and refinement alternatives. APHIS/AC Policy 12 clarifies that APHIS/AC adopts this framework and considers "alternatives" to include "some aspect of replacement, reduction, or refinement in pursuit of the minimization of animal pain and distress consistent with the goals of the research."[16] U.S. Government Principles III–V and the *Guide* (pp. 4–5) call for similar consideration, invoking Russell and Burch's language. An investigator searches for alternatives *to* animal research by looking for non-animal models that would replace the use of sentient animals, and for alternatives *in* animal research by searching for ways to reduce animal numbers and to reduce animal pain and distress. (See 16:4.)

12:13 What is meant by replacement alternatives?

Opin. Replacement alternatives are conceptually the most straightforward: find ways to generate research data without using sentient animals at all. Candidates for consideration include studying cells in tissue culture (*in vitro* techniques), microorganisms, computer simulations, making better use of human epidemiological data and human volunteers, or using inanimate models.[5] It can include replacing sentient animals, such as mammals, with animals presently thought to be much less sentient, such as insects or protozoa. It might include using slaughterhouse material, or cadavers of animals euthanized for purposes other than research or teaching. A paradigmatic example is the production of monoclonal antibodies in cell culture, as opposed to growing them in mouse ascites fluid. Though animals may still be the source of the cells in culture, the use of whole live sentient animals in a potentially painful way is replaced with the in vitro technology.

More problematic is the attempt to replace more sentient animals with less sentient animals (this author strongly resists the terms *higher animal* and *lower animal* or to reference to a *"phylogenetic scale,"* as none of these terms have precise biological definitions. Despite this, "lower on the phylogenetic scale" is the term in the *Guide* [p. 5]). Are frogs, fish, and octopi all reliably known to be less sentient than mammals? Among the mammals, are there any data that rodents are significantly less sentient than dogs, swine, or even primates? For the purposes of this chapter, *sentience* is defined as the capacity to experience pain and suffering. It is more than simply the ability to sense stimuli (even noxious stimuli) and less than the complex self-awareness and self-consciousness possessed by humans. There is no biological reason to assume that it is an all-or-nothing trait, possessed equally, in full, by all animals that possess it. Rather, it exists on a continuum across species, with monkeys, apes, and dolphins, as far as we know, more sentient than fish or squid.[17] Among mammals, while some species are clearly more self-conscious and more intelligent than others, the author is less confident stating that rodents differ in sentience from dogs or swine or even primates and cetaceans. (See 13:3.)

Surv. When considering the use of a "less sentient" animal as an animal model, which of the following statements is applicable to your IACUC? (check as many as appropriate)

• Not applicable	43/297(23.7%)
• We have no definition of "sentient"	106/297 (35.7%)
• Of a lower phylogenetic class (i.e., mammal > bird > reptile > amphibian > fish > invertebrate	95/297 (32.0%)
• Of a smaller size (e.g., a mouse is less sentient than a rat)	13/297 (4.4%)
• Of lesser innate intelligence (e.g., monkey is more sentient than a mouse	3/297 (1.0%)
• We consider all mammals as equally sentient	29/297 (9.8%)
• We consider all vertebrates as equally sentient	39/297(13.1%)
• I don't know	8/297 (2.7%)
• Other	2/297 (~1%)

12:14 What is meant by *reduction alternatives*?

Opin. Reduction includes all the efforts to lower the numbers of animals used. This often means rethinking statistical tests to use just the number necessary for statistically valid result. Reduction attempts may rely on refining the study, as when use of healthier, more genetically homogeneous animals lower in-group variability, allowing smaller number of animals in each group. Reduction efforts may rely on careful planning of statistical analyses, use of power analyses or other methods for projecting the minimum number of animals needed, and use of historical controls, when appropriate.[18] When concurrent controls are required (as is most typical), it means carefully thinking how many control groups are needed. Sharing tissues across several projects when euthanizing animals is another form of reduction.

12:15 What is meant by *refinement alternatives*?

Opin. Refinement alternatives are the most varied, as they comprise all the myriad ways to rethink animal care and use to reduce the potential for pain or distress. Some examples are listed below. The author recommends this exercise for investigators seeking refinements in their studies and uses it when reviewing IACUC protocols: First, conduct a birth-to-death welfare inventory: visualize every step in the animal's life from the initial point of contact (birth at the facility or arrival into the facility) until the animal's euthanasia or departure from the institution. List every reasonable potential source of significant pain or distress, whether related to housing or to the experimental procedures. Each of those potential sources of pain or distress should then be addressed, whether in a targeted literature search, consultation with peers who have published on the methodology, conducting pilot studies or consultation with veterinarians and animal care specialists and with as much creative thinking as the investigator can muster.

Some Refinement Alternatives to Reduce Pain or Distress in Animal Research

- Choice of experimental endpoints that precede onset of disease or mortality
- Improved use of anesthetics and painkillers
- Housing social animals in compatible groups
- Using flexible tethers to replace rigid restraint devices
- Replacing open surgery with endoscopic techniques
- Providing supportive veterinary care
- Maintaining infection-free animal colonies
- Designing cages that allow animals to dig, run, climb, hide
- Training animals to cooperate with research procedures
- Frequent monitoring of body weight or other indicators of well-being
- Using positive reinforcement in behavioral studies
- Killing animals using the least painful methods

Source: From Carbone, L., *What Animals Want: Expertise and Advocacy in Laboratory Animal Welfare Policy*, Oxford University Press, New York, 2004. With permission.

12:16 How do the "three Rs" of alternatives complement or conflict with one another?

Reg. The *Guide* (p. 5) states that "Refinement and reduction goals should be balanced on a case-by-case basis, and also that "Principal investigators are strongly discouraged from advocating animal reuse as a reduction strategy, and reduction should not be a rationale for reusing an animal or animals that have already undergone experimental procedures especially if the well-being of the animals would be compromised."

Opin. Replacement, reduction, and refinement are not mutually exclusive concepts. Replacing some of the animals in a study reduces the overall numbers. Refining sample collection techniques may reduce the potential for pain as well as the number of animals needed. For example, if an assay requiring 1000 μl of mouse blood can be replaced with one that requires 10 μl, it may become possible to study a small cohort of animals sequentially, rather than euthanizing mice for each data time point needed (1000 μl cannot generally be collected from a mouse on a non-euthanasia basis). Studying the smaller cohort sequentially may result in less intersubject variability and thus may lower the statistical need for animals at each time point of data collection. Moreover, 10 μl can be collected via the facial, tail, or leg veins in a less invasive procedure than the collection of 1000 μl would require.

On the other hand, attention must be paid to the risk of increasing welfare costs when pursuing alternatives. If an investigator reduces animal numbers by more intensively studying a smaller cohort, those animals may be at risk of undergoing more procedures per animal, and that may jeopardize their welfare. This is a concern noted in the *Guide* (p. 5). There are times when refinement can be more important than reduction, by spreading animal welfare costs over more animals.

Efforts to switch species can also lead to challenges. Replacing dogs with mice, for instance, can greatly increase the number of animals needed (for example, studies that require large volumes of blood collection at several time points may require a separate mouse for each time point sample, whereas a single dog could provide samples over several time points), and procedures that can be easily performed with hand restraint and minimal pain in dogs (again, blood collection is a good example) may require anesthesia and more invasive techniques in smaller subjects. There are many reasons to replace dogs, or even monkeys, with mice or rats (cost, public relations concerns, regulatory issues, training requirements, health and safety issues, as well as issues of experimental design). This author remains skeptical that mice are sufficiently less sentient than larger mammals for this concern alone to drive the replacement effort, especially when animal numbers and the invasiveness of procedures will dramatically increase.

12:17 How can the IACUC evaluate the investigator's search for and consideration of alternatives?

Reg. Under the AWA (§2143,a,3,B) and AWAR (§2.31,d,1,ii), the IACUC is required to assure that "the principal investigator has considered alternatives to procedures that may cause more than momentary or slight pain or distress" and "has provided a written narrative description of the methods and sources, e.g., the Animal Welfare Information Center used to determine that alternatives were not available."

APHIS/AC Policy 12 expands on this requirement, stating that APHIS/AC "continues to recommend a database search as the most effective and efficient method for demonstrating compliance with the requirement to consider alternatives to painful/distressful procedures," and that "when a database search is the primary means of meeting this requirement, the narrative should include:

1. The name(s) of the databases searched (due to the variation in subject coverage and sources used, one database is seldom adequate);
2. The date the search was performed;
3. The period covered by the search; and
4. The search strategy (including scientifically relevant terminology) used."

Policy 12 also includes provision for conditions in which database searching is not the primary route for researching alternatives:

> However, in some circumstances, (as in highly specialized fields of study), conferences, colloquia, subject expert consultants, or other sources may provide up-to-date information regarding alternatives in lieu of, or in addition to, a database search. Sufficient documentation, such as the consultant's name and qualifications and the date and content of the consult, should be provided to the IACUC to demonstrate the expert's knowledge of the availability of alternatives in the specific field of study. For example, an immunologist cited as a subject expert may or may not possess expertise concerning alternatives to in vivo antibody production.

Importantly, Policy 12 states: "If a database search or other source identifies a bona fide alternative method (one that could be used to accomplish the goals of the animal use proposal), the IACUC may and should ask the PI to explain why an alternative that had been found was not used. The IACUC, in fact, can withhold approval of the study proposal if the Committee is not satisfied with the procedures the PI plans to use in his study."

Note that this requirement applies not just to the initial review of a study, but to review of any proposed major modification to an ongoing study. Policy 12 further states, "Animal Care would normally expect the principal investigator to reconsider alternatives at least once every 3 years, consistent with the triennial de novo review requirements of the Public Health Service Policy on Humane Care and Use of Laboratory Animals (IV,C,5)."

The *Guide* (p. 25) is less detailed on this topic. It specifies that justification of animal numbers should be statistically justified, whenever possible. The availability of refinements and replacements are two of the topics that "should be considered in the preparation of the protocol by the researcher and its review by the IACUC."

Opin. Institutions should comply with APHIS/AC Policy 12, especially for AWA-regulated species, but should also recognize its limitations. The recommendation to use "a database search as the most effective and efficient method" for complying with the requirement to consider alternatives assumes that most of the relevant information is found in published, indexed articles. This author does not share that assumption. Very useful information can be found through consultation with others performing similar work and with veterinarians and animal care specialists who focus on laboratory animals. That said, the author believes it would be the rare project in which consultation with experts were the sole or even primary route to alternatives and would advise most investigators to include the electronic resources described later. The IACUC cannot thoroughly review literature searches if the only information it receives are the names of the databases, the key words, and the dates of the search.

Surv. 1 What does your IACUC accept as an appropriate means of searching for alternatives to painful or distressful procedures? (check as many boxes as appropriate)

- Not applicable 7/296 (2.4%)
- A single database search 120/296 (40.5%)
- At least two database searches 176/296 (59.5%)
- Consultation with experts in the field 142/296 (48.0%)
- Attendance at scientific meetings or study sections 64/296 (21.6%)
- Other 14/296 (4.7%)

"Other" responses varied, including veterinary consultation, investigators' experience and expertise, and submission of data.

Surv. 2 If a database search is typically used by researchers/educators at your institution to help find alternatives to painful/distressful procedures, what does your IACUC require for a "written narrative description of the methods and sources … used to determine that alternatives were not available?"

- Not applicable 11/295 (3.7%)
- Dates of search(es), database(s) searched, and search terms used 88/295 (29.8%)

• Dates of search(es), database(s) searched, search terms used, plus a description of what the search(es) produced	60/295 (20.3%)
• Dates of search(es), database(s) searched, search terms used, a description of what the search(es) produced, and rationale for rejecting some/all of the alternatives that the search produced	130/295 (44.1%)
• Other	6/295 (2%)

12:18 What sources, resources, and methods are useful when searching for alternatives to animal use?

Reg. The AWAR (§2.31,d,ii) and APHIS/AC Policy 12 mention the Animal Welfare Information Center (AWIC) as a useful resource, but do not mandate its use.

Opin. An information search that fully meets the intent of the AWAR requires several approaches, some of them almost opposites. Assuring, for example, that a proposed study of a particular cytokine's activity in intestinal lymphoid tissue does not unnecessarily duplicate other work requires a targeted search in the relevant immunological literature, with some strategy (such as conference attendance) for learning of abstracts, posters, and other evidence of work not yet published. That same immunological literature might provide some information relevant to the search for replacements to whole live mice for the study. However, if the proposed work requires a surgical approach to the intestines, then a far more general search on mouse surgery, anesthesia, and analgesia is warranted, one that should hardly be confined to the immunological literature.

There are dozens of American and international Internet-based resources for conducting a search for alternatives. Some of those of the most general interest are described in the following; specific disciplines also have specialized resources.

The 1985 AWA amendment that mandated IACUCs and the search for alternatives also established the Animal Welfare Information Center (AWIC) within the National Agriculture Library. AWIC maintains a Web site with many useful links (http://awic.nal.usda.gov). Its staff periodically updates and publishes "quick bibliographies"—literature searches on select topics of fairly general interest, such as "rodents." The site also includes information and tutorials, backed up by staff information specialists available for consultation on procedures to conduct literature searches.

For biomedical research, PubMed is certainly the most utilized database search engine. It is maintained by the National Library of Medicine and is accessed on the Web at http://www.ncbi.nlm.nih.gov/pmc/. Its search capabilities include "old Medline" as well as in-process citations not yet available in Medline.

Agricola is the National Agriculture Library's article citation database (http://agricola.nal.usda.gov/). Its agriculture and animal health focuses complement PubMed, and several journals directly related to animal welfare, not indexed in PubMed, are indexed in Agricola.

Altweb (http://altweb.jhsph.edu) is a project of the Center for Alternatives to Animal Testing at the Johns Hopkins Bloomberg School of Public Health. It has searches that can be conducted through its online interface (such as the Altweb Anesthesia/Analgesia Database), as well as links to several other databases.

The University of California Center for Animal Alternatives (http://www.lib.ucdavis.edu/dept/animalalternatives/mission.php) shares some features of Altweb and AWIC. It has some searches on specific topics available, in which the center librarian has set up a search template that the investigator can then run as a live search via the Internet. Most of these search templates are on fairly general topics (such as analgesia, blood collection, or euthanasia) and can be run through both PubMed and Agricola. The site also provides useful guidance on conducting an alternatives search.

The author finds great value in direct communication, especially for refining experimental procedures. Laboratory animal veterinarians and other professionals have access to the American Association for Laboratory Animal Science's Compmed list-serve and archives (http://www.aalas.org/). Online discussions allow shared anecdotal information on research animal care and use, and even shared animals and animal tissues, none of which would be retrieved through a literature search engine such as PubMed.

Investigators developing new experimental techniques should be encouraged not simply to read the Materials and Methods section of published papers, but to contact and visit laboratories conducting the procedure directly, to learn the subtle hands-on knowledge required to perform a procedure or maintain a particular line of animals competently or other applied information.

Finally, general commercial search engines (such as Yahoo, Bing, and Google, including Google Scholar) and medically focused commercial sites (such as Embase: http://www.embase.com) also are very useful at times, especially for identifying individuals and institutions that have experience relevant to particular animal models.

12:19 Must the IACUC assure that the animal use activity does not unnecessarily duplicate previously conducted work? If so, how is this done?

Reg. The AWAR (§2.31,d,1,iii) require that "the principal investigator has provided written assurance [to the IACUC] that the activities do not unnecessarily duplicate previous experiments." The *Guide* (p. 26) lists unnecessary duplication of experiments as a topic the investigator should consider in preparation of the protocol for IACUC review. Neither the AWAR nor the *Guide* requires that the IACUC make this assessment directly.

Opin. The IACUC must not solely accept the investigator's claim that the proposal does not entail unnecessary duplication but apparently should review how the investigator knows this to be true. Though the search strategy for non-duplication differs from the search for alternatives, a simple approach to documenting compliance is to include a statement on the investigator's signed protocol application that the work does not entail unnecessary duplication; in addition, the documentation for an alternatives search should include information on how that search would uncover unnecessary duplication.

The key word here is *unnecessary*. Duplication and replication are essential components of hypothesis-driven research and are permitted if they serve a valid scientific purpose, such as confirming another's work or validating laboratory assay systems. To determine whether the duplication is or is not necessary, the IACUC needs to evaluate the justification provided by the investigator who proposes the duplication.

12:20 Is the use of animals in teaching protocols considered necessary duplication?

Opin. There are no clear federal laws, regulations, or policies pertinent to this question. A genuine consideration of alternatives, not a simple statement that the protocol is for teaching, is warranted. The important question is not whether duplication (i.e., teaching a new cohort of students using procedures employed in previous semesters) is necessary, but whether animal use, especially invasive animal use, is necessary. A careful alternatives search on this topic would include consideration of published studies on the educational effectiveness of live animal versus other teaching methods and consultation with faculty at other institutions who are employing alternative methods. Though animal laboratories may be retained in the curriculum, refinements and reductions may be appropriate. These could include videotaping some animal laboratories to show future generations of students, requiring preliminary skills development with nonliving models prior to handling live animals, or delegating the most challenging part (e.g., pithing or anesthetizing an animal) to an experienced professor rather than an inexperienced roomful of students.

12:21 Since the AWAR definition of an animal includes dead animals (see 12:1), should the IACUC request a justification for the use of dead animals?

Reg. The PHS Policy does not require IACUC review and approval of the use of dead animals unless they are killed for the purpose of being used in PHS-supported activities (see NIH/OLAW FAQ A.3). In the latter case, the IACUC is really reviewing the use of live animals being euthanized for research, teaching, or testing and the focus is on animal care and handling before and during euthanasia, not after. The AWAR are less straightforward. While the statutory definition of *animal* includes dead animals used or intended for use in research, the definition of a *research facility* is one using or intending to use live animals (AWAR §1.1, Research Facility). Therefore, reference to committee review of use of animals would seem to apply solely to use of live animals (AWAR §2.31).

Opin. From a legal point of view, justification does not appear to be required. It is an excellent application of alternatives to make maximal use of euthanized animals by using them as teaching cadavers or distributing needed tissues to several investigators, rather than requiring each to purchase animals separately. IACUC review could slow the process and increase the burden of in-house tissue sharing with no improvement in animal welfare and should not be required. A middle position between total oversight and no oversight would be to allow in-house tissue sharing without case-by-case IACUC oversight but to require oversight if dead animals are being received through purchase or donation from outside the institution. Health and safety oversight is necessary if the cadaveric material comes from animals likely to be harboring zoonotic infections (e.g., macaques, sheep, and wild animals) or from animals on infectious disease or other hazardous studies. (See 8:7; 14:28.)

12:22 Should the IACUC request a justification for euthanasia of animals?

Reg. The AWAR (§1.1 Euthanasia; §2.31,d,1,i; §2.31,d,1,v), PHS Policy (IV,C,1,c; IV,C,1,g) and the *Guide* (pp. 123–124) require the IACUC to review the proposed method of euthanasia and stress the need to minimize pain and distress. Methods likely to cause animal pain or distress must be explicitly approved by the IACUC. Beyond this, there is no special justification required for killing animals (though the numbers of animals killed must be justified, as must the numbers of animals handled and used in any manner, including those used and later retired or adopted). While euthanasia need not be justified to the IACUC, refusal to euthanize animals that experience untreatable pain or distress does require very special justification. The AWA (§2413,a,3,C,v) requires that "that the withholding of tranquilizers, anesthesia, analgesia, or euthanasia when scientifically necessary shall continue for only the necessary period of time." U.S. Government Principle VI states, "Animals that would otherwise suffer severe or chronic pain or distress that cannot be relieved should be painlessly killed at the end of the procedure or, if appropriate, during the procedure."[19]

12:23 Should the IACUC request justification for the use of animals in field studies?

Reg. The AWAR (§1.1, Field Study) state that a field study is "a study conducted on free-living wild animals in their natural habitat. However, this term excludes any study that involves an invasive procedure, harms, or materially alters the behavior of an animal under study." Under the AWAR (§2.31,d,1) animals involved in some field studies are exempt from IACUC review but require the same standards as laboratory-based research if invasive techniques will be performed or the animals' behavior is altered. PHS Policy does not distinguish between field studies and laboratory studies; therefore, field studies require IACUC review and approval if covered under an Animal Welfare Assurance (see NIH/OLAW FAQ A.6). The *Guide* (p. 32) lists field investigations as one of the eight topics that are "the most common requiring special IACUC consideration" and calls attention to special welfare concerns (including return to wild or other disposition), to permitting requirements, and to human health and safety concerns. (See 15:9–15:12.)

Opin. Field investigations vary from hands-off observation to capturing, holding, marking, surgically altering or tissue collection. If invasive procedures occur in field studies, the AWAR exemption does not apply and they must be reviewed. "Materially altering" animal behavior is left undefined, including duration of any alteration. Studies that require capture and brief restraint for marking or tagging animals may not be invasive, but they may affect animal behavior (especially if anesthesia is used), and the marking or tagging method may itself affect the animals in various ways. Vertebrate animals in PHS funded field studies are not excluded from IACUC consideration in institutions covered by the PHS Policy, regardless of the invasiveness or behavioral effects of the study. For consistency, it is a good policy to grant the same review of justifications for field studies as for non-field studies. IOs should also remember that many tagging systems identify the institution placing the tag, and will want a record that the study was duly reviewed if animals with problems are later found and identified as institutional research subjects. This author would recommend that the IACUC, not just the investigator, determine whether a particular field study might pose significant risk to animals to warrant IACUC review.

12:24 Is a requirement by a funding or regulatory agency to use animals, or to use animals in particular ways, considered adequate justification?

Reg. APHIS/AC Policy 12 states:

> The rationale for federally-mandated animal testing (for example, testing product safety/efficacy/potency) should include a citation of the appropriate government agency's regulation and guidance documents. Mandating agency guidelines should be consulted since they may provide alternatives (for example, refinements such as humane endpoints or replacements such as the Murine Local Lymph Node Assay) that are not included in the Code of Federal Regulations. If a mandating agency-accepted alternative is not used, the IACUC must review the proposal to determine adequate rationales have been provided, and pain and discomfort limited to that which is unavoidable.[16]

NIH/OLAW has posted guidance clarifying the relative roles of Scientific Review Groups (SRGs) serving NIH funding agencies, and IACUCs. Briefly, if the SRG raises a specific "concern" about the vertebrate animal section of a grant application the NIH informs the applicant of that concern. The applicant must resolve the concern in conjunction with the NIH Program Officer. Once that happens the PI is expected to convey to the IACUC any modifications related to animal usage that may have emanated from the NIH review.[25]

Opin. Adjudication as to what is proper use of animals is not dictated *solely* by granting or regulatory agencies. The institution, through its IACUC, must determine whether the justification to use animals is appropriate, and the measure that IACUCs use is *scientific* justification. If a granting or regulatory agency requires the use of animals, or a particular species or strain, their reasons for doing so could be submitted to the IACUC for its consideration. However, the IACUC must base that decision on the parameters previously noted, not just on the policy of the granting agency. Other details of a protocol may also be mandated either by a granting agency or by a regulatory agency that will review the data. For example, approval of veterinary vaccines may require submission of vaccine trial data that use death as an end point. Though efficacy may be well demonstrated long before mortality, or even before significant morbidity, animal lives would be wasted on such a study if the efficacy data would not be accepted by the Department of Agriculture, which licenses such vaccines. Approving such a study therefore means approving potentially painful procedures when alternatives have been rejected not for scientific but for regulatory reasons.

There may be occasions when the IACUC finds itself in a bind, asked to approve agency-mandated procedures that the IACUC considers unjustified. An IACUC might strike a balance in which it approves some animal use, but also initiates its own correspondence to persuade the agency in question to change its mandate or, at least, better explain and justify it. In such a situation, it seems the IACUC may have little choice but to defer to the agency requirements, even if it disagrees with the agency on the justification.

Surv. Is a federal requirement to use animals for a safety test considered to be adequate justification for your IACUC to approve animal use?

- Not applicable 82/296 (27.7%)
- Yes 58/296 (19.6%)
- No 40/296 (13.5%)

- It depends on the "requirement" because some requirements are actually suggestions — 76/296 (25.7%)
- I don't know — 31/296 (10.5%)
- Other — 9/296 (3.0%)

12:25 Are there any federal requirements dictating that animals or certain species of animals must be used in certain forms of research or product safety testing?

Reg. The AWA and the HREA do not include mandates for animal testing of any sort. The Food and Drug Administration (FDA) and the Environmental Protection Agency (EPA) are the two federal agencies most responsible for the review of product safety testing. Testing requirements are complicated by the wide range of medicines, medical devices, environmental chemicals, and other products they regulate. For the most part, these agencies provide general guidance on the types of testing data required for licensing a product. The FDA's guidance on exploratory Investigational New Drug (IND) applications does not explicitly mandate animal use prior to approval to proceed with human studies, and indeed, emphasizes that there is flexibility in preclinical requirements depending on the proposed IND. The guidance goes on to provide details on what animal use would be typical; for example, for repeat-dose toxicology studies. The rat is the usual species chosen for this purpose, but other species might be selected. In addition to studies in a rodent species, additional studies in non-rodents, most often dogs, can be used to confirm that the rodent is an appropriately sensitive species.[20] For specific models, the FDA states what is customarily done, again without specifically mandating when alternative species or studies will suffice. For evaluating cardiovascular devices, for example, they write: "While the selection of an appropriate animal model is left to the sponsor, FDA has observed that dogs comprise the bulk of research involving cardiac pacing and models in which heart failure is necessary. FDA has also observed that small ruminants serve as a model for prosthetic heart valves and artificial circulatory support devices, and that small swine represent the majority of other cardiovascular research animals."[21] The FDA gives guidance on use of rabbits in pyrogen testing.[22] The FDA also recognizes consensus standards on product safety testing, in part to harmonize with other nations' regulatory practices.[22]

Opin. Individuals not already thoroughly familiar with testing data required for a particular type of product or device will find value in consultation with the relevant regulatory agency before initiating animal studies. Guidelines vary with the model and the agency and are under constant revision.

12:26 What is the position of the FDA and the Food, Drug and Cosmetics (FD&C) Act relative to using animals for cosmetics testing?

Reg. The FDA has stated:

> The FD&C Act does not specifically require the use of animals in testing cosmetics for safety, nor does the Act subject cosmetics to FDA premarket approval. However, the agency has consistently advised cosmetic manufacturers to employ whatever testing is appropriate and effective for substantiating the safety of their products. It remains the responsibility of the manufacturer to substantiate the safety of both ingredients and finished cosmetic products prior to marketing.

> Animal testing by manufacturers seeking to market new products may be used to establish product safety. In some cases, after considering available alternatives, companies may determine that animal testing is necessary to assure the safety of a product or ingredient. FDA advocates that research and testing derive the maximum amount of useful scientific information from the minimum number of animals and employ the most humane methods available within the limits of scientific capability... We also believe that prior to use of animals, consideration should be given to the use of scientifically valid alternative methods to whole-animal testing... FDA supports the development and use of alternatives to whole-animal testing as well as adherence to the most humane methods available within the limits of scientific capability when animals are used for testing the safety of cosmetic products.[26]

12:27 Should an IACUC approve development of an animal model if another well-established model exists, simply because the investigator has no experience with the established model?

Opin. The IACUC should evaluate each protocol on the basis of how well the investigator has justified the use of whichever species, strain, or construct is proposed. One issue is the appropriateness of the animal model. Crucial questions would include whether development of a new model required large numbers of animals or invasive work solely to validate the model, the costs to animal welfare in the investigator's learning to use the established model, and the likelihood—a question of scientific merit—of the new model's relating to the established literature sufficiently that results can be accepted through a peer review process.

12:28 Since many people believe that cephalopods are sentient and because they have a well-developed nervous system, does an IACUC have the authority to include such animals under its purview? What about other invertebrates?

Opin. Neither the AWAR nor PHS Policy covers the use of invertebrate animals. The *Guide* (p. 2) "generally" defines laboratory animals as vertebrate animals but states, with regard to invertebrates such as cephalopods that it "establishes general principles and ethical considerations that are also applicable to these species." The *Guide*'s bibliographies include reference material on invertebrates. There are no restrictions that either APHIS/AC or NIH/OLAW has placed on IACUCs related to oversight of animal care and use activities that exceed the scope of their regulations or policies. Therefore, an American institution can confer upon its IACUC the authority to oversee the care and use of invertebrates but is not presently required to do so. (See 31:4.)

12:29 Must the IACUC review and approve the use of animals destined to be used as a food source for other animals being used in IACUC-approved studies?

Opin. Studies of animal predation that involve feeding live or euthanized animals to others are, of course, research projects that must receive IACUC review and approval. Beyond that, the AWAR and PHS Policy do not explicitly call for IACUC approval for animals maintained as dietary items for other animals. NIH/OLAW has published guidance that in such a case, if the animals are not covered through the IACUC's protocol review process, the IACUC should oversee this use as "a covered component of the institutional program of animal care and use."[23] Also, see NIH/OLAW FAQ D.16. (See 8:33.) In addition to the welfare of the prey species, the *Guide* (pp. 84–85) cautions that their suitability as part of a complete diet, and their continuous availability must be ensured. It is essential to rule out their potential as infectious vectors as well.

References

1. National Institutes of Health, Office of Laboratory Animal Welfare. 2012. Frequently Asked Questions. http://grants.nih.gov/grants/olaw/faqs.htm.
2. Moorman, S.J. 2001. Development of sensory systems in zebrafish (Danio rerio). *ILAR J.* 42:292–298.
3. Carbone, L. 2011. Pain in laboratory animals: the ethical and regulatory imperatives. *PLoS One.* 6(9):e21578.
4. National Air and Space Administration. 2001. NASA Principles for the Ethical Care and Use of Animals. http://www.nal.usda.gov/awic/pubs/IACUC/nasa.htm#pri.
5. Carbone, L. 2004. *What Animals Want: Expertise and advocacy in laboratory animal welfare policy.* New York: Oxford.
6. Association for the Assessment and Accreditation of Laboratory Animal Care International. 2012. Frequently Asked Questions: Harm-Benefit Analysis. http://www.aaalac.org/accreditation/faq_landing.cfm#B3.
7. Dresser, R. 1989. Developing standards in animal research review. *J. Am. Vet. Med. Assoc.* 194: 1184–1191.
8. Stafleu, F.R. 1994. The ethical acceptability of animal experiments as judged by researchers. PhD diss. Universiteit Utrecht.
9. Stafleu, F.R., B.D. Baarda, F.R. Heeger et al. 1994. The influence of animal discomfort and human interest on the ethical acceptability of animal experiments. In Welfare and Science: Proceedings of the Fifth Symposium of the Federation of European Laboratory Animal Science Associations. Brighton, UK: Royal Society of Medicine Press.
10. U.S. Department of Agriculture, Animal and Plant Health Inspection Service. 2011. Animal Care Resource Guide. Policy 11. Painful and distressful procedures. http://www.aphis.usda.gov/animal_welfare/policy.php?policy=11.
11. U.S. Department of Agriculture, Animal and Plant Health Inspection Service. 2013. *Animal Welfare Inspection Guide,* Appendix A. http://www.aphis.usda.gov/animal_welfare/downloads/Inspection%20Guide%20-%November%202013.pdf.
12. Committee on Recognition and Alleviation of Pain in Laboratory Animals. 2009. The *Recognition and Alleviation of Pain in Laboratory Animals,* 270. Washington, D.C.: National Academies Press.
13. National Institutes of Health, Office of Laboratory Animal Welfare. 1997. OPRR Reports, Number 98-01, Production of monoclonal antibodies using mouse ascites method. http://grants.nih.gov/grants/olaw/references/dc98-01.htm.
14. Richardson, C.A. and P.A. Flecknell. 2005. Anaesthesia and post-operative analgesia following experimental surgery in laboratory rodents: Are we making progress? *Altern. Lab. Anim.* 33:119–127.
15. Russell, W.M.S. and R.L. Burch. 1959. *The Principles of Humane Experimental Technique.* 238. London: Methuen & Co. Ltd.
16. U.S. Department of Agriculture, Animal and Plant Health Inspection Service. 2011. Animal Care Resource Guide, Policy 12. Written narrative for alternatives to painful procedures. http://www.aphis.usda.gov/animal_welfare/policy.php?policy=12.
17. DeGrazia, D. Taking Animals Seriously. 1996. 302. Cambridge: Cambridge University Press.

18. Festing, M.F. and D.G. Altman. 2002. Guidelines for the design and statistical analysis of experiments using laboratory animals. *ILAR J.* 43:244–258.
19. U.S. Interagency Research Animal Committee. May 20, 1985. Principles for the utilization and care of vertebrate animals used in testing, research, and training. *Federal Register.*, 20864–20865.
20. U.S. Food and Drug Administration. 2006. Guidance for industry, investigators, and reviewers: Exploratory IND studies. http://www.fda.gov/downloads/Drugs/GuidanceComplianceRegulatoryInformation/Guidances/ucm078933.pdf.
21. U.S. Food and Drug Administration. 2010. Guidance for industry and FDA Staff: General considerations for animal studies for cardiovascular devices. http://www.fda.gov/MedicalDevices/DeviceRegulationandGuidance/GuidanceDocuments/ucm220760.htm.
22. U.S. Food and Drug Administration. 2007. Guidance for industry and FDA Staff—Recognition and use of consensus standards. http://www.fda.gov/MedicalDevices/DeviceRegulationandGuidance/GuidanceDocuments/ucm077274.htm.
23. Garnett, N.L. 2003. A word from OLAW. *Lab Anim. (N.Y.)* 32(10):19.
24. Willems, R.A. and J.A. Nelson. 2004. The top ten tips for completing the USDA annual report. *Lab Anim. (N.Y.)* 33(8):25.
25. National Institutes of Health, Office of Laboratory Animal Welfare. 2010. Notice NOT-OD-10-128, Clarification on the roles of NIH Scientific Review Groups (SRG) and Institutional Animal Care and Use Committees (IACUC) in review of vertebrate animal research. http://grants.nih.gov/grants/guide/notice-files/not-od-10-128.html.
26. U.S. Food and Drug Administration. 2006. Cosmetics: Animal testing. http://www.fda.gov/Cosmetics/ProductandIngredientSafety/ProductTesting/ucm072268.htm.

13

Justification of the Number of Animals to Be Used

Ed J. Gracely

Introduction

Choosing an appropriate number of subjects is an important part of any research project. A proper sample size is essential to obtaining valid results and to minimizing the number of individuals exposed to the potential risks and harms of research. For this reason, the IACUC is interested in ensuring that a sufficient but not excessive number of animals are used. A study with too few animals is ethically problematic because the animals are being subjected to potentially painful procedures or loss of life with relatively little likely benefit to the advancement of knowledge. On the other hand, a study using more animals than are truly needed is also problematic, because it unnecessarily exposes some of them to the same harms.

This chapter discusses key issues related to IACUC review of animal numbers. The existing regulations and policy are based on the use of performance standards empowering IACUCs to use professional judgment in weighing questions about animal numbers. Although IACUCs differ on many of these issues, experience, common sense, and the results of an informal survey all help provide direction. The survey results emanate from a questionnaire sent primarily to academic institutions.

13:1 What do the AWAR and PHS Policy state with respect to justifying the number of animals requested in the protocol?

Reg. The AWAR (§2.31,e,1; §2.31,e,2) state that the proposal to use animals must include "identification of the species and the approximate number of animals to be used" and "a rationale for involving animals and for the appropriateness of the species and numbers of animals to be used." U.S. Government Principle III states, "The animals selected for a procedure should be of an appropriate species and quality and the minimum number required to obtain valid results." The PHS Policy (IV,D,1,a; IV,D,1,b) also states that all applications and proposals that are submitted to the PHS and that involve animals must include "identification of the species and approximate number of animals to be used" and "rationale for ... appropriateness of the species and numbers used."

The *Guide* (p. 25) states that "whenever possible, the number of animals and experimental group sizes should be statistically justified (e.g., provision of a power analysis...)." This is the most specific statement of how animal numbers should be justified in any of these documents. The *Guide* also notes that, "...hypothesis testing, sample size, group numbers, and adequacy of controls can relate directly to the prevention of unnecessary animal use or duplication of experiments." Similar guidelines are provided by the Worksheet for Review of the Vertebrate Animal Section (VAS) at the NIH/OLAW,[1] which states that "Estimates for the number of animals to be used should be as accurate as possible. Justification for the number of animals to be used may include considerations of animal availability, experimental success rate, inclusion of control groups and requirements for statistical significance; cite power calculations where appropriate." And, elsewhere in the document, "Provide justification for … Number of animals to be used (cite power calculations, if appropriate) with specific justification for large numbers of animals."

NIH/OLAW FAQ F.2[2] adds a bit more detail. "Although the PHS Policy does not explicitly require a mechanism to track animal usage by investigators, it does require that proposals specify a rationale for the approximate number of animals to be used and be limited to the appropriate number necessary to obtain valid results. This implicitly requires that institutions establish mechanisms to document and monitor numbers of animals acquired and used, including any animals that are euthanized because they are not needed. Monitoring should not exclude the disposition of animals inadvertently or necessarily produced in excess of the number needed or which do not meet criteria (e.g., genetic) established for the specific study proposal. Institutions have adopted a variety of administrative, electronic, and manual mechanisms to meet institutional needs and PHS Policy requirements."

Opin. The phrase "approximate number" appears in both the AWAR and the PHS Policy, without further explanation. Needless to say, all animal number estimations are approximate in some sense. Animals die, procedures fail, unexpected events occur, and new research directions emerge. Furthermore, it is always possible to amend a protocol to deal with such eventualities or with data more variable than had been anticipated, thus requiring additional animals per group. If the wording in the law and the policy is interpreted as primarily intended to encompass these sorts of uncertainties, then that wording is consistent with requiring an initial estimate that is specific, but understood implicitly to be approximate.

It is possible that the regulations may have been intended to allow a broader type of flexibility (such as "from 200 to 250 animals will be required"). As this type of request is not prohibited, if the IACUC finds it to be reasonable and scientifically justified, then it may approve an approximate number of animals.

NIH/OLAW FAQ F.2[2] makes clear that researchers cannot decide to monitor and record only animals that are actually used in studies. Animal welfare regulations (and ethical considerations) apply to any animal produced in the process of doing the research, even if it has the wrong genotype or gender, or is merely surplus. The IACUC is responsible for ensuring that these animals are accounted for and humanely disposed of if they are not used.

13:2 What constitutes sufficient justification for a requested number of animals?

Reg. (See 13:1.)

Opin. The AWAR and PHS Policy do not address this issue. Thus, it is left to the institution to develop appropriate policies. The IACUC must be able to determine how all of the requested animals will be utilized, which research questions will potentially be answered by the number to be used, and why those questions could not be answered with fewer animals or no animals at all. Describing animal use is generally straightforward, except in studies that are open ended and have evolving goals. In the latter instance, the investigator should attempt to provide a sequence of events and a desired end point. A justification of numbers based solely on an ongoing need for a certain number of animals per year, without an explanation of what is likely to be accomplished in each specific period is unacceptable.

An interesting set of three comments by researchers and veterinarians on the justification of numbers for an open-ended study has been published.[3–5] Four invited commentators (in three articles) responded to a scenario in which an established scientist planned a 3-year series of studies requiring a total of 5500 mice. The project had already been funded (pending IACUC approval) by the NIH. The discussion centered on the fact that the justification of numbers is brief for such a large number of animals, being basically a statement of animal needs per week multiplied by 3 years × 52 weeks = the total number. To provide some context, note that an APHIS/AC reviewer who read this chapter before publication agreed with this chapter's author: "This justification would not be adequate for AWAR purposes. A scientific justification is needed; i.e., an explanation of why x number of animals per week is needed." Interestingly, only one of the responses to the scenario (the last one) took a similar position. One commentator was fairly satisfied with the justification, especially given that the NIH grant review panel was satisfied. This respondent wanted only a few examples of the kinds of studies to be done. The second comment (written by two respondents) listed a variety of questions, mainly about ways the researcher might reduce the required animals or better justify the weekly details (for example, could more cells be obtained from each animal by using improved techniques?). It was not clear whether this pair of commentators would have accepted the unspecified research to be done if these more practical questions had been satisfactorily answered. The third respondent took objection to the whole process, arguing that if the researcher could not explicitly lay out the studies to be done over 3 years, then approval should be requested only for that series that could be planned in advance, with amendments to request more animals as needed. Clearly this is a topic that generates strong and diverse feelings!

Justifying the specific animal numbers for individual research questions is often difficult as well and there is no perfect way to do it. Previous research, while sometimes helpful, is only a crude guide to the number of animals needed in the present study. Even statistical power analysis requires assumptions (such as the magnitude of effects worth looking for) that are often subjective or that must themselves be estimated from other data.

This author believes that sample size justification should be seen as a means to ensure that researchers and IACUCs are aware of the issues and are attempting to prevent serious miscalculations. A moderate amount of detail is sufficient to accomplish this. Consider three specific approaches, each of which may be adequate when it is applicable:

- Report of a power analysis, with enough information provided to show that the researcher knows how to analyze the data and use power analysis techniques. Question 13:4 lists key considerations of a fairly thorough power analysis. A briefer approach may be acceptable, but this author's experience suggests that attempts at an abbreviated power analysis often produce descriptions that are garbled or lack key information.

- Citation of previous research, with sufficient information provided to indicate that the previous research is similar enough in concept and methodology to the present proposal as to make it a reasonable model for sample sizes in the latter.
- Derivation of animal numbers from material needs, when the study has no statistics (e.g., histological characterization of tissues), with a clear indication why the specific amount of material is needed, and why the number of animals requested is appropriate to provide that amount of material.

Survey question 1 asks whether IACUCs believe in the importance of justifying animal numbers. Fortunately, virtually all of them do, and 90% are willing to critique the content of those justifications. A smaller percentage believes that scientific reviewers should take that responsibility, not the IACUC.

Surv. 1 Does your IACUC regard reviewing sample size (animal numbers) justification as part of its responsibility?

• Not applicable	3/294 (1%)
• Yes, we review animal number justifications and may critique their content.	263/294 (89%)
• Yes, but we primarily assure that a justification is provided. We let scientific reviewers verify the justification if there is a grant or contract.	24/294 (8%)
• No. It is a scientific question outside of the purview of the IACUC.	3/294 (1%)
• Our decision to review or not depends on the funding source or species of animal being used.	0/294 (0%)
• Other	1/294 (0.3%)

Survey questions 2 and 3 focus on instances in which inferential statistics are not relevant because the researcher is determining sample size from the amount of material needed for histological or biochemical analysis. The majority (59%–70%) of the institutions for whom the questions applied indicated an intermediate level of requirements in these situations.

Surv. 2 Some protocols may not require any statistical analysis (e.g., histological analysis). When presented with such a protocol, what is the minimum your IACUC will accept for sample size justification?

• Not applicable	23/295 (8%)
• Few details, such as a statement that "n" animals are needed to provide enough tissue for all histological analyses	108/295 (37%)
• Simple calculation of the total amount of tissue needed and the amount available from one animal	88/295 (30%)

- A more detailed justification of the number of histological observations required and the amount of tissue available from one animal — 73/295 (25%)
- Other — 3/295 (1%)

Surv. 3 For studies in which a particular cell type (e.g., a dendritic cell) or a cell product (e.g., an enzyme) is required, what is the minimum your IACUC will accept as an animal number justification

- Not applicable — 37/296 (12%)
- Limited details, such as "n" animals are needed to provide enough material for all the assays to be performed — 76/296 (26%)
- Simple calculation, such as the amount of product needed for all analyses and the amount that can be acquired from one animal — 115/296 (39%)
- A detailed explanation of why a particular amount of product is required combined with the amount that can be acquired from one animal — 68/296 (23%)
- Other — 0/296 (0%)

Survey questions 4–8 focus on studies that include some statistical analyses as an integral component. About half of the committees (56%) routinely request a power analysis (survey 4), and another 14% sometimes do. Of those protocols with inferential statistics to justify the numbers (presumably including a power analysis), committees run the gamut of details expected. For this item, over 10% of the committees responded "not applicable." Including them in the denominator, the modal requirement (49%) was a simplified description of the power analysis (survey 5). Most committees (70%) will accept a justification based on the PI's experience (survey 6), with the majority of those requiring details of study comparability. A full 88% will accept a justification based on previous published work (not necessarily by the PI, survey 7). They also mostly want details of comparability as part of the justification. However, only a minority of IACUCs check the references provided (survey 8). Since so little specific guidance is given in the regulations, and even the *Guide* (p. 25) only recommends the use of statistical methods (interpreting "should be" as weaker than "must be") the onus is on the individual IACUCs to determine when sufficient justification has been provided.

Surv. 4 When the experimental design allows, does your IACUC request a statistical justification (i.e., power analysis) for the number of animals being requested?

- Not applicable — 4/294 (1.4%)
- Yes, almost always — 165/294 (56.1%)
- Only sometimes — 42/294 (14.3%)
- Not usually — 35/294 (11.9%)
- We always leave the choice of how to justify animal numbers to the investigator — 47/294 (16%)
- Other — 1/294 (0.3%)

Surv. 5 If an investigator uses inferential statistics to justify the number of animals in a particular protocol, what is the minimum statistical information your IACUC would usually accept?

- Not applicable 33/295 (11%)
- Details (e.g., test to be used, standard deviation, effect size, alpha, power, expected differences), plus the results of the power analysis. 55/295 (18.6%)
- A basically correct, brief statement of a power analysis with enough information to indicate competence in performing the analysis 145/295 (49.2%)
- We will accept most anything that looks like a statistical response 24/295 (8.1%)
- A simple statement that a power analysis has been done 32/295 (10.9%)
- Other 6/263/295 (2%)

Surv. 6 Would your IACUC routinely accept sample size justification in a statistical-type study that is based on a statement of previous experience of the investigator, and nothing else?

- Not applicable 15/295 (5%)
- Yes, if details of study comparability are provided 159/295 (54%)
- Yes, if that is what the investigator chooses to provide 46/295 (15.6%)
- No 70/295 (23.7%)
- Other 5/295 (1.7%)

Surv. 7 Does your IACUC accept sample size justification in a statistical-type study that is based on references to similar work in published literature?

- Not applicable 13/294 (4.4%)
- Yes, if adequate details are provided to demonstrate comparability 210/294 (71.4%)
- Yes, if that is what the investigator chooses to provide 49/294 (16.7%)
- No 18/294 (6.1%)
- Other 4/294 (1.4%)

Surv. 8 If your IACUC accepts sample size justifications based on references to similar work in the published literature, does any person check to see if the literature citation(s) actually supports the contention of the animal user?

- Not applicable 31/294 (10.5%)
- Yes, usually 84/294 (28.6%)
- Yes, occasionally 75/294 (25.5%)
- Yes, but rarely 39/294 (13.3%)
- No 46/294 (15.7%)
- I don't know 16/294 (5.4%)
- Other 3/294 (1.0%)

Survey 9 asks an interesting question: what to do if a study is almost statistically significant, albeit not quite, and the PI asks for a few more animals to bump it up to being significant. This is, of course, a problematic situation since adding a few extra subjects to a non-significant result can distort the p-value and lead to an increased risk of type 1 (false positive) error. Sample size should normally be decided in advance, and stuck to. If you have a non-significant result and think it would be significant with a bigger sample size, a new study is called for, not a bit-by-bit increase in N until you get what you want. Unfortunately, few IACUC's seem to understand this basic principle. The exception to this rule is with well-planned sequential study designs, in which stopping rules are established a priori.

In human subjects research it is common to design long studies with large sample sizes using sequential stopping rules. The researchers plan to perform interim analyses at pre-defined points and to stop the study if the results merit that. In some cases, the data are analyzed after every new subject is completed. Sequential testing is a specific procedure, with its own appropriate modifications to data analysis. For example, if the study is not stopped early, the analysis of the final dataset may require a smaller p-value to attain significance than would otherwise have been needed, in order to compensate for the multiple "looks" along the way.

Recent publications by Douglas Fitts, notably reference 12, provide an easily-applied mechanism for using such methods in small studies. For example, suppose a researcher does a preliminary power analysis which suggests 10 animals per group and decides that 6 per group is the smallest number worth trying, while 18 per group is the largest he or she would even consider. The study proceeds by doing the analysis after 6 animals have been completed in each group, and then after 1 more animal has been added to each group in turn. The trick is that the p-value for significance at each step is required to be less than or equal to 0.0165 (the new alpha level). Interestingly, if the p-value is 0.24 or greater at any step, the study also stops. The details of the significance and non-significance cutoffs vary with other decisions—see the reference for details. Such methods can be powerful and may on the average use fewer animals than a single fixed sample-size model, but the PI must be prepared to continue to the end if neither stopping criterion is reached.

Surv. 9 How does your IACUC approach an amendment request to add additional animals to a study for the purpose of reaching statistical significance if the original number approved fell just short of reaching significance? Check as many boxes as appropriate. (Note: Because of this, percentages add to >100%.)

• Not applicable	42/296 (14.2%)
• We request statistical justification for the additional number requested	132/296 (44.6%)
• We usually approve a request if it is for 5% or fewer animals	85/296 (28.7%)
• We are more strict about approving animals covered by the Animal Welfare Act	24/296 (8.1%)
• We consider any such request to be a major amendment	64/296 (21.6%)

- We typically consider such a request to be a minor amendment — 58/296 (19.6%)
- We consider such a request to be scientifically inappropriate and do not allow it — 7/296 (2.4%)
- Other — 13/296 (4.4%)

Finally, survey question 10 addresses the question of replication within the same protocol. About half the respondents who found the question applicable said that they "often" allow replication, slightly more often for in vivo than in vitro studies.

Surv. 10 Does your IACUC allow for the repeat of an experiment (as part of the same protocol) in order to demonstrate experimental repeatability prior to publication? Check as many boxes as appropriate. (Note: Because of this, percentages add to >100%.)

- Not applicable — 85/294 (28.9%)
- Often yes, for in vitro studies using animal tissues — 104/294 (35.4%)
- Often yes, for in vivo studies — 123/294 (41.8%)
- Not often for in vitro studies — 18/294 (6.1%)
- Not often for in vivo studies — 55/294 (18.7%)
- Other — 31/294 (10.5%)

13:3 Relative to the number of animals used, are there different regulatory requirements for different species?

Reg. Definitions of an animal relative to the PHS Policy and the AWAR are provided in 12:1. U.S. Government Principle III, the *Guide* (p. 25), and the AWAR (§2.31,e,2) all require a justification for the appropriate species proposed for the animal use activity.

Opin. The definition of an animal varies with the species that are included under the purview of the IACUC but the definitions make no distinctions in terms of animal number justification. For example, there is no wording that distinguishes different phylogenetic levels of equally sentient animals in terms of animal number justification. (See 12:13.)

APHIS/AC Policy 12,[6] which is meant to help interpret the AWAR requirement for an alternatives search, clearly states that less sentient species (e.g., insects) should be used if possible. This guideline does not appear to have implications for decisions among more sentient species, such as mammals and birds. Researchers sometimes speak of wanting to use the "lowest phylogenetic level" that will answer a question, but unless there is the possibility of using a species (such as an insect or worm) that is clearly less subject to pain than higher animals, there is nothing in the regulations to mandate this, nor to suggest that more justification for a particular number of subjects is required for "higher" species than for "lower" ones among sentient animals. (See 12:8.)

One distinction of interest involves the study of endangered species as covered under the Endangered Species Act of 1973[7] and other relevant legal documents such as the Convention on International Trade in Endangered Species of Wild Flora and Fauna (CITES) treaty.[8] It seems likely that the total number of such animals to be studied, as well as the number to be euthanized, would be subject to stricter standards for endangered species than would otherwise prevail for other animals. Any IACUC dealing with such a situation will have to review all of the relevant documents.

Other than endangered species, there are few species-specific distinctions relative to justifying the number of animals proposed for use. From ethical, humanitarian, and regulatory standpoints, it is as important to justify the use of 100 mice properly as it is to justify the use of 100 dogs. Invertebrates are not included in the definition of an animal noted above but it seems reasonable that certain apparently sentient and intelligent invertebrate groups (such as octopuses) be given the same consideration as vertebrates. Naturally, the appropriate species for the research must be used. (See 12:4; 12:8; 12:13.)

Is there a level of animal intelligence that renders killing animals less justifiable? Is it important to provide more justification for the utilization or euthanasia of ten apes compared to ten mice? There are no such distinctions in the regulations (at least in terms of justifying the numbers of animals used) but implicit in all animal number justifications is the idea that the knowledge gained must be adequate to offset the impact on the individual animals involved. Each IACUC will need to determine whether highly intelligent species require special consideration in maintaining this balance. The statistical analysis is not affected, but the decision as to whether the research should occur at all may be.

13:4 What are the key considerations in using a statistical power analysis to justify animal numbers?

Opin. There are several key considerations when a researcher wants to perform and present a fully usable and appropriate power analysis. The researcher should have a thorough understanding of the appropriate statistical analysis for the design or must be willing to seek out and collaborate with someone who does. Much basic science research that uses statistics involves several groups or conditions and often a number of time points. Such designs require advanced analytic techniques, such as the analysis of variance (ANOVA) and specific follow-up tests (such as the Tukey test). In some instances the major question involves the interaction of two independent variables, as tested in a two-way ANOVA. These analytic decisions have important implications for the power analysis. For example, a researcher with a complex design who estimates power merely for a two-group comparison may reach a substantially incorrect conclusion. Often, a thoughtful analyst can reduce a complex question to one for which sample sizes can be estimated with simpler methods, but this process requires a good understanding of the statistical issues.

The study design must be spelled out in sufficient detail for IACUC (or other) reviewers to understand it and to see how the power analysis is configured to match the analysis that will be employed. The researcher must present and justify the specific parameters and assumptions required by the power analysis. Typically, this will involve estimates of effect sizes (such as the mean difference between two groups) and estimates of variability, such as the standard deviation. Analytic methods have different requirements that must be determined before a power analysis can be run. Prior experience, the theory underlying the research, and pilot data may be useful.

13:5 What role can a biostatistician play in evaluating the justification of the number of animals requested?

Opin. Biostatisticians are trained to perform the kind of calculations used to determine sample size. They can advise the researcher how to analyze the data properly and can help to write the analysis and animal number sections of the protocol. To be most helpful, biostatisticians also should be familiar with the IACUC and its task. They should understand the audience that their comments will reach and understand enough of the science to be able to recognize what is central and what is not. Having a statistician either on the committee or available to help researchers could be very useful. Since the development of statistical methods and the performance of power analyses are both fairly lengthy tasks, IACUCs should not expect a statistician (whether a member of the committee or not) to perform these tasks routinely without making appropriate collaborative, and possibly financial, arrangements.

13:6 What exactly is a pilot study? How do pilot studies differ from other studies in animal number justification?

Opin. A pilot study is one involving a small number of animals (rarely more than ten) used to demonstrate that a technique can work or to estimate the variability in the data before performing a statistical analysis. A pilot study does not require the extent of sample size justification required by "full" studies. Nevertheless, a pilot study is part of an IACUC protocol and, therefore, does require a justification that is sufficient to indicate why that number of animals is needed to achieve the results of the pilot. If pilot studies are to be exempted from power analysis or other relatively stringent methods for sample size justification, then common sense suggests that they be narrowly defined and limited to a few special cases. Every institution should develop and disseminate specific guidelines, or an SOP, concerning what will be accepted as a pilot study.

It is important to note that pilot studies are much more useful for estimating variability in the data than for determining the likely magnitude of effects. The latter requires a study much larger than most pilots. Furthermore, if a pilot study is used to estimate magnitudes of differences or to determine whether or not a particular manipulation is worth pursuing further, then the pilot animals should not be used in the succeeding full study (see 13:8). If pilot study animals serve only to estimate variability or to verify that a technique works in your laboratory (e.g., that you can actually get reliable results on a particular piece of equipment), it may be acceptable to include the pilot study animals in the full study. (See 9:39.)

13:7 Can a study using four groups of ten animals each be considered a pilot study?

Opin. In the author's opinion it is difficult to see how so many animals could be considered a pilot study or why a pilot study would require four groups. Variability can normally be adequately estimated with one group (or, at most, two groups) of animals. Viability of a technique also can typically be tested on one or two groups.

Every possible situation cannot be anticipated. For example, consider a researcher who has four technically difficult manipulations, each of which must be shown to work before a larger study can be undertaken utilizing all four. Conceivably, a pilot study with four groups could be justified here.

13:8 If animals are requested both for a pilot study and for the full study to follow, should the IACUC approve both at the same time?

Opin. Assuming the number of animals used in the pilot study is justified (see 13:6) and until the results of the pilot study are known, the IACUC should not approve the entire number of animals requested.

The pilot may show that the basic technique is flawed and the whole study should not progress. In other instances, a pilot has as its purpose the estimation of variability in the data upon which a power analysis will be based. Until these estimates are in hand, any number of animals requested for the full study must be considered unjustified. Granting some kind of "conditional" approval pending the results of the pilot study sends the wrong message and may be confusing, yet many institutions require some preliminary report from the investigator before the full study is implemented. PIs could misunderstand this as in effective "approval but for a technicality," which may not be the intent. It is, in this author's opinion, best to grant approval only for the number of animals fully and initially justified. (See 9:39.)

Survey 1 asks about the transition from pilot to full study. This could be tricky, because the pilot might change the number required, or even indicate that the entire procedure is not workable and must be revised. As noted, most IACUC's (53%) take this uncertainty into account by withholding either approval or release of the animals until results of the pilot are available. This percentage might actually be higher, since 30% said that responses vary depending on the protocol, and 8% said that the question was not applicable.

Surv. 1 If a pilot study is approved by your IACUC, how does your IACUC proceed with approving the total number of animals needed for the larger study?

• Not applicable	23/296 (7.8%)
• We approve all the animals at once	21/296 (7.1%)
• We approve all the animals at once but only allow the investigator to order the animals for the pilot study. Once the pilot study is reviewed by the IACUC we may release the remaining animals to the investigator	24/296 (8.1%)
• We only approve animals for the pilot study and do not approve (via amendment) the remaining animals until the IACUC evaluates the pilot study results	134/296 (45.3%)
• Our decisions vary, based on specific protocols	90/296 (30.4%)
• Other	4/296 (1.4%)

Survey question 2 asks about a related issue: What if the number of animals needed in the second part of a study is dependent on the earlier part, even if neither is a "pilot" per se. 38% reported that they require an amendment to approve the remaining animals. But 63% (combining first and third applicable responses) allow the investigator to estimate or "best guess the numbers," and, presumably, approve them all in advance (unless the estimate turned out to be badly wrong, I assume). This question allowed more than one answer, so some respondents may have required an estimate but not necessarily approved all the animals from it until initial results were in hand.

Surv. 2 If the number of animals required for the latter part of a study is dependent upon findings from an earlier part of the same study (on the same protocol), how does your IACUC approach this situation? Check as many boxes as applicable.

- Not applicable 25/295 (8.5%)
- We require an estimate of the number of animals needed based on the most likely outcome of the first part of the study 151/295 (51.2%)
- We approve the initial number needed then the investigator submits an amendment with a justification for the remaining number needed 113/295 (38.3%)
- We allow the investigator to make a best guess of the number needed without additional justification 31/295 (10.5%)
- Other 9/295 (3.1%)

13:9 Can the IACUC demand that a pilot study be done?

Opin. The IACUC has the right and responsibility to require an adequate justification of the number of animals requested (see 13:1). If a researcher is unable to provide that, the IACUC may withhold its approval. If a pilot study can break the impasse, it may be an appropriate choice for both the IACUC and the investigator.

The IACUC, by exercising its responsibility to assure that the number of animals to be used is appropriate, may withhold approval of a project until a pilot study is completed. If a pilot study is agreed to, it requires IACUC approval (see 13:6–13:8). There is a brief reference in NIH/OLAW FAQ D.11[2] suggesting that IACUCs may be able to demand a pilot, without details, "Whether proposed by investigators or required by the IACUC, pilot studies require review and approval by the IACUC in accordance with the PHS Policy." (See 9:39.) Since the IACUC will only grant full approval after the pilot study is completed (if it follows the preceding advice) it will necessarily need to receive a report on the pilot, with a revised request for the remaining animals.

A survey question on this topic was not asked for this edition of *The IACUC Handbook* but in the second edition survey[9] 69% of 162 respondents agreed that IACUCs could require a pilot study, 24% had no policy, and only 6% said no.

13:10 Can an IACUC withhold approval of a protocol that requests fewer animals than the IACUC believes are needed for a valid study?

Reg. (See 13:1.)

Opin. Yes, if the researcher does not adequately justify the use of that number. The primary question is whether or not the sample size justification itself is adequate (see 13:1). Section 13.2, above discusses methods to adequately justify sample size. If the justification is invalid or inappropriate, the researcher should be asked to improve it. If the researcher fails to do this acceptably, then approval of the protocol is withheld until the justification is satisfactory to the IACUC. The use of too few or too many animals can be equally deleterious. A researcher who requests fewer animals than one might have expected is not excused from showing how that number is sufficient for research purposes.

A survey question on this topic was not asked for this edition of *The IACUC Handbook*; however, in the second-edition survey[9] 69% of 161 respondents agreed that their IACUC "would withhold approval" of a protocol that appears to be requesting too few animals, 24% said they would not, and 7% replied "other."

13:11 When justifying the number of animals requested, should fetuses and neonates be counted as vertebrate animals?

Reg. The AWAR and PHS Policy do not address this issue in detail. However, NIH/OLAW has provided guidance on this issue in FAQ A.4 and A.5[2] that state: "Although avian and other egg-laying vertebrate species develop backbones prior to hatching, OLAW interprets the PHS Policy as applicable to their offspring only after hatching." This statement makes clear that hatched chicks should be counted. It would be reasonable to infer that the same logic would apply to neonate mammals. Thus, it seems that born and hatched vertebrates should all be counted. The handling of fetuses is less clear (see *Opinion*).

Opin. Fetuses should be counted in almost all cases; at least those advanced enough to feel pain or distress. This opinion is based on ethics, rather than the law and associated regulations, but behaving in a way that makes sense ethically is probably also good legal practice. The purpose of the IACUC is to help assure high standards of animal welfare in the institution. If a fetus can feel pain, then its welfare matters and it should become an IACUC concern. Nevertheless, it is difficult to develop comprehensive and systematic guidelines for preventing fetal pain, because so little is known about many species and their precise neural development. (See 8:11; 14:19; 16:34; 16:35.)

If late-stage fetal or neonatal animals are directly the subjects of research studies, accounting for them follows fairly naturally from the sample size justification itself. The justification of animal numbers in such instances will be critically determined by the number of fetuses needed.

A survey question (14:19) suggested that many IACUCs do not "count" fetal animals until they are born and sometimes do not count even born animals until weaned. However, the intent of the respondents was not clear. Perhaps those committees that only count weaned animals are referring to breeding protocols in which animals are not in any way utilized for research until they reach adulthood. This author can understand why an IACUC might not want to count every fetus or neonate in a breeding situation when only those that survive to adulthood are used in experiments. Nevertheless, this author would still prefer to see all late-stage fetuses and born animals tallied and included. Suppose a researcher had a fragile strain of rat with 80% mortality rate in the postnatal period. An ethical analysis of the value of this research against the number of animals used would have to consider all of the animals produced, not just the 20% that survive to be manipulated. In addition, the possibility that those that died would have suffered from their disorder before succumbing to it would also be relevant, and would only be salient if the total number were reported.

The *Institutional Animal Care and Use Committee Guidebook*[10] section on breeding does not discuss fetuses but suggests that suckling animals be subdivided into those to be used in any way in the research versus those merely to be euthanized before weaning. Both groups are then counted and reported, as recommended here.

13:12 To reduce the number of animals used, how might an animal be used in more than one biomedical research, teaching, or testing protocol?

Reg. The AWAR (§2.31,d,1,x) and the *Guide* (p. 30) limit the situations under which animals can be subjected to a second major survival surgery or related operative procedure. This happens primarily in instances of scientific necessity and is clarified in APHIS/AC Policy 14 and NIH/OLAW FAQ F9.[2] (See 16:19; 16:52 and Chapter 18.)

Opin. Any sacrificed animal should be utilized in as many studies as can benefit from the use of its tissues. It is appropriate to do so any time several protocols can use an animal without subjecting it to any additional pain or distress. A second use of a live animal in a non-painful or non-distressful study is also appropriate and can reduce the total number of animals involved in research. For example, after a survival surgery study in which animals need not be euthanized, a simple non-stressful behavioral study might be appropriate.

A real problem concerns the reuse of an animal in a painful or distressing way after it has already been used once in this way. Federal guidelines focus on surgery but common sense suggests that the painful or distressful reuse of an animal that previously was subjected to a distressful or painful procedure requires careful justification by the investigator and careful consideration by the IACUC.

13:13 How else can animal numbers be reduced?

Opin. It is worthwhile to consider ways that animal numbers can be reduced carefully. A table of possible methods for reducing the number of animals used is provided in the *Institutional Animal Care and Use Committee Guidebook*[10]. Some of what is recommended is already customary practice for most researchers, such as standardized procedures; use of "healthy, genetically similar animals"; and good postoperative care to minimize losses. Other suggestions include sharing of tissue (as noted in 13:12), several terminal procedures per animal, appropriate choice of control groups, and maximization of the use of information with good statistical software.

Morton[11] describes some additional methods. For example, he notes that if a certain drug would be considered as too toxic for use if even one of six subjects had severe adverse reactions, it would make sense to test the six in sequence and then stop the study if one of them, in fact, had such a reaction. He notes that in certain kinds of research (such as testing antibiotics for efficacy in an animal model prior to use in a Phase I human trial), there may be little purpose in a strict significance level (e.g., setting the cutoff for significance at $p = 0.01$) because any findings must still be replicated in humans.

Morton[11] notes that pilot studies can help to prevent problems and the data may be able to be used in the main study. Nevertheless, as noted in 13:6, one needs to be cautious with this concept. If pilot studies are used merely to test procedures and estimate variance, it is true that one can use the data in the full study. If, however, pilot studies are used to determine, for example, which drugs from a large set of drugs have an effect, one needs independent replication of the results with entirely new data. Anyone proposing to mix pilot data with full study data should consult a statistician.

13:14 What role can veterinarians and animal facility personnel play in reducing the number of animals used?

Opin. The IACUC has a general responsibility to oversee the appropriate use of animals in biomedical research. It closely interacts with the AV and the animal care staff. Veterinarians and other animal facility personnel have responsibilities that generally include assuring that animals are obtained from high-quality vendors, reviewing the health status of incoming animals on the basis of a vendor's reports, assuring quarantine and testing of animals when appropriate, and facilitating the re-derivation of animal strains when appropriate. They also are responsible for developing the entire program of preventive medicine and often oversee all animal care operations. The use of healthy, well-treated, and well-maintained animals helps decrease the numbers of animals needed by decreasing the need to repeat studies.

Veterinarians and animal facility personnel also can assist with preoperative, operative, and postoperative planning to help assure animal survival, further decreasing the numbers needed.

References

1. National Institutes of Health. Worksheet for Review of the Vertebrate Animal Section (VAS). http://grants.nih.gov/grants/olaw/VASchecklist.pdf. Accessed September 5, 2013.
2. National Institutes of Health, Office of Laboratory Animal Welfare. 2013. Frequently asked questions. PHS Policy on Humane Care and Use of Laboratory Animals. http://grants.nih.gov/grants/olaw/faqs.htm#useandmgmt_2.
3. Silverman, J. 2004. Animal use for in vitro work: How much justification is enough? *Lab Animal (N.Y.)* 33(5):15.
4. Doyle, R. 2004. Give some examples. *Lab Animal (N.Y.)* 33(5):15.
5. Matthews, M. 2004. A little short. *Lab Animal (N.Y.)* 33(5):16.
6. U.S. Department of Agriculture, Animal and Plant Health Inspection Service. 2011. Animal Care Policy Manual, Policy 12. Considerations of alternatives to painful/distressful procedures. http://www.aphis.usda.gov/animal_welfare/policy.php?policy = 12.
7. U.S. Fish and Wildlife Service. 2013. *Endangered Species Act (P.L. 93-205, 87 Statute 884)*. http://www.fws.gov/endangered/laws-policies/index.html.
8. Convention on International Trade in Endangered Species. www.cites.org. Accessed February 22, 2013.
9. Gracely, E.J. 2007. Justification of the number of animals to be used. In *The IACUC Handbook, 2nd edition*, eds. J. Silverman, M. Suckow and S. Murthy, Chapter 13. Boca Raton: CRC Press.
10. Applied Research Ethics National Association/Office of Laboratory Animal Welfare. 2002. *Institutional Animal Care and Use Committee Guidebook, 2nd edition, 98*, 130–133. Bethesda: National Institutes of Health.
11. Morton, D. 1998. The importance of non-statistical design in refining animal experiments. *ANZCCART News* 11(2, insert):1–17. http://www.adelaide.edu.au/ANZCCART/publications/ImportanceNonStat_T5Arch.pdf.
12. Fitts, D. 2010. Improved stopping rules for the design of efficient small-sample experiments in biomedical and biobehavioral research. *Behav. Res. Meth.* 42:3–22.

14

Animal Acquisition and Disposition

Michael J. Huerkamp, Larry Iten, and David R. Archer

Introduction

Millions of animals are acquired and used for scientific research annually in the U.S. Because animal acquisitions for research and the eventual disposition of these animals are under the stringent regulation of federal law and the oversight of funding agencies, there is a lot at stake. If animals are not acquired and used in accordance with appropriate laws and standards, institutions risk punitive measures and/or public relations dilemmas. Complicating these issues are myriad details related to the potential sources of animals or research specimens (e.g., farms, stockyards, commercial breeders, USDA Class B dealers, foreign import, pet shops, wild-caught, donations or loans from citizens, institutional breeding colonies, reanimation from sperm or embryo cryopreserved repositories, transfer from other research projects, interstate movement, and abattoirs), the type of research for which animals are acquired, the animal species to be used, stage of development/maturation, and eventual disposition (e.g., euthanasia, slaughter for food, donation to raptor rehabilitation program, and adoption). After animals are acquired, the tracking and accounting of their use are arguably among the major administrative challenges at a research institution. Consequently, in order to act in a legal, fair, consistent, and rational manner, it is important for IACUCs to be cognizant of the many issues related to acquisition, records, tracking, and disposition of research animals.

14:1 What are the usual and ordinary sources of animals for research?

Opin. The most conventional way to obtain animals for research is to purchase them from outside the institution from a commercial vendor, licensed dealer, or farm. Within an institution, sources include breeding or stock colonies and those transferred from or exchanged with another research project. Other sources, usually requiring a certain level of justification, include pet shops, private pets (such as those used in veterinary clinical studies with owner consent) (see 8:31; 15:8; 14:42), and those transferred from other research institutions. In certain geographic areas, dogs and cats may be available from animal control programs, but this availability is highly variable, depending on locality. Animals may also be studied in, or collected from, the wild.

In some instances, scientists may require tissues, rather than live animals, for research. As an alternative to purchasing animals to be immediately euthanized, scientists may obtain biological materials from slaughtered livestock, captured wildlife, harvested marine animals, or, perhaps, transfer from a colleague who has euthanized animals for another purpose.

It is important to understand that the breadth of species defined as *animal* varies with the regulating body and does not necessarily adhere to the classic biological definition. PHS Policy (III,A) and the *Guide* (p. 2) include all vertebrates in their definition of animal, while the AWAR (§1.1, Animal) specifically excludes most mice and rats, birds bred for research, horses not used specifically for research purposes, and livestock or poultry used for food, fiber, or agricultural research. (See 12:1.)

14:2 What are the legal requirements governing procurement of animals and tissues for research?

Reg. The AWAR requires that animals used for biomedical research

- Be acquired by lawful means and in compliance with record-keeping requirements (§2.35,b).
- Have a rationale for involving animals, and for the appropriateness of the species to be used (§2.31,e,2). Have appropriate identification of the species and the approximate number of animals to be used (§2.31,e,1).

Opin. For animals used for food, fiber, and in agricultural research, the AWAR do not apply, but recommendations are given in the *Guide for the Care and Use of Agricultural Animals in Agricultural Research and Teaching.*[1] For research funded by entities of the PHS, the salient features of the AWAR are reinforced by the PHS Policy (IV,D,1,a; IV,D,1,b), U.S. Government Principle III, and the *Guide* (pp. 11–35, 106–107).

The provisions in the *Guide*, including those that relate to justifying and tracking animal use, are also used by AAALAC in accrediting research institutions.[2] While the AWAR excludes poikilothermic animals and typical laboratory rats and mice from these considerations, all vertebrates are covered under PHS Policy and, by extension, AAALAC. Consequently, in most research institutions in the U.S., federal regulations, AAALAC standards, and granting agencies require that vertebrate animals be acquired lawfully, used judiciously, and disposed of appropriately. Non-PHS-funded research may be included in PHS requirements if an institution, via its Animal Welfare Assurance with NIH/OLAW, indicates that it will include activities funded by sources other than the PHS. Institutions may exclude non-PHS-funded or agricultural animal activities from its Assurance, yet still empower the IACUC to oversee the use of animals in those areas. Areas that are not directly supported by the PHS, but are functionally, programmatically or physically in an area that could affect PHS supported activities would not be considered excluded from the Assurance. In the case of field studies, local, state-provincial, federal-national, or international laws, regulations and permit requirements likely will pertain to scientific collecting, transport, possession, sale, purchase, barter, exportation, and importation of specimens or parts thereof, or other activities involving native or nonnative species of animals whether endangered or not[3] and fall under IACUC purview minimally for mammals regulated under the AWAR (§1.1, Animal).

The determination of the use of tissue versus the use of animals would depend on how the tissue is obtained. Specifically for PHS funded research, if animals are purposely obtained to collect tissue, that would be considered use of vertebrate animals and the requirements of the PHS Policy, including an Assurance and IACUC approval, would apply. If tissues are provided as extra material from another IACUC approved study that would not be considered animal use under the PHS Policy.

Also impacting research animal acquisition and disposition are regulations and standards covering importation of animals and safety and efficacy of biologicals,[4] endangered and threatened species,[5,6] the safety of the American food supply,[7] the use of infectious agents (biohazards, bloodborne pathogens),[8–11] radioisotopes (usually state regulated), and chemicals and certain toxins.[12] Finally, beyond these bounds, an institution must also act and set policy in ways that serve to protect its good reputation.

14:3 Is a land grant college that uses animals only in agricultural research and training required to follow the AWAR, PHS Policy, or both?

Reg. The AWAR (§1.1, Animals, §2.1,a,3,vi) and the *Guide* (pp. 32–33) specifically exclude livestock bought, sold, or used in agriculture production, agricultural research, or agricultural training. PHS Policy also does not normally cover animals acquired and studied for agricultural purposes; however, it does apply if the agricultural research and teaching component are part of a larger institution with an institutional Animal Welfare Assurance that states that all institutional entities will adhere to PHS Policy. This becomes a germane issue when PHS-funded research is done on institutional farms or other agriculture-related facilities.

Opin. Regardless of the category of research, institutions are expected to provide oversight of all research animals and ensure that pain and distress are minimized (*Guide* p. 33). Whether the research is interpreted as falling within the realm of biomedical or agricultural research, and the concomitant standards for care and use, is a decision best made by the IACUC in consideration of the goals of the investigator and the wellbeing of the study subjects and not necessarily as defined by the setting (i.e., farm versus laboratory) for the holding of animals or conduct of experiments (*Guide* p. 33). Where animals are used in research strictly for agricultural purposes, the *Guide* does not apply. For animals maintained in a farm setting, the *Guide for the Care and Use of Agricultural Animals in Research and Teaching* becomes the applicable guidance document.[1] In order to preserve relations with the USDA and a good reputation and to obtain and maintain AAALAC accreditation, any institution doing agricultural research with food and fiber species should endorse the principles in the agricultural guide.[1] The standards of the *"Ag Guide,"* similar to the *Guide*, advocate roles for the IO, AV and a fully authorized IACUC and promote thorough protocol review and semiannual inspections of housing and study areas. Adherence to standards in that document also promotes good science, enhances animal wellbeing, maintains credibility with the public, provides standardization of research conditions across institutions (and hence reduces variability), and ultimately protects the institutional privilege of conducting research with animals.

14:4 Since neither PHS Policy nor the AWAR address the research use of amphibians, reptiles, fish, and other ectotherms in any detail, should the use of these species in research be monitored by the IACUC?

Reg. PHS Policy (III,A) does cover the use of all vertebrates in research including reptiles, amphibians, and fish and the *Guide* (pp. 77–88) addresses husbandry of aquatic species in research to a significant degree covering areas of environmental conditions and enrichment, water quality, life support systems, habitat enclosure characteristics, behavior and social management, husbandry, sanitation, identification and record-keeping.

Opin. Although the PHS Policy is intentionally broad in scope and does not prescribe specifics about the care and use of any species, it assigns that task to the IACUC and allows for professional judgment. Many of the principles it advocates generally can be adapted to animal care and use programs for animals of any phylogenetic class. Recommendations for specific species are often available from organizations that have an interest in the appropriate care and use of these species in laboratory and field studies.[13–18] It is clear, however, that individual requirements for these three classes of vertebrates, which contain more than 28,000 species, cannot be addressed in a single set of guidelines. Consequently, NIH/OLAW recommends that the advice of experts be obtained to design and develop studies and suitable housing and care procedures for species not commonly used in research.[19,20]

14:5 For the purpose of IACUC review, are birds used in scientific research covered by the AWAR or PHS Policy?

Reg. Although wild-caught birds are covered by the AWAR, those bred for research are not covered by the AWA as a consequence of the Farm Security and Rural Investment Act of 2002 amending the definition of an animal in the AWA (AWAR, §1.1, Animal). (See 12:1.) Historically, all birds were excluded from the AWA.[21] The PHS Policy covers the use of all vertebrates in research, and this category obviously includes birds (PHS Policy III,A).

Opin. As is the case for other vertebrate species, PHS Policy assigns oversight for the care and use of birds to the IACUC and then allows for professional judgment. There are numerous scenarios in which birds may be used in research. Depending upon specific need, birds may be acquired from the wild or from domestic commercial or private sources. Commercial purveyors of birds may provide them for regulated (research) use or for non-research purposes. At our institution, purchased society finches and chicks and trapped wild songbirds are used in neuroscience experiments. Canaries and zebra finches may be acquired from pet stores for similar use. Broad spectra of wild indigenous species are captured for sample collection in studies of enzootic ectoparasitism and West Nile virus prevalence. Domestic chickens, under some circumstances, have proved useful for antibody production, and quail may be used in cardiac research. Where avian species are not commonly used in research, NIH/OLAW recommends that expert advice be obtained to design and develop studies and suitable housing and care procedures.[19,20]

14:6 Are there any special legal requirements regarding accounting and tracking of specific species used in research?

Reg. The federal government specifically requires that detailed records be kept on individual dogs and cats acquired and used in research (AWAR §2.35,b,1–§2.35,b,8) and that annual use of regulated species be reported by pain/distress category (AWAR §2.36,b,5–§2.36,b,8). The species required to be reported are listed under Animal in §1.1, Definitions of the AWAR (see 14:1). Federal and state laws generally require special permits for the use of endangered species in captivity or in field research. Specific record keeping can aid the researcher in proving that animals were born in captivity and in compliance with applicable laws.

Opin. Annual tracking of animal use[22] is a potentially valuable management resource. Such data can be used to show trends in program size and changes in focus that may be useful in program justification and projections. Such information may also be requested by institutional administrators or public officials and may be helpful in clarifying facts about animal research for the public. Many institutions, ours, for example, require accounting of animal use in an annual report.

14:7 Should investigators be allowed to acquire, but not use, animals prior to IACUC approval of the protocol?

Opin. It is generally not advisable to allow advance purchase, capture, or acquisition by other means for several reasons. First, there may be unintentional use or temptation to use or to prepare to use the animals without appropriate authorization. The conduct of any animal activities that have not been reviewed and approved by the IACUC is considered by APHIS/AC and NIH/OLAW as noncompliance, and for NIH/OLAW, it must be reported to that office.[23] Second, if animals are acquired prior to consideration of a research proposal and the proposed use of the animals is subsequently rejected by the IACUC or the number of animals approved by the IACUC is reduced from the original request, this situation may present problems in managing the population and would constitute violation of legal requirements to use the minimal number of animals necessary (see 14:15). An exception to this might be fish maintained and studied in large numbers, those kept at marine or freshwater experiment stations for prospective use, and those trapped in the wild (see 14:8; 14:25).

14:8 Is it necessary to have IACUC approval for animals that are just being maintained (no research being performed)?

Opin. In addition to animals directly used in research, NIH/OLAW and APHIS/AC implicitly require that institutions establish mechanisms to monitor and document all animals acquired or produced.[22] This includes any animals that are kept for breeding purposes, nonhuman primates awaiting assignment in core pools at primate centers, dogs and cats in centralized institutional conditioning programs, gift rodents secured in quarantine programs, and even animals held in abeyance in the case of a lapsed or suspended protocol. Without IACUC approval or at the expiration of a protocol, animal use is not valid. Continuation of animal activities beyond the expiration or use of animals without approval is a serious violation of the PHS Policy and AWAR and is reportable to NIH/OLAW.[23] In the case of an expired protocol, especially where the likelihood exists that it will be legally renewed or the animals will be transferred to another protocol, rather than the wasteful and internecine approach of animal depopulation, the IACUC should establish a mechanism for temporary oversight and management of these animals. Even in the case of fish caught and held for prospective use, a husbandry and holding protocol is appropriate. While other mechanisms may be contemplated to manage these circumstances generating a protocol would be consistent with standard practices (see 14:7).

14:9 What institutional entity should procure animals for research?

Opin. One of the critical responsibilities of an IACUC is to assure that the institution lawfully procures appropriate animals for research. This is best accomplished through the institution's animal resource program or other appropriately designated office.[24] Veterinarians involved in the animal resource program must have specialized training in laboratory animal medicine (AWAR §2.31,b,3,i; PHS Policy IV,A,3,b,1; *Guide* p. 15) and usually have the expertise to evaluate animal sources and identify those that are appropriate for research given the institution, its resources, and its needs.

It is well within the jurisdiction of the IACUC, especially in consideration of its responsibility to ensure an adequate program of veterinary care and a safe work environment, to provide guidance to the animal resources program concerning the species, legal sources, genotypes, and microbiological background of animals acquired for research use.

In smaller academic or industrial programs with part-time or consultant veterinary staffs, the IACUC may well want or need to have considerable input into this process. Coordination of veterinary care, vendor approval, and a centralized purchasing system will aid the tracking of animal use as discussed in 14:6; 14:10; 14:11; 14:13; and 14:14.

14:10 Should the IACUC become involved in approving an "acceptable" animal vendor?

Opin. The loss of animals or research to preventable infectious diseases has ramifications for the institutional requirement to ensure that research is not unnecessarily duplicative (AWAR §2.31,d,1,iii; *Guide* p. 26).[1] Consequently, it is within the purview of the IACUC to work with the professional veterinary staff to identify acceptable animal health standards. In most instances, IACUCs defer to the professional judgment of the veterinary staff to identify vendors that meet appropriate health and legal standards. Often, this is the most effective means of addressing this obligation. Nevertheless, when rodent use is high, when certain animals are scarce, or when funds are inadequate, veterinarians may be under pressure to permit the entry of animals with suboptimal health status into the institution. In this situation, it may be worthwhile for the IACUC, as the de facto representative of the community of scientists, to work with the veterinary staff to develop and use standards that are in the best interests of the community of scientists within the institution.

14:11 What constitutes an acceptable animal vendor? What procedures can the IACUC or animal resources program use to assess vendor quality?

Opin. In general, animal vendors should have a practice of producing consistently healthy animals of a specific genotype and health status in compliance with applicable laws, statutes and regulations, and institutional needs. Under the AWAR

(§2.40,a) vendors are required to have a program of veterinary care. Specifically, the AWAR require licensing of any of the following that supply warm-blooded animals, other than common laboratory rats and mice, to research institutions:

- Pet shops (§2.1,a,3,i)
- Breeders selling more than 25 dogs or cats annually that were born and raised on their premises (§2.1,a,3,iv)
- Persons or commercial enterprises deriving more than $500 annually from such sales (§1.1, Dealer; §2.1,a,3,ii)
- Any person or entity (Class B Dealer) selling any dog or cat not born and raised on its premises (§2.1,a,3,iv)

Additionally, institutions should make a dedicated effort to ensure that all transactions and related procedures involving animal procurement are done in a lawful manner (*Guide* pp. 106–107). Other considerations are whether the vendor has an adequate program of veterinary care and a history of consistently meeting consumer needs and expectations. Ideally, vendors should be AAALAC accredited or provide assurance that they meet standards for their industry (i.e., health quality, animal care, customer service) and applicable federal laws. For rodent vendors, the health status of the animals should be documented through a regular program of health surveillance and verified by a rational means that includes sampling technique, sampling strategy, appropriate diagnostic tests, sufficiently frequent testing, a reliable laboratory, and prompt reporting of changes in health status. In most instances, the evaluation and consideration of animal vendors are handled within the program of veterinary medical care, with IACUC oversight.

14:12 What precautions should the IACUC and animal resources program require if animals of unknown health status must be acquired?

Opin. Acquisition of animals of unknown health status should be limited to those instances when the animals are only available from one source. Nevertheless, this process still has several risks. The primary threat is the transmission of infectious diseases to pathogen-free animals. In addition to causing otherwise preventable pain and distress, research may be compromised or invalidated and therefore deemed unnecessarily duplicative (see 14:10). Consequently, the IACUC should ensure that the institution has the components of a program of veterinary medical care that provide for the stabilization, isolation, and health characterization of animals of an undefined health status (*Guide* pp. 109–112).[1]

Quarantine programs should be suitably long to detect infection through the expression of clinical signs or developing antibody titers or permit the incubation or subclinical stage of a disease to become manifested via culture, the presence of parasites and/or their eggs, pathogen nucleic acids, or pathognomonic lesions and thereby minimize the risk of introduction of pathogens into established colonies. Where bacteriology or polymerase chain reaction (PCR) technology can be employed, results can often be obtained quickly; however, responsive antibody production may not be detectable for 1 to 3 weeks or more, depending on the amount and timing of the pathogen inoculum, husbandry factors, and characteristics of the host. Some mice, for example, are purported to have poor, uneven or low antibody production in response to *Mycoplasma pulmonis* infection.[25] Along with health surveillance program attributes, mouse age and genotype profoundly influence mouse parvovirus shedding and detection.[26] Although investigators may feel a stringent quarantine and testing program is a hindrance to their research, data are readily compromised when using potentially infected animals.

Health background information from trustworthy sources should be sufficient to enable the veterinarian to establish appropriate quarantine parameters (e.g., duration, diagnostic procedures, therapeutic or prophylactic interventions, etc.) and to enable the physiological, nutritional and behavioral adaptation of new arrivals to any social or environmental novelty.

An effective quarantine program should also provide for the rederivation of diseased animals by surgical or other appropriate means and other relevant services such as pathogen testing for cells and cell lines that may potentially affect the health of animals within the program.

14:13 Should the IACUC attempt to track the number of all animals acquired or only those used under approved protocols?

Opin. Although neither the PHS Policy nor the AWAR explicitly require an institutional mechanism to track animal usage by investigators under IACUC approved activities, both require that applications to the IACUC specify and include a rationale for the approximate number of animals proposed to be used (AWAR §2.31,e,1; §2.31,e,2; PHS Policy IV,D,1,b; *Guide* p. 25). These provisions implicitly require that institutions establish mechanisms to monitor and document the number of animals acquired or produced and used in approved activities.[22] This includes any number of animals that are kept for breeding purposes or culled prior to research use and not subjected to any experimental manipulations.[27] Tracking is particularly important for institutions that maintain large breeding colonies of nonhuman primates where animals produced from this population may be held for lengthy periods before assignment to a research protocol. Institutions that use dogs and cats must maintain acquisition and disposition records in compliance with federal law (AWAR §2.35,b) and should invoke steps to verify that pets are not being received (*Guide* p. 106). These resources may also operate, depending on the institutional need, by maintaining pools of unassigned animals and allotting them to research protocols only after a suitable stabilization period. In some cases, however, tracking of animal numbers may be difficult or counterproductive (see 14:15).

14:14 How should numbers of animals acquired and used be tracked by the IACUC?

Reg. The AWAR (§2.36,b,5–§2.36,b,8) require that the common names and numbers of warm-blooded animals (other than common laboratory rats and mice) used for teaching, research, experiments, or tests or held for all such purposes except applied production agriculture research be included in the annual report to APHIS/AC. NIH/OLAW only requires reporting of average daily inventories of animals, by species, in new and renewal Animal Welfare Assurances.[28]

Opin. It is implicitly required that institutions establish mechanisms to monitor and document the number of animals acquired and used in approved activities (see 14:13).[22] The most effective approach combines a methodology for tabulating animal use with a policy regarding the accuracy or precision of the process. Methods of tabulation can be either manual or automated. The use of commercial animal ordering software and census tabulation by bar code scanning or radiofrequency identification (RFID) allows for most integrated animal resources programs to track, with relative ease, the number of animals purchased from commercial sources and imported into the institution through a quarantine program. This process provides accurate figures with relatively little investment in time and personnel. Admittedly, accounting for the production of rodents bred and weaned on site may be more challenging. When acquisitions are linked to IACUC approval numbers, investigators, depending upon the capabilities of the technology, can be automatically informed when their use has reached a preset percentage (e.g., 80%–90%) of the animals approved on that protocol and with a request to provide a specific justification if it is anticipated that the number of animals ultimately required will exceed the number approved.[19] Small institutions that use limited numbers of animals may find it most effective to maintain a hardcopy log of each IACUC approved activity, merely subtracting the number of animals acquired for each order or bred and added to the census from the number approved, with verbal or written notification to the investigator as the number of animals approved is approached.[19]

With respect to the precision of the tabulation process, IACUCs in aggregate historically have approached this consideration with some lack of unity.[29,30] While most committees attempt to track all animal use with precise accuracy, others attempt to track use with approximate accuracy and apparently roughly one in twenty do not do anything at all (see survey 1, below). More flexibility is often found in practices concerning rodents and ectotherms than those concerning nonrodent mammals. The major reason for this diversity in accounting essentially relates to discrepancies in tracking and reporting requirements between the AWAR and PHS Policy. Species not regulated by the AWAR (e.g., rats, mice, birds, fish) are often accounted for by using approximate means (PHS Policy IV,D,1,a), appropriate amounts (*Guide* p. 12), or the minimum number to obtain valid results (U.S. Government Principle III) rather than prescribed precise figures. Even when a species may not be regulated under the AWAR, where the experimental use is highly invasive or has considerable potential for significant pain or distress, it may, at the discretion of the IACUC, be subjected to higher scrutiny and require precise justification and tracking.

Institutions that track and report the approximate number of certain species used in research may be in full compliance with the AWAR and PHS Policy. However, many internal and external factors (including financial management, public opinion, and the spirit of the law) support an accurate accounting of the animals used in research. Whatever approaches and processes an institution chooses, it must satisfy the PHS Policy requirement that the number of animals used be limited to the appropriate number necessary to obtain valid results.[22,31] The AWAR (§2.31,e,2) also has a requirement to provide a rationale for the appropriateness of the number of animals to be used.

Surv. 1 Does your IACUC (or an administrative office) track the number of animals acquired for research with precise or approximate accuracy?

- We track all animals acquired with precise accuracy 193/295 (65%)
- We track all animals acquired with approximate accuracy 70/295 (24%)

- We do not track the number of animals acquired 18/295 (6%)
- Not applicable 11/295 (4%)
- Other 3/295 (1%)

Surv. 2 In cases in which your IACUC tracks the number of animals acquired for research with only approximate accuracy, what species are involved? Check all applicable responses.

- Mice 84/244 (34%)
- Rats 77/244 (32%)
- Rabbits 43/244 (18%)
- Fish 42/244 (17%)
- Guinea pigs 34/244 (14%)
- Amphibians 32/244 (13%)
- Hamsters 31/244 (13%)
- Birds 31/244 (13%)
- Vertebrates used in field research 28/244 (11%)
- Not applicable 138/244 (57%)
- Other (including swine, nonhuman primates, reptiles, ferrets, cattle, bats) 17/244 (7%)

14:15 What factors prevent the precise prediction and accounting of the number of animals acquired and used for research?

Opin. Although precise accounting of animals is not always required (see 14:14), investigators must still satisfy the requirement that the number of animals used be limited to the appropriate number necessary to obtain valid results (AWAR §2.31,e,1–§2.31,e,2; PHS Policy IV,A,1,g; PHS Policy IV,D,1,a; U.S. Government Principles III). This requirement often produces conflicting pressures, on the one hand to justify animal numbers (including the use of statistical analysis and power calculations (see Chapter 13), on the other to account for the inherent unpredictability of science. Foremost, it is fundamentally difficult to forecast animal requirements accurately for the 3- to 5-year funded lifetime of all but the smallest and simplest endeavors, especially given that results and their impact on subsequent research direction cannot typically be predicted. This difficulty has a virtually exponentially greater effect in sizable, active laboratories with multiple projects and numerous technicians, graduate students, and postdoctoral fellows carrying out various aspects of research. These conditions can often lead to discrepancies between the number of animals approved and the number necessary for the completion of the work. In addition to changes in research direction or increases in experimental complexity, failed experiments, unforeseen technical challenges, unexpected disease, and other complications may cause more animals to be used than originally expected (or requested). Given these circumstances, it is not reasonable to expect scientists to be able to project the exact number of animals needed to do the work. It is reasonable, however, to require a valid estimation and, where additional animals must be added to enable ongoing work to continue via the protocol modification process, to expect a logical explanation of why previous approximations were incorrect or why more animals are needed.

The real challenge in animal tracking is found in large breeding colonies of prolific animals such as ectotherms (e.g., fish), chickens, or rodents. Dealing with the sheer volume of animals produced can be a full-time job for animal caretakers and users, and attempts to count the animals accurately are often overwhelming. Activities involving the production of genetically manipulated animals are of low yield but require the production of large numbers of animals, the vast majority of which are unsuitable for research.[27] (See 14:18.) Avian and rodent breeding colony accounting may be facilitated if the populations are managed by a centralized animal resources program, but there are trade-offs in this practice such as insulation of researchers from their animals and risks of inequity between production and demand.

Additional difficulties are presented when animals are used or collected in the field or bred in fisheries or when animals are transferred from one investigator to another. In large-scale pond and tank production systems, where fish are propagated or held for subsequent research projects or used in production aquaculture research, there may be tens to hundreds of thousands of fish involved.[32] In such studies, it is not ordinarily possible to count the precise number of animals used and an estimation of the number of fish acquired and maintained should be sufficient. Additionally, handling of fish, simply for purposes of counting, may be distressful and counterproductive to the maintenance of good health.[32] As such, IACUCs should consider whether the precise tabulation of fish is critical to the experiment itself or important to realistic minimization of fish pain or distress, or whether it constitutes an otherwise groundless exercise. In the context of protocol review involving fish, it is often reasonable for the IACUC to review the general number of animals involved, the rationale for that number, and justification for any vagueness.

While it is not outside the realm of possibility to count every single animal in an enormous population, the cost–benefit ratio is not conducive to the practice, and doing so arguably would have little impact on the humane care or use of animals except in identifying situations where a disease, environmental variability, or genetic contamination may be influencing the numbers of pups born and weaned. Although of less significance, receiving procedures may not account for extra animals included in a shipment to be counted against ceilings approved by the IACUC (see 14:21).

Other accounting challenges include instances in which animals may be used in more than one study or where animals are kept for several years for the same study. Animal use in more than one study usually is of one of two types. First are studies in which the animal is killed for a specific purpose but tissues are taken for another study (discussed in 14:30). Next are studies in which the longevity or rarity of the animal, the duration (e.g., aging) or innocuous nature of the primary study (e.g., behavioral), or the need to study disease pathogenesis or response to treatment over a lengthy period would allow for the animals to be used in long-term or justified sequential or serial experiments. The latter scenarios would usually involve nonhuman primates or higher vertebrates; thus it is especially important that each use of an animal is justified in an application and that multiple experiments do not jeopardize the quality of the data or unduly impact the welfare of the animal. For reporting purposes, animals should only be included once except when specific procedures are required to be reported. For purposes of annual reporting of species regulated by the AWAR, animals are to be counted only once, regardless of the number of studies in which they are used, and those used in more than one protocol should be counted in the most painful or distressful category.[4,30] When animals such as dogs, cats, or monkeys are studied for several years or alternate between periods of study and disuse, they should be counted annually.[33]

In the context of the overall research program, scientists often fail to see precise accounting as valuable and thus do not commit precious human resources to the activity. There may also be a disincentive to accurate reporting if it brings the risk of extra time and effort required to amend protocols and add more animals. Peer review groups of most funding agencies do not micromanage projects down to details such as animal numbers. Instead, most rely on the powerful influence of funding, an influence that is often underestimated by the public and largely ignored by the vocal opponents of research. IACUCs, in turn, may stringently enforce what has been conveyed as a loose regulatory standard for animal number approximation (AWAR §2.31,e,1–§2.31,e,2; PHS Policy IV,A,1,g; PHS Policy IV,D,1,a; U.S. Government Principles III). Sometimes, an "us and them" relationship can develop between investigators and IACUCs and can lead to less than optimal accounting of animal usage. Generally, IACUCs need to communicate their role to investigators in an effective way and be responsive to the sometimes rapidly changing needs of the investigators. An effective mechanism is the institutional training program (particularly for new investigators), supported by periodic workshops or communiqués to the established community of scientists explaining the relevant laws and the obligation of the investigators.

Widely available, especially for many larger programs, are technological innovations that save labor and enhance process consistency, such as electronic census acquisition by bar code scanning or RFID. The use of these technological tools ordinarily forces institutions to address the issue of accounting for carryover animals as old protocols and associated cage cards expire, renewed protocols are initiated, and replacement cage cards must be produced. The identification or estimation of the number of carryover animals will be integral to the accuracy of this process and must be considered in the renewal along with any new animals to be added. Otherwise, investigators will run the risk of exhausting the number of animals approved by the IACUC before the anticipated conclusion of the project or the IACUC risks approving substantially more animals than necessary.

The deployment of census management technology is not a cure-all; however, as there remain limitations related to most products and their applicability and possibly explaining why 24% of IACUCs surveyed using tracking systems find them to be difficult to use, not used effectively, or inaccurate (see 14:38). Despite all these considerations, the expectation of most organizations with research oversight responsibility is that there will be a good faith effort made in specifying and tracking the number of animals used.[22,28]

14:16 Should the number of animals generated by breeding colonies be tracked?

Reg. The AWAR requires reporting of the species and number of regulated animals that are bred, conditioned, or held for use (§2.36,b,8). NIH/OLAW implicitly requires that institutions have mechanisms and processes in place to document and monitor the numbers of animals acquired and used including those produced from in-house breeding colonies.[22]

Opin. Breeding animals for research purposes, stated simply, constitutes research. Consequently, animals produced by breeding colonies should be counted. The expectation of legal, regulatory, and accrediting bodies is that research facilities make a good faith effort at counting not only the number of progeny from these colonies that are used directly in research, but the total number of animals produced.[27] A previous survey from 2007 showed that IACUCs consistently, and in overwhelming numbers, require the tabulation of rodents and other mammalian species produced from in-house colonies.[30] When a breeding colony is a centralized core, or supports multiple projects (or even individual investigators), animal production should be tracked by a dedicated breeding protocol. In the case of dogs and cats, federal law requires precise accounting at the time of weaning (AWAR §2.35,b; §2.38,g,3; §2.50,d). For most other species, progeny are likewise counted at weaning and debited against a numeric ceiling approved by the IACUC and include both those designated for use in research as well as those that will not be used.[27] Regrettably, the unavoidable production of unusable offspring is an integral part of the research enterprise.[27] (See 14:18.)

14:17 What information is useful and how should the IACUC address the acquisition of new research animals from breeding colonies?

Opin. The IACUC's role for oversight regarding breeding colonies includes ensuring that the need for a breeding colony has been established on the basis of scientific or animal welfare concerns; that the procedures for care, genetic authenticity, genotyping, and use employed in the breeding colony are consistent with the *Guide* and evaluated and approved by the IACUC on a regular basis (i.e., as part of both protocol review and the semiannual program review); and that there is a mechanism for tracking animals.[24,27,34,35] To review SOPs for breeding colonies, the IACUC will need information about colony management. These practices should be briefly described in the investigator's animal protocol and justification provided for departure from any standard institutional practices.

To reduce the potential for unnecessary animal wastage and assure the validity of data derived from experiments using animals from the colony, the IACUC should assess the specific mouse breeding colony management knowledge and skills of persons managing the colonies. Special emphasis should be placed on this individual's background in rodent genetics, breeding practices, genetic monitoring, and understanding of nomenclature.

Examples of other information that might prove useful to the committee include the typical number of breeders and number of young per cage, the breeding system (including number of females per male or continuous versus interrupted mating), the intended weaning age, methods for identification of individual animals, and even the breeding scheme (e.g., cross–intercross, backcrossing). This sort of information is especially important with polygamous or continuous mating systems when there is the prospect of large numbers of pups from several litters to be present in any cage. The estimated number of animals that are kept for breeding purposes and not subject to any experimental manipulations should be part of the animal protocol.[27] Intensively-bred rodent colonies must be managed in ways that demonstrate successful production outcomes and preserve animal well-being.[36]

Determining which animals to include in the estimated number of animals on an animal protocol can be especially daunting to the investigator and the IACUC in the absence of IACUC-developed guidelines. Studies involving genetic analysis are particularly animal intensive. It is possible for the investigator to estimate the number of animals required, but difficult for the IACUC to evaluate this estimate in the absence of experience.[27] A reasonable estimate is for the requirement of about 1200 mice to map a single gene with recessive inheritance and full penetrance derived from a nucleus of 10 to 12 monogamously mated mice.[27] Similar numbers of mice may be required for quantitative trait loci analysis using the F_2 progeny derived from an initial cross of four to six pairs. Using "speed" congenic technology, up to 750 mice may be necessary to derive a congenic strain, providing the homozygous mutant is fertile.[27] Even in the straightforward case of established colonies in which all offspring consistently meet criteria to be suitable for research, the exact number of animals can only be approximated because it is impossible to predict the exact number and sex of offspring. After founder transgenic or "knock-out" mice have been identified, between 80 and 100 mice may be needed to maintain and characterize a line and assure heritability under the assumptions of five breeder pairs per line, regular breeder replacement prior to senescence, no unusual infertility, and adequate numbers of weanlings for genotyping and phenotyping.[27]

In the preparation of a protocol, the estimated number of animals should clearly distinguish among breeders, young that cannot be used in experiments because they are of the wrong genotype or gender, and animals that will ultimately be subject to experimental manipulations. If a study requires fertilized one-cell eggs, embryos, or fetuses, the protocol should indicate the number of eggs, embryos, or fetuses that are required for proposed studies.[27] The estimated number of experimental animals may be limited to the number of female animals that are mated and euthanized or surgically manipulated to collect the required eggs, embryos, or fetuses. In this situation, males might be listed as breeders if they are not subject to any experimental manipulation. Likewise, when breeding colony animals will be used as sources of tissues, cells, or other biological materials, the total number to be used for experiments should be estimated on the basis of the typical amount of biological material needed for an experiment, the approximate number of animals necessary to yield the material, and the number of experiments projected.

The accounting of animal production can be done by animal care personnel recording births and weaning. On a large scale, this is an intrusive and labor-intensive process often best done electronically. As a new cage of mice is populated with weanlings removed from a production cage, either the exact number of mice or an average (typically 3.25–3.5 mice/cage) can be deducted from the maximal number of mice approved from the protocol. Regardless of approach, the process must support the mandate to the IACUC to limit the number of animals produced necessary to support the work and not exclude those animals produced in excess of need, either inadvertently or necessarily, or genetic culls.[28] Litter sizes also should be assessed with sufficient frequency and in a manner ensuring that space recommendations of the *Guide* are routinely heeded.[28] As an alternative, investigators report to the designated office, at regular intervals, the number of animals born, weaned, or used in studies.[24] This report can be tallied against the numbers in the approved protocol.

14:18 Is there a percentage of unused breeding colony animals that are euthanized (without having been used as breeding stock or experimental subjects) that should be tolerated by the IACUC?

Opin. Substantial numbers of animals may be required to maintain a breeding colony. For example, while trying to establish a breeding colony for a new mutant rodent or ectotherm model, the investigator will also be working to determine phenotype, to identify affected physiological system(s), and to define inheritance pattern in addition to breeding founder and subsequent generational animals. Under these circumstances, the production of unused animals that are subsequently euthanized is almost inescapable. This is a significant issue for commercial and core rodent production colonies that must operate at a production excess (approximately 30% or more) in order to meet unexpected increases in market demand or that produce animals at an excess when demand unexpectedly takes a downturn. A number of user factors such as newly funded projects, changing experimental initiatives, failure to plan, cessation of standing orders, and periods of relative research dormancy alone or collectively may have a dynamic impact on production. For animals from a production colony, matching supply with need is a moving target, and, therefore, production cannot reasonably be expected to match demand.

The case of creating genetically unique animals by using homologous or nonhomologous recombination presents many examples of animals that are produced and euthanized without further use. In the creation of a transgenic (nonhomologous insertion) mouse using the pronuclear microinjection technique, only 10%–30% of the mice born will carry the transgene; thus, 70%–90% cannot be used.[37] Although most transgenic founders will subsequently transmit the foreign gene in 50% of their offspring, approximately 20%–30% are mosaic and transmit the gene at a lower frequency (i.e., 5%–10%).[37,38] Consequently, in the breeding of transgenic founder animals anywhere from 50% to 95% of the offspring will not carry the gene of interest, will have no research value in many instances, and must be euthanized.

In the case of genetically unique mice or breeding of small colonies used by a single investigator during periods when progeny may not be needed for experiments and where sperm or embryo cryopreservation techniques may not be available, mating must continue in order to perpetuate the colony. When this happens, the offspring that are not selected as breeders must be euthanized. In managing inbred animals by a single-line system, genetic divergence is prevented by euthanizing breeders, their progeny, and their grand-progeny that cannot be traced within a minimal number of generations to a common ancestor. As one might imagine, in a large production operation involving multiple genotypes of animals, the euthanasia of large numbers of animals results.

For these reasons, it is recommended that IACUCs permit animals to be produced from breeding colonies in reasonable excess of projected or historical need.

14:19 Should unweaned animals (particularly rodents), or animals used or harvested *in utero*, be accounted for in the protocol review process?

Opin. If painful procedures are to be performed on animals, including those that are immature but at a stage of development able to perceive pain (i.e., late *in utero* or neonatal), it is clearly within the jurisdiction and responsibility of the IACUC to address this issue (*Guide* pp. 25–26). (See 8:11; 13:11; 16:35.) If suckling rodents will be subject to any manipulation, such as thymectomy, toe clip, ear notch for identification, tail tip excision for genotyping, or behavioral tests, the estimated number of manipulated sucklings must be included in the number of animals used.[27] Essentially all IACUCs adhere to this doctrine and, although no specific guidance exists, most also oversee the use of fetuses in research (see survey below). NIH/OLAW clarifies that institutions must appropriately monitor and document numbers of all animals acquired (through breeding or other means) and used in approved activities and to include genetic culls or those produced in excess of need.[28]

Surv. 1 Does your IACUC ever consider a mammalian fetus to be a vertebrate animal subject to the oversight of your IACUC?

- Not applicable 95/295 (32%)
- Yes, at any fetal age 70/295 (24%)
- Yes, after a certain fetal age (species dependent) 46/295 (16%)
- No 71/295 (24%)
- Other 13/295 (4%)

Surv. 2 At what age does your IACUC begin to count neonatal mice in the animal census?

- Not applicable 72/292 (25%)
- At or very soon after birth (i.e., within a day of birth) 74/292 (25%)
- Within 48–72 hours after birth 16/292 (5%)
- Within 4–5 days after birth so as not to disturb the litter 24/292 (8%)
- At weaning 97/292 (33%)
- Other 9/292 (3%)

14:20 Should preweanling rodents be counted and debited from approved protocols especially if procedures are done, and at what age?

Opin. If suckling rodents will be subject to any manipulation, such as surgery, permanent physical identification, biopsy for genotyping, or behavioral tests, the estimated number of manipulated sucklings must be included in the number of animals used.[27] While this expectation seems reasonable, it may be difficult to manage logistically, especially may add confusion to the process of monitoring and tracking animal use, and may be highly associated with the size of the mouse population and census and other animal accounting processes. Be that as it may, NIH/OLAW stresses that "institutions need to appropriately monitor and document numbers of animals acquired (through breeding or other means) and used in approved activities" and that such practices "not exclude the disposition of animals inadvertently or necessarily produced in excess of the number needed or which do not meet criteria (e.g., genetic) established for the specific study proposal."[28] In particular, it is difficult to tabulate precise numbers of rodents born without disturbing them at critical times at or around birth and during rearing. The animal resources program's engaging in this process adds disruption and expense to the process without any apparent animal welfare benefit. While it is possible for the research staff to report these numbers to the animal resources program or IACUC office as they are generated by the manipulative event, a high degree of dedication to the process of all parties is required. After protocol review and depending upon the accuracy of information conveyed from the IACUC, there is a risk for the person responsible for tracking to mix, confuse, and interchange the total number of animals born, the subset of any used for a distressful manipulation, the accounting of weanlings, and any parental animals that may be purchased (see 14:19 and 14:21). In institutions where it is applicable, approximately 75% of IACUCs attempt to tabulate the precise number of rat or pups born (see survey below). At our institution the use of fetuses or animals up to the age of weaning is not tracked unless there are research activities involving pain or distress, but, rather, the adult breeding animals needed to produce the necessary number of offspring. Clearly, if painful procedures are to be done on animals, including presumably young ones capable of pain perception or experiencing of distress, it is within the aegis of the IACUC to address this issue (*Guide* pp. 25–26).

Surv. With reference to indicating the number of pups born in a rat or mouse litter.

- We try to count the individual animals 154/293 (53%)
- We estimate the number of pups in the litter 44/293 (15%)
- Not applicable 83/293 (28%)
- Other 12/293 (4%)

14:21 A commercial animal vendor sends two more rats than were ordered. How should the two extra animals be accounted for in the IACUC's and PI's records?

Opin. On one hand, it may not be necessary to tabulate the extra rodents given that the animals were not ordered, were received unexpectedly, and could not reasonably be considered by using statistical analysis or other means in the formulation of a protocol. Conversely, the animals do represent a windfall that could be used in manners or for purposes outside IACUC oversight. If the approved use involves highly invasive procedures, considerable pain or distress by design, or species of interest for any variety of reasons (e.g., regulated by the AWAR), the IACUC may be particularly interested in regulating any access to individual animals in excess of the IACUC-reviewed and approved figures. In the end, the housing density, census tabulation methodology, and protocol debiting procedures at the operative level in the animal research facility are likely to determine whether the extra animals are counted. For example, at our institution the rat census is done by counting each individual animal. Each rat has a corresponding individual cage card, and electronic census methods are used. Therefore, the extra rats would be debited from the protocol when new cage cards were made at the time of initial housing immediately after receipt. If the census is done by the cage, as for mice at our institution, extra mice in a shipment may or may not be counted. Capture of these animals for accounting purposes would hinge upon the research instructions for housing density and whether additional cage cards must be generated to accommodate all of the cages. In the context of the overall numbers of animals received by the typical investigator or acquired annually by the institution, these extra animals, included in shipments to buffer against any incidental mortality, are generally insignificant. The authors recommend that the IACUC work with the institutional animal resources program to develop a written policy and procedure regarding the disposition of animals received unexpectedly.

14:22 An IACUC approves the use of one litter of rodent pups (approximately ten pups) that will be born of one dam. What can the IACUC do to prevent a person from purchasing a total of 11 animals, rather than just the one dam?

Opin. The most important considerations for the IACUC are to define when an animal is considered to be "acquired" and, in the context of protocol review and approval, to limit the approved number of animals to those that have been "acquired." Generally, this translates into the age or experimental conditions at which the institution's SOP dictate it is to be counted on the census. Under practical conditions and especially where the census is substantial, rodents are often not tabulated until weaned. By extension, it is also critical for these definitions and expectations to be clearly conveyed to the animal purchasing party (commonly, a clerk or animal order specialist affiliated with the animal resources program).

The NIH/OLAW mandate to document all animals produced, even if not used scientifically, including those culled and euthanatized before weaning, and whether or not painful procedures are done,[28] creates a possible conundrum. Institutions seemingly must commit resources to document every animal acquired and particularly all unweaned rodents, whether or not used in actual research. The alternative is to use human and technological resources rationally to apply constraints that in the end allow only the acquisition of the number of animals approved by the IACUC to conduct valid experiments. For example, when timed-pregnant female rats are purchased as the source of pups to be used in a study, to prevent excessive ordering it is necessary to segregate the number of sucklings projected for experimental use from the number of dams estimated to produce them, and to convey the appropriate information to the purchasing entity. If 1000 nursing rat pups of either sex have been justified for use in acute experiments and 100 outbred, pregnant female rats have been estimated to produce that number, the total number of rats approved to be *purchased* for the study would be 100 (the timed-pregnant females). If the IACUC was to approve 1100 rats (1000 + 100) and convey this to the purchasing entity (and even injudiciously to the investigator), this might create an error whereby up to 1100 timed-pregnant females could be ordered. The requirement to monitor, document, and scientifically justify the acquisition of all 1100 rats required by the study can be met in the context of the IACUC protocol review. After all, the pivotal regulatory requirements to approve a certain number of animals for a study are somewhat conflicting and ambiguous. The various charges are to consider the approximate (AWAR §2.31,e,1; PHS Policy IV,D,1,a), appropriate (AWAR §2.31,e,2; PHS Policy IV,D,1,b), or minimal (PHS Policy IV,A,1,g) number of animals needed. Arguably, this is not necessarily the precise number. As a facet of protocol review, and in an attempt to meet these requirements, it may be helpful to separate the number of adult animals to be acquired, those to be subsequently born at the institution, and, of those born, those to be used scientifically while still nursing (not ordinarily counted if rodents) and those ultimately to be weaned (customarily tabulated).

The authors caution institutions to weigh the pros and cons of tasking someone with the requirement of tabulating animals as they are born and debiting them promptly from an IACUC-imposed limitation. Such intrusive activity seemingly creates an animal welfare threat (e.g., litter cannibalism or abandonment), may disrupt normal behavior (e.g., nursing care, sleeping), may add variability to the environment, and may not provide much of a benefit beyond the rational and comprehensive consideration of animal numbers done in the context of IACUC review. Beyond these considerations, there is also an annoyance risk in this process for the investigator. For example, assume that the IACUC approves 100 male and female rats and some of the females are pregnant. Their progeny are not only born, but also weaned and counted. In this situation the scientist will be in a deficit (over the approved limit) situation that must be administratively reconciled. These undesirable predicaments can be obviated when the policies and procedures are clear and understood by all parties.

14:23 An investigator uses avian embryos in his research. What is the responsibility of the IACUC, if any, in overseeing this activity?

Opin. Although PHS Policy does not discuss considerations attendant to embryonated bird eggs (or mammalian fetuses, for that matter), nonvertebrate animals (chick embryos within eggs) have the potential to develop into hatched vertebrate animals. While embryonal states of avian species develop vertebrae at a stage in their development prior to hatching, NIH/OLAW has interpreted *live vertebrate animal* to apply to avians only after hatching.[39,40] However, the risk of eggs hatching and producing chicks (requiring food, water, proper housing, and veterinary care and placing them under the purview of PHS Policy) dictates that IACUCs consider developing policies for different aged avian embryos, newly hatched birds, the point at which bird embryos are considered vertebrate animals, and interventions that address the care or euthanasia of animals that hatch unexpectedly.[40–42] For chickens, the last 3 days of incubation (incubation days 18–21) represent the last stage of embryo development and coincide with the chick's drawing the yolk sac into the body and having sufficient pulmonary maturation to handle oxygen and carbon dioxide exchange.[43] During this period, some chicks may hatch normally and some prematurely hatched chicks may survive outside the egg with little additional care. (See 8:11; 16:35.)

14:24 Zebrafish are born with their yolk sac attached. At what stage of development are they considered vertebrate animals under the PHS Policy? (See 12:1.)

Reg. The PHS Policy has been interpreted by NIH/OLAW to apply to fish and amphibians at the time of hatching.[44]

Opin. As larval forms of fish and amphibians have vertebrae, PHS Policy applies to the offspring of egg-laying vertebrates upon hatching. For zebrafish larvae, for example, this typically occurs 3 days post-fertilization. This definition should cover all reproductive strategies found in fish, including oviparous and viviparous species.[45] Fish are not regulated by the AWAR.

14:25 How, and with what precision, should animals be accounted for when used in field studies?

Reg. The AWAR (§1.1, Field Study) grants an exemption to field studies in which free-ranging wild animals are observed in their natural environment and without the involvement of invasive procedures or intrusions that harm or materially alter behavior. Where federal funds are used to support field studies, PHS Policy and the U.S. Government Principles, as specified in the *Guide* (pp. 199–200), dictate that the approximate number of animals to be captured be addressed.

Opin. Although the AWAR grants an exemption to observational field studies, few such studies can be done without interventions or some degree of invasiveness.[46] Given the variety of procedures that could be done in the field, the associated stress and risk of untoward outcomes associated with capture, and variation in researcher experience and skills, virtually any procedure, even simple handling, has the potential to cause harm. Because the tabulation of animal numbers revolves around the issue of invasiveness, it is important for the IACUC and involved researchers to address the specific procedures, potential refinements, and qualifications and training of personnel. After appropriate deliberation, it should be possible to define the relative invasiveness of the field study. If the study is determined to involve more than detached observation, the IACUC and investigator must address not only the acquisition of animals by capture, but any allowable rate of attrition related to capture and the experimental interventions. IACUCs also should be aware that the federal and state wildlife permitting processes are usually rigorous and involve an assessment of hazard of the procedures and overall risk to the population to be studied. However, this does not necessarily include the implications for individual animals. Investigators should define the potential number and species of targeted and untargeted species that might be captured. Additionally, the protocol should address, and the IACUC should consider, the merit of allowing investigators to use excess animal numbers. The numbers of animals required in studies of wildlife will vary greatly depending on study design, species' life-history characteristics, and questions posed. The needs for behavioral studies range from the capture of only a few animals where the focus is on a specific behavior to an entire population to mark all individuals.[3] Genetic, taxonomic, ecological, and other studies require a minimum sample size for statistical analyses. Too few animals might not allow the investigator to address research questions with sufficient scientific robustness and, subsequently, will result in a waste of animals if the results cannot be applied to test a hypothesis.[3] Statistical tools, such as power analysis, however, still have utility for field studies in estimating the number of animals required to obtain statistical significance for a given level of variance and a minimum difference between samples.

It is reasonable for the IACUC to review and approve a field study protocol with only a general description of animal numbers provided there are a rationale for that number and explanation for any vagueness.

While in small-scale, well-defined experiments in the laboratory using only a few fish or birds or wild-caught rodents, it is possible to limit animal acquisition to a specific number, this is not ordinarily the situation in the field. For example, in the capture of fish, netting of birds, or live trapping of small mammals, IACUCs should appreciate that while investigators can define the number and identity of targeted species in the field required to meet their research objectives, it is not reasonable to expect them to know how many will actually be captured or caught. This is especially true in survey work and in other types of exploratory investigations.[3] If captured targeted and accidental species are of such rarity as to be unlikely to be encountered, or if the captured animals are highly stressed and have a strong likelihood of dying upon release, some argue that it is irrational not to use them to advance the knowledge base to benefit society.[32] In the situation in which capture has an unexpected high yield, this serendipitous increase in animal numbers may provide a more robust database for evaluating specific hypotheses. Finally, the IACUC and field investigators should establish a mechanism to gain IACUC approval for protocol modification expeditiously from the field. A DMR process that allows for rapid review and approval or allowing a broad yet justifiable range for number of animals to be collected would be best practices. Anticipating and describing the rare occurrences in the protocol allows for their review and approval and is preferable to addressing a noncompliance situation after the fact.

The investigator should also provide assurance to the IACUC that collection permits necessary for the use of wild mammals have been issued for the proposed project and copies of permits might be requested by a prudent IACUC.[3] Many states require an annual report as a condition of permitting. These reports can be submitted to the IACUC to satisfy the APHIS/AC and NIH/OLAW annual reporting requirements of the number of animals used. (Animals that are part of field studies, as defined in the AWAR §1.1, Field study, are not required to be listed in the APHIS/AC annual report.) Investigators should additionally obtain permission of the owner, operator, or manager of any private property to be used or of the management authority for public lands involved as a site before commencing fieldwork thereon.[3]

14:26 Should the IACUC allow for additional animals to be acquired for approved studies to compensate for animal losses due to random deaths, experimental error, husbandry-caused mortality, surgical complications, and the like?

Opin. It is reasonable to assume that not all animals that begin on a study will be alive at the end, as a result of causes other than those related to the experiment. In a breeding colony of healthy C57Bl/6J mice, for example, the mortality rate of adult mice in otherwise good health and not subjected to experimental manipulation has been documented at 0.13% per week.[47] At our institution, the tracking of mouse mortality by our animal resources program from 2001 to 2011 and for an approximate average daily census over that period of 42,500, the mortality rate on a weekly basis was at least 0.21% +/− 0.05% (mean +/− 1 standard deviation). This was for an uncharacterized and mixed population of inbred, outbred, mutant, and variously crossbred mice used for breeding, aging, invasive and noninvasive experimental purposes, and other applications. The number most likely represents an under-estimate to some degree as it comprised only mortality reported by animal care personnel and, with rare exception, not those animals found by researchers. Separate from research factors, the husbandry program can be a source of unanticipated morbidity or mortality such as from cage flooding or dehydration if, for example, an automated watering system fails. The risk of not allowing compensation for some degree of unplanned attrition is that data will be collected from a population with insufficient animals, statistical analysis will lose robustness, some opportunities to arrive at significant conclusions will be forfeited, and animals may be used wastefully. The jeopardy in allowing overly excessive animal numbers may be preventable animal pain and distress and undue loss of animals that may be undetected by the IACUC for a considerable period.

The important consideration for the IACUC is to determine a "normal" or "acceptable" rate of spontaneous mortality for the genotype and that associated with a given procedure or aspect of research. Excluding iatrogenic causes, attrition ordinarily from natural causes should not be very high and in the range of 5%–7% annually for many populations of healthy, non-aged adult inbred mice[38,47] and substantially less for non-inbred species higher on the phylogenetic scale. However, in situations in which the animal phenotype predisposes to a general lack of vigor or even death, and inexperienced persons are being trained in a new procedure, mortality above nominal rates may be observed. Especially with surgical models, IACUCs are often asked to allow for additional numbers of animals to account for technical mistakes, experimental error, and other complications.

Mortality related to surgery may be influenced by the genetic background of the animal, the anesthetic regimen, competency of the surgeon, programs of intraoperative and postoperative care, and the invasiveness of the surgical procedure. High-risk procedures such as thoracotomy, bowel transplant, and three-quarter and five-sixth nephrectomy models will have ostensibly higher associated death rates than uncomplicated procedures such as embryo transfer, orchiectomy, or cranial implant. Mortality should not be very high in surgery, however, when well-trained and experienced persons use healthy animals and established procedures that meet contemporary veterinary standards including appropriate asepsis and postoperative care. In the absence of readily available published or internal data applicable to the model, a reasonable expectation for mortality in association with surgery in such cases might be 5% or less. When highly invasive, painful, or distressful experiments are planned, especially for species regulated by the AWAR, the IACUC may choose to exercise appropriate oversight by approving only the minimal quantity of animals rationalized as being necessary. Additional animals might only be allowed to be added by amendment, after IACUC review. Most IACUCs proceed with no guidelines regarding unforeseen losses. Instead, the rationale for extra animals is considered on a case-by-case basis in the context of protocol review. When IACUCs allow for a mortality "cushion" to be built into experiments, the general practice is to allow for a 10% or lower mortality rate without additional review.

Given that it is reasonable for new technical employees (not to mention students engaged in research in academic institutions) to be learning procedures, the IACUC might prudently focus on training by initially allowing a number of animals to be used for nonsurvival training preparations rather than artificially increasing the sizes of experimental groups.

Surv. What percentage of animals does your IACUC typically approve for unexpected deaths or related research problems?

- Not applicable 27/296 (9%)
- We handle each request on a case-by-case basis 212/296 (72%)
- Ten percent or less 48/296 (16%)
- More than ten percent 1/296 (~0%)
- Typically whatever the investigator requests 8/296 (3%)
- Other 0/296 (0%)

14:27 How can the IACUC prevent the excessive acquisition or breeding of animals beyond approved limits?

Opin. Many institutions have automated systems that will alert an appropriate individual when an investigator has reached a preset percentage (e.g., 80%–90%) of the number of animals approved for a specific project and can prevent ordering of animals in excess of the number approved.[24] The circumstances are more problematic when animals are produced to excess from an in-house breeding colony. In many instances, it may be not only impractical, but draconian, to cease all breeding immediately, euthanatize all animals that are in excess of the IACUC allowance, separate all breeders, and the like. It is incumbent upon the IACUC to articulate a clear policy and work closely and collegially with the investigators and animal resources program to nurture a culture of compliance. In the rare and unfortunate case of recalcitrance, the IACUC can resort to the options of protocol suspension or denial of access to facilities and notification of the proper regulatory authorities and funding entities.

14:28 Is IACUC approval needed for the use of animal tissues acquired from a slaughterhouse?

Reg. Ordinarily, neither the AWAR nor the PHS Policy protocol review requirements apply to dead animals in the context of research, including those obtained from abattoirs.[48] (See 8:7; 12:21; 14:51.)

Opin. The use of shared tissues and slaughterhouse material is an effective application of the "three Rs" and is encouraged where scientifically appropriate.[49] IACUC approval is not required if the collection takes place postmortem and as a byproduct of the commercial enterprise.[49] (If AWA regulated species are euthanized specifically for the purposes of a research study, this is regulated under the AWA.) The use of dead animals or parts of animals in research, however, is an area for which a clear institutional policy can help prevent serious misunderstandings and possible compliance problems.[49] It is in the institution's best interest to ensure that issues such as occupational health and safety, potential disease transmission to colony animals, institutional and personal liability, and related matters are addressed. A few IACUCs delegate oversight for this to some combination of the AV, safety officer, and IACUC Chair or other official designee of the committee. Whatever the mechanism, it is recommended that the IACUC and investigators maintain appropriate documentation, and institutions tailor their policy to meet their needs.[49] If the samples are obtained antemortem and are not incidental to slaughter (e.g., blood), or if they dictate the slaughter procedures in any way, then a protocol may be necessary, depending on the nature of the study. Examples of scenarios of when a protocol is likely to be needed include where biological fluids (e.g., blood, urine, milk, saliva, semen) or tissue specimens or biopsies are collected from animals prior to death or where live fetuses are obtained.

PHS grant applicants using shared animal tissues or slaughterhouse materials are advised to specify the origins of the tissues when describing their proposed use as appropriate in an application.[49] Otherwise, any reference to the use of animal tissues in the application is likely to trigger questions related to IACUC review. Therefore, an explanation in the application that the tissues will be harvested from dead animals at a slaughterhouse will help prevent complications in the peer review process.

Surv. Does your IACUC review the use of animal tissues acquired from a slaughterhouse?

• Not applicable	115/295 (39%)
• We have no policy regarding that type of acquisition	54/295 (18%)
• IACUC review is not required	79/295 (27%)
• A formal IACUC review is required	22/295 (7%)
• We require notification via a notice of intent, memorandum of understanding, informal review by an IACUC member, or other such means of notification	23/295 (8%)
• Other	2/295 (1%)

14:29 What should be required by the IACUC if a PI wishes to use blood or other body parts from an animal that was killed at a slaughterhouse, animal control institution, veterinary clinic or for reasons otherwise not related to research?

Opin. It is recommended that IACUCs have a pro forma policy and procedure established for this sort of eventuality. Many IACUCs find it effective to have short, one-page forms for the purpose of specifically documenting animal tissue acquisition and use. A useful mechanism is to have a form for tissues from any source including abattoirs, intramural projects in which animals are euthanized and tissues are made available to others, or extramural transfers. Completed forms can be given designated member review by the committee chairperson or designee, acting under the auspices of the IACUC, although this is not required. (See 14:28, 14:51.)

14:30 Is IACUC approval needed for the use of animal tissues acquired from animals euthanized as part of an unrelated experiment? For example, without having an IACUC-approved protocol, can a scientist remove body parts immediately after euthanasia from an animal on a colleague's IACUC protocol?

Opin. NIH/OLAW considers the use of shared tissues to be an effective application of reduction, refinement, and replacement and encourages this practice.[49] Assuming the animals and the method of euthanasia were covered under an approved IACUC protocol, IACUC approval is not needed.[49] Likewise, the harvest of tissues from dead animals itself is not considered research under the AWA and therefore is not a regulated activity.[49] However, if animals are killed for purposes of obtaining or using their tissues or other materials or if there must be distinct antemortem manipulation necessary to obtain the tissues, IACUC review and approval are required.[48] Many institutions choose to require protocol review for all activities involving living or dead animals. This places the decision-making control at the IACUC level, and where the practice may best serve the interests of the institutions for a variety of regulatory and nonregulatory reasons (e.g., public relations, liability, occupational health and safety).[49] Committees that elect not to oversee this activity generally do so as long as the tissues collected do not alter the approved procedures in any way (including addition of extra animals), that the animals were used originally for the research of others (and not for the individual receiving the tissue), that hazard assessment showed no risk of injury or infection for personnel, and that the tissues were collected after the animal was dead. In these instances it is important for investigators and IACUCs to avoid the temptation to misuse the dead animal exemption to circumvent policy and regulation. For example, the idea that the sole determining issue is whether or not whole animals are involved (making oversight of tissue acquisition unnecessary since tissues do not constitute whole animals) is incorrect.[49] IACUCs should consider this issue and have a policy and practice. Some committees may abdicate this responsibility because the animal is dead or only parts are taken, but this has the potential for creating problems if safeguards or understandings are not in place to ensure personnel safety and oversight of the ultimate use(s) of the animal. In institutions where IACUC approval is required (see 14:28) the process often involves the use of a brief form and a designated member review by the IACUC Chair, AV, or other designee. In all cases, a "paper trail" is recommended to show that an institution has applied the appropriate standards to the acquisition, use, and disposition of animals.[49] Likewise, to prevent peer review complications, PHS grant application procedures should be assiduously observed.

Surv. Does your IACUC approve the use of tissues acquired from dead animals that were euthanized as part of an unrelated experiment?

- Not applicable 63/295 (21%)
- Yes 124/295 (42%)
- No 93/295 (32%)
- I don't know 9/295 (3%)
- Other 6/295 (2%)

14:31 Are there reasons why IACUCs should review the acquisition of animal tissues acquired from animals euthanized as part of an unrelated experiment?

Opin. Assuming the animals and the method of euthanasia were covered under an approved protocol, IACUC approval is generally not needed.[49] However, approval is obviously necessary if the investigator receiving the tissue, in the course of meeting research needs, causes more animals to be used or different procedures to be performed on live animals (see 14:30). Review or policy establishment by the IACUC should also be considered when there is a risk of pathogen transmission from one facility to another within a research institution. For example, a number of rodent viruses may be transferred in tissues. Direct or indirect contact with such tissues may infect previously disease-free rodents in another animal research facility, invalidate the research in which the animals were used, and cause additional animals to be used unnecessarily as a replacement for those infected. Occupational health and safety issues also dictate IACUC consideration of tissue transfer and use. Some rodent pathogens, such as lymphocytic choriomeningitis virus, are transmissible to humans. Tissues from nonhuman primates, wild-caught animals, livestock, and unconditioned dogs or cats may also serve as a source of pathogens for humans. Additionally, there may be experiments done that could lead to public relations challenges for the institution, such as research using fetal tissue that may not be consistent with the institutional mission or philosophy, or research on tissues acquired from animals euthanized at an animal shelter.

14:32 Without having an IACUC-approved protocol, can a scientist remove body parts immediately before euthanasia from an anesthetized animal on a different scientist's IACUC protocol?

Reg. NIH/OLAW and APHIS/AC stipulate that any procedure involving antemortem manipulation of live animals must be subject to protocol review.[48]

Opin. In the context of the tissue-receiving investigator, it should be noted that NIH/OLAW has issued guidance that identifies a change, including addition of a species, as an example of a "significant change" requiring prior IACUC approval.[50] Likewise, IACUCs must have administrative mechanisms to ensure that all participating personnel are adequately trained and qualified, participate appropriately in the occupational health and safety program, and meet all other applicable criteria.[51] Acquisition of body parts from a live animal by a person not approved to be engaged in the procedure represents potentially serious noncompliance with the AWAR, PHS Policy, and federal granting procedures. NIH/OLAW has indicated its determination that conducting unauthorized animal research activities is a serious and reportable violation.[23] This position is also consistent with the NIH Grants Policy Statement (Rev. 03/01) that identifies the substitution of one animal model for another (or presumably addition of an animal model where none existed previously) as an example of "change of scope" likely to require prior NIH (grants management officer) approval.[52,53] The signature block on the face page of the PHS form 398 grant application requires the institution to declare that it meets and will maintain compliance with all applicable terms and conditions of award. Consequently, IACUCs should have clearly articulated and well-known policies in this regard. When the situation of tissue sharing has been encountered by IACUCs, the majority require an IACUC review, but interestingly, survey of IACUC Chairs (below) shows a noteworthy minority of 9% of IACUCs where joint use of an anesthetized subject may apply (down from 16% in 2007)[54] are apparently not aware of this need or, despite the gravitas if Accredited and/or Assured, proceed in apparent defiance of the requirement. Given that prudent sharing of animals is encouraged as a means of reducing the number of animals used in research,[49,55] it may not be necessary for the second scientist to have an independent protocol. Nevertheless, the "piggy-backed" scientist should be added to the personnel list and meet credentialing qualifications, and the IACUC should specifically review any new or modified procedures for the existing protocol.

Surv. Does your IACUC require its approval for the use of tissues acquired from an anesthetized animal that is about to be euthanized (i.e., before the animal is dead) to be used in an unrelated experiment?

- Not applicable 40/295 (14%)
- Our IACUC would require its approval before tissues could be removed 221/295 (75%)
- Our IACUC would not require its approval before tissues could be removed 24/295 (8%)
- I do not know 9/295 (3%)
- Other 1/295 (<1%)

14:33 Is IACUC approval needed for acquisition of animal tissues from abroad?

Opin. Investigators who collaborate with scientists in other countries may receive animal tissues from these foreign sources and may sometimes go abroad to collect samples from captive wild animals maintained in zoological collections and in research colonies. Just as animal-related activities supported by the PHS require review by the IACUC of the domestic awardee institution, foreign institutions that serve as performance sites must have a Foreign Assurance on file with NIH/OLAW.[19,56,57] NIH/OLAW considers institutions whose scientists are engaged in such collaborative work accountable for the animal-related activities from which they receive animals or animal parts. When a foreign institution holds a PHS Assurance, it also is expected to conduct the study in accordance with the applicable host nation's policies and regulations. In the specific case of sample collection, the review should take into account the species involved, the nature of the specimen, and the degree of invasiveness of the procedure and should give appropriate consideration to the use of anesthetics and analgesics. When samples are to be obtained directly by citizens of a foreign country for subsequent shipment, the recipient PHS supported investigator should determine the proposed methods of collection and present that information to his or her IACUC for review. Prior to sample collection, and regardless of whether specimens are obtained by an awardee institution's investigator directly or by persons in a foreign country, NIH/OLAW strongly recommends that each awardee institution consult other agencies of the U.S. government concerning importation requirements.[19,58] Depending on the species involved and the nature of the specimen, the following may be of assistance: the U.S. Fish and Wildlife Service, Department of the Interior (for compliance with the International Convention on Trade in Endangered Species of Fauna and Flora [CITES]),[6] USDA/APHIS (regarding potential animal pathogens),[4] and Centers for Disease Control and Prevention (concerning importation of nonhuman primates and potential pathogens of human beings).[8]

14:34 Using PHS grant monies, a research institution purchases surgically modified mice from a commercial vendor. Must the vendor have an NIH/OLAW Assurance? (See 18:13.)

Reg. The PHS Policy is applicable to all PHS-supported activities involving animals, whether the activities are performed at a PHS agency, an awardee institution, or another institution.[59]

Opin. The determining issue is whether the surgery is conducted in response to a specific custom surgery request or whether the animals were previously surgically modified and available for sale "off the shelf" (e.g., from a catalog) before the PI's need was anticipated.[59] If an investigator requests that a specific custom surgical procedure or procedures be performed on an animal for use in activities funded by the PHS, then the organization or person who conducts the surgical procedure(s) is considered to operate from a performance site and must either have on file with NIH/OLAW an approved Animal Welfare Assurance or be included as a component of the applicant organization's Assurance.[59] If the animals are commercially available as surgically modified models, such as in the case of hypophysectomized rats or mice with implanted devices, then the supplier is not considered a performance site and is not required to file an Assurance with NIH/OLAW.

14:35 Under what conditions can the IACUC permit the use of animals purchased from a retail pet store?

Reg. The AWAR require licensing of pet shops that supply warm-blooded animals, other than birds and domesticated laboratory rats and mice, for research purposes (§1.1, Retail Pet Store; §2.1,a,3,i).

Opin. As a general rule, this is a practice that should be avoided. It may be acceptable for ectotherms, such as fish; for rare types of rodents or rabbits; or for certain birds or reptiles when such animals cannot be acquired from a more traditional or conventional source. Where regulated species (see 14:1) are to be obtained from a pet shop, the institution has an obligation to obtain the informed consent of pet shop management and to procure the animals in a lawful manner (*Guide* pp. 106–107). The purchase of animals from pet shops should also fall under the auspices of the IACUC because of the potential for abuse and the risk to the institutional reputation should this practice become known to the public. Animals that are obtained from pet stores may not be acclimated to a laboratory environment and are generally of unknown genetic and health backgrounds. With respect to the latter, the veterinary staff should be called upon to evaluate the health status of the animals and quarantine, isolate and otherwise manage these animals to protect other research colonies.

Surv. Does your IACUC permit the acquisition of research animals from a pet store or pet supplier?

- Yes, we do — 25/296 (8%)
- Yes, but only for species not regulated by the Animal Welfare Act regulations — 25/296 (8%)
- No, pet stores are prohibited as a source of animals — 154/296 (52%)
- This question has not been encountered by our IACUC — 55/296 (19%)
- Not applicable — 26/296 (9%)
- Other — 11/296 (4%)

14:36 Are there legal requirements that must be met for purchase of animals from a commercial farm?

Reg. See 12:1 for the APHIS/AC and PHS Policy definition of an animal. Regulated farm animals in research include the following as per APHIS/AC Policy 17:

- Farm animals used to manufacture or test biologics for nonagricultural or nonproduction animals, or humans. This includes biologics that are produced or tested for possible use in either agricultural or nonagricultural species, such as multispecies rabies vaccines.
- Farm animals that are used as models for human subjects or nonagricultural animals (e.g., using calves to develop an artificial heart for humans).
- Farm animals used for biomedical teaching; that is, the training of human or veterinary medical personnel in medical methods and procedures, such as surgery, diagnostic techniques, anesthesia and analgesia.

Animals that are exempt from the AWAR include:

- Farm animals used to manufacture or test veterinary biological products intended for use in the diagnosis, treatment, or prevention of diseases in agricultural animals.
- Farm animals used in agricultural teaching, such as farm or ranch management procedures (e.g., hoof trimming, shearing), handling practices and breeding techniques.

APHIS/AC Policy 8 clarifies when a dealer's license is required for selling hoofed stock:

- Sells animals only for regulated purposes such as biomedical research, exhibition or as pets.
- Sells the majority of their domesticated farm hoof stock (sheep, cattle, goats, pigs, llamas) for regulated purposes and more than 10 animals are sold for regulated purposes in a 12-month period.
- Sells more than 10 wild hoof stock (such as deer, bison, or elk) for regulated purposes in a 12-month period or one or more exotic animals such as a zebra, hippopotami, ibex, camel, giraffe, etc.

Opin. APHIS/AC regulates farm animals produced and used for biomedical research purposes (APHIS/AC Policy 17) and their sale (APHIS/AC Policy 8). The research facility becomes responsible at the point they assume custody of the animal. Therefore, if an employee of the research facility goes to a livestock market or farm and purchases livestock, then that person, as an agent of the research facility, must satisfy the AWAR transportation requirements (AWAR §3.136–§3.142) and the institution must satisfy health and husbandry standards (AWAR §3.125–§3.133).

14:37 Can animals be transferred between studies that have IACUC-approved protocols?

Opin. Provided certain requirements are met, it is permissible to transfer animals between studies. Survey data from 2000,[60] 2007[30] and this volume show this to be a universal and standard practice. Unrestricted trading of animals between research protocols, however, could lead to unauthorized use, defeat attempts at tracking animal utilization, put legal records pertaining to dogs and cats in jeopardy (AWAR §2.35,c,1–§2.35,c,2), and be potentially confusing to veterinarians who try to determine the ownership of animals requiring medical care. As such, transfer procedures should include informed IACUC consent for the transaction; approval by the IACUC of the use of the species in the research; a means to assure that multiple, unrelated major survival surgical procedures are prevented; and a mechanism to credit animals to one protocol and transfer them to another via a debit transaction. IACUCs may choose to discriminate in terms of the stringency of process control. For example, the process could be managed through an IACUC administrative function or by the animal resources program on behalf of the IACUC or deferred to the involved scientists. Important considerations in this regard might be whether or not the species is AWA regulated or the degree of invasiveness of the intended use. For those species that are not regulated under the AWA, such as rats and mice bred for use in research, and fish, IACUCs might allow less restricted transfer. The technological innovation of census management by bar code scanning allows transfer procedures by crediting of the animals to the donor's account and debiting the exchanged animals from the total approved for the recipient collaborator.

Surv. Does your IACUC permit animals to be transferred between studies having IACUC-approved protocols?

- Not applicable 32/294 (11%)
- Yes, we permit transfers 226/294 (76%)
- No, we do not permit transfers 25/294 (9%)
- Other 11/294 (4%)

14:38 Under what circumstances or conditions may animals be transferred between studies and how can the IACUC track this activity?

Opin. At institutions where there are breeding colonies of animals supporting more than one research project, there must be a mechanism to transfer animals from the breeding colony to IACUC-approved research protocols. An example of the latter is an institution that has a core facility for the production of genetically engineered animals. Those of a desired genotype are produced under one IACUC-approved activity (the core) and are then transferred to the IACUC-approved activity of a specific scientist for research use. Other cases involve simple transfers in which one investigator receives a few breeders of one-of-a-kind genetically manipulated mice from the breeding colony of a colleague. Nonhuman primates used in operant conditioning or behavioral paradigms present another example. These animals are usually purchased from a commercial vendor and are quarantined in the animal research facility. During the quarantine phase (or shortly thereafter) the investigator may become aware of behavioral considerations that render the animal less appropriate for the intended research and more appropriate for use by a colleague. In this instance, the investigators may wish to exchange animals. The same may be true for dogs or cats obtained for use in research involving surgical procedures. In this circumstance, demeanor and size may lead scientists to trade animals between acute and chronic studies, or where the animals have anatomical or other attributes that are better suited for certain studies than others. Two investigators, one with expertise in a specific experimental manipulation and a second with animals of a unique genotype, may collaborate by transferring the animals from the original owner to the party doing the manipulation.

The most important consideration is that the IACUC employs a mechanism to ensure that the total number of animals used and the species of animal are appropriate in terms of the approved protocol and that there are not multiple major operative procedures in unrelated research projects (AWAR §2.31,d,1,x,A; *Guide* p. 30). The IACUC should develop some guidelines for review of historical research use, the transfer of animals between approved protocols, and the subsequent monitoring of wellbeing and outcomes if this is a circumstance that might occur repeatedly. An important consideration in setting guidelines is obtaining the surgical history of the animals from either the investigator or veterinary staff. This is necessary to prevent animals from being unintentionally subjected to multiple major survival operative procedures. When applicable, the IO must submit a request to APHIS/AC and receive approval in order to allow a regulated animal to undergo multiple survival surgical procedures in unrelated experiments (*Guide* p. 30; APHIS/AC Policy 14).[61] The movement of animals from one protocol to another ideally should involve debiting or crediting the number of animals against those approved for the study. IACUCs often refer these activities to the animal resources program, which often has the staffing and computer resources to handle the task. Some IACUCs, however, find it advantageous to require the PI to inform the committee of the practice by memorandum or other mechanism.

Surv. Assume animals are bred on Protocol A and then used on Protocols B and C. Does your IACUC have an effective system of tracking the number of animals born and then transferred to the other protocols?

• We do not have a tracking system	34/296 (11%)
• We have a tracking system but it is not very accurate	28/296 (9%)
• We have a reasonably accurate tracking system	136/296 (46%)
• We have a reasonably accurate tracking system, but it is difficult to use or is not used effectively	14/296 (5%)
• I don't know	5/296 (2%)
• Not applicable	73/296 (25%)
• Other	6/296 (2%)

14:39 If a protocol is being renewed, how can the IACUC ensure that it is not unintentionally renewing the original number of animals requested?

Opin. This is an important issue from the perspective of preventing unnecessarily duplicative experiments and providing a justification of the total number of animals used in research. An example of this situation is useful for illustrative purposes.

An investigator is funded for a 5-year study involving 30 macaques. He purchases them over the first 3 years of the study and then submits a renewal after 3 years, requesting 30 macaques. How can the IACUC be sure that the 30 animals requested on the renewal are new monkeys, and not just a relisting of those already on the study and still in the investigator's colony?

As an extension of this problem, what if the investigator originally was approved for 30 animals but only acquired 20 by the time of the 3-year renewal? If ten animals are now requested in the renewal, how can the IACUC determine whether these are the remaining ten from the original approval, or ten more above the original request of 30 (for a total of 40)?

These problems can be addressed with careful formulation of the renewal and modification requests. IACUC review may be facilitated when requests are supported by progress reports of animal usage on the previous iteration. For an annual renewal (end of first or second year for PHS Policy regulated species) the application should not include animal numbers, as the request has already been approved for the first 3 years. The annual accounting of animals for APHIS/AC should be monitored at the level of acquisition and then once annually for individual animals kept over multiple years.

Requests for additional animals before the end of the original protocol should be handled in a different manner than new or 3-year renewals and specifically indicate that additional animals are required. One mechanism to help eliminate the issues raised in the examples is to close out the original protocol regardless of the number of animals used or not used and to consider the continuation as an essentially new proposal, with a new set of experiments and requirement for justification of animal numbers. Obviously there are circumstances, as in the example, where further questions, such as the following, should be asked:

- Is this a new protocol or a 3-year renewal?
- If the protocol is a renewal, what is the previous protocol number?
- If the protocol is a renewal, are there *existing* animals that are already assigned to the study?
- If so, how many? (Do not include this number in your answer to the next question.)
- Justify the number and use of animals specific to this application.

Many IACUCs do not consider this issue with meticulous detail. Instead, they rely on faith that good science is not going to be repetitious or wasteful, that the vast majority of scientists can be trusted, that economic factors (as much as anything else) have a limiting effect on animal use and dictate judiciousness, and that the IACUC application will show that the research requires a certain number of animals and is novel and important.

When an animal resources program performs census management using electronic technology, it dictates downstream processes for the IACUC. Computerization of the practice typically requires that animals residing under an expiring protocol be managed in one of two ways at the point where the new protocol becomes active. Once the expiring protocol is deactivated, the animals are debited from the original number approved under the old protocol and considered to have been expended, *ipso facto*, as a result of termination. They then are immediately issued a new cage card (with new protocol and billing account numbers), considered as a new acquisition, and debited from the new IACUC protocol number. A second option is to transfer the animals from the expiring protocol to the new one, treating them as an internal receipt. To reduce confusion during consideration of the 3-year renewal, some IACUCs find it useful for the remaining animals to be added to the new protocol to be listed or described on a supplementary document separate from the renewal application.

In summary, special attention should be paid to the type of application (new or renewal), the numbers of animals used, the number requested, and the number to be carried over from the previous protocol. Although it is beyond the jurisdiction of individual IACUCs, in the opinion of the authors, many of these problems could be alleviated if periods covered by IACUC protocols and extramural grants could be matched in duration.

Surv. Assume a study is being renewed for an additional period of time. What actions does your IACUC take to assure that animals that were not used during the original approval period are being included in a request for additional animals to be used during the renewal period? More than one response was possible.

- Our IACUC does not have policy on this matter — 81/294 (28%)
- If there are any "leftover" animals from the initial approval they must be transferred to the renewed protocol and counted as part of the number of animals approved for the renewed protocol — 138/294 (47%)
- Animals carried over to the renewed protocol are added to the total number of animals to be used via a separate line item. They are then debited from the grand total when the renewal is activated (such as when updated cage cards are made) — 23/294 (8%)
- The expired protocol is closed out and any remaining animals are deleted from the census. They can be added onto the renewal via a special form but they are debited from the total — 24/294 (8%)
- Carryover animals are specifically identified in the renewal. No additional consideration is given except that only the total number of new animals needed over the next 3 years is divulged to those who purchase the animals — 29/294 (10%)
- Investigators are encouraged to identify carryover animal numbers, but doing so is not enforced — 15/294 (5%)
- Not applicable — 31/294 (11%)
- Other — 5/294 (2%)

14:40 What guidelines can the IACUC establish for the use of endangered species in research?

Opin. The use of endangered species in research must conform with all federal and state laws and international conventions to which the U.S. subscribes, including the Lacey Act, the Endangered Species Act, the Marine Mammal Protection Act, and the CITES treaty. It is desirable for IACUC guidelines to investigators to repeat this requirement.

14:41 Should institutional programs related to veterinary care (such as a rodent sentinel program, quarantine programs involving diagnostic sample collection from animals, necropsy service, rodent colonies for raptor rehabilitation programs, and therapeutic surgical interventions) be approved via the protocol review system?

Reg. The AWAR (§2.31,a) require that all animal activities for regulated species be reviewed by the IACUC. PHS Policy (IV,B,1) and the *Guide* (p. 9) state that the entire "program" must be reviewed by the IACUC.

Opin. PHS Policy applicability is not limited to research; it also includes all activities involving animals, such as testing and teaching. NIH/OLAW has determined that although animals used as sentinels, breeding stock, chronic donors of blood and blood products and for other like needs may not be part of specific research protocols, their use for these purposes contributes significantly to the institutional research program and constitutes activities involving animals.[62,63] Consequently, the IACUC must receive and approve protocols, have appropriate systems to monitor the use of animals prior to the commencement of such activities, and then perform reviews at the appropriate intervals.[62] The authors also interpret this to include rodent colonies used as a source of food for raptors in rehabilitation programs (see 8:33), animals used for training purposes, core rodent breeding services, and those held on expired protocols. All clinical activities (e.g., diagnostic and therapeutic interventions) and the necropsy service involving existing research or instructional animals ordinarily are under the supervision of the AV, should be covered under the semiannual institutional program assessment, and do not require approval under the protocol review system. (See 27:13.)

Surv. At your institution, which of the following programs require an IACUC-approved protocol? More than one response was acceptable.

- Our rodent sentinel program 188/280 (67%)
- We do not have a rodent sentinel program 76/280 (27%)
- Animal (mammal) quarantine program 129/280 (46%)
- We do not have an animal quarantine program 78/280 (28%)
- Rodent surgical rederivation program 95/280 (34%)
- We do not have a rodent surgical rederivation program 95/280 (34%)
- Animal adoption program 35/280 (13%)
- We do not have an animal adoption program 192/280 (69%)

14:42 Does the IACUC need to approve studies that use privately owned animals, such as those in a clinical trial in a school of veterinary medicine? (See 14:50: 29:64; 15:7; 15:8.)

Reg. The AWAR (§1.1) include all regulated species used, or intended for use, in research, teaching, and testing. The PHS Policy (III,A) includes any live vertebrate animals used or intended for use in research, research training, experimentation, or for related purposes (if the work is PHS funded or the institution voluntarily includes all animals under its Animal Welfare Assurance). Neither the AWA, AWAR, nor PHS Policy distinguishes between privately owned animals and those owned by the institution.

Opin. The determining factor as to whether PHS Policy and the AWAR are applicable is whether the veterinary engagement with the animal is in response to the requirements for research or in the usual care of a patient.[64–66] Given that the animals are enrolled in a clinical trial in this scenario, the IACUC must provide oversight. The law and regulations give the IACUC room to deliberate justifiable exemptions, particularly where the use is not covered in the AWAR or PHS Policy (see bulleted examples that follow). Ultimately this involves doing what is appropriate by all parties and documenting the exemption and its rationale. If the treatment plans and monitoring elected by the client-owners are not influenced by the study, then the research use would be incidental and may not fall under the AWAR or PHS

Policy.[66] In this case, some entity would need to determine whether or not IACUC oversight applied and the IACUC likely would be in the most qualified position to decide. Beyond that, it is strongly recommended for the protection of the institution that the IACUC have a written policy with respect to the research use of client-owned animals and, at the least, an informed consent is obtained from the client as the representative of the animal (the subject).

As noted in the Regulatory comments to this question, when federal oversight is mandated, neither the AWAR nor PHS distinguishes privately owned animals from those owned by the institution. A number of factors are to be considered, including but not limited to the following:

- The specific type of work
- Species to be used
- The need to board or retain the animals for a period
- The nature and resources of the facility
- Involvement of federal funding
- Requirement of IACUC review by a nonfederal granting agency
- The content of the institutional Animal Welfare Assurance

Pets used in clinical trials are privately owned animals recruited from volunteers and, as such, usually remain the property of the owners, who maintain responsibility for their upkeep, housing, and transportation. However, once these animals appear on institutional property, as noted previously, rules should be established as to how the animals are managed and temporarily housed. Such rules should take into consideration facility operations, informed consent, institutional liability, public safety, and animal welfare. When client-owned animals are donated to the institution, the full range of compliance issues takes effect. At those veterinary schools where the IACUC does not review the research use of client-owned animals (e.g., does not have an NIH/OLAW Assurance), teaching hospitals typically have policies governing the conduct of research. Where the veterinary school has an Animal Welfare Assurance or the species to be used is covered by the AWAR or the Assurance of the parent university, the IACUC clearly has full oversight responsibility. In situations not covered by an Assurance or the AWAR, the proposed research may be reviewed by the department and then a hospital board or committee akin to an institutional review board for human subject experimentation. In practice, the overwhelming preponderance of IACUCs elect to oversee the experimental use of client-owned animals. (See 8:31.)

Surv. With reference to veterinary clinical trials at a veterinary or agricultural school: Does your IACUC approve these clinical trials?

- Not applicable — 253/288 (88%)
- No, a separate clinical trials committee approves these trials — 1/288 (<1%)
- Yes, the IACUC approves these trials because there is no other committee to do so — 4/288 (1%)
- Yes, the IACUC approves these trials because they are considered to be research — 29/288 (10%)

- A clinical trials committee approves the clinical trials but IACUC approval may be required in certain cases — 4/288 (1%)
- I don't know — 2/288 (<1%)
- Other — 1/288 (<1%)

14:43 An institution, a technical college, does not own or house animals but in the veterinary technology program, students encounter animals in classroom demonstrations of noninvasive procedures and during externships. Is the IACUC required to follow the AWAR, PHS Policy, or both, and is it required to monitor the acquisition and disposition of the animals?

Reg. The AWAR apply to the use of animals in teaching including those that might be encountered in a veterinary technology program such as dogs, cats, guinea pigs, hamsters, or rabbits (§1.1, Animals).

Opin. Although not applicable where such animal use is supported by tuition or other institutional funds, for purposes of consistency and simplicity all instructional use of animals should be reviewed by the IACUC.[67]

The American Veterinary Medical Association requires that the instructional use of animals in accredited veterinary technology programs, regardless of funding source, species, or nature of use, be under the authority of an IACUC.[68] This logically extends even to instances when student-owned animals are used for the classroom demonstration of noninvasive procedures such as physical restraint for examination, vital signs recording, and positioning of conscious patients for radiography and that remain on the premises for less than 12 hours. Under this scenario at the teaching institution, it is within the purview of the IACUC to monitor and track the numbers of animals procured or participating, the frequency of use, and the eventual disposition of the animals. When students work with client-owned animals during an externship at a community veterinary private practice establishment for purposes of acquiring practical skills in specimen collection, medication administration, surgery assistance, or postoperative care, the institution has a stake in assuring that these experiences are meaningful and that the students will be appropriately supervised, mentored, and trained.

Techniques learned in this environment will be carried into subsequent careers. While the level at which oversight occurs is a matter of institutional decision, it is logical to assign the responsibility to the IACUC. In this case, the IACUC should at least consider whether or not these experiences should be covered by protocols.

14:44 An institution's veterinarians or physicians occasionally are asked to provide veterinary services to a pet at no or reduced cost. How should the IACUC address these situations?

Opin. Although research institutions have great resources and clinicians with unique skills, engaging in these activities is a legal "slippery slope" fraught with risk. There is great potential for both institutional and veterinary liability should something go awry.

All states have veterinary practice acts that regulate the practice of veterinary medicine in the state and protect veterinary private practice as a business. When unlicensed specialists, including physicians, undercut fees or provide veterinary services for individuals who otherwise would seek care from private practitioners, the institution and medical professional assume a legal risk.

Some institutions and individuals believe that they can protect themselves by rendering free service or requiring that clients sign consent forms. Nevertheless, the judicial system most likely would not enforce provisions of consent forms in which clients agree not to hold an institution or individual liable for the negligent treatment of an animal.[69] Nor will any consent by a client to have a veterinary procedure done by a non-veterinarian protect the non-veterinarian from prosecution for practicing veterinary medicine without a license.[69]

In some instances, veterinarians and institutions must be pragmatic. Acutely ill or dying animals on or near institutional property, such as stray cats in end-stage disease, dogs that have been hit by cars, or injured birds, may require humane intervention and cooperation with local animal control authorities. Whatever the reason, veterinarians are bound by oath to relieve animal suffering, but as noted previously, must also act in ways that protect research animals from those who have undefined health status. Consequently, institutions may want to develop policies that address situations in which emergencies involving animals that are not university property require prompt and humane intervention. The unwritten policy of the authors' institution is to stabilize the condition of animals that require critical care with the understanding that they will be transported by a county animal control officer immediately to a local veterinary hospital for subsequent care.

14:45 A laboratory animal facility is associated with a human hospital. What policy, if any, should the IACUC have relative to housing Seeing Eye dogs or other animals that might accompany patients admitted into the hospital?

Opin. At the authors' institution, with a large medical center and several affiliated hospitals, Seeing Eye dogs and animals involved in pet-facilitated therapy programs are permitted in the wards and public areas of the hospital. These issues fall under the jurisdiction of hospital policy and practice and not under the oversight of the IACUC as the animals are privately owned and not used for research, testing, teaching, or training. Under the Americans with Disabilities Act, hospitals, as public accommodations, are expected to provide reasonable access to services and facilities (health care) for those who use service animals[42] but are not required to provide care or supervision while they are on site.[70] Likewise, at our medical center, pet-facilitated therapy is made available to patients through a relationship with a bonded, outside provider. The provider ensures that the privately owned animals participating in the program are of the appropriate age, size, and disposition; are acclimated to patients and the hospital environment; and are in apparent good health (including parasite control and full immunizations) as certified annually by a contract veterinarian.

14:46 What are the legal requirements regarding the disposition of research animals?

Opin. Animals used in research may be disposed of by euthanasia, adoption, retirement (e.g., endangered species), safe and responsible return to production agriculture (food and fiber species), sale for slaughter, release to the wild, or transfer to another institution or research project. In most instances, animals used in research are euthanized. Considerations when disposing of these animals include the experimental history (e.g., method of euthanasia and exposure to chemicals, toxins, radiation, infectious agents, recombinant DNA, human tissues, and drugs) and may involve burial in a landfill, incineration, alkaline hydrolysis with discharge into the sewage system, or rendering into animal feeds. In some instances, particularly in the case where radioisotopes have been given to an animal, the carcass must be held until the radioactivity decays to safe levels.

14:47 Can animals used in research, teaching, or testing be adopted as pets once they have been used in an IACUC-approved project? What is an appropriate mechanism for facilitating this process?

Opin. Provided the animals are emotionally and physically healthy with appropriate socialization and have not been disfigured or disabled by the research, adoption of animals that make suitable pets is appropriate at the discretion of the institution.[71] This practice, as have many others, may have legal ramifications. Guidelines for adoptions should be formally developed in consultation with legal counsel. Important considerations relevant to adoption include adoption to a reliably suitable and permanent home, consent of the researcher, approval of the request by the IACUC, and confirmation of the good health of the animal by a veterinarian. There may be other restrictions, such as limiting adoptions to employees and students of the institution and not to the general public. Some public institutions do not permit adoption *per se* as adoption entails a change in ownership that involves divestiture of public property. Pragmatic approaches to circumvent this challenge may include permitting animals to leave under the status of a long-term loan or to allow retired animals to remain in possession of the laboratory with care provided by staff. This may lead, however, to questions of whether such "institutional pets" are covered in a protocol and how the program is funded. At the author's institution, healthy animals are released after the approval process and the completion of relevant records (i.e., USDA Veterinary Services Form 18-6: Record of Disposition of Dogs and Cats). Copies of all correspondence related to adoptions and the permanent medical records of adopted animals are maintained in the animal resources program office.

Surv. Does your IACUC permit the adoption of animals as pets once their research use has been completed?

- Not applicable 52/295 (18%)
- Never 111/295 (38%)
- Sometimes 111/295 (38%)
- I don't know 10/295 (3%)
- Other 11/295 (3%)

14:48 Should the IACUC permit farm animal species that have been used in teaching, testing, or research to be sold for slaughter? What guidelines would be useful to the IACUC in this regard? (See 18:14.)

Opin. Domestic livestock, usually raised for food or fiber, are also commonly used in research, teaching and testing. In some instances, it may be appropriate to return live animals that have been used for research or training purposes to production agriculture or to render carcasses into livestock feed. However, if this is done, it should be done with caution and with full awareness of the potential for institutional and individual liability.

When animals used in research are subsequently slaughtered for human or animal food, the major concern is adulteration of the meat by residues (e.g., antibiotics, hormones, toxins, radioisotopes, carcinogens) or infectious agents that would render the meat or its food products unwholesome. The agencies that regulate slaughter are the USDA's Food Safety and Inspection Service (for federally inspected facilities) or state agencies for state inspected slaughter facilities. Endangerment of the nation's food supply by failure to adhere to food safety standards constitutes a felony in the U.S. that could be referred to the appropriate legal jurisdiction. Given these potential legal ramifications, an institutional policy on the sale, slaughter, consumption, or rendering of research animals is strongly encouraged. The policy should be developed with appropriate legal counsel and should be consistent with the institution's Animal Welfare Assurance. Veterinary care of animals sent to market or slaughter and drug withdrawal periods must be consistent with the Animal Drug Use Clarification Act of 1994 (AMDUCA).[7] In the event that the animals are given a new animal drug as defined by the Food and Drug Administration (FDA),[72] no meat, eggs, or milk from those animals may be processed for human food without FDA approval. If animals are being slaughtered at the research facility for subsequent consumption, the method of euthanasia must be stated in the approved protocol and should be in compliance with the AWAR (§1.1, Euthanasia) and the American Veterinary Medical Association's guidelines on euthanasia.[42] For animals that have had invasive surgical procedures, the general practice should be for the level of surgical asepsis and postoperative care to be consistent with current veterinary standards for the species. Where IACUCs allow this practice, FDA restrictions are adhered to and the animals must have no lingering evidence of manipulation such as the presence of a cannula or other permanent implant.

Surv. Does your IACUC permit farm animals that have been used for teaching, testing, or research to be sold for slaughter or released back to production agriculture?

- Not applicable 162/292 (55%)
- This has not been discussed by our IACUC 32/292 (11%)
- Yes, we always allow slaughter or release back to production agriculture 7/292 (2%)
- Yes, but under specified circumstances 32/292 (11%)
- No, we do not allow these animals to be slaughtered or released back to production agriculture 57/292 (20%)
- Other 2/292 (~1%)

14:49 Is IACUC approval required (either formally or informally) to donate surplus or culled rodents to zoos or similar institutions to be used as a food source for reptiles or birds?

Reg. The *Guide* (p. 73) stipulates that conventional, biologic, and hazardous waste, including carcasses, should be removed and disposed of regularly and safely including rendering hazardous wastes innocuous by an appropriate means and in accordance with all applicable regulations.

Opin. The AWAR and PHS Policy are silent with respect to the disposal of culled or euthanized research animals with the exception of requirements related to humane euthanasia practices (AWAR §2.31,d,v; §2.31,e,5; PHS Policy IV,C,1,c). The AWAR technically would apply only to rodent species other than rats of the genus *Rattus* and mice of the genus *Mus* (e.g., hamsters, voles, squirrels; §1.1 Animal). Be that as it may, the diversion of carcasses from research sources to raptor rehabilitation programs, zoological parks, other exhibitors, or as a food source for the reptile pets of employees may represent a cost-effective and beneficial arrangement particularly for those recipient operations with expenses functioning on slim financial margins and, in the case of employee pets, as a component of the facility biosecurity program (i.e., preventing employees from maintaining rodent colonies at home of a lesser health status). Recipients may become highly dependent upon the arrangement and, superficially, the public relations benefits might seem to be positive for all of those involved. The authors recommend, however, that these undertakings be entered with mutual caution. Cells immortalized with viral agents, cells carrying viral genomic material, tumorigenic human cells, and cell lines of human or nonhuman primate origin inoculated into animals present potential hazards beyond the research environment particularly if disposed via digestion by other animals. Disposal of potentially hazardous carcasses by alimentation would not be consistent with the conditions of the *Guide* nor applicable bloodborne pathogen and biosafety standards.[9–11] With respect to genetically engineered mice, regardless of how harmless or inconsequential the insert or targeted mutation, one must be cognizant of the public relations challenge created by the term "transgene" and the difficulties in prevailing in a 15-second sound bite debate on the news with a partially informed concerned party. Institutions clearly have an obligation to create internal processes ensuring that pathogenic materials, chemicals, radioactivity, and other known or potential hazards properly stream for disposal as hazardous waste or are rendered into a harmless state. The simplest solution, where surplus or culled rodents are to be donated, is to establish programs allowing only the release for this purpose of animals of wild-type genetic background (in this day and age, a small fraction of any census) known to be free of hazards. This topic is highly complex and potentially dynamic from the aspect of safety regulation with exceptions and exemplars for safe diversion of carcasses for food use beyond the limited purvey of the treatment of this subject here. As such and at a minimum, responsible institutional review is recommended with the IACUC ordinarily the best body to coordinate this.

Surv. Is your IACUC's approval required (either formally or informally) to donate surplus or culled rodents to zoos or similar institutions to be used as a food source for reptiles or birds of prey?

- Not applicable 75/296 (25%)
- We have no policy on this topic 65/296 (22%)
- We do not allow such donations 70/296 (24%)
- We allow such donations, but only of dead animals 16/296 (5%)
- We allow such donations of dead or live animals 7/295 (2%)
- We allow such donations, but only of dead animals known to be free of potential hazards (e.g., infectious organisms, transgenes, radioactivity, etc.) 43/296 (15%)

- We allow such donations of live or dead animals known to be free of potential hazards (e.g., infectious organisms, transgenes, radioactivity, etc.) 15/296 (5%)
- Other 5/296 (2%)

14:50 Is the IACUC required to approve the use of privately owned animals that are used for teaching general physical examinations at a veterinary school?

Reg. The AWAR (§1.1) apply in the case of regulated species used for teaching purposes. The PHS Policy (III,A) includes any live vertebrate animals used or intended for use in research, research training, experimentation, biological testing, or for related purposes. Neither the AWA, AWAR, nor PHS Policy distinguishes between privately owned animals and those owned by the institution.

Opin. The discriminating factors to consider as to whether PHS Policy and the AWAR apply are the species, source of funding for the activity, and whether the supervised teaching of veterinary skills and engagement with the animal is truly for instructional purposes only and does not involve research. The AWAR would apply for instructional activities with regulated species and IACUC oversight would be necessary given the use of the patients for training purposes (see 12:1). The PHS Policy would not apply given that experimentation is not a component of the activity and financial support most likely would derive from tuition and client fees and ordinarily would not be derived from PHS. For additional discussion, see 14:42.

14:51 Do IACUCs have to approve the use of cadaver animals obtained from a pound?

Reg. The AWAR include in the definition of "Animal" both live subjects and dead specimens of regulated species (§1.1 Animal) including those obtained from pounds or shelters (§1.1 Pound or Shelter, Random Source). The *Guide* (p. 106) requires that all animals be acquired lawfully. APHIS/AC Policy 20 clarifies the licensing of sales of dead animals.[73]

Opin. The inclusion of both live and dead specimens of regulated species under the definition of "Animal" in the AWAR includes those obtained from pounds or shelters (§1.1 Pound or Shelter, Random Source) and presumably any dog and/or cat cadavers; however, research facility record-keeping requirements under the AWAR pertain only to live animals (§2.35,b, Recordkeeping requirements). While the *Guide* requires lawful acquisition of animals, its definition of laboratory animal (p. 2) carried with it the practical interpretation and associated historical application that attention would be directed to cases of live animals. APHIS/AC Policy 20 requires the licensing of distributors of dead animals or parts only if the dealers or persons are engaged in commerce and seemingly would exclude pounds as a regulated source providing no money changed hands. While it is safe to conclude that the IACUC would not be required to approve the acquisition of cadaver animals donated from a pound, the involvement of a financial transaction may have different implications. IACUCs may want to carefully review and consider situations of cadaver donation from pounds given the animal welfare, public relations, and other implications related to the care, handling, and humane euthanasia practices at the source. (See 14:28.)

References

1. Committees to revise the Guide for the Care and Use of Agricultural Animals in Research and Teaching. 2010. *Guide for the Care and Use of Agricultural Animals in Agricultural Research and Teaching,* 3rd edition, 1–9. Champaign, IL: Federation of Animal Science Societies.
2. Association for Assessment and Accreditation of Laboratory Animal Care International. AAALAC now using Three Primary Standards. http://www.aaalac.org/about/guidelines.cfm. Accessed July 14, 2013.
3. Sikes, R.S., and W.L. Gannon. 2011. Guidelines of the American Society of Mammalogists for the use of wild mammals in research. *J. Mammal.* 92:235–253.
4. U.S. Department of Agriculture, Code of Federal Regulations, Animal and Plant Health Inspection Service—Veterinary Services, Title 7; Part 3714 (b)(2), Washington, D.C.
5. U.S. Department of the Interior, Code of Federal Regulations, Permits for Scientific Purposes, Enhancement of Propagation or Survival, or for Incidental Taking of Endangered Wildlife, Title 50; Part 17, Washington, D.C.
6. U.S. Department of the Interior, Code of Federal Regulations, Endangered Species Convention, Title 50; Part 23, Washington, D.C.
7. U.S. Department of Health and Human Services, Code of Federal Regulations, Extralabel Drug Use in Animals, Title 21; Part 530, Washington, D.C.
8. U.S. Department of Health and Human Services, Code of Federal Regulations, Foreign Quarantine; Importations; Dogs and cats; Turtles, tortoises, and terrapins; Nonhuman primates; etiologic agents, hosts and vectors, Title 42; Part 71; Subpart F, Washington, D.C.
9. U.S. Department of Labor, Title 29; Part 1910.1030, Occupational and Health Standards, Bloodborne Pathogens, Washington, D.C.
10. U.S. Department of Health and Human Services. 2009. *Biosafety in Microbiological and Biomedical Laboratories,* 5th edition, Appendix H: Working with Human, NHP, and other Mammalian Cells and Tissues.
11. McCully, R.E. 1994. Letter re: Standard Interpretation of OSHA Bloodborne Standard 29CFR1910.1030. https://www.osha.gov/pls/oshaweb/owadisp.show_document?p_table=INTERPRETATIONS&p_id=21519&p_text_version=FALSE%20.
12. Occupational Health and Safety Administration, Code of Federal Regulations, Occupational Safety and Health Standards; Toxic and Hazardous Substances; Chemical Hygiene Plan, Title 29; Part 1910; Subpart Z, Washington, D.C.
13. Committee on Standards, Institute of Laboratory Animal Resources. 1974. *Amphibians: Guidelines for the Breeding, Care and Management of Laboratory Animals*: Washington, D.C.: National Academies Press. http://www.nap.edu/catalog.php?record_id=661.
14. Pough, F.H. 1992. Recommendations for the care of amphibians and reptiles in academic institutions. *ILAR J.* 33(4):S5–S18.
15. Schaeffer, D.O., K.M. Kleinow, and L. Krulisch. 1992. *The Care and Use of Amphibians, Reptiles, and Fish in Research.* Betheseda: Scientists Center for Animal Welfare.
16. American Society of Ichthyologists and Herpetologists, American Fisheries Society, American Institute of Fisheries Research Biologists. 1987. Guidelines for use of fishes in field research. *Fisheries.* 13(2):16–23.
17. Herpetological Animal Care and Use Committee. American Society of Ichthyologists and Herpetologists. 2004. Guidelines for use of live amphibians and reptiles in field and laboratory research, 2nd edition. http://www.research.fsu.edu/acuc/policies_Guidelines/ASIH_HACC_GuidelinesAmphibians.pdf.
18. DeTolla, L.J., S. Sriniva, B.R. Whitaker et al. 1995. Guidelines for the care and use of fish in research. *ILAR. J.* 37:159–173.

19. Potkay, S., N.L. Garnett, J.G. Miller et al. 1997. Frequently asked questions about the Public Health Service Policy on Humane Care and Use of Laboratory Animals. *Contemp. Top. Lab. Anim. Sci.* 36:47–50.
20. National Institutes of Health, Office of Laboratory Animal Welfare. 2013. Frequently Asked Questions D.17, What guidelines should IACUCs follow for fishes, amphibians, reptiles, birds, and other nontraditional species used in research? http://grants.nih.gov/grants/olaw/faqs.htm-proto_17.
21. U.S. Department of Agriculture. 2011. Animal Care Policy Manual, Policy #17, Regulation of Agricultural Animals. http://www.aphis.usda.gov/animal_welfare/policy.php?policy=17.
22. National Institutes of Health, Office of Laboratory Animal Welfare. 2013. Frequently Asked Questions F.2, Is the IACUC responsible for tracking animal usage? http://grants.nih.gov/grants/olaw/faqs.htm-useandmgmt_2.
23. National Institutes of Health, Office of Laboratory Animal Welfare. 2005. Notice NOT-OD-05-034, Guidance on prompt reporting to OLAW under the PHS Policy on Humane Care and Use of Laboratory Animals. http://grants.nih.gov/grants/guide/notice-files/NOT-OD-05-034.html.
24. Applied Research Ethics National Association/Office of Laboratory Animal Welfare. 2002. *Institutional Animal Care and Use Committee Guidebook*, 2nd edition, 153. Bethesda: National Institutes of Health.
25. Jacoby, R.O., J.G. Fox, and M. Davisson. 2002. *Biology and Diseases of Mice*, 2nd edition. New York: Academic Press.
26. Macy, J.D., G.A. Camero, P.C. Smith et al. 2011. Detection and control of mouse parvovirus. *J. Am. Assoc. Lab. Anim. Sci.* 50:516–522.
27. Applied Research Ethics National Association/Office of Laboratory Animal Welfare. 2002. Institutional Animal Care and Use Committee Guidebook. 2nd edition, 130–133. Bethesda: National Institutes of Health.
28. Wigglesworth, C. 2006. A word from OLAW. *Lab Anim (NY)*. 35:17.
29. Huerkamp, M.J., and D. Archer. 2007. Animal Acquisition and Disposition. In *The IACUC Handbook*, 2nd edition, eds. J. Silverman, M.A. Suckow, and S. Murthy, 188. Boca Raton: CRC Press.
30. Huerkamp, M.J., and D. Archer. 2007. Animal Acquisition and Disposition. In *The IACUC Handbook*, 2nd edition, eds. J. Silverman, M.A. Suckow, and S. Murthy, 189–225. Boca Raton: CRC Press.
31. U.S. Government Principles for the Utilization and Care of Vertebrate Animals Used in Testing, Research, and Training. http://grants.nih.gov/grants/olaw/references/phspol.htm-USGovPrinciples. Accessed July 15, 2013.
32. Borski, R.J., and R.G. Hodson. 2003. Fish research and the institutional animal care and use committee. *ILAR J.* 44:286–294.
33. U.S. Department of Agriculture. Guidelines for Preparing USDA Annual Reports and Assigning USDA Pain & Distress Categories. http://oacu.od.nih.gov/ARAC/documents/USDA_Reports.pdf. Accessed July 15, 2013.
34. Potkay, S. and W.R. DeHaven. 2000. Common areas of noncompliance. *Lab Anim (NY)*. 29(5):32.
35. Wolff, A., N. Garnett, S. Potkay et al. 2003. Frequently asked questions about the public health service policy on humane care and use of laboratory animals. *Lab Anim (NY)*. 32(9):33–36.
36. Association for Assessment and Accreditation of Laboratory Animal Care International. 2013. Frequently Asked Questions D.1. Animal Environment, Housing and Management: Trio Breeding. http://www.aaalac.org/accreditation/faq_landing.cfm#Ctrio.
37. Nagy, A., M. Gertsenstein, K. Vinterstein et al. 2003. *Manipulating the mouse embryo: A laboratory manual*. 3rd edition, 152. Plainview, NY: Cold Spring Harbor Laboratory Press.
38. Wilkie, T.M., R.L. Brinster, and R.D. Palmiter. 1986. Germline and somatic mosaicism in transgenic mice. *Dev. Biol.* 118:9–18.

39. National Institutes of Health, Office of Laboratory Animal Welfare. 1991. The Public Health Service Responds to Commonly Asked Questions. *ILAR News* 33(4):68–70. http://grants2.nih.gov/grants/olaw/references/ilar91.htm-1.
40. National Institutes of Health, Office of Laboratory Animal Welfare. 2013. Frequently Asked Questions A.4., Does the PHS Policy apply to live embryonated eggs? http://grants.nih.gov/grants/olaw/faqs.htm-App_4.
41. Saif, Y.M., and W.L. Bacon. 1996. Simply stated. *Lab Anim (NY)*. 25(5):22.
42. American Veterinary Medical Association. 2013. AVMA guidelines for the euthansia of animals: 2013 edition. https://www.avma.org/KB/Policies/Documents/euthanasia.pdf.
43. Swayne, D.E. 1996. Preventative Measures. *Lab Anim (NY)*. 25(5):22.
44. National Institutes of Health, Office of Laboratory Animal Welfare. 2013. Frequently Asked Questions, A.5, Does the PHS Policy apply to larval forms of amphibians and fish? http://grants.nih.gov/grants/olaw/faqs.htm-a5.
45. Williams, B. 1999. Wildlife Research and the IACUC. *AWIC Bulletin*. 10:1–2.
46. Mulcahy, D.M. 2003. Does the Animal Welfare Act apply to free-ranging animals? *ILAR J*. 44:252–258.
47. Reeb-Whitaker, C.K., B. Paigen, W.G. Beamer et al. 2001. The impact of reduced frequency of cage changes on the health of mice housed in ventilated cages. *Lab Anim (NY)*. 35:58–73.
48. National Institutes of Health, Office of Laboratory Animal Welfare. 2013. Frequently Asked Questions, A.3, Does the PHS Policy apply to use of animal tissue obtained from dead animals? http://grants.nih.gov/grants/olaw/faqs.htm-App_3.
49. Garnett, N., and W.R. DeHaven. 1997. OPRR and USDA/Animal Care response on applicability of the Animal Welfare Regulations and the PHS Policy to dead animals and shared tissues. *Lab Anim (NY)*. 26(3):21.
50. National Institutes of Health, Office of Laboratory Animal Welfare. 2013. Frequently Asked Questions D.9, What is considered a significant change to a project that would require IACUC review? http://grants.nih.gov/grants/olaw/faqs.htm-proto_9.
51. National Institutes of Health, Office of Extramural Research. 2003. Notice NOT-OD-03-046, Revised guidance regarding IACUC approval of changes in personnel involved in animal activities. http://grants.nih.gov/grants/guide/notice-files/NOT-OD-03-046.html.
52. National Institutes of Health, Office of Extramural Research. 2011. NIH Grants Policy Statement, Part II: Terms and Conditions, Subpart A, Change in Scope. http://grants.nih.gov/grants/policy/nihgps_2011/nihgps_ch8.htm.
53. National Institutes of Health, Office of Laboratory Animal Welfare. 2013. Frequently Asked Questions B.13, Does OLAW expect the IACUC to notify NIH when there is a change in an animal activity supported by PHS funds? http://grants.nih.gov/grants/olaw/faqs.htm-b13.
54. Huerkamp, M.J., and D. Archer. 2007. Animal Acquisition and Disposition. In *The IACUC Handbook*, 2nd edition, eds. J. Silverman, M.A. Suckow, and S. Murthy, 211. Boca Raton: CRC Press.
55. Brown, P. 2007. A word from OLAW. *Lab Anim (NY)*. 36:14.
56. National Institutes of Health. 2012. Notice NOT-OD-12-081, Change in criteria for renewal of animal welfare assurances for foreign institutions. http://grants.nih.gov/grants/guide/notice-files/NOT-OD-12-081.html.
57. National Institutes of Health, Office of Laboratory Animal Welfare. 2013. Frequently Asked Questions D.13, If an animal activity will be performed outside of the U.S. (either by a foreign awardee or by a foreign institution as a subproject for a domestic awardee), is the awardee's IACUC required to review and approve that activity? http://grants.nih.gov/grants/olaw/faqs.htm-proto_13.
58. National Institutes of Health, Office of Laboratory Animal Welfare. 2013. Frequently Asked Questions D.15, Is IACUC approval required for the collection of samples in foreign countries from captive wild animals or research colonies? http://grants.nih.gov/grants/olaw/faqs.htm-proto_15.

59. National Institutes of Health, Office of Laboratory Animal Welfare. 2013. Frequently Asked Questions A.2, Does the PHS Policy apply to the production of custom antibodies or to the purchase of surgically modified animals? http://grants.nih.gov/grants/olaw/faqs.htm-a2.
60. Huerkamp, M.J., and D. Archer. 2007. Animal Acquisition and Disposition. In *The IACUC Handbook,* 2nd edition, eds. J. Silverman, M.A. Suckow, and S. Murthy, 191. Boca Raton: CRC Press.
61. U.S. Department of Agriculture. 2011. Policy 14, Major survival surgery dealers selling surgically-altered animals to research. http://www.aphis.usda.gov/animal_welfare/policy.php?policy=14.
62. Potkay, S., N. Garnett, J.G. Miller et al. 1995. Frequently Asked Questions about the Public Health Service Policy on Humane Care and Use of Laboratory Animals. *Lab Anim (NY).* 24(9):24–26.
63. National Institutes of Health, Office of Laboratory Animal Welfare. 2013. Frequently Asked Questions D.16, Is IACUC approval required for the use of animals in breeding programs, as blood donors, as sentinels in disease surveillance programs, or for other non-research purposes? http://grants.nih.gov/grants/olaw/faqs.htm-proto_16.
64. National Institutes of Health, Office of Laboratory Animal Welfare. 2013. Frequently Asked Questions A.8, How can the IACUC determine if activities involving privately owned animals constitute veterinary clinical care or research activities? http://grants.nih.gov/grants/olaw/faqs.htm-a8.
65. National Institutes of Health, Office of Laboratory Animal Welfare. 2013. Frequently Asked Questions A.7, Does the IACUC need to approve research studies that use privately owned animals, such as pets? http://grants.nih.gov/grants/olaw/faqs.htm-a7.
66. Brown, P., and C. Gipson. 2010. A word from OLAW and USDA. *Lab Anim (NY).* 39(3):68.
67. Applied Research Ethics National Association/Office of Laboratory Animal Welfare. 2002. *Institutional Animal Care and Use Committee Guidebook,* 2nd edition, 142. Bethesea: National Institutes of Health.
68. American Veterinary Medical Association. Committee on Veterinary Technician Education, Activities. 2013. *Accreditation Policies and Procedures of the AVMA Committee on Veterinary Technician Education and Activities (CVTEA).* https://www.avma.org/ProfessionalDevelopment/Education/Accreditation/Programs/Pages/cvtea-pp.aspx.
69. Tannenbaum, J. 1998. Personal Communication.
70. U.S. Department of Justice, Code of Federal Regulations, Nondiscrimination on the Basis of Disability by Public Accomodations and in Commercial Facilities; Service Animals, Title 28; Part 36, Washingon, D.C.
71. National Institutes of Health, Office of Laboratory Animal Welfare. 2013. Frequently Asked Questions F.11, Can IACUCs authorize the adoption of research animals as pets after the animals are no longer needed for study? http://grants.nih.gov/grants/olaw/faqs.htm-useandmgmt_11.
72. U.S. Department of Health and Human Services. Code of Federal Regulations, New Animal Drugs for Investigational Use, Title 21; Part 511, Washington, D.C.
73. U.S. Department of Agriculture. Animal Care Policy Manual. 2011. Policy 20, Licensing Sales of Dead Animals. http://ww.aphis.usda.gov/animal_welfare/policy.php?policy=20.

15

*Animal Housing, Use Sites, and Transportation**

Robert M. Bigsby and Robin Crisler

Introduction

This chapter focuses on examples of acceptable and non-acceptable housing or holding locations for animals used in research, teaching, and testing. The AWA and AWAR,[1] the *Guide*,[2] and the PHS Policy[3] are the primary references utilized. There are few absolutes when considering animal housing and use sites. Professional judgment, institutional policy, and protocol goals and needs must all be considered when determining what is and is not an acceptable location for the care and use of animals.

IACUC Chairs from throughout the United States and Canada responded to a questionnaire about animal housing and use sites. Their responses are tabulated and summarized where appropriate to provide further insight into the breadth of policies currently in practice.

15:1 What is the definition of an animal facility?

Reg. The AWAR (§1.1) do not provide a definition for "animal facility," however animal housing is described within separate definitions for each of the following terms: *housing facility, indoor housing facility, outdoor housing facility, sheltered housing,* and *study area*. For example, a housing facility is any land, premise, shed, barn, building, trailer, or other structure housing or intended to house animals.

PHS Policy (III,B) defines an *animal facility* as "any and all buildings, rooms, areas, enclosures, or vehicles, including satellite facilities, used for animal confinement, transport, maintenance, breeding, or experiments inclusive of surgical manipulation."

The *Guide* (pp. 135–136) states that an animal facility consists of functional areas for animal housing, care, and sanitation; receipt, quarantine, and separation of animals, separation of species, or isolation of individual projects; and storage. Some facilities may encompass space for surgery, intensive care, necropsy, irradiation, diet preparation, procedural space, imaging, clinical treatment, containment or barrier facilities, supply receiving, cage washing, carcass storage, facility administration, training, locker rooms, break rooms and maintenance areas. (See 24:12.)

Opin. The authors' working definition of an animal facility is an integrated concept that is inclusive of the many programmatic aspects that relate to animal housing, care,

* This revised Chapter 15 builds on the contributions of Dr. Cynthia Gillett, author of this chapter in the 1st and 2nd editions of *The IACUC Handbook*.

and use in an institutional setting. However, for the purposes of this chapter, the term *animal facility* shall refer to the primary housing location and support space for animals and animal studies overseen by the IACUC.

15:2 What are the circumstances and length of time that animals may be kept at a location other than their primary housing site?

Reg. There are two basic scenarios in which this can occur: for research purposes (lab, surgical suite, testing area) or for medical reasons (veterinary or surgical treatment area). In both cases if housed greater than 12 hours these will be considered *study areas* as per the AWAR §1.1 definition and should be inspected by the IACUC as per AWAR §2.31,c. The AWAR (§2.31,c,2) require that any location where animals are housed for more than 12 hours be inspected by the IACUC at least once every 6 months.

PHS Policy (III,B) states that "a satellite facility is any containment outside of a core facility or centrally designated or managed area in which animals are housed for more than 24 hours." Nevertheless, PHS Policy (II) requires compliance with the AWA where applicable. Therefore, the PHS Policy definition of a *satellite facility* and the duration of stay requirements may be applied to species that are not regulated by the AWAR, such as typical laboratory rats and mice (and see NIH/OLAW FAQ E.1).[4,5] However, the more restrictive AWA regulation (§2.31,c,2) requiring IACUC semiannual inspection of locations where animals are housed for more than 12 hours must be applied to all AWA-regulated species.

The *Guide* (p. 25) states that the IACUC is responsible for oversight of animal use areas; however, it does not define that term. It also states (p. 134) that "Animals should be housed in facilities dedicated to or assigned for that purpose, not in laboratories merely for convenience. If animals must be maintained in a laboratory to satisfy the scientific aims of a protocol, that space should be appropriate to house and care for the animals and its use limited to the period during which it is required."

Opin. The authors' interpretation of the term *animal use area*, as stated in the *Guide* (pp. 14, 25), is that it implies research procedural areas such as a laboratory or testing site. Neither the *Guide*, the AWAR, nor the PHS Policy prohibits removal of animals from the animal holding facility for procedures or for housing elsewhere, such as in an investigator's laboratory or a satellite facility. Additionally, NIH/OLAW indicates "Institutions are responsible for oversight of all animal-related activities regardless of how long or where the activity occurs." See NIH/OLAW FAQ E.1.[4] However, doing so may trigger additional regulatory oversight responsibility of the IACUC or bring into play individual institutional requirements. Although one could apply different time standards on the basis of whether or not an AWAR-regulated species is involved, applying the most restrictive criterion (12 hours) will result in compliance in all situations.

Surv. For how long does your IACUC allow animals to be removed from their primary housing site without an IACUC-approved exception to federal regulations? (More than one response may have been relevant for individual respondents.)

- Not applicable 84/296 (28%)
- Up to 12 hours for species covered by the AWA 46/296 (16%)
- Up to 12 hours for all species 15/296 (39%)

• Up to 24 hours for species covered only by the PHS Policy	30/296 (10%)
• Up to 24 hours for all species	31/296 (10%)
• Greater than 24 hours for all species	4/296 (1.4%)
• I don't know	20/296 (6.8%)

15:3 What are examples of acceptable circumstances for keeping animals outside the primary animal facility?

Reg. The *Guide* (pp. 134–135) refers to a centralized animal facility as one in which support, care and use areas are adjacent to the animal housing space, while in a decentralized system animal housing and use may occur in space that is not solely dedicated to animal care and support, e.g., laboratories, treatment or imaging facilities, or surgical suites physically separated from animal housing/support areas. The *Guide* (p. 135), states "The opportunity for exposure to disease agents is much greater in these (decentralized) situations and special consideration should be given to biosecurity, including transportation to and from the site, quarantine before or after use of the specialized research area, and environmental and equipment decontamination."

Opin. The AWAR and PHS Policy, while providing definitions and oversight recommendations for animal facilities (including satellite facilities; see 15:1), do not provide examples of acceptable circumstances or rationale for keeping animals outside the primary animal facility. However the *Guide* (pp. 146–150) does provide examples of circumstances where decentralization of animal housing and use may be preferred, such as facilities for imaging or irradiation, or laboratories for behavioral studies.

As the *Guide* indicates, there are other considerations when housing animals outside a centrally managed facility. For reasons of disease control, some facilities have a policy stating that if animals leave the facility they may not be returned. This is particularly applicable to microbiologically defined (specific pathogen free) rodents. The method and route of transportation to the study area also must be monitored to minimize potential for disease transmission and to ensure conformance with transportation guidelines. Facility construction, environmental monitoring, security, consistency of animal care, and veterinary availability should also be considered when housing outside of the primary facility.

Surv. What are some circumstances that are typically acceptable to your IACUC for keeping animals outside of the primary animal facility? (More than one response may have been relevant for individual respondents.)

• Not applicable	79/296 (27%)
• Biocontainment requirements	95/296 (32%)
• Needed research equipment cannot be placed within the animal facility	189/296 (64%)
• Behavioral or similar testing would be affected by daily movement of animals to and from their usual holding room	140/296 (47%)
• Postoperative monitoring	100/296 (34%)
• Need for frequent observations or sampling	104/296 (35%)
• Other	25/296 (8.4%)

Some of the survey responses for the "other, please specify" category noted that animals may be kept outside the primary housing facility in instances where adequate scientific justification had been provided and evaluated by the IACUC on a case-by-case basis; where a specialized housing environment was required to meet species-specific requirements or research needs; temporary housing for imaging or terminal procedures; during building construction, heating, ventilation and air conditioning (HVAC) system failures, power outages, emergency or natural disaster; and instances where investigators had allergies to other species in the primary facility.

15:4 Are there regulatory considerations regarding the actual act of transportation that the IACUC must consider when animals are moved from their primary housing site to a secondary site?

Reg. The AWAR (Part 3, Standards) have extensive transit regulations for each regulated species. They include primary enclosures, primary conveyances, animal identification, food and water requirements, care in transit, handling, sanitation, ventilation, shelter, and temperature. The AWAR (§2.38,f) also cover the transit of animals as part of an adequate "handling" consideration, which implies animal handling during transportation is included as an item the IACUC must consider, though this is not explicitly stated. PHS Policy II requires compliance with the AWAR, where applicable, and PHS Policy III defines an animal facility to include vehicles used for transport.

The *Guide* (pp. 107–109) states: "All transportation of animals … should be planned to minimize transit time and the risk of zoonoses, protect against environmental extremes, avoid overcrowding, provide food and water when indicated, and protect against physical trauma." In addition, "special considerations may be necessary for transporting animals during certain phases of their life or in certain conditions, such as pregnant, perinatal, and geriatric animals; animals with preexisting medical conditions (e.g., diabetes mellitus); and animals surgically prepared by the supplier." Importantly, for the sake of a controlled experimental environment, the transportation process "should provide an appropriate level of animal biosecurity," that is, measures should be taken "to control known or unknown infections in laboratory animals" (*Guide* p. 109). "Transportation of animals in private vehicles is discouraged because of potential animal biosecurity, safety, health, and liability risks for the animals, personnel, and institution."

Transport of animals that have been infected with an infectious agent or ones that are carrying grafted tissue that is suspected to be infected with such an agent is subject to national and international rules regulating transport of hazardous materials. The website for the Centers for Disease Control and Prevention (CDC)[6] contains links to relevant regulatory agencies.

Opin. The AWAR on transport of animals are designed to ensure humane animal care. These regulations should be consulted when planning intrafacility and interfacility institutional animal transfers. While animal transport is not required to be directly reviewed and approved by the IACUC, transport activities should be reviewed during the semiannual program evaluation to ensure appropriate procedures are being followed. Since they can be considered animal use areas, transport vehicles used routinely by the institution should be included in the semiannual facility inspection, assessing sanitation and climate control systems just as with any other animal housing unit. Animal transport issues also should be raised during protocol review when justification for off-site housing is being requested. International transport is outside the scope of this chapter; however all applicable national and international standards and laws should be followed.

Biosafety concerns are paramount when considering transport of animals carrying infectious agents or tissues (human or animal xenografts) that cannot be reasonably considered to be non-infectious. According to the Dangerous Goods website for the International Air Transport Association (IATA):[7] "A live animal that has been intentionally infected and is known or suspected to contain an infectious substance must not be transported by air unless the infectious substance contained cannot be consigned by any other means. Infected animals may only be transported under terms and conditions approved by the appropriate national authority." Although there are guidelines for packaging tissue specimens containing potentially hazardous material, there are no readily available guidelines for securing a live animal for such transport and the reader is advised to contact the CDC and IATA for more information.

Surv. 1 Does your IACUC allow small animals (e.g., rodents) to be transported from one animal housing site to another by private vehicles?

Note: Past survey responses from the 2nd edition of The IACUC Handbook[8] (p. 230) indicated about 8% of survey respondents were allowed to transport larger animals (e.g., dogs, monkeys, pigs) between housing sites in private vehicles with IACUC-approval.

• Not applicable	77/296 (26%)
• No	115/296 (39%)
• Yes, with IACUC approval	93/296 (31%)
• Yes, and IACUC approval is not necessary	8/296 (2.7%)
• Other	3/296 (1.0%)

Surv. 2 If your IACUC allows the use of privately owned passenger vehicles for animal transport, what environmental conditions have been established for that purpose? (More than one response may have been relevant for individual respondents.)

• Not applicable	171/280 (61%)
• IACUC approval is required	79/280 (28%)
• Cages must be secured within the vehicle (e.g., via seat belts)	55/280 (20%)
• Vehicles must maintain a temperature that is appropriate for the species	82/280 (29%)
• Even if approved by the IACUC we discourage this practice as there may be animal allergens or other factors left behind from the transported animals	39/280(14%)
• Other: (Please specify)	10/280 (3.6%)

Note: Some of the survey responses for the "other, please specify" category indicated that use of specialized transport containers, environmentally controlled vehicles, and a separate animal holding area inside the vehicle were required. Other causes for use of private vehicle included for conduct of wildlife research, and in cases of catastrophic emergency.

15:5 Are there other practical considerations besides regulatory requirements the IACUC must consider regarding animal transport?

Reg. The AWAR (§2.38,f,1) state that handling of all animals shall be done as expeditiously and carefully as possible in a manner than does not cause trauma, overheating, excessive cooling, behavioral stress, physical harm, or unnecessary discomfort. The *Guide* (p. 107) indicates that transportation should provide for animal biosecurity, minimize zoonotic risks, protect from environmental extremes, avoid overcrowding, provide for animals' needs and comfort, and protect people and animals from trauma.

Opin. The practical, nonregulatory day-to-day considerations that can arise during intra- and interinstitutional animal transport tend to be situational and are usually best handled by discussions with the laboratory animal facility professional staff that oversee the transportation processes. Examples include situations such as whether or not a box of mice can be taken outside to transport them to a laboratory in another building; whether or not a cardboard box can be used to transport a rabbit to the laboratory; or which elevator can be used to transport animals to a laboratory. The answers to these and similar issues depend on factors such as climate, time of year, facility design, and institutional policies. The IACUC, during the course of its semiannual facility inspection or through reports of concern about transportation, may find evidence of unacceptable transportation practices (e.g., a soiled transport enclosure, inappropriate elevator or corridor use) that will need to be addressed with the PIs and laboratory animal care staff.

It is advisable that an institution have a transportation policy or SOP to describe processes, provide continuity, and to help establish expectations between investigators and animal facility staff. This policy should be reviewed by the IACUC as part of the program evaluation even though aspects of the policy may contain nonregulatory elements. Guidelines for transport of animals that have been administered biologic, chemical or radiologic hazards should be established with the help of safety experts and the appropriate institutional committees (Institutional Biosafety Committee, Radiologic Safety Committee, Environmental Health and Safety). Exposure of persons to allergens through the transportation process must also be considered. Individuals who do not have direct animal contact may not be covered by the institution's Occupational Health and Safety Program, yet still need information about the risks of laboratory animal allergy and instructions on what to do should they experience symptoms. Transportation practices should be designed to reduce and limit exposure of all persons that may be exposed to airborne allergens along the transport route, such as the general public and employees in laboratories adjacent to animal use areas.

15:6 If animals are being housed in a research laboratory setting, are there species differences to consider?

Reg. The housing requirements for various APHIS/AC-regulated species are addressed in Part 3 of the AWAR. These standards apply if the animals are approved by the IACUC to be "housed" in a laboratory. The IACUC may make exceptions to these standards if there is scientific justification to do so (AWAR §2.31,c,3). Likewise, PHS Policy (IV,C,1,d) requires the same standards of care for animals housed in or away from the primary care facility, ensuring that "living conditions of animals will be appropriate for their species and contribute to their health and comfort." The *Guide* (pp. 41–103) provides additional examples of species-specific environmental considerations such as temperature, illumination, noise, social needs, environmental enrichment, space, husbandry, and sanitation. (See 15:11.)

Opin. The need for laboratory housing should be protocol driven and not motivated by convenience or economics (*Guide* p. 134). Not all laboratory facilities will be structurally or environmentally appropriate for housing all types of animal species. The suitability of laboratory housing and the expertise and training of scientists to provide nonmedical care must be carefully evaluated on a case-by-case basis by the IACUC. A comprehensive husbandry plan for the animals should be submitted to the IACUC for review. The IACUC should keep records of approved laboratory housing sites and perform, or delegate, periodic inspections of these areas when animals are present, in addition to the scheduled semiannual facility inspections.

Surv. How does your IACUC inspect animal housing and use sites if they are at geographically distant locations? That is, locations that would require unique travel plans were the IACUC to do an on-site inspection. (More than one response may have been relevant for individual respondents.)

- Not applicable 160/295 (54%)
- At a geographically distant institution with a collaborating institution, we depend on that institution's IACUC 72/295 (24%)
- At a geographically distant field site we ask nearby colleagues to inspect the site 9/295 (3.0%)
- At a geographically distant field site we depend on videos or similar means to evaluate the site 20/295 (7.0%)
- We inspect all sites by ourselves, whether near or distant 55/295 (19%)
- We do not inspect our geographically distant sites 22/295 (7.0%)
- Other 9/295 (3.0%)

15:7 May animals be housed at sites not belonging to an institution? Under what circumstances?

Reg. The AWAR (§2.38,i) state, "If any research facility obtains prior approval of the APHIS AC Regional Director, it may arrange to have another person hold animals, *provided* that

1. The other person agrees, in writing, to comply with the regulations in this part and the standards in part 3 of this subchapter, and to allow inspection of the premises by an APHIS official during business hours;
2. The animals remain under the total control and responsibility of the research facility; and
3. The Institutional Official agrees, in writing, that the other person or premises is a recognized animal site under its research facility registration."

APHIS/AC suggests that APHIS Form 7009 should be used for approved off-premises housing sites. The APHIS/AC Regional Office can be contacted for help and guidance on this topic should questions arise.

PHS Policy (IV,A) requires that awardee institutions and other participating institutions that receive PHS support have an approved Assurance on file with NIH/OLAW. NIH/OLAW has stated that vendors that supply custom antibodies to investigators, or perform a surgery conducted in response to a specific request, must either have an Animal Welfare Assurance on file with NIH/OLAW or be included in the receiving research institution's Assurance statement (see NIH/OLAW FAQ A.2.[9] Additionally, if AWAR-regulated species, such as rabbits, are used to produce antibodies at the vendor's site or are in other research studies, then the vendor must be registered with APHIS/AC as a research facility. APHIS/AC Policy 10 clarifies the need for licensing or registration of a facility using regulated species for antibody production.[10] (See 8:10; 19:6.) The *Guide* does not address this issue.

Opin. The legal necessity for the noninstitutional site to be APHIS/AC registered must be evaluated along with a separate verification that an Assurance is on file with NIH/OLAW if one is legally required, as outlined above. AAALAC accreditation status for the noninstitutional site may also be highly desirable. These considerations are dependent on the primary institution's status and on the individual housing situation. Affiliate Veterans Administration Medical Center facilities are an example of facilities where housing animals at a noninstitutionally owned site might be routinely permitted. In the latter example, this should be listed as a "site" of the research institution. Institutions may out-source rodent quarantine or embryo rederivation to commercial vendors specializing in those services. AAALAC follows animal ownership in terms of defining who is responsible for animals at an off-site program for purposes of accreditation. During the accreditation process, AAALAC considers the outside contractor's accreditation standing as well as that of the institution seeking accreditation.[11]

Surv. Under what circumstances would your IACUC approve the housing of animals at sites not belonging to your institution? (More than one response may have been relevant for individual respondents.)

Responses to the "other, please specify" category included: during serious facility emergencies or disasters, during construction, for field research, for specialized procedures, or specific research requirements.

• Not applicable	79/294 (27%)
• If we would run out of general housing space	32/294 (11%)
• Need for specialized housing (e.g., for nonhuman primates)	61/294 (21%)
• Collaborative research arrangements	178/294 (61%)
• I don't know	11/294 (3.7%)
• Other: (Please specify)	27/294 (9.2%)

15:8 Are there any circumstances under which animals may be housed in a person's home or on a private farm?

Reg. Both NIH/OLAW and APHIS/AC will consider the use of the animal, the housing, and the animal's ownership to determine the appropriate level of regulatory oversight that is required. As far as the use of the animal in concerned, APHIS/AC specifically does not regulate pets or animals owned by private citizens in a veterinarian–client relationship because the owner's consent is present and clinical work is outside of the original intent of the AWA. Animals in a veterinarian–client relationship are not regulated even when they are being treated at a research institution. In addition, as per the definition of animal in the AWA (§2132,g), APHIS/AC has no regulatory authority over animals used in agricultural research that are housed on private farms.

In a guidance statement provided for this question, APHIS/AC (with NIH/OLAW concurrence) indicated that pet and client owned animals participating in animal study proposal for educational and teaching purposes (e.g., as part of a veterinary technical program) are regulated during the activity. Although the pet owner's home would not be inspected, the animal study protocol utilizing these animals must be reviewed and approved by the IACUC.

In the event an institution elects to house its animals used in biomedical research in a private home or private farm for greater than 12 hours, the home or farm would be considered a study area (AWAR §1.1, Study Area) and therefore subject to IACUC inspection. Under the PHS Policy (III,B) the same is true for animals housed in a private home or farm for more than 24 hours. With regard to PHS policy, pets and livestock used in research must be covered under an IACUC-approved protocol and the institution must have an NIH/OLAW-approved Assurance covering all performance sites (see related PHS information in 15:7).

Opin. The origins and intent of the AWARs are rooted in the need for protection of pets from theft and re-sale, especially for the unauthorized sale of stolen animals for use in research institutions (AWA §2131), which was a national concern first documented in the November 29, 1965 issue of *Sports Illustrated* magazine regarding Pepper, a Dalmatian dog, and helped pave the way for Congress to enact the AWA in 1966.

Prior approval by the IACUC for private housing is considered by surveyed respondents to be essential. The authors feel that IACUC-approval is needed for these animals, even when PHS funding is not supporting the research. When animals remain privately owned, as opposed to institutionally owned, they most often are housed in that person's home or farm (e.g., blood donor animals or veterinary clinical studies on companion and farm animals).

Surv. Under what circumstances would your IACUC allow animals to be housed in a person's home or on a private farm?

- Not applicable; issue has not arisen — 74/147 (50%)
- None — 53/147 (36%)
- Certain animals may be housed at a private farm or home with IACUC approval — 16/147 (11%)
- Other — 4/147 (2.7%)

15:9 Do all wildlife or field studies need to be reviewed by the IACUC?

Reg. The AWAR definition of *animal* (see 12:1) includes most mammals; therefore, wild mammals are subject to IACUC protocol review. However, the regulations specifically exempt the IACUC from the requirement to review field studies that are conducted on free-living wild animals in their natural habitat and which do not involve an invasive procedure, any harm to the animal, or any material alteration of the behavior of an animal under study (AWAR §1.1; §2.31,d,1).

PHS Policy II requires compliance with the AWA, as applicable, and (III,A) defines an *animal* as any live, vertebrate animal. Therefore, all vertebrate animals used in field studies fall under the PHS Policy. The *Guide* (p. 32) states that "Investigations may involve the observation or use of nondomesticated vertebrate species under field conditions" and that "...the basic principles of humane care and use apply to the use of animals living in natural conditions." The *Guide* (p. 32) indicates that field studies need assessment for issues including zoonotic disease and safety of humans and the study animals.

Opin. If an institution's NIH/OLAW Assurance covers all vertebrate animals, not just those funded by PHS, the IACUC must review wildlife field studies of all vertebrates that include activities other than simple unobtrusive observation. It is the authors' opinion that applying uniform policies to all animals involved in the institution's animal care and use program regardless of the type of research or teaching activity involved promotes consistent oversight.

APHIS/AC, through consultation with the authors of this chapter, has indicated that it is implied through the AWA (§2143,a,7,A) regarding "professionally acceptable standards governing the care, treatment, and use of animals...", that the *Guide for the Care and Use of Laboratory Animals*, the *Guide for the Care and Use of Agricultural Animals in Research and Teaching*, as well as peer-reviewed taxon-specific journals and publications are considered acceptable standards of care to APHIS/AC. These resources may be helpful to IACUCs when evaluating wildlife or field studies.

15:10 If animals are wild-trapped, are there conditions the IACUC should place on that activity in terms of type of trap, frequency of checking of the trap, and euthanasia of injured or to-be-collected animals?

Reg. In accordance with the AWAR (§2.31,d,1) field investigations which involve invasive procedures, any harm to an animal, or any material alteration of the behavior of an animal under study, requires IACUC review. However a *field study*, as per the definition in the AWAR (§1.1) is exempt because it does *not* harm or materially alter the behavior of an animal under study. AWAR §2.38,f,1 indicates: "Handling of all animals shall be done as expeditiously and carefully as possible in a manner that does not cause trauma, overheating, excessive cooling, behavioral stress, physical harm, or unnecessary discomfort." The IACUC may make exceptions to these standards if there is scientific justification to do so (AWAR §2.31,c,3).

The USDA under AWA §2145 is required to consult and cooperate with other Federal Agencies, States, and political subdivisions in regard to welfare of animals used in research. As a result it is implied that IACUCs have a responsibility to ensure compliance with the appropriate entity in regards to permits, handling, and transportation.

The *Guide* (p. 32) points out that field investigations may require international, state, and/or local permits and that the permitting agencies may require a review of the scientific merit of the proposed study. The *Guide* states: "When species are removed from the wild, the protocol should include plans for either a return to their habitat or their final disposition, as appropriate." In addition, potential occupational health and safety issues should be reviewed by the appropriate institutional committee(s), "with assurances to the IACUC that the field study does not compromise the health and safety of either animals or persons in the field."

Opin. To help ensure appropriate animal care and use, the IACUC is required to review field studies that involve animal trapping. Thus, protocols must provide sufficient detail on methods to be used (e.g., trapping, tagging, collaring, blood collections, euthanasia) and frequency of observations, and a contingency plan for animals hurt in the collection process. Appendix E of the *Institutional Animal Care and Use Committee Guidebook*[12] describes local, national and international regulations requiring permits for conduct of field studies. *The Guidebook* and publications of several professional societies provide guidelines to the capture and care of wild animals.[12–18]

15:11 Should field sites be included in the semiannual IACUC inspections?

Reg. The AWAR (§2.31,c,2) state that "animal areas containing free-living wild animals in their natural habitat need not be included in such inspections." This statement agrees with the APHIS/AC *Animal Welfare Inspection Guide* to facility inspections.[19] Neither the *Guide* nor the PHS Policy addresses the issue of inspecting field sites; however, PHS Policy II requires compliance with the AWA as applicable. Although the PHS Policy does not specifically discuss wild animals or field studies, the definition of *animal* in PHS Policy III,A applies to all live vertebrate animals. Therefore, under strict interpretation field sites could be considered animal study areas by NIH/OLAW and require semiannual inspection by the IACUC. NIH/OLAW clarifies in their FAQ E.4[4] that "While semiannual IACUC inspections of field study sites are not required and in many circumstances are impractical, IACUCs should be apprised of the circumstances under which studies are conducted so that they can consider risks to personnel and impact on study subjects. This may be partially accomplished by written descriptions, photographs, or videos that document specified aspects of the study site. The IACUC should also ensure that appropriate permits are in place."(See 15:6.)

Opin. It is the authors' opinion that the question of whether field study sites are to be included in the IACUC semiannual inspection must be decided on a case-by-case basis. Certainly, field study sites that include housing enclosures should be inspected, as the animals involved are no longer truly free-living and are dependent on adequate monitoring and oversight. The IACUC should try to inspect such sites when they are actually in use. However, practical issues must also be considered. If the field site is geographically distant, sending an inspection team may not be feasible. It is recommended that IACUCs should be sensitive to public concerns regarding the trapping, restraint, and capture of animals under wildlife and field conditions. IACUC monitoring of these studies may help avoid instances of negative public perception.

Surv. Does your IACUC include field sites in its semiannual inspections? (More than one response may have been relevant for individual respondents.)

- Not applicable 143/295 (48%)
- No 87/295 (29%)
- Always if close by 32/295 (11%)
- Always, using "surrogate inspections" if not close by 26/295 (8.8%)
- Yes, but only if the protocol includes an invasive procedure 3/295 (1.0%)
- I don't know 2/295 (0.7%)
- Other 17/295 (5.8%)

15:12 Are animals that are used in field studies included on the APHIS/AC annual report?

Reg. The AWA (§2143 a,7,A) requires research facilities to annually report that the provisions of the Act along with professionally acceptable standards governing the care and use of animals in research have been followed. The AWAR (§2.36) defines this as the Annual Report and provides greater detail on the requirements.

A regulated species, in accordance with the definition of the word animal (AWAR §1.1), not only includes free-living and wild animals but also includes free-living rats (*Rattus*) and mice (*Mus*), because they were *not* bred for research. As a result, all animals used in field investigations that *do not* meet the definition of a field study in AWAR §1.1 must be listed on the APHIS/AC Annual Report. It is a regulated activity to perform research on APHIS/AC-regulated animals in the field when the work fails to meet the AWAR's definition of field study. In contrast, animal use that falls under the AWAR's definition of field study (§1.1) need not be included on the APHIS/AC annual report because this is not a regulated activity. (See 15:11.)

Opin. Including the animals on the APHIS/AC annual report was occasionally not done by some survey respondents. However, the institution and the IACUC should provide a mechanism for reporting (on the APHIS/AC annual report) the AWAR-regulated species used in field investigations. For example, the IACUC can send each investigator who has an approved protocol involving research in the field an annual request for summary information on the number and species used during the reporting period (October 1 to September 30). These numbers can then be compiled for the institutional annual report, which is due before December 1 of each year.

Special note: The authors wish to highlight that the use of the terminology 'field study' and 'field investigation' are not interchangeable and have entirely different meanings in the regulatory context of the AWAR's definitions in section §1.1. This wording is often confused and if incorrectly used would make a distinct difference in an institution's regulatory compliance. Similarly, if the term 'wildlife research' is used, this needs to be further described as either a 'field study' under the AWAR's definition, or a regulated activity conducted in the field.

15:13 With reference to housing requirements, should the IACUC make an attempt to distinguish proactively between biomedical research and agricultural research when reviewing proposals that use large farm animals?

Reg. The definition of an *animal* in the AWAR (§1.1) excludes "farm animals, such as, but not limited to, livestock or poultry used or intended for use as food or fiber, or livestock or poultry used or intended for use for improving animal nutrition, breeding, management, or production efficiency, or for improving the quality of food or fiber." (See 12:1.)

The *Guide* (pp. 32–33) states that "The use of agricultural animals in research is subject to the same ethical considerations as for other animals in research … The protocol, rather than the category of research, should determine the setting (farm or laboratory). Housing systems for farm animals used in biomedical research might or might not differ from those in agricultural research; animals used in either biomedical or agricultural research can be housed in cages, stalls, paddocks, or pastures."

PHS Policy II does not make reference to a particular type of research. It states that it is applicable to all PHS-conducted or supported activities involving animals. The HREA (§495a), on which the PHS Policy is based, does refer to "animals to be used in biomedical and behavioral research," a reference that, by inference, might be construed as not applicable to agricultural research. NIH/OLAW FAQ A.1 indicates "There are many valid reasons for institutions to perform program oversight institution-wide using uniform and consistent standards for animal care and use. Institutions must implement the PHS Policy for all PHS supported activities involving animals, and must ensure that any standards that might not be consistent with PHS Policy do not affect or pose risks to PHS supported activities." As stated in the *Institutional Animal Care and Use Committee Guidebook*[12] (p. 121), "OLAW advises institutions that uniform and consistent standards are an essential ingredient in a quality animal care and use program. Public perception of a potential double standard should also be considered." This admonition is elaborated on in NIH/OLAW FAQ A.1.

Opin. The AWAR language has become the working definition of agricultural research. The unstated converse is that farm animals used in studies where the goal is the advancement of biomedical science are regulated by the AWAR. Therefore, the practical reason why an IACUC might wish to distinguish between biomedical and agricultural research is that the latter need not be included in AWAR requirements such as annual reporting and search for alternatives and need not conform to recommendations in the *Guide* with particular reference to the space and physical plant recommendations that are focused on biomedical research animals. (See 15:14.) It is the authors' opinion that applying uniform policies to all animals involved in the institution's animal care and use program regardless of the type of research or teaching activity involved promotes consistent oversight; however, decisions on housing farm animals used in biomedical research under conditions typical of an agricultural setting should be made by the IACUC on a case-by-case basis.

APHIS/AC, through consultation with the authors of this chapter, has indicated that it is implied through the AWA (§2143,a,7,A) regarding "professionally acceptable standards governing the care, treatment, and use of animals…", that the *Guide for the Care and Use of Laboratory Animals*, the *Guide for the Care and Use of Agricultural Animals in Research and Teaching*, as well as peer-reviewed taxon-specific journals and publications are considered acceptable standards of care to the USDA.

15:14 What criteria are used to distinguish biomedical from agricultural research? How are the standards different for biomedical research housing sites versus agricultural research housing sites?

Reg. (See 15:13.)

Opin. It is clear that farm animals such as sheep, cows, goats, swine, and horses used in biomedical research are covered by both the AWAR and the *Guide*. It is also clear that the same animals used for research on livestock production are not ordinarily regulated by the AWAR. PHS-funded agricultural projects (which would be covered by the *Guide*) are very unlikely, given that the focus of PHS funding is biomedical research.

What is not clear are the circumstances under which farm animals used in biomedical research must be housed according to the *Guide*'s recommendations as compared to when they may be housed in more farm-like settings (such as paddocks, pastures, and barns). The *Guide* (p. 33) states that the housing system chosen should be protocol driven. The authors interpret that to mean that farm animals used in studies in which the minimization of variables is vital to the study's outcome (e.g., transplant research) should be housed in the more environmentally controlled setting of a traditional laboratory animal facility. Conversely, farm animals intended for use in studies for which each animal's environment need not be finely controlled (e.g., antibody production) may be housed in farm-type settings. As stated in the *Guide* (p. 33), decisions on the type of enclosure and density of housing for farm species should be made with due consideration to both the researcher's goals and the animal's well-being. For animals housed at an accredited facility in the agricultural setting, AAALAC's perspective on this topic is that "…the housing and care for farm animals should meet the standards that prevail on a high-quality, well-managed farm. The collective professional judgment of the responsible oversight body (i.e., IACUC, Ethics Committee), principal investigator and veterinarian should determine which standard(s) applies best with regard to the care and welfare of agricultural animals, based on a performance approach in the context of the requirements of the study and the species used. The rationale for making this determination should be documented."[10]

Surv. 1 What criteria does your IACUC use to distinguish biomedical from agricultural studies? (More than one response may have been relevant for individual respondents.)

Note: A few IACUCs have an agricultural animal subcommittee, or a separate IACUC for wildlife and agricultural research. Other survey responses indicated that if the end goal benefited another species, the research was considered biomedical.

• Not applicable	211/295 (72%)
• If the end goal is to benefit human health, we consider it biomedical research	50/295 (17%)
• If the end goal is to benefit production practices, we consider it agricultural research	35/295 (12%)
• We look at the funding, e.g., NIH funding means biomedical research while the Pork Producers Council means agricultural research	17/295 (5.8%)
• We have no specific criteria	36/295 (12%)
• Other: (Please specify)	3/295 (1.0%)

Surv. 2 Which standards does your IACUC use for agricultural research housing sites? (More than one response may have been relevant for individual respondents.)

• Not applicable	241/294 (82%)
• Those of a well-managed farm	15/294 (5.1%)
• Those in the *Guide for the Care and Use of Agricultural Animals in Research and Teaching*	45/294 (15%)
• Housing standards are protocol driven	9/294 (3.0%)
• I don't know	1/294 (0.3%)
• Other	3/294 (1.0%)

15:15 Is it acceptable to house different animal species in the same holding room?

Reg. The AWAR (§2.33,b,1 – §2.33,b,2) state that adequate veterinary care standards require appropriate facilities for housing animals and appropriate methods for prevention of disease. In regard to keeping in the same animal holding/housing room, the AWAR (§2.33,b,5) indicate adequate veterinary care allows for care in accordance with currently accepted veterinary medical and nursing procedures.

The AWAR (§1.1) define a *primary enclosure* as "any structure or device used to restrict an animal or animals to a limited amount of space, such as a room, pen, run, cage, compartment, pool, or hutch." The AWAR do not address housing of different species in the same holding room but do address housing of different species in the same primary enclosure.

The AWAR (§3.33,b) prohibit housing hamsters in the same primary enclosure with any other animal species. The same is true for guinea pigs. Housing rabbits in a primary structure with other species is prohibited, "unless required for scientific reasons" (AWAR §3.58,a). Furthermore, the AWAR (§3.33,c; §3.58,b) require the separation of hamsters, guinea pigs, and rabbits from same-species animals that are under quarantine or treatment for a communicable disease so as to minimize dissemination of such disease. This separation requirement also applies to other species that may be susceptible to contracting the disease from the hamster or guinea pig, or rabbit. The AWAR (§3.7,d) prohibit cohabitation of dogs or cats with other species in the same primary enclosure unless they are compatible. Nonhuman primates may not be housed with other species of primates or animals unless they are compatible, do not prevent access to food, water, or shelter by individual animals, and are not known to be hazardous to the health and well-being of each other (§3.81,a,3). For all other regulated species the AWAR (§3.133) state that, "Animals housed in the same primary enclosure must be compatible. Animals shall not be housed near animals that interfere with their health or cause them discomfort."

PHS Policy is silent on the subject of mixing different species in the same primary enclosure or holding room, although PHS Policy II requires compliance with the AWAR, where applicable.

The *Guide* (p. 52) states that "Housing should provide for the animals' health and well-being while being consistent with the intended objectives of animal use." The *Guide* (p. 111) also states that "Physical separation of animals by species is recommended to prevent interspecies disease transmission and to eliminate the potential for anxiety and physiologic and behavioral changes due to interspecies conflict.... it may also be acceptable to house different species in the same room, for example, two species that have a similar pathogen status and are behaviorally compatible."

Opin. Professional judgment usually dictates that mixing of certain species in the same room or same primary enclosure is inappropriate to their life histories (e.g., cats and dogs), behavioral characteristics, or predator-prey relationships. The potential for disease and parasite transmission is an additional consideration regarding housing different species in the same room or enclosure. As examples of recommended separate housing by species, the *Guide* (pp. 111–112) notes that rats and mice should be housed separately where risk of interspecies transmission with disease-causing bacteria or viruses may exist, and nonhuman primates should be separated by geographical origin due to differing viral disease susceptibility. Our opinion (supported by AAALAC)[11] is that the practice of cohousing differing species in the same room or primary enclosure should be specifically reviewed and approved (or have approval withheld) by the IACUC on a case-by-case basis for all of the factors mentioned above.

Surv. Which species do you house together (in separate cages) in the same holding room? (More than one response may have been relevant for individual respondents.)

- Not applicable, we work with only one species 19/294 (6.4%)
- Not applicable, we never mix species in the same holding room 152/294 (52%)
- Whatever species the veterinary or animal care specialists allow us to mix 29/294 (9.9%)
- Rodent species that are kept in microisolator-type cages 63/294 (21%)
- Amphibians and reptiles (same or different species) 36/294 (12%)
- Farm species 13/294 (4.4%)
- Farm species and dogs 3/294 (1.0%)
- Multiple nonhuman primate species 5/294 (1.7%)
- Other: (Please specify) 20/294 (6.8%)

Note: Several respondents commented that mixing species is a veterinary, not an IACUC decision whereas others felt it was an IACUC decision that also required concurrent veterinary recommendation. Institutions not responding to housing certain species together may have not done so because they do not house that species at all or they have not encountered a need to house species together. Housing different species of fish together was the most common response in the "other" category, followed by birds. Other reasons given for co-housing different species included for post-operative housing in an Intensive Care Unit during anesthetic recovery, and under ABSL-2 conditions with special circumstances.

References

1. Title 9 - Animals and Animal Products, Subchapter A – Animal Welfare, Code of Federal Regulations, http://www.gpo.gov/fdsys/pkg/CFR-2009-title9-vol1/xml/CFR-2009-title9-vol1-chapI-subchapA.xml. Accessed September 28, 2012.
2. Committee for the Update of the Guide for the Care and Use of Laboratory Animals. 2011. *Guide for the Care and Use of Laboratory Animals*, 8th edition. Washington, D.C.: National Academies Press.
3. National Institutes of Health, Office of Laboratory Animal Welfare. *Public Health Service Policy on Humane Care and Use of Laboratory Animals.* http://grants.nih.gov/grants/olaw/references/phspol.htm. Accessed September 28, 2012.
4. National Institutes of Health, Office of Laboratory Animal Welfare. 2013. Frequently Asked Questions. PHS Policy on Humane Care and Use of Laboratory Animals. http://grants.nih.gov/grants/olaw/faqs.htm.
5. National Institutes of Health, Office for Protection from Research Risks. 1991. The Public Health Service responds to commonly asked questions. *ILAR News* 33(4):69. http://grants.nih.gov/grants/olaw/references/ilar91.htm#2.

6. Centers for Disease Control and Prevention. 2012. 12th CDC International Symposium on Biosafety. http://www.cdc.gov/biosafety/.
7. International Air Transport Association. 2013. Infectious substances. https://www.iata.org/whatwedo/cargo/dgr/Pages/infectious_substances.aspx.
8. Silverman, J., M.A. Suckow and S. Murthy. 2007. *The IACUC Handbook*, 2nd edition. Boca Raton: CRC Press.
9. Potkay, S., N.L. Garnett, J.G. Miller et al. 1995. Frequently asked questions about the Public Health Service Policy on Humane Care and Use of Laboratory Animals. *Lab Anim. (NY)* 24(9): 24. Also see Office of Laboratory Animal Welfare. 1995. OPRR Reports, #95-02. http://grants.nih.gov/grants/olaw/references/dc95-3.htm and. Accessed September 28, 2012.
10. U.S. Department of Agriculture, Animal and Plant Health Inspection Service. 2011. Animal Welfare Policy 10. Specific Activities Requiring a License or Registration. http://www.aphis.usda.gov/animal_welfare/policy.php?policy=10.
11. Association for Assessment and Accreditation of Laboratory Animal Care International. 2012. Frequently Asked Questions D.5, Multiple animal species in a housing room. http://www.aaalac.org/accreditation/faq_landing.cfm#C. Also see Association for Assessment and Accreditation of Laboratory Animal Care International. Selecting the appropriate standard(s) for the care and use of agricultural animals. http://www.aaalac.org/accreditation/positionstatements.cfm#ag. Accessed October 25, 2012.
12. Applied Research Ethics National Association/Office of Laboratory Animal Welfare. 2002. *Institutional Animal Care and Use Committee Guidebook, 2nd edition*. Bethesda: National Institutes of Health. http://grants.nih.gov/grants/olaw/GuideBook.pdf.
13. Use of Fishes in Research Committee. 2004. *Guidelines for the Use of Fishes in Research*. 2004. Joint publication of the American Fisheries Society, the American Society of Ichthyologists and Herpetologists, and the American Institute of Fisheries Research Biologists. Bethesda: American Fisheries Society. http://fisheries.org/docs/policy_useoffishes.pdf.
14. Herpetological Animal Care and Use Committee. 2004. *Guidelines for the Use of Live Amphibians and Reptiles in Field and Laboratory Research, 2nd edition*. American Society of Ichthyologists and Herpetologists. www.asih.org/files/hacc-final.pdf.
15. Sikes, R.S., W.L. Gannon, and the Animal Care and Use Committee of the American Society of Mammalogists. 2011. Guidelines of the American Society of Mammalogists for the use of wild mammals in research. *J. Mammal.* 92:235-253. http://www.mammalogy.org/articles/guidelines-american-society-mammalogists-use-wild-mammals-research-0.
16. Braun, C.E. (ed). 2005. *Techniques for Wildlife Investigations and Management, 6th ed*. Bethesda: The Wildlife Society.
17. Captive Care Committee. 2007. *IPS International Guidelines for the Acquisition, Care, and Breeding of Nonhuman Primates, 2nd ed*. International Primatological Society. http://www.internationalprimatologicalsociety.org/docs/IPS_International_Guidelines_for_the_Acquisition_Care_and_Breeding_of_Nonhuman_Primates_Second_Edition_2007.pdf.
18. The Ornithological Council. 2010. Guidelines to the Use of Wild Birds in Research, 3rd edition. http://www.nmnh.si.edu/BIRDNET/documents/guidlines/Guidelines_August2010.pdf. Accessed April 22, 2013.
19. U.S. Department of Agriculture, Animal and Plant Health Inspection Service. 2013. Animal Welfare Inspection Guide. http://www.aphis.usda.gov/animal_welfare/downloads/Inspection%20Guide%20-%20November%202013.pdf.

16

Pain and Distress

Alicia Z. Karas and Jerald Silverman

Introduction

Pain or other forms of distress are important considerations in laboratory animals. While there is a great diversity of views, it is clear that for many, minimization of pain and distress improves acceptability of use of animals in research and testing.[1–3] The concept of minimization of harm to animals was described in Russell and Burch's classic text, *The Principles of Humane Experimental Technique*,[4] and has been subsequently popularized as the "3 Rs" of animal experimentation. These are: Reduction of the number of animals used, Replacement of animals with nonsentient alternatives (or with human subjects), and Refinement of experimental design to minimize pain and distress. Refinements may include efforts to reduce stress or pain such as training or acclimation of animals and investigators, altered experimental methods, provision of social contact, supportive care, environmental enhancement, increased monitoring for pain and distress in order to intervene or end an experiment, etc. Over time, as we have begun to better understand the assessment and impact of pain and distress, refinement has become an increasingly important ethical aspect of animal research. In addition, the alleviation of unnecessary pain and distress may improve the quality of scientific data obtained. The manner in which experimental animals are handled, housed, and fed has an impact on physiological processes such as development, toxicology, oncology, infection, cognition, or cardiovascular status; these and other examples abound in the literature.[5–22] Pain and other stressors have been demonstrated to result in production of cytokines and activation of an endocrine state of catabolism, translating into an overall spectrum of effects such as decreased immune responses and healing rates, increases in tumor metastasis retention after surgery, and the development of chronic pain states.[23–29] Severe, uncontrolled stress is well known as both a trigger for disease and an impediment to recovery. When the goal of an experimental intervention is to improve disease outcome, the effect of unnecessary stress on subjects is concerning if it might obscure the impact of a studied treatment.[30] Pain and distress may represent unrecognized experimental variables that can have significant effects on research results, and the reduction of painful or distressful experiences can potentially benefit both the welfare of the animals used and the scientific output from the project.

One of the most common arguments against incorporation of refinements is that it will risk "changing the model." Fears of changing long established models often fail to take into consideration that some models have experienced (perhaps to a greatly under-appreciated degree) dramatic changes over a period of time as laboratory animal husbandry, medicine, and even strain/stock variations have been introduced. One example is that many rodent experimental surgery protocols have been modified over the past two decades, from use of anesthesia based on fixed dose injectable regimens (e.g., pentobarbital or ketamine-xylazine combinations) to isoflurane inhalation. Equally concerning is that the outcome of an experiment may hinge to a substantial extent on the way that animals are handled, as Chesler and colleagues[31] demonstrated: the variable having the greatest influence on a very "objective" measurement (hotplate tail flick in the mouse) was the identity of the individual experimenter, trumping the effects of strain, sex, season, etc. Inevitably, differences in models occur within and between laboratories; attempts to standardize conditions may falter simply by the nature of who is doing the experiment. There are currently even arguments proposed *against* rigorous standardization of environmental conditions for a given model in terms of improving and understanding external validity.[32] A major impediment is that it may be impossible to know enough detail about how animals are used: many of what could be important factors are not routinely specified in published manuscripts.[33–36] It is anticipated that changes in editorial guidelines for scientific journals will bring about improvements in characterization and documentation of all exposure factors in order to help interpret differences between results, or putatively, to boost confidence in the robustness of the findings of animal models.[34,35]

Many IACUCs find themselves in situations where an investigator indicates that "refinement 'X' cannot be used because it will interfere with (or change) the model." There may be a tendency for the IACUC to award the benefit of the doubt to the investigator, based on a theoretical justification rather than actual knowledge or experience. In this case, the question is whether the IACUC should simply accept the *theory* that it will impact the model—if the deleterious impact on the model cannot be backed up by specific evidence. After all, one would not accept the fact that a new treatment will work because *in theory* it should. The IACUC must thus weigh the available specific evidence and try to strike a balance between a concern for animal welfare and the requirements of a particular research project. Over time, the acceptability of withholding refinements due to theoretical, undocumented potential effects on the model will continue to evolve. Very few of us can recall a time where successful justification was made for withholding anesthesia altogether for surgery because it could interfere with a study goal. Instead, the current practice is that the best type of anesthesia alternative is used and the compromise taken into account as an unavoidable limitation. Quite a few other experimental animal techniques have gone out of favor as well, as evidence for their distress causing nature has gathered, yet science continues to progress.

Similarly, in many instances we also lack the scientific evidence to inform us about the "true" experiences of experimental animals: about the extent of pain or distress that will be caused by a study protocol, or the efficacy of a technique to relieve or prevent pain or distress, or what constitutes a meaningful positive element. A rapidly expanding animal welfare science literature can contribute much to a thoughtful analysis of whether adoption of refinements will have the desired benefit if individuals with relevant expertise are consulted. The value of laboratory animal welfare studies cannot be overestimated when decisions are required about the impact of procedures. But, as with claims of theoretical interference with model, the IACUC often must decide, with no evidence to consult, whether the addition of a new proposed technique to improve the welfare of animal subjects might have a positive benefit, and be therefore worth the time and cost. Simple acceptance of the *theoretical* benefit without specific evidence carries the risk of not actually helping animals and being more resource intensive, but unless it actually causes harm to animals or compromises data, one might argue that in such a case, the benefit of the doubt could be given to the animals. Fortunately, when measures are adopted that are intended to reduce pain or distress, it is possible to evaluate whether a change for the better was actually achieved, and if not deemed worthwhile, to reconsider.

Another argument against use of refinements, or enhanced monitoring, is that the process may be time consuming and involve additional cost. Many preclinical rodent models of disease do not report supportive care in the form of thermal, fluid or nutritional measures (and therefore we should assume that these are not part of the model), while supportive measures would unconditionally be used in the target population (e.g., people with medical conditions). Supportive measures, by their very nature, also tend to help to reduce the amount of distress suffered by the patient and would arguably as well, do so for the animal subjects. Supportive or palliative measures might require some frequency of monitoring and care above and beyond the daily cage check. While the incorporation of critical care that is equivalent to human medical care may be impossible from a size or species point of view, some basic refinements might be achievable. It is important to acknowledge that an investigator would not be expected to shy away from additional time, effort, or cost to attend to details that would make the experiment itself successful, and that reluctance to improve animal welfare because of the work involved may inadvertently harm the external validity of the model in addition to posing an ethical dilemma. Sometimes, the situation arises where some degree of stress or distress is fundamental to the model's success: for example, when a degenerative condition does not develop rapidly enough when animals are given access to exercise wheels. If this is the case, then the balance of welfare against experimental goal will shift, but any data that support a finding that refinement *cannot* be used in the model is still of critical importance to the field and should be reported.

It follows that a consideration of methods for minimizing or eliminating pain and distress should be of central concern to IACUCs for legal, ethical, and scientific reasons. But looking forward to the future, one additional concept must be mentioned. In the period since the last version of this text, the definition of good animal welfare has continued to evolve. "Good welfare" originally meant that animals were in a state of good health and productivity and that measures were taken to reduce or remove negative influences on their physical and (more recently) mental state. The current paradigm for good welfare has begun to consider that animals should be given access to elements of "positive" valence—such as companionship, comfort and "things to do."[37] Buchanan-Smith and colleagues[38] have proposed an enhanced definition of refinement; one that incorporates measures to "maximize(s) well-being" using a "proactive approach to promote positive aspects of welfare." In a system that is currently geared towards merely meeting the legal or standard requirements, it seems unlikely that IACUCs will push investigators to balance the unavoidable negative aspects of laboratory animal welfare by considering ways to add positive elements during the lifespan of an animal. However, as society will very likely continue to evolve in its attitude toward what is acceptable use of experimental animals, the scientist and the IACUC member should be aware of the changing paradigms and relationships of "good welfare" and "necessary harm."

16:1 What is the difference between stress and distress?

Reg. Distress is defined in the *Guide* (p. 121) as an aversive state in which an animal fails to cope or adjust to various stressors with which it is presented. The AWA and AWAR do not provide a definition of distress.

Opin. It is very important to understand that there is a difference between stress and distress. The most recent report from the *Committee on Recognition and Alleviation of Distress in Laboratory Animals*[39] stated that the adaptive *stress response* and a *state of distress* can be differentiated on the basis of "good scientific evidence" but that, "distress remains a complex and still poorly understood phenomenon." Many attempts have been made to define and differentiate stress and distress, and these vary in their approach.[39,40] *The Committee on Recognition and Alleviation of Distress in Laboratory Animals*[39] wrote that based on current scholarly thinking "stress denotes a real or perceived perturbation to an organism's physiological homeostasis or psychological well-being." The role of the stress response is to make the organism able to adapt to a given situation, theoretically to regain a state of homeostasis, or balance, which will allow it to function well. The ability to adapt physiologically or psychologically to a disturbance is thus of benefit to the individual animal, however, the end result is not always to return the animal to its exact state prior to the stressful event.[41] In addition, it is commonly emphasized that the events that are *stressors*, and the *condition* of mounting a stress response are not always unpleasant. As an example, strenuous play may trigger a stress response but could be highly rewarding and enjoyable to the animal. Some recent work has questioned the current definitions of stress and stressor and made the case that the condition of *stress* is present only when the individual cannot either predict a stimulus or control its environment (i.e., adaptation is not possible); otherwise the body's physiologic response to perturbations resulting in the constellation of events that is known as "the stress response" is simply that—a response.[40]

When the effort required to adapt to some stressors, or the aversive nature of the stressor, increases and reaches a point where the animal fails to be able to adapt adequately, this is considered by some authors to represent distress. In other words, it is in the "failing to cope with" either severe, sustained or multiple combined stressors that the animal's state becomes that of distress. Stressors often cause a low level unpleasant affective or emotional state (e.g., thirst) and the degree of distress caused by water deprivation will increase as the animal's affective and physiologic states become overwhelmed. Unless manifested by overt external physical signs or behaviors, states of distress may be difficult to recognize in laboratory animals—i.e., it might be difficult to determine when a mouse becomes distressed by water deprivation, but renal failure and hypovolemic shock are states that can be measured. McMillan[42] gives a simple explanation for why certain things (such as hypoxia, hunger, nausea, urinary bladder distention, or being too cold) cause an increasingly unpleasant state; which is that they represent a threat to the animal's survival. He also proposes that when distress is severe, the animal is "increasingly compelled to focus on the threat and not on matters less relevant to survival." This may be one key to detecting when distress is overwhelming—to note when the animal cannot deal with anything outside its own body. It is also important to recognize that a poor physiologic state does not always mean that distress is present: an animal may have cancer or other disease for some time and not be "aware" or affected by its body's efforts or inability to cope and while the animal may thus be said to have a relatively poor welfare state from the standpoint of health, it is not (yet) distressed. Judgments and conclusions made during an ethical review process would ideally be based on objective criteria, such as percentage of weight loss, or radiographic signs of a tumor. But following the example above, merely seeing an animal reach a measureable threshold (objective criterion) does not mean that distress is present, and distress may be experienced by an animal in situations where we cannot (or do not) assess their mental state. While under general anesthesia an animal cannot consciously perceive the presence of a negative state, and thus distress is prevented by loss of consciousness.

Pain is only one of a number of aversive states that can cause distress, although the regulations governing laboratory animal use tend to favor scrutiny of protocols for pain and its alleviation due to their wording. Pain is complicated as a cause of distress; the degree of noxiousness of a stimulus might differ between individual animals, between species, and may depend on the situation. For example, being housed on wire cage bottoms might be more distressful for rodents that have injured footpads than for normal rodents, and standing on a hot surface is likely to be more painful for a mouse than for a horse. It might be said that an aversive state has dimensions of time and magnitude and becomes "distress" when the product of the two exceeds a certain threshold. Thus, weeks of chronic exposure to a low level aversive situation, such as barren housing, may not cause as much distress as weeks of increasing pain from tumor growth. In any case, the IACUC will have to take responsibility for estimating the degree of averseness of a condition, which in some cases has been studied scientifically (e.g., by preference and motivation studies) and which in other cases may require a subjective interpretation (e.g., this is what a human would feel like in an equivalent state), in order to guide attempts at refinement.[43]

16:2 What is the difference between pain and nociception?

Reg. *Pain* is not directly defined in the AWAR, but a *painful procedure* (§1:1, Painful Procedure) is defined as one "that would reasonably be expected to cause more than slight or momentary pain or distress in a human to which that procedure is applied." The example provided is pain in excess of that caused by injections or other minor procedures. The *Guide* (p. 120) states that pain is a complex experience that typically results from stimuli that damage or have the potential to damage tissue. *Nociception* is not defined in either the AWAR or PHS Policy. The definition of a *painful procedure* in U.S. Government Principle IV (which is incorporated into the PHS Policy) and the *Guide* (p. 120) is similar to the AWAR definition.

Opin. Nociception represents only the detection of certain stimuli by sensory nerves and the transmission of signals to the central nervous system. In other words, nociception represents basically the way that noxious (potentially tissue damaging) stimuli are "wired" through the spinal cord and brain. Noxious input of a sufficient magnitude will cause depolarization of sensory nerve endings; examples of such stimuli types are heat or cold, pressure, and chemical irritants. The signal travels through the sensory nerve tracts to the spinal cord and potentially to the brain. Noxious stimuli may concomitantly trigger both involuntary (e.g., reflex limb withdrawal, autonomic nervous responses) and voluntary (purposeful) aversive movements. Pain is the subjective experience of nociceptive signal input. The nociceptive signals arrive at different levels of the brain (depending on the phylogenetic level of the species) and may be "processed" or "interpreted" there as being painful. When the overall interpretation of the stimulus is that it is painful, an affective–motivational response occurs, involving prior experience and contextual information, and thus the *pain experience* depends upon many factors.

Although reflex motor and autonomic responses may occur in the unconscious or anesthetized animal in response to noxious stimuli, they are not typically defined as pain because the animal is "unaware" of the stimulus (the neurologic input–output situation is that of nociception). The administration of anesthetics to the point where both voluntary and involuntary movements are suppressed, and autonomic responses are blunted in response to a noxious input, is taken to be sufficient to eliminate the experience of pain for the duration of the anesthesia. Upon recovery to a conscious state it is assumed that at this point the animal is capable of "experiencing" or perceiving pain.

Pain in humans is described as having both a sensory and an emotional component. For example, the International Association for the Study of Pain (IASP) definition is as follows: "Pain is an unpleasant sensory and emotional experience associated with actual or potential tissue damage or described in terms of such damage."[44] Defining animal pain is difficult because of uncertainties relating to the emotional (affective) component of pain in non-humans. This has been avoided in some definitions by interpreting pain in relation to its effects on animal behavior. For example: "Pain in animals is an aversive sensory experience that elicits protective motor actions, results in learned avoidance, and may modify species-specific traits of behavior, including social behavior."[45]

Nociception and pain are said to serve the organism; protecting it from damage. One neuroscientist's view of pain is that it evolved as part of the function of homeostasis (ensuring that the body regulates its own well-being).[46] Water and salt balance, for example, are regulated and involve internal physiologic compensations as well as behavioral drives. Similarly, pain is argued to have evolved to reflect the condition of the body arising from a tissue injury. Nevertheless, as a sensation, pain can vary dramatically. "The behavioral drive that we call pain usually matches the intensity of the sensory input but it can vary under different conditions, and can become intolerable or, alternatively, disappear, just as hunger or thirst."[46] Regardless of whether the reader agrees that animals experience emotions, for familiar animal species it can be argued that pain does exist: it has the behavior-motivational component, it can vary depending on the circumstances of what else is happening at the time, and if the experience continues with sufficient magnitude, it can lead to adverse health consequences.

A useful elucidation of nociception, pain, pain assessment, pain recognition and treatment is given in the Institute of Laboratory Animal Resources report *Recognition and Alleviation of Pain in Laboratory Animals*.[47] Definitions, importance, recognition, and treatment goals for pain are also reviewed in the American College of Veterinary Anesthesiologists position paper on pain in animals.[48]

16:3 What should be the minimal expectations of the IACUC related to the relief of pain and distress?

Reg. The IACUC, under the AWA (§2143,a,3,A; §2143,a,3,C) and AWAR (§2.31,d,1,iv,A–C) must assure that procedures causing more than slight pain or distress will be performed with appropriate sedatives, analgesics, or anesthetics unless certain specific criteria are met (e.g., IACUC-approved scientific justification to withhold analgesia for only the necessary period of time). They must involve the AV or the AV's designee in their planning and not use paralytics without anesthesia. Animals experiencing severe or chronic pain or distress that is not alleviated must be euthanized during or at the end of the study. The AWAR (§2.31,d,1,ii) require consideration of alternatives to procedures that may cause more than momentary pain or distress to animals and require a written narrative description of the methods and sources used to determine that alternatives were not available. APHIS/AC Policy 11 is quite clear: "The [IACUC] is responsible for ensuring that investigators have avoided or minimized discomfort, distress and pain to the animals." APHIS/

AC Policy 12 states that a database search remains the most effective and efficient method for demonstrating compliance with this requirement. (See 16:4.) To aid researchers with their database searches, the AWA (§2143,e) specified the establishment of the National Agricultural Library (NAL) to serve as a resource on the 3 R's to help minimize pain and distress.

The PHS Policy (IV,C,1,a–IV,C,1,c) is worded similarly to the AWAR relative to avoiding or minimizing pain and distress, the need to use appropriate means to alleviate pain and distress, and the need to euthanize animals during or at the end of a procedure if severe pain or distress cannot be alleviated. PHS Policy (IV,D,1,d) notes that applications and proposals to the PHS must contain a description of the procedures that are designed to assure that discomfort and injury to animals will be limited to that which is unpreventable in the conduct of scientifically valuable research, and the proper drugs will be used where indicated to minimize animal pain and discomfort. U.S. Government Principles IV-VI largely reiterate the AWAR and PHS Policy.

The *Guide* (p. 120), U.S. Government Principle IV, and the AWAR (§1.1, Painful Procedure) assume that procedures that cause pain in humans will probably cause pain in nonhuman animals. The *Guide* (p. 121) states that the selection of analgesia and anesthesia for animals should reflect appropriate veterinary judgment. It also notes (p. 120) that the recognition of pain in different species is a key to its prevention or alleviation. The *Guide* (pp. 122–123) warns that some drugs, such as sedatives, anxiolytics, and neuromuscular-blocking agents, are not analgesics or anesthetics.

The authors note that APHIS/AC (via its veterinary medical officers and regional offices) and NIH/OLAW are always available for consultation on regulatory issues affecting the IACUC review of protocols involving pain or other forms of distress.

Opin. Pain and distress, when significant to the animal, are undesirable and also form the basis for much public disquiet about the use of animals in research. The role of the IACUC is to thoroughly review all protocols for elements that might lead to pain or distress, and to assess the appropriateness of methods designed to minimize these. In doing so, there should be scrutiny of the "search for alternatives." Alternatives in this context range from replacement of animals altogether, to the refinement of methods of animal treatment so that less total pain or distress are experienced (see "3Rs" in the Introduction to this chapter). The investigator may not be cognizant of the many ways that an animal's experience can be affected. Perhaps it is in this arena that the IACUC will have its most important oversight, reviewing not only the proposed use of pain or distress alleviation and monitoring strategies, but also the training and competency of those involved in the procedures. For example, an inexperienced investigator may handle an animal inappropriately and cause significant pain or distress that could have been prevented by proper training. It also is likely that a person skilled in surgical procedures will carry out a project using fewer animals (because of fewer technical failures), cause less postoperative pain (because of less tissue trauma), and effect a more rapid recovery (because of shorter anesthesia time) than an inexperienced or less competent investigator. The total amount of pain or distress from all aspects of animal use, including not only experimental procedures themselves but also such things as marking/identification and general husbandry procedures, must also be considered.

For an IACUC to decide whether the PI has sufficiently considered pain and distress alternatives, all potential sources of painful and nonpainful distress should have been identified, and alleviation or prevention techniques sought. Although the regulations use broad terms such as *appropriate* (drugs) or *unavoidable* (discomfort) (AWAR §2.31,d,iv,A; §2.31,e,4; PHS Policy IV,C,1,b), some methods to reduce pain or distress might be possible but extremely unfeasible (e.g., requiring years of surgical training prior to permitting a new surgeon to operate). In many instances it may be difficult or impossible to find a resource that offers advice on how to reduce pain or distress. Also, the minimization of distress caused by stressors that might contribute to pain, such as by providing additional bedding, more easily reachable food or water, fluid therapy, or other improvements in comfort and homeostasis (*Guide* p. 122) are frequently overlooked forms of refining animal use. This may occur because the wording of the AWAR can lead scientists to be primarily focused on anesthetics, analgesics or tranquilizers as refinements. The IACUC can play an important role in overcoming this limitation by specifically asking for this type of relief of pain or distress to be included in the protocols they review, by requiring it in training of PIs, and reviewing the need for supportive care with veterinary and animal facility staff through the AV.

Finally, the IACUC should ensure that animals are observed appropriately for their species-specific condition at regular intervals, using generally agreed upon criteria (e.g., body condition scoring, active behaviors such as food consumption, lameness, and attitude) in both the normal state and during experimental conditions in which pain or distress is expected. Monitoring of animal well-being can ensure that measures to reduce pain or distress are utilized when needed. The frequency of observations should usually be greater when a significant abnormal state is anticipated. All of these factors should be integrated into the IACUC's expectations when assessing the efforts an investigator intends to make in order to reduce pain and distress.

16:4 What alternatives must be considered for protocols that will involve pain or distress? How can this be done?

Reg. The AWAR (§2.31,d,1,ii) require the PI to consider alternatives to procedures that may cause more than momentary or slight pain or distress to the animal. The PI must provide a written description of the methods and sources used to determine that alternatives are not available. This requirement is clarified in APHIS/AC Policy 12, which provides examples of sources that can be used (e.g., Medline, the Animal Welfare Information Center (AWIC) of the National Agricultural Library). AWIC offers training in performing acceptable searches, and will perform searches upon request (http://awic.nal.usda.gov/).

Policy 12 states that the minimal written narrative should include the database(s) or other sources used, the date of the search, the years covered by the search, and the keywords or search strategy used. More importantly, Policy 12 notes that reduction, replacement, and refinement (see 16:3), not solely animal replacement, are generally considered part of the entire concept of "alternative methods" and should be addressed as part of the search for alternatives.

The *Guide* (p. 5) states that consideration should be given to the use of inanimate systems or nonvertebrate animals. U.S. Government Principle III, which is part of the PHS Policy, states that mathematical models, computer simulation, and in vitro biological systems should be considered. (See 12:12–12:18.)

Opin. From 16:3 it can be seen that all protocols that have the potential to cause pain and distress should be reviewed with reference to reduction, replacement, and refinement. This requires significant effort of both the IACUC and the PI. The PI should at least explain, in some detail, how an animal is to be handled and justify their use of animals. They should indicate the literature sources they have reviewed to show that there is no practical alternative to animal use. From each step outlined in the protocol, handling or experimental details can be identified and used as keys to search for alternatives. SOPs for a variety of common actions can be developed by investigators at an institution, in concert with the veterinary experts, so that uniform expectations are available to the IACUC. Deviations from SOPs must be reviewed with respect to whether they will produce an acceptable level of well-being. As the number of databases on alternatives grows, reference to searches on these can be included on an IACUC application. (See 16:23.)

From 16:3 it can be seen that all protocols that have the potential to cause pain and distress should be reviewed with reference to reduction, replacement, and refinement. This requires significant effort of both the IACUC and the PI. The PI should at least explain, in some detail, how an animal is to be handled and justify their use of animals. They should indicate the literature sources they have reviewed seeking alternatives and, if no alternatives can be found, explain that in their justification for the use of animals. From each step outlined in the protocol, handling or experimental details can be identified and used as keys to search for alternatives. SOPs for a variety of common actions can be developed by investigators at an institution, in concert with the veterinary experts, so that uniform expectations are available to the IACUC. Deviations from SOPs must be reviewed with respect to whether they will produce an acceptable level of well-being. As the number of databases on alternatives grows, reference to searches on these can be included on an IACUC application.

Investigators also should demonstrate to the IACUC an appropriate experimental design that maximizes the information gained and minimizes the number of animals used. It is helpful to indicate whether expert advice from a statistician has been sought, particularly when the experimental design is complex (*Guide* pp. 26–27). A useful practical approach to estimating numbers of animals required has been published.[49] Finally, a range of options are available for reducing pain and distress, and those appropriate to the investigation should be considered. For example, it might be possible to conduct a study entirely while the animal is under general anesthesia, reduce pain by using appropriate analgesics, or as technology expands, to use less invasive methods of collecting data so that the overall numbers of animals are reduced. It also may be possible to limit the distress experienced by animals, or the duration of that distress, by requiring defined end points for a study (e.g., placing upper limits on total tumor burden in studies of carcinogenesis or defining a set of criteria that will be used to determine when an animal should be removed from study and humanely killed). The *Guide* (p. 26) recommends the use of a pilot study to evaluate pain or distress that might be experienced from procedures that have not been previously encountered. (See 16:60.)

16:5 Since the actual assessment of pain cannot be made until after a procedure is performed, is it appropriate for the IACUC to attempt to assess and categorize the potential for pain and distress when first reviewing a protocol?

Reg. The AWA (§2143,a,7,A) and the AWAR (§2.36) require an annual report for regulated species, encompassing animal use activity in testing, research, and teaching from October 1 through September 30 (the USDA fiscal year). The annual report

(AWAR §2.36,b,5–§2.36,b,7) categorizes the number of animals used on the basis of the level of pain or distress they experienced. Category C (on APHIS/AC Annual Report form 7023) is used for no pain or distress and no use of pain-alleviating drugs, Category D is for pain or distress alleviated by drugs, and Category E is for pain or distress not alleviated by drugs as a consequence of research needs. The PHS Policy has no requirement for the categorization of pain and distress. (See 16:2 for the definition of a *painful procedure*.)

Opin. Many IACUCs use the APHIS/AC pain and distress categories for all animals, not only those regulated by the AWAR. Nevertheless, the AWA requirement applies to its regulated species only. The wording of AWAR §2.36,b,5–§2.35,b,7 is in the past tense, suggesting that the information provided to APHIS/AC in the annual report should be based on the pain or distress actually experienced by the animal, not solely on an educated estimate made by the IACUC or the PI prior to the initiation of the study. However, Appendix A.4 of the APHIS/AC *Animal Welfare Inspection Guide*[50] indicates that when an IACUC protocol is reviewed the committee must decide on the appropriate pain/distress category in order to allow for an appropriate search for alternatives to painful or distressful procedures if pain or distress is anticipated. While prospective estimates of the number of animals that will fall into each of the pain or distress categories can simplify the work of the IACUC, the reality is that research needs and experiences often change during the course of a study. If the IACUC uses a prospective system of assigning pain or distress categories, the IACUC should have an established process that allows a PI (or the IACUC) to re-assess and change the categorization of some or all animals, retrospectively, on the basis of animals' actual experience. The latter can be assessed during the annual review process.

Although the degree of pain cannot be determined for any particular individual animal until after a procedure has been performed, it is frequently possible to make an informed estimate of the likely degree of pain that may be caused. If the procedure, or a similar procedure, has been carried out previously at the institution, then this can be used to help predict the likely consequences. If the technique or species/strain is new to the institution, then colleagues at other institutions where the procedure has been performed should be consulted. If at all possible, individuals who actually performed or cared for animals should be asked for input. Extrapolations of pain experienced by humans undergoing similar procedures can act as a rough guideline. There should be few occasions when some indication cannot be obtained as to the likely pain and distress consequences. This initial assessment should certainly be sufficient to determine, for example, what level of analgesic use would be appropriate, and what type of aftercare the animal might require. In all instances, measures for the control of pain require monitoring of animals to ensure they are effective.

16:6 Can the IACUC approve protocols in which animals will experience pain or distress not relieved by the use of anesthetics or analgesics? Under what circumstances?

Reg. If an investigator can justify, in writing, that withholding anesthesia or analgesia is required for scientific reasons and will only continue for the necessary period, then the IACUC can potentially approve such an activity (AWAR §2.31,d,1,iv,A).

The PHS Policy (IV,C,1,b) has similar wording. The *Guide* (p. 27) indicates that for studies with the potential for unalleviated pain or distress the IACUC is obligated to weigh the objectives of the study against animal welfare concerns.

Opin. IACUCs may frequently be asked to approve protocols in which animals experience pain or distress that, due to experimental needs, cannot be alleviated by the use of anesthetics or analgesics. For example, in studies of pain pathophysiology and treatment, animals are subjected to noxious stimuli or tissue injury and may experience pain, but analgesic administration may interfere with the protocol to such an extent that its use would invalidate the data obtained. Pain and distress may also occur in animal models such as those involving neoplasia, chronic organ failure, infectious diseases, toxicity testing, and a wide range of other circumstances. In many of these studies, alleviation of pain and distress by pharmacological means may not be possible because of interactions between the drugs used and the research protocol. However, methods to reduce pain or disease-related disability, such as providing environmental support (e.g., non-drug therapy such as petting an animal, use of soft surface for resting) may act to reduce overall distress (*Guide* p. 122). Because the APHIS/AC annual report for AWA regulated species is available to the public, some facilities have elected to mention these non-drug methods of alleviating pain or distress in that report. (See 16:3; 16:4.)

When pain or distress is unpreventable, IACUCs should ask the PI whether the potentially serious impact of pain or distress upon animal physiological processes will detrimentally influence the study's results. Pain and stress can cause metabolic and immunological alterations in animals (see Introduction to this chapter). If a control group does not experience the equivalent amount of pain or stress of the experimental group, then results of certain types of studies might potentially be biased to a greater degree than if both groups were treated with analgesics. Another consideration is whether the benefit of the information or results outweighs the total degree of animal distress imposed by the study (*Guide* p. 27). Whenever there is unalleviated pain or distress, efforts must be made to reduce the number of animals used in such a study, replace the study with alternative techniques that do not cause pain, and refine the study design so that pain and distress are minimized. Although these basic tenets should be applied to all protocols, they clearly are of particular importance in circumstances in which significant pain and distress can be anticipated. (See 16:10; 16:13; 17:37.)

16:7 If the AV and a PI differ on the pain or distress potential for a particular procedure, is it the responsibility of the AV, IACUC, PI, or all to provide supporting evidence for their point of view?

Reg. With reference to surgery, APHIS/AC Policy 3 notes that all animal activity proposals must specify details concerning the relief of postoperative pain and distress. The specific details must be approved by the AV or his/her designee. Policy 3 also states that the AV retains the authority to change post-procedural care as necessary to ensure the comfort of the animal but this change must be made with the approval of the IACUC if the veterinarian intends to alter post-procedural care for the remainder of the animals. Therefore, when this occurs, it is likely that the PI, with the advice of the veterinarian, will need to submit a modification of the protocol to the IACUC for review and approval.

Under the AWAR the AV provides guidance to the PI relative to the use of anesthesia and analgesia (§2.33,b,4; §2.31,d,1,iii,B) while the IACUC maintains its responsibility to assure studies are performed with proper sedation, anesthesia, or analgesia (AWAR §2.31,d,1,iv,A). The PHS Policy itself is silent on the authority of the AV to change post-procedural care; however, the *Guide* (p. 114) notes that the AV and the PI should make all efforts to work together to determine the appropriate treatment or course of action. (See 16:8.) In the event of a pressing health problem where consensus between the AV and the PI cannot be resolved, the AV must have the authority to institute appropriate measures to relieve severe pain or distress (*Guide* p. 114).

Opin. Veterinarians and veterinary care staff are professionals trained to understand species differences in behavior and husbandry needs. Whether or not specifically trained in pain medicine, they have the educational or practical background to be able to provide informed opinions. In some instances, the veterinarians, veterinary technicians, and animal care staff are in a position to observe animals more often or with greater expertise than the PI's personnel. Some PIs also have expertise in animal husbandry or behavior or may have acquired substantial experience over the course of time. If the PI can offer evidence to the IACUC about the impact of a study on animal well-being, then this should be taken into account. However, as the science of animal welfare advances and other improvements in research refinement are published, veterinarians and animal welfare scientists may be able to contribute new knowledge to the assessment of the potential for pain or distress and to suggest remedies for their alleviation. Any decision made by an IACUC must incorporate input from all of those involved in animal care and experimentation.

16:8 A PI provides, in writing, a scientific rationale for withholding the provision of analgesia during and after a painful procedure. Is the IACUC obliged to accept this rationale?

Reg. The AWAR (§2.31,d,1,iv,A) require the use of analgesia during a procedure that causes more than momentary pain or distress unless the PI provides the IACUC written scientific justification for not providing it. The PHS Policy (IV,C,1,b) has the same wording. The AWA (§2143,a,6,A) and the AWAR (§2.31,a) state that except as authorized by law or by the AWAR the IACUC cannot prescribe methods or set standards for the design, performance, or conduct of actual research or experimentation by a research facility. The HREA (§495,a,2,B) states that the guidelines established in the PHS Policy "shall not be construed to prescribe methods of research" but also indicates that this statement does not override compliance with the PHS Policy (HREA §495,b,1). Both the AWAR (§2.31,c,6) and PHS Policy (IV,B,6) provide the IACUC with the authority to approve the use of animals in research or other animal activities.

Opin. It has been argued that the scientific justification presented to the IACUC must have some substance; otherwise it is no better than no justification at all.[51] The same reference stated that if a logical justification could be provided, even if "reasonable minds may differ as to whether the justification suffices, the IACUC must defer to the investigator." In contrast, both APHIS/AC and NIH/OLAW responded that it is their interpretation "that the AWA and PHS Policy do not require an IACUC to approve a proposed project against its will."[52] It has been argued that the scientific justification presented to the IACUC must have some substance; otherwise it is no better than no justification at all.[51] The same reference stated that if a logical justification could be provided, even if "reasonable minds may differ as to whether the justification suffices, the IACUC must defer to the investigator." In contrast, both APHIS/AC and NIH/OLAW responded that it is their interpretation "that the AWA and PHS Policy do not require an IACUC to approve a proposed project against its will."[52] This is reaffirmed in NIH/OLAW FAQ D.6 which states that IACUCs must confirm that the protocol is consistent with the *Guide* unless a scientific justification for a departure is presented and is acceptable to the IACUC.

Any justification to withhold analgesia should be clearly reasoned and, whenever possible, documented by laboratory data or literature references. Generic statements, such as "Analgesia will interfere with the collection of useful data" may be correct, but the IACUC should not accept such statements at face value. (See 16:7.)

16:9 What are some typical criteria an IACUC or an investigator can use to determine whether an animal is in pain or distress?

Reg. The *Guide* (pp. 120–121) provides examples of some species-specific manifestations of pain (such as nasal porphyrin discharge in rodents) and quotes U.S. Government Principle IV which states that unless the contrary is known, it should be considered that conditions causing pain in humans also may cause pain in other animals. The AWAR (§1.1, Painful procedure) uses similar wording.

Opin. Attempts to recognize pain or distress in animals rely on understanding the potential causes of pain or distress, observing animals using a combination of behavioral and physiological variables, and looking for deviations over time from normality. In the case of pain, for example, an animal may change its spontaneous behavior so that it becomes less active, decreases grooming or nest building activity, and reduces its food and water consumption. It also may change its responses to handling, for example, by exhibiting increased aggression. Conversely, some animals may become apathetic and unresponsive to handling. Body weight changes may occur as a result of acute or chronic pain. The criteria to be used for recognition depend on the species, the type of pain, and the degree of socialization, but there is one constant: all of them currently are based on behavior. There is no purely objective way to assess "clinical" pain or distress in animals, and so the only way to evaluate them is to observe behavior or representations of behavior (such as nest quality or utilization of enrichments). Species specific signs of pain have been reported, and the literature continually expands.[47,53,54] Similar guidance regarding the detection of distress is available.[39]

The key point to note is that signs of pain and distress may be difficult to detect and their recognition almost always requires a detailed knowledge of the normal behavior of the animal species but also behavior within the confines of the laboratory enclosure. Recognition also requires that sufficient time be allocated for observation of the animal, and, in some circumstances, it may be necessary to observe the animal's behavior in such a way that it is unaware of the presence of the observer (e.g., by using a video camera).

Spontaneous behavior can be studied in species that are not accustomed to being handled. Nevertheless, enriched housing may be necessary because full behavioral repertoires may not be evident in barren environments, even if the animal is not experiencing overt pain or distress. In socialized, easily handled species such as the dog or goat, palpation of surgical wounds may elicit evasive behaviors (flinching, glance at the site, vocalizations). If the administration of an analgesic or other therapy restores behavioral indices toward the normal, then unalleviated pain or distress may have been present.

16:10 An IACUC protocol indicates that a procedure is moderately painful to its canine subjects for approximately 1 hour, then never repeated. During that time, pain-relieving drugs cannot be used, but an alternative pain relief mechanism will be used. That is, calming music, a darkened room, and human petting of the animal will constantly occur. The technique has not been previously attempted. Should the IACUC consider this to be alleviated or unalleviated pain?

Reg. (See 16:2.) for the definition of a *painful procedure* and 16:3 for related regulatory information. For painful procedures in which appropriate anesthetic, analgesic, or tranquilizing drugs would have adverse effects on procedures, results, or the interpretation of teaching, research, experimentation, or tests, the AWAR (§2.36,b,7) specifically require an explanation of the procedures and reasons such drugs were not used. The *Guide* (p. 122) notes that in addition to anesthetics, analgesics, and tranquilizers, nonpharmacological control of pain and proper nursing support is often effective. (See 16:6.)

Opin. The determination of whether pain was or was not alleviated is an outcome assessment. When the likely outcome is not known from a review of the literature or from other sources, a pilot study may prove helpful (*Guide* p. 26). (See 16:60.)

Evidence for benefit of a variety of non-pharmacologic methods of pain relief can be found in the literature; substantiation depends on the modality, situation and species. Social contact can offer relief of anxiety, and pain may be worse in anxiety.[47] Practical experience in clinical veterinary practice has shown that many socialized animals, especially dogs and cats, respond positively to human presence and touch. The mechanisms behind the beneficial effects of social contact and touch based therapies have been investigated and may involve the production of oxytocin or other neuropeptides that modulate pain.[55,56] The question is whether the proposed strategies of music, dim light and petting would be considered sufficient to deem pain "alleviated," and the answer depends on the response of the animal—does its behavior indicate that it is coping adequately with the moderate pain?

Obtaining a positive response to such contact depends critically upon the previous experience of the animal and the nursing skills of the personnel involved. If an animal has not been adequately socialized to accept (and welcome) human contact, then attempts to provide reassurance may be counterproductive and increase the animal's distress. If such a technique is to be considered as a replacement for conventional pain-alleviating techniques, then it is important that the IACUC require that the personnel involved along with their previous relationship with the animals used be specified. The personnel should have experience with the technique, and ideally the technique should be familiar to the animals (acclimation to restraint, devices, etc.). When considering whether the justified restriction on use of drugs to reduce pain is acceptable, the IACUC might ask the question of whether physiologic changes that accompany the non-pharmacologic pain relief are substantially different than those that might occur with use of a drug. Finally, it is appropriate to consider other, non-pharmacological methods of pain relief, such as application of heat or cold, acupuncture, or transcutaneous nerve stimulation. (See 16:11; 16:12 for a discussion of decerebration as a method of non-pharmacologic pain relief.)

16:11 Should a decerebrated cat be placed in APHIS/AC annual report pain/distress Category D (alleviated pain or distress) or E (unalleviated pain or distress)? (See 16:12.)

Reg. The AWA (§2143,a,3,A) and the AWAR (§2.31,e,4; §2.36) refer to the use of drugs to alleviate pain. (See 16:5.) Neither the PHS Policy nor the *Guide* requires categorization of pain or distress; however, the PHS Policy requires absolute compliance with the AWAR for regulated species (PHS Policy IV,C,1).

Opin. An experimentally decerebrated animal cannot feel pain because of the inactivation of the cerebral cortex and thalamus. The decerebration is performed with traditional pharmacologic anesthesia, suggesting that APHIS/AC Category D is appropriate. In reality, the issue of pain is moot, since the animal cannot perceive any potential pain because the pain recognition centers of the brain have been inactivated or physically removed. Thus, if there is nociception but no perception of pain (see 16:2) there is no need to further alleviate pain and category D is appropriate as a consequence of the initial use of general anesthesia. The *Guide* (p. 122) recognizes that non-pharmacologic control of pain is often effective. The issue of decerebration has been discussed in the literature and it has been recommended that properly performed decerebration be considered the equivalent of continuous general anesthesia.[57]

16:12 Can mammals that have a cerebral cortex feel pain if made decerebrate?

Opin. When addressing this question it is important to differentiate between pain and nociception (see 16:2). Nociception is the response to damaging or potentially damaging stimuli, whereas pain has a subjective component and is interpreted by most humans (and other mammals) as an unpleasant experience.[44,47] It is widely accepted that the presence of a functioning forebrain is required for the perception (or more explicitly, *for the experience*) of pain, so decerebration should remove that capacity. Since thalamic structures are believed to play a role in pain perception, it is usually considered that removal of the forebrain, including the thalamic nuclei, ensures that pain cannot be perceived. The position in regard to decorticate animals is less clear. Comparison with humans leads one to presume that such animals have no capacity for pain perception. However, given the higher levels of organized behavior shown by decorticate or decerebrate animals of some species (e.g., rats), some caution is required in making these extrapolations. (See 16:11.)

16:13 Can nonmammalian vertebrates lacking or having only a primitive cerebral cortex (e.g., frogs) feel pain?

Opin. Since the capacity to experience the subjective, unpleasant component of pain in humans appears to rely on the presence of a functioning cerebral cortex, it is often assumed that animals with a less well developed cortex have less capacity to feel pain. Comparison of the frog with humans is simply a more extreme comparison than the more frequent extrapolation of a rat or mouse to humans. In both instances, we have little insight into the nature of the experience of pain in the animal, nor of its significance to the individual. We have adopted an approach that presumes an animal with a certain level of central nervous system development can experience pain, and research involving these species should be conducted in a way that reduces the likelihood of causing pain or distress.

It seems unlikely that frogs experience pain in the same manner as humans, but since we can demonstrate nociception in amphibians[58] and since their degree of cerebral development cannot be said to preclude the possibility of pain perception, we should assume that these species can experience pain. What is more problematic about the assumption that they experience pain is that pharmacologic alleviation of pain in nonmammalian species is poorly studied (although studies of nociception in amphibians and less cerebrally complex animals are available) and clinical signs of pain are not well characterized.[47]

16:14 Can ketamine alone or with xylazine be considered adequate anesthesia for major surgery, such as abdominal surgery in laboratory rodents, rabbits, or cats?

Opin. Ketamine can cause immobilization, lack of awareness, amnesia, and analgesia. This drug has been in use for decades; over time we have come to appreciate many of its beneficial properties. This drug continues to be used in humans for specific indications as a sedative or anesthetic. Its clinical use in both veterinary and human medicine has another relatively new major facet, which is in treatment of acute and chronic pain, both of visceral and somatic origin, with "subanesthetic" concentrations of ketamine.[59–63] There is controversy about whether it can act as a true analgesic (i.e., to decrease the pain of an incision) versus its use as an antihyperalgesic (i.e., to prevent postoperative pain); the latter is well established and depends on the situation. However, as current thinking is that the anesthetic action of ketamine "renders central nervous system centers unable to receive or process sensory information"[59] it may be considered part of an anesthetic protocol in certain cases. Ketamine in combination with xylazine and Telazol® has been deemed satisfactory anesthesia for feline neuters for over a decade.[64] Moreover, as inhibitory effects on pain have been demonstrated with pre- or intraoperative administration of ketamine, its use in combination with other agents for anesthesia may result in improved surgical analgesia compared to some other modern anesthetics (e.g., propofol or isoflurane).[59–61,65] Recent research and clinical experience will refute older concepts of the drug which may be held by many veterinarians.

The degree of analgesia and of immobilization from ketamine may vary considerably between species.[66] Ketamine alone continues to be regarded as inadequate anesthesia for abdominal surgery in rodents, rabbits, cats, or indeed any species.[67,68] The degree of muscle relaxation is poor and this often renders ketamine alone unsuitable even for superficial procedures. The addition of injectable drugs with sedative effects such as xylazine, dexmedetomidine, acepromazine, opioids, or diazepam greatly improves the quality of anesthesia. In general, combining ketamine with drugs that have analgesic and sedative properties (e.g., xylazine, dexmedetomidine, +/– opioids) produces more effective surgical conditions than combinations such as ketamine with acepromazine or ketamine with diazepam.[66] When combined with xylazine (or other alpha 2 agonists), ketamine provides surgical anesthesia in many species, including rodents, rabbits, and cats.[69,70] The degree of anesthesia provided by ketamine combined with other sedatives/

analgesics may therefore be sufficient for abdominal surgery in rodents, rabbits, cats and other species; however, this will depend on the situation.[71] Disadvantages include the fact that fixed-dose combinations given as an intravenous or intramuscular/intraperitoneal bolus injection may provide anesthesia of insufficient duration, and also may have deleterious effects on physiologic parameters without additional supportive measures, and the adequacy of the surgical plane of anesthesia varies with species, strain/stock, dose, administration route, or procedure itself.

16:15 Is chloralose considered to be an anesthetic or a hypnotic?

Opin. Chloralose is a hypnotic, as are many other agents that are used to anesthetize animals (e.g., pentobarbital). It is often assumed, incorrectly, that because a drug is hypnotic it cannot be used to provide surgical anesthesia. The definition of *hypnosis* is "a condition of artificially induced sleep, or a state resembling sleep, resulting from moderate depression of the central nervous system from which the patient is readily aroused."[71] Although drugs classed as hypnotics produce this effect as the dosage is increased, progressively greater depression of the central nervous system occurs, so that animals progress from sedation to hypnosis to general anesthesia. Some hypnotics cause such severe cardiovascular depression at the doses necessary to achieve general anesthesia that they cannot be used in this way without risking the death of the animal. Others, such as chloralose and the barbiturates, can be given at doses sufficient to produce general anesthesia in many species. Different species, and different strains of the same species, vary in their response to these drugs, and in some circumstances the depth of anesthesia will be inadequate for surgery. Provided the anesthetic depth is assessed (e.g., by evoking a response to a painful stimulus such as a toe pinch), dosages of hypnotics can be adjusted as necessary or additional analgesia can be provided by drugs such as morphine or fentanyl.[72]

Chloralose is not a modern anesthetic agent. It gained favor and is still commonly used in certain types of studies (e.g., cardiovascular, respiratory and neurovascular physiology, neurotransmission) because of its reputation for creating less interference with the processes being studied than other anesthetic agents and because it can produce long-lasting immobility. However, in the past, chloralose was often compared to pentobarbital, and newer anesthetic techniques may provide acceptable hypnosis or anesthesia without the potential for pain or distress that chloralose has been cited to cause. Certain methods of administering chloralose may cause injury and subsequent pain or other morbidity during the postoperative period. Problems cited for chloralose include adynamic ileus and intestinal injury when administered intraperitoneally, phlebitis associated with extravascular accidental injection, and stressful recovery.[73] In many types of studies, alternative modern anesthetic techniques, conscious trained animals, and decerebrate preparations have been shown to be equal or superior methods of immobilization.[74–78] There are still many instances in which use of chloralose, alone or in combination, is the most suitable technique for a measurement. Anesthetics have variable effects on experimental results, and an investigator should carefully scrutinize the literature for newer, less noxious anesthetics or conscious or decerebrate techniques, rather than relying on traditional beliefs.[79,80] If, for example, an author uses modern anesthetics in one species and chloralose in another to measure the same parameters, it could be assumed the modern anesthetic would be appropriate for both species.[81] Many experts recommend that chloralose not be used for survival surgery, and in addition, pharmaceutical grade chloralose is not commercially available.[66,73,82,83]

16:16 Should the IACUC request special safety precautions when urethane is used as an anesthetic?

Opin. Urethane is mutagenic and carcinogenic,[84] and, if it is to be used as an anesthetic, appropriate precautions should be adopted to prevent safety hazards to personnel. The exact nature of the precautions may vary, but they generally mirror those required to ensure safe handling and use of carcinogens. The IACUC should consider utilizing the services of the institution's biosafety committee. (See 16:17 and Chapter 20.) One consideration is that pharmaceutical grade urethane is not commercially available.

16:17 Should the use of urethane be allowed as anesthesia for recovery surgery?

Opin. Urethane has been used for the long-lasting immobility, anesthesia, and reported cardiovascular stability that it produces. A good recent review of its useful and adverse effects is available.[83] Hypertonic solutions may cause tissue damage and lead to adverse health consequences when it is given intraperitoneally and recovery can be very prolonged.[83] Both the availability of newer injectable and inhalant anesthetic techniques and many disadvantageous features of urethane suggest that it should be used rarely if at all, and only with careful consideration of alternative methods.[73,74,77,82] Personnel safety is a major consideration and exposure to urethane may cause neoplasia in some species (see 16:16).

16:18 Should the IACUC approve the use of a nonpharmaceutical grade drug for anesthesia or analgesia if pharmaceutical grade alternatives are available?

Reg. The AWAR does not comment directly on the use of nonpharmaceutical grade drugs; however, the AWA (§2143,a,3,A) states that adequate veterinary care with the appropriate use of analgesic or tranquilizing drugs is required. The AWAR (§2.33,a) also require adequate veterinary care. APHIS/AC Policy 3, which is meant to provide guidance on the AWAR, states, "Investigators are expected to use pharmaceutical-grade medications whenever they are available, even in acute procedures. Non-pharmaceutical-grade chemical compounds should only be used in regulated animals after specific review and approval by the IACUC for reasons such as scientific necessity or non-availability of an acceptable veterinary or human pharmaceutical-grade product. Cost saving alone is not an adequate justification for using nonpharmaceutical-grade compounds in regulated animals." *Note that this wording in APHIS/AC Policy 3 will be undergoing revision in the near future but the intent will remain the same.* PHS Policy (IV,C,1) requires compliance with the AWA for regulated species.

The HREA (§495,a,2,A) requires the appropriate use of analgesics and anesthetics. PHS Policy (IV,C,1,b) states that painful procedures will be performed with appropriate analgesia or anesthesia (unless otherwise justified). The *Guide* (p. 31) notes that pharmaceutical grade chemicals should be used, when available, for all animal-related procedures.

It should be noted that NIH/OLAW allows for the use of non-pharmaceutical grade pentobarbital as a substitute for the pharmaceutical grade drug as the current price of the latter is so excessive as to render the drug unavailable. This is a special case decision by NIH/OLAW and does not apply to other drugs at the time of this writing (April 2013 and October 2013). See Comment 24 in the transcript of the NIH/OLAW online seminar on the use of non-pharmaceutical grade chemicals and other substances in animals (http://grants.nih.gov/grants/olaw/120301_seminar_transcript.pdf).

Opin. The preparation of nonpharmaceutical grade drug solutions can lead to inconsistent product quality and the potential for adverse effects, including tissue damage, pain, or lack of sterility or of a sufficient clinical effect. For this reason, their use may constitute less than adequate veterinary care. NIH/OLAW and APHIS/AC have clarified that a non-pharmaceutical grade anesthetic agent can be used when scientifically necessary, appropriately justified, and approved by the IACUC.[85] The IACUC should ensure that investigators have adequately considered the utility of alternative pharmaceutical grade drugs, and on an ongoing basis, that when use of nonpharmaceutical grade drugs is approved, that these solutions are prepared, stored and used with appropriate consideration of sterility, purity, stability and related factors. A more detailed discussion about the use of non-pharmaceutical quality drugs or chemicals is found in 27:16 and a PHS Position Statement[86] and NIH/OLAW FAQ F.4.

Surv. Does your IACUC allow investigators to use reagent grade drugs for anesthesia (such as tribromoethanol?)

• Not applicable	44/295 (14.9%)
• Yes, almost always	8/295 (2.7%)
• Yes, but only with a science-based explanation	110/295 (37.3%)
• No, (unless the non-pharmaceutical grade drug itself is being studied)	127/295 (43.0%)
• Other	6/295 (2.0%)

"Other" responses included: Only when pharmaceutical grade is unavailable; used non-pharmaceutical grade sodium pentobarbital due to experimental needs.

16:19 The AWA, the AWAR, and the *Guide* have restrictions on performing multiple major survival surgeries on the same animal. (See 16:22.) Are there any such restrictions on performing multiple painful procedures on the same animal?

Reg. There are no specific prohibitions on performing multiple painful procedures on the same animal, if approved by the IACUC, provided that that procedure is not a major operative one where an animal that had a major operative procedure for research was transferred to a different protocol in which another major operative procedure is needed. The latter case would be a Special Circumstance which requires IACUC approval and approval from APHIS/AC (AWAR §2.31,d,1,x,C). PHS Policy (IV,D,1,d) notes that applications and proposals to the PHS must contain a description of the procedures that are designed to assure that discomfort and injury to animals will be limited to that which is unpreventable in the conduct of scientifically valuable research and that proper drugs will be used where indicated to minimize animal pain and discomfort. U.S. Government Principle IV speaks of avoiding or minimizing discomfort, distress, and pain when consistent with sound scientific practices. A goal of the *Guide* (p. xiii) is to assist institutions in the humane care of laboratory animals. NIH/OLAW FAQ F.9 directly addressed multiple major survival surgical procedures, indicating that the practice is acceptable if

1. Included in and essential components of a single research project or proposal
2. Scientifically justified by investigator

Or necessary for clinical reasons.

The AWAR (§2.31,d,1,i) state that procedures involving animals will avoid or minimize discomfort, distress, and pain to the animals. The AWAR (§2.31,d,1,ii) also state that the PI must consider alternatives to procedures that may cause more than momentary or slight pain or distress. Further, the AWAR (§2.31,e,4) note that the procedures to be used should be designed to limit pain and discomfort to that which is unavoidable for the conduct of scientifically valuable research.

Opin. It appears that the intent of the federal regulations restricting multiple major survival surgeries is to reduce the total amount of pain and distress caused to an individual animal and to prevent animals from being transferred at the end of one study to another study where the animal may be subjected again to major survival surgery. Therefore, it seems prudent that studies in which animals are potentially subjected to multiple painful procedures should be reviewed by IACUCs using a paradigm similar to that used for multiple major survival surgeries (AWAR §2.31,d,1,x; *Guide* p. 30). (See 16:22.) A review for this purpose is implied in the *Guide* (p. 30) which states that "some procedures characterized as minor may induce substantial post-procedural pain or impairment and should similarly be scientifically justified if performed more than once in a single animal." If a procedure causes significant tissue damage (e.g., skin burns, osteoarthritis, or nerve damage) but is not a major surgery, there is still a potential that the animal is left with a chronic state in which pain or distress thresholds may be lowered and subsequent uses might result in more total pain than if another animal were used. On the other hand, a distinction should be made between *potentially* painful procedures and those that involve *unalleviated* pain. Techniques to reduce pain and distress include local and systemic analgesics, anesthetics, and operant conditioning/acclimation. Other modalities can be used alone or in combination and should be aggressively pursued if they will not specifically interfere with the study.

If a study requires that an animal be subjected to repeated painful or distressful manipulations, then the investigator should outline for the IACUC the impact, magnitude, and duration of the pain; its potential effects on experimental outcome; methods to monitor the animal's ability to adapt to or recover from the event; and reasons why the pain cannot be relieved, if this is proposed. (See 16:6.) Without such information any attempt to set a limit on the frequency with which such an event can be repeated is arbitrary. In the course of a particular study, repetition of procedures in which unalleviated pain is caused may be necessary. NIH/OLAW, in reviewing 16:19, has added that "Allowing transfer of an animal to another study in which additional major painful procedures are performed does not comply with the AWAR regulations or the PHS Policy. OLAW does not condone the use of such a transfer to circumvent guidance and would consider this noncompliant with PHS Policy and U.S. Government Principles." (See 16:22.)

Surv. 1 If an animal is transferred from one study to another, does your IACUC have a means of determining whether prior major survival surgery was performed on that animal?

- Not applicable 71/295 (24.1%)
- Yes 171/295 (58.0%)
- No 4/295 (1.4%)
- No, unless the issue is specifically brought to the committee's attention 42/295 (14.2%)
- Other 7/295 (2.4%)

Surv. 2 If an animal is transferred from one study to another, does your IACUC have an effective means of determining if an animal experienced significant pain or distress on the original study?

- Not applicable 66/296 (22.3%)
- No 7/296 (2.4%)
- No, unless the issue is specifically brought to the committee's attention 64/296 (21.6%)
- Yes 152/296 (51.4%)
- Other 7/296 (2.4%)

16:20 Is it a good idea to have an institution-wide policy that some form of analgesia must be provided for any major survival surgical procedure?

Reg. The AWAR (§2.31,d,1,iv,A) state that procedures involving more than momentary pain or distress will be performed with appropriate sedatives, analgesics, or anesthetics unless their withholding is justified, in writing, by the PI. APHIS/AC Policy 3 states that all animal activity proposals involving surgery must provide specific details of pre- through post-procedural care and relief of pain and distress. U.S. Government Principle IV speaks of avoiding or minimizing discomfort, distress, and pain when consistent with sound scientific practices.

Opin. Analgesics, and often more than one class of analgesic drug and non-drug approaches to preventing pain, should be used for any potentially painful survival surgical procedure, whether major or minor. (See 16:2–16:4, 16:9.) We concur with APHIS/AC Policy 3 that the drug (or other method of analgesia) to be used should be specified in the IACUC protocol and readily available for use. The type and amount of analgesia will depend on the degree of pain that is caused. Published estimates of the degree of pain expected from various types of surgery[54,87,88] (as well as for any disease, injury or condition for that matter) can be used to help decide on the amount, type, and combinations of analgesics to be used. In many cases, the amount of pain an animal will be expected to perceive is comparable to that felt by humans in the similar situation (AWAR §1.1, Painful Procedure; U.S. Government Principle IV; *Guide* p. 120). In order to get an idea of the incidence and magnitude of pain that a surgery or condition causes in humans, the authors find it instructive to consult the human clinical literature. Studies of analgesic interventions for a particular procedure (e.g., hip replacement surgery) or a disease state (e.g., ulcerative colitis) commonly include a description of the severity of pain. To best comply with both the regulatory and ethical aspects of humane animal care, the institution should have a policy stating that analgesia for any survival surgery or other pain inducing procedure or disease state should be provided to the animal, and that the type of analgesia should be commensurate with the degree of expected pain, unless adequate scientific justification is provided for not doing so.

16:21 Is the use of an analgesic necessary after embryo transfer in mice?

Reg. (See 16:3; 16:20.)

Opin. For many years, the successful production of offspring via embryo transfer in mice was said to be inhibited by use of certain anesthetics or analgesics, and thus analgesia was considered "contraindicated." Embryo recipient mice therefore were subjected to unalleviated pain that was justified as being "necessary." Embryo transfer surgery is expected to cause some degree of postsurgical pain. The degree of pain from a flank incision and subsequent visceral manipulations depends upon surgical finesse, the type of anesthetic agent used, and other factors. There are few studies investigating the amount of pain that this commonly performed procedure causes in mice; at least one indicated that it had an impact on behavior,[89] and the equivalent surgery in humans would require analgesia. In the authors' experience, mice given no postoperative analgesia lose body weight after bilateral flank ovariectomy, an indication that this surgery has an impact on them.[90] Furthermore, while some anesthetic agents (e.g., ketamine, xylazine) possess analgesic effects that may modestly persist into the postoperative period, others (e.g., isoflurane, pentobarbital) do not. In recent years, several studies of use of a variety of analgesics alone or in combination do not support a reduction in viability of transferred embryos in mice.[91–94] Although these studies are limited in terms of mouse strain/stock and repeatability, in combination with lack of data supporting a deleterious effect of analgesia after embryo transfer in mice, and the rationale and limited data indicating that mice do experience pain after embryo transfer, indicate that at least one single dose of perioperative analgesia is warranted and is not detrimental. (See 16:14.)

Surv. Does your IACUC require the use of appropriate analgesia after embryo transfer surgery in mice?

- Not applicable 193/287 (67.3%)
- We do not require an analgesic 6/287 (2.1%)
- We do not require an analgesic but we encourage its use 5/287 (1.7%)
- We usually require the use of an analgesic 78/287 (27.2%)
- Other 5/287 (1.7%)

16:22 AWA-regulated animals arrive at an animal facility, already having undergone major surgery (under anesthesia) at the vendor's company. The animals healed well before arrival, and there was no apparent pain or distress upon arrival. There will be no painful or distressful procedures when used at the purchaser's institution. On the APHIS/AC annual report form (Form 7023), are these animals placed in Column C (no pain or distress) or Column D (pain or distress alleviated by drugs)?

Reg. The AWAR (§2.31,d,1,x) and the *Guide* (p. 30) restrict multiple survival surgeries to those circumstances described in 18:11. (See 16:5 for general considerations of pain categories.)

Opin. On the APHIS/AC annual report form (Form 7023) of the receiving facility these animals should be listed in Column C because no surgery was performed by the facility. The vendor, however, would need to place the animals in Column D on its annual report. If these animals were to be transferred to another protocol at the receiving facility where a major operative procedure is to be done, the facility must get approval from APHIS/AC (as special circumstance study) as well as IACUC approval because of the first surgery AWAR (§2.31,d,1,x). (See 18:13.)

Vendors usually have their own APHIS/AC registration; in those circumstances the vendor's annual report states the appropriate pain or distress category for the animal while at the vendor's facility. If the vendor does not have an independent APHIS/AC registration, the receiving institution must arrange to have the vendor's animal use operations covered under the receiving institution's registration. How the animals in the study are categorized at the receiving institution should refer to the specific way that the animals are to be used there. In the scenario presented there was no apparent pain upon arrival. If there is no need from that point onward to administer analgesics or anesthetics to minimize pain or distress, then the study animals are placed in column C. An institution might choose to base its animal procurement decisions on whether or not a vendor claims to use analgesics at the time of the surgery.

16:23 When an IACUC reviews a literature search strategy for alternatives to painful or distressful procedures, must each potentially painful or distressful procedure be listed as a keyword in the search?

Reg. (See 16:4.)

Opin. (See 16:4.) For an investigator to consider adequately whether the proposed use of laboratory animals will be conducted by using currently available techniques (i.e., refined), a search of each potentially painful or distressful procedure should be done. This type of search is challenging but can be aided by use of resources such as the USDA's Animal Welfare Information Center,[95] or the Animal Welfare Institute database for refinement.[96] In addition to looking for nonsentient animal models (which may be found to be unsuitable), the PI is strongly urged to search for methods by which animal well-being can be monitored or improved. For example, researchers in unrelated disciplines may have developed behavioral assessment tools or otherwise refined animal use methods that can be applied to the new protocol as it is developed. In order to most fully reduce the amount of pain or distress caused, each type of painful or distressful exposure should be listed as search terms. Importantly, use of the word *'alternatives'* as a search term may unduly limit the number of useful hits in a literature search, whereas using terms such as *severity, assessment,* crossed with terms such as *model* and *animal,* and with *pain, nociception,* or *illness,* may reveal established techniques for assessment of pain, disability, or humane end points. It should also be noted that such papers may not have as their primary focus the reduction of pain or distress, and that the methods section of the papers will have to be examined for the necessary details.

Surv. Have animal users at your institution performed literature searches that resulted in the implementation of alternatives to painful or distressful procedures that were in the original IACUC application? Check as many boxes as appropriate.

- Not applicable 21/295 (7.1%)
- No 62/295 (21.0%)
- Yes, but infrequently for basic research needs 132/295 (44.7%)
- Yes, and somewhat often for basic research needs 22/295 (7.5%)
- Yes, but infrequently for educational needs 26/295 (8.8%)
- Yes, and somewhat often for educational needs 13/295 (4.4%)
- I don't know 52/295 (17.6%)

16:24 Is a measured decrease in food or water consumption a reasonable indicator of pain in laboratory rodents?

Reg. (See 16:3.)

Opin. Numerous factors can influence food and water consumption in rodents and pain is only one of these. If a rodent reduces its food and water intake, pain should be included as a possible cause along with oral or dental disorders, infectious disease, changes in environment, alteration in the husbandry regimen, and other stressors. Nevertheless, in the postoperative period, immediate changes in food and water consumption have been cited to be indicators of postoperative pain.[97–100] As rodents are often group-housed, measuring individual food and water consumption can be difficult. It is often more convenient to record body weight. It is important to note that many studies are carried out on growing animals, so establishing average weight gain for a few days before an operative procedure is necessary. After surgery, rats and mice almost invariably show a small fall in body weight (between 5% and 15%, depending upon the nature of the surgical procedure) as a consequence of reduced food and water consumption. In some cases, perioperative analgesia will attenuate this fall in bodyweight compared with untreated control animals, thus it seems reasonable to conclude that some of the reduction is due to postsurgical pain.[97–100] Some authors have reported that use of buprenorphine in both operated and control rodents may result in weight loss, thus additional data about the relationship of body weight changes to pain are needed.[99,100]

Although measuring body weight is said to be useful as an indicator of pain or distress, it is essentially a retrospective index of the amount of pain or discomfort that was produced over the preceding 12 to 24 hours. Thus, weight loss probably indicates that unalleviated pain or distress has already occurred, making it a more useful barometer to guide changes in future animal surgeries. Behavioral assessment, if carried out effectively in "real time," allows dose rates of analgesics to be adjusted, depending on the animals' responses. (See 16:25.)

16:25 What percentage of weight loss over what period might indicate that an animal is experiencing pain or distress?

Opin. (See 16:24.) Loss of weight does not necessarily indicate that an animal is experiencing pain and distress, but something is amiss that might lead to distress. Chronic weight loss, particularly if unexpected, should be investigated carefully. In many institutions, limits are placed on weight loss, on the basis of the assumption that a fall in weight must reflect some adverse occurrence and some degree of distress. Brief periods of inappetence will cause weight loss of 5–10% in rodents and this can happen within a 24-hour period after surgery in mice. Once weight loss exceeds 20%, general loss of body condition and fat or muscle mass becomes clinically apparent and it would be reasonable to expect animals to feel distressed. Most guidelines are based on consensus, not on published data, and it is advisable to examine the issue on a case-by-case basis. It is important to establish why the weight loss is occurring. Is this due, for example, to lowered feed intake, increased metabolism, or decreased absorption of nutrients? Is the weight loss an unpreventable consequence of the research protocol, or could supplemental feeding, use of analgesics, antianxiety drugs, environmental changes, etc., prevent or ameliorate it? If cachexic states associated with disease or cancer are not the focus of the research, then supportive measures may allow animals to survive longer, making long term studies feasible. It also is important to compare an animal's weight with that of an untreated, normally growing control. Finally, it is worth noting that adults of several species of laboratory animals when fed *ad libitum* become obese. This emphasizes the importance of adopting a reasoned, logical approach to interpreting the significance of weight loss in an animal. Determining whether animals that are losing weight are distressed may need to take into account whether the animal is behaving relatively normally.

Surv. When weight loss is accepted by your IACUC as one indicator that an animal may be in pain or distress, what percentage of weight loss is typically considered to be significant (as compared to control animals or a previous maximum weight)? Assume the average recovery time for any procedure that might have been performed has passed.

- Not applicable 32/294 (10.9%)
- Usually 5–10% 34/294 (11.6%)
- Usually 10–15% 77/294 (26.2%)
- Usually 15–20% 78/294 (26.5%)
- Greater than 20% 47/294 (16.0%)
- I don't know 26/294 (8.8%)

16:26 What are useful guidelines for the IACUC to consider when the use of neuromuscular blocking (NMB) agents is requested? (See 16:27.)

Reg. The AWAR (§2.31,d,1,iv,C) state that NMB agents (paralytics) should not be used without anesthesia. The PHS Policy (U.S. Government Principle V) states that surgical or other painful procedures should not be performed on unanesthetized animals paralyzed by chemical agents. The *Guide* (p. 123) states that when these agents are used, it is recommended that the appropriate amount of anesthetic first be defined on the basis of results of a similar procedure that used the anesthetic without a blocking agent.

Opin. NMBs prevent voluntary muscle activity so that an animal can no longer respond to painful or other stimuli by moving, including eye position changes and respiratory efforts. It is essential to preclude the very real possibility that the level of anesthesia is inadequate and the animal is aware of the surgical procedure being performed ("awareness"), which is potentially painful and distressful. Lack of movement in response to a surgical stimulus is the most commonly used method to assess the adequacy of anesthetic depth; when using NMBs as part of an anesthetic protocol, new methods of assessment and monitoring are required. In the management of anesthesia for human patients, NMBs are often used to facilitate endotracheal intubation and major surgery, and estimates of "awareness" are based on postsurgical reports. Current reviews indicate that approximately 1–4 of every 1000 patients can recall intraoperative events that occurred while he or she was under anesthesia.[101,102] Put into perspective, these human cases were anesthetized by anesthetists who had postgraduate professional training, whereas most animals are anesthetized by technicians or other individuals who have far less training. Thus the use of NMBs in research animals risks causing unnecessary pain or distress when inadequate anesthetic monitoring is used.

Before examining what constraints might reasonably be placed on the use of these drugs, the IACUC should first establish why it is thought necessary to use them. In some instances, investigators may simply have taken a human anesthetic protocol and decided to use it in their animal model. This is often inappropriate, since the rationale for NMB use in humans often is not applicable to other animals.[103] Human (M.D.) surgeons are accustomed to operating under conditions of little or no muscle tone; veterinary surgeons are not. It is not necessary, for example, to use an NMB to allow assisted (mechanical) ventilation in nonhuman animals. Veterinary anesthesiology textbooks discuss mechanical ventilation during anesthesia, and these chapters do not mention use of neuromuscular blockers. In a discussion of respiratory management of anesthetized animals, three options for preventing spontaneous breathing are listed:

1. Slight hyperventilation
2. Achieving adequate anesthetic levels
3. Use of NMB

Option 1 is listed as the easiest.[104] Thus, NMBs are not necessary for mechanical ventilation and the IACUC should not accept this as the sole reason for requiring NMBs. Neither are NMBs needed to produce sufficient muscle relaxation for the majority of surgical procedures. It is true that electrosurgery and cautery cause skin and muscle to twitch in the absence of NMBs, which can make the limb or head move slightly for brief periods. Nevertheless, this is only critical if the structures being accessed (e.g., eye, brain, spinal cord) are sensitive to tiny movements. Under those conditions neuromuscular blockade is only necessary during specific portions of surgery (not, for example, for incision of skin and cranium). For the most part, patient movement can be prevented by ear bars and other securing measures. Certain intraocular surgeries may require a central eye position and an NMB may be needed to facilitate this. However, human intraocular surgery is performed on awake patients using a local anesthetic block of the orbit. Presumably, in the anesthetized animal, a skillful orbital local block will assure the central eye position as well as provide excellent preemptive analgesia and there will be no need for NMBs. If, however, NMBs are required, then the following points should be considered:

- The investigators should have experience with anesthesia and surgery in the species used and should be using a familiar anesthetic regimen that is effective for causing insensibility to pain and lack of awareness *on its own for that surgery* in the absence of an NMB.
- The NMB should not be administered until after the anesthetic has reached a stable level and surgery has commenced. If practical, the NMB should be allowed to wear off periodically, so that somatic reflex responses can be assessed before additional doses of NMB are administered.
- The heart rate and blood pressure should be monitored and elevations (15–20% or more) of either in response to painful stimuli indicate the need for additional anesthesia. It is important to note that this monitoring technique is not fully reliable and awareness in humans can occur without major changes in these variables.[105] With some anesthetic regimens (e.g., a volatile anesthetic such as isoflurane), monitoring the electroencephalogram can be of value, but this technique also can prove unreliable.[57] (See 18:1.)
- If inhalant anesthesia is used, an end tidal agent monitor (which monitors the exhaled anesthetic gas concentration as an indication of the central nervous system concentration) is extremely useful to monitor and document that the animal's anesthetic level was in the range in which insensibility is typically present. End tidal agent measurements are currently only feasible in larger animals (rabbit and larger).
- The IACUC should scrutinize the anesthetic records of cases in which NMBs are used to determine whether there is documentation of all anesthetic techniques and indication that proper humane use of these agents occurred.

16:27 If an NMB is used along with an anesthetic, what might the IACUC request to help assure that the animal is anesthetized and not simply immobilized?

Reg. (See 16:26.)

Opin. Some common situations in which NMBs are used include neurophysiological and imaging studies. In many of these studies, a very light plane of anesthesia must be maintained to minimize interference between the anesthetic regimen and the study protocol. It is precisely in these circumstances that inadvertent production of inadequate anesthesia is most likely. Some investigators have suggested that after completion of surgery, the surgical wounds can be infiltrated with local anesthetic, general anesthesia discontinued, and the animal immobilized because of the effects of the NMB drug (although it must receive ventilatory support). This proposal raises two concerns. First, it is difficult to ensure the adequacy of local anesthetic blockade—either initially or later when the NMB has been administered. Second, even if the animal is pain free, it may become distressed if exposed to other stimuli (e.g., odors, sounds) and is unable to react because it is paralyzed. General anesthesia is used in animals not only to provide insensibility to pain but also to induce unconsciousness, which helps prevent the distress caused by physical restraint and experimental manipulations. A paralyzed, conscious animal, even if pain free, is likely to experience considerable distress. If NMBs are to be used during surgery or painful procedures, then the monitoring techniques that can be applied are those described in 16:26.

16:28 Is the use of a local anesthetic to perform a minor procedure considered to be alleviation of pain by anesthesia?

Reg. (See 16:3.)

Opin. Yes. Local anesthetics offer a valuable alternative to general anesthesia for workers carrying out a range of different procedures. In some species, the use of regional anesthesia produced by local anesthetics can be considered the method of choice (e.g., a paravertebral block to carry out abdominal surgery in cattle).[71,106] Provided the administration is carried out competently, pain can be prevented either by local infiltration of drugs such as bupivacaine, infiltration around nerve trunks, or epidural or intrathecal administration. Experience with both humans and other animals suggests that these techniques are useful, provided any distress caused by physical restraint or other (nonpainful) procedures can be controlled.[107] The duration of local anesthetic action varies with the agent and additional local anesthetic or other analgesic administration may be necessary to maintain alleviation of pain. In addition, local anesthetics can cause momentary but significant pain on injection. Techniques to reduce this pain involve the use of sedatives or tranquilizers (which by reducing anxiety or the level of consciousness can act to raise the pain threshold), buffering of acidic solutions, use of smallest needle possible, and other methods.

16:29 What records related to the use of anesthetics and analgesics should be maintained and reviewed by the IACUC?

Reg. The AWAR (§2.35,a; §2.35,f) require that all records and reports directly relating to proposed activities involving animals (i.e., those elements of research, testing, or teaching procedures that involve the care and use of animals [AWAR §1.1, Activity]) and proposed significant changes in ongoing activities reviewed and approved by the IACUC are to be maintained for at least 3 years after the completion of the activity. PHS Policy (IV,E,2) has the same requirement. APHIS/AC Policy 3 refers to maintaining animal health records for at least 1 year after the disposition or death of the animal; nevertheless, the policy notes that a longer retention period may be needed to comply with other applicable laws or policies.

Although anesthesia and analgesia records are not specifically required by the AWAR except for marine mammals (AWAR §3.110) the concept of adequate veterinary care (AWAR §2.33; *Guide* pp. 12, 14, 105) strongly implies the need for appropriate anesthesia and analgesia records. The *Guide* (p. 115) states that medical records are a key element of the veterinary care program.

Opin. Records of analgesic use should vary, depending upon the type and complexity of the procedure. As a minimum, the anesthetic drugs or analgesic drugs, dosage, time, route of administration, and effect should be recorded, together with details about the experimental animal (age, weight, sex, strain, etc.). Increasingly, provision of this information will also be required by journals.[34,35] If surgical procedures are to be carried out, the investigator should note the adequacy of anesthetic depth and the way this was assessed. If additional doses of drugs are given, the time and route of administration and the effect should be recorded. (See 27:21.)

The researchers also should record the duration of anesthesia, the recovery time to sternal recumbency, any morbidity, any unexpected adverse effects (e.g., vomiting), and when full recovery commenced (normal activity, feeding, drinking, etc.) and indicate how this assessment was made. As discussed in 16:26 and 16:27, it may be desirable to keep a continuous recording of heart rate and blood pressure, and this is mandatory if neuromuscular blocking agents are used. These data permit a critical review of anesthetic practices and should be considered an essential part of any scientific protocol, because interactions between anesthetic complications and the study objectives can readily occur.

When using analgesics, the drug, dose, and route of administration should be recorded. Any additional doses given also should be noted. It may be considered helpful to link this information with the pre- and post-procedure observation of variables such as body weight and clinical appearance.

In many instances it is difficult to discern whether a medical condition has emanated from study or nonstudy causes and therefore it is prudent to maintain all medical records for at least 3 years after the euthanasia or other disposition of the animal.

Surv. Does your IACUC require investigators to maintain anesthesia and analgesia usage records (other than notes on cage cards) for rats and mice? This refers to animal/cage identification, dates, dosages, drug used, effects, and any other items considered important by your IACUC.

- Not applicable 29/290 (10.0%)
- No 51/290 (17.6%)
- Yes, always 139/290 (47.9%)
- Sometimes, depending on the details of the procedure 69/290 (23.8%)
- Other 2/290 (0.7%)

16:30 Should the use of anesthetics and analgesics be based on a strict dosage schedule or on varying dosages as determined by sound clinical judgment?

Reg. The AWAR (§2.33,b,4) require the AV to guide the PI and other personnel in the appropriate use of anesthesia, analgesia, and immobilization. The AWAR (§2.33,b,5) give the AV additional authority, stating that adequate pre- and post-procedural care shall be in accordance with current established veterinary medical and nursing procedures. APHIS/AC Policy 3 states that the AV has the authority to alter post-operative care due to unexpected pain or distress but the IACUC must approve a significant change to the protocol if the AV requests to alter post-operative care for the remaining animals.

PHS Policy (IV,A,3,b,1) states that a veterinarian will have direct or delegated authority and responsibility for activities involving animals. The *Guide* (p. 120) notes that an integral component of veterinary medical care includes effective programs for the alleviation of pain. PHS Policy (IV,A,1) requires institutions to use the *Guide* as a basis for developing and implementing an animal use program.

Opin. Anesthetics and analgesics should be given at least initially according to dose schedules. These are determined from the scientific literature but are often modified because of variations in response by different strains of animals. Administering a standard dose of anesthetic may produce the desired effect, but it also can cause animals to appear to be too deeply anesthetized or inadequately anesthetized. After assessing the response to an anesthetic for the particular strain, age, and sex of animal to be used in a study, a more appropriate dose schedule can be developed. The variation can be substantial. For example, when using the duration of unconsciousness as an indicator of anesthetic efficacy, doubling of sleep time can be observed in different strains of mice.[108]

It is relatively easy to adjust anesthetic drug doses to suit particular groups of animals, but more difficult to do this with analgesics because of our limited ability to assess postoperative pain. There is contention about whether a PI can propose to "give analgesics as needed" to treat pain. "As needed" dosing strategies are effective in humans, only when they can self-administer strong analgesics by means of patient-controlled analgesia devices, or by oral dosing, as they can determine when additional analgesics are needed and can titrate the dose accordingly. "As needed" analgesic dosing strategies in animals suffer from two major limitations: First, whether it will be possible at all to detect significant pain in the individual animal, and second, whether timing of the assessments will be frequent enough that lengthy periods of substantial pain are prevented. (See 16:54.) If pain assessment is not performed skillfully, then no amount of vigilance will suffice for preventing pain in that animal.

For nonrodent mammalian species, the literature and experience with preventive and multimodal analgesic techniques are rapidly expanding. Many authors recommend the administration of analgesics prior to incision as part of a balanced anesthetic technique, with the continuation of analgesia during the postoperative period. Fixed dose administration for 12–24 hours postoperatively, combined with monitoring for pain between doses, extending the duration of treatment for procedures that are felt to cause moderate or severe pain, and using modalities that overlap in terms of analgesic coverage are advocated in many current veterinary textbooks on pain management.[47,88,109] In our opinion, a problem with fixed dose schedules is that they may lead to insufficient pain relief. In fact, to be evaluated properly, an animal's pain should ideally be assessed in the interval between doses and just prior to redosing in order to detect whether there are adequate duration and magnitude of effect. An argument for the use of a fixed dose schedule, with monitoring and ability to intervene between doses for the first 24–48 hours post-surgery, is that this method rarely leads to complications of excessive analgesic side effects. It also has the benefit of assuring that if pain assessment skills are limited, the animal will be given at least a baseline of coverage for pain.

A general inadequacy of postoperative pain relief in humans, among other factors, was often ascribed to use of rigid dosing schedules with no inherent flexibility.[110] Individual animals will vary in their response to surgery. Therefore, dose rates obtained from the literature are a helpful starting point, but every attempt should be made to assess the adequacy of analgesia in each individual animal, in case a particular animal requires more analgesia than others. Even if retrospective measurements such as body weight (see 16:24; 16:25) are used, this permits variation of the "standard" dose in subsequent studies. In some species (e.g., dog, goat) complex behaviors, human–animal interactions, or animal–animal interactions are relatively well understood, and these behaviors can be used as baselines for postoperative observation to construct scales for the quantization of pain. Evaluation of pain in rodent and other "prey" species by observing behavior is more challenging: animals may be inactive because of photoperiod, observer presence, or pain. This is an area in need of study and educational efforts.

One major barrier to the provision of analgesia in the immediate postoperative period is that the dosing interval for opioid analgesics is shorter than the period animals are typically left without monitoring or treatment (e.g., overnight). This should be recognized, and if longer-duration techniques can be used (e.g., transdermal fentanyl patch) or the overlap of long-duration agents (e.g., nonsteroidal anti-inflammatory drugs) is not sufficient for analgesic throughout the night, then overnight monitoring and treatment, or other scheduling modifications, will be necessary in order to meet regulatory and ethical requirements.

Surv. Does your IACUC require overnight monitoring of animals when a procedure has been performed and additional analgesia or observations are likely to be needed during the night?

- Not applicable 58/296 (19.6%)
- No 16/296 (5.4%)
- Not usually 44/296 (14.9%)
- Yes, usually 115/296 (38.9%)
- Yes, but only for species covered by the Animal Welfare Act 3/296 (1.0%)
- Yes, but only if requested by our veterinarian and approved by the IACUC 46/296 (15.5%)
- Yes, but only for Animal Welfare Act covered species and when requested by our veterinarian and approved by the IACUC 6/296 (2.0%)
- Other 8/296 (2.7%)

16:31 If the use of anesthesia might impair the survival ability of an animal to be released in the wild as part of a field study, can the IACUC appropriately approve procedures that use either no anesthesia or anesthetic regimens that do not produce complete pain alleviation?

Reg. Neither the AWAR nor the PHS Policy has pain relief exemptions for animals used in field studies. A field study is defined in the AWAR (§1.1) as any study conducted on free-living wild animals in their natural habitat. It excludes a study that involves an invasive procedure or harms or materially alters the behavior of an animal under study. The PHS Policy does not have a specific definition of a field study but notes that veterinary and IACUC input may be needed for field studies involving anesthesia (*Guide* p. 32). When anesthesia is required for a field study the IACUC must assure that the proposed study is in accord with the *Guide*.[111]

Opin. The use of animals in field studies poses particular difficulties, since it is usually intended that the animal will survive the study and resume its normal activities. Anesthesia may impair this process. However, the introduction of reversible anesthetic regimens has greatly improved management of wildlife anesthesia. Similarly, in smaller species, use of modern inhalational agents (e.g., isoflurane) can result in very rapid recovery with few significant aftereffects. The use of potent inhalational agents such as isoflurane in simple induction chambers is not generally recommended, as dangerous concentrations of anesthetic are produced.[66] However, in field conditions at moderate to low environmental temperatures (<15°C) it is possible to use them, provided the animals are observed closely for signs of overdose. Finally, the use of local anesthetics can be considered as a means of minimizing an animal's experiencing of pain. If all these options have been considered and rejected as impracticable or ineffective, then an ethical judgment must be made as to whether the aims of the project outweigh the (presumably momentary) pain or distress caused by the required manipulations. Alternatively, if the project seeks only to obtain tissue samples from an animal, a judgment must be made as to whether the procedure should be allowed without adequate anesthesia and the animal allowed to recover, or the procedure carried out with anesthesia and the animal euthanized rather than released. (See 15:10; 16:8.)

16:32 What is the maximal practical length of time that the IACUC should allow a rat to be deprived of food or water? How do these suggestions change for a mouse, dog, nonhuman primate, or other common species?

Reg. The AWAR (§2.38,f,2,ii) state that short-term withholding of food or water from animals is allowed when specified in an IACUC-approved activity that includes a description of monitoring procedures. APHIS/AC Policy 11 uses food or water deprivation beyond that necessary for normal presurgical preparation as an example of a procedure that may cause more than momentary or slight distress. PHS Policy (IV,A,1) requires institutions to follow the *Guide* (pp. 30–31), which addresses food or fluid restrictions. The *Guide* (pp. 30–31) specifically notes that restriction for research purposes should be the least restriction necessary to achieve the scientific objective while maintaining animal well-being. The *Guide* (p. 31) also indicates that records should be established to monitor physiologic or behavioral indices, including criteria for temporary or permanent removal of an animal from the experimental protocol. NIH/OLAW FAQ F.19 also discusses food and fluid restrictions. (See 16:58.)

Opin. Food and water deprivation can be carried out for many reasons, and attempting to establish maximal periods of deprivation without reference to the aims of a particular study is undesirable. For example, a project might be judged so important in terms of its potential benefits that very prolonged periods of food deprivation could be sanctioned if this were considered a necessity. At the extreme, an IACUC might consider whether any project could justify withdrawal of food until an animal dies as a result of this procedure.

Often, only relatively minor periods of food and water deprivation are required, for example, to ensure an empty stomach or gastrointestinal tract or to induce a catabolic state in the animal. In each study, it is important to determine the minimal period of deprivation needed to achieve the desired objective. Often 16- or 24-hour periods are chosen, as these are a convenient interval for removal of the food or water at the end of a working day, followed by use of an animal the following morning. In the case of small rodents, this period may be excessive and may induce unnecessarily severe effects because of the high metabolic rate of these smaller species. It has been shown that when presented with a limited quantity of food (e.g., half the amount normally consumed overnight), rats eat normally until the food is exhausted. Thus, by reducing the amount of food placed in their food hopper, effective food removal (e.g., from 3:00 A.M.) can be achieved. This is sufficient to produce an empty stomach.[112] Investigators should note, of course, that coprophagy occurs in these and other species, so complete food deprivation can only be achieved by combining fasting with the use of an anal cup to prevent ingestion of feces.[113]

A second aspect of the animal's normal biological processes also should be considered. Some species (e.g., rats) normally feed only in the dark phase of their photoperiod. If food is withdrawn overnight and a procedure carried out the next day, and if that procedure causes adverse effects, the rat may not eat the following night or the next day. As a result, 48 hours of fasting may have inadvertently been produced. This may have consequences for the particular study, as well as for the welfare of the animals concerned.

When dealing with larger species, longer periods of food withdrawal might be needed and, generally, are better tolerated than in smaller animals. If food is being withdrawn solely to reduce the risk of vomition during induction or recovery after anesthesia, a period of 8 to 16 hours is appropriate for dogs, cats, ferrets, or nonhuman primates, but unnecessary in rabbits and rodents as these latter species do not vomit. Withholding of water is not recommended for more than a few hours in monogastric species. Withholding of food in pigs can be helpful in reducing the volume of gut contents for abdominal surgery and in ruminants a minimum of 12 hour withholding of food and water may reduce some of adverse consequences of anesthesia, such as regurgitation, and rumenal tympany.[66]

16:33 Is it appropriate to use electric shock to stimulate animals to run or walk on a treadmill?

Reg. PHS Policy (IV,C,1,a) requires that procedures with animals avoid or minimize discomfort, distress, and pain, consistently with sound research design. The AWAR (§2.31,d,1; §2.31,d,1,i) state that unless acceptable justification for a departure from the standards of care in the AWAR is presented in writing, procedures involving animals will avoid or minimize discomfort, distress, and pain to the animals. APHIS/AC Policy 11 cites inescapable noxious electrical shock as a procedure that may cause more than momentary pain or distress. The AV must be consulted on procedures which cause greater than slight or momentary pain or distress (AWAR§2.31,d,1,iv,B; AWAR §2.33,b,4).

Opin. When using any conditioning stimulus, it is desirable to use a reward system rather than a mild or moderate noxious stimulus such as electric shock (*Guide* p. 31). When examining a proposal for the latter, an IACUC should require evidence to show that an aversive stimulus is the only technique that can be used to produce the required behavior in the animal. If it is concluded that aversive stimuli are the only practical conditioning stimuli, then the least aversive and potentially tissue-damaging stimulus used to motivate animals should be chosen (e.g., a puff of air, sound, or vibration) over one that is overtly pain producing.

16:34 Can a mammalian fetus feel pain? If so, at what age can a rodent fetus be presumed to feel pain? Should it be included as a vertebrate animal when the IACUC considers the number of animals requested for the study?

Reg. (See 12:1; 13:11.)

Opin. The concern relating to fetal and neonatal pain has emerged largely from studies in human infants. Prior to the mid-1980s, many procedures were undertaken on human infants without effective anesthesia or analgesia. A series of studies indicated that human neonates can experience pain,[114,115] and investigations in neonatal rodents have shown not only that pain can (or at least responses to noxious stimuli) be demonstrated, but that these experiences produce long-term changes in the nervous system. The stage at which these abilities become functional during fetal and neonatal development is still under debate, and case-by-case interpretation may be needed as the extent of mammalian neurophysiological development differs, depending on the species. In rodents and other species, anatomical studies suggest that *nociceptive* responsiveness is present in the second half of gestation, but as reviewed, nociception does not equal pain perception (see 16:2).

Anesthesia is being advocated for human fetuses over the age of 20 weeks.[115–118] It is debatable whether this anatomical development translates into a capacity to experience pain. Some authors have put forth the suggestion that consciousness begins at birth and that the experience of pain requires consciousness; others disagree.[76,115,119–121] Neurologic maturation, indicated by EEG "differentiation" may dictate the point at which pain can be perceived.[120,122,123] Some recent veterinary reviews suggest that on the basis of available evidence, the more neurologically advanced at birth the fetus of a given species is, the more likely it is to feel pain prior to birth, but that in the case of rodents and rabbits, and other species with poor maturation at birth, pain may not be perceived until after birth for hours to days.[120,122–124] It seems appropriate to be concerned that some mammalian fetuses have some capacity to experience pain in late stages of gestation and adapt research protocols accordingly. The European Union Directive, issued in 2010, states (p. 39) that pain should be considered in mammalian fetuses from the last 1/3 of gestation.[125] It also seems safest to include the fetus when considering the number of animals to be approved in an IACUC submission. Of course, practical considerations should be noted in that an investigator cannot reasonably be expected to predict exactly how many fetal animals in a multiparous species may be present in each pregnant female. Reviews of pain in the fetus and neonate may be of value to IACUC members.[120–124] (See 8:11; 13:11; 14:19; 17:31.)

16:35 Can unhatched avian embryos be presumed to feel pain? Is IACUC approval needed for the use of avian embryos?

Reg. NIH/OLAW has interpreted the term *live vertebrate animal* to apply to avians (e.g., chick embryos) only after hatching.[126] See also NIH/OLAW FAQ A.4. Birds bred specifically for research are not currently regulated under the AWAR.

Opin. Answering this question is difficult when dealing with avian species, since we have a very limited knowledge of nociception and pain perception in these animals. Although it seems illogical to conclude that the capacity to experience pain emerges at the instant of hatching, we cannot as yet determine the stage of development at which this capacity is sufficiently well developed to warrant concern.[127] In the United Kingdom, arbitrary limits were set by legislation enacted in 1986, and this pragmatic approach may be the best way forward for IACUCs: that is, acknowledge that pain may occur, acknowledge our uncertainty as to the stage of development at which this occurs, and set some initial guidelines that can be reviewed as our understanding of avian neurobiology improves. This same approach would allow studies on early embryonic stages to proceed without IACUC approval, but would require approval once a particular developmental stage has passed. (See 14:23; 16:34.)

16:36 Is hypothermia an acceptable form of anesthesia for fetal or neonatal homeothermic animals?

Opin. Hypothermia produces immobility and apparent insensibility in neonates and fetuses and has become a well-established means of "anesthetizing" neonatal rodents. At low temperatures nerve conduction slows and may be blocked, and depression of body systems can produce unconsciousness. Nevertheless, to date no convincing studies of neonatal central nervous system responses to noxious stimuli during hypothermia have been carried out. In addition, it has been suggested that since rewarming from hypothermia is associated with pain in humans,[128] the technique may be undesirable even if it produces a state of insensibility. Perhaps a more constructive approach is to examine the alternative anesthetic techniques available for use in neonates. Because volatile and injectable anesthetics can be used to produce safe and effective anesthesia,[129,130] it seems reasonable for an IACUC to ask why it is necessary to use a questionable technique when more acceptable alternatives are available. Investigators who have used all of these techniques often find the use of conventional anesthesia more convenient, especially for prolonged surgical procedures, since the neonates do not need to be maintained on an ice pack for the duration of surgery.

Surv. Does your IACUC typically accept hypothermia as an acceptable form of anesthesia for neonatal mice or rats? Check as many boxes as appropriate.

• Not applicable	75/295 (25.4%)
• Yes, up to 5 days of age	40/295 (13.6%)
• Yes, up to 10 days of age	7/295 (2.4%)
• Yes, up to 14 days of age	2/295 (0.7%)
• Yes, and requires an explanation as to why other anesthetics are not acceptable	37/295 (12.5%)
• No	142/295 (48.1%)
• Other	4/295 (1.4%)

16:37 Is hypothermia an acceptable form of anesthesia for poikilothermic animals?

Reg. PHS Policy (IV,C,1,b) and the AWAR (§2.31,d,iv) state that procedures causing more than slight or momentary pain or distress will be performed with appropriate sedation, analgesia, or anesthesia unless otherwise justified in writing by the PI.

Opin. Similar concerns to those described in 16:36 relate to the use of hypothermia as a means of "anesthesia" in poikilotherms. As with mammalian neonates, well-established alternative anesthetic regimens are available and it, therefore, seems unnecessary to use a questionable technique. A review of hypothermia for reptiles and amphibians, although not recent, may be of interest to IACUC members.[131]

16:38 A PI proposes to administer a small amount of nonirritating fluid daily for 1 month by intraperitoneal injection into hamsters. The injection is performed by a skilled person and the sites will be rotated. Is this APHIS/AC pain/distress Category C or Category E?

Reg. (See 16:5 for a discussion of the APHIS/AC pain/distress categories.) The *Guide* (p. 26) states that "If little is known about a specific procedure, limited pilot studies… conducted under IACUC oversight, are appropriate." (See 13:6; 16:60.)

Opin. The essential question is: does the procedure cause momentary pain or distress, or does it cause more substantial pain? The investigator and IACUC are unlikely to be able to answer this question at the outset. Secondary questions include: on what basis can the fluid be deemed "nonirritating"? In other words, has histopathologic exam of the peritoneum been performed and does the animal indicate by behavioral cues that the effect of the injection procedure is transient? Also, what is the potential for complications—the risk of penetrating an abdominal organ, which would be associated with greater pain or illness? These questions can only be answered in a prospective manner; by observing of animals after a single and then after multiple days of injections, by body weight monitoring, and by comparisons to age-matched (and ideally sham-handled) animals. The degree to which a relatively minor noxious stimulus can lead to pain depends on whether inflammation occurs and on the "emotional" state of the animal. In a recent poster presentation it was reported that 30 daily intraperitoneal injections into CD-1 mice resulted in no difference in the complete blood count and behaviors, and there were minimal focal histopathological lesions from the needle punctures that did not appear to be clinically significant.[132] Following animals by daily monitoring of behavior and weight can aid in detecting the magnitude of the procedure's impact on them; use of operant conditioning (provision of a food reward) may aid in reducing some of the averseness by means of exerting a descending modulation of nociception to a less painful experience. Once this information is known, future categorization can be done with more certainty, and the initial study can be retrospectively characterized as C or E.

16:39 A PI proposes to administer a small amount of nonirritating fluid daily for 1 month by oral gavage to unanesthetized and untranquilized hamsters. The gavage is performed by a skilled person. Is this APHIS/AC pain/distress Category C or Category E?

Reg. (See 16:5; 16:38.)

Opin. Oral gavage has the potential to cause irritation and inflammation of the esophagus, which might be expected to result in lower weight gain and food consumption. Body weight curves, posture and swallowing behavior during voluntary oral intake are likely to be very indicative of the presence of this pain. If there is a well trained and experienced person performing the gavage, Category C would be appropriate. (See 16:38.)

16:40 An animal is given general anesthesia for several hours to permit a prolonged but noninvasive imaging procedure to be performed. Should this type of restraint be considered to be a distressful procedure for which anesthesia was used to relieve the distress?

Reg. (See 16:5 for pertinent regulations that refer to the relief of either pain *or* distress.) (See 16:1–16:2 for definitions of *distress*.)

Opin. APHIS/AC Policy 11 includes immobility as an example of a procedure that may lead to more than momentary or slight distress in a conscious animal. However, if the animal is anesthetized for the duration of the procedure, the potential distress has been alleviated by the appropriate use of a drug, as noted in the regulations. Assuming appropriate IACUC approval has been gained and no other experimental manipulations are involved, this would be reported as Column D on the APHIS/AC annual report for species regulated by APHIS/AC because restraint may cause pain or distress. However, if the positioning of the animal during the imaging procedure is such that it causes tissue injury, then additional monitoring would detect signs of this and require amelioration by use of analgesics or prevention by altering the method of positioning.

16:41 An inhalation anesthetic is used to restrain an animal briefly for a painless procedure (such as an intranasal inoculation). Does this fall under APHIS/AC annual report Category C or D?

Reg. (See 16:5. See 16:1 and 16:2 for definitions of distress.)

Opin. (See 16:42.) The primary advantage of using sedatives or anesthetics to immobilize animals for the performance of nonpainful procedures is that the drugs are used in lieu of more forceful or injurious physical restraint, which would likely be painful or distressful to the animal. Therefore, the IACUC must evaluate the rationale presented by the PI for using an anesthetic. For example, if the IACUC agrees with the PI that the extent of physical restraint needed to make an accurate intranasal inoculation is likely to distress the animal more than momentarily, then the use of an anesthetic (or perhaps a sedative) to reduce the need and consequences of that extreme restraint is appropriate. Alternately, the PI might argue that anesthesia is required to have the animal deeply inhale the intranasal inoculation and that other means of assuring lung delivery would be more distressful or painful. With either explanation, the IACUC must agree with the stated need. Although using an anesthetic or sedative might be considered a refinement of the experimental technique, it does not change the fact that regulated animals which receive drugs for the purpose of alleviating pain or distress are required under the AWAR to be placed in Category D.

To help assure there is no confusion as to when animals can receive general anesthesia yet still be included in Category C, APHIS/AC has provided the following clarification:

> The IACUC has the latitude to consider anesthetizing or tranquilizing an animal for radiographs, MRI, or similar procedures as Category C if it determines that only minimal pain or distress is involved with the procedure. If, however, the IACUC determines that any needed restraint process is itself painful or very distressful requiring medications (sedatives, tranquilizers, anesthetics) for alleviation, then Category D is appropriate.

16:42 What limits should the IACUC place on chronic restraint?

Reg. The PHS Policy (IV,C,1,a; IV,C,1,b; U.S. Government Principle IV) and the AWAR (§2.31,d,1,i; §2.31,d,1,ii) note that the IACUC should determine that a proposed activity prevents or minimizes discomfort and distress, and that alternatives to painful or distressful procedures have been considered. The *Guide* (p. 29) is more specific. It states that prolonged restraint (including chairing of nonhuman primates) should be avoided unless it is scientifically essential and is approved by the IACUC. It suggests the use of less restrictive systems, such as a backpack-fitted infusion pump for nonhuman primates and free-stall housing for farm animals. The *Guide* (pp. 29–30) provides advice for restraint such as: restraint time should be the minimum required to accomplish the research objectives; animals should be acclimated to the restraint device; animals that fail to acclimate should be removed from the study; restraint devices should not be used as no more than a convenience in handling animals, and so forth. Specific requirements for the use of restraint devices in nonhuman primates can be found in the AWAR (§3.81,d).

Opin. Chronic (prolonged) restraint can be required for a variety of purposes but is usually needed when administering compounds or sampling body fluids via implanted catheters or obtaining continuous recording of physiological variables. The need for prolonged physical restraint has been reduced by the development of harness and swivel devices that allow an animal some degree of movement, and implantable telemetry devices or ambulatory infusion systems that allow complete freedom of movement. While these devices are refinements that may reduce the need for restraint and any accompanying pain and distress potential, there are some caveats. The weight of, and the surgical procedure required to insert telemetry implant devices, can inflict a metabolic burden or postoperative pain. Tethers and metabolism cages allow freedom of movement but require animals to be isolated from conspecifics and kept in barren environments. Boredom and social isolation can also cause distress. The effects of environmental stressors on many experimental variables, including toxicity of substances, prenatal developmental biological processes, and cardiovascular, neurologic, and immunologic parameters have been well documented, and indeed, restraint methods are used as an experimental tool to study the effects of stress.[8,10,12,14,133–136]

When considering protocols that require restraint (or any of the refinements mentioned), an IACUC should question whether physical restraint is needed or whether an alternate approach can be adopted. This must be balanced by an appreciation that these alternate systems may still cause some pain, distress, or discomfort to an animal, both after implantation and during the earlier conditioning period. If a study can be completed with (for example) acclimated physical restraint of 1 or 2 hours, this might be preferable to the use of a tether system. An IACUC member may easily understand the discomfort of prolonged positioning without the ability to change posture by trying to sit very still in the chair during a meeting.

As the period of restraint increases, the balance between different systems changes. It is important to determine whether the animal can be readily trained to accept physical restraint. Measurement of stress-sensitive indices, such as blood glucose concentration, heart rate, or behavioral stereotypy, suggests that animals can be acclimated or conditioned to accept restraint in slings or other restraint devices.[137,138] Problems arise, however, when physical restraint is carried out for prolonged periods in such a manner that the animal is unable to eat, drink, or carry out any normal behaviors, such as grooming.

The issue of restraint, therefore, is not simply one of "how long," but critically depends on the method used, the prior experience of the animal, and the availability of alternative, less stressful systems.

16:43 A research project with hamsters requires 0.5 ml of blood once a week for 6 months. Postorbital blood collection, under anesthesia, is approved by the IACUC. Even with a skilled technician, is it reasonable for the IACUC to assume pain or distress will occur over time from so many collections?

Reg. Neither the AWAR nor the PHS Policy directly addresses postorbital blood collection. (See 16:3–16:5; 16:44.)

Opin. Abnormal behavior indicative of pain was studied in rats after a single postorbital puncture under anesthesia. Using intensive monitoring, the duration of the impact of the procedure appeared to be only about 5 hours.[139,140] Repeated tissue injury may cause altered threshold to pain sensitivity, and repeated sampling might cause increased pain over time. The severity of impact on an individual animal might be measured by alterations in body condition score, body weight, and of course, gross appearance of the eye. Alternating between the left and right eyes would reduce the number of times a single eye was sampled, lowering the overall potential for induction of a painful state.

16:44 Does postorbital blood collection from rodents normally require anesthesia, analgesia, or tranquilization?

Reg. Neither the AWAR nor the PHS Policy directly addresses postorbital blood collection.

Opin. (See 16:43). Postorbital (retroorbital) blood sampling in rodents can provide moderate quantities of blood quickly and easily and when carried out expertly seems to cause a minimum of long-term complications. When it is carried out less competently, injuries to the globe can occur and these may be severe.[141] The procedure itself may be no more painful than peripheral venipuncture (e.g., of the tail vein). Nevertheless, many research units only carry out the technique with anesthesia, both to reduce any pain and to minimize inadvertent injury should the animal struggle during the procedure. At least one study concluded that anesthesia of rats with either carbon dioxide or isoflurane did not confer benefits to rats having periorbital blood sampling, as the stress of the anesthesia added to the recovery.[140] The procedure is undoubtedly stressful, and perhaps a more appropriate question to ask is, Why is the retroorbital plexus to be used, rather than alternative techniques such as the facial vein or saphenous vein?[140,142–144] (See 16:4; 16:5; 16:43.)

16:45 Relative to pain or distress, how many episodes of blood collection by the postorbital sinus or postorbital plexus method should the IACUC reasonably allow on a single animal?

Opin. A reasonable response to this question is for an IACUC to require a report on the incidence of complications (corneal abrasions, retrobulbar hemorrhage or abscessation, damage to the globe, etc.) produced by the technique. Similar assessments of sequelae to other methods of venipuncture can lead some IACUCs to conclude that repeated orbital sinus puncture is acceptable, while others may conclude that it should never be used. Very few studies have accurately assessed the incidence of complications associated with this procedure, and all guidelines appear based on a compromise between an investigator's wishes and an IACUC's unease at the effects on the animal. The authors urge the IACUC to require follow-up (including postmortem histologic evaluation of the globe and periorbital tissues) when allowing use of this technique. It then is possible to make an assessment based on data rather than opinion. (See 16:5; 16:43; 19:9; 19:10.)

16:46 For performing procedures such as Southern blots or polymerase chain reactions, can the tail of a rat or mouse be clipped without causing more than momentary pain to the animal? If yes, what length of tail on mice?

Opin. Removal of a small portion of the tail has become an established procedure in many institutes that work with transgenic animals. There is little doubt that removing the tip of a sensitive structure causes pain. A number of studies have been done to examine the short or long term effects on rodents and whether anesthesia or analgesia is effective.[144–153] The conclusions vary but there is clearly evidence for an effect that is more than momentary; however, handling of animals also causes short term effects on behavior and physiology. Removal of longer portions (25 mm) caused long term (at 3 weeks) sensitivity compared to smaller portions (2.5 mm).[149,150] Anesthesia without lasting analgesic effects may have more detrimental impact on mice than performing the procedure without anesthesia.[144,146,147,153] A consensus may shortly begin to emerge as the results of studies of mice of varying age and of technique (i.e., amount of tail removed) accumulate. These studies have used an assortment of assessment methods, such as physiological, behavioral and persistent pain hypersensitivity as well as histological evaluation. Currently, different institutions vary in their opinion as to whether the procedure requires anesthesia and whether anesthesia should be local or general. One laboratory carries out the procedure with brief general anesthesia (with isoflurane) and provides postprocedure analgesia (with carprofen). An alternative technique described by Meldgaard and associates,[152] which uses swabs of buccal cells, could be attractive in that it replaces a surgical procedure with potential for pain. However, a subsequent study found that the physiologic and behavioral changes that result from both invasive (i.e., cutting tissue) and non-invasive DNA sampling methods (i.e., hair or mucosal swabs) did not differ, nor did any sampling method differ from those of animals who were simply restrained.[153]

Surv. 1 Does your IACUC permit tail clipping as a means of obtaining tissue for genotyping?

• Not applicable	94/292 (32.2%)
• No	14/292 (4.8%)
• Yes, but only 2 mm or less	64/292 (21.9%)
• Yes, but only 5 mm or less	87/292 (29.8%)
• Yes, but only 1 cm or less	17/292 (5.8%)
• Yes, and <1 cm is acceptable	6/292 (2.1%)
• Other	10/292 (3.4%)

Surv. 2 If your IACUC permits tail clipping for genotyping, at what age do you allow this to be done *without* local or general anesthesia? Assume no more than a 3 mm sample will be removed.

- Not applicable 115/291 (39.5%)
- Less than 7 days of age 23/291 (7.9%)
- Less than 10 days of age 24/291 (8.3%)
- Less than 17 days of age 9/291 (3.1%)
- Less than 21 days of age 72/291 (24.7%)
- Less than 28 days of age 18/291 (6.2%)
- At any age 11/291 (3.8%)
- Other 19/291 (6.5%)

Surv. 3 Does your IACUC require the use of alternatives to tail clipping for genotyping? Check all appropriate boxes.

- Not applicable 110/291 (37.8%)
- We do not require an alternative 150/291 (51.5%)
- We require the use of alternatives, including cheek scrapes 7/291 (2.4%)
- We require the use of alternatives, including saliva analysis 7/291 (2.2%)
- We require the use of alternatives, including ear punch tissue 20/291 (6.9%)
- We require the use of alternatives, including hair analysis 5/291 (1.7%)
- We require the use of alternatives, including toe clips 5/291 (1.7%)
- Other 12/291 (4.1%)

"Other" included project dependent decisions and age-related decisions.

16:47 What are the important issues and what are some general guidelines for the IACUC to consider relative to methods of animal identification?

Reg. The *Guide* (p. 75) states that toe clipping, as a method of identification of small rodents, should be used only when no other individual identification method is feasible although it may be the preferred method for neonatal mice up to 7 days of age. Although the *Guide* lists other methods of identification (e.g., tattooing, ear notching) that can potentially be painful or distressful to an animal, only toe clipping is singled out for additional discussion. Fin clipping, identification tags, etc. can be used to identify aquatic species (*Guide* p. 87). The AWAR (1:1, Painful Procedure) define a *painful procedure* as one that causes more than momentary pain or distress (e.g., as can occur from a needle prick). PHS Policy (IV,C,1,a; IV,C,1,b) has similar wording and the same intent. Identification requirements for dogs and cats are noted in the AWAR (§2.38,g). (See 16:48.)

Opin. Accurate and reliable identification of laboratory animals is essential for most research projects. Methods should

- Be reliable
- Cause minimal pain or distress to an animal

- Be simple to use
- Be standardized, so that the identification system can be readily interpreted by all concerned
- Not impact the well-being of non-captive animals

A further factor is the cost associated with the technique, which assumes greater significance when large numbers of animals must be identified. The IACUC should balance the consequences of failure of an identification system (loss of animals from a study and the need to repeat the investigation using additional animals) against the impact on the animal of the marking technique. For example, the PI should consider whether the technique would impact the ability to survive in the animal's environment. These issues are often debated when an investigator wishes to use relatively low-cost (in economic terms) techniques such as ear punching or toe clipping, rather than more expensive methods such as microchip implantation. Aside from concerns that the former methods are less reliable (fighting can result in loss of further portions of the ear), methods of physical marking result in a minor mutilation of the animal that may cause more than momentary pain. Checking the identification of an animal may cause stress from physical restraint. In contrast, the placement of microchips might be considered to cause only momentary pain and identification can be carried out more easily, often without the need to restrain the animal. Recently, however, a few studies have supported that toe clipping in very young mice does not cause appreciable indications of persistent pain and that a microchip transponder was more aversive than toe clipping.[153–155] The authors of one study suggested that toe clipping might be an alternative that allows identification of animals but also provides a sample for DNA testing. The fact that two such studies have appeared in the literature only recently argues for literature review to take place on a frequent basis to be able to take advantage of the most current knowledge to guide animal care. The impact of a variety of identification procedures in rodents is reviewed in a recent publication.[150]

16:48 Are there any circumstances under which the IACUC would allow a toe to be clipped as a means of animal identification or to obtain tissue for analysis by Southern blot or polymerase chain reaction?

Reg. (See 16:47.)

Opin. A similar approach to that of 16:47 must be taken to this question. Investigators should state why they consider it necessary to remove a piece of tissue if less invasive methods are available. These issues may seem relatively trivial when weighed alongside questions of end points in carcinogenesis studies or subjecting of animals to major survival surgery, but they should be addressed in a way that recognizes our general concern to reduce to a minimum the pain or distress caused to animals used in biomedical research. The pain and distress may be minor, but the numbers of animals involved can be considerable.

Surv. 1 Does your IACUC permit toe clipping of young rats or mice (10 days of age or younger) as a means of animal identification? Check as many boxes as appropriate.

- Not applicable 71/293 (24.2%)
- No, we do not permit toe clipping as a means of identification 128/293 (43.7%)

- Yes, with adequate scientific justification 71/293 (24.2%)
- Yes, with or without justification 9/293 (3.1%)
- Yes, but only up to a pre-specified age limit that is based on available literature 28/293 (9.6%)
- No 142/293 (48.5%)

Surv. 2 Does your IACUC permit toe clipping of young rats or mice as a means of collecting tissue for genotyping?

- Not applicable 100/294 (34.0%)
- No, we do not 104/294 (35.4%)
- Yes, with adequate scientific justification 75/294 (25.5%)
- Yes, with or without adequate scientific justification 10/294 (3.4%)
- Other 5/294 (1.7%)

16:49 What justification should an IACUC expect if an investigator insists that the death of an animal or the death of 50% of her or his animals is the most appropriate end point for the study?

Reg. Neither the AWAR nor the PHS Policy addresses the use of the lethal dose 50% (LD_{50}) test. The U.S. Food and Drug Administration does not require LD_{50} test data to establish levels of toxicity (Federal Register, 53: 39650–39651, 1988). A search for alternatives to potentially painful or distressful procedures is required under the AWAR §2.31,d,1,ii.

Opin. There are two issues to consider here. First is the justification for the use of death as an end point, and second is the use of a 50% mortality criterion. The use of death as an end point has been regarded as essential in some investigations, but rapid developments are being made in this field and it is an area in which both investigators and the IACUC should make particular efforts to keep abreast of the current literature. The usual reason offered for selecting death as an end point is the difficulty of reliably differentiating animals that will die from those that will recover, despite their showing severe clinical signs of illness or toxicity. Surrogate humane end points have been proposed as a means to intervene when death can reliably be predicted from clinical signs or test results. In toxicity studies, death signifies a "cutoff," the dose was lethal, and by setting a 50% criterion, the toxic dose is "defined." In infectious disease, disease prevention, or therapeutic studies, death indicates a lethal infectious burden or failure of therapy. Often, these studies involve species for which detailed monitoring is not typically performed (e.g., rodents). If there is not a peracute onset of morbidity, animals gradually develop progressively more severe abnormalities, commencing with mild depression of normal activity and signs associated with lack of grooming activity (e.g., ruffled fur coat) and finally progressing to coma and death. When the clinical signs include subjectively distressing changes such as convulsions, weakness, cachexia, or severe dyspnea, there is pressure to euthanize an animal rather than allow further deterioration of its condition. As previously mentioned, the difficulty is the fear that some of these animals may recover and early euthanasia might invalidate a test result.

Several constructive suggestions have been proposed to reduce the need to use death as an end point in studies. There is also an alternative to using a strict 50%

lethality criterion for acute toxicity studies that can reduce animal numbers used by 60–70%.[156] Furthermore, it is important to keep in mind that the results of a given test or exposure may be markedly affected by the handling or environment of the animals (see 16:41). This means that careful attention to laboratory conditions, training of personnel, and animal monitoring will play a vital role in experimental validity. Because rodents are typically not given a detailed health evaluation with the frequency that their rapid metabolic rate and comparatively shorter life span dictate, a conclusion that the agent, disease, or intervention caused death could be erroneous. Instead, the animals may have died as a result of weakness and inability to eat. Recognition of what is actually happening physiologically is missed when death is the only criterion evaluated; more detailed tracking of clinical signs or parameters might lead to better information, thereby increasing the discriminatory power of the experiment. Guidelines on the management of animals being used for the assessment of novel antibacterial or antifungal agents have been proposed by representatives of organizations involved in this type of investigation.[157]

In some circumstances, simple clinical indices such as the development of profound hypothermia can be used to predict death reliably. There remain many studies, however, in which no validated criteria have been devised. In these circumstances, an IACUC should carefully assess whether criteria can be developed during the progress of the study. In many instances, failure to develop these criteria may be due to infrequent or inadequate observations of the animals. At other times critical events may occur when personnel are not usually available, thereby precluding detailed observation. There seems no doubt that progress in refining end points has only occurred as a result of PIs carefully evaluating their own particular models. It may be that even after careful assessment, no progress is made and animals must be allowed to die if the aims of the study are not to be defeated. It is essential, however, for an IACUC to be proactive, require genuine attempts to develop nonlethal end points, and not allow investigators to use death simply because it is an unambiguous and easy criterion they can apply. Finally, the adoption of nonlethal end points can have beneficial effects on a study because they allow blood and tissue samples to be taken that may enable added data to be obtained.

16:50 A mammal is born with a genetic defect that leads to an abnormal but nonlethal health condition very soon after birth (e.g., an inability to use its hind limbs). It adapts as well as can be expected to that condition and requires no significant additional levels of husbandry or veterinary care. Should the IACUC consider this condition as nonpainful and nonstressful (Category C on the APHIS/AC annual report) or as unalleviated pain or distress (Category E on the APHIS/AC annual report)?

Opin. There are two possibilities for classifying such studies, and the IACUC should judge each individual circumstance as it arises. In the example, it is assumed that animals do not require additional levels of husbandry to supplement grooming of the affected limbs and there is also no risk of traumatic damage and self-inflicted wounds. Another example is congenital blindness or deafness. Although animals may have no apparent abnormality on clinical examination and require no special husbandry, one might consider that the lack of ability to move about or sensory deprivation causes an increase in the barrenness of the environment to that animal, and therefore a form of deprivation distress. On the other hand, one could argue that since we are unable to detect any apparent adverse effect, except by neurological examination, we should classify this as nonpainful or nondistressful, and therefore Category C. It also may be necessary to reconsider the classification as the animal ages and all aspects of the life cycle are taken into account.

16:51 When a condition such as loss of sight or Parkinson's Disease is acutely induced in a laboratory animal should this be considered distressful to that animal?

Reg. (See 16:1, 16:2, and 16:60 for definitions of pain and distress.) US Government Principle IV states that unless the contrary is established, procedures that cause distress in humans should be considered as potentially being able to cause distress in other animals. Similar wording is used in the AWAR (§1.1, Painful procedure). (See 16:9.)

Opin. Induced conditions in laboratory animals may have different clinical presentations depending on the species of animal, the method used to induce the condition, the rapidity of onset, associated research procedures, and other variables. For example, the induction of Parkinson's Disease with MPTP (1-methly-4-phenyl-1,2,3,6-tetrahydropyridine) in nonhuman primates typically is presented as a clinically more severe condition than when the same chemical is used to induce the same disease in the mouse. The determination as to whether or not an animal is distressed by the induced condition can be subjective but as with the definition of pain (and pain is a form of distress) it is helpful to consider if the same condition would be distressing if it occurred in a human. Thus, if a human became acutely blinded, even if there was no associated pain, would that person be distressed? If a person became acutely Parkinsonian, even with the best of supportive care, would that person be distressed? If, in the opinion of the IACUC, the answer is yes to either or both of these questions then it is reasonable to assume that an animal would also be distressed. (See 16:50.)

16:52 Cats are used to teach tracheal intubation techniques. They are anesthetized, recovered, and used again. What is a reasonable number of times, and at what intervals, might a cat be used for this purpose? How does the IACUC determine this?

Opin. It is more appropriate to say that an animal can be used as often as required for such a procedure, provided it shows no behavioral changes that increase in severity over time; is treated appropriately for and allowed to recover from any pain or disability that might result from the procedure; continues to feed, drink, and grow normally; and does not develop clinical abnormalities. Also, the handling of the animal under anesthesia is critical, as the amount of trauma caused in one session can range from minimal to substantial. Animals should always be thoroughly examined for evidence of laryngeal, oral, or other trauma at the end of the procedure. Familiarizing an animal to the procedure and providing rewards for appropriate behavior and an enriched environment can lead to improved welfare of the cats (there can be no reason not to keep such animals in groups and in pens, rather than individual cages). This welfare enhancement could not easily be achieved when animals are used on a few occasions, euthanized, and a naive animal used for the next series of manipulations. The availability of non-sentient alternatives (in this case models for training intubation) should be monitored closely and a training method, which requires that the effort be "duplicated" for each new group of trainees, that involves a degree of pain or distress should be closely compared to models as they become available.

16:53 A drug to be used in a research project is known to cause tremors or mild seizures that occur once and last for approximately 2 minutes. Should the IACUC classify this as unalleviated distress? In this example, if seizures occur frequently, should the IACUC classify this as unalleviated distress?

Reg. The AWAR (§2.36,b,7) specifically require an annual report indicating painful or distressful procedures for which the use of appropriate anesthetic, analgesic, or tranquilizing drugs would adversely affect the procedures, results, or interpretation of the teaching, research, experiments, surgery, or tests.

Opin. Many drugs produce mild neurological side effects, and when these include seizures, often particular concern over the welfare of the animals being used arises. It is generally perceived that seizure activity can cause postictal pain, particularly when marked muscle spasm or other bodily injury occurs. In the case of seizures that cause a loss of consciousness, there is no evidence that the seizure itself is painful, although we must also consider that anxiety or fear may occur in the subject or in animals cohoused with others who might react to the seizure episode. It is the opinion of APHIS/AC[158] that seizures such as this constitute unalleviated distress.

As in other areas of research using animals, we have no clear evidence regarding the degree of pain or distress caused by seizures. There is evidence that in humans, some seizure activity may occur while the person is consciously experiencing the event, reports of rare cases where individuals report severe pain during the seizure, and a relatively high reported incidence of post seizure headache.[159,160] It could be argued that if the seizure is of a type that results in loss of consciousness and is not associated with potentially distressing aftereffects, then it should not be considered to produce unalleviated distress. There is no way to assess whether an animal is aware during a seizure, thus the IAUC may want to classify it as unalleviated distress unless detailed observations of the animal during and after the seizure appear to indicate that it is unaffected.

16:54 A PI performs a research procedure and states that he will euthanize an animal with a pentobarbital overdose as soon as any pain or distress is detected. Under those circumstances, would the IACUC place those animals in the APHIS/AC annual pain and distress report Category C or D?

Reg. (See 16:5; 16:62.) See 16:1–16:2 for definitions of *pain* and *distress.*

Opin. In this question, pentobarbital will be used for euthanasia, not anesthesia. The central issue is the timing of observations relative to the onset of pain or distress. If, for example, the animal is continuously observed until it shows the first signs indicative of *slight or momentary pain or distress* and then is immediately euthanized, the study might be placed in Category C. This is because Category C is reserved for animals that experience no more than slight or momentary pain or distress and the pain or distress in the example never progressed beyond the "slight or momentary" stage. The fact that euthanasia was performed does not change the

APHIS/AC category to Category D (pain or distress alleviated by drugs) because it should be reported to APHIS/AC the most severe category of pain or distress *actually experienced* by the animal. In this case, it was only slight or momentary (Category C).

In a real world situation animals are rarely observed with such rigor and (as another example) an animal that is observed only twice a day might experience 12–18 hours of *more than* slight pain before the pain is noted by an observer. If this happens, and the observer notes more than slight pain that will not be alleviated due to experimental needs, Category E is appropriate because it is reserved for unalleviated pain or distress that is more than just slight or momentary. However, if the pain is then alleviated by the use of drugs, Category D should be used for that animal. (See 16:30.)

16:55 Should an IACUC allow cardiac puncture without prior anesthesia?

Reg. Cardiac puncture as a means to effect euthanasia by exsanguination (unless there is prior anesthesia, stunning, or sedation) is considered unacceptable by the AVMA euthanasia guidelines because of anxiety associated with extreme hypovolemia.[161] PHS Policy (IV,C,1,g) requires compliance with the AVMA *Guidelines for the Euthanasia of Animals* unless the IACUC approves a deviation. APHIS/AC Policy 3 states that the method of euthanasia must be consistent with the current AVMA euthanasia guidelines. (See 17:2.)

There are no specific regulatory statements concerning cardiac puncture for purposes other than euthanasia, assuming the volume of blood removed is sufficiently small as not to cause hypovolemia. Personnel performing cardiac puncture must be appropriately trained (AWAR §2.31,d,1,viii; U.S. Government Principle VIII; PHS Policy IV,C,1,f).

Opin. The authors know of no evidence to indicate that direct cardiac puncture, in and of itself, is painful when used for clinical purposes (sampling of blood) or delivery of certain types of anesthetic euthanasia solution (e.g., pentobarbital). The entry point on the body surface is likely to be the factor that influences pain—pericardial fluid removal (pericardiocentesis) is accomplished in patients by administering a local anesthetic intercostally with a small gauge needle prior to inserting a longer, larger gauge needle. This is done to decrease the pain of inserting a needle as the intercostal nerves are easily impinged upon, causing marked pain. Clinically, pericardiocentesis is performed under conscious patients with echocardiographic and electrocardiographic guidance, to prevent accidental contact with critical structures inside the thorax. Rather than being sensitive to a needle prick, visceral structures (myocardium and pericardium) are innervated by sensory fibers that are responsive to distention, inflammation, or ischemia. Cardiac puncture can also be accomplished transdiaphragmatically. Insertion of a needle in this manner would be expected to cause no more pain than an intraperitoneal injection. Rapid removal of blood (exsanguination) might cause cardiac ischemia, which can be associated with moderate to severe pain. Removal of a small sample of blood may not be painful per se, but contact with a coronary artery can lead to pericardial hemorrhage and death. Thus, direct cardiac puncture might be useful only when other approaches to sampling blood do not exist. If cardiac puncture is used to deliver euthanasia solution, then there should be a justification for why it cannot be preceded by a technique that will reduce the pain of intercostal injection, or by peritoneal injection, and the administration of intracardiac or intravenous potassium chloride would be prohibited due to the painfulness of this method. If the investigator proposes to use this technique in an unanesthetized or unsedated animal to sample blood, the anatomical approach and justification should address these concerns.

16:56 Is rabbit "hypnosis" a suitable form of restraint for painful procedures?

Opin. *Hypnosis* is one of many terms designating an immobile, unresponsive state that can be induced in numerous species, referred to as an *immobility response* (IR). IR can be triggered by various physical and psychological manipulations (predator threat, violent struggle, thoracic compression, restraint by neck skin "scruffing," supine position).[162] Reviews of reports in animals and a study of IR in rabbits indicate that between-individual responses are variable and may depend on conditioning or maturation.[163] Evidence exists for analgesia in some species (or an increased threshold to noxious stimulus, but animals can be roused from the IR state by stronger stimuli, and the duration of an IR is not predictable.[162] Commonly performed restraint techniques, such as application of a twitch to horses or restraint by neck skin pinch, can produce a degree of immobile tolerance to a brief stimulus, and thus can be considered preferable to more injurious or strenuous restraint. Nevertheless, because of the lack of predictability, induction of an IR in order to have an animal tolerate painful procedures lasting more than a few moments is not recommended. Also, those intending to use IR for momentary restraint must be prepared to accommodate a significant percentage of animals that do not develop the IR.

16:57 Is tattooing of animals a painful procedure?

Reg. The AWAR (§1.1, Painful procedure) defines a painful procedure as one that would reasonably be expected to cause more than slight or momentary pain or distress in a human being to which that procedure was applied, that is, pain in excess of that caused by injections or other minor procedures. The AWAR (§2.36,b,5) indicates that tattooing should be categorized on its annual report as a procedure involving no pain, distress, or use of pain-relieving drugs.

Opin. Based on the experiences of humans, the size, location, and manner of application of a tattoo are all factors in determining whether the application of a particular tattoo is painful. All permanent methods of tattooing cause a certain amount of tissue damage and it is therefore likely that a certain amount of pain will occur although it may be minimal and momentary. The use of new sterile needles, nontoxic pigments, disinfected skin, and anesthesia (if appropriate) all contribute to lessening any immediate or subsequent pain from tattooing. Experience suggests that the pain caused by tattooing of rodents, particularly neonatal animals, is usually minimal. The impact of a variety of identification procedures in rodents is reviewed, and a comparison of three types, including tattooing, in adult rats is reported in a recent paper.[150]

16:58 Are food or fluid restrictions considered pain or distress?

Reg. The AWAR (§1.1, Painful procedure) defines a painful procedure as one that would reasonably be expected to cause more than slight or momentary pain or distress in a human being to which that procedure was applied, that is, pain in excess of that caused by injections or other minor procedures. APHIS/AC Policy 11 indicates that food and/or water deprivation or restriction beyond that necessary for normal presurgical preparation *may* cause more than momentary or slight distress. The *Guide* (pp. 30–31) states that when the availability of food or water is regulated animals should be closely monitored to ensure that food and fluid intake meet their nutritional needs and that written records should be kept for each animal to document daily food and fluid consumption, hydration status, and any behavioral and clinical changes. (See 16:32.)

Opin: NIH/OLAW, in a position statement, has written "The IACUC must evaluate the level of restriction and potential adverse consequences in regulating food or fluid. The IACUC must also evaluate the methods for assessing the health and well-being of animal activities that involve regulation of food or fluid. The IACUC has the authority to approve scientific justifications for departures from the recommendations in the *Guide*. For instance, using scheduled access to food or fluid sources may be justified by describing procedures based on performance standards that assure adequate maintenance of hydration, body weight, and behavioral and clinical health. (See also 16:32). It may be necessary to monitor both food and fluid intake if regulation of one influences consumption of the other."[164] NIH/OLAW FAQ F.19 has similar wording.

Many studies report the withholding of water for physiologic studies. There is no question that dehydration causes distress that increases with time. Long established standard paradigms may need to be challenged and alternatives actively sought. A brief search of methods sections of papers using water deprivation in rats indicates that water deprivation in rats may be imposed for periods of 24–96 hours. One refinement study reports that rats exhibit signs of dehydration (lethargy, hunched posture, miosis and decreased skin elasticity) after 18–22 hours of water restriction.[165]

Food deprivation is also a cause of distress, and the effects will depend on the species and environmental conditions.

16:59 Is death at the end of a normal lifespan considered a painful experience?

Reg. NIH/OLAW Notice NOT-OD-05-034[166] states that under certain conditions mortality of animals may *not* have to be reported to NIH/OLAW, based on the IACUC's consideration of the surrounding circumstances.

Opin. Examples provided in the above NIH/OLAW notice include

- Death of animals that have reached the end of their natural life spans
- Death of neonates when husbandry and veterinary medical oversight of dams and litters was appropriate
- Death from spontaneous disease when appropriate quarantine, preventive medical, surveillance, diagnostic and therapeutic procedures were in place and followed
- Death from manipulations that fall within parameters described in an IACUC-approved protocol

Death, even at the end of a normal lifespan, may involve pain or other distress. If an animal has a sudden event and dies, there is possibly no pain or distress experience. Most animals that die spontaneously do so as a result of organ dysfunction, cancer, or lack of ability to maintain homeostasis and there is a high likelihood that pain or distress will occur. Intervention with euthanasia would ideally be used to prevent moderate to severe pain. However, particularly in small laboratory animals, monitoring of parameters such as body weight or body condition score, mobility, evidence of nest behavior, etc., would be required.

Surv. If an untreated control animal dies a so-called "natural death" due to old age, what USDA/AC pain/distress category is assigned by your IACUC (assume a USDA/AC category is to be used)?

- Not applicable 82/295 (27.8%)
- Category B – if the animal was just being held and not used for experimentation 83/295 (28.1%)
- Category C – if the animal was part of a study or for teaching and the study/teaching resulted in minimal or no pain/distress 86/295 (29.2%)
- Category E – unalleviated pain/distress due to experimental needs 6/295 (2.0%)
- I don't know 31/295 (10.5%)
- Other 7/295 (2.4%)

16:60 Can the IACUC suggest pilot studies to determine the extent of pain or distress when they are not known?

Reg. Although pain or distress is not formally defined in the AWAR, (see 16:2 for the definition of a painful procedure and 16:1 for a discussion of distress.) The *Guide* (p. 120) defines pain as a complex experience that typically results from stimuli that damage or have the potential to damage tissue. The *Guide* (p. 121) defines distress as an aversive state in which an animal fails to cope or adjust to various stressors with which it is presented. The *Guide* (p. 26) states that "if little is known about a specific procedure, limited pilot studies, designed to assess both the procedure's effects on the animals and the skills of the research team and conducted under IACUC oversight, are appropriate." NIH/OLAW in FAQ D.11 has stated: "Pilot studies may be appropriate to determine the technical feasibility of larger studies or to make initial assessments of the effect of procedures on animals."

Opin. Pilot studies, or the conduct of the experiment in a limited number of animals (e.g., 2–6) will allow the PI, veterinary staff and IACUC to determine whether monitoring and treatment strategies are adequate to maintain well-being. This may involve careful physical exam by a veterinarian, and revision of methods if unanticipated problems occur. (See 13:6.)

16:61 If during a study an animal covered by the AWAR accidentally fractures a leg bone and the animal receives appropriate veterinary treatment, which APHIS/AC annual report pain and distress category should the animal be assigned to?

Reg. The AWAR (§2.36,b,1) requires an annual report to assure, in part, that professional standards of treatment have been used including the use of anesthetic, analgesic, and tranquilizing drugs when required by the IACUC. Animals used are placed into one of three pain and distress categories (AWAR §2.36,b,5–7). The AWAR

(§2.33,b,2) requires adequate veterinary care to treat diseases and injuries. PHS Policy IV,C,1,e requires medical care for animals by a qualified veterinarian. (See 16:5.)

Opin. Although the AWAR (§2.36,b,5–§2.36,b,7) require an annual report categorizing regulated species used during experimentation, teaching, or testing into one of three categories, this categorization does not apply to accidental injuries to an animal. The APHIS/AC *Animal Welfare Inspection Guide,* Appendix A[50] states that in a circumstance analogous to that described herein, the animal should be reported in the pain category appropriate to its experiences in the study. The accident does not affect the reporting category. Therefore, if the animal did not experience any pain or distress as part of the approved study it would be reported in category C (no pain or distress or use of pain relieving drugs) even though the leg fracture may have caused initial pain and anesthesia and analgesics may have been used during its treatment.

16:62 If an animal covered by the AWAR experiences unanticipated pain as part of a research procedure and the pain is treated in a timely manner, which of the APHIS/AC annual report categories for pain and distress is appropriate to use?

Reg. (See 16:61.)

Opin. The key phrase in the above example is "in a timely manner," which is the wording used in examples provided by APHIS/AC.[50] Although neither the AWAR nor the APHIS/AC Inspection Guide[50] define that phrase it is reasonable to assume it implies that the animal is successfully treated as soon as the pain is recognized and that observations have been made with a frequency approved by the IACUC. With those caveats in place it is suggested by the APHIS/AC Inspection Guide to assign Category D to the animal (pain or distress alleviated by the use of drugs).

16:63 If an animal covered by the AWAR experiences unanticipated pain during a research procedure and it is euthanized in a timely manner, which of the APHIS/AC annual report categories for pain and distress is appropriate to use?

Reg. Animals that cannot have chronic pain or distress alleviated will be euthanized at the end of the procedure or, if appropriate, during the procedure (AWAR §2.31,d,1,v; PHS Policy IV,C,1,c). (See 16:61.)

Opin. As noted in 16:62, the key phrase is "in a timely manner" and the phrase assumes that the pain is recognized during observations made within the frequency approved by the IACUC and the animal is euthanized very soon thereafter. The APHIS/AC *Consolidated Inspection Guide*[50] suggests that under such a circumstance Category D (pain or distress alleviate by the use of drugs) is appropriate. Technically, euthanasia should be accomplished by the use of a drug to comply with the wording of Category D, however, it is the opinion of the authors that non-drug means of euthanasia complying with the AVMA euthanasia guidelines[161] likely would be found acceptable to APHIS/AC. The *Guide* (p. 28) emphasizes the need for careful monitoring of animals to help assure animal well-being. *It should be noted that the Consolidated Inspection Guide is undergoing revision and may provide updated guidance in the future.*

16:64 An animal on a study experiences unanticipated pain from an infection that is entirely unrelated to the study. The pain cannot be treated due to interference with the goals of the study. Nevertheless, the animal can be treated with antibiotics. Which of the APHIS/AC annual report categories for pain and distress is appropriate to use?

Reg. See 16:61. The AWAR (§2.33,b,2) requires adequate veterinary care to treat diseases and injuries. PHS Policy (IV,C,1,e) requires medical care for animals by a qualified veterinarian.

Opin. Under the circumstances described in the above example Category E is most appropriate (unalleviated pain or distress because pain relieving drugs would have adversely affected the procedure, results, or interpretation of the study). It was the needs of the study, not the infection that dictated the withholding of analgesia. Even if the antibiotics that were used eventually resolved the infection and mitigated the pain, it is unlikely that the pain was alleviated in a timely manner (see 16:62). This interpretation is aligned with an example provided by APHIS/AC in its *Consolidated Inspection Guide.*[50] *It should be noted that the Consolidated Inspection Guide is undergoing revision and may provide updated guidance in the future.*

16:65 Does the use of sterile thioglycollate medium, given intraperitoneally for the collection of peritoneal macrophages or neutrophils, cause a painful peritonitis?

Opin. Often called a thioglycollate peritonitis model, the purpose is to attract inflammatory cells to the site of sterile inflammation. The procedure typically involves a single intraperitoneal injection of sterile 4% thioglycollate medium and subsequent euthanasia of the animal to collect cells from the peritoneal cavity. Although non-sterile peritonitis is often a painful condition in humans, numerous empirical observations of mice used for this sterile peritonitis procedure suggest that there is no pain experienced during the relatively short (typically 48 hours) period of time prior to euthanasia.

Surv. If your IACUC approves a study with the injection of sterile thioglycollate medium to induce "thioglycollate peritonitis" for macrophage harvest, what USDA/AC pain/distress category is assigned (assume that your IACUC assigns such categories).

- Not applicable 179/290 (61.7%)
- Category C – minimal or no pain/distress 13/290 (4.5%)
- Category D – Pain/distress alleviated by drug use because our IACUC requires an analgesic to be used 43/290 (14.8%)
- Category E – Unalleviated pain/distress due to experimental requirements 48/290 (16.6%)
- Other 7/290 (2.4%)

References

1. Lund, T.B., M.R. Morkbak, J. Lassen et al. 2012. Painful dilemmas: A study of the way the public's assessment of animal research balances costs to animals against human benefits. *Public Underst. Sci.* doi:10.1177/0963662512451402.
2. Nuffield Council on Bioethics Report. 2005. The ethics of research involving animals, pp. 35–36 http://www.nuffieldbioethics.org/sites/default/files/The%20ethics%20of%20research%20involving%20animals%20-%20full%20report.pdf.

3. Swami, V., A. Furnham, and A.N. Christopher. 2008. Free the animals? Investigating attitudes toward animal testing in Britain and the United States. *Scand. J. Psychol.* 4: 269–276.
4. Russell, W.M.S., and R.L. Burch. 1959. *The Principles of Humane Experimental Technique.* London: Methuen.
5. Antoni, M.H., S.K. Lutgendorf, S.W. Cole et al. 2006. The influence of bio-behavioural factors on tumour biology: Pathways and mechanisms. *Nat. Rev. Cancer.* 6:240–248.
6. Bowers, S.L., S.D. Bilbo, F.S. Dhabhar et al. 2008. Stressor-specific alterations in corticosterone and immune responses in mice. *Brain Behav. Immun.* 2:105–113.
7. Caso, J.R., J.C. Leza, L. Menchen. 2008. The effects of physical and psychological stress on the gastro-intestinal tract: Lessons from animal models. *Curr. Mol. Med.* 8:299–312.
8. Cory-Slechta, D.A., M.B. Virgolini, M. Thiruchelvam et al. 2004. Maternal stress modulates the effects of developmental lead exposure. *Environ. Health Perspect.* 112:717–730.
9. Dalm, S., V. Brinks, M.H. van der Mark et al. 2008. Non-invasive stress-free application of glucocorticoid ligands in mice. *J. Neurosci. Meth.* 170:77–84.
10. Damon, E.G., C.H. Hobbs, F.F. Hahn. 1986. Effect of acclimation to caging on nephrotoxic response of rats to uranium. *Lab. Anim. Sci.* 36:24–27.
11. Engler, H., L. Dawils, S. Hoves et al. 2004. Effects of social stress on blood leukocyte distribution: The role of alpha- and beta-adrenergic mechanisms. *J. Neuroimmunol.* 156:153–162.
12. Grootendorst, J., E.R. De Kloet, S. Dalm et al. 2001. Reversal of cognitive deficit of apolipoprotein E knockout mice after repeated exposure to a common environmental experience. *Neurosci.* 108:237–247.
13. Hurst, J.L. and R.S. West. 2010. Taming anxiety in laboratory mice. *Nat. Methods.* 7:825–826.
14. Kerr, L.R., R. Hundal, W.A. Silva et al. 2001. Effects of social housing condition on chemotherapeutic efficacy in a Shionogi carcinoma (SC115) mouse tumor model: Influences of temporal factors, tumor size, and tumor growth rate. *Psychosom. Med.* 63:973–984.
15. Kiank, C., P. Koerner, W. Kebler et al. 2007. Seasonal variations in inflammatory responses to sepsis and stress in mice. *Crit. Care Med.* 35:2352–2358.
16. Kiank, C., B. Holtfreter, A. Starke et al. 2006. Stress susceptibility predicts the severity of immune depression and the failure to combat bacterial infections in chronically stressed mice. *Brain Behav. Immun.* 20:359–368.
17. Leussis, M.P. and S.C. Heinrichs. 2006. Routine tail suspension husbandry facilitates onset of seizure susceptibility in EL mice. *Epilepsia.* 47:801–804.
18. Lupien, S.J., B.S. McEwen, M.R. Gunnar et al. 2009. Effects of stress throughout the lifespan on the brain, behaviour and cognition. *Nat. Rev. Neurosci.* 10:434–445.
19. Renoir, T., T.Y. Pang, A.J. Hannan. 2012. Effects of environmental manipulations in genetically targeted animal models of affective disorders. *Neurobiol. Dis.* doi:10.1016/j.nbd.2012.04.003.
20. Sternberg, W.F. and C.G. Ridgway. 2003. Effects of gestational stress and neonatal handling on pain, analgesia and stress behavior in adult mice. *Physiol. Behav.* 78:375–383.
21. Van Loo, P.L., N. Kuin, R. Sommer et al. 2007. Impact of living apart together on postoperative recovery of mice compared with social and individual housing. *Lab. Anim.* 41:441–455.
22. Voikar, V., A. Polus, E. Vasar et al. 2005. Long-term individual housing in C57BL/6J and DBA/2 mice: Assessment of behavioral consequences. *Genes Brain Behav.* 4:240–252.
23. Beilin, B., Y. Shavit, E. Trabekin et al. 2003. The effects of postoperative pain management on the immune response to surgery. *Anesth. Analg.* 97:822–827.
24. Ben-Eliyahu, S., G.G. Page, R. Yirmiya et al. 1999. Evidence that stress and surgical interventions promote tumor development by suppressing natural killer cell activity. *Int. J. Cancer.* 80:880–888.
25. Christian, L.M., J.E. Graham, D.A. Padgett et al. 2006. Stress and wound healing. *Neuroimmunomodulat.* 13:337–346.
26. Desborough, J.P. 2000. The stress response to trauma and surgery. *Br. J. Anaesth.* 85:109–117.
27. Dhabhar, F.S. 2009. Enhancing versus suppressive effects of stress on immune function: Implications for immunoprotection and immunopathology. *Neuroimmunomodulat.* 16:300–317.
28. Glaser, R. and J.K. Kiecolt-Glaser. 2005. Stress-induced immune dysfunction: Implications for health. *Nat. Rev. Immunol.* 5:243–251.

29. Perkins, F.M. and H. Kehlet. 2000. Chronic pain as an outcome of surgery: A review of predictive factors. *Anesthesiol.* 93:1123–1133.
30. Stuller, K.A., B. Jarrett and A.C. DeVries. 2012. Stress and social isolation increase vulnerability to stroke. *Exper. Neurol.* 233:33–39.
31. Chesler, E.J., S.G. Wilson, W.R. Lariviere et al. 2002. Identification and ranking of genetic and laboratory environment factors influencing a behavioral trait, thermal nociception, via computational analysis of a large data archive. *Neurosci. Biobehav. Rev.* 26:907–923.
32. Richter, S.H., J.P. Garner, C. Auer et al. 2010. Systematic variation improves reproducibility of animal experiments. *Nat. Methods* 7:167–168.
33. Hooijmans, C.R., M. Leenaars and M. Ritskes-Hoitinga. 2010. Gold standard publication checklist to improve the quality of animal studies, to fully integrate the three Rs, and to make systematic reviews more feasible. *ATLA-Altern. Lab. Anim.* 38:167–182.
34. Institute for Laboratory Animal Research. 2011. *Guidance for the Description of Animal Research in Scientific Publications*. Washington, D.C.: National Academies Press.
35. Kilkenny, C., W.J. Browne, I.C. Cuthill et al. 2010. Improving bioscience research reporting: The ARRIVE Guidelines for reporting animal research. *PLoS Biol.* 8:e1000412. doi:10.1371/journal.pbio.1000412.
36. Wever, K.E., T.P. Menting, M. Rovers et al. 2012. Ischemic preconditioning in the animal kidney, a systematic review and meta-analysis. *PLoS ONE.* 7:e32296. doi:10.1371/journal.pone.0032296.
37. Mellor, D.J., E. Patterson-Kane and K.J. Stafford. 2009. *The Sciences of Animal Welfare*. Hoboken: Wiley-Blackwell.
38. Buchanan-Smith, H.M., A.E. Rennie, A. Vitale et al. 2005. Harmonising the definition of refinement. *Anim. Welfare.* 14:379–384.
39. Committee on Recognition and Alleviation of Distress in Laboratory Animals. 2008. *Recognition and Alleviation of Distress in Laboratory Animals*. Washington, D.C.: National Academies Press.
40. Koolhaas, J.M., A. Bartolomucci, B. Buwalda et al. 2011. Stress revisited: A critical evaluation of the stress concept. *Neurosci. Biobehav. Rev.* 35:1291–1301.
41. Korte, S.M., B. Olivier and J.M. Koolhaas. 2007. A new animal welfare concept based on allostasis. *Physiol. Behav.* 92:422–428.
42. McMillan, F.D. 2003. A world of hurts—Is pain special? *J. Amer. Vet. Med. Assoc.* 223:183–186.
43. Orlans, F.B. 1987. Review of experimental protocols: Classifying animal harm and applying "refinements." *Lab. Anim. Sci.* 37:50–56.
44. International Association for the Study of Pain. 2011. IASP taxonomy. http://www.iasp-pain.org/AM/Template.cfm?Section=Pain_Definitions.
45. Zimmerman, M. 1984. Neurological Concepts of Pain, its Assessment and Therapy. In *Neurophysiological Correlates of Pain*, ed. Bromm, B. Amsterdam: Elsevier.
46. Craig, A.D. 2003. A new view of pain as a homeostatic emotion. *Trends Neurosci.* 26:303–307.
47. Committee on Recognition and Alleviation of Pain in Laboratory Animals. 2009. *Recognition and alleviation of pain in laboratory animals*. Washington, D.C.: National Academy Press.
48. American College of Veterinary Anesthesiologists. 1998. Position paper on the treatment of pain in animals. *J. Am. Vet. Med. Assoc.* 213:628–630.
49. Khamis, H.J. 1997. Statistics and the issue of animal numbers in research. *Contemp. Topics Lab. Anim. Sci.* 36:54–59.
50. U.S. Department of Agriculture. Animal and Plant Health Inspection Service. 2013. *Animal Welfare Inspection Guide*. http://www.aphis.usda.gov/animal_welfare/downloads/Inspection%20Guide%20-%20November%202013.pdf.
51. Francione, G.L. and A.E. Charlton. 1998. Bound by law. *Lab. Anim. (NY)*. 27(3):20–21.
52. Garnett, N.L. and W.R. DeHaven. 1998. A word from the government. *Lab Anim. (NY)*. 27(3):21.
53. Morton, D.B. and P.H.M. Griffiths. 1985. Guidelines on the recognition of pain, distress and discomfort in experimental animals and an hypothesis for assessment. *Vet. Rec.* 116:431–436.
54. Karas, A.Z., P.J. Danneman and J.M. Cadillac. 2008. Strategies for assessing and minimizing pain. In *Anesthesia and Analgesia in Laboratory Animals*, ed. R. Fish, P.J. Danneman, M. Brown, et al. San Diego, Academic Press.

55. Lund, I., L-C. Yu, K. Uvnas-Moberg et al. 2002. Repeated massage-like stimulation induces long-term effects on nociception: Contribution of oxytocinergic mechanisms. *Eur. J. Neurosci.* 16:330–338.
56. So, P.S., J.Y. Jiang and Y. Qin. 2008. "Touch therapies for pain relief in adults. *Cochrane Database Syst. Rev.* 4. Art.No.: CD006535. doi: 0.1002/14651858.CD006535.pub2.
57. Silverman, J., N.L. Garnett, S.F. Giszter et al. 2005. Decerebrate mammalian preparations: Unalleviated or fully alleviated pain? A review and opinion. *J. Am. Assoc. Lab. Anim. Sci.* 44:34–36.
58. Stevens, C.W. 2004. Opioid research in amphibians: An alternative pain model yielding insights on the evolution of opioid receptors. *Brain Res. Rev.* 46:204–215.
59. Aroni, F., N. Iacovidou, I. Dontas et al. 2009. Pharmacological aspects and potential new clinical applications of ketamine: Reevaluation of an old drug. *J. Clin. Pharm.* 49:957–964.
60. Granry, J.-C., L. Dube, H. Turroques et al. 2000. Ketamine: New uses for an old drug. *Curr. Opin. Anaesthesio.* 13:299–302.
61. Persson, J. 2010. Wherefore ketamine? *Curr. Opin. Anaesthesiol.* 23:455–460.
62. Olesen, A.E., T. Andresen, L.L. Christrup et al. 2009. Translational pain research: Evaluating analgesic effect in experimental visceral pain models. *World J. Gastroenterol.* 15:177–181.
63. Strigo, I.A., G.H. Duncan, M.C. Bushnell et al. 2005. The effects of racemic ketamine on painful stimulation of skin and viscera in human subjects. *Pain* 113:255–264.
64. Williams, L.S., J.K. Levy, S.A. Robertson et al. 2002. Use of the anesthetic combination of tiletamine, zolazepam, ketamine, and xylazine for neutering feral cats. *J. Am. Vet. Med. Assoc.* 220:1491–1495.
65. Himmelseher, S. and M.E. Durieux. 2005. Ketamine for perioperative pain management. *Anesthesiol.* 102:211–220.
66. Flecknell, P.A. 1996. *Laboratory Animal Anesthesia*, 2nd edition. London: Academic Press.
67. Banknieder, A.R., J.M. Phillips, K.T. Jackson et al. 1978. Comparison of ketamine with the combination of ketamine and xylazine for effective anesthesia in the rhesus monkey (Macaca mulatta). *Lab. Anim. Sci.* 28:742–745.
68. Boschert, K., P.A. Flecknell, R.T. Fosse et al. 1996. Ketamine and its use in the pig: Recommendations of the Consensus Meeting on Ketamine Anaesthesia in Pigs, Bergen 1994. *Lab. Anim.* 30:209–219.
69. Flecknell, P.A. 1997. Medetomidine and atipamezole: Potential uses in laboratory animals, *Lab Anim. (N.Y.)* 26(2):21–27.
70. Green, C.J., J. Knight, S. Precious et al. 1981. Ketamine alone and combined with diazepam or xylazine in laboratory animals: A 10-year experience, *Lab. Anim.* 15:163–170.
71. Thurmon, J.C., W.J. Tranquilli and G.J. Benson. 1996. *Lumb and Jones' Veterinary Anesthesia*, 3rd ed. Baltimore: Williams & Wilkins.
72. Rubal, B. and C. Buchanan, C. 1986. Supplemental chloralose anesthesia in morphine premedicated dogs. *Lab. Anim. Sci.* 36:59–64.
73. Silverman, J. and W.W. Muir. 1993. A review of laboratory animal anesthesia with chloral hydrate and chloralose. *Lab. Anim. Sci.* 43:210–216.
74. Ching, M. 1984. Comparison of the effects of althesin, chloralose-urethane, urethane, and pentobarbital on mammalian physiologic responses. *Can. J. Physiol. Pharmacol.* 62:654–657.
75. Faber, J.E. 1989. Effects of althesin and urethane-chloralose on neurohumoral cardiovascular regulation. *Am. J. Physiol.* 256(3 Pt. 2):R757–765.
76. Momosaki, S., K. Hatano, Y. Kawasumi et al. 2004. Rat-PET study without anesthesia: Anesthetics modify the dopamine D1 receptor binding in the rat brain. *Synapse* 54:207–213.
77. Theodorsson, A., L. Holm and E. Therodorsson. 2005. Modern anesthesia and preoperative monitoring methods reduce pre- and postoperative mortality during transient occlusion of the middle cerebral artery in rats. *Brain Res. Protoc.* 14:181–190.
78. Yuan, C.S. 1988. Effects of chloralose-urethane anesthesia on single-axon reciprocal Ia IPSPs in the cat. *Neurosci. Lett.* 94:291–296.
79. Wagner, P.G., F.L. Eldridge and R.T. Dowell. 1991. Anesthesia affects respiratory and sympathetic nerve activities differentially. *J. Auto. Nerv. Syst.* 36:225–236.

80. Zaugg, M., E. Lucchinetti, D.R. Spahn et al. 2002. Differential effects of anesthetics on mitochondrial K_{ATP} channel activity and cardiomyocyte protection. *Anesthesiol.* 97:15–23.
81. Savarese, J.J., M.R. Belmont, M.A. Hashim et al. 2004. Preclinical pharmacology of GW280430A (AVA430A) in the rhesus monkey and in the cat. *Anesthesiol.* 100:835–845.
82. Lukas, V.S. 1994. Animal use in toxicity evaluation. In *Anesthetic Toxicity,* ed. S.A. Rice and K.J. Fish, Chapter 2. New York: Raven Press.
83. Meyer, R.E. and R.E. Fish. 2008. Pharmacology of injectable anesthetics, sedatives, and tranquilizers. In *Anesthesia and Analgesia in Laboratory Animals*, ed. R. Fish, P.J. Danneman, M. Brown, et al. San Diego, CA: Academic Press.
84. Field, K.J. and C.M. Lang. 1988. Hazards of urethane (ethyl carbamate): A review of the literature. *Lab. Anim.* 22:255–262.
85. National Institutes of Health, Office of Laboratory Animal Welfare. 2012. Use of Non-Pharmaceutical-Grade Chemicals and Other Substances in Research with Animals. http://grants.nih.gov/grants/olaw/120301_seminar_transcript.pdf.
86. National Institutes of Health. Office of Laboratory Animal Welfare. 2011. Position Statements. http://grants.nih.gov/grants/olaw/positionstatement_guide.htm.
87. Matthews, K.A. 2000. Pain assessment and general approach to management. *Vet. Clin. N. Am.-Small* 30:729–755.
88. Gaynor, J.S. and W.W. Muir. 2002. Acute pain management, a case based approach. In *Handbook of Veterinary Pain Management*, eds. J.S. Gaynor and W.W. Muir. 346. St. Louis: Mosby.
89. Goecke, J.C., H. Awad, J.C. Lawson et al. 2005. Evaluating postoperative analgesics in mice using telemetry. *Comp Med.* 55:37–44.
90. Karas, A.Z., Unpublished observations, 2005.
91. Parker, J.M., J. Austin, J. Wilkerson et al. 2011. Effects of multimodal analgesia on the success of mouse embryo transfer surgery. *J. Am. Assoc. Lab. Anim. Sci.* 50:466–470.
92. Goulding, D.R., P.H. Myers, E.H. Goulding et al. 2010. The effects of perioperative analgesia on litter size in Crl:CD1(ICR) mice undergoing embryo transfer. *J. Am. Assoc. Lab. Anim. Sci.* 49:423–426.
93. Krueger, K.L. and Y. Fujiwara. 2008. The use of buprenorphine as an analgesic after rodent embryo transfer. *Lab Anim. (N.Y.).* 37:87–90.
94. Smith, J.C., T.J. Corbin, J.G. McCabe et al. 2004. Isoflurane with morphine is a suitable anaesthetic regimen for embryo transfer in the production of transgenic rats. *Lab. Anim.* 38:38–43.
95. National Agriculture Library, Animal Welfare Information Center. http://awic.nal.usda.gov/ Accessed Feb. 4, 2013.
96. Animal Welfare Institute. Additional Resources. http://awionline.org/content/additional-resources-4 Accessed Feb. 4, 2013.
97. Liles, J.H. and P.A. Flecknell. 1993. A comparison of the effects of buprenorphine, carprofen and flunixin following laparotomy in rats. *J. Vet. Pharm. Ther.* 17:284–290.
98. Liles, J.H. and P.A. Flecknell. 1993. The effects of surgical stimulus on the rat and the influence of analgesic treatment. *Br. Vet. J.* 149:515–525.
99. Clark, M.D., L. Krugner-Higby, L.J. Smith et al. 2004. Evaluation of liposome-encapsulated oxymorphone hydrochloride in mice after splenectomy. *Comp. Med.* 54:558–563.
100. Blaha, M.D. and L. Leon, 2008. Effects of indomethacin and buprenorphine analgesia on the postoperative recovery of mice. *J. Am. Assoc. Lab. Anim. Sci.* 47:8–19.
101. Sebel, P.S., T.A. Bowdle, M.M. Ghoneim. 2004. The incidence of awareness during anesthesia: A multicenter United States study. *Anesth. Analg.* 99:833–839.
102. Samuelsson, P., L. Brudin and R. Sandin. 2007. Late psychological symptoms after awareness among consecutively included surgical patients. *Anesthesiol.* 106:26–32.
103. Marsch, S.C.U. and W. Studer. 1999. Guidelines to the use of laboratory animals: What about neuromuscular blockers? *Cardiovasc. Res.* 42:565–566.
104. Paddleford, R. Pulmonary dysfunction. 1996. In *Lumb and Jones Veterinary Anesthesia,* 3rd edition, ed. J.C. Thurmon, W.J. Tranquilli and J.G. Benson, 774. Baltimore: Williams and Wilkins.

105. Whelan, G. and P.A. Flecknell. 1992. The assessment of depth of anesthesia in animals and man. *Lab. Anim.* 26:153–162.
106. Hall, L.W. 1991. *Veterinary Anesthesia*, 9th edition. London: Balliere Tindall.
107. Lemke, K.A. and S.D. Dawson. 2000. Local and regional anesthesia. *Vet. Clin. North Am. – Small.* 30:839–857.
108. Lovell, D.P. 1986. Variation in pentobarbitone sleeping time in mice. 1. Strain and sex differences. *Lab. Anim.* 20:85–90.
109. Pascoe, P.J. Perioperative pain management. 2000. *Vet. Clin. North Am.- Small* 30:917–932.
110. Smith, G. Postoperative pain. 1984. In *Quality of Care in Anesthetic Practice*, ed. J.N. Lunn, 164–192. London: McMillan Press.
111. National Institutes of Health, Office of Laboratory Animal Welfare. 2012. Frequently asked questions A.6, Does the PHS Policy apply to animal research that is conducted in the field? http://grants2.nih.gov/grants/olaw/faqs.htm#App_6.
112. Vermeulen, J.K., A.D. Vries, F. Schlingmann et al. 1997. Food deprivation: Common sense or nonsense? *Anim. Technol.* 48:45–54.
113. Waynforth, H.B. and P.A. Flecknell. 1992. *Experimental and Surgical Techniques in the Rat*. London: Academic Press.
114. Anand, K., W.G. Sippell and A. Aynsley-Green. 1987. Randomised trial of fentanyl anesthesia in preterm babies undergoing surgery: Effects on the stress response. *Lancet.* 329(8524):243–248.
115. Anand, K.J. 2000. Effects of perinatal pain and stress. *Prog. Brain Res.* 22:117–129.
116. Anand, K.J.S. and M. Maze. 2001. Fetuses, fentanyl and the stress response: Signals from the beginnings of pain? *Anesthesiol.* 95:823–825.
117. Houfflin-Debarge, V., A. Delelis, S. Jaillard et al. 2005. Effects of nociceptive stimuli on the pulmonary circulation in the ovine fetus. *Am. J. Physiol. Regul. Integr. Comp. Physiol.* 288: R547–553.
118. Senat, M.V., C. Fischer, C. and Y. Ville. 2002. Funipuncture for fetocide in late termination of pregnancy. *Prenatal Diagnosis.* 22:354–356.
119. Mellor, D.J. 2010. Galloping colts, fetal feelings, and reassuring regulations: Putting animal-welfare science into practice. *J. Vet. Med. Educ.* 37:94–100.
120. Mellor, D.J. and T.J. Diesch. 2006. Onset of sentience: The potential for suffering in fetal and newborn farm animals. *Appl. Anim. Behav. Sci.* 100:48–57.
121. Lowery, C.L., M. P. Hardman, N. Manning et al. 2007. Neurodevelopmental changes of fetal pain. *Semin. Perinatol.* 31:275–282.
122. Diesch, T.J., D.J. Mellor, C.B. Johnson et al. 2007. Responsiveness to painful stimuli in anaesthetised newborn and young animals of varying neurological maturity (wallaby joeys, rat pups and lambs). Proc. 6th World Congress on Alternatives & Animal Use in the Life Sciences August 21–25, Tokyo, Japan. *AATEX* 14(Special Issue):549–552.
123. Murrell, J.C., D.J. Mellor and C.B. Johnson. 2008. Anaesthesia and analgesia in the foetus and neonate. In *Anesthesia and Analgesia in Laboratory Animals*. ed. R. Fish, P.J. Danneman, M. Brown et al. Waltham: Academic Press.
124. White, S.C. 2012. Prevention of fetal suffering during ovariohysterectomy of pregnant animals. *J. Am. Vet. Med. Assoc.* 240:1160–1163.
125. The European Parliament and the Council of the European Union. 2010. Directive 2010/63/EU of the European Parliament and of the Council of 22 September 2010 on the protection of animals used for scientific purposes. *Official Journal of the European Union* L276:33–79.
126. National Institutes of Health, Office for Protection from Research Risks. 1991. The Public Health Service responds to commonly asked questions. *ILAR News* 33(4):68.
127. Mellor, D.J. and Diesch, T. J. 2007. Birth and hatching: Key events in the onset of awareness in the lamb and chick. *New Zeal. Vet. J.* 55:51–60.
128. Davis, K.D. and G.E. Pope. 2002. Noxious cold evokes multiple sensations with distinct time courses. *Pain* 98:179–185.
129. Danneman, P.J. and T.D. Mandrell. 1997. Evaluation of five agents/methods for anesthesia of neonatal rats. *Lab. Anim. Sci.* 47:386–395.

130. Park, C.M., K.E. Clegg, C.J. Harvey-Clark et al. 1992. Improved techniques for successful neonatal rat surgery. *Lab. Anim. Sci.* 42:508–513.
131. Martin, B.J. 1995. Evaluation of hypothermia for anesthesia in reptiles and amphibians. *ILAR J.* 37:186–190.
132. Davis, J.N., C.L. Courtney, R.G. Hunter et al. 2012. Behavioral, clinical, and pathologic changes seen in mice receiving repeated intraperitoneal injections. Poster presented at the annual meeting of the American Association for Laboratory Animal Science, Minneapolis.
133. Colomina, M.T., M.L. Albina, D.J. Sanchez et al. 2001. Interactions in developmental toxicology: Combined action of restraint stress, caffeine, and aspirin in pregnant mice. *Teratology* 63:144–151.
134. Cordero, M.I., J.J. Rodriguez, H.A. Davies et al. 2005. Chronic restraint stress down-regulates amygdaloid expression of polysialylated neural cell adhesion molecule. *Neuroscience.* 133:903–910.
135. Costa, A., A. Smeraldi, C. Tassorelli et al. 2005. Effects of acute and chronic restraint stress on nitroglycerin-induced hyperalgesia in rats. *Neurosci. Lett.* 383:7–11.
136. King-Herbert, A.P., T.W. Hesterburg, P.P. Thevenaz et al. 1997. Effects of immobilization restraint on Syrian golden hamsters. *Lab. Anim. Sci.* 47:362–366.
137. Hassimoto, M., T. Harada and T. Harada. 2004. Changes in hematology, biochemical values, and restraint ECG of rhesus monkeys (*Macaca mulatta*) following 6-month laboratory acclimation. *J. Med. Primatol.* 33:175–186.
138. Natelson, B.H., S.L. Hoffman and N.A. Cagin. 1980. A role for environmental factors in the production of digitalis toxicity. *Pharmacol. Biochem. Behav.* 12:235–237.
139. Van Herck, H., V. Baumans, H.A.G. Boere et al. 2000. Orbital sinus blood sampling in rats: Effects upon selected behavioral variables. *Lab. Anim.* 34:10–19.
140. Toft, M.F., M.H. Petersen, N. Dragsted et al. 2006. The impact of different blood sampling methods on laboratory rats under different types of anaesthesia. *Lab. Anim.* 40:261–274.
141. Van Herck, H., V. Baumans, C.J.W.M. Brandt et al. 1998. Orbital sinus blood sampling in rats as performed by different animal technicians: The influence of technique and expertise. *Lab. Anim.* 32:377–386.
142. Heimann, M., H.P. Kasermann, R. Pfister et al. 2009. Blood collection from the sublingual vein in mice and hamsters: A suitable alternative to retrobulbar technique that provides large volumes and minimizes tissue damage. *Lab. Anim.* 43:255–260.
143. Heimann, M., D.R. Roth, D. Ledieu et al. 2010. Sublingual and submandibular blood collection in mice: A comparison of effects on body weight, food consumption and tissue damage. *Lab. Anim.* 44:352–358.
144. Holmberg, H., M.K. Kiersgaard, L.F. Mikkelsen et al. 2011. Impact of blood sampling technique on blood quality and animal welfare in haemophilic mice. *Lab. Anim.* 45:114–120.
145. Hankenson, F.C., L.M. Garzel, D.D. Fischer et al. 2008. Evaluation of tail biopsy collection in laboratory mice (*Mus musculus*): Vertebral ossification, DNA quantity, and acute behavioral responses. *J. Am. Assoc. Lab. Anim. Sci.* 47:10–18.
146. Morales, M.E., and R.W. Gereau. 2009. The effects of tail biopsy for genotyping on behavioral responses to nociceptive stimuli. *PLoS One.* 4:e645.
147. Hankenson, F.C., G.C. Braden-Weiss, and J.A. Blendy. 2011. Behavioral and activity assessment of laboratory mice (Mus musculus) after tail biopsy under isoflurane anesthesia. *J. Am. Assoc. Lab. Anim. Sci.* 50:686–94.
148. Arras, M., A. Rettich, B. Seifert et al. 2007. Should laboratory mice be anaesthetized for tail biopsy? *Lab. Anim.* 41:30–45.
149. Sørensen, D.B., C. Stub, H.E. Jensen et al. 2007. The impact of tail tip amputation and ink tattoo on C57BL/6JBomTac mice. *Lab. Anim.* 41:19–29.
150. Kasanen, H.E., H-M. Voipio, H. Leskinen et al. 2011. Comparison of ear tattoo, ear notching and microtattoo in rats undergoing cardiovascular telemetry. *Lab. Anim.* 45:154–159.
151. Zhuo, M. 1998. NMDA receptor-dependent long term hyperalgesia after tail amputation in mice. *Eur. J. Pharmacol.* 349:211–220.

152. Meldgaard, M., P.J. Bollen, and B. Finsen. 2004. Non-invasive method for sampling and extraction of mouse DNA for PCR. *Lab. Anim.* 38:413–417.
153. Cinelli, P., A. Rettich, B. Seifert et al. 2007. Comparative analysis and physiological impact of different tissue biopsy methodologies used for the genotyping of laboratory mice. *Lab. Anim.* 41:174–184.
154. Castelhano, M.J., N. Souas, F. Ohl et al. 2010. Identification methods in newborn C57BL/6 mice: A developmental and behavioural evaluation. *Lab. Anim.* 44:88–103.
155. Schaefer, D.C., I.N. Asner, B. Seifert et al. 2010. Analysis of physiological and behavioural parameters in mice after toe clipping as newborns. *Lab. Anim.* 44:7–13.
156. Rispin, A., D. Farrar, E. Margosches et al. 2002. Alternative methods for the median lethal dose (LD50) test: The up-and-down procedure for acute oral toxicity. *ILAR J.* 43:233–243.
157. Acred, P., T.D. Hennessey, J.A. MacArthur-Clark et al. 1994. Guidelines for the welfare of animals in rodent protection tests: A report from the Rodent Protection Test Working Party. *Lab. Anim.* 28:13–18.
158. DeHaven, W.R., personal communication, 1999.
159. Cavanna, A.E. M. Mula, S. Servo et al. 2008. Measuring the level and content of consciousness during epileptic seizures: The Ictal Consciousness Inventory. *Epilepsy & Behav.* 13:184–188.
160. Förderreuther, S., A. Henkel, S. Noachtar et al. 2002. Headache associated with epileptic seizures: Epidemiology and clinical characteristics. *Headache.* 2:649–655.
161. American Veterinary Medical Association. 2013. AVMA Guidelines for the Euthanasia of Animals: 2013 edition. https://www.avma.org/KB/Policies/Documents/euthanasia.pdf.
162. Carli, G. 1992. Immobility responses and their behavioral and physiologic aspects, including pain: A review. In *Animal Pain*. ed. C.E. Short and A. Van Poznak, 543–554. New York: Churchill Livingstone.
163. Danneman, P.J., W. J. White, W.K. Marshall et al. 1988. An evaluation of analgesia associated with the immobility response in laboratory rabbits. *Lab. Anim. Sci.* 38:51–57.
164. National Institutes of Health, Office of Laboratory Animal Welfare. Position Statements: OLAW Responds to Concerns Regarding Adoption of the Guide for the Care and Use of Laboratory Animals: Eighth Edition http://grants.nih.gov/grants/olaw/positionstatement_guide.htm#nonpharma]05-034. Accessed Feb. 4, 2013.
165. Kulick, L.J., D.J. Clemons, R.L. Hall et al. 2005. Refinement of the urine concentration test in rats. *Contemp. Topics Lab. Anim. Sci.* 44:46–49.
166. National Institutes of Health, Office of Laboratory Animal Welfare. 2005. Notice NOT-OD-05-034, Guidance on prompt reporting to OLAW under the PHS Policy on Humane Care and Use of Laboratory Animals. http://grants.nih.gov/grants/guide/notice-files/NOT-OD-05-034.html.

17

Euthanasia

Peggy J. Danneman

Introduction

This chapter addresses questions on the topic of euthanasia. The chapter is organized into eight sections: General Guidelines and IACUC Policy; Endpoints; Inhalant and Other Chemical Methods; Physical Methods; Adjunctive Methods—Stunning, Pithing, Exsanguination; Pre-Natal and Neonatal Rodents; Field Studies; and Other Euthanasia-related Questions. In addition to the *Guide*, PHS Policy, and AWAR, the major 'regulatory' (see 17:2) reference for this chapter is the 2013 *AVMA Guidelines for the Euthanasia of Animals*. This newest iteration of the AVMA Guidelines was just released just prior to the third edition of *The IACUC Handbook* going to press and some of the questions in this chapter reflect issues related to the application of the 2007 *AVMA Guidelines on Euthanasia* that were clarified or changed in the 2013 iteration. Note also that, at the time this third edition of *The IACUC Handbook* went to press, NIH/OLAW was in a public comment period concerning the new AVMA *Guidelines*, and updated guidance may be posted near the time the Handbook is published. The reader is urged to consult the most recent guidance from NIH/OLAW and APHIS/AC prior to making decisions regarding acceptable techniques and policies for euthanasia. The Canadian Council on Animal Care (CCAC) also has published several guideline documents that address various issues related to euthanasia. Although U.S. and other non-Canadian readers are not required to adhere to the CCAC guidelines, these documents include some excellent information that should be of interest to all readers. Where pertinent, this information is summarized in the Regulatory and/or Opinion parts of this chapter.

General Guidelines and IACUC Policy

17:1 What are some general guidelines for euthanasia?

Reg. Euthanasia of experimental animals may be performed for many reasons, including procurement of tissues or blood as part of the experimental design. Animals also may be euthanized to alleviate otherwise untreatable pain or distress. Both the AWAR (§2.31,d,1,v) and PHS Policy (IV,C,1,c) require investigators to euthanize animals that would otherwise experience severe or chronic pain or distress that cannot be relieved by other means. In such instances, animals should be euthanized at the end of the procedure or, if appropriate, during the procedure. The

Guide (p. 123) further indicates that animal use protocols should specify endpoint criteria that will "enable a prompt decision by the veterinarian and the investigator to ensure that the endpoint is humane and, whenever possible, the scientific objective of the protocol is achieved."

Opin. The term euthanasia is derived from Greek and means "easy death." According to definitions in the AWAR (§1.1, Euthanasia), the *Guide* (p. 123), and the *AVMA Guidelines for the Euthanasia of Animals*,[1] euthanasia involves killing in a humane manner that causes rapid unconsciousness and death with little or no pain or distress. These goals are met by selecting an agent or technique that is appropriate for the situation, taking into account the species of animal and its age and temperament, the availability of appropriate equipment for restraint, and the environment in which the euthanasia will be performed. It also is important that the person performing the euthanasia be compassionate, gentle, fully trained and technically proficient in the technique to be used, and experienced in the humane restraint of the species to be euthanized. Because other animals may be distressed by visual, auditory, or olfactory signals from the animal being euthanized, it has been considered preferable to perform the procedure in an area where other animals are not present.[1–3] However, some studies suggest that animals of at least some species (e.g., rats) are no more distressed by witnessing the violent death of conspecifics than they are by witnessing procedures such as routine cage changing.[4–7] Perhaps for this reason, the *AVMA Guidelines for the Euthanasia of Animals*[1] fall short of making a general recommendation to avoid having other animals present during euthanasia; instead, it advises against this practice when dealing with "sensitive" species.

Aside from the all-important humane issues, three broad, overlapping criteria should be considered when selecting a method of euthanasia.

Regulatory: The method should be in compliance with all relevant regulations and guidelines. Deviations from these regulations and guidelines are permitted under certain circumstances and should be approved on the basis of sound written justification for scientific or medical reasons (*Guide* p. 123; AWAR §2.31,d,1,xi; PHS Policy IV,C,1,g). It should be noted that some states and Canadian provinces also have specific laws related to euthanasia.

Human: The method should take into account personnel and management issues, including the qualifications and training of the personnel administering euthanasia; the need to minimize emotional distress in human participants and observers by proper attention to the aesthetic implications of the method; and attention to the health and safety of humans (and other animals).

Scientific: The method should take into account the potential effect of the method on the scientific objectives of the research or planned postmortem evaluations. Any method of euthanasia should be viewed as having the potential to interfere with scientific objectives and it is important to choose the method with care if any data will be collected from the animal after death.

Other criteria to consider are the reliability of the method in producing a rapid and humane death, the relative likelihood that an apparently dead animal might recover following disposal, the additional time or procedures that may be required to assure that an animal will not recover after disposal, the expense and availability of drugs or equipment, the potential for human drug abuse, the ability to maintain euthanasia equipment in proper working order, safety for predators or scavengers that may eat the carcass, and potential environmental impacts of the method itself or animal's carcass following disposal.[1,3,8]

17:2 The AVMA Guidelines for the Euthanasia of Animals are not recognized as law. Must the IACUC use them when determining the appropriateness of the proposed method of euthanasia?

Reg. It is true that the AVMA euthanasia guidelines are presented as guidelines, not legal requirements. However, the AVMA Panel on Euthanasia (p. 8) acknowledges that the Guidelines will be used by some "as part of regulatory structures designed to protect the welfare of animals used for human purposes."[1] PHS Policy (IV,C,1,g) specifically states that the IACUC must determine that methods of euthanasia in a proposed research project are consistent with the AVMA euthanasia guidelines "unless a deviation is justified for scientific reasons in writing by the investigator." While this wording allows for deviation from the *AVMA Guidelines for the Euthanasia of Animals* under specific circumstances, it clearly indicates that IACUCs are required to give the Guidelines careful consideration when evaluating a proposed method of euthanasia. In NOT-OD-02-062, NIH/OLAW emphasizes that any approved deviation from AVMA Guidelines must be project specific and that IACUCs may not issue blanket waivers of applicable AVMA recommendations.[9] APHIS/AC Policy 3 states that "the method of euthanasia should be consistent with the current AVMA Guidelines on Euthanasia…" and does not address the possibility of IACUC-approved deviations.[10] However, the AWAR (§2.31,d,1,xi) state that "Methods of euthanasia used must be in accordance with the definition of the term set forth in 9 CFR part 1, Sec. 1.1 of this subchapter, unless a deviation is justified for scientific reasons, in writing, by the investigator."

Opin. For the most part, adherence to the AVMA Guidelines should present little challenge to the IACUC, but it is possible that situations will arise where judgments must be made within the framework or spirit of the Guidelines and regulations. The Guidelines are based on scientific evaluation of the pertinent literature, but published findings and/or conclusions regarding a particular euthanasia method are not always consistent, nor is all pertinent information necessarily to be found in the published literature. In such situations the IACUC may need to make its own determination regarding the best approach. In consultation with the AV, it must apply knowledge of the literature and other pertinent facts as they relate to the species in question along with sound professional judgment to achieve the goal of humane euthanasia. The IACUC also may encounter situations where an investigator wishes to use a method of euthanasia that is not addressed in the AVMA Guidelines. The potential for such a situation is acknowledged within the Guidelines, which go on to state that a veterinarian who is experienced with the species in question should apply both professional judgment and knowledge of clinically acceptable techniques to the situation. Finally, situations may arise where a deviation from the Guidelines is necessitated by scientific or clinical objectives. Both PHS Policy (IV,C,1,g) and the *Guide* (p. 123) state that the IACUC may approve a method of euthanasia that is inconsistent with the AVMA Guidelines provided that the investigator has justified the method for scientific (PHS Policy and *Guide*) or medical (*Guide*) reasons.

In response to a survey of IACUC Chairs conducted for the 2nd edition of *The IACUC Handbook*, most respondents (154 of 163 responses) indicated that they either always required following the AVMA recommendations (31%) or required following the AVMA recommendations unless scientific justification was provided for a deviation (63%).[11]

17:3 Is the performance of euthanasia considered pain relief when assigning APHIS/AC pain categories?

Opin. This question is not addressed directly in the AWAR or APHIS/AC Animal Care Policies. In the AWAR (§2.36,d,6–7), it is specifically the use of "appropriate anesthetic, analgesic, or tranquilizing drugs" that distinguishes animals placed in Category D from those in Category E. However, this question is directly addressed in APHIS/AC's *Animal Welfare Inspection Guide*.[12] In Appendix A (Guidance on Category E for Research Facilities), a hypothetical example is presented that involves euthanasia of an animal on study after it is recognized that the animal is experiencing unexpected pain or distress due to the research procedures. The guidance provided indicates that this animal would be reported in Category D.[12]

17:4 Can the AV or a DVM with delegated authority demand or perform euthanasia of an animal that s/he believes is in significant pain or distress?

Reg. Both the AWAR (§1.1, Attending Veterinarian) and PHS Policy (IV,A,3,b) state that the AV must have direct or delegated authority for activities involving animals at the facility, and the AWAR (§2.33,a,2) require that the AV have appropriate authority to ensure the provision of adequate veterinary care. However, neither the AWAR nor PHS Policy directly addresses the authority of the AV to euthanize an animal that s/he believes is experiencing significant pain or distress. The *Guide* (p. 114) is more specific in this regard, stating that the AV has the authority to intervene, even perform euthanasia, to relieve pain or distress "if the responsible person (e.g., investigator) is not available or if consensus between the investigator and veterinary staff cannot be reached concerning treatment."

Opin. The regulations and guidelines clearly state that the AV is responsible for the health and welfare of animals at the institution. Less clearly acknowledged is the role of the AV in supporting animal research at the institution. These two roles—animal protector and research facilitator—must be balanced, a goal that sometimes requires walking a fine line. The AV must protect animal health and welfare and must remain vigilant to assure that his/her professional judgment is not compromised by institutional interests that are unrelated to the responsible conduct of research.[13] Nevertheless, every effort must be made to avoid unnecessarily compromising the use of the animal in research. Upon encountering an animal that is experiencing unanticipated significant pain or distress (i.e., not addressed in the approved animal use protocol), the first thought should be to identify an effective treatment that is consistent with the goals of the study. Only if this is not possible, or the treatment fails, should an intervention that is inconsistent with experimental objectives be considered. Among the most difficult situations is one in which the IACUC-approved protocol clearly states that animals will experience significant pain or distress prior to reaching the experimental endpoint. It would not be appropriate for the AV to intervene and euthanize an animal that was exhibiting the expected signs of pain or distress but had not yet reached the IACUC-approved experimental endpoint.

17:5 If an animal is to be euthanized for clinical reasons that are unrelated to the research project (e.g., spontaneous disease, severe injury), must the method of euthanasia be one that was approved during review of the research protocol by the IACUC?

Opin. There are no specific guidelines that address this question. In the author's opinion, the primary concern should be the welfare of the animal to be euthanized, although the objectives of the research should be taken into account also. If the method of euthanasia is critical to research objectives, it would be desirable to euthanize the animal in accordance with the approved protocol. However, if the method of euthanasia is not critical to research objectives, or if use of the method of euthanasia specified in the protocol is contraindicated by the animal's clinical condition, the method used should be the one that would provide the most humane death for the animal. In any case, the method chosen should be consistent with recommendations of the *AVMA Guidelines for the Euthanasia of Animals*.[1]

Surv. If an animal is to be euthanized for clinical reasons that are unrelated to ongoing research (e.g., spontaneous disease, severe accident), which method of euthanasia listed below can be used? Check all that apply.

• Not applicable	10/295 (3.4%)
• Only the method specified in the IACUC protocol	143/295 (48.5%)
• Any method normally accepted by the IACUC (e.g., carbon dioxide, pentobarbital)	111/295 (37.6%)
• The method requested by the Principal Investigator	35/295 (11.9%)
• The method approved by our veterinarian	176/295 (59.7%)
• I don't know	1/295 (0.3%)
• Other	3/295 (1.0%)

Responses in this survey showed both similarities to, and differences from, responses in a previous survey conducted for the 2nd edition of *The IACUC Handbook*.[11] In both surveys, approximately 60% of respondents indicated that any method approved by the veterinarian was acceptable for euthanasia of animals for non-research-related reasons. However, in contrast to the present survey, where over 48% of respondents indicated that the only method(s) that could be used was the one(s) specified in the protocol, only 12% of respondents in the previous survey said that, in the absence of an IACUC-approved exception, the method used had to be one specified in the protocol.

17:6 Is it acceptable to euthanize animals in the presence of other animals of the same species?

Reg. There are no regulatory prohibitions against euthanizing animals in the presence of conspecifics, although the *AVMA Guidelines for the Euthanasia of Animals* notes that, for sensitive species, it is desirable to avoid having other animals present when individual animals are being euthanized.[1]

Opin. It has long been believed that vocalizations, behaviors, or odors released by animals undergoing frightening or stressful experiences could cause distress in other animals, particularly animals of the same species. For this reason, it is often recommended that other animals not be present during euthanasia, especially euthanasia of a conspecific.[1,2,14] There is some evidence that rats are aroused, and perhaps stressed, when exposed to conspecifics that are undergoing stressful procedures. De Laat *et al.* observed that rats exposed to animals that had been stressed by transportation were more behaviorally aroused than rats exposed to animals that had not been transported.[15] In a series of studies, Sharp and colleagues[4–7] observed that heart rate and/or blood pressure were elevated in rats that witnessed other rats undergoing procedures such as decapitation, necropsy, routine cage change, restraint, vaginal lavage, or subcutaneous or tail vein injections. The animals also showed increases in heart rate and blood pressure when exposed to urine from stressed rats or dried rat blood. The magnitude of the cardiovascular changes observed in these studies was influenced by housing density and gender. In most cases, cardiovascular responses to witnessing decapitation were generally no greater than, and sometimes less than, the responses to witnessing procedures such as routine cage change. While these studies suggest that rats at least are stressed by witnessing a particularly violent euthanasia procedure, the evidence does not support the conclusion that witnessing euthanasia of conspecifics is any more stressful to other animals than is witnessing many other, more benign (at least from the human perspective) manipulations.

Surv. Does your IACUC allow for the euthanasia of animals in the presence of animals of the same species (either in visual or auditory range) that are not being euthanized?

- Not applicable 9/295 (3.1%)
- Yes we do for all species 33/295 (11.2%)
- Yes we do, but only for some species 57/295 (19.3%)
- No we do not 169/295 (57.3%)
- I don't know 17/295 (5.8%)
- Other 10/295 (3.4%)

Responses in this survey differed somewhat from those in a previous survey conducted for the 2nd edition of *The IACUC Handbook*.[11] The differences reflect an apparent upsurge in awareness of this issue as a potential concern. In the previous survey, 28% of respondents indicated that they had never considered this matter, compared with only 9% in the current survey who responded that the question was not applicable to their organization or that they did not know their IACUC's policy on this issue. Among those who had a policy to address this issue, the percentage of respondents indicating that they do not allow animals to be euthanized in the presence of conspecifics increased from 40% in the previous survey to 57% in the present survey. This corresponded to a decrease—from 15% in the previous survey to 11% in the current survey—in the number of respondents indicating that they allowed animals of any species to be euthanized in the presence of conspecifics. Interestingly, there was a sharp increase—from 8% in the previous survey to 22% in the present survey—in the percentage of respondents who indicated that they permitted euthanasia of some (but not other) species in the presence of conspecifics.

17:7 Can tissues that are removed post-euthanasia be used for research that has not been approved by the IACUC?

Reg. The AWAR pertain to both live and dead animals (AWAR §1.1, Definitions) and dead animals or parts that are sold for research may also be regulated under the AWAR.[16] While the PHS Policy covers only live vertebrate animals, IACUC approval is required if: 1) the animals will be killed specifically for the purpose of obtaining or using their tissues; or 2) use of the tissues will require project-specific ante-mortem manipulation of the animals.[17]

Opin. As noted above, there are specific circumstances under which IACUC approval for the use of tissues from dead animals might be mandated under federal regulations or requirements. However, even under circumstances where IACUC approval might not be a regulatory requirement, it is generally appropriate for the IACUC to review all aspects of the proposed use of animals in research. The IACUC is expected to make a determination that studies using animals are scientifically valuable and consistent with the U.S. Government Principles. Principle II states that "procedures involving animals should be designed and performed with due consideration of their relevance to human or animal health, the advancement of knowledge, or the good of society." If the objectives of the study will be met in part through the use of tissues collected post-mortem, the IACUC must review these post-mortem uses to fully understand the research and the contribution of animals to it. Consider a situation where it might be possible to make use of tissues from an animal that otherwise would be discarded following its use in another study. This would be a desirable means of reducing the total number of animals used, provided that previous use of the animal did not render the tissues unacceptable for the proposed post-mortem study. For example, it would be important to verify that the animal had not been treated with any substances that might have resulted in contamination of the tissues with residues that would interfere with the post-mortem use or pose a human safety risk. In the author's opinion, the IACUC should at least be notified prior to the 'reuse' of tissues, even if they would simply have been discarded—this is essential if the IACUC is to be fully informed regarding all uses of animals at the institution.

17:8 Is perfusion under anesthesia considered euthanasia or a nonsurvival surgical procedure?

Reg. Perfusion fixation is not addressed in the federal regulations, but it is addressed as an adjunctive method of euthanasia in the *AVMA Guidelines for the Euthanasia of Animals*.[1] Provided that animals are fully anesthetized prior to perfusion, this technique is viewed as a fully acceptable adjunctive method of euthanasia.

Opin. Perfusion, or perfusion fixation, is a technique for replacing the blood in the animal's body with tissue fixative. Typically this is done by initial infusion of saline to remove the blood, followed by infusion of the fixative. Death may occur relatively early or late in this process. Animals must be deeply anesthetized prior to starting the procedure.[1] Sedation would not be adequate for perfusion fixation, which involves pain or distress related to replacement of blood with fixative and, in most cases, some degree of surgical manipulation.

Whether to classify perfusion under anesthesia as a method of euthanasia or a nonsurvival surgical procedure has been a topic of debate. Because preparation of an animal for perfusion often requires opening the thorax or abdomen to provide access to the heart, many IACUCs have classified the procedure as nonsurvival surgery. However, preparation for perfusion may involve nothing more than cut-down to a major vessel or simple venipuncture. Furthermore, if the animal is deeply anesthetized prior to starting the procedure, even if surgical preparation is involved, perfusion fixation meets the definition of euthanasia in the AWAR (§1.1, Definitions), AVMA Guidelines,[1] and *Guide* (p. 123), i.e., rapid loss of consciousness followed by death without pain or distress. On this basis, many IACUCs have classified the procedure as euthanasia, particularly if there is minimal or no surgical preparation involved. With the publication of the 2013 *AVMA Guidelines for the Euthanasia of Animals*, the AVMA Panel on Euthanasia has taken a definite stand in this debate by classifying perfusion as an adjunctive method of euthanasia. Thus, for most circumstances, it is most consistent with current guidelines for IACUCs to classify perfusion under anesthesia as euthanasia. The only exception might be a situation where the animal is subjected to a surgical procedure, at the termination of which it is perfused prior to regaining consciousness.

Surv. Does your IACUC consider liver perfusion (or another perfusion procedure) as a primary form of euthanasia or is it considered nonsurvival surgery? NOTE: This survey was performed prior to the publication of the 2013 *AVMA Guidelines for the Euthanasia of Animals*, at a time when the then current 2007 *AVMA Guidelines for Euthanasia* did not address perfusion under anesthesia.

- Not applicable 97/296 (32.8%)
- Euthanasia 31/296 (10.5%)
- Nonsurvival surgery 103/296 (34.8%)
- Nonsurvival surgery if a body cavity is opened, euthanasia if a peripheral vein is used 28/296 (9.5%)
- I don't know 34/296 (11.5%)
- Other 3/296 (1.0%)

Responses in this survey differed from those in a previous survey conducted for the 2nd edition of *The IACUC Handbook*.[11] In the previous survey, more respondents classified this procedure as euthanasia (38%) than as nonsurvival surgery (25%), whereas the reverse was true in the current survey (10% euthanasia vs. 35% nonsurvival surgery). There was less difference between the two surveys in the percentage of respondents who classified it as nonsurvival surgery if a body cavity was opened (12% in the previous survey vs. 9% in the current survey). The percentage of respondents answering "not applicable" or "I don't know" was much higher in the current survey (44%) than in the previous survey (23%), possibly suggesting that perfusion may be used less commonly now than it was a few years ago.

17:9 Should the IACUC review methods used to collect and remove escaped or verminous rodents from the animal facility?

Reg. Neither PHS Policy nor the AWAR address this question directly. The *Guide* (p. 74) states that any trapping methods used to control rodent pests should be humane, and also notes that live-catch traps require frequent observation and humane euthanasia after capture. The *AVMA Guidelines for the Euthanasia of Animals* state in general terms that the same standards for humane euthanasia should be applied to killing wild or feral animals.[1] Wild-caught animals should be handled and euthanized in a manner that causes the least amount of stress. These Guidelines specifically address the acceptability of kill traps for euthanasia. These traps do not always cause death in a manner consistent with the criteria for euthanasia established by the AVMA Panel on Euthanasia and use of live traps followed by other methods of euthanasia is preferred. If used, kill traps should be checked daily and designed to kill captured animals swiftly; if a trapped animal is found alive, it should be killed quickly and humanely.

Opin. Nothing in the *Guide* or *AVMA Guidelines for the Euthanasia of Animals* specifically dictates IACUC involvement in determining methods for controlling escaped or verminous rodents. However, PHS Policy and the AWAR require the IACUC to review and approve the animal care and use program, of which pest control is an essential aspect. The IACUC, in close consultation with the AV, is also well qualified to ensure that the provisions of the *Guide* and AVMA Guidelines are followed with respect to humane euthanasia of wild/feral animals in general and trapping of wild/feral pests in particular. There is an added element of ethical obligation for the welfare of escaped rodents. It is ironic that the status of research rodents can change so dramatically—from valued, even pampered, to scorned—when they escape. Nonetheless, the point could be made that the IACUC's responsibility to protect the welfare of these animals should continue despite this change in status.

17:10 If a trained person performs euthanasia improperly, resulting in injury to an animal or disposal of an animal that is not dead, what actions should the IACUC take?

Reg. NIH/OLAW,[9] CCAC,[18] the *Guide* (p. 124), and the *AVMA Guidelines for the Euthanasia of Animals*[1] emphasize that death must be verified after euthanasia and before disposal of the animal. This is viewed as particularly important when using carbon dioxide (CO_2) for euthanasia, as animals may appear dead while in the chamber, but recover when exposed to room air.[9] In addition, PHS Policy (IV,B,4), the AWAR (§2.31,c,4), and the *Guide* (p. 23) stipulate that it is the responsibility of the IACUC and the IO to review and/or investigate animal welfare concerns at the institution. Both the AWAR (§2.31,d,7) and PHS Policy (IV,C,7) require that a report be submitted to APHIS/AC and/or NIH/OLAW if an activity is suspended by the IACUC as a result of such an investigation. NIH/OLAW views failure to ensure death of animals after euthanasia as a serious noncompliance with PHS Policy and requires PHS-Assured institutions to report the incident promptly—whether or not the incident results in suspension of an activity—with a full explanation of circumstances and actions taken.[9,19] Readers are encouraged to consult NIH/OLAW's Reporting Noncompliance webpage at http://grants.nih.gov/grants/olaw/reporting_noncompliance.htm for more information on this topic.

Opin. Improper performance of euthanasia, leading to injury of the animal or recovery of the animal following disposal, is a serious matter. The regulations and other guidelines are unambiguous in their requirements that: 1) death must be confirmed following euthanasia; and 2) animal welfare concerns must be reviewed by the IACUC. Disposal of a live animal after a botched attempt at euthanasia clearly would constitute a failure to confirm death, and either injury to an animal during a failed attempt at euthanasia or disposal of a live animal would fall within the definition of an animal welfare concern. Furthermore, failure to ensure death is identified by NIH/OLAW as an example of serious noncompliance with PHS Policy.[9,19] In a situation where an animal recovers following disposal, the following actions would be appropriate: investigate the situation to determine why it happened and what can be done to prevent it from happening again; and, for PHS-Assured institutions, report the incident to NIH/OLAW. A report need be made to APHIS/AC only if the corrective action includes suspension of an activity involving a covered species (see Chapter 29). Reporting to NIH/OLAW might not be necessary following a failed attempt at euthanasia provided that the failure is recognized immediately and the animal is killed promptly, before it regains consciousness. If for some reason the animal were not euthanized immediately, it would most certainly require a report to NIH/OLAW. Either way, this situation would qualify as an animal welfare concern, requiring some level of review to determine the appropriate corrective and preventive actions. Retraining might be in order to prevent future occurrences. Retraining should focus on: proper performance of the technique in question; recognition of cessation of vital signs in the species in question; and/or effective application of an adjunctive method of euthanasia, e.g., thoracotomy, decapitation, cervical dislocation, or exsanguination.[1,2,8,20]

Surv. If a trained person performs euthanasia improperly, resulting in the injury or disposal of an animal that is unconscious but not dead, what actions would your IACUC typically take? Check as many boxes as appropriate.

- Not applicable 14/295 (4.7%)
- No action for a first-time offense 14/295 (4.7%)
- We typically take no action, even for repeated offenses 1/295 (0.3%)
- We may give a verbal reprimand 116/295 (39.3%)
- We may give a written reprimand 174/295 (59.0%)
- We may provide training 174/295 (59.0%)
- We would notify NIH/OLAW for covered species 158/295 (54.6%)
- We would notify USDA/Animal Care for covered species 123/295 (41.7%)
- I don't know 12/295 (4.1%)
- Other 16/295 (5.4%)

Endpoints

17:11 What is the difference between an experimental endpoint and a humane endpoint?

Reg. The *Guide* (p. 27) defines these terms as follows: "The experimental endpoint of a study occurs when the scientific aims and objectives have been reached. The humane endpoint is the point at which pain or distress is prevented, terminated or relieved in an experimental animal." The *AVMA Guidelines for the Euthanasia of Animals* distinguish between euthanasia as an act undertaken to serve the animal's best interests vs. the use of humane techniques to induce the most rapid and painless and distress-free death possible, while further noting that these mandates are codependent.[1] The Guidelines also state that, in research situations, animal welfare considerations are balanced against the merits of the experimental design and merits of the research. In these situations, the IACUC must apply the principles of refinement, replacement, and reduction, and ensure a respectful death for research animals.

Opin. The goal for all studies should be to identify an endpoint that is humane and supports the goals of the research. If animals are terminated before scientific objectives are achieved, the value of the study is compromised and animal lives—not to mention time and resources—are wasted or used to poor end. However, if animals are allowed to suffer beyond the point where scientific objectives have been achieved, their suffering is unnecessary and the situation must be defined as inhumane. Ideally the endpoint should be selected before the study starts, and it should be clearly defined so that everyone can agree when it is reached. This process should involve the investigator, the AV, and the IACUC. In many, perhaps even the majority, of studies, identification of an appropriate endpoint is not difficult. However, for some studies—those involving cancer, infectious disease, or trauma are examples—identifying an endpoint that is both humane and scientifically sound is more of a challenge. If the investigator has significant experience with the model, s/he may be able to use existing data from previous studies to identify an appropriate endpoint. If not, a thorough review of the literature may reveal endpoints that proved useful for similar studies. In cases where there are no previous or published data upon which a decision can be based, a pilot study may be performed for the purpose of identifying a suitable endpoint. In any of these situations, further refinement of the endpoint may be possible over time.

Surv. Which description below is closest to how your IACUC typically approaches or defines a humane endpoint?

- Not applicable — 7/295 (2.4%)
- The point in a study in which an animal is killed — 8/295 (2.7%)
- The point in a study in which an animal is to be killed if pain or distress cannot be alleviated — 61/295 (20.7%)
- The point in a study in which pain or distress must be alleviated by some means, including euthanasia — 115/295 (38.0%)
- We consider humane endpoints to be protocol specific and we do not set specific criteria or definitions — 102/295 (34.6%)
- Other — 2/295 (0.7%)

17:12 What is the moribund condition? Is it an acceptable endpoint in biomedical research studies?

Opin. *Moribund* has been variously defined as "being in the state of dying; approaching death", "in a dying state", and "dying; at the point of death."[21–23] In studies where meeting scientific objectives requires knowing that a subject will die within a predictable timeframe, euthanasia of the animal when it becomes moribund may be viewed as a more humane alternative to allowing the animal to die a natural death. The question is whether an animal is likely to experience significant pain or distress in the interval between being recognized as moribund and natural death. Depending on the animal's level of conscious perception at the point where it is judged to be moribund, euthanasia at this point might or might not prevent additional suffering.

It must be assumed that prompt euthanasia would prevent suffering in animals that are judged to be moribund but that still appear conscious and responsive. But what about animals that appear unconscious? Studies of human patients show that apparent unconsciousness may not be correlated with a complete lack of perception, suggesting that even apparently comatose animals might benefit from euthanasia in lieu of progression to natural death.[24] However, the same studies suggest that a significant percentage of moribund, unconscious animals would experience no further suffering if left to die naturally. Thus, as a general rule, it is desirable to euthanize animals prior to the point where they enter a terminal comatose state. Recognizing that moribund animals might be past the point of suffering, the Canadian Council on Animal Care advises performing detailed observations of the animals to help set pre-moribund endpoints.[3] Similarly, ILAR's Committee on Pain and Distress in Laboratory Animals cautions that, unless euthanasia would interfere with experimental goals, animals should be killed before they become moribund.[2]

In a survey conducted for the 2nd edition of *The IACUC Handbook*, the question was asked whether the moribund condition was a potentially acceptable endpoint.[11] Over half (74%) of the survey takers responded that their IACUC's considered it to be potentially acceptable, but most (91%) of those indicated that it was acceptable only with scientific justification. A substantial number (21%) indicated that the moribund condition was never acceptable as an endpoint.

Surv. With reference to the word "moribund," which description below is closest to how your IACUC interprets its meaning?

- Not applicable 12/294 (4.1%)
- Moribund means near death; the animal is not conscious 30/294 (10.2%)
- Moribund means near death; the animal may or may not be conscious 182/294 (61.9%)
- Moribund means very ill but nothing more than that 29/294 (9.9%)
- Moribund means any animal experiencing pain or distress 13/294 (4.4%)
- I don't know 26/294 (8.8%)
- Other 2/294 (0.7%)

17:13 In situations in which natural death is not an acceptable endpoint, what clinical signs can be used to determine the appropriate time for euthanasia?

Reg. Federal regulations provide no specific guidelines as to which clinical signs might be most useful in determining an appropriate humane endpoint. The *Guide* (p. 27) states only that the PI should identify in the study protocol an endpoint that is both humane and scientifically sound. The Canadian Council on Animal Care provides the same general guidance along with some more specific suggestions regarding physiological and behavioral parameters that could be assessed as indicators of pain and/or distress associated with deterioration in the animal's condition.[3] The CCAC also provides some more detailed guidelines for selecting appropriate endpoints in cancer, toxicology, infectious disease, or pain research, as well as studies involving monoclonal antibody production.[3]

Opin. Signs that can be used to determine the appropriate time for euthanasia fall into two general categories: (1) those indicating pain, distress, and or deterioration in clinical condition in an animal in which effective treatments are contraindicated or have failed; and (2) those that predict an irreversible progression to the outcome required to meet scientific objectives, e.g., death. General indicators of pain, distress and overall clinical condition include: body weight; body condition score; physiological indices such as heart rate, respiration, and body temperature; external appearance; unprovoked behavior; and behavioral responses to external stimulation. The value of these indicators is increased when each is evaluated according to a precisely defined scoring system, and combining the individual scores into a cumulative score provides a means of assessing the animal's overall deviation from normal.[3,25] Species-specific signs of chronic pain or distress also may be useful in identifying early humane endpoints.[26,27]

Signs that predict an irreversible progression to a scientifically valid outcome may be more variable depending on the research area or condition being studied. Potentially useful signs can be identified based on a literature search, previous experience with the model, and/or a pilot study. Before being adopted as definitive endpoints, potential endpoints should be evaluated in studies designed to demonstrate their reliability in predicting the desired experimental outcome (e.g., pilot studies). Humane endpoints that have proven useful as valid scientific endpoints include: decrease in body condition score with or without changes in body weight and/or appearance or behavior for cancer research;[24,28,29] weight loss and/or changes in body temperature for infectious disease research;[24,30] and loss of muscular control or decrease in body temperature for vaccine potency testing.[31]

Surv. Which of the following methods has your IACUC approved in the past to help determine when an animal is to be euthanized? Check as many as appropriate.

- Not applicable — 17/295 (5.8%)
- A group of nonspecific clinical signs (e.g., ruffled fur, vocalizations, weight loss) — 220/295 (74.6%)
- A group of clinical signs, each having a numerical value, and euthanasia is performed when a certain sum is reached — 108/295 (23.6%)
- A disease severity index (e.g., when paralysis is first noticed) — 138/295 (46.8%)
- Body condition scoring — 127/295 (43.1%)
- Study-specific signs developed by the investigator, with or without input from the veterinarian or IACUC — 164/295 (55.6%)
- Other — 13/295 (4.4%)

Inhalant and Other Chemical Methods

17:14 What are the most appropriate conditions (e.g., concentration, flow rate) for euthanasia of adult rodents with carbon dioxide?

Reg. The *AVMA Guidelines for the Euthanasia of Animals*[1] view CO_2 as conditionally acceptable for euthanasia of those species in which aversion or distress can be minimized. A gradual-fill procedure, in which 10%–30% of the chamber volume is displaced per minute, is recommended. The flow of CO_2 into the chamber must be continued for at least 1 minute after respiratory arrest. It is not acceptable to place conscious animals in a chamber that has been pre-filled with 100% CO_2. When gradual-fill methods cannot be used, the animal should be rendered unconscious prior to immersing it into 100% CO_2. CO_2 and CO_2 mixtures must be supplied in a "precisely regulated and purified form... typically from a commercially supplied cylinder or tank." Additional requirements include "an appropriate pressure-reducing regulator and flow meter or equivalent equipment with demonstrated capability for generating the recommended displacement rates for the size container being utilized." Mixing the CO_2 with oxygen is not recommended. These guidelines are similar to those of the Canadian Council on Animal Care, which maintains that CO_2 is not an ideal method for humane killing in any species and that it should be used as the sole agent for euthanasia only under circumstances where other methods are impractical.[18] NIH/OLAW also discourages prefilling the euthanasia chamber and emphasizes the importance of using only compressed CO_2 in cylinders.[9]

Opin. The acceptability of CO_2 as an agent for euthanasia of adult rodents has been called into question in recent years. Even amongst those who support the use of CO_2 for this purpose, there is controversy regarding the conditions that are most likely to ensure a humane death. The most controversial issues are: whether or not to prefill the chamber with CO_2 before adding animals; the rate of introduction of CO_2 into the chamber once animals are present; and whether or not to dilute the CO_2 with oxygen. While acknowledging that high concentrations of CO_2 could be distressful to some animals, both the 2000 *Report of the AVMA Panel on Euthanasia*[32] and 2007 *AVMA Guidelines on Euthanasia*[33] concluded that CO_2 was acceptable for euthanasia of laboratory rodents. Prefilling the chamber was viewed as a desirable method for hastening loss of consciousness and the optimal flow rate was one that displaced at least 20% of the chamber volume per minute. These recommendations, now obsolete, are in sharp contrast to those made in the 2013 *AVMA Guidelines for the Euthanasia of Animals*,[1] which are summarized above. All three AVMA reports concluded that there was no apparent advantage to combining oxygen with the CO_2 and the 2013 report recommends against this practice.

There is evidence that animals experience significant pain and/or distress when exposed to high concentrations of CO_2.[34–37] This is consistent with research indicating that carbon dioxide, which forms carbonic acid in a moist environment (e.g., the nasal mucosa), specifically excites small nerve fibers that transmit sensations of pain.[38,39] For this reason, it has been recommended that conscious animals should not be exposed to CO_2 concentrations greater than 70%.[36] On the other hand, unconsciousness and death occur more quickly when animals are exposed to high concentrations of CO_2. At least in part because animals collapse more quickly when exposed to initial high concentrations of CO_2, some observers argue that prefilling the chamber with CO_2 before introducing animals may reduce anxiety and struggling;[2] others suggest that pain and distress can be reduced if the chamber is not prefilled.[36,37,40,41] Some observers noted no differences in the behavior of rats euthanized in prefilled versus non-prefilled chambers.[42,43] As noted above, the *AVMA Guidelines for the Euthanasia of Animals: 2013 Edition* views the immersion of conscious animals into 100% CO_2 as unacceptable.[1] NIH/OLAW recommends prefilling "only under circumstances in which such use has not been shown to cause distress."[9] The American College of Laboratory Animal Medicine goes further, recommending that the chamber not be prefilled "because inspiration of high concentrations of CO_2 is both aversive and painful."[20]

Several sources suggest the use of low gas flow rates that lead to a gradual increase in the concentration of CO_2 in the chamber.[1,2,18,20,40,44] Two reasons are cited for this recommendation:

- The turbulence and loud hissing associated with the rapid influx of gas into the chamber appear distressing to animals.
- Slowly rising carbon dioxide concentrations allow the animal to lose consciousness prior to asphyxiation and with less pain and distress than they might experience if exposed high concentrations of CO_2 while still fully conscious.

A flow rate that would displace 10%–30% of the chamber volume is recommended.[1,18,20,44]

The benefit, if any, of diluting CO_2 with oxygen has been an ongoing topic of debate. The addition of oxygen has been observed to reduce distress,[45] provide no benefit,[42] and even potentially increase distress[46] in animals euthanized with CO_2. The overall conclusion is that there is no convincing evidence that mixing oxygen with CO_2 provides a distinct benefit,[1,20] and the 2013 *AVMA Guidelines for the Euthanasia of Animals* recommends against this practice.[1] Data from studies of 'air hunger' (an uncomfortable urge to breathe) in humans reinforce this position. This highly unpleasant, even frightening, condition can be induced experimentally in subjects rendered mildly hypercapnic by exposure to CO_2 concentrations of 7%–8%.[47–49] Although the phenomenon of air hunger has not been demonstrated conclusively in animals, there is evidence to suggest that the physiological basis for air hunger is well developed in small rodents.[50,51] The addition of oxygen would likely do nothing to minimize any sensation of air hunger experienced during exposure to the high concentrations of CO_2 needed for euthanasia, but would likely prolong the time to unconsciousness and death; it is not recommended.[1]

Surv. If your IACUC allows the use of carbon dioxide for the euthanasia of small rodents, which of the following statements are true? Check all that apply.

• Not applicable	41/294 (13.9%)
• The euthanasia chamber must be filled slowly with animals already in it	103/294 (35%)
• The euthanasia chamber must be rapidly filled with animals already in it	52/294 (17.7%)
• We require the use of a gas regulator on all CO_2 valves	187/294 (63.6%)
• We require the use of a gas regulator on some CO_2 valves	8/294 (2.7%)
• I'm not sure about some or any of the above statements	46/294 (15.6%)

17:15 For euthanasia of rodents with carbon dioxide, are there any guidelines regarding proper use of the chamber?

Reg. Both the *AVMA Guidelines for the Euthanasia of Animals*[1] and NIH/OLAW[9] address general considerations for placing animals in the chamber prior to euthanasia with CO_2. Both emphasize the importance of avoiding overcrowding and also of minimizing distress that might be caused by mixing animals of different species or animals that are otherwise incompatible or simply unfamiliar with each other. If necessary, animals should be restrained or separated to avoid harm to themselves or others. The AVMA Guidelines further emphasize the need to keep chambers clean "to minimize odors that might cause distress in animals subsequently euthanized."

Opin. For practical purposes, the density in the chamber must not exceed the point where the CO_2 can circulate freely, ensuring that all animals will be equally exposed. From an animal welfare perspective, it should be possible—at least in theory; the animals may have other ideas!—for all animals to have all four feet on the floor at the same time, i.e., no animal should be forced due to lack of space to stand partially or completely on top of other animals. It is preferable to load the animals into the chamber immediately prior to administration of euthanasia. However, if it will be necessary to hold animals in the chamber for any interval prior to euthanasia, density should be maintained below the point where the temperature, humidity, and levels of oxygen and CO_2 within the chamber would be adversely affected.

As noted by both NIH/OLAW and the AVMA Guidelines, density is only one of the factors that should be taken into consideration to protect animal welfare during CO_2 euthanasia.[1,9] Equally important are concerns related to social compatibility and other potential stressors to which animals might be subjected by humans seeking to complete an unpleasant task as quickly and efficiently as possible. Laboratory mice and rats exhibit evidence of fear, stress, and possibly distress in response to a variety of environmental conditions and procedures in the vivarium, including noises, pheromones and other odors, handling, and transportation.[1,52,53] Mice in particular also exhibit signs of stress when combined with unfamiliar conspecifics; depending on sex, strain, and other factors, they may or may not behave aggressively toward each other in such situations.[54–57] Stress due to these factors can be minimized by: not mixing animals from different social groupings; transporting and euthanizing animals in the home cage; transporting them on carts that roll smoothly and quietly; and cleaning the euthanasia chamber between groups of animals.[1,9,20]

17:16 Is carbon dioxide an appropriate agent for euthanasia of rabbits?

Reg. The *AVMA Guidelines for the Euthanasia of Animals* list CO_2 as acceptable with conditions for the euthanasia of adult rabbits, noting that prior sedation will minimize distress associated with CO_2 euthanasia in this species. The Guidelines also observe that animals of the genus *Oryctolagus* may have prolonged survival times following CO_2 exposure.[1] In contrast, the Canadian Council on Animal Care does not view CO_2 as either fully or conditionally acceptable for euthanasia of rabbits.[18]

Opin. CO_2 is viewed by many as an effective, but often less preferred, method of euthanasia for rabbits.[1,2,58,59] Green[59] and the AVMA Guidelines,[1] for example, note that it is a useful agent but that it appears to cause apprehension or distress in rabbits. In the author's opinion, CO_2 is an acceptable agent for euthanasia of sedated rabbits, but other methods are preferable. When rabbits are exposed to high concentrations of this gas, particularly at first, they tend to struggle and kick, and the potential for injuring themselves or an animal handler is significant. This is a common response of rabbits to any stimulus that causes pain or fear and is probably not indicative that rabbits experience more discomfort than other animals exposed to CO_2. If scientific or other circumstances necessitate the use of CO_2 to euthanize a conscious rabbit, it is essential that the animal be sedated and properly restrained (e.g., in a commercial rabbit restrainer) to prevent it from kicking. The rabbit should never be placed loose in a euthanasia chamber, where the animal could injure itself if it were to panic during the procedure. If the rabbit is still conscious, the CO_2 should be introduced slowly into the chamber, and appropriate precautions should be taken to assure that the animal is dead prior to disposal.

Surv. Does your IACUC allow the use of carbon dioxide as a euthanasia agent for adult rabbits?

- Not applicable 130/294 (44.2%)
- Fully acceptable 8/294 (2.7%)
- Acceptable only with scientific justification 16/294 (5.4%)
- Acceptable only if the rabbit is anesthetized or sedated prior to euthanasia 11/294 (3.7%)
- Not acceptable for rabbits 105/294 (35.7%)
- I don't know 24/294 (8.16%)

17:17 Is nitrogen or argon—alone, combined, or mixed with carbon dioxide—an appropriate agent for euthanasia of rodents or rabbits?

Reg. The AVMA *Guidelines for the Euthanasia of Animals* and Canadian Council on Animal Care view nitrogen and argon as unacceptable as sole agents for euthanizing most mammals, including rodents and rabbits.[1,18] However, the AVMA Guidelines indicate that high concentrations of nitrogen or argon (>98%) may be used to kill laboratory rodents after they have been rendered unconscious by other acceptable means, further noting that prolonged exposure may be required to ensure death.

Opin. Nitrogen and argon are inert gases that are available commercially in compressed gas cylinders. A mixture of argon (70%) and carbon dioxide (30%) is also commercially available for use in welding. These agents, alone or in combination, have been used to euthanize animals of several species, including dogs,[60] rats,[61] mink,[62] swine,[63,64] turkeys,[65] and chickens.[66–69] Disturbing reactions, including convulsions, were seen in some species after induction of unconsciousness. Animals of most species would willingly enter hypoxic chambers and overt signs of distress were seldom observed in conscious animals. However, only poultry and swine were willing to remain in hypoxic chambers to the point of unconsciousness. In contrast, rats invariably showed signs of distress when exposed to mixtures of nitrogen and air.[61] Similarly, several investigators[37,70,71] noted that rats and mice were reluctant to enter chambers filled with argon or that they left chambers with gradually rising argon concentrations prior to losing consciousness. Nonetheless, Leach *et al.*[37] concluded that rodents appeared to find 100% argon less aversive than either carbon dioxide or a mixture of carbon dioxide and argon.

Surv. Does your IACUC approve the use of other gases or gas mixtures (such as argon, nitrogen) alone or with carbon dioxide for small rodent euthanasia? Check as many boxes as appropriate.

- Not applicable 110/294 (37.4%)
- We have no policy 68/294 (23.1%)
- We allow argon alone 0/294 (0%)
- We allow nitrogen alone 0/294 (0%)

- We allow various mixtures of argon, nitrogen, and carbon dioxide — 2/294 (0.75)
- We do not allow the use of other gases alone or in mixture — 105/294 (35.7%)
- I don't know — 18/294 (6.1%)

Physical Methods

17:18 What type of justification is needed if an investigator states that a small rodent is to be euthanized by cervical dislocation?

Opin. The *AVMA Guidelines for the Euthanasia of Animals* views cervical dislocation of small rodents and rabbits as acceptable with conditions.[1] Language in the 2007 Guidelines[32] that this technique should be used in research settings only when scientifically justified has been removed from the 2013 Guidelines. Nonetheless, methods of euthanasia must be reviewed by the IACUC, which may require justification for use of this technique vs. one that is classified by the AVMA Guidelines as fully acceptable. The best justification would be provision of data—from the PI's own laboratory or the published literature—showing how each acceptable method would interfere with specific scientific objectives using the model in question, and also showing that cervical dislocation would not interfere. In lieu of such data on each acceptable method, a plausible rationale might be presented for extrapolating data demonstrating the undesirable effects of one particular method to other similar methods, e.g., from isoflurane to all inhalant anesthetics. In the absence of any existing data, the IACUC has the option of either: 1) accepting a plausible explanation of how fully acceptable methods would be likely to (or might) adversely affect scientific objectives; or 2) requiring the PI to perform a study to generate data. In deciding whether to require the PI to perform a pilot study, the IACUC would need to assess whether the sacrifice of additional animal lives would outweigh the value of the information that would be gained. In the author's opinion, the most humane option would be to approve the use of cervical dislocation on the basis of likely inference with the research and provide whatever training is necessary to ensure the proficiency of the personnel who will perform the procedure. If personnel are skilled in the technique, cervical dislocation is a humane method of euthanasia.[1]

Surv. What type of justification does your IACUC request if an investigator states that a small rodent is to be euthanized by cervical dislocation? Check as many boxes as appropriate.

- Not applicable — 47/296 (15.9%)
- We require no additional justification — 28/296 (9.5%)
- A general statement that other means of euthanasia would interfere with the study — 39/296 (13.2%)
- A specific statement that other means of euthanasia would interfere with the study but references are not required — 83/296 (28%)

- A specific statement that other means of euthanasia would interfere with the study AND includes one or more published references or laboratory data — 126/296 (42.6%)
- I don't know — 2/296 (0.7%)
- Other — 14/296 (4.7%)

17:19 Is cervical dislocation an acceptable method of euthanasia for large rats (> 200 gm) or rabbits (> 1 kg)?

Reg. The *AVMA Guidelines for the Euthanasia of Animals* classify manual cervical dislocation as a conditionally acceptable method for euthanasia of rabbits and for mice and rats weighing less than 200 g.[1] In contrast to the 2007 iteration of the *AVMA Guidelines on Euthanasia*,[33] the 2013 Guidelines do not limit application of this technique to smaller rabbits, although the observation is still made that this procedure is physically more difficult with larger rabbits (and rats) due to their increased cervical muscle mass. The current Guidelines emphasize the importance of assuring technical competence in personnel who will perform this procedure on conscious mature rabbits, and they state that commercially available equipment designed to aid in this procedure should be evaluated for effectiveness. The Canadian Council on Animal Care classifies cervical dislocation as conditionally acceptable for rodents and rabbits in general, but further specifies that commercial dislocators/luxators must be used on heavier rats (>200 g) and rabbits (> 2 kg).[18]

Opin. It is generally acknowledged that it is physically challenging to perform manual cervical dislocation on larger, more heavily muscled animals. As a result, there is a high potential for severely injuring, without killing, such an animal. The likelihood of this outcome can be significantly reduced by use of a mechanical dislocator, and is only with use of this equipment that the CCAC views cervical dislocation as conditionally acceptable for larger rats or rabbits.[18] The AVMA Guidelines do not recognize any conditions under which cervical dislocation of conscious mature rats might be acceptable, but consider the procedure conditionally acceptable for conscious mature rabbits.[1] While not requiring the use of a mechanical dislocator for larger rabbits, the AVMA Guidelines advise institutions to evaluate the effectiveness of such equipment. An IACUC in the U.S. would have to grant an exception from AVMA Guidelines before an investigator could euthanize a mature rat (>200 g) by cervical dislocation. In the author's opinion, such an exception might be considered—with excellent scientific justification—if the skill of the person performing the technique were unquestioned, if a mechanical dislocator were to be employed, and/or if the animal were to be heavily sedated or anesthetized for the procedure. These same conditions must apply for euthanasia of mature rabbits by cervical dislocation to be performed in accordance with AVMA Guidelines.

Surv. Does your IACUC permit euthanasia of large rats (>200 grams) or rabbits (>1 kg) by cervical dislocation? Check as many boxes as appropriate.

- Not applicable — 66/296 (22.3%)
- Yes, for rats — 34/296 (11.5%)
- Yes, for rabbits — 0/296 (0%)
- Yes, for rabbits or rats but only if first anesthetized or sedated — 27/296 (9.1%)

- No, for rats 156/296 (52.7%)
- No, for rabbits 161/296 (54.4%)
- I don't know 20/296 (6.8%)
- Other 6/296 (2.0%)

When this question was asked in a survey conducted for the 2nd edition of *The IACUC Handbook*,[11] the percentage of respondents indicating that their IACUC did not permit cervical dislocation of large rats (55%) was similar to that in the current survey (53%). However, whereas 54% of respondents in the current survey would not allow cervical dislocation of large rabbits, only 34% of respondents in the previous survey disallowed this procedure on large rabbits. A small percentage of respondents (1.9%) in the previous survey specified that cervical dislocation of large rabbits was permitted if the personnel were skilled, and 14% (vs. 9% in the current survey) indicated that cervical dislocation of larger animals was permitted if the animals were anesthetized first. Also, 7.5% of the respondents in the first survey stated that they had no policy on this issue.

17:20 Must a moribund mouse be anesthetized prior to euthanasia by decapitation or cervical dislocation?

Reg. Anesthesia per se is not required prior to decapitation or cervical dislocation,[1,18] although the *AVMA Guidelines for the Euthanasia of Animal*s specifies that animals must be unconscious or anesthetized prior to cervical dislocation in situations where the operator has not demonstrated technical proficiency.[1] On the other hand, the Guidelines (pp. 16, 80, 83) specifically and repeatedly equate the moribund condition with unconsciousness or anesthesia, suggesting that the Guidelines would not require anesthesia of a moribund animal prior to cervical dislocation even by a less technically proficient operator.

Opin. The AVMA Guidelines require that small rodents be unconscious or anesthetized prior to cervical dislocation by a person whose skill is unproven. This is to protect the welfare of the animal, which might be left seriously injured but not dead following a bungled attempt at cervical dislocation. This is less of a concern with decapitation, which is technically (though, for most people, not psychologically) easier to perform correctly, but concerns for both human safety and animal welfare dictate that animals be rendered unconscious prior to decapitation by untrained individuals. Could an untrained operator perform either cervical dislocation or decapitation as safely and humanely on a moribund mouse as on an anesthetized mouse? While a moribund animal is by definition near death, it is not necessarily unconscious (see 17:11) and might still be able to experience pain or distress. It might even be able to struggle, thereby posing a human safety risk during decapitation by an untrained individual. It is best to assume that the moribund mouse would be able to experience pain and distress, and to assign euthanasia by decapitation or cervical dislocation to a skilled operator. If for any reason the procedure must be performed by an inexperienced individual, the mouse should be deeply sedated or anesthetized beforehand.

When asked in a survey conducted for the 2nd edition of *The IACUC Handbook* whether their IACUC would require anesthesia of a moribund mouse prior to cervical dislocation or decapitation, many people responded that this question was not applicable to their situation or that it had never arisen.[11] Among those who had policies on this issue, 14% indicated that their IACUC would require anesthesia, 14% that their IACUC would not require anesthesia, and 14% indicated that they would require scientific justification for withholding anesthesia, 7% said they would require anesthesia only if the mouse were still conscious.

17:21 Is it acceptable to use scissors, rather than a guillotine, for decapitation of adult mice?

Opin. The *AVMA Guidelines for the Euthanasia of Animals*[1] notes that "sharp blades"—presumably scissors—can be used for decapitation of neonatal rodents, but the use of scissors for decapitation of adult mice is not addressed by the Guidelines or in any of the U.S. regulations pertaining to euthanasia. The ILAR Committee on Pain and Distress in Laboratory Animals[2] states that decapitation of rodents may be accomplished by using a guillotine or heavy shears. The Canadian Council on Animal Care does not specifically address the use of scissors for decapitation, but may be alluding to—and tacitly approving?—their use in the following statement: "blades should be kept very sharp and guillotines should be well maintained and cleaned between uses."[18] In the author's opinion, humane euthanasia of adult mice using heavy, well-sharpened scissors is possible, provided that the operator is highly skilled in this technique. However, this procedure is exceedingly distasteful for most people, and this can lead all but the most resolute and experienced operator to perform the procedure less quickly and cleanly than is required for humane euthanasia. For this reason, use of scissors for decapitation is best limited to euthanasia of neonates or for adult mice that have been heavily sedated or anesthetized. People who wish to use this approach to euthanize fully conscious adult mice should be especially skilled and experienced.

When asked in a survey conducted for the 2nd edition of *The IACUC Handbook* whether their IACUC would consider the use of scissors as acceptable for decapitation of adult mice, 63% of the respondents answered 'no.'[11] Only 4% responded that their IACUC considered this technique fully acceptable, but an additional 8% indicated that it would be acceptable if the mouse were sedated or anesthetized beforehand.

17:22 When reviewing a protocol that involves euthanasia of mice by cervical dislocation or decapitation, how should the IACUC determine that personnel performing these techniques are properly trained?

Reg. PHS Policy (IV,C,1,f), the AWAR (§2.31,d,1,viii), and the *Guide* (pp. 16–17, 124) emphasize that personnel performing procedures on animals must be properly trained. In all instances, euthanasia procedures are explicitly or implicitly included in these requirements (see Chapter 21). The *AVMA Guidelines for the Euthanasia of Animals* and Canadian Council on Animal Care specifically emphasize the need for personnel to be properly trained prior to performing euthanasia by decapitation or cervical dislocation.[1,18]

Opin. Although PHS Policy (IV,C,1,f), the AWAR (§2.31,d,1,viii), and the *Guide* (p. 15) specify that the IACUC is responsible for assuring that personnel are appropriately trained and that training programs are effective, none of these documents specifies the basis upon which the IACUC would determine that a particular individual has been properly trained. The *AVMA Guidelines for the Euthanasia of Animals* state that inexperienced persons should practice on dead or anesthetized animals until they are proficient in performing these procedures properly and humanely.[1] The Canadian Council on Animal Care also emphasizes the importance of assuring that personnel are trained in the humane performance of these procedures, as well as in the safe use of equipment for decapitation.[18] As for who should provide the training, the AWAR (§2.33,b,5) places guidance on euthanasia under the program of veterinary care and §2.32 makes "the research facility" responsible for providing training. The *Guide* (p. 16) states only that it is an institutional responsibility to assure that training is provided. The ILAR Committee on Pain and Distress in Laboratory Animals is bit more specific, maintaining that the IACUC and AV should provide for the training and supervision of personnel who will perform euthanasia. Training by the AV or IACUC-approved institutional training program provides the IACUC with the highest degree of certainty that personnel are properly trained. An alternative approach is to have the PI or other skilled individual perform the training. In this case, direct observation of the trainee's skill by the AV, institutional trainer, or other experienced member of the IACUC is desirable prior to approving him/her to perform the procedure unsupervised.

Ideally, the skills of trained personnel would be reassessed or reapproved on a regular basis—every 2–5 years, for example. Trained personnel would be evaluated in a less formal manner as part of the post-approval monitoring process. Should any concerns arise regarding the proficiency of a particular operator, it would be incumbent upon the IACUC to require an assessment of their skill by the AV, institutional trainer, or other IACUC-approved individual.

Surv. When reviewing a protocol that involves the euthanasia of mice by cervical dislocation or decapitation, how does your IACUC determine that personnel performing the procedure are properly trained to do so? Check as many boxes as appropriate.

- Not applicable 51/296 (17.2%)
- We rely on the investigator to provide assurance that personnel are properly trained 152/296 (51.4%)
- We ask the veterinary staff to make appropriate observations 123/296 (41.6%)
- We have a post-approval monitoring program that periodically checks techniques 81/296 (27.4%)
- We require pre- or post-approval training 85/296 (28.7%)
- I don't know 1/296 (0.3%)
- Other* 9/296 (3.0%)

*"Other" responses included: 4 respondents indicated that some degree of monitoring, review or certification was required; 1 respondent indicated that the veterinarian would determine proficiency if any concerns arose about the operator's proficiency; 3 respondents indicated that training was provided by institutional trainers or the vivarium staff; and 1 respondent indicated that only the PI was "allowed to do this." One respondent specified that annual sharpening of guillotines was required.

Adjunctive Methods—Stunning, Pithing, Exsanguination (and see 17:8)

17:23 An investigator states that she has many years of experience euthanizing rats by swinging them rapidly by the tail and hitting their heads against the edge of a table. Should the IACUC allow this procedure, if properly performed, as an acceptable form of euthanasia?

Reg. The *AVMA Guidelines for the Euthanasia of Animals* acknowledges that manually applied blunt force trauma to the head can be an effective and humane method of euthanasia when it is performed correctly on animals with thin craniums.[1] However, humane concerns can arise when it is performed repeatedly or on an inappropriate subject, and it can be aesthetically displeasing to human personnel. As a result, the AVMA Guidelines recommend that those who use this technique actively search for alternative methods. Concussion (crushing blow to the head) is recognized by the Canadian Council on Animal Care as conditionally acceptable under emergency conditions only.[18]

Opin. Cranial trauma created by swinging the animal and hitting the back of its head on the edge of a table is an old method that was used primarily to euthanize small laboratory animals with thin craniums; it is seldom mentioned in the more recent literature. However, this technique has been described as a humane method of euthanasia or as a method of rendering animals unconscious prior to exsanguination or cervical dislocation in small animals.[14,41,59,72] Waynforth and Flecknell emphasize that the procedure must be performed correctly and that training should take place using deeply anesthetized animals.[41] The need for training of personnel is similarly emphasized by the *AVMA Guidelines for the Euthanasia of Animals*, which also emphasize the need to monitor personnel for proficiency and assure that they are aware of the aesthetic implications of the technique.[1] Stunning by hitting the animal's head on the edge of a table requires a high level of concentration in addition to considerable technical proficiency. It takes only a small deviation in the force or angle of the blow to result in severe injury rather than instant death. Inconsistency in the performance of the technique is more likely as personnel become fatigued after performing the procedure repeatedly. Furthermore, swinging the animal by the tail is likely to produce distress, even if collision of the head were to produce rapid death. Therefore, it is questionable whether this particular method of stunning would qualify as euthanasia, as defined in the AWAR. In the author's opinion, the IACUC should allow use of this method only when it is well justified and the committee is satisfied that other, preferable alternatives cannot be used.

Surv. Does your IACUC accept stunning of rodents, if properly performed, as an acceptable form of euthanasia?

- Not applicable 66/295 (22.4%)
- We have no policy on stunning 61/295 (20.7%)
- Acceptable without scientific justification 0/295 (0%)
- Acceptable only with scientific justification 10/295 (3.4%)
- Not acceptable 157/295 (53.2%)
- Other 1/295 (0.3%)

When asked the same question in a survey conducted for the 2nd edition of *The IACUC Handbook*[11] a higher percentage of the respondents replied that their IACUC had no policy on stunning (38.6% in the previous survey vs. 20.7% in the current survey), but a lower percentage responded that their IACUC viewed stunning as not acceptable (44.4% in the previous survey vs. 53.2% in the current survey). 14% of respondents in the previous survey indicated that stunning was acceptable with scientific justification, demonstration of proficiency and/or use of a second method to assure death.

17:24 Is pithing an acceptable method for euthanasia of conscious frogs?

Reg. The *AVMA Guidelines for the Euthanasia of Animals* strongly discourages the use of pithing for euthanasia of frogs and other amphibians unless the animal is anesthetized or otherwise unconscious; personnel must be properly trained in the technique.[1] Similarly, the Canadian Council on Animal Care indicates that pithing may be used to render frogs brain dead, but the animals must be at a surgical plane of anesthesia for the procedure.[18]

Opin. Simple pithing—destruction of the brain only—does not result in the death of the frog (the heart remains beating, respirations continue, and many reflexes remain intact) and is not viewed as an effective or acceptable method of euthanasia. Double pithing, which involves destruction of the brain followed immediately by destruction of the spinal cord, is required to assure death. Pithing requires considerable technical proficiency; if the person performing the technique is not skilled, the animal could experience considerable pain and suffering.

In a survey conducted for the 2nd edition of *The IACUC Handbook*, respondents were asked whether their IACUC viewed pithing as an acceptable method of euthanasia for conscious frogs.[11] Most of those who responded (63%) indicated that the question was not applicable to their institution, that the issue had never arisen at their institution, or that their institution had no policy. Another 15% answered that pithing was acceptable only with scientific justification, and a nearly equal number (14%) answered that it was acceptable only if the animal was anesthetized. Only 5% considered it acceptable without justification.

17:25 Is sedation (e.g., with diazepam or acepromazine) an acceptable alternative to anesthesia for pretreatment of animals prior to exsanguination?

Reg. The *AVMA Guidelines for the Euthanasia of Animals* state that exsanguination may be used as an adjunctive method to ensure death in animals that have been stunned or are otherwise unconscious.[1] However, the Guidelines also indicate that if exsanguination will be used to obtain blood products, sedation is an acceptable pretreatment alternative to anesthesia or stunning. The Canadian Council on Animal Care is less equivocal, stating that animals must be unconscious prior to exsanguination and that sedation is inadequate to ensure unconsciousness.[18]

Opin. While there is no reason to believe that animals would experience significant pain associated with competently performed exsanguination via a needle or cannula placed in a peripheral vein, there is concern that animals would experience distress associated with hypovolemia-induced anxiety.[1,2] It is not clear whether sedation is adequate to eliminate this distress. In the author's opinion, anesthesia should be required unless compelling scientific justification is presented to explain why anesthetic drugs, but not a particular sedative drug, would interfere with scientific objectives. There is no question that an animal should be anesthetized prior to an exsanguination procedure that involves creation of a surgical incision. Exsanguination by cardiac puncture also requires anesthesia.[73–75]

When asked in a survey conducted for the 2nd edition of *The IACUC Handbook* whether sedation was an acceptable alternative to anesthesia for pretreatment prior to exsanguination, responses were as follows: 24% answered that their IACUC had no formal policy; 28% answered that sedation was not an acceptable alternative; 2% answered that sedation was an acceptable alternative; 16% answered that sedation was permitted only with scientific justification; and 27% said that this question was not applicable to their institution.[11]

17:26 Must a moribund rabbit be anesthetized prior to euthanasia by exsanguination?

Reg. See 17:20 and 17:25.

Opin. This question hinges in part on whether a moribund rabbit would be capable of perceiving pain and distress. As discussed in 17:11 and 17:19 it cannot be assumed that a moribund animal is incapable of perceiving pain and distress and it is generally best to proceed under the assumption that a moribund animal could experience pain and distress. Were the rabbit healthy and fully alert, anesthesia would be advisable prior to exsanguination (see 17:25). However, there is an additional issue to consider in the case of a moribund animal. An animal near death might be unable to tolerate a general anesthetic and could die before or early in the exsanguination procedure, possibly limiting the amount of blood that could be collected. This could have a negative impact on the research if it were necessary to collect a large quantity of blood for use in the research project. In such a situation, the need to minimize distress in the rabbit would have to be balanced with the requirements of the study. Provided that the blood were to be collected via a needle or cannula placed in a peripheral vein, light anesthesia or sedation might be more appropriate than deep anesthesia. The justification would have to be more compelling to consider waiving the need for anesthesia prior to cardiac puncture, and even a moribund animal should be anesthetized prior to making any kind of surgical incision.

When asked in a survey conducted for the 2nd edition of *The IACUC Handbook* whether it would be necessary to anesthetize a moribund rabbit prior to exsanguination, respondents answered as follows: 17% indicated that their IACUC had no policy on this issue; 29% indicated that anesthesia would be required; 8% indicated that anesthesia would be required only if the rabbit were still conscious; 1% indicated that anesthesia was required only if the procedure involved penetration of a body cavity; 3% indicated that anesthesia might not be required if it might result in the death of the rabbit prior to blood collection; and 41% indicated that this question was not applicable to their institution.[11]

17:27 How should death be ensured prior to disposal of animals killed by methods such as carbon dioxide or inhalant anesthetics?

Reg. See 17:23.

Opin. It is generally recognized as essential that death be verified before disposal of an animal that has been euthanized.[1,2,8,18] This is especially important when animals are euthanized by techniques that can cause loss of consciousness or cessation of visible respirations well before death occurs, as these animals are more prone to regain consciousness after disposal. The potential for regaining consciousness following disposal is a particular concern with CO_2 asphyxiation.[2] Maintaining the animal in the CO_2-rich environment for a period of time after the animal appears dead can help to minimize this potential, as can observation of the animal for a few minutes after re-exposure to room air.[1,2,9,20] Death can be verified by examining the animal for cessation of a combination of vital signs, including heartbeat, respiration, and corneal reflex,[1,2,20] but the only sign that definitively confirms death is rigor mortis.[1] Another way to assure that the animal is dead is to follow up with a physical method once the animal appears dead or unconscious. Reliable follow up methods include decapitation, cervical dislocation, thoracotomy, or exsanguination.[1,2,9,20]

Surv. 1 Does your IACUC have a policy of "double kill"? That is, using a second method of euthanasia to assure a euthanized animal is truly dead.

• Not applicable	17/295 (5.8%)
• Yes, we have such a policy	152/295 (51.5%)
• Yes, we have such a policy but only for certain animal species	30/295 (10.2%)
• No, we do not have such a policy	90/295 (30.5%)
• Other	6/295 (2.0%)

Surv. 2 If your IACUC has a "double kill" policy requiring a physical method to assure that an animal is dead, which of the following methods are acceptable? Check as many boxes as appropriate.

• Not applicable	93/277 (33.6%)
• Bilateral thoracotomy	123/277 (44.4%)
• Decapitation	130/277 (46.9%)
• Cervical dislocation	144/277 (52.0%)
• Exsanguination	140/277 (50.5%)
• Removal of a vital organ	78/277 (28.2%)
• Other*	9/277 (3.2%)

*In addition to the methods listed above, one person each responded that their IACUC would accept the following methods for confirming death: cessation of breathing for 2 minutes; perfusion; observation of respiration and/or heart rate; and anesthesia. Two people were specific regarding the species for which double kill is required; one noted that double kill is required only for small animals in which death might be difficult to verify otherwise, and the other noted that a double kill method is used for frogs. One respondent stated that they require confirmation of death but that they do not require that it be by double kill, another clarified that they accept unilateral (vs. bilateral) thoracotomy, and a third specified that cranial concussion (stunning) must be followed by pithing or severing of the spinal cord.

Peri-Natal and Neonatal Rodents

17:28 What agents or techniques are most appropriate for euthanasia of neonatal altricial rodents (e.g., rats and mice)?

Reg. The *Guide* (p. 124) states that methods such as injection of a chemical agent, cervical dislocation, or decapitation should be considered as alternatives to CO_2 for euthanasia of neonatal rodents. While CO_2 is not ruled out as a potentially acceptable method, it is noted that neonatal rodents will require longer exposure times than adults to ensure death. The *AVMA Guidelines for the Euthanasia of Animals* echo these concerns, noting that neonatal mice may take up to 50 minutes to die from CO_2 exposure.[1] Intraperitoneal (IP) injection of pentobarbital is the recommended method for euthanasia pre-weanling small mammals. Conditionally acceptable methods include: injectable dissociative anesthetics combined with α-adrenergic receptor agonists or benzodiazepines; nonflammable inhalant anesthetic gases; CO_2; hypothermia; immersion in liquid nitrogen; decapitation; and cervical dislocation.

Opin. There are few published guidelines that specifically address euthanasia of neonatal animals. The few specific recommendations that do exist emphasize the fact that newborn animals are typically more resistant to hypoxia and more capable of coping with high environmental CO_2 than are older animals.[1,18,20] For this reason, it is often recommended either that inhalant agents (e.g., inhalant anesthetics, CO_2, carbon monoxide) not be used to euthanize neonates[2] or that these agents be used only when the prolonged exposure times necessary to ensure death can be employed.[1,18,20,76–78] Klaunberg and colleagues found that CO_2 was more effective than halothane for euthanasia of neonatal mice up to 7 days of age, and that CO_2 produced more rapid euthanasia of older preweanlings than either halothane or intraperitoneal pentobarbital.[76] Other methods that have been specifically identified as acceptable or conditionally acceptable for euthanasia of neonatal rats and mice include: decapitation, cervical dislocation, anesthetic overdose (prolonged exposure times may be needed with inhalant anesthetics), and immersion in liquid nitrogen, or anesthesia followed by perfusion with chemical fixatives.[1,28,20] Hypothermia can be used instead of injectable or inhalant anesthetics to anesthetize pups <6 days of age prior to euthanasia by another method.[20]

When choosing a method of euthanasia for a neonate, it should be kept in mind that neonates, including neonatal altricial rodents, appear to be at least as capable of perceiving pain as adult animals.[18,20,79–83]

In the author's opinion, the most humane methods of euthanasia for neonatal mice or rats ≤10 days of age are: IP injection of pentobarbital, cervical dislocation, or decapitation with heavy, sharp scissors. Sedation or anesthesia is not necessary prior to decapitation or cervical dislocation if the procedure is performed correctly by a skilled operator (see 17:20). Decapitation and cervical dislocation are aesthetically upsetting for many people, so CO_2 asphyxiation may be a more desirable alternative under circumstances where pentobarbital is not available or practical. Prolonged exposure of the pups to CO_2 (and maybe also to inhalant anesthetics) is essential, as is verification of death prior to disposal.

When asked in a survey conducted for the 2nd edition of *The IACUC Handbook* what agents or techniques were acceptable for euthanasia of neonatal rats and mice, the most common responses were carbon dioxide, alone or followed by a physical method (66%), and decapitation, with or without anesthesia (64%).[11] Other responses included: anesthetic overdose (54%), hypothermia followed by a physical method (28%), cervical dislocation (20%), and hypothermia alone (6%).

17:29 Is carbon dioxide an appropriate agent for euthanasia of neonatal rodents? Are conditions for its use different for neonates than for adults?

Reg. Noting that neonatal rodents are resistant to the hypoxia-inducing effects of CO_2 and that they therefore require longer exposure times for euthanasia with this gas, the *Guide* (p. 124) states that alternative methods should be considered, e.g., injection with chemical agents, cervical dislocation, or decapitation. The *AVMA Guidelines for the Euthanasia of Animals* recommends IP pentobarbital as the preferred method for euthanasia of unweaned small mammals, but views CO_2 as conditionally acceptable for this purpose.[1]

Opin. As noted above, the *AVMA Guidelines for the Euthanasia of* Animals view CO_2 as not preferred but conditionally acceptable for the euthanasia of small mammals.[1] The need for a prolonged period of exposure is emphasized. The Canadian Council on Animal Care and ACLAM Task Force on Rodent Euthanasia also address the use of CO_2 in neonatal rats and mice; both define neonates as animals up to 10 days of age.[18,20] CCAC classifies CO_2 as only conditionally acceptable even for adult rodents and emphasizes that the time to death may be considerably longer in neonates.[18] The need for prolonged exposure times is equally emphasized by ACLAM, which also stresses the need to verify that the pups are dead prior to disposal.[20] In fact, time to death may exceed 35 minutes in neonatal rats and 50 minutes in neonatal mice exposed to 100% CO_2.[77,78] Resistance to CO_2 decreases steadily during the first 10 days after birth, but even 10-day-old rats and mice require longer exposure times than adults of the same species.[76–78] During the first 7–8 days of life, inbred mice took longer than outbred mice to die following CO_2 exposure;[77] this difference was not demonstrated for rats.[78]

In the author's opinion, CO_2 is an acceptable but not preferred method of euthanasia for neonatal rats and mice ≤6 days of age (see 17:28). If CO_2 will be used to euthanize neonates of these species, prolonged exposure times are essential. To ensure death, recommended exposure times are 45 minutes for rat pups up to 6 days of age and 10–20 minutes for older pre-weanlings. For mice, recommended exposure times are 60 minutes for pups up to 6 days of age and 15–20 minutes for older pre-weanlings. As with adult animals euthanized with CO_2, it is essential to verify that the pups are dead prior to disposal.

Surv. If your IACUC allows the use of carbon dioxide for the euthanasia of mice, are there different requirements for neonates?

- Not applicable 80/295 (27.1%)
- We have no different requirements 41/295 (13.9%)
- We require a longer exposure time for neonates 106/295 (35.9%)
- We recommend but do not require a longer exposure time for neonates 12/295 (4.1%)
- I don't know 31/295 (10.5%)
- Other 25/295 (8.5%)

17:30 Is decapitation with sharp scissors an acceptable method for euthanasia of 1–2 day old rats or mice?

Reg. The *AVMA Guidelines for the Euthanasia of Animals* note that 'sharp blades' can be used in place of a guillotine for decapitation of neonatal rodents.[1] The Canadian Council on Animal Care does not mention the use of scissors for decapitation of neonates, but does note that 'sharp, well-maintained' scissors are acceptable for the decapitation of rodent fetuses.[18]

Opin. Decapitation is viewed as a conditionally acceptable method of euthanasia for rats and mice of any age.[1,18] While guillotines are generally recommended for decapitation of adults (see 17:21), scissors are an acceptable alternative for neonates and late term fetuses.[1,18,20] In the author's opinion, decapitation with heavy, sharp scissors is a preferred method of euthanizing neonatal rats and mice (see 17:28). If scissors are used for this purpose, it is essential that they be kept clean and sharp. It is also essential that the operator be experienced in this procedure and able to decapitate the pup with a single, swift, decisive cut.

Surv. Does your IACUC approve decapitation with a sharp scissors as an acceptable method of euthanasia of 1–2 day old rats or mice? Assume IACUC approval of decapitation as the method of euthanasia.

- Not applicable 73/296 (24.7%)
- Yes, it is acceptable 109/296 (36.8%)
- Yes, it is acceptable but only if the animal is sedated 42/296 (14.2%)
- No, it is not acceptable 50/296 (16.9%)
- I don't know 17/296 (5.7%)
- Other* 5/296 (1.7%)

*One respondent indicated that use of scissors was acceptable for mice but not rats, and another indicated that decapitation was acceptable for neonatal rats and mice but that the IACUC had no policy regarding the use of scissors per se. A third person noted that pups must be anesthetized by hypothermia prior to the procedure, and another stated that the IACUC would have to review a request to decapitate using scissors on a case by case basis. The fifth respondent answered that the issue had never arisen at their institution.

17:31 If a pregnant mouse or rat dies or is euthanized, must additional steps be taken to ensure death of the fetuses?

Reg. The *Guide* (pp. 123–124) states that special consideration should be given to euthanizing fetuses depending on species and gestational age, but provides no further guidance on what this special consideration might entail. The *AVMA Guidelines for the Euthanasia of Animals* state that no additional steps need be taken to ensure the death of rodent fetuses following death of the dam because mammalian fetuses are unconscious during pregnancy and birth and that they cannot suffer while dying in utero following the death of the dam from any cause.[1] The Canadian Council on Animal Care takes a different stand on this issue: no additional steps would be needed to ensure death of fetuses during the first two trimesters of gestation; but, during the last trimester, the fetuses must be euthanized separately following euthanasia of the dam.[18]

Opin. Based on neural development, many have concluded that it is plausible, even likely, that humans and other mammals, including rats and mice, are able to perceive pain prior to parturition.[84–89] Although some question remains whether the capacity to experience pain truly exists before birth,[90,91] there is accumulating support for the belief that human fetuses are able to experience pain during the third trimester of pregnancy, as early as 26 weeks of gestation.[87–89] The evidence is not as robust for rodents, but rat and mouse fetuses exhibit neural and behavioral responses to painful and other types of somatosensory stimulation during late gestation, as early as embryonic day 17 in the rat.[84–86,92,93] On the basis of this evidence, it is now widely recommended that late term rat and mouse fetuses be handled in a manner that minimizes the potential for pain or distress. To quote the Canadian Council on Animal Care "during the final third of gestation, fetuses should be given the same ethical considerations as apply to the fully mature animal."[18] While the AVMA Guidelines have taken a sharply different stand on this issue, many believe that additional steps should be taken following the death of the mother to protect the welfare of the fetuses. In the author's opinion, it is preferable to err on the side of caution until this issue can be decided more definitively. Appropriate methods of euthanasia include those recommended for neonatal rats and mice (see 17:28). As to what 'late term' means in terms of gestational age, it is applied to fetuses as young as embryonic day 15.[94] There are no data or recommendations suggesting that special precautions are needed to minimize pain or distress in fetuses during the first 2 trimesters of gestation.

Field Studies

17:32 Is it acceptable to use inhalant gases to euthanize small rodents in the field?

Reg. The *Guide* (p. 32) is largely silent on this question except to comment that, in general, issues associated with euthanasia in the field are similar, if not identical, to those associated with euthanasia in the laboratory. The AWAR are similarly silent except in APHIS/AC Policy 3, which addresses the use of gunshot for certain situations encountered in the field. Note however, that where a field study is not exempted from the AWAR, §2.31,d,1,xi would apply; i.e., "Methods of euthanasia used must be in accordance with the definition of the term set forth in 9 CFR part 1, Sec. 1.1 of this subchapter, unless a deviation is justified for scientific reasons, in writing, by the investigator."

The *AVMA Guidelines for the Euthanasia of Animals* describe the use of inhalant anesthetics, CO_2, carbon monoxide (CO), and other inert gases as conditionally acceptable for the euthanasia of free-ranging wild rodents, noting that the same conditions must be met for use of these agents as would apply for their use with domestic animals.[1] The Canadian Council on Animal Care also views CO_2 or CO as acceptable agents for use in the field, but emphasizes the need for a closed chamber and attention to personnel safety.[95] The CCAC also endorses the use of volatile anesthetic gases, except in the case of species that are able to hold their breath for long periods of time; however ether and nitrous oxide are not recommended for euthanasia of wildlife.[95]

Opin. Techniques of euthanasia that are preferred in the laboratory also may be applicable in the field. However, the *Guide* (p. 32), *AVMA Guidelines for the Euthanasia of Animals*, APHIS/AC Policy 3, and CCAC all observe that there are situations in which it may be necessary to approach euthanasia in the field differently than in the laboratory.[1,10,95] Different methods may be mandated because of feasibility, human safety, animal welfare, or animal safety. Research objectives may limit the use of some methods or agents for wildlife species, but humane concerns must over-ride mere convenience when selecting a method.[1] Provided that proper equipment is available and precautions are taken to protect personnel from harm, inhalant gases are acceptable for euthanasia of small rodents in the field. Note however that the CCAC specifically advises against the use of nitrous oxide or ether for this purpose as these gases are combustible and/or explosive and have potential for human toxicity and abuse.[95]

Surv. Does your IACUC allow the use of inhalant gases as a primary means of euthanasia for small rodents captured in field research? Check as many boxes as appropriate.

- Not applicable 216/296 (73%)
- Yes, we allow the use of isoflurane 52/296 (17.6%)
- Yes, we allow the use of diethyl ether 3/296 (1%)
- Yes, we allow the use of other inhalant gases 21/296 (7.1%)
- No, we do not allow the use of inhalant gases 9/296 (3%)
- I don't know 11/296 (3.7%)
- Other 3/296 (1%)

17:33 Is thoracic compression an acceptable method of euthanasia for wild rodents trapped in field studies?

Reg. The use of thoracic compression as a method of euthanasia is not addressed in the AWAR, PHS Policy, or the *Guide*. The *AVMA Guidelines for the Euthanasia of Animals* state that, under rare circumstances, thoracic compression might be used to euthanize animals that are deeply anesthetized or otherwise unconscious, or as a final confirmatory procedure to ensure death in cases where the animal's status is uncertain.[1]

Opin. In contrast to the *AVMA Guidelines for the Euthanasia of Animals*, the American Society of Mammalogists (ASM) views thoracic compression as an acceptable alternative for euthanasia of small mammals under circumstances where use of injectable pharmaceutical agents or inhalant gases may be stressful for the animals or pose risks to investigators.[96] ASM notes that thoracic compression has been used effectively for decades by practicing mammalogists.[96] Sikes *et al.* note that thoracic compression appears to be painless and that it maximizes the use of the carcass.[96]

Surv. Does your IACUC consider thoracic compression an ethical form of euthanasia for small rodents trapped in field research?

- Not applicable 195/293 (66.6%)
- We have no policy 38/293 (13%)
- Acceptable with scientific justification 14/293 (4.8%)
- Acceptable without scientific justification 3/293 (1%)
- Not acceptable 42/293 (14.3%)
- Other 1/293 (0.3%)

17:34 Sometimes animals are unintentionally injured during field studies. Is it necessary for the investigator to euthanize these animals?

Reg. Except for situations in which live, wounded animals are found in kill traps, the issue of how to deal with animals that are injured during field studies is not specifically dealt with in the federal regulations or policies. However, the AWAR (§2.33,b,2) require "The use of appropriate methods to prevent, control, diagnose, and treat diseases and injuries." This requirement also applies to animals in those field studies that are not exempted from regulation (based on the AWAR (§1.1) definition of "field study"). Kill traps must be checked at least once daily and any trapped animals that are not dead must be killed quickly and humanely.[1] The Canadian Council on Animal Care addresses this issue more broadly, stating that the investigator must euthanize any animal in the field that is suffering unrelievable pain or distress as a result of capture, handling, or experimental manipulations.[95]

Opin. Both the AVMA Panel on Euthanasia and CCAC emphasize that challenges encountered during field research do not reduce the ethical obligation of the responsible individual to reduce pain and distress during the performance of euthanasia.[1,95] This obligation is echoed by the American Society of Mammalogists, which further states that protocols involving fieldwork should explicitly describe circumstances and methods for euthanizing injured or distressed animals, even for studies where injury or death is not an anticipated outcome.[96] The *AVMA Guidelines for the Euthanasia of Animals* address in more general terms the performance of euthanasia to improve animal welfare.[1] For an animal that no longer enjoys good welfare—defined as feeling well, functioning well, and able to perform innate behaviors—the humane alternative is to give it a good death. While this discussion in the Guidelines centers primarily around companion animals, it could apply equally well to wild animals used in field studies. In the author's opinion, investigators have an ethical obligation to humanely kill any animals that would otherwise experience unrelieved pain or distress, or would be more vulnerable to predation, as a result of the investigative team's actions.

Surv. Does your IACUC have a field research policy that injured animals (e.g., from investigator trapping) be euthanized?

- Not applicable 175/295 (59.3%)
- We do not have such a policy 48/295 (16.3%)
- We have such a policy 49/295 (16.6%)
- We have such a policy but there are some IACUC approved exceptions 13/295 (4.4%)
- I don't know 4/295 (1.4%)
- Other 6/295 (2%)

Other Euthanasia-Related Questions

17:35 Which methods of euthanasia are acceptable for euthanizing zebrafish?

Reg. The *AVMA Guidelines for the Euthanasia of Animals* lists several methods as acceptable for the euthanasia of fish in general.[1] Of these, the following are acceptable for zebrafish in research facilities: immersion into buffered tricaine methane sulfonate (MS222, tricaine, TMS), buffered benzocaine, quinaldine sulfate, or 2-phenoxyethanol; and rapid chilling in an ice-water slurry. In addition, rapid chilling followed by immersion in a dilute sodium hypochlorite or calcium hypochlorite solution is acceptable for zebrafish embryos and larvae. Methods that are conditionally acceptable for zebrafish in research settings include: immersion in CO_2-saturated water, eugenol, isoeugenol, or clove oil; decapitation followed by pithing; cranial concussion followed by pithing; and maceration. The Canadian Council on Animal Care (CCAC) guidelines are similar in some respects but different in others (see below).

Opin. Both the *AVMA Guidelines for the Euthanasia of Animals* and CCAC guidelines on euthanasia recommend immersion in tricaine methane sulfonate (MS222, tricaine, TMS) or benzocaine as preferred methods for euthanizing fish.[1,18,97] Other methods that are viewed as acceptable or conditionally acceptable by both AVMA and CCAC are immersion in clove oil and maceration.[1,18] Cranial concussion also is listed as conditionally acceptable by both guidelines, but the AVMA Guidelines further specify that this must be followed by pithing.[1,18] Listed as acceptable or acceptable with conditions by AVMA but not addressed by CCAC are: immersion in quinaldine sulfate or 2-phenoxyethanol; rapid chilling in an ice-water slurry; rapid chilling followed by immersion in a dilute sodium hypochlorite or calcium hypochlorite solution is acceptable (for zebrafish embryos and larvae); immersion in CO_2-saturated water, eugenol, or isoeugenol; and decapitation followed by pithing. Conversely, immersion in etomidate is listed as acceptable by CCAC[18] but is not addressed by AVMA, and exsanguination under anesthesia is listed as conditionally acceptable by CCAC[96] but is not addressed by AVMA. The biggest difference in these two sets of guidelines is in relation to metomidate, immersion in which is listed as an acceptable method of euthanasia by CCAC,[18] but which is viewed as an unacceptable method by AVMA.[1]

Surv. Which methods of euthanasia are generally acceptable to your IACUC for euthanizing zebrafish? Check as many boxes as appropriate.

• Not applicable	147/294 (50.0%)
• An overdose of MS-222 (tricaine methane sulfonate)	113/294 (38.4%)
• Rapid immersion in ice water	28/294 (9.5%)
• Decapitation	26/294 (8.8%)
• Sodium pentobarbital	11/294 (3.7%)
• Benzocaine HCl	11/294 (3.7%)
• Clove oil	8/294 (2.7%)
• Carbon dioxide	5/294 (1.7%)
• I don't know	33/294 (11.2%)
• Other	3/294 (1.0%)

17:36 Sea hares (*Aplysia*) are not a regulated species under the AWAR and PHS Policy. Is it necessary for the IACUC to review methods of euthanasia for sea hares used in research?

Opin. As invertebrates, sea hares (also known as sea slugs) are not regulated under the AWAR or PHS Policy. AAALAC states that invertebrates used in research are covered as part of an accredited program "where they are relevant to the unit's mission."[98] However, their use in research does not necessarily require formal IACUC review and approval. The level of IACUC oversight, and interest in review of the research by AAALAC site visitors, may vary depending on the species of invertebrate, with greater oversight expected for higher level species such as octopi than for lower level species such as sea slugs. Recognizing the importance of invertebrates in research, as display animals, and even as companion animals, the 2013 iteration of the *AVMA Guidelines for the Euthanasia of Animals* includes a section on euthanasia of invertebrates.[1] The Guidelines acknowledge the uncertainty regarding the ability of invertebrates to experience pain, distress or compromised welfare, but assume that a conservative and humane approach to the care of any animal is warranted and expected by society. Guidelines also exist in the published literature for provision of anesthesia and analgesia for invertebrates.[99,100]

In the author's opinion, IACUC review of activities involving invertebrates should, at a minimum, assure that: the housing and care of the animals will be appropriate for their species; the maintenance and research use of these animals will not pose any threat to the health or safety of humans or other animals; and methods of euthanasia of these animals will be consistent with the AVMA Guidelines unless an exception is justified. In the case of a lower level invertebrate such as *Aplysia*, IACUC oversight might consist of nothing more than review of a letter from the investigator describing the intended research use of the animals, the provisions for housing and care of the animals, any appropriate precautions to prevent escape of the animals or other unintended impact on humans or other animals, methods that will be taken to prevent unnecessary pain or distress in the animals, and the method of euthanasia that will be used at the end of the study.

17:37 On what basis might an IACUC approve use of a method of euthanasia that is not reviewed in the *AVMA Guidelines for the Euthanasia of Animals*?

Opin. It is acknowledged within the report itself that circumstances may arise that are not covered by the *AVMA Guidelines for the Euthanasia of Animals*.[1] In such situations, it is recommended that an appropriate method be chosen by a veterinarian who has experience with the species to be euthanized and who will apply professional judgment based on the animal's size, physiology, and behavior. Within a research setting, the method may actually be proposed by the investigator, but the IACUC should rely heavily on the advice of the AV (or other veterinary expert) in deciding whether to approve the method. Additional pertinent information on the method may be found in the published literature or in euthanasia guidelines developed by other professional bodies, e.g., the Canadian Council on Animal Care or International Council for Laboratory Animal Science (ICLAS). Insights might also be gained from colleagues at other institutions who are familiar with the method in question. As approval is being deliberated, all members of the IACUC should consider whether the method will satisfy the minimum requirements for a humane death: rapid loss of consciousness without pain, distress, or fear; and reliable progression to death following loss of consciousness. Availability of personnel with adequate expertise to apply the method in a skillful manner should also be taken into account, as should potential effects on the health, safety and emotional responses of human personnel and other animals (see also 17:1).

References

1. American Veterinary Medical Association. 2013. *AVMA Guidelines for the Euthanasia of Animals: 2013 Edition*. https://www.avma.org/KB/Policies/Documents/euthanasia.pdf.
2. National Research Council. 1992. Euthanasia. In *Recognition and Alleviation of Pain and Distress in Laboratory Animals: A Report of the Institute of Laboratory Animal Resources Committee on Pain and Distress in Laboratory Animals*, 102–116. Washington, D.C.: National Academy Press.
3. Canadian Council on Animal Care. 1998. *Guidelines on: Choosing an Appropriate Endpoint in Experiments Using Animals for Research, Teaching and Testing*. Ottawa: Canadian Council on Animal Care.
4. Sharp J.L., T.G. Zammit, T.A. Azar et al. 2002. Stress-like responses to common procedures in male rats housed alone or with other rats. *Contemp. Topics Lab. Anim. Sci.* 41(4):8–14.
5. Sharp J.L., T.G. Zammit. T.A. Azar et al. 2002. Does witnessing experimental procedures produce stress in male rats? *Contemp. Topics Lab. Anim Sci.* 41(5): 8–12.
6. Sharp J.L., T.G. Zammit, T.A. Azar et al. 2003. Are "by-stander" female Sprague-Dawley rats affected by experimental procedures? *Contemp. Topics Lab. Anim. Sci.* 42(1):19–27.
7. Sharp J.L., T.G. Zammit, and Lawson D.M. 2002. Stress-like responses to common procedures in rats: Effect of the estrous cycle. *Contemp. Topics Lab. Anim. Sci.* 41(4):15–22.
8. Committee for the Update of the Guide for the Care and Use of Laboratory Animals. 2011. *Guide for the Care and Use of Laboratory Animals*, 123. Washington, D.C.: National Academy Press.
9. National Institutes of Health. 2002. Notice NOT-OD-02-062, PHS Policy on Humane Care and Use of Laboratory Animals Clarification Regarding Use of Carbon Dioxide for Euthanasia of Small Laboratory Animals. http://grants.nih.gov/grants/guide/notice-files/NOT-OD-02-062.html.
10. U.S. Department of Agriculture, Animal and Plant Health Inspection Service. 2011. Animal Care Policy Manual. Policy 3, Veterinary Care. http://www.aphis.usda.gov/animal_welfare/policy.php?policy=3
11. Danneman, P.J. 2007. Euthanasia. In *The IACUC Handbook, 2nd Edition*. ed. J Silverman, M.A. Suckow and S. Murthy, 287–320. Boca Raton: CRC Press.
12. U.S. Department of Agriculture, Animal and Plant Health Inspection Service. 2013. *Animal Welfare Inspection Guide*. http://www.aphis.usda.gov/animal_welfare/downloads/Inspection%20Guide%20-%20November%202013.pdf.
13. American College of Laboratory Animal Medicine. Public Statements: Position Statement on Adequate Veterinary Care. http://www.aclam.org/Content/files/files/Public/Active/position_adeqvetcare.pdf. Accessed June 2012.
14. Canadian Council on Animal Care. 1993. Euthanasia. In *Guide to the Care and Use of Experimental Animals*, Vol. 1, ed. E.D. Olfert, B.M. Cross and A.A. McWilliams, 141–153. Ottawa: Canadian Council on Animal Care.
15. de Laat, J.M., G. van Tintelen, A.C. Beynen. 1989. Transportation of rats affects behaviour of non-transported rats in the absence of physical contact. *Z Versuchstierkd.* 32:235–237 (preliminary communication).
16. U.S. Department of Agriculture, Animal and Plant Health Inspection Service. 2011. Animal Care Policy Manual. Policy 20. Licensing Sales of Dead Animals. http://www.aphis.usda.gov/animal_welfare/policy.php?policy=20.
17. National Institutes of Health, Office of Laboratory Animal Welfare. 2013. Frequently Asked Questions A.3, Does the PHS Policy apply to the use of animal tissue or materials obtained from dead animals? http://grants.nih.gov/grants/olaw/faqs.htm#App_3.
18. Canadian Council on Animal Care. 2010. *CCAC Guidelines on: Euthanasia of animals used in science*. Ottawa: Canadian Council on Animal Care.
19. National Institutes of Health, Office of Laboratory Animal Welfare. 2005. Notice NOT-OD-05-034, Guidance on Prompt Reporting to OLAW under the PHS Policy on Humane Care and Use of Laboratory Animals. http://grants.nih.gov/grants/guide/notice-files/NOT-OD-05-034.html.

20. American College of Laboratory Animal Medicine. 2005. Public Statements: Report of the ACLAM Test Force on Rodent Euthanasia. http://www.aclam.org/Content/files/files/Public/Active/report_rodent_euth.pdf.
21. Merriam Webster's Collegiate Dictionary. 1995. 10th Edition, Springfield, MA: Merriam-Webster.
22. Dorland's Illustrated Medical Dictionary. 1981. 26th Edition, Philadelphia: W.B. Saunders.
23. Stedman's Medical Dictionary. 2006. 28th edition. Philadelphia: Lippincott Williams & Wilkins.
24. Toth, L.A. 1997. The moribund state as an experimental endpoint. *Contemp. Topics Lab. Anim. Sci.* 36:44–48.
25. Morton, D.B. 2000. A systemic approach for establishing humane endpoints. *ILAR J.* 41:80–86.
26. Danneman, P.J. 1997. Monitoring of analgesia. In *Anesthesia and Analgesia in Laboratory Animals*, eds. D.F. Kohn, S.K. Wixson, W.J. White et al., 83–103. New York: Academic Press.
27. Carstens, E. and G.P. Moberg. 2000. Recognizing pain and distress in laboratory animals. *ILAR J.* 41:62–71.
28. Wallace, J. 2000. Humane endpoints and cancer research. *ILAR J.* 41:87–93.
29. Paster, E.V., K.A. Villines and D.L. Hickman. 2009. Endpoints for mouse abdominal tumor models: Refinement of current criteria. *Comp. Med.* 48:234–241.
30. Olfert, E.D. and D.L. Godson. 2000. Humane endpoints for infectious disease animal models. *ILAR J.* 41:99–104.
31. Hendriksen, C.F.M. and B. Steen. 2000. Refinement of vaccine potency testing with the use of humane endpoints. *ILAR J.* 41:105–113.
32. American Veterinary Medical Association. 2000. 2000 Report of the AVMA Panel on Euthanasia. *J. Am. Vet. Med. Assoc.* 218:669–696.
33. American Veterinary Medical Association. 2007. AVMA Guidelines on Euthanasia. http://www.avma.org/issues/animal_welfare/euthanasia.pdf.
34. Thurauf, N, I. Friedel, C. Hummel et al. 1991. The mucosal potential elicited by noxious chemical stimuli with CO_2 in rats: Is it a peripheral nociceptive event? *Neurosci Lett.* 128:297–300.
35. Peppel, P. and F. Anton. 1993. Responses of rat medullary dorsal horn neurons following intranasal noxious chemical stimulation: Effects of stimulus intensity, duration, and interstimulus interval. *J Neurophysiol.* 70:2260–2275.
36. Danneman, P.J., S. Stein, and S.O. Walshaw. 1997. Humane and practical implications of using carbon dioxide mixed with oxygen for anesthesia or euthanasia of rats. *Lab. Anim. Sci.* 47:376–385.
37. Leach, M.C., V.A. Bowell, T.F. Allan et al. 2002. Aversion to gaseous euthanasia agents in rats and mice. *Vet Rec.* 52:249–257.
38. Steen, K.H., P.W. Reeh, F. Anton et al. 1992. Protons selectively induce lasting excitation and sensitization to mechanical stimulation of nociceptors in rat skin, in vitro. *J Neurosci.* 12:86–95.
39. Thurauf, N, W. Ditterich and G. Kobal. 1994. Different sensitivity of pain-related chemosensory potentials evoked by stimulation with CO2, tooth pulp event–related potentials, and acoustic event–related potentials to the tranquilizer diazepam. *Br. J. Clin. Pharmacol.* 38:545–555.
40. Britt, D.P. 1986. The humaneness of carbon dioxide as an agent of euthanasia for laboratory rodents. In *Euthanasia of unwanted, injured, or diseased animals or for educational or scientific purposes*, 19–31. Hertfordshire, U.K.: Universities Federation for Animal Welfare.
41. Waynforth, H.B. and P.A. Flecknell. 1992. Miscellaneous techniques. In *experimental and surgical technique in the rat*, 313–340. London: Academic Press.
42. Hewett, T.A., M.S. Kovacs, J.E. Artwohl et al. 1993. A comparison of euthanasia methods in rats, using carbon dioxide in prefilled and fixed low rate filled chambers. *Lab. Anim. Sci.* 43:579–582.
43. Smith, W. and S.B. Harrap. 1997. Behavioural and cardiovascular responses of rats to euthanasia using carbon dioxide gas. *Lab. Anim.* 31:337–346.
44. Burkholder, T.H., L. Niel, J.L., Weed et al. 2010. Comparison of carbon dioxide and argon euthanasia: Effects on behavior, heart rate, and respiratory lesions in rats. *J. Am. Assoc. Lab. Anim. Sci.* 49:448–453.
45. Coenen, A.M.L., W.H.I.M. Drinkenburg, R. Hoenderken et al. 1995. Carbon dioxide euthanasia in rats: Oxygen supplementation minimizes signs of agitation and asphyxia. *Lab. Anim.* 29:262–268.

46. Ambrose, N., J. Wadham, and D. Morton. 2000. Refinement of euthanasia. In *Progress in the reduction, refinement and replacement of animal experimentation*, ed. M. Balls, A.M. Van Zeller, and M.E. Halder, 1159–1171. Amsterdam: Elsevier.
47. Banzett, R.B., R.W. Lansing, K.C. Evans et al. 1996. Stimulus-response characteristics of CO_2-induced air hunger in normal subjects. *Resp. Physiol.* 102:19–31.
48. Banzett, R.B. and S.H. Moosavi. 2001. Dyspnea and pain: Similarities and contrasts between two very unpleasant sensations. *Am. Pain Soc. Bull.* 11:6–8.
49. Moosavi, S.H., E. Golestanian, R.W. Binks et al. 2002. Hypoxic and hypercapnic drives to breathe generate equivalent levels of air hunger in humans. *J. Appl. Physiol.* 94:141–154.
50. Fisher, J.T. 2009. The TRPV1 ion channel: Implications for respiratory sensation and dyspnea. *Respir. Physiol. Neurobiol.* 167:45–52.
51. Schimitel, F.G., G.M. de Almeida, D.N. Pitol et al. 2012. Evidence of a suffocation alarm system within the periaqueductal gray matter of the rat. *Neuroscience* 200:59–73.
52. Balcombe, J.P, N.D. Barnard and C. Sandusky. 2004. Laboratory routines cause animal stress. *Contemp. Top. Lab. Anim. Sci.* 43:42–51.
53. Castelhano-Carlos, M.J. and V. Baumans. 2009. The impact of light, noise, cage cleaning and in-house transport on welfare and stress in laboratory rats. *Lab. Anim.* 43:311–327.
54. Wimer, R.E. and J.L. Fuller. 1968. Patterns of behavior. In *Biology of the Laboratory Mouse, 2nd Edition*, ed. E. L. Green, 629–653. New York: Dover Publications, Inc.
55. Marchlewska-Koj, A. and M. Zacharczuk-Kakietek. 1990. Acute increase in plasma corticosterone level in female mice evoked by pheromones. *Physiol. Behav.* 48:577–580.
56. Van Loo, P.L.P, L.F.M. Van Zutphen and V. Baumans. 2003. Male management: Coping with aggression problems in male laboratory mice. *Lab. Anim.* 37:300–313.
57. Pletzer, B., W. Klimesch, K. Oberascher-Holzinger et al. 2007. Corticosterone response in a resident-intruder-paradigm depends on social state and coping style in adolescent male Balb-C mice. *Neuro. Endocrinol. Lett.* 28:585–590.
58. Bivin, W.S. 1994. Basic biomethodology. In *The Biology of the Laboratory Rabbit*, ed. P.J. Manning, D.H. Ringler and C.E. Newcomer, 83. San Diego: Academic Press.
59. Green C.J. 1979. Euthanasia. In *Animal anesthesia*, 237–241. London: Laboratory Animals Ltd.
60. Herin, R.A., P. Hall, and J.W. Fitch. 1978. Nitrogen inhalation as a method of euthanasia of dogs. *Am. J. Vet. Res.* 39:989–991.
61. Hornett, T.D. and A.P. Haynes. 1984. Comparison of carbon dioxide/air mixture and nitrogen/air mixture for the euthanasia of rodents: Design of a system for inhalation euthanasia. *Anim. Technol.* 35:93–99.
62. Enggaard Hansen, N., A. Creutzberg and H.B. Simonsen. 1991. Euthanasia of mink (Mustela vison) by means of carbon dioxide (CO_2), carbon monoxide (CO) and nitrogen (N_2). *Br. Vet. J.* 147:140–146.
63. Raj, A.B. 1999. Behaviour of pigs exposed to mixtures of gases and the time required to stun and kill them: Welfare implications. *Vet. Rec.* 144:165–168.
64. Raj, A.B.M and N.G. Gregory. 1996. Welfare implications of the gas stunning of pigs. 2. Stress of induction of anaesthesia. *Anim. Welf.* 5:71–78.
65. Raj, M. and N.G. 1994. An evaluation of humane gas stunning methods for turkeys. *Vet. Rec.* 135:222–223.
66. Raj, A.B.M., N.G. Gregory and S.R. Wotton. 1991. Changes in the somatosensory evoked potentials and spontaneous electroencephalogram of hens during stunning in argon-induced anoxia. *Br. J. Vet. Res.* 147:322–330.
67. Raj, A.B.M. and P.E. Whittington. 1995. Euthanasia of day-old chicks with carbon dioxide and argon. *Vet. Rec.* 136:292–294.
68. Webster, A.B. and D.L. Fletcher. 2001. Reactions of laying hens and broilers to different gases used for stunning poultry. *Poult. Sci.* 80:1371–1377.
69. Gerritzen, A., B. Lambooij, H. Reimart et al. 2004. On-farm euthanasia of broiler chickens: Effects of different gas mixtures on behavior and brain activity. *Poult. Sci.* 83:1294–1301.
70. Niel, L. and D.M. Weary. 2007. Rats avoid exposure to carbon dioxide and argon. *Appl. Anim. Behav. Sci.* 107:100–109.

71. Makowska, I.J., L. Niel, R.D. Kirkden et al. 2008. Rats show aversion to argon-induced hypoxia. *Appl. Anim. Behav. Sci.* 114:572–581.
72. Clifford, D.H. 1984. Preanesthesia, anesthesia, analgesia, and euthanasia. In *Laboratory Animal Medicine,* eds. J.G. Fox, B.J. Cohen and F.M. Loew, 528–563. New York: Academic Press.
73. Waynforth, H.B. and P.A. Flecknell. 1992. Methods of obtaining body fluids. In *Experimental and surgical technique in the rat,* 68–99. London: Academic Press.
74. Adams, R.J. 2002. Techniques of experimentation. In *Laboratory Animal Medicine,* 2nd Edition, ed. J.G. Fox, L.C. Anderson, F.M. Loew et al., 1005–1045. San Diego: Academic Press.
75. Hayward, A.M., L.B. Lemke, E.C. Bridgeford et al. 2007. Biomethods and surgical techniques. In *The mouse in biomedical research, 2nd Edition, volume 3: Normative biology, husbandry, and models,* eds. J.G. Fox, M. Davisson, F.W. Quimby et al., 437–488. San Diego: Academic Press.
76. Klaunberg, B.A., J. O'Malley, T. Clark, T et al. 2004. Euthanasia of mouse fetuses and neonates. *Contemp Top. Lab. Anim. Sci.* 43:29–34.
77. Pritchett, K., D. Corrow, J. Stockwell et al. 2005. Euthanasia of neonatal mice with carbon dioxide. *Comp. Med.* 55:275–281.
78. Pritchett-Corning, K.R. 2009. Euthanasia of neonatal rats with carbon dioxide, *J. Am. Assoc. Lab. Anim. Sci.* 48:23–27.
79. McLaughlin, C.R., A.H. Lichtman, M.S. Fanselow et al. 1990. Tonic nociception in neonatal rats. *Pharmacol. Biochem. Behav.* 36:859–862.
80. McLaughlin, C.R. and W.L. Dewey. 1994. A comparison of the antinociceptive effects of opioid agonists in neonatal and adult rats in phasic and tonic nociceptive tests. *Pharmacol. Biochem. Behav.* 49:1071–1023.
81. Blass, E.M., C.P. Cramer and M.S. Fanselow. 1993. The development of morphine-induced antinociception in neonatal rats: A comparison of forepaw, hindpaw, and tail retraction from a thermal stimulus. *Pharmacol. Biochem. Behav.* 44:643–649.
82. Guy, E.R. and F.V. Abbott. 1992. The behavioral response to formalin in preweanling rats. *Pain* 51:81–90.
83. Fitzgerald, M. 1994. Neurobiology of fetal and neonatal pain, in *Textbook of Pain,* ed. P.D. Wall and R. Melzack, 153. London: Churchill Livingstone.
84. Yi, D.K. and G.A. Barr. 1997. Formalin-induced c-fos expression in the spinal cord of fetal rats. *Pain* 73:347–354.
85. Barr, G.A. 1998. Maturation of the biphasic behavioral and heart rate response in the formalin test. *Pharmacol. Biochem. Behav.* 60:329–335.
86. Barr, G.A. 2011. Formalin-induced c-fos expression in the brain of infant rats. *J. Pain.* 12:263–271.
87. Lee, S.J., H.J. Ralston, E.A. Drey et al. 2005. Fetal pain: A systematic multidisciplinary review of the evidence. *J. Am. Med. Assoc.* 294:947–954.
88. Rokyta, R. 2008. Fetal pain. *Neuro. Endocrinol. Lett.* 29:807–814.
89. Lowery, C.L., M.P. Hardman, N. Manning et al. 2007. Neurodevelopmental changes of fetal pain. *Semin. Perinatol.* 31:275–282.
90. Mellor, D.J., T.J. Diesch, A.J. Gunn et al. 2005. The importance of 'awareness' for understanding fetal pain. *Brain Res. Brain Res. Rev.* 49:455–471.
91. Murrell, J.C., D.J. Mellor, and C.B. Johnson. 2008. Anesthesia and analgesia in the foetus and neonate. In *Anesthesia and Analgesia in Laboratory Animals, 2nd Edition,* ed. R.E. Fish, M.J. Brown, P.J. Danneman and A.Z. Karas, 593–608. San Diego: Academic Press.
92. Smotherman, W.P. and S.R. Robinson. 1985. The rat fetus in its environment: Behavioral adjustments to novel, familiar, aversive, and conditioned stimuli present in utero. *Behav. Neurosci.* 99:521–530.
93. Coppola, D.M., L.C. Millar, C.J. Chen et al. 1997. Chronic cocaine exposure affects stimulus-induced but not spontaneous behavior of the near-term mouse fetus. *Pharmacol. Biochem. Behav.* 58:793–799.
94. National Institutes of Health, Office of Animal Care and Use. 2013. Guidelines for the Euthanasia of Rodent Fetuses and Neonates. http://oacu.od.nih.gov/ARAC/documents/Rodent_Euthanasia_Pup.pdf

95. Canadian Council on Animal Care. 2003. *Guidelines on: The care and use of wildlife*. Ottawa: Canadian Council on Animal Care.
96. Sikes, R.S. and W.L. Gannon and the Animal Care and Use Committee of the American Society of Mammalogists. 2011. Guidelines of the American Society of Mammalogists for the use of wild mammals in research. *J. Mammal.* 92:235–253.
97. Canadian Council on Animal Care. 2005. *Guidelines on: The care and use of fish in research, teaching and testing*. Ottawa: Canadian Council on Animal Care.
98. Association for the Assessment and Accreditation of Laboratory Animal Care, International. 2013. Frequently asked questions, Animals included in the AAALAC International accredited "unit." http://www.aaalac.org/accreditation/faq_landing.cfm#A2
99. Cooper, J.E. 2011. Anesthesia, analgesia and euthanasia of invertebrates. *ILAR J.* 52(2):196–204.
100. Lewbart, G.A. 2012. *Invertebrate Medicine, 2nd edition.* Ames: John Wiley & Sons, Ltd.

18

Surgery

Lester L. Rolf Jr.

Introduction

The IACUC and AV have the responsibility to help ensure the humane care and use of animals undergoing surgery and to help ensure that individuals who perform that surgery are appropriately qualified and trained. This chapter addresses several major considerations of surgery using animals, such as anesthesia, aseptic technique, and the perioperative care associated with the conduct of survival and non-survival animal surgery in biomedical research and teaching environments.

18:1 What information regarding survival surgical procedures should be included by the investigator in the IACUC protocol? How detailed a description is necessary?

Reg. There are numerous regulatory requirements that relate to survival surgical procedures. In the broadest sense, they revolve around adequate veterinary care (see Chapter 27; AWAR §2.33; PHS Policy IV,C,1,e). Specific areas of concern for the IACUC include the following:

- Use of appropriate anesthesia and analgesia (AWAR §2.31,d,1,iv,A)
- Use of aseptic technique (AWAR §2.31,d,1,ix)
- Conduct of multiple survival surgeries (AWAR §2.31,d,1,x; Animal Care Policy 14[1])
- Qualifications of surgical personnel (AWAR §2.31,d,1,viii)
- Perioperative care (AWAR §2.31,d,1,ix)
- Whether a major operative procedure or a non-major operative procedure will be performed (AWAR §2.31,d,1,ix)
- Description of the procedures to be performed (AWAR §2.31,d,2; §2.31,e,3; PHS Policy IV,C,2; IV,D,1) and compliance, as applicable, with the AWA required (PHS Policy, II; IV,C,1)

Opin. An investigator should provide sufficient surgical procedure detail in the IACUC protocol for the committee to confirm that acceptable surgical techniques are proposed and the AV can evaluate the perioperative care program pertaining to adequate veterinary care needed.[2] In the author's opinion, incision location, chronic instrumentation and implants (when applicable), and the method of wound closure (including size and type of suture) should be included with the description of surgical procedures. In addition, the description of the method of anesthesia should include preanesthetic, anesthetic, and analgesic drugs and doses, routes, and frequency of administration. It is helpful for the protocol to include an appropriate range of agents, routes, dosages, frequencies, procedures, etc., that will allow flexibility and enable the research team to use appropriate professional judgment and still provide enough detail for the IACUC to determine that the proposed care is humane and appropriate. If neuromuscular blocking agents are to be used, the agent, dosage, and administration route of the agonist and reversal agent should be listed. Procedures that may cause more than momentary pain or distress to the animals should not involve use of neuromuscular blocking agents unless anesthesia is also used (AWAR §2.31,d,1,iv,C). When neuromuscular blocking agents are used, it is appropriate to require the PI to document a lack of pain response, including parameters which are not affected by neuromuscular blockade (such as heart rate) from the animal subjects during the procedure. This places the AV and IACUC in a much more enlightened position with regard to these study paradigms. (See 16:26.)

Surv. Which of the following items relating to major survival surgical procedures are typically requested by your IACUC to be included on a protocol application? Check as many boxes as appropriate.

• Not applicable	42/292 (14.4%)
• Incision location	228/250 (91.1%)*
• Incision length	171/250 (68.4%)*
• Instruments to be used	127/250 (50.8%)*
• Type of suture material to be used	196/250 (78.4%)*
• Method of wound closure (e.g., in layers, one suture for all)	228/250 (91.2%)*
• When (or if) skin sutures or clips will be removed	206/250 (82.4%)*
• I don't know	1/250 (0.6%)*

* To calculate percentage of responses, the number of "Not applicable" responses was subtracted from the total to more accurately note the degree to which responders contend with these issues.

In the second Edition of *The IACUC Handbook*[3] respondents were asked for information regarding instrumentation and implants, with 68% reporting a requirement for such information in their protocol format. The current survey, using the term *"instruments to be used,"* provoked a reduced response of only 51%. In this author's opinion, this was probably interpreted to indeed mean "instruments" rather than unique devices that might be implanted for monitoring physiological signs; provide continuous access to blood sampling; allow for incremental bone elongation, etc. Further, in the second edition of *The IACUC Handbook* 48% of respondents indicated that the type of suture material to be used was a requirement of the protocol form. This number has increased in the current survey, with 78% of respondents in the current survey indicating this to be a requirement. Similarly, 91 and 82%, respectively of the current respondents reported that the method of wound closure and the use of wound clips *in lieu* of skin sutures was required information on their protocol forms. In the second edition of *The IACUC Handbook* these figures were 57 and 71%, respectively.

18:2 What information related to pharmaceuticals used during surgery should be included in the IACUC protocol?

Opin. With respect to pharmaceuticals administered as part of a surgical procedure, it is this author's opinion that an adequate description would include all preoperative sedatives and anticholinergic agents as well as preemptive analgesics administered prior to anesthetic induction and maintenance, with either a volatile anesthetic or parenteral (intravenous, intraperitoneal, or intramuscular) medication. Further, planned use of non-pharmaceutical grade compounds should be indicated and appropriately justified (APHIS/AC Policy 3; *Guide* p. 31)

The anesthetic agent and its inhaled concentration (if a volatile agent is used) or dosage and route of administration (if a parenteral agent is used) should be indicated and evaluated by the AV and the IACUC. If delivered by drip and/or constant infusion, the concentration and infusion rate, or at least a range, should be included. A range of concentrations for volatile agents and the carrier gas should be detailed. Reversal agents, if applicable, their dosage and route of administration should also be included in the protocol. Institutional web sites providing tables of possible agents from which PIs may "pick and choose" may add ease to writing of protocols, but it is the responsibility of the IACUC, especially with the input from the AV, for fine tuning any medication protocols to maximize safety of the patient while under anesthesia, and a quick and uneventful recovery following anesthesia. It also is useful for the protocol to include a range of dosages, routes of administration, and appropriate agents such that the research team has flexibility to make changes based upon professional judgment during the course of the project.

If neuromuscular blocking agents will be used as part of the proposed research, the drug, dosage and route of administration must be included in the protocol. Further, the protocol should include a discussion regarding how the depth of anesthesia will be determined to ensure that the patient remains pain-free while paralyzed.

If significant blood loss is likely to occur and/or extracorporeal perfusion will be used during the surgical maneuver, the use of heparin, protamine, calcium and potassium electrolytes and electrolyte fluids needs to be described in the protocol. Further, it should be clear whether the perfusionist or the anesthetist will be responsible for adjustment of fluids and medications during the course of the surgical event. Indecision in this regard or confusion during the surgical maneuver can be disastrous for the animal patient.

Surv. What information is typically requested of an investigator performing a survival surgery procedure? Check as many boxes as appropriate.

• Not applicable	32/294 (10.9%)
• Pre-anesthetic drugs used	217/262 (82.8%)*
• Anesthetic drugs used	259/262 (98.9%)*
• Analgesic drugs used	254/262 (96.9%)*
• The dose, frequency, and route of administration of anesthetic and analgesic drugs	255/262 (97.3%)*
• The dosage, frequency, and route of administration of all other drugs and chemicals used	245/262 (93.5%)*

- The name (e.g., lactated Ringer's), dosage, and route of fluid therapy (if used) — 198/262 (75.6%)*
- I don't know — 7/262 (2.7%)*
- Other — 0/262 (0.0%)

* To calculate percent of responses, the number of "Not applicable" responses was subtracted from the total to more accurately note the degree to which responders contend with these issues.

18:3 What information with respect to qualifications of the surgeon should be included in the IACUC protocol?

Opin. With respect to surgical qualifications, aseptic technique, and monitoring of the animal, the IACUC is given responsibility for implementation of SOPs, policies and training programs that will assure that animal well-being and humane treatment are priorities during every surgical procedure, survival or non-survival. Often this responsibility is met through employees specifically trained in certain aspects of general animal care and more specifically perioperative and operative procedures. In many cases such individuals are veterinarians, physician-surgeons, or technicians who have been tutored in specific surgical procedures by veterinarians and PIs. It is important that technical skills are kept current and that the well-being of the animal patient is always of primary concern. Personnel, including veterinarians and surgeons, may require species-specific and procedure-specific training before being allowed to conduct survival or non-survival animal surgery. It is the responsibility of the IACUC to ascertain that only qualified individuals are allowed to perform surgery on animals. Having the AV or facility staff monitor new procedures and personnel; acknowledging publications in which personnel are identified as having performed specific procedures; and mentoring of personnel in laboratories where the proposed surgery is routinely performed are but a *few* of the means available to the IACUC to ensure competency in the surgical theater.

Surv. For major survival surgical procedures, which of the following information items are requested from an investigator on his/her IACUC protocol application? Check as many boxes as appropriate.

- Not applicable — 46/293 (15.7%)
- Length of the surgeon's experience as a surgeon — 144/247 (58.3%)*
- Length of the surgeon's experience with the procedure in the species being used — 196/247 (79.4%)*
- Qualifications of others directly involved in surgery (e.g., the anesthetist) — 193/247 (78.1%)*
- Methods of intra-operative monitoring of the animal — 223/247 (90.3%)*
- The methods of aseptic technique that will be followed — 221/247 (89.5%)*
- The frequency of postoperative observations until the animal is recovered from general anesthesia (the definition of recovery is to be acceptable to the IACUC) — 232/247 (93.9%)*
- I don't know — 1/247 (0.40%)*

* To calculate percentage of responses the number of "Not applicable" responses was subtracted from the total to more accurately note the degree to which responders contend with these issues.

In the previous edition of *The IACUC Handbook*[3] aspects of the survey shown above were covered under the query, "*How much detail should be provided in the IACUC protocol relative to perioperative care*?" Perioperative care included pre-operative, intra-operative and post-operative monitoring, care and recovery of the surgical patient. The earlier survey was more focused on identification of personnel who would be associated with the procedure as surgeon and anesthetist. The current survey was directed more to the qualifications for the specific surgical procedure to be performed and qualifications of all the support personnel, including the anesthetist. Support staff, surgeons, perfusionists, anesthetists, radiology technicians, etc., should be identified by name and their qualifications for specific procedures.

Almost 90% of current responders from institutions where survival surgery is conducted identified the requirement to spell out the way in which aseptic technique would be implemented. This author assumes this to mean conditions such as the surgical attire; the preparation of the skin after hair removal, if applicable; display of sterilized instruments on instrument tables covered with sterile cloth; draping of the patient, etc., and all surgery is to be conducted in a dedicated operating room or area with appropriate ventilation. In the earlier survey, only 67% of respondents indicated this was required to be on their institution's IACUC protocol.

18:4 Is a cutaneous biopsy considered a survival surgical procedure?

Reg. The *Guide* (p. 117) identifies skin biopsy as one of a number of examples of a *minor surgical* procedure, describing a minor surgical procedure as one in which there is minimal tissue trauma, no penetration of a body cavity and little or no physical impairment from such a procedure. The PHS Policy does not specify what constitutes minor surgery, though major survival surgery is characterized as surgery which "penetrates and exposes a body cavity or produces substantial impairment of physical or physiologic functions or involves extensive tissue dissection or transection."[4] Similarly, the *Guide* (p. 117) and NIH/OLAW FAQ F.13 define major survival surgery as one which "penetrates and exposes a body cavity, produces substantial impairment of physical or physiologic functions, or involves extensive tissue dissection or transection..." AWAR §1.1 provides a definition of a major operative procedure as "any surgical intervention that penetrates and exposes a body cavity, or any procedure which produces permanent impairment of physical or physiologic function." In addition, AWAR §2.31,d,1,ix classifies procedures as either major operative procedures or non-major operative procedures. In conventional veterinary clinical practice, a skin biopsy would usually be performed at least with a sedative if not a local anesthetic; might require suturing or at least repair with tissue adhesive and would, therefore, be considered a surgical procedure.

Opin: There is a wide spectrum of procedures done under the umbrella of skin biopsy, particularly in human dermatology, where very superficial layers of the epidermis can be removed for histological evaluation, usually with the patient anesthetized using only local anesthesia. However, some of these procedures can reach the dermis. Despite the overall minimal nature of such procedures, all of the attendant risks, though reduced, are those that would be associated with penetration into a body cavity. That is, there is the risk of infection; risk of non-healing and requirement for debridement and/or a secondary surgical repair; risk of hemorrhage and nerve damage and depending upon the area and reasons for biopsy, the need for prophylactic antimicrobial use. Because living tissue has been incised, a cutaneous biopsy would be considered a minor survival surgical procedure.

18:5 What information regarding *nonsurvival* surgical procedures should be included by the investigator in the IACUC protocol?

Reg. The *Guide* (p. 118) indicates that while not all techniques that should be followed for survival surgery would need to be followed for non-survival procedures, that "...at a minimum, the surgical site should be clipped, the surgeon should wear gloves, and the instruments and surrounding area should be clean... "

Opin. As with survival surgical procedures (see 18:1) an investigator should provide sufficient detail in the IACUC protocol when describing nonsurvival surgical procedures so that the committee can confirm that acceptable surgical techniques are proposed and the AV can evaluate the preoperative and intraoperative management pertaining to adequate veterinary care. The method of anesthesia described should include preanesthetic and anesthetic drugs, ranges of dosages, routes and frequency of administration. If neuromuscular blocking agents are to be used, the agent, dosage range, and administration route of the agonist and reversal agent should be listed. When neuromuscular blocking agents are used, it is appropriate to require the PI to document lack of pain response, including parameters such as heart rate, which are not affected by neuromuscular blockade, from the animal subjects during the procedure. This places the AV and IACUC in a much more enlightened position with regard to these study paradigms. (See 16:26.) Some IACUCs choose to be more restrictive than current regulations (i.e., *Guide* p. 118; AWAR §2.31,d,1,ix) with issues such as appropriate surgical attire, dedicated space requirements, and use of aseptic procedures. When this occurs, the IACUC should clearly define such requirements to the investigator.

Surv. For *non-survival* surgical procedures, what information is requested from an investigator? Check as many boxes as appropriate.

• Not applicable	28/295 (9.5%)
• Pre-anesthetic and anesthetic drugs, dosages, and route of administration	255/267 (95.5%)*
• The anticipated length of the procedure	88/267 (33.0%)*
• Whether or not aseptic technique will be used	161/267 (60.3%)*
• If used, the dose and route of neuromuscular blocking agents	198/267 (74.2%)*
• If used, the precautions to assure an animal is not awake during the use of neuromuscular blocking agents	206/267 (77.2%)*
• I don't know.	2/267 (0.7%)*
• Other	5/267 (1.9%)*

* To calculate percent, the number of "Not applicable" responses was subtracted from the total to more accurately note the degree to which responders contend with these issues.

18:6 How detailed a description is necessary in an IACUC protocol that contains *nonsurvival* surgical procedures?

Opin. The description should be adequate to assure the IACUC that, in fact, normal procedures are not abbreviated and that conduct of the surgery will be done with the same care and attention that is applied in survival procedures.

Some procedures require less detail than others, particularly those that do not involve extensive intraoperative management. For example, a tissue harvest conducted alone or, perhaps, after some physiological measurements, might only require a "bare bones" surgical description, such as "Midline laparotomy is performed in surgically prepped and draped animals. The renal artery and veins are ligated at the pelvis and the kidney extirpated. The animal is euthanized as described in item…"

A more detailed description might be warranted when the intraoperative time and care are likely to be substantial, such that failure to provide detailed attention would likely result in pain or distress to the animal. Examples include a series of physiological measurements such as blood pressure, pressure volume curves of the ventricular chambers, and transmural pressures, in which maintenance of vascular cannulas is involved. In such cases, the physiological stability of the anesthetized animal is required for reliable data. Also, because the potential for hemorrhage and the need for volume replacement exist, the IACUC requires more detailed information.

18:7 If an animal is anesthetized and then undergoes open-chest terminal perfusion, should the procedure be counted as euthanasia or a non-survival surgical procedure?

Reg. The *Guide* (p. 118) simply describes a non-survival surgical procedure as one in which the animal patient is euthanized before recovery from anesthesia. Euthanasia is the act of humanely killing animals by methods that induce rapid unconsciousness and death without pain or distress. Unless a deviation is justified for scientific or medical reasons, methods should be consistent with the *AVMA Guidelines for the Euthanasia of Animals.*[5] APHIS/AC Policy 3 defines acute terminal procedures as those in which an animal is anesthetized for the purpose of conducting a research procedure, such as surgery, and the animal is euthanized without ever regaining consciousness.[6]

Opin. In the author's opinion, this particular procedure could be classified as either survival or non-survival surgery. The final semantics depends upon the protocol description of the ways in which the animal is to be terminated and "why" this particular method is chosen. Importantly, the criteria for euthanasia have been met as part of the procedure, with the animal undergoing terminal perfusion under anesthesia.

If experimental physiological measurements were made as part of the procedure (e.g., blood pressure, ventricular contractile force, FEV_1; Purkinje fiber conduction velocity, etc.) and then the animal was perfused, it might be considered a non-survival surgical procedure. If in fact, the only thing done was to perfuse the heart and lung so that tissue could be taken for histology or other analytical work, it might be called a tissue harvest and could be considered as a euthanasia procedure rather than a non-survival surgery. (See 17:8.)

18:8 Can the IACUC classify surgical procedures as minor versus major? By what criteria? How can this classification be useful to the IACUC when reviewing protocols?

Reg. The AWAR (§1.1) define a *major operative procedure* as "any surgical intervention that penetrates and exposes a body cavity or any procedure that produces permanent impairment of physical or physiological functions." The AWAR classify survival surgical procedures as either major or non-major operative procedures (§2.31,d,1,ix). The *Guide* (p. 117) states that surgical procedures are categorized as major or minor. A *major survival surgery* is defined as surgery that "penetrates and exposes a body cavity, produces substantial impairment of physical or physiologic functions, or involves extensive tissue dissection or transection." The *Guide* (p. 117) offers as a general guideline that procedures such as laparotomy, thoracotomy, joint replacement, and limb amputation are examples of major surgical procedures. The *Guide* (p. 117) defines *minor survival surgery* as surgery that "does not expose a body cavity and causes little or no physical impairment" and suggests that procedures such as wound suturing, peripheral vessel cannulation, percutaneous biopsy, routine agricultural animal procedures such as castration and most procedures routinely done on an 'outpatient' basis in veterinary clinical practice are examples of minor survival surgery. PHS Policy (IV,A,1) requires institutions to use the *Guide* as a basis for their animal care and use program. (See 26:5.)

Opin. Classification of surgical procedures by an IACUC as major or minor is based on the preceding regulatory definitions. This classification is useful to an IACUC when reviewing a protocol because it is the basis for the IACUC's determination of the type of surgical facility and the degree of aseptic technique required (see 18:17–18:23). Major survival surgery on nonrodents must be performed only in facilities designed, operated, and maintained for that purpose. Minor survival surgery (non-major operative procedures as defined by APHIS/AC) and all surgery on rodents do not require separate dedicated facilities; however, aseptic technique must be used (AWAR §2.31,d,1,ix; *Guide* p. 118). Further, for animals to be used in multiple major survival surgeries, such use must be scientifically justified in the protocol and approved by the IACUC (AWAR §2.31,d,1,x,A; *Guide* p. 30; NIH/OLAW FAQ F.9). There is no regulatory limitation to multiple *minor* survival surgical procedures. Nevertheless, an IACUC should use professional judgment to limit the number of minor surgical procedures performed on an animal.[7]

18:9 Should laparoscopy be considered a major operative procedure (major survival surgery)?

Reg. (See 18:8.) The AWAR (§1.1) define a *major operative procedure* as "any surgical intervention that penetrates and exposes a body cavity or any procedure that produces permanent impairment of physical or physiological functions." The *Guide* (p. 117) states that a major survival surgery is one that penetrates and exposes a body cavity or produces substantial impairment of physical or physiologic functions.[4] NIH/OLAW FAQ F13 provides general guidance to the IACUC about the classification of laparoscopic procedures as major or minor surgery.

Opin. The regulatory definition of major survival surgery is not presented with respect to the extent of entry or exposure of a specific body cavity. If the body cavity has been entered, a criterion has been met for defining the procedure as a major survival surgery. With laparoscopic procedures, there are usually at least two incision sites, one for the operative scope and one for the viewing scope. Depending upon the procedure, there may be more. The welfare of the animal patient should be the focus. While the laparoscopic technique minimizes trauma, it also places the animal at increased risk of hidden hemorrhage. Additionally, the repeated entry and withdrawal of the scopes in this technique and their awkward handling increase the likelihood of unnoticed contamination by surgeon or assistants. Further, laparoscopic equipment is more difficult to clean and sterilize, thereby increasing the risk of bacterial contamination, despite reduced wound size.

Surv. Does your IACUC usually consider a laparoscopic procedure into the abdomen or thorax to be a major or minor operative procedure?

- Not applicable 79/294 (26.9%)
- Generally a major procedure 150/294 (51.0%)
- Generally a minor procedure 9/294 (3.1%)
- We evaluate the overall invasiveness of the procedure before determining if it is major or minor 50/294 (17.0%)
- I don't know 6/294 (2.0%)
- Other 0/294 (0.0%)

18:10 Does your IACUC usually consider a craniotomy (via one or more burr holes) to be a major or minor operative procedure

Reg. The AWAR (§1.1) define a major operative procedures as "any surgical intervention that penetrates and exposes a body cavity or any procedures that produces permanent impairment of physical or physiological functions." The *Guide* (p. 117) states that surgical procedures are categorized as major or minor. A *major survival surgery* is defined as surgery that "penetrates and exposes a body cavity, produces substantial impairment of physical or physiologic functions, or involves extensive tissue dissection or transection" and offers that procedures such as laparotomy, thoracotomy, joint replacement, and limb amputation are examples of major surgical procedures. The *Guide* (p. 117) defines *minor survival surgery* as surgery that "does not expose a body cavity and causes little or no physical impairment" and suggests that procedures such as wound suturing, peripheral vessel cannulation, percutaneous biopsy, routine agricultural animal procedures such as castration and most procedures routinely done on an 'outpatient' basis in veterinary clinical practice are examples of minor survival surgery. PHS Policy (IV,A,1) requires institutions to use the *Guide* as a basis for their animal care and use program and NIH/OLAW recognizes the authority of the IACUC to determine whether specific manipulations used in research are major operative procedures (NIH/OLAW FAQ F13).

Opin. For uniformity and clarity, the NIH/OLAW position seems congruent with the majority opinion of laboratory animal veterinarians and AVs. APHIS/AC has left this decision to the discretion of the IACUC. A request to classify a craniotomy, as well as any other manipulation involving penetration into a body cavity, as a minor surgical procedure would need to be evaluated by the IACUC with input from the PI and the AV. For instance, a common clinical procedure might puncture the thorax or abdomen with cannulated needles to withdrawn fluid without exposing the corresponding body cavity. This is a relatively common clinical practice and in both instances the surrounding tissues usually seal the opening. Thus, it might be argued that the body cavity has not been significantly exposed and the procedure qualifies as a minor surgical procedure. In a similar vein, percutaneous puncture with a closed endoscope into the coelomic cavity of an avian species for the purpose of sexing, with the closure of the skin around the exiting endoscope, might be considered a minor procedure. In this regard, it is much more difficult to envision entering the cranium, even with a very fine trephine, without exposing the cavity, at least momentarily. The risks of significant hemorrhage are greater in the cranium than a similar penetration into either the chest or abdomen. In this author's opinion, the greater risk of hemorrhage, as well as the more severe consequences related to physical tissue damage and/or bacterial contamination of the brain warrant the classifying craniotomy as a major survival surgery.

Surv. Does your IACUC usually consider craniotomy (via one or more burr holes) to be a major or minor operative procedure?

• Not applicable	95/295 (32.2%)
• We generally consider it to be a major procedure	153/295 (51.9%)
• We generally consider it to be a minor procedure	12/295 (4.1%)
• We evaluate the size of the burr hole(s) before deciding on major or minor	1/295 (0.3%)
• We evaluate the overall invasiveness of the procedure before deciding on major or minor	25/295 (8.5%)
• I don't know	9/295 (3.0%)
• Other	0/295 (0.0%)

Note that nearly 1/3 of respondents indicated that craniotomy is not performed at their institutions; and, of those institutions where craniotomy is performed, most (52%) indicate that it is regarded as major survival surgery.

18:11 Under what circumstances can an individual animal be used in multiple survival surgical procedures?

Reg. Although performing multiple major survival surgical procedures on an individual animal is discouraged, the *Guide* (p. 30) states that they may be permitted if scientifically justified by the investigator and approved by the IACUC; if necessary for clinical reasons; or if the surgeries are included in and essential components of a research project or protocol. The AWAR (§2.31,d,1,x,A-C) and NIH/OLAW[7] permit multiple major survival operative procedures (and see NIH/OLAW FAQ F.9 for guidance on multiple survival surgeries on the same animal):

- When scientifically justified by the investigator in writing
- When needed as a routine veterinary procedure or to protect the health and well-being of an animal
- When other special circumstances are authorized by the Administrator, APHIS, USDA, on a case-by-case basis
- APHIS/AC Policy 14 provides additional clarification. It states that a second major survival operative procedure must not be performed on an animal in a separate animal study activity. It goes on to state that in order to comply

> with the intent of the AWA, animals surviving major operative procedures in one animal study activity must be identified in a manner that effectively precludes their use in additional animal study activities involving major survival operative procedures.[1]

Regulations that address multiple survival surgical procedures on an individual animal have exemption criteria where it is allowed. According to the AWA (§2143,a,3,D) exemptions are authorized by the Secretary of Agriculture. However, as specified by the AWAR (§2.31,d,1,x) requests are sent to the Administrator, APHIS/USDA. The Secretary of Agriculture has delegated this responsibility to the APHIS Administrator, who has further delegated this to the Deputy Administrator of Animal Care. According to APHIS/AC Policy 14,[1] requests are made to the appropriate Animal Care Regional Director, who forwards them to the Animal Care Assistant Deputy Administrator for review and recommendation to the Deputy Administrator. The IO of the research facility should make the exemption request to the appropriate Animal Care Regional Director, who, as noted, will forward it to the Animal Care Deputy Administrator.

NIH/OLAW[7] permits multiple major survival operative procedures on individual animals in the following circumstances:

- The procedures are included in and essential components of a single research project or proposal;
- The multiple surgical procedures are scientifically justified by the investigator; or
- The multiple surgical procedures are necessary for clinical reasons.

Opin. Examples of special circumstances that may potentially justify performing multiple surgical procedures on an animal include procedures that are related components of a research project, involve conservation of scarce animal resources, or are needed for clinical reasons. The *Guide* (p. 30), APHIS/AC Policy 14[1] and interpretation of PHS Policy[8] do not consider cost savings alone as an adequate justification for performing multiple major survival surgeries on a single animal. (See 16:22; 18:14–18:15.)

18:12 How many survival surgical procedures to harvest oocytes from Xenopus frogs are acceptable for a single animal?

Reg. There are no explicit regulations or guidelines from NIH/OLAW, APHIS/AC or the *Guide* regarding this specific procedure. Indeed, the AWAR (§1.1, Animal) exclude amphibians as a regulated species per the definition of "animal." However, guidance from NIH/OLAW and the *Guide* as described above (18:11) for multiple survival surgery might have import to this issue. In this regard, the surgeries must be scientifically justified; part of one distinct study and protocol; be done under general anesthesia; be attentive to using the best aseptic techniques possible under the circumstances; and provide appropriate perioperative monitoring.

Opin. In the author's opinion, multiple surgical oocyte collection should be allowed in these species, based primarily on existing experiences of those who have been doing this procedure for years with minimal apparent impact on the well-being of the frogs. There appears to be rapid postoperative healing of the surgical site.

Proficiency through experience can limit the incision length and return the animal to normal eating and behavior quite quickly. The fact that subsequent harvests result in good quality eggs from a frog that maintains or increases weight during collection intervals, suggests that physiologic recovery is quite rapid. These latter facts strongly suggest that perioperative pain and distress may be more subdued compared to that seen with mammalian species experiencing abdominal surgery. Reproductive tissues (ovaries) are not removed in this procedure and almost no manipulation or mechanical disturbance of other coelomic viscera are required, minimizing postoperative scarring, adhesions, and attendant pain. The IACUC may wish to consider that an adequate rest period before the next surgical collection is stated and that the two sides of the abdominal cavity might be alternated for surgery to access the oocytes.

Surv. When performing routine oocyte collections on *Xenopus* frogs, how many survival surgical procedures does your IACUC typically allow on any one animal?

• Not applicable	201/293 (68.6%)
• One surgery	10/92 (10.9%)*
• Two surgeries	11/92 (12.0%)*
• Three surgeries	13/92 (14.1%)*
• Four surgeries	5/92 (5.4%)*
• Five surgeries	9/92 (9.8%)*
• Six surgeries	5/92 (5.4%)*
• More than six surgeries	2/92 (2.2%)*
• I don't know	37/92 (40.2%)*

* To calculate the percentage of respondents the number of "Not applicable" responses was subtracted from the total to more accurately note the degree to which responders contend with this issue.

Relatively few of the respondents are from institutions where *Xenopus* oocyte harvest is part of the biomedical research endeavor. Curiously, nearly 50% of respondents who allow multiple surgeries, there appears to be a conservative approach to the number of survival harvests before the final terminal harvest. The survey does not specify if this number is inclusive of the terminal harvest but this author has assumed that. Two-thirds of these respondents are from institutions where the IACUC has limited total harvests to three or less.

18:13 An animal is surgically modified by the vendor to meet a research need (e.g., ovariectomy of a rat). Must that vendor have its own PHS Assurance statement if the animals are sold to an institution and used in a project supported by PHS funds? Should the IACUC obtain a copy of that Assurance statement?

Opin. Purchase of animals that have been surgically modified by a vendor and are available for general sale is covered by PHS Policy if the surgery is conducted in response to a specific, custom request.[9] If an institution purchases an animal to be used in a PHS-supported study, and the animal has been surgically modified by the breeder in advance of sale, as a readily available "off-the-shelf" catalog item, then the vendor is not required to have his or her own Assurance on file with NIH/OLAW. On the other hand, if an investigator subgrants or subcontracts with a supplier to produce surgically-modified animals according to specification for a PHS-supported study, the supplier must have on file an Animal Welfare Assurance with NIH/OLAW and must provide verification of IACUC approval for the animal activity. (NIH/OLAW FAQ A.2 discusses the applicability of the PHS Policy to the purchase of surgically modified animals.) If both institutions have Assurances, the IACUCs may choose which IACUC will review the protocol for the animal activities being conducted. It is recommended that if an IACUC defers protocol review to another IACUC, documentation of the review should be maintained by both committees.[9] The institution must obtain the supplier's Assurance number by contacting the supplier or from the NIH/OLAW website having a list of Assured institutions (http://grants.nih.gov/grants/olaw/assurance/300index.htm). Obtaining a copy of the Assurance statement would be a local decision made at the institutional level. (See 8:10; 16:22.)

18:14 Should the IACUC permit a farm animal to recover from major surgery so that it can be sold for slaughter once it has fully recovered? (See 14:48.)

Reg. The AWAR (§2.31,d,1,ix) and the *Guide* (pp. 115–120) do not prohibit recovery from a first major surgery provided that all related requirements are met. However, PHS Policy (IV,C,1,a) states that "procedures with animals will avoid or minimize discomfort, distress, and pain to the animals, consistent with sound research design." This is similarly stated in the AWAR (§2.31,d,1,i). The *Guide for the Care and Use of Agricultural Animals in Research and Teaching*[10] (p. 10) is similarly specific with allowance given (1) when the surgeries are justified and approved by the IACUC; and (2) if needed to ensure or maintain the health of the animal.

Opin. Assuming that the investigator's research or educational needs can be met with nonsurvival surgery, anesthesia without recovery supports the welfare interests of the animal by fully preventing potential pain and discomfort rather than *minimizing* potential postoperative pain with analgesics. If meat from the slaughtered animal was destined for human consumption, the IACUC has to ensure that the animal received only FDA-approved medications and drug withdrawal periods were met. Ultimately, the IACUC's awareness of the public's expectations and perceptions regarding the use of animals in research and education should take precedence in making this decision. Use of animals in research is held to a higher ethical standard than many other uses. The IACUC needs to be aware of all Federal, State, and local laws regarding sending research animals into the food chain (for either human or animal consumption). The potential negative public perception of returning farm animals to the food chain after use in research or education might be a basis for the IACUCs to discourage and decide not to permit this practice, particularly given the negative publicity that might result in the event of an adverse event following entry of such meat into the food chain.

18:15 Should the IACUC permit a hamster, ovariectomized as a customer service by the vendor, to undergo another major survival surgical procedure? For regulatory compliance purposes, should this be considered two major survival surgical procedures on the same animal?

Reg. The AWAR (§2.31,d,x,A–§2.31,d,x,C) stipulate that no animal should be used in more than one survival major operative procedure unless justified for scientific reasons, required as part of the veterinary care of the animal, or otherwise exempted by the APHIS/AC Administrator as a Special Circumstance. (See 18:11.) Further clarification is offered by APHIS/AC Policy 14[1] which indicates that animals undergoing a major survival operative procedure by the vendor must be identified as such to prevent their use in another major survival operative procedure. In the event another surgery is required for scientific purposes, APHIS/AC can be contacted with a request for Special Circumstance considerations (AWAR §2.31,d,1,x,C). The *Guide* (p. 30) also clearly identifies the need for scientific justification if more than one major survival surgery is to be performed on an animal.

Opin. Ovariectomy, performed according to current veterinary practices, would have to be classified as a major surgery, since a body cavity has been entered, even when entered laparoscopically (see 18:9). Further, the removal of the ovaries would result in permanent physiological changes to the animal; thus, it would be logical for the IACUC to regard this situation as one in which multiple major survival surgical procedures are being conducted. The IACUC may permit, as an exception based on scientific merit, multiple major survival surgical procedures as described above.

18:16 A major surgical procedure is performed by a veterinarian for a spontaneous medical condition that is unexpected and unrelated to any ongoing research. Have conditions been established for prohibiting a subsequent survival major operative procedure unless scientifically justified and IACUC approval is given for "multiple major operative procedures from which the animal recovers"?

Reg. This condition is specifically addressed by the AWAR (§2.31,d,1,x,B) and is clarified in APHIS/AC Policy 14[1] which indicates that an animal that "has a major operative procedure as part of a facility's veterinary care program (unrelated to research), or as an emergency surgery, may still be used in a research proposal that requires a major survival operative procedure." While not specifically stated, the assumption is that the emergency procedure was such that the procedure did not interfere with the potential use of the animal in a dedicated study.

Opin. It is the responsibility of the AV and the PI to advise the IACUC on the medical and scientific suitability of these animals for further study. The *Guide* (p. 30) suggests that additional consideration could be given to the species availability or other unique features of the animal that might make it a valuable resource for additional studies involving major survival surgery, though such use should be critically reviewed by the IACUC.

18:17 If an investigator dedicates a part of a laboratory bench top for performing major survival surgery on a hamster, is this considered as being compliant with the AWAR?

Reg. The AWAR (§1.1, Animal) specifically include hamsters as a regulated species. However, the AWAR (§2.31,d,1,ix) indicate that major operative procedures performed on rodents do not require dedicated facilities, though aseptic technique is required. The *Guide* (p. 144) indicates that space dedicated to surgery and related activities when used for this purpose and managed to minimize contamination from other activities within the room being conducted at the same time is acceptable for rodent surgery. NIH/OLAW has provided similar guidance on this topic in FAQ F.7, What are the requirements for conducing rodent survival surgery?

Opin. Conducting hamster survival surgical procedures on a laboratory bench is acceptable as long as there is no competing activity in the area at that time. Further, use of this area is compliant as long as conditions minimize contamination from other activities within the laboratory. The AWAR exception for rodents should not be interpreted to exempt the surgeon from the use of aseptic technique, nor from being appropriately garbed. Further, as the hamster is a regulated species, adequate intraoperative and postoperative records of observations, analgesic or other medication administration, pain assessments, and notes on surgical healing should be maintained.

18:18 What is aseptic technique?

Reg. The *Guide* (p. 118) states that aseptic technique includes:

- Preparation of the patient, with body hair removal and disinfection of the operative site
- Preparation of the surgeon, such as wearing of appropriate surgical attire, face masks, and sterile surgical gloves
- The use of sterilized instruments, supplies, and implanted materials
- The use of operative techniques to reduce the likelihood of infection

The *Guide* (p. 118) also states that "aseptic technique is used to reduce microbial contamination to the lowest possible practical level. No procedure, piece of equipment, or germicide alone can achieve that objective: aseptic technique requires the input and cooperation of everyone who enters the surgery area. The contribution and importance of each practice varies with the procedure."

Opin. By definition, surgical *aseptic technique* is the performance of a surgical procedure in a manner to minimize exposure of the patient to pathogenic organisms.[11] (See 26:20.)

18:19 Is aseptic technique necessary for performing survival surgery in animals?

Reg. The AWAR (§2.31,d,1,ix) require survival surgery to be performed on all regulated animals using aseptic procedures, which include surgical gloves, sterile instruments; and aseptic techniques. The *Guide* (p. 118) states that "aseptic technique should be followed for all survival surgical procedures."

Opin. An IACUC has a responsibility to ensure use of acceptable standards of aseptic technique. Semiannual inspections of areas where surgery is performed can help ensure this standard. (See Chapter 26.) Performing survival surgery using aseptic technique also is necessary to achieve a satisfactory surgical outcome with reduced risk of infection. Aseptic technique is used to reduce microbial contamination of a surgical wound and exposed tissues to the lowest possible practical level (*Guide* p. 118). To use less than optimal aseptic technique potentially increases bacterial contamination and subsequent risk of infection, which can compromise an animal's health postoperatively. In addition, lack of clinical evidence of postoperative infection does not rule out clinically unapparent infection. A study performed in rats showed that infection can be clinically unapparent and yet cause adverse physiologic and behavioral responses that may affect research results and may not be recognized.[12]

18:20 Are the standards for aseptic technique different for rodents compared to other mammals?

Reg. Neither the AWAR (§2.31,d,1,ix) nor the *Guide* (p. 144) require a dedicated surgical facility for major survival rodent surgery as compared to non-rodent surgery. The AWAR (§2.31,d,1,ix) state that all survival surgery must be performed using aseptic procedures, including surgical gloves, masks, sterile instruments, and aseptic technique. The *Guide* (p. 118) states that "aseptic technique should be followed for all survival surgical procedures" and makes no distinction in this regard for rodents versus non-rodent mammals.

Opin. Because surgery involving rodents is often of a minor nature and may involve surgery on multiple individuals during a single surgery session, modifications in standard aseptic techniques are often followed and experience has shown that such practices are typically successful, including use of one sterile instrument pack for up to five rodents incorporating techniques to maintain sterility between animals, and optional wearing of a mask with cap and sterile gown.[13,14] Although modifications in standard aseptic techniques may be necessary or desirable for rodents, the performance standards to prevent or minimize exposure of the patient to pathogenic organisms to reduce the likelihood of infection must be met and the well-being of the animals should not be compromised.[15]

While the rationale for less stringent regulations governing rodent surgery appears obvious, there are reasons to be more restrictive, on an institutional basis, than required. Many laboratories work with both regulated and nonregulated species. The "lower-key" atmosphere of rodent surgery can spill over into surgical activities involving surgical work with regulated species. If one is doing five sequential rabbit or guinea pig surgeries, five separate sterilized instrument packs will be required. It may not be unreasonable for an IACUC to have this same standard for rodents, even though it is not required by regulations. If a particular IACUC is more restrictive than the regulations, these additional requirements should be clearly defined so that the PI can perform studies in a compliant fashion.

18:21 Does aseptic technique need to be followed for nonsurvival surgery?

Reg. The AWAR (§2.31,d,1,ix) state that, "Major operative procedures on non-rodents will be conducted only facilities intended for that purpose which shall be operated and maintained under aseptic conditions." The area to be used should be clean, free of clutter, and prepared by acceptable veterinary sanitation practices that would be used in a standard examination/treatment room. The *Guide* (p. 118) does not require using aseptic technique for nonsurvival surgery; however, "the surgical site should be clipped, the surgeon should wear gloves, and the instruments and surrounding area should be clean."

Opin. The extent to which an investigator should exceed the minimal regulatory requirements for nonsurvival surgery and apply any or all of the components of aseptic technique as described in 18:18 depends on the experimental protocol and the surgery being performed. Professional judgment is necessary to evaluate the probability of virulent bacterial contamination and subsequent host responses that would invalidate research results. Slattum and colleagues[15] reported results demonstrating a need for aseptic technique when performing nonsurvival surgery. They found that gram-negative bacteremia and septic shock developed in dogs during nonsurvival cardiopulmonary studies performed without using aseptic technique. Laboratory-prepared nonsterile intravenous solutions were found to be contaminated with gram-negative bacteria. Bacteremia and septic shock ceased to occur after initiating some components of aseptic technique such as using sterile commercial saline solution and other sterile intravenous injectables, disinfecting equipment and instruments, and using sterile gloves. (See 26:4.) If nonsurvival surgery is conducted in a dedicated "survival" surgery area, then aseptic technique should be used or the surgical room be returned to an appropriate level of cleanliness before it is used for majantigenor survival surgery

18:22 What level of aseptic technique should the IACUC require for field surgery involving wildlife?

Reg. The AWAR (§2.31,d,1,ix) state that survival surgery conducted at field sites does not require dedicated facilities but must be performed using aseptic procedures. The *Institutional Animal Care and Use Committee Guidebook*[16] suggests that aseptic practices are a best practice when performing survival surgery on wildlife in field studies. The *Guide* (p. 118) indicates that general principles of aseptic technique should be followed for all survival surgical procedures, with no distinction made between those performed in surgical facilities versus those performed under field conditions. The *Guide* (p. 115) also indicates that standard techniques may need to be modified under conditions such as field surgery, and in such cases, a close assessment of surgical outcomes may require criteria other than morbidity and mortality.

Opin. An experimental animal surgical facility environment may be unnecessary when performing survival surgery on wildlife if it is not needed to improve animal well-being or surgical outcome as measured by lack of postoperative complications, improved surgical survival, or minimized pain and stress. Taking free-living wildlife to a dedicated facility to perform surgery could, in fact, be detrimental to the well-being and survival of the animals, though settings typical of clinical veterinary practice may also be suitable for field surgery involving wildlife.[7] Nevertheless, aseptic technique must be used to prevent or minimize exposure of the animal to pathogenic organisms. This includes preparation of the operative site (including hair removal and disinfection); surgical mask; sterile surgical gloves, instruments, supplies, and implanted materials; and use of operative techniques to reduce the likelihood of infection (*Guide* p. 118). Steam sterilization or autoclaving is the preferred method of surgical instrument sterilization. However, chemical "sterilization" using liquid chemicals with appropriate contact time may be necessary in some field surgery settings. Ultimately, professional judgment must be used to optimize the circumstances for the environment where the surgery is performed and the aseptic techniques used.

18:23 What level of aseptic technique should be required for survival surgery performed on non-mammalian aquatic species?

Reg. As described in the *Guide* (p. 118) general principles of aseptic technique should be followed for all survival surgical procedures. There are no specific regulations for aseptic technique with regard to aquatic species.

Opin. The NIH Intramural Research Program provides internal guidelines[17] for aseptic technique during *Xenopus* oocyte collection that may be useful best practices:

"Surgeries should be done as aseptically as practical including the use of sterilized instruments and powderless gloves. Instruments should be sterilized by autoclaving or using a glass-bead sterilizer. The use of cold sterilants should be avoided so that these potentially toxic chemicals are not inadvertently introduced into the surgical site or onto permeable amphibian skin." To this point, the description is quite like that for aseptic technique typically used for rodents or for non-survival surgical procedures. The NIH intramural guidelines also state, "The use of surgical drapes and preparation of the surgical site remains controversial for aquatic species. The use of a sterile drape and preparation of the surgical site with dilute povidone iodine solution has been recommended. The use of these chemical agents may disrupt the normal skin flora of the patient and the constant mucous production of *Xenopus* skin makes any sterilization effort transient. When chemical surgical preps are used, they should be limited to the immediate area around the incision site and should only be solutions, not scrubs containing soaps or detergents."

"Similarly, arguments have been presented regarding the use of surgical drapes. Drapes may be useful to keep mucus from getting on instruments and suture material, and can be moistened to keep skin from drying during surgery. However, amphibian skin can be easily damaged and paper drapes that become wet pose no barrier to bacteria. NIH veterinarians report that the incidence of clinical complications following surgical oocyte harvesting is rare." It is also worth noting here that the skin of fish is also protected with antimicrobial rich mucus as well as scales that need to be protected for rapid healing and tissue repair.

In this author's opinion, the surgeon should be wearing sterile, powderless, surgical gloves and surgical drapes need not be used. The author is unaware of requirements by any research institution for survival surgery on non-mammalian aquatic species to be performed in a surgical suite. As with surgeries for rodents, it is advisable to perform such surgery in at least a dedicated space, since such an area keeps a focus on the surgery, needs of the patient and the proper environment for conducting the surgery with expediency given the less than ideal conditions.

A dilemma exists with respect to the use of organic iodine (povidone iodine) as a skin antiseptic as a means to not only improve aseptic conditions, but also to reduce the possible contamination of harvested oocytes from *Xenopus*. Any potential benefit must be balanced against the possibility of chemical harm to the skin. It is not clear at this point that any antiseptic prevents infections and/or enhances wound healing as part of surgical harvest of *Xenopus* oocytes, but until shown otherwise, it might be assumed that use of such antiseptics offer benefit to the animal and to the harvested oocytes. Alternatively, one could express the egg mass through a dilute povidone solution the way neonates are treated for cesarean derived rodents, followed with a saline rinse.

18:24 Would less than optimal aseptic technique be acceptable if a PI's records indicated excellent surgical success with a lack of postoperative infections?

Reg. The AWAR (§2.31,d,1,ix) require that all survival surgery on regulated species be performed using aseptic technique. The *Guide* (p. 118) similarly indicates that aseptic technique should be followed for all survival surgical procedures.

Opin. No. In this author's opinion, the performance of survival surgery using less than optimal aseptic technique indicates that although the IACUC and AV reviewed and approved the procedures, the review was not thorough enough or the investigator was inadequately trained in aseptic technique. In other words, the IACUC did not fulfill its responsibility to assure use of acceptable standards of aseptic technique.

Aseptic technique is used to reduce microbial contamination of a surgical wound and exposed tissues to the lowest possible practical level[14] (*Guide* p. 118). To use less than optimal aseptic technique potentially increases bacterial contamination and subsequent risk of infection which would compromise the animal's health. Lack of clinical evidence of postoperative infection does not rule out clinically unapparent infection. A study performed in rats showed that infection can be clinically unapparent and cause unrecognized, adverse physiologic and behavioral responses that can affect research results.[14]

Of course, it is hard to argue with success. Under some circumstances, less than optimal aseptic technique may be acceptable as long as the animal's health is not compromised and the procedures are approved by the IACUC; however, the IACUC should be cautioned not to overly focus on the product rather than the process. The IACUC should regularly review approved aseptic techniques vis-à-vis current acceptable standards of proper veterinary care.

18:25 Should the IACUC demand that aseptic technique be used when surgery is performed on farm animals in the field (e.g., standing rumenotomy performed in a barn)?

Reg. The AWAR (§2.31,d,1,ix), which apply to farm animals used in biomedical research, state that survival surgery conducted at field sites does not require dedicated facilities but must be performed using aseptic procedures. PHS Policy (IV,A,1) requires compliance with standards for survival surgery as outlined in the *Guide* when using farm animals in biomedical research. The *Guide* (p. 115) recognizes that modification of standard techniques might be necessary when performing field surgery; however, animal well-being should not be compromised. When modifications are implemented, a thorough assessment of surgical outcomes should be performed to ensure that appropriate procedures are followed. Surgical outcome assessment may require other criteria in addition to clinical morbidity and mortality.

When using farm animals in agricultural research, standards for survival surgery, as outlined in the *Guide for the Care and Use of Agricultural Animals in Agricultural Research and Teaching* (*Ag Guide*),[10] should be applied.[10] The *Ag Guide* (p. 21) states that "major survival surgeries should be performed in facilities designed and prepared to accommodate surgery, and standard aseptic surgical procedures should be employed." Aseptic surgical procedures include use of cap, mask, gown, gloves, and sterile instruments as well as appropriate operative site preparation and draping. Minor survival surgical procedures that do not expose a body cavity and cause little or no physical impairment (e.g., wound suturing and peripheral vessel cannulation) may be performed under less stringent conditions if performed in accordance with standard veterinary practices. Therapeutic and emergency surgeries (e.g., cesarean section, bloat treatment, and displaced abomasum repair) are sometimes required in agricultural situations that are not conducive to rigid asepsis. However, every effort should be made to conduct minor and emergency survival surgeries in a sanitary and aseptic manner.

Opin. When farm animals in the "field" require elective major survival surgery, it should be performed in facilities designed and maintained for surgery using appropriate aseptic procedures including cap, mask, sterile gown and surgical gloves, sterile instruments, and aseptic technique (see 18:18). Therefore, an elective rumenotomy should not be performed in a barn. Minor survival procedures may be performed under less stringent conditions than major procedures but require sterile instruments and aseptic technique. When therapeutic and emergency surgeries do not allow transport to dedicated surgical facilities, the facilities used should be clean and methods of aseptic technique used to prevent or minimize exposure of the animal to pathogenic organisms to reduce the likelihood of infection.

18:26 How can the IACUC assure that personnel are qualified to perform surgical procedures?

Reg. The PHS Policy (IV,C,1,f) and AWAR (§2.31,d,1,viii) place responsibility with the IACUC to determine that personnel performing surgical procedures are qualified and trained in those procedures. NIH/OLAW further describes application of the PHS Policy by stating that institutions must "ensure that research staff members performing experimental manipulation, including anesthesia and surgery, are qualified through training or experience to accomplish such procedures humanely and in a scientifically acceptable fashion…[18]

The *Guide* (pp. 115–116) recognizes that personnel performing surgery on research animals have a wide range of educational backgrounds and might require various levels and kinds of training to ensure that good surgical technique is used. Research institutions should also perform continuing and thorough assessments of surgical outcomes to ensure that appropriate procedures are followed. (See 21:36.)

Opin. The IACUC's goal is to assess surgical competency before IACUC approval of a protocol. The IACUC should review an individual's education, training, certification, and experience for assessment of general surgical competency and qualifications to perform the specific surgical procedure. To assist the AV and IACUC in developing appropriate training programs, the Academy of Surgical Research (ASR) developed and published training guidelines for research surgery commensurate with a person's formal education and training background.[19] The ASR training guidelines can be used when evaluating a person's educational background to determine what an individual may already know. For example, a physician trained in a surgical specialty may need less training to perform surgery on animals within his area of surgical expertise. However, he or she may require training in interspecies variations of anatomy, anesthesia, analgesia, and postoperative care methods. The need for formal training might be waived if the surgeon participates in a multidisciplinary team approach to perform the specific surgical protocol and includes additional experienced personnel qualified to work with animals.

To assure a surgeon's competence, the IACUC may require that a laboratory animal veterinarian observe or assist with at least the first surgery. Assistance could continue with subsequent procedures until competency is achieved and a specific surgical procedure is predictable to the satisfaction of the veterinarian. Although there is no single credential to ensure competency, the best credential is a documented record of previous successful performance of the proposed surgical procedure on the specified species, demonstrating minimal operative and postoperative complications. With such documentation, there should be no need for additional training. After IACUC approval of a surgical protocol, continuing assessment of surgical outcomes should be performed. Participation and input from a laboratory animal veterinarian, animal care staff, surgical technicians, and the investigator are needed to assure ongoing use of appropriate procedures, to address complications and evaluate their rate of occurrence, and to initiate necessary corrective changes.

18:27 What is meant by perioperative care?

Opin. Perioperative care encompasses all events associated with a surgical procedure. A perioperative care program comprises three overlapping components.[20]

- Preoperative planning and management
- Intraoperative care
- Postoperative care

Detailed descriptions of the perioperative care program components have been published.[20] Preoperative planning should include

- Identification of members of the multidisciplinary surgical team, all of whom should provide input into presurgical planning
- Roles and training needs of personnel
- Equipment and supplies required
- Facilities for conducting procedures

An anesthetic protocol should be developed, including anesthetic agents, techniques, and methods of anesthetic monitoring to be used. Planning of the surgical procedure should include aseptic techniques to be used (see 18:18) and assessment of indications for perioperative antibiotics. A postoperative care plan should be outlined. Preoperative management should include

- A preoperative animal-health assessment with a physical examination and laboratory examination if indicated
- A period of stabilization to a new environment for animals before undergoing surgical procedures
- Preoperative fasting of a specified duration if indicated for the species to be used
- Administration of preoperative medications or antibiotics

Components of intraoperative care include

- Monitoring of anesthetic level and vital organ function
- Provision of vital organ support such as parenteral fluid administration, supplemental oxygen, and maintenance of body temperature
- Proper surgical technique, which comprises:
  - Gentle tissue handling
  - Effective hemostasis

- Maintenance of sufficient blood supply to tissues
- Asepsis
- Accurate tissue apposition
- Proper use of surgical instruments
- Appropriate use of monitoring equipment
- Expeditious performance of the surgical procedure

The postoperative period can be divided into three overlapping phases: recovery from anesthesia, acute postoperative care, and long-term postoperative care. Frequent assessment of thermoregulation and cardiovascular and respiratory function is required during anesthetic recovery and acute postoperative care. Additional care may include

- Monitoring of the surgical incision
- Thermal support to combat hypothermia
- Parenteral fluid administration to maintain hydration
- Administration of analgesics for postoperative pain
- Administration of prophylactic antibiotics and other drugs

Long-term postoperative care after anesthetic recovery and adequate physiologic stabilization requires at least once-daily monitoring until sutures are removed (usually at 10–14 days postoperatively) and any postoperative complications are resolved. Monitoring of vital signs, hydration, feed and fluid intake, feces and urine output, attitude and activity, surgical wound condition, body weight, and signs of postoperative pain and infection should be performed. Special diets, analgesics, bandaging, antibiotics, and other medications may be indicated. After suture removal, postoperative care required depends upon the species and surgical procedure. For example, chronic catheters or other partially exteriorized implants require ongoing monitoring and care. Monitoring of body weight should be scheduled throughout the postoperative period.

18:28 What level of perioperative care is appropriate for rodents when compared to nonrodents?

Opin. The general components of the perioperative care programs (see 18:27) are the same for rodents and nonrodents. Nevertheless, some elements of the programs differ. For rodent surgery, the surgical "team" may be reduced to one person, who serves as surgeon, anesthetist, surgical technician, and scrub nurse. When one person performs surgery on multiple animals at one sitting, as frequently occurs in rodent surgery, careful presurgical planning is required to assure availability of all supplies and equipment required to perform surgery and support necessary modifications in standard aseptic technique. The preoperative animal-health assessment should include vendor-supplied colony health testing (e.g., serology) and a visual examination rather than a physical examination of each animal when received.[7] Serologic testing of sentinel animals in the facility or other health surveillance results also may be useful.

Although sophisticated methods for intraoperative monitoring of anesthetized rodents are available when scientifically required, such methods might not be practical or possible in many research situations. Simply observing chest wall movement to determine respiratory rate and palpating the apical pulse through the chest wall may be sufficient to assess cardiovascular and respiratory stability.[21] Procedures used in larger animal species for intraoperative vital organ support such as intravenous fluid therapy can be difficult to use in rodents and may require other routes of administration of fluids or other strategies for supporting circulating blood volume.

The same intensity of monitoring and supportive care commonly provided for larger animals during recovery from anesthesia and acute postoperative care requires modification for rodents. In addition, performing surgery on multiple animals at one sitting, which frequently occurs, requires developing methods of supportive care for the anesthetic recovery of multiple animals simultaneously.

18:29 How much detail should be provided in the IACUC protocol relative to perioperative care?

Opin. In this author's opinion, the perioperative care detail that should be provided in the IACUC protocol includes the following:

- Names, qualifications, and responsibilities of participating personnel including the surgeon, the person who will be administering and monitoring anesthesia, and the person who will perform postoperative care
- Location where surgery will be done
- Duration of fasting, with rationale if greater than 24 hours
- Perioperative medications and antibiotics; dose, volume, route, and frequency of administration; duration of treatment
- Intraoperative monitoring and methods to be used to assess adequate anesthesia level
- Surgical anesthesia monitoring system if neuromuscular blocking agents are administered
- Aseptic techniques to be used
- Frequency of animal monitoring during anesthetic recovery and postoperative period
- Postoperative monitoring and care
- Indication of compliance with postoperative monitoring and care guidelines in the institution's IACUC policy manual
- Indication of expected health changes or possible postoperative complications and description of methods of monitoring and care
- Postoperative analgesics to be given; agent, dose, route, and frequency of administration; criteria for determining need for analgesics

18:30 Can expired pharmaceuticals be used for *survival surgery*?

Reg. APHIS/AC Policy 3[6] states that the use of expired medical materials such as pharmaceuticals "on regulated animals is not considered to be acceptable veterinary practice and is not consistent with adequate veterinary care." Policy 3 specifies that expired anesthetics, analgesics, and emergency drugs cannot be used on any regulated animals for major survival operative procedures. NIH/OLAW states that, "Euthanasia, anesthesia and analgesia agents should not be used beyond their expiration date, even if a procedure is terminal."[22]

Opin. In this author's opinion, expired drugs and fluids of any kind should not be used in animals undergoing survival surgery. The use of expired anesthetics, analgesics, euthanasia, and emergency drugs should not be allowed regardless of the procedure intended. Loss of pharmacological integrity may lead to insufficient pain relief or other unexpected anesthetic problems. It is fundamental to proper veterinary care that the user has confidence in the administered drugs; this is best assured by use of non-expired pharmaceuticals.

18:31 Is the use of expired pharmaceuticals acceptable for nonsurvival surgery?

Reg. APHIS/AC Policy 3 states that medical materials, other than anesthetics, analgesics, and emergency drugs, can be used beyond their 'used by' date if "their use does not adversely affect the animal's well-being or compromise the validity of the scientific study."[6] Facilities should have a written policy addressing the use of expired drugs in nonsurvival surgery or require that their intended use be described in the investigator's IACUC protocol submitted for approval.[6] NIH/OLAW states that, "Euthanasia, anesthesia and analgesia agents should not be used beyond their expiration date, even if a procedure is terminal."[22]

Opin. Without written documentation of safety and efficacy from the manufacturer and IACUC approval of an investigator's written request for their use in nonsurvival surgery, use of expired drugs and fluids should not be allowed for animals undergoing nonsurvival surgery.

18:32 Should the IACUC request information on the type of suture materials that will be used?

Opin. The PI should describe the type of suture material and method of wound closure on the protocol form to assure that they are appropriate for the species and the surgical incision. To facilitate wound healing, the type of suture material used to close the skin should be chosen to minimize tissue reaction and potential for wound infection. For example, to produce a lesion on healthy skin, an inoculum of 10^7 or more *Staphylococcus aureus* organisms is required, but in the presence of a silk suture, the required inoculum is reduced to less than 10^3 bacteria.[23] The capillary action of braided suture material and the inherent difficulty in keeping a wound clean can combine to increase the probability of infection at the surgical site.[23] Appropriate wound closure using methods that bury sutures should be considered when animals may bite or pick at exposed skin sutures.

18:33 Should the IACUC request information about when sutures (or clips) will be removed, if that is necessary?

Opin. The appropriate time for suture, skin staple, or clip removal might be provided to investigators in an institutional IACUC policy manual, which should contain guidelines on postoperative care. Rather than request that the PI restate when sutures or clips will be removed, a written indication on the protocol form that the PI has read and will comply with the postoperative care guidelines provides adequate assurance that timely suture removal will be done. If written IACUC guidelines are not available to the PI, the IACUC should request that the PI state on the protocol form when sutures or clips will be removed.

18:34 Should the IACUC request information on how rodents or other small animals will be kept warm during and after surgery?

Reg. The AWAR (§2.33,b,4–5) require appropriate care in accordance with current accepted veterinary medical and nursing practices. The *Guide* (pp. 26, 120) notes the importance of postoperative care, including thermoregulation.

Opin. Because of the large surface area-to-body mass ratio of rodents and other small animals, heat loss resulting in hypothermia is likely to occur without adequate thermal support during surgery and recovery from anesthesia. Maintenance of normal body temperature significantly reduces anesthesia-related cardiovascular and respiratory disturbances (*Guide* p. 119) and is often critical to the successful recovery of rodents from anesthesia. Methods to combat hypothermia can be suggested to investigators in an institutional IACUC policy manual, which should contain guidelines for rodent surgery and postoperative care. Perioperative observations and support are often considered to be important aspects of species-specific training, including that for rodents. If the PI is expected to follow institutional guidelines with respect to body temperature maintenance, the protocol form could have an item in which the PI can indicate his or her intent to comply with these guidelines. This endorsement provides adequate assurance that appropriate thermal support will be provided. If written IACUC guidelines describing methods to combat hypothermia are not available to the PI, the IACUC should request that the PI state, on the protocol form, how rodents or other small animals will be kept warm during and after surgery.

18:35 Should the IACUC request information about the frequency of postoperative observations?

Opin. The frequency of postoperative observation is determined by the nature of the surgical procedure and the stage of recovery.[7] Guidelines for minimal frequency of postoperative observation could be provided to investigators in an institutional IACUC policy manual for postoperative care. A written indication on the protocol form, by the investigator, that she or he has read and will comply with these guidelines provides adequate assurance that appropriate postoperative observation frequency will be carried out. However, if a surgical procedure necessitates more frequent observations than the guidelines recommend, the customized monitoring frequency and circumstances requiring monitoring should be described in the protocol. If there are no written IACUC guidelines available to the PI, the IACUC should request that the PI provide information about the frequency of postoperative observations on the protocol form.

18:36 When is an animal considered to be sufficiently recovered to return it to its home cage?

Opin. Many biomedical scientists believe that an animal should remain in a recovery area until it has recovered from anesthesia, physiological parameters are adequately stabilized, and it has regained normal ambulatory and protective behaviors before being moved to its home cage. The caveat to this is that application and practicality are at least species and size dependent. Physiological stability and normal ambulatory behavior might be very different for an animal that has undergone a thoracotomy versus one that has undergone arthroscopy. At some institutions, continuous observation by caregivers is required until the animal can be extubated, the core temperature approaches normothermia, and the patient remains in sternal recumbency.

Preemptive analgesia may cause a postoperative animal to appear lethargic. Unlike a tranquilized animal, however, one treated with narcotic or nonnarcotic analgesics will respond to non-pain stimulation (e.g., retching). The ability to stay sternal is associated with the ability to raise the head as needed and allows the animal to cope with postoperative vomition should it occur. The ability to raise the head is also an added safeguard when there is need to place water bowls that are shallow enough to prevent accidental immersion of the nostrils.

18:37 May animals that have undergone surgery be recovered in their home cage, or is there a need for a separate recovery area?

Reg. The *Guide* (p. 145) states that "a postoperative-recovery area should provide the physical environment to support the needs of the animal during the period of anesthetic and immediate postsurgical recovery and should be sited to allow adequate observation of the animal during this period." The species and type of surgical procedure will dictate the type of caging and support equipment required, which should be designed to support physiologic functions, such as thermoregulation and respiration.

Opin. During anesthetic recovery, an animal should be in an area that is warm, safe, quiet, comfortable, and appropriate for the needs of the individual animal and species. Dedicated recovery rooms are recommended for species such as rabbits, cats, dogs, and swine. A separate recovery area provides the necessary physical environment, ambient temperature control, and equipment for monitoring and supportive care to manage complications that may occur during recovery. Animals can be housed individually in appropriately sized cages designed to prevent injury to occupants. Individual housing also prevents potential trauma from cage mates. In addition, recovery areas can be located to facilitate the appropriate frequency of monitoring an animal by personnel.

Small ruminants that are too large to fit into recovery cages and large farm animals can recover in their own pen or stall if provisions are made for adequate observation, warmth, and animal safety. The added advantage of conspecific odors, sounds, and visual contact may enhance the animal's recovery. Foam- or air-filled padding placed on the floor in secure, warm areas are also useful for these species. The additional space afforded by a pen or stall also allows for assistance with changing position and moving larger species that may be unable to move themselves or may begin thrashing about during recovery. Regardless of the location, the criteria noted in 18:36 also apply: monitoring of the animal until sternally recumbent, a core body temperature approaching normothermia, and extubation of the animal.

Rodents often recover from anesthesia in the same laboratory where surgery is performed. Provisions to support body temperature are frequently necessary as thermal support is often critical to the successful recovery of rodents from anesthesia. Recovering rodents should be housed individually to prevent injury by cage mates. Although continual monitoring may not be possible, the animals should be observed frequently until recovered from anesthesia and adequately stabilized before being returned to their home cages.

18:38 Investigators keep relative to surgery and the perioperative period?

Reg. The AWAR (§2.33,b,5; §2.35,f) indicate that adequate pre- and post-procedural care is expected and that all records and reports must be maintained for at least three years past the end of the activity; hence, it might be inferred that medical and surgical records are of import. APHIS/AC Policy 3[6] requires appropriate post-operative record keeping in accordance with accepted veterinary procedures for covered species. The *Guide* (p. 115) states that appropriate medical records documenting animal well-being and tracking animal care and use.

Opin. The specific information requested in medical records is detailed in 26:10 and 27:21. The *Institutional Animal Care and Use Committee Guidebook*[16] suggests that as best practice "an intra-operative anesthetic monitoring record should be kept and included with the surgeon's report as part of the animal's records. This record should be available to the personnel providing postoperative care. Postoperative records at a minimum should reflect that the animal was observed until it was extubated and had recovered the ability to stand. These should be supplemented by records evaluating the animal's recovery, administration of analgesics and antibiotics, basic vital signs, monitoring for infection, wound care, and other medical observations."

Preoperative records should include a health profile and medical and vaccination history (when applicable) provided by the vendor and subsequently supplemented by the clinical care staff during quarantine or with annual updates, as appropriate. Examples of additional information include vaccinations for dogs or cats, quarterly or semiannual blood work, and tuberculosis testing and serological profiles in nonhuman primates. A physical exam to assess an animal's health status should be performed. A preanesthetic evaluation should include records of body weight, body temperature, heart rate, and respiratory rate to establish baseline data, recognizing that these values might be influenced by preoperative or preanesthetic medications. The need for diagnostic radiographic tests and laboratory evaluation of blood, urine, and feces will depend on the animal species, research protocol, and health history of the animal or colony. Preoperative administration of medications or antibiotics should be documented.

An anesthetic record should document anesthetic administration, monitoring of anesthetic depth (including reflexes), and frequent assessment of the physiologic status of an animal including body temperature and cardiovascular and respiratory function.[24] The extent of record keeping and monitoring sophistication depends upon the potential complications associated with the anesthetic regimen, the surgical procedure, and the animal species. Minimal documentation should include vital signs, core body temperature, heart rate, and respiratory rate, recorded at an average interval of every 15 minutes, and the anesthetic administered, dose, route, and time of administration.

The postoperative period can be divided into three overlapping phases: recovery from anesthesia, acute postoperative care, and long-term postoperative care.[7]

The extent of record keeping and intensity of monitoring during the postoperative period depend upon the species and the surgical procedure. Generally, the most intensive and frequent monitoring is required during recovery from anesthesia, since it is then that the animal is most vulnerable to the combined adverse effects of anesthesia and hypothermia. Frequent assessment of thermoregulation and cardiovascular and respiratory function should continue during anesthesia recovery and acute postoperative care until the animal is adequately stabilized. Vital signs, including body temperature, pulse rate and rhythm, respiratory rate and rhythm, and oxygen saturation, should be recorded approximately every 15 minutes during anesthesia recovery with less frequent monitoring possible as an animal stabilizes. Documentation of clinical observations including mucous membrane color, capillary refill time, pain assessment, and surgical wound condition is recommended. Administration of analgesics, antibiotics, parenteral fluids, and other medications should be recorded and should include the drug name, dosage, route, and time given. A brief description of the surgery should be recorded with the postoperative care records or as a separate surgeon's report, with a notation about complications and/or unexpected need to deviate from the approved surgical plan.

During long-term postoperative care after anesthetic recovery and adequate physiologic stabilization, monitoring should continue at least once a day until suture removal (usually at 10–14 days postoperatively) and any postoperative complications are resolved. Vital signs, hydration, feed and fluid intake, feces and urine output, attitude and activity assessment, pain assessment, surgical wound condition and care, and monitoring for postoperative infection should be recorded. All postoperative care and medications administered as described with anesthesia recovery should be documented. After suture removal, frequency and parameters monitored depend on the species and surgical procedure. For example, chronic catheters or other partially exteriorized implants require documentation of ongoing monitoring and care. Monitoring of body weight should be scheduled throughout the postoperative period. (See 26:10–26:15.)

18:39 What is considered health record maintenance "consistent with current veterinary practice" and how are these records used?

Reg. The AWAR (§2.33,b,5; §2.35,f) indicate that adequate pre- and post-procedural care is expected and that all records and reports must be maintained for at least three years. APHIS/AC Policy 3[6] requires appropriate postoperative record keeping in accordance with accepted veterinary procedures for covered species. (See 18:38.) The *Guide* (p. 115) states that, "Medical records are a key element of the veterinary care program and are considered critical for documenting animal well-being as well as tracking animal care and use at a facility."

Opin. The animal health record describes a problem and observations, action plan, and resolution or lack thereof, whether alone or superimposed on other problems. It is a communication tool that allows for a variety of trained persons to interact with the patient in a meaningful way and have others on the care team recognize exactly what was done, why it was done, and why an anticipated outcome may or may not have been realized. It is a solid mechanism for communication between the clinical care staff and similarly trained physician/surgeons. In the research environment, especially with novel animal models, the reason for unexpected deaths can frequently be resolved by consulting a well-documented animal health record. The well-organized and documented animal health record serves to put everyone on the same page and is critical for good research. (See 26:10–26:15; 27:21–27:25.)

References

1. U.S. Department of Agriculture, Animal and Plant Health Inspection Service. 2011. Policy 14, Major Survival Surgery Dealers Selling Surgically-Altered Animals to Research. http://www.aphis.usda.gov/animal_welfare/policy.php?policy=14.
2. Silverman, J. 1994. Protocol review: Whose responsibility is it? *Lab Anim. (NY)* 23(2):22.
3. Rolf, L. A. 2007. Surgery. In *The IACUC Handbook,* 2nd edition, eds J. Silverman J., M.A. Suckow, S. Murthy, 321–343. Boca Raton: CRC Press.
4. National Institutes of Health. Office of Laboratory Animal Welfare. Sample animal study proposal. http://grants.nih.gov/grants/olaw/sampledoc/oacu3040-2.htm#surg. Accessed March 28, 2013.
5. American Veterinary Medical Association. 2013. *AVMA guidelines for the euthanasia of animals: 2013 edition.* https://www.avma.org/KB/Policies/Documents/euthanasia.pdf.
6. U.S. Department of Agriculture, Animal and Plant Health Inspection Service. 2011. Policy 3, Veterinary Care. http://www.aphis.usda.gov/animal_welfare/policy.php?policy=3.
7. Brown, M.J., P. T. Pearson, and F. N. Tomson. 1993. Guidelines for animal surgery in research and teaching. *Am. J. Vet. Res.* 54:1544–1549.
8. Potkay, S., N.L. Garnett, J.G. Miller et al. 1997. Frequently asked questions about the Public Health Service Policy on Humane Care and Use of Laboratory Animals. *Contemp. Top. Lab. Anim. Sci.* 36(2): 47–50.
9. National Institutes of Health, Office of Laboratory Animal Welfare. 2013. Frequently Asked Questions, A.2, Does the PHS Policy apply to the production of custom antibodies or to the purchase of surgically modified animals? http://grants.nih.gov/grants/olaw/faqs.htm#App_2.
10. Committees to revise the Guide for the Care and Use of Agricultural Animals in Research and Teaching. 2010. *Guide for the Care and Use of Agricultural Animals in Research and Teaching,* 3rd Edition. Champaign, IL: Federation of Animal Science Societies.
11. Committee on Educational Programs in Laboratory Animal Science, National Research Council. 1991. Survival surgery and postsurgical care. In *Education and Training in the Care and Use of Laboratory Animals: A Guide for Developing Institutional Programs,* 61. Washington, D.C.: National Academy of Science Press.
12. Bradfield, J.F., T.R. Schachtman, R.M. McLaughlin et al. 1992. Behavioral and physiologic effects of inapparent wound infection in rats. *Lab. Anim. Sci.,* 42:572–578.
13. Hoogstraten-Miller, S.L., and Brown, P.A. 2008. Techniques in aseptic rodent surgery. *Curr. Protoc, Immunol.* Chapter 1:Unit 1.12.1-1.12-14.
14. Sharp, P., and Villano, J. 2012. *The Laboratory* Rat, 2nd edition. Boca Raton: CRC Press.
15. Slattum, M.M., L. Maggio-Price, R.F. DiGiacomo et al. 1991. Infusion-related sepsis in dogs undergoing acute cardiopulmonary surgery. *Lab. Anim. Sci.* 41:146–150.
16. Applied Research Ethics National Association/Office of Laboratory Animal Welfare. 2002. *Institutional Animal Care and Use Committee Guidebook,* 2nd edition, chapter C.3.g. Bethesda: National Institutes of Health.
17. National Institutes of Health, Office of Animal Care and Use. 2010. Guidelines for egg and oocyte harvesting in *Xenopus laevis.* http://oacu.od.nih.gov/ARAC/documents/Oocyte_Harvest.pdf.
18. National Institutes of Health, Office of Laboratory Animal Welfare. 2013. Frequently Asked Questions, G.1, What kind of training is necessary to comply with PHS Policy, and how frequently should it be provided? http://grants.nih.gov/grants/olaw/faqs.htm#instresp_1.

19. Academy of Surgical Research. 2009. Guidelines for training in surgical research with animals. *J. Invest. Surg.* 22:218–225.
20. Brown, M.J., and J.C. Schofield. 1994. Perioperative care. In *Essentials for Animal Research: A Primer for Research Personnel*, eds. B.T. Bennett, M.J. Brown and J. C. Schofield, 79. Washington, D.C.: National Agricultural Library.
21. Gaertner, D. J., Hallman, T. M., Hankenson, F. C. et al. 2008. Anesthesia and analgesia for laboratory rodents. In *Anesthesia and Analgesia in Laboratory Animals*, eds. R. E. Fish, M. J. Brown, P. J. Danneman, and A. Z. Karas, 239. New York: Academic Press.
22. National Institutes of Health, Office of Laboratory Animal Welfare. 2012. Frequently Asked Questions, F.5, May investigators use expired pharmaceuticals, biologics, and supplies in animals? http://grants.nih.gov/grants/olaw/faqs.htm#useandmgmt_5.
23. McCurnin, D.M. and R.L. Jones. 1993. Principles of surgical asepsis. In *Textbook of Small Animal Surgery*, 2nd edition, ed. D.H. Slatter, 114. Philadelphia: W.B. Saunders.
24. Brown, M.J. 1994. Principles of anesthesia and analgesia. In *Essentials for Animal Research: A Primer for Research Personnel*, eds. B.T. Bennett, M.J. Brown and J.C. Schofield, 39. Washington, D.C.: National Agricultural Library.

19

Antigens, Antibodies, and Blood Collection

Harold F. Stills Jr.

Introduction

Modern biologic research techniques often require the production of specific polyclonal and monoclonal antibodies as an essential component of the research protocol. Production of both polyclonal and monoclonal antibodies may require the immunization of animals, therein presenting the IACUC with the dilemma and duty of evaluating the immunization procedures and schedules with respect to potential animal pain and distress. Published guidelines from a number of sources are available to assist the IACUC in developing policies and procedures for animal use in antibody production.[1–6] Although in vitro alternatives to the use of live animals in the production of antibodies are often appropriate (see 19:7), this chapter primarily focuses on those circumstances in which the IACUC approves the use of live animals for antibody production.

Production of a high-quality antibody requires an appropriate stimulation of the immune system. The initial immune activating event involves processing of the antigen by antigen-presenting cells (APCs), which are primarily the macrophages and the dendritic cells, such as the Langerhans cells in the epidermis of the skin. Activation of these cells, along with the recruitment of additional cells for antibody production, is often enhanced by the addition of various biologically active compounds (primarily targeting toll-like receptors) in an adjuvant. Protecting the antigen from rapid degradation in the body and permitting the slow release of the antigen to the APCs are other functions often performed by an adjuvant. The end results of the use of an adjuvant are both an increase in the desirable antibody response and the production of a substantial inflammatory response with undesirable tissue destruction and potential pain and distress.

Of the available adjuvants in use today, none has proved more effective or controversial than Freund's complete adjuvant (FCA).[8–15] First described in 1942,[7,8] it was originally composed of paraffin oil, mannide monooleate, and killed mycobacteria. The presently available formulations of FCA, however, differ dramatically from the historic formulation in using a much purer and less toxic mineral oil and producing significantly fewer severe inflammatory lesions.[9,10] Injected into animals, FCA produces a chronic granulomatous inflammatory reaction at the injection site. Granulomas often may be detected in draining lymph nodes, spleen, kidney, and other organs where microdroplets of the emulsion have been distributed by the vascular or lymphatic systems. Intradermal injections routinely produce ulcerations as early as 12 to 14 days post injection, which persist for 8 weeks or longer.[11] The microscopic lesions produced by injection of FCA are primarily those of granulomatous inflammation and focal necrosis.[11–17]

19:1 What type of justification should be required for the use of adjuvants in antibody production?

Reg. (See 19:2.)

Opin. The dramatic inflammatory reaction historically generated by the injection of the original formulation of Freund's complete adjuvant (FCA) has restricted its use to research antibody production. This has resulted in refinements in the original FCA composition designed to reduce the resulting inflammation, as well as the development of and efforts to encourage the use of other adjuvants that purport to produce similar antibody responses without the granulomatous lesions associated with FCA. Unfortunately, many of the alternative adjuvants selected to replace FCA produce severe inflammatory reactions.[12,14,18–20] All adjuvants, by nature of their mode of action, induce an inflammatory reaction. Since many of the compounds for which antibodies are required are poorly immunogenic, adjuvants are essential. The choice of any adjuvant should be based upon a sound scientific rationale that considers the specific immunogen, the species being immunized, the route of administration, the impact on the animal's well-being, and the desired antibody.[4,21]

Surv. Does your IACUC require justification for the use of Freund's Complete Adjuvant in mice? (Check as many boxes as applicable)

• Not applicable—we do not use Freund's Complete Adjuvant in mice	113/294 (38.4%)
• Yes, when used by any route	157/294 (53.4%)
• Yes, if we were to approve subcutaneous use	11/294 (3.7%)
• Yes, if we were to approve intradermal use	9/294 (3.1%)
• Yes, if we were to approve intramuscular use	7/294 (2.4%)
• Yes, if we were to approve intraperitoneal use	9/294 (3.1%)
• Yes, if we were to approve intradermal use in the foot pad	8/294 (2.7%)
• I don't know	14/294 (4.8%)

19:2 Is the use of FCA a painful or distressful procedure?

Reg. APHIS/AC Policy 11 lists the use of FCA under "procedures that may cause more than momentary or slight pain."[22] That same policy goes on to state that FCA used for antibody production "may cause a severe inflammatory reaction depending on the species and route of administration." PHS Policy (U.S. Government Principle IV) and the *Guide* (p. 120) require the minimization of discomfort, distress, and pain in concert with good science. Additionally, the *Guide* (p. 26) requires that the investigator and the IACUC evaluate the "impact of the proposed procedures on the animal's well-being" in the preparation and approval of the animal use protocol.

Opin. There is much disagreement in the literature regarding the pain and distress elicited by the use of FCA. Numerous reports have failed to document any pain or distress in animals injected with FCA[14,16,17,23] or in tuberculin-negative humans.[24,25] In mice and other rodents, however, the intraperitoneal injection of FCA and other adjuvants has been associated with the formation of granulomas, fibrous adhesions, abdominal fluid distension and clinical signs indicative of pain and distress.[1,13,14,26–29] The use of intraperitoneal FCA in rats as a model of acute inflammation[29–31] is an additional indication of the potential for pain and distress. The use of analgesics with FCA immunization in mice has been recommended based upon behavioral evaluations and a lack of detected impact on immunization efficacy.[32] In the author's experience, uncomplicated FCA–antigen emulsion intradermal and subcutaneous injections are not characterized by tenderness, irritation, or any other indication of greater than slight or momentary pain or distress in the animal. Feed consumption and body weights are unaffected by uncomplicated FCA–antigen injections.

The choice of immunogen, regardless of the choice of adjuvant, may induce an autoimmune condition that is associated with pain and distress. The adjuvant-induced model of arthritis is a prime example and has even been recommended as a model of chronic clinical pain.[33]

Surv. Which USDA/AC pain classification does your IACUC usually assign to projects using Freund's complete adjuvant for antibody production? Check as many boxes as appropriate.

- Not applicable 121/295 (41.0%)
- Category C (no pain/distress) 17/295 (5.8%)
- Category D (pain/distress that will be alleviated with drugs) 60/295 (20.3%)
- Category E (no relief from pain/distress due to experimental needs) 67/295 (22.7%)
- It varies with the species 15/295 (5.1%)
- It varies with the site of inoculation 32/295 (10.9%)
- I don't know 26/295 (8.8%)
- Other 7/295 (2.4%)

19:3 Should the amount of killed mycobacteria in the Freund's Complete Adjuvant (FCA) be limited?

Opin. Formulations of FCA containing no more than 0.1 mg/ml of dry mycobacterial cell mass have been recommended as a method to reduce the associated inflammatory response.[11] Nevertheless, commercially available formulations of FCA generally range from 0.4 to 1.0 mg/ml of dry mycobacterial cell mass.

Surv. Does your IACUC limit the amount of mycobacteria in Freund's Complete Adjuvant?

- Not applicable 121/294 (41.2%)
- There is no limit specified 77/294 (26.2%)
- Our limit is 0.5 mg/ml 8/294 (2.7%)
- Our limit is up to 1.0 mg/ml 4/294 (1.4%)
- I don't know 78/294 (26.5%)
- Other 6/294 (2.0%)

19:4 What procedures should be required by the IACUC to minimize the potential for contamination of injection sites during the injection of adjuvant–antigens?

Opin. Using aseptic procedures for antigen–adjuvant injections, including ensuring sterile preparation of the antigen–adjuvant mixture, clipping the injection site, and sterile scrubbing of the injection site with an antiseptic, has been recommended.[1,3,4,16,17] The injection of viable contaminating bacteria with an adjuvant–antigen mixture often results in the formation of an abscess with the influx of neutrophils, exudation, hyperemia, and pain. Additionally, the clipping of the animal's fur permits a more thorough examination of the injection sites post-immunization.

Surv. What procedures does your IACUC require for minimizing contamination of injection sites? More than one response per institution is possible.

- Not applicable 100/293 (34.1%)
- We have no policy 54/293 (18.4%)
- We require a sterile preparation of antigen–adjuvant mixture 77/293 (26.4%)
- We require clipping of the animal's fur prior to injection 61/293 (20.8%)
- We require swabbing the area with a disinfectant prior to injection 80/293 (27.3%)
- We require a surgical (or near surgical) prep of the area prior to injection 16/293 (5.6%)
- I don't know 40/293 (13.7%)
- Other 4/293 (1.5%)

19:5 What type of justification or information should be required by the IACUC for that committee to approve a schedule of antigen–adjuvant injections?

Opin. The administration of booster injections prior to the optimal time not only increases the potential for pain and distress but also may be detrimental to the ultimate antibody level and affinity.[1,16,34–37] Animals mounting a robust immune response may develop an arthus reaction (an acute inflammatory and painful reaction) at the booster injection site. The antigen, quantity of antigen, adjuvant used, and injection route are all important factors in determining the optimal schedule for immunization and, optimally, the individual animal's antibody titer should be followed to determine when the antibody level has plateaued and booster injections are indicated.[38]

Surv. Does your IACUC require investigators to monitor serum titers before booster injections are given? (Check as many boxes as appropriate.)

- Not applicable 112/292 (38.4%)
- Yes, we require monitoring 24/292 (8.2%)
- No, we do not require monitoring 90/292 (30.8%)
- It depends on the antigen used 30/292 (10.3%)
- I don't know 40/292 (13.7%)
- Other 5/292 (1.7%)

19:6 What oversight actions should an IACUC consider if an investigator has contracted with a commercial laboratory for the production of custom antibodies in animals?

Reg. The AWAR (§2.31,d,1,ii; §2.31,d,1,iii) and the PHS Policy (IV,C,1,a) require that PIs provide assurances that, in the case of painful animal procedures, alternatives have been searched for and found unacceptable and that the proposed activities are not duplicative. PHS Policy requires that any institution using animals with PHS funds, even if the funds are awarded by subcontracting or subgranting, must either have an approved Animal Welfare Assurance or be included as a component under the primary institution's Assurance. If both institutions have approved Assurances, it is only necessary that one of the institutions' IACUCs review and approve the specific activity. IACUCs may choose which IACUC will review the protocol for the animal activities being conducted. Verification of IACUC-approval for the production of antibodies must be project-specific. It is recommended that if an IACUC defers protocol review to another IACUC, documentation of the review should be maintained by both committees. The subcontracting or subgranting institution, however, retains accountability for "providing effective oversight mechanisms to ensure compliance with the PHS Policy."[39,40]

For PHS funded projects involving the production of custom antibodies, the NIH Grants Policy applies (NIH GPS Part II, 4.1.1.3). Under consortium (subaward) agreements in which the grantee collaborates with another organization, the grantee, as the primary recipient of NIH grant funds, is accountable for the performance of the project. Animal welfare requirements that apply to grantees also apply to subcontracts and subprojects. Each institution in the consortium must have an Assurance (Domestic, Interinstitutional or Foreign)..

Opin. The contracting of antibody production is a difficult problem for many institutions. The availability of numerous commercial contracting companies along with the many alliances between peptide-generating and antibody-producing companies makes institutional oversight a nightmare. IACUCs must be diligent in their attempts to identify potential situations and be proactive in the education of PIs as to the necessity of notifying and working with the IACUC when contracting antibody production from a commercial firm. (See 8:10; 16:22; 18:13.)

The IACUC's duty in reviewing the contracted facilities protocols and procedures for antibody production protocols is less well defined. Some IACUCs require a complete review of contracted protocols in addition to the review by the IACUC of the contracted facility. Others limit their requirements to the documentation of IACUC approval from the contracted PHS Assured facility, while others do no review or have no requirements. The availability of a commercially available and acceptable antibody is a clear and definite example of an alternative to additional animal use and duplicative procedures that is typically not evaluated by commercial production facilities. Since the contracting institution retains partial accountability, it would seem prudent that IACUCs should review contracted antibody production protocols to assure that the procedures and policies involved adhere to those of the contracting institution and that the acceptability of available commercial antibodies has been evaluated.

Surv. If an investigator uses a commercial laboratory to produce antibodies using Public Health Service funds, what information is required by your IACUC? (Check as many boxes as appropriate.)

- Not applicable 127/295 (43.1%)
- We ask for a copy of the laboratory's PHS Assurance approval and its USDA registration (for USDA covered species) 71/295 (24.1%)
- We ask to see that laboratory's IACUC approval letter for raising antibodies 45/295 (15.3%)

• We review the laboratory's IACUC-approved protocol	24/295 (8.1%)
• We have no policy	67/295 (22.7%)
• I don't know	18/295 (6.1%)
• Other	8/295 (2.7%)

19:7 Should additional justifications be requested by the IACUC for the production of monoclonal antibodies using the mouse ascites method?

Reg. NIH/OLAW issued a Frequently Asked Question directing that IACUCs "critically evaluate the proposed use of the mouse ascites method… IACUCs must determine that (i) the proposed use is scientifically justified; (ii) methods that avoid or minimize discomfort, distress, and pain (including in vitro methods) have been considered; and (iii) the latter have been found unsuitable."[41]

Opin. The development of practical and reasonably priced in vitro methods for hybridoma growth, coupled with evidence that the mouse ascites method causes discomfort, has led many IACUCs to require that in vitro methods be utilized. Published comparisons of the in vitro methods and the ascites method have shown the results to be highly dependent upon the in vitro system used and the specific hybridoma.[42–48] In the author's opinion, there are few justifications for the use of the ascites method for monoclonal antibody production. Nearly all hybridomas can be propagated in vitro and the levels of monoclonal antibody production can be increased by the use of specifically designed growth chambers, hollow-fiber bioreactor systems, and other means. In all instances, the burden of proof should be upon the PI to show that multiple in vitro methods have been attempted and were unsuccessful prior to the IACUC's approving the ascites method of monoclonal antibody production.

A report [49] commissioned by the NIH summarizes the findings of the National Academy of Sciences as to the scientific necessity for producing monoclonal antibodies by the mouse method, including ways to minimize any pain or distress that might be associated with that method. The report also discusses regulatory considerations for the mouse method and summarizes the current stage of development of tissue culture methods.

Surv. What justifications are typically acceptable to your IACUC for permitting the in vivo production of monoclonal antibodies by the mouse ascites method? (Check as many boxes as appropriate.)

• Not applicable	155/294 (52.7%)
• We do not request any special justification	13/294 (4.4%)
• In vitro production is too expensive	5/294 (1.7%)
• The investigator tried in vitro methods and was not successful	75/294 (25.5%)
• The literature indicates in vitro methods unsuccessful and in vivois required	70/294 (23.8%)
• The structure or molecular weight of the protein suggests in vitro production will fail	31/294 (10.5%)
• I don't know	35/294 (11.9%)
• Other	16/294 (5.4%)

19:8 Should the IACUC limit the number of allowable peritoneal taps for the collection of ascitic fluid?

Opin. The process of removing the excessive peritoneal fluid from mice used in monoclonal antibody production requires the use of a large-gauge needle and, often, some anesthesia. Many institutions limit the total number of peritoneal taps (with the last tap performed post mortem) while other institutions place limits on the animal's physical condition.[3] All hybridoma lines, based to a large degree upon the plasmacytoma fusion partner, behave somewhat differently when growing in the animal's peritoneal cavity. Certain hybridomas are extremely invasive, resulting in bloody peritoneal fluid on the first tap. The tumor masses that hybridomas form in the peritoneal cavity may be either disseminated or solid and singular.[50] Likewise, the dynamics of monoclonal antibody production from specific hybridomas also varies.[51] The practice of placing absolute limitations on the number of taps is inappropriate as it fails to recognize the differences between hybridoma lines; however, there is little reason for the maximal number of taps to exceed three.

Surv. What limit does your IACUC place on the number of peritoneal taps when using the mouse ascites method for monoclonal antibody production?

- Not applicable 159/295 (53.9%)
- We have no limits 10/295 (3.4%)
- One tap only 21/295 (7.1%)
- Two to three taps 45/295 (15.3%)
- More than three taps 2/295 (0.7%)
- I don't know 58/295 (19.7%)

19:9 What is the limit for an acceptable volume of blood withdrawal from a single collection?

Opin. Numerous studies have associated severe hemodynamic changes and hemorrhagic shock with blood losses greater than 30% of the total blood volume.[52,53] Smaller volume losses in rats (15%–20% of the blood volume) reduce cardiac output by nearly 50%.[54–56] Some authors have recommended a maximal withdrawal of 15% of blood volume,[57,58] while others have reported acceptable results with blood collections of up to 25% of the total blood volume in mice.[59] In the author's opinion a single blood collection of up to 20% of the total blood volume is acceptable.

Surv. What is the maximum blood volume withdrawal that your IACUC accepts as reasonable for a single withdrawal from a mouse? (Check as many boxes as applicable.)

- Not applicable 42/294 (14.3%)
- We have no policy 36/294 (12.2%)
- No more than 7.5% of the animal's total blood volume 50/294 (17.0%)
- No more than 10% of the animal's total blood volume 111/294 (37.8%)
- No more than 15% of the animal's total blood volume 25/294 (8.5%)
- More than 15% of the animal's total blood volume 6/294 (2.0%)
- I don't know 23/294 (7.8%)
- Other 15/294 (5.1%)

19:10 What limits should be set if repeated blood sampling is a component of the protocol?

Reg. The AWAR (§2.31,d,1,i), the PHS Policy (IV,C,1,a), and the *Guide* (p. 12) all require that procedures used on animals be chosen to avoid or minimize discomfort, distress, and pain to the animals consistent with sound research design.

Opin. While standard hematology tests (hematocrit and hemoglobin concentration) may be used in larger animals to monitor the effects of multiple blood withdrawals upon an individual animal's well-being, they are rarely used in mice and rats. IACUCs are therefore left to prospectively evaluate the impact of proposed multiple blood collections upon animal well-being and to set limitations. Several studies with repeated blood collections have reported that collection of 250–500 µl daily from adult mice (approximately 10%–20% of total blood volume) results in a significant drop in the hematocrit and a profound reticulocytosis.[60,61] The findings of several authors[57,62] suggest that weekly blood collection volumes exceeding 7.5% of the estimated blood volume results in significant hematologic changes supports a recommendation to limit weekly collections to under 10% of the total blood volume whenever maintaining hematologic parameters is necessary. If scientifically necessary, weekly blood collections of up to 15% of total blood volume in males and up to 25% in females has been reported to be sustainable in mice.[59] (See 16:45.)

Surv. 1 What is the maximum amount of blood your IACUC typically permits for once weekly blood collection in adult mice? (Check more than one box if appropriate.)

• Not applicable	47/293 (16.0%)
• We have no policy	47/293 (16.0%)
• 10% of the estimated blood volume	112/293 (38.2%)
• 15% of the estimated blood volume	13/293 (4.4%)
• 20% of the estimated blood volume	7/293 (2.4%)
• 25% of the estimated blood volume	0/293 (0%)
• I don't know	34/293 (11.6%)
• Other	39/293 (13.3%)

Surv. 2 What is the minimum length of time until the next blood draw that is typically accepted by your IACUC when the maximum allowable amount of blood is withdrawn from an adult mouse?

• Not applicable	55/294 (18.7%)
• We have no policy	45/294 (15.3%)
• One week	73/294 (24.8%)
• Two weeks	51/294 (17.4%)
• Three weeks	14/294 (4.8%)
• More than three weeks	14/294 (4.8%)
• I don't know	34/294 (11.6%)
• Other	8/294 (2.7%)

19:11 What should the IACUC require regarding orbital sinus (retro-orbital, postorbital) and other blood collection procedures in rodents?

Reg. The AWAR (§2.31,d,1,i), the PHS Policy (IV,C,1,a), and the *Guide* (p. 12) all require that procedures used on animals be chosen to avoid or minimize discomfort, distress, and pain to the animals consistent with sound research design. The AWAR (§2.31,d,1,iv,A) and the PHS Policy (IV,C,1,b) also require that procedures that may cause more than momentary or slight pain or distress to the animals will be performed with appropriate sedation, analgesia, or anesthesia unless scientific justification is provided by the investigator.

Opin. Orbital sinus or retro-orbital (postorbital) blood collection has been a standard technique in mice and rats that is not without potential problems. Several studies have described the tissue trauma resulting from orbital sinus bleeding[63–65] and other studies have described behavioral changes resulting from orbital sinus bleeding.[66,67] It is clear from the literature[68,69] and personal experience that the level of trauma associated with the orbital sinus bleeding technique is highly correlated with the skill of the technician performing the procedure. The variability in results and the potential for serious tissue trauma with resulting pain and distress has led many IACUCs[70] and professional organizations[62] to require the use of anesthesia or sedation for orbital sinus blood collections. Unless the technician performing the procedure has demonstrated the requisite skill and experience to repeatedly perform orbital sinus blood collections without trauma to the animal, it is the author's opinion that anesthesia should be required.

Surv. 1 Does your IACUC usually require anesthesia or sedation for postorbital blood collection in mice?

• Not applicable	90/293 (30.7%)
• No anesthesia or sedation is required	27/293 (9.2%)
• Sedation alone is sufficient	22/293 (7.5%)
• We require general anesthesia	106/293 (36.2%)
• Topical ocular anesthesia only is sufficient	19/293 (6.5%)
• I don't know	20/293 (6.8%)
• Other	9/293 (3.1%)

Opin. Many IACUCs recommend alternative sites to the orbital sinus blood collection procedure. These alternative sites may include the facial vein (submandibular vein, superficial temporal vein, linguofacial vein), the sublingual vein, the femoral vein, the saphenous vein, the tail vein, and others. Each of these sites has advantages and disadvantages with the site chosen often determined by the familiarity of the staff with the procedure. The facial vein has become a popular alternative to orbital sinus blood collection permitting rapid blood collection without anesthesia in mice. This procedure, however, is not without potential complications with poor blood sample quality, tissue trauma, and reduced weight gains in comparison to the sublingual collection method.[71,72]

Surv. 2 Does your IACUC allow blood collection from the facial vein (submandibular vein, superficial temporal vein, linguofacial vein)?

• Not applicable	106/292 (36.3%)
• Yes, mice only	61/292 (20.9%)
• Yes, rats only	9/292 (3.1%)
• Yes, rats and mice	72/292 (24.7%)
• No, we do not allow that route to be used for blood collection	31/292 (10.6%)
• Other	13/292 (4.5%)

References

1. Canadian Council on Animal Care. 2002. Guidelines on: Antibody Production. http://www.ccac.ca/Documents/Standards/Guidelines/Antibody_production.pdf.
2. Grumstrup-Scott, J. and D. Greenhouse. 1988. NIH intramural recommendation for the research use of complete Freund's adjuvant. *ILAR News* 20:9.
3. Jackson, L.R. and J.G. Fox, J.G. 1995. Institutional policies and guidelines on adjuvants and antibody production. *ILAR J.* 37:141–152.
4. Leenaars, M. and C.F.M. Hendriksen. 2005. Critical steps in the production of polyclonal and monoclonal antibodies: Evaluation and recommendations. *ILAR J.* 46:269–279.
5. Leenaars, P.P., C.F.M. Hendriksen, W.A. de Leeuw et al. 1999. The production of polyclonal antibodies in laboratory animals. *ATLA-Altern. Lab. Anim.* 27:79–102.
6. Workman, P., P. Twentyman, F. Balkwill et al. 1998. United Kingdom Co-Ordinating Committee on Cancer Research (UKCCCR) Guidelines for the Welfare of Animals in Experimental Neoplasia (Second Edition). *Br. J. Cancer* 77:1–10.
7. Freund, J. 1956. The mode of action of immunologic adjuvants. *Bibl. Tuberc.* 10:130–148.
8. Freund, J. and D. McDermott, 1942. Sensitization to horse serum by means of adjuvants. *P. Soc. Exp. Biol. Med.* 49:548–553.
9. Stewart-Tull, D. 1998. The future potential for the use of adjuvants in human vaccines. In *biomedical science and technology: Recent developments in the pharmaceutical and medical sciences*, ed. A.A. Hincal and H.S. Kas, 129–136. New York: Plenum Press.
10. Stewart-Tull, D.E.S. 1995. Freund-type mineral oil adjuvant emulsions. In *The theory and practical application of adjuvants*, ed. D.E.S. Stewart-Tull, 1–19. New York: John Wiley and Sons.
11. Broderson, J.R. 1989. A retrospective review of lesions associated with the use of Freund's adjuvant. *Lab. Anim. Sci.* 39:400–405.
12. Leenaars, P.P., C.F. Hendriksen, A.F. Angulo et al. 1994. Evaluation of several adjuvants as alternatives to the use of Freund's adjuvant in rabbits. *Vet. Immunol. Immunopathol.* 40:225–241.
13. Leenaars, P. P., C.F. Hendriksen, M.A. Koedam et al. 1995. Comparison of adjuvants for immune potentiating properties and side effects in mice. *Vet. Immunol. Immunopathol.* 48:123–138.
14. Leenaars, P.P., M.A. Koedam, P.W. Wester et al. 1998. Assessment of side effects induced by injection of different adjuvant/antigen combinations in rabbits and mice. *Lab. Anim.* 32:387–406.
15. Stills, H.F., Jr. 1994. Polyclonal antibody production. In *The Biology of the Laboratory Rabbit*, 2nd edition, ed., P.J. Manning, D.H. Ringler and C.E. Newcomer, 435–448. San Diego: Academic Press.
16. Stills, H.F., Jr., Polyclonal Antibody Production. 2012. In *The Laboratory Rabbit, Guinea Pig, Hamster, and Other Rodents*, ed. M.A. Suckow, K.A. Stevens and R.P. Wilson, 259–274. Amsterdam: Elsevier.

17. Stills, H.F., Jr. and M.Q. Bailey. 1991. The use of Freund's complete adjuvant. *Lab Anim. (N.Y.)* 20(4):25–30.
18. Hendriksen, C.F.M. and J. Hau, J. 2003. Production of polyclonal and monoclonal antibodies. In *Handbook of Laboratory Animal Science,* 2nd edition, ed. J. Hau and J. Van Hoosier, 391. Boca Raton: CRC Press.
19. Johnson, D.K. 1994. Adjuvant comparison in rabbits. *Sci. Anim. Care* 5(2):2–4.
20. Johnson, D.K. 1994. Adjuvant comparison in rabbits. In *Rodents and Rabbits: Current Research Issues*, eds. S.M. Niemi, J.S. Venable and H.N. Guttman, 77–80. Washington, D.C.: Scientists Center for Animal Welfare.
21. Stills, H.F., Jr. 2005. Adjuvants and antibody production: Dispelling the myths associated with Freund's complete and other adjuvants. *ILAR J.* 46:280–293.
22. U.S. Department of Agriculture, Animal and Plant Health Inspection Service. 2011. Animal Care Policy 11, Painful and Distressful Procedures. http://www.aphis.usda.gov/animal_welfare/policy.php?policy=11.
23. Smith, D.E., M.E. O'Brien, V.J. Palmer et al. 1992. The selection of an adjuvant emulsion for polyclonal antibody production using a low-molecular-weight antigen in rabbits. *Lab. Anim. Sci.* 42:599–601.
24. Chapel, H.M. and P.J. August. 1976. Report of nine cases of accidental injury to man with Freund's complete adjuvant. *Clin. Exper. Immunol.* 24:538–541.
25. Hughes, L.E., R. Kearney and M. Tully. 1970. A study in clinical cancer immunotherapy. *Cancer* 26:269–278.
26. Giffen, P.S., J. Turton, C.M. Andrews et al. 2003. Markers of experimental acute inflammation in the Wistar Han rat with particular reference to haptoglobin and C-reactive protein. *Arch. Toxicol.* 77:392–402.
27. Lipman, N.S., L.J. Trudel, J.C. Murphy et al. 1992. Comparison of immune response potentiation and in-vivo inflammatory effects of Freund's and RIBI adjuvants in mice. *Lab. Anim. Sci.* 42:193–197.
28. Toth, L.A., A.W. Dunlap, G.A. Olson, G.A. et al. 1989. An evaluation of distress following intraperitoneal immunization with Freund's adjuvant in mice. *Lab. Anim. Sci.* 39:122–126.
29. Wanstrup, J. and H.E. Christensen. 1965. Granulomatous Lesions in Mice Produced by Freund's Adjuvant; Morphogenesis and Phasic Development. *Acta Pathol. Mic. Sc.* 63:340–354.
30. Geisterfer, M. and J. Gauldie. 1996. Regulation of signal transducer, GP13O and the LIF receptor in acute inflammation in vivo. *Cytokine* 8:283–287.
31. Olivier, E., E. Soury, J.L. Risler, Smith, F. et al. 1999. A novel set of hepatic mRNAs preferentially expressed during an acute inflammation in rat represents mostly intracellular proteins. *Genomics* 57:352–364.
32. Kolstad, A.M., R.M. Rodriguiz, C.J. Kim et al. 2012. Effect of pain management on immunization efficacy in mice. *J. Amer. Assoc. Lab. Anim. Sci.* 51:448–457.
33. Besson, J.-M.R. and G. Guilbaud. 1988. The arthritic rat as a model of clinical pain?: Proceedings of the international symposium on the arthritic rat as a model of clinical pain, held in Saint-Paul de Vence, France, 6–8 June 1988. Amsterdam: Elsevier.
34. Herbert, W.J. 1968. The mode of action of mineral-oil emulsion adjuvants on antibody production in mice. *Immunology* 14:301–318.
35. Hu, J.G., A. Ide, T. Yokoyama et al. 1989. Studies on the optimal immunization schedule of the mouse as an experimental animal. The effect of antigen dose and adjuvant type. *Chem. Pharm. Bull.* 37:3042–3046.
36. Hu, J.G. and T. Kitagawa. 1990. Studies on the optimal immunization schedule of experimental animals. VI. Antigen dose-response of aluminum hydroxide-aided immunization and booster effect under low antigen dose. *Chem. Pharm. Bull.* 38:2775–2779.
37. Hu, J.G., T. Yokoyama, and T. Kitagawa. 1990. Studies on the optimal immunization schedule of experimental animals. IV. The optimal age and sex of mice, and the influence of booster injections. *Chem. Pharm. Bull.* 38:448–451.

38. Hanly, W.C., J.E. Artwohl and B.T. Bennett. 1995. Review of polyclonal antibody production procedures in mammals and poultry. *ILAR J.* 37:93–118.
39. Potkay, S., N.L. Garnett, J.G. Miller et al. 1995. Frequently asked questions about the public health service policy on humane care and use of laboratory animals. *Lab Anim. (N.Y.)* 24:24–26.
40. Wolff, A., N. Garnett, S. Potkay et al. 2003. Frequently asked questions about the Public Health Service Policy on Humane Care and Use of Laboratory Animals. *Lab Anim. (N.Y.)* 32(9):33–36.
41. National Institutes of Health, Office of Laboratory Animal Welfare. 2013. Frequently Asked Questions F8, When institutions collaborate, or when the performance site is not the awardee institution, which IACUC is responsible for review of the research activity? http://www.grants.nih.gov/grants/olaw/faqs.htm#proto_8.
42. Bruce, M.P., V. Boyd, C. Duch et al. 2002. Dialysis-based bioreactor systems for the production of monoclonal antibodies—alternatives to ascites production in mice. *J. Immunol. Meth.* 264(1–2):59–68.
43. Jackson, L R., L.J., Trudel, J.G., Fox, and N.S., Lipman. 1996. Evaluation of hollow fiber bioreactors as an alternative to murine ascites production for small scale monoclonal antibody production. *J. Immunol. Meth.* 189(2):217–231.
44. Lipman, N.S. and L.R. Jackson. 1998. Hollow fibre bioreactors: An alternative to murine ascites for small scale (< 1 gram) monoclonal antibody production. *Res. Immunol.* 149:571–576.
45. Marx, U. 1998. Membrane-based cell culture technologies: A scientifically and economically satisfactory alternative to malignant ascites production for monoclonal antibodies. *Res. Immunol.* 149:557–559.
46. Peterson, N.C. and J.E. Peavey. 1998. Comparison of in vitro monoclonal antibody production methods with an in vivo ascites production technique. *Contemp. Topics Lab. Anim. Sci.* 37(5):61–66.
47. Trebak, M., J.M. Chong, D. Herlyn et al. 1999. Efficient laboratory-scale production of monoclonal antibodies using membrane-based high-density cell culture technology. *J. Immunol Meth.* 230:59–70.
48. Valdes Veliz, R. 2002. Alternative techniques to obtain monoclonal antibodies at a small scale: Current state and future goals. *Biotecnologia Aplicada* 19:119–131.
49. National Research Council (U.S.). Committee on Methods of Producing Monoclonal Antibodies. 1999. *Monoclonal antibody production.* Washington, D.C.: National Academies Press. http://www.ncbi.nlm.nih.gov/books/NBK100199/.
50. Jackson, L.R., L.J. Trudel, J.G. Fox et al. 1999. Monoclonal antibody production in murine ascites - I. Clinical and pathologic features. *Lab. Anim. Sci.* 49(1):70–80.
51. Jackson, L.R., L.J. Trudel, J.G. Fox et al. 1999. Monoclonal antibody production in murine ascites - II. Production characteristics. *Lab. Anim. Sci.* 49(1):81–86.
52. Kaushansky, K. and W.J. Williams. 2010. *Williams Hematology,* 8th edition. New York: McGraw-Hill Medical.
53. Noble, R.P., M.I. Gregersen, P.M. Porter et al. 1946. Blood volume in clinical shock. 2. The extent and cause of blood volume reduction in traumatic, hemorrhagic, and burn shock. *J. Clin. Invest.* 25(2):172–183.
54. Ploucha, J.M. and G.D. Fink. 1985. Systemic hemodynamics during hemorrhage in the conscious rat and chick. *Fed. Proc.* 44:1349–1349.
55. Ploucha, J.M. and G.D. Fink. 1986. Hemodynamics of hemorrhage in the conscious rat and chicken. *Am. J. Physio.* 251:R846–R850.
56. Sapirstein, L.A., E.H. Sapirstein, and A. Bredemeyer. 1960. Effect of hemorrhage on the cardiac output and distribution in the rat. *Circ. Res.* 8:135–148.
57. Mitruka, B.M. and H.M. Rawnsley. 1977. *Clinical Biochemical and Hematological Reference Values in Normal Experimental Animals.* New York: Masson Publishing.
58. Raabe, B.M., J.E. Artwohl, J.E. Purcell et al. 2011. Effects of weekly blood collection in C57BL/6 mice. *J. Amer. Assoc. Lab. Anim. Sci.* 50:680–685.
59. Holm, T.M., A. Braun, B.L. Trigatti et al. 2002. Failure of red blood cell maturation in mice with defects in the high-density lipoprotein receptor SR-BI. *Blood* 99:1817–1824.

60. Macleod, J.N. and B.H. Shapiro. 1988. Repetitive blood-sampling in unrestrained and unstressed mice using a chronic indwelling right atrial catheterization apparatus. *Lab. Anim. Sci.* 38:603–608.
61. Diehl, K.H., R. Hull, D. Morton et al. 2001. A good practice guide to the administration of substances and removal of blood, including routes and volumes. *J. Appl. Toxicol.* 21:15–23.
62. McGuill, M.W. and A.N. Rowan. 1989. Biological effects of blood loss: Implications for sampling volumes and techniques. *ILAR News* 31(4):5–20.
63. McGee, M.A. and R.R. Maronpot. 1979. Harderian gland dacryoadenitis in rats resulting from orbital bleeding. *Lab. Anim. Sci.* 29:639–641.
64. Vanherck, H., V. Baumans, N.R. Vandercraats et al. 1992. Histological changes in the orbital region of rats after orbital puncture. *Lab. Anim.* 26(1):53–58.
65. Le Net, J.E.L., D.P. Abbott, P.R. Mompon et al. 1994. Repeated orbital sinus puncture in rats induces damages to optic nerve and retina. *Vet. Pathol.* 31:621.
66. van Herck, H., V. Baumans, H.A.G. Boere et al. 2000. Orbital sinus blood sampling in rats: Effects upon selected behavioural variables. *Lab. Anim.* 34:10–19.
67. van Herck, H., V. Baumans, C. Brandt et al. 2001. Blood sampling from the retro-orbital plexus, the saphenous vein and the tail vein in rats: Comparative effects on selected behavioural and blood variables. *Lab. Anim.* 35:131–139.
68. Everds, N.E., Hematology of the Laboratory Mouse. 2007. In *The Mouse in Biomedical Research: Normative Biology, Husbandry, and Models,* 2nd edition, eds. J. G. Fox, S. W. Barthold, M. T. Davisson et al., 133–170. Amsterdam: Elsevier.
69. van Herck, H., V. Baumans, C. Brandt et al. 1998. Orbital sinus blood sampling in rats as performed by different animal technicians: The influence of technique and expertise. *Lab. Anim.* 32:377–386.
70. Taylor, R., K.E. Hayes and L.A. Toth. 2000. Evaluation of an anesthetic regimen for retroorbital blood collection from mice. *Contemp. Topics Lab. Anim. Sci.* 39(2):14–17.
71. Heimann, M., H.P. Kaesermann, R. Pfister et al. 2009. Blood collection from the sublingual vein in mice and hamsters: A suitable alternative to retrobulbar technique that provides large volumes and minimizes tissue damage. *Lab. Anim.* 43:255–260.
72. Heimann, M., D.R. Roth, D. Ledieu et al. 2010. Sublingual and submandibular blood collection in mice: A comparison of effects on body weight, food consumption and tissue damage. *Lab. Anim.* 44:352–358.

20

Occupational Health and Safety

Robin Lyn Trundy and Susan Stein Cook*

Introduction

The use of animals for research purposes has always been directly or indirectly linked to human health concerns. Animal allergens, zoonotic diseases, and physical injuries are only some of the issues that need to be addressed in a general occupational health program associated with animal research. Although regulations and guidelines dealing with human health and protection are a valuable resource for establishing a comprehensive occupational health program, their scope and level of detail will always be limited. Fortunately, most current health and safety regulations are performance oriented, focusing on the goal (i.e., a safe work environment) rather than describing how to get there. This allows institutions to tailor their safety programs to protect employees in the most effective way.

General Occupational Health and Safety

20:1 What is the general responsibility of the IACUC toward ensuring a safe work environment for persons working with laboratory animals?

Reg. The need to protect the health and safety of personnel involved in animal-based research is addressed in the *Guide* (p. 17) and the PHS Policy (IV,A,1,f). The PHS Policy (IV,A,1) requires institutions to use the *Guide* as the basis for their institutional program for activities involving animals. The *Guide* states that institutions must establish an occupational health and safety program (OHSP) as part of the overall animal research program. The *Guide* (p. 17) specifically references the National Research Council publication *Occupational Health and Safety in the Care and Use of Research Animals*[1] to be used as a tool for establishing a comprehensive OHSP. While the main responsibility of the OHSP lies with the institution, the IACUC also has to assume at least partial responsibility for the program as it is specifically charged with the oversight and evaluation of the animal care and use program (*Guide* p. 17). The institutional OHSP should identify potential hazards

* The authors thank Stefan Wagener, Robert J. Ceru and Kristin Erickson for their contributions to previous editions of *The IACUC Handbook*. The authors also thank Ben Edwards for his contribution to this edition of the chapter.

in the work environment and conduct a critical assessment of the associated risks (*Guide* p. 18).

Opin. The applicable referenced guidelines and regulations provide limited detail specifically addressing the IACUC's responsibility for occupational health and safety. Therefore, the institution must manage and assign responsibilities covering the various programs required by the *Guide*. The quality of the OHSP will be impacted significantly by the quality of interaction of the IACUC with relevant institutional groups responsible for functions such as the research program, environmental health and safety, and occupational health services. Institutions are required to address OHSP issues and assign responsibilities to various groups or individuals inside the institution for management and oversight of the different components. For example, occupational health requires the input of medical professionals, while physical, chemical, biological, and radiological hazards are best addressed by specific committees or safety professionals with expertise in these areas. Ideally, the IACUC assumes the role of a facilitator, interacting with the various programs, committees, and individuals to ensure the existence of a comprehensive OHSP and compliance with the *Guide*.

20:2 What is the general responsibility of the IACUC toward ensuring a safe work environment for persons having access to an animal facility but not working with laboratory animals (e.g., maintenance, clerical personnel, visitors)?

Reg. As stated in 20:1, the institution at large is primarily responsible for the establishment of an OHSP. The PHS Policy (IV,A,1,f) requires a health program for personnel who have frequent contact with animals, as well as all persons who work in laboratory animal facilities. According to the *Guide* (pp. 18–19) "The extent and level of participation of personnel in the OHSP should be based in the hazards posed by the animals and materials used, the exposure intensity, duration, and frequency; to some extent, the susceptibility of the personnel; and the history of occupational illness and injury in the particular workplace."

In addition, the National Research Council publication *Occupational Health and Safety in the Care and Use of Research Animals*[1] (pp. 12, 17) as referenced by the *Guide*, suggests including not only animal caretakers, technicians, students, volunteers, investigators, and veterinarians, but also facility maintenance personnel, housekeepers, security, and other staff who must perform job duties in animal research environments.

Opin. Ideally, the IACUC facilitates the close interaction among institutional groups dealing with employee health and safety as it relates to animal research facilities. For example, the institution's environmental health and safety (EHS) office oversees the health and safety of maintenance personnel working in areas with known chemical and biological hazards. The institutional occupational health group and the medical professionals should identify any necessary surveillance criteria. The assistance of these and other health and safety groups should be requested by the IACUC to perform risk assessments, project review, and health and safety recommendations.

20:3 Does the IACUC have a responsibility to ensure a safe working environment for research technicians, investigators, and students who use animals outside the animal facility?

Reg. The PHS Policy (IV,A,1,f) is applicable to all institutional programs involving animals, not only in animal facilities. Therefore, the OHSP must address all risks to investigators, technicians, and students associated with PHS-supported animal activity, regardless of where it occurs. The risks associated with unusual experimental conditions, such as those encountered in field studies or wildlife research, should also be addressed under the hazard identification and risk assessment responsibilities of the OHSP (*Guide* p. 18).

Opin. The IACUC works in concert with the institution's occupational health and safety professionals to foster a safe work environment and ensure safe project design.

General and specific regulatory health and safety requirements (i.e., Occupational Safety and Health Administration [OSHA] regulations), however, do apply to worker protection inside and outside a facility. Hazards associated with animal research outside a facility can be of great concern. One example is field research involving wild animals. The incorrect use of equipment to trap or immobilize wild animals can pose a significant physical hazard to animals and humans. Electrical and chemical hazards can also be present. Zoonotic diseases, such as rabies or hantavirus pulmonary syndrome can be life threatening. Field sites may be located in areas with possible seasonal risk of exposure to vector borne diseases (e.g., Lyme Disease, West Nile Virus). Prudent practices include risk assessment of all animal use protocols by the knowledgeable institutional safety groups or individuals to assure and maintain worker safety outside the facility.

20:4 What elements should the OHSP include?

Reg. The *Guide* (pp. 17–23) outlines an occupational health and safety program, as required in the PHS Policy (IV,A,1,f). This was summarized by NIH/OLAW[2] as follows: Minimally, the program should include

- Pre-placement medical evaluation
- Identification of hazards to personnel and safeguards appropriate to the risks associated with the hazards
- Appropriate testing and vaccinations
- Training of personnel regarding their duties, any hazards, and necessary safeguards
- Policies and facilities that promote cleanliness
- Provisions for treating and documenting job-related injuries and illnesses
- Facilities, equipment, and procedures should be designed, selected, and developed to reduce the possibility of physical injury or health risk to personnel
- Good personal hygiene practices, prohibiting eating and drinking, use of tobacco products, and application of cosmetics and/or contact lenses in animal rooms and laboratories
- Personal protective equipment (PPE)

Opin. The minimal components outlined above are necessary for an effective OHSP. Safety professionals and occupational health professionals should work collaboratively with veterinary and animal care management staff to address all of these components. While different groups may be responsible for administration of certain components, it is important for the groups to communicate openly with the other groups when developing policies and procedures to build a cohesive program that truly supports the safety needs of the end users.

20:5 What circumstances may require additional medical surveillance activities beyond standard occupational health examinations for those who work with animals?

Reg. Some OSHA standards require a specific medical surveillance activity if an employee is exposed to a hazard at levels or durations above those that are considered to be safe. Examples include the Occupational Noise Exposure Standard,[3] the Respiratory Protection Standard,[4] and the Bloodborne Pathogens Standard.[5]

Under the Occupational Noise Exposure Standard[3] employees who are exposed to noise levels at or above 85 decibels for an 8-hour time-weighted average (TWA) must have baseline and annual audiometric testing through the occupational health provider. These levels may be exceeded in some facilities, as a result of noise from animals or equipment.

The Respiratory Protection Standard[4] requires that any employee who is required to wear a respirator for protection against an airborne hazard on the job be medically evaluated before wearing a respirator; additional ongoing surveillance requirements are determined by the occupational health provider. This standard may apply to the use of certain chemical products in the research facility and for work with airborne infectious agents.

The Bloodborne Pathogens Standard[5] requires that employees who are exposed to certain human-derived materials (which include human cells under most circumstances) be included in the bloodborne pathogens exposure control program. These personnel must be offered the hepatitis B vaccination and receive specific post-exposure services if an employee sustains an exposure to such materials through a splash to the eyes, nose, or mouth; absorption through broken skin; or a cut with a contaminated object. In the animal research environment, personnel who handle human cells during the introduction of these materials into an animal model or harvest these materials from the animal model are likely to be affected by these requirements. The *Guide* (p. 22) also states, "Vaccination is recommended if research is to be conducted on infectious diseases for which effective vaccines are available. More specific recommendations are available in *Biosafety in Microbiological and Biomedical Laboratories, 5th edition* (BMBL)."[6]

Opin. In addition to the OSHA requirements, an institution should ensure that the occupational health–medical surveillance activities include additional components to protect personnel if infectious disease research is under way in the animal research environment or laboratories. The BMBL[6] is considered a *de facto* standard by many agencies and organizations. This document includes agent summary statements under Section VIII for environments where an infectious disease risk is present. Additionally, the document provides information on modes of infection, laboratory safety and containment recommendations, vaccinations, medical surveillance information, and considerations for work with specific agents and biological materials. Subsections include: bacterial agents, fungal agents, parasitic agents, rickettsial agents, viral agents, arboviral agents and related zoonotic viruses, toxins, and prion diseases.

20:6 Should the IACUC establish safe working rules for an animal facility or laboratory?

Reg. The IACUC has no specific regulatory mandate for establishing safe working rules unless they pertain specifically to animal health and care and are based on the *Guide* or other relevant animal use and care regulations. (See 20:1.)

Opin. "Safe working rules" (or administrative controls) are only one component of hazard control prescribed by OSHA standards. For all recognizable occupational hazards, employers need to develop a hazard control program that employs several methods to minimize employee exposure to the hazard. These methods (with animal research examples) include

- Procedural change to eliminate the hazardous component, (e.g., use of computer models or cell culture in place of in vivo studies)
- Procedural change to substitute the hazardous component for a less-hazardous component (e.g., use of a comparable non-aggressive species instead of an aggressive species)
- Use of engineering controls to isolate the hazard (e.g., physical or chemical restraint methods, sharps containers, biosafety cabinets and fume hoods)
- Administrative controls to support the application of hazard control methods (e.g., training, task-specific practices, medical surveillance)
- Use of personal protective equipment when exposure to the hazard cannot be eliminated through the use of the other methods (e.g., hand, face and eye protection; disposable gowns)

While administrative controls certainly support the execution of tasks in a safe manner, they need to include all elements of hazard control. The IACUC should be vigilant to assure that hazard control programs (not just "safe working rules") are available for animal research activities; it is ultimately the institution's responsibility (as the employer) to establish and maintain these programs.

20:7 What types of hazards are relevant to animal research settings?

Opin. Animal research settings involve unique hazards related to working with animals, but such common workplace hazards as cuts, burns, slips, trips or falls also occur. In addition, the research might involve unique physical, chemical, biological, or radiological hazards. An overview of relevant physical, chemical, biological, and protocol-related hazards has been published in Chapter 3 of the document entitled *Occupational Health and Safety in the Care and Use of Research Animals.*[1]

In general, physical hazards in the animal research environment can include bites, scratches, and kicks; sprains and strains caused by moving of equipment; operational hazards involving electricity, machinery, and noise; or protocol-induced hazards such as radioactivity or the use of lasers. Chemical hazards can be directly related to a specific agent or procedure. Examples include chemicals used for processing tissues and cleaning or disinfecting research equipment. The accidental inhalation of waste anesthetic gases provides another route of chemical exposure. Research protocols might require the application of toxins, carcinogens, and other hazardous chemicals. Biological hazards, such as infectious pathogens, can be introduced through naturally occurring or experimentally infected research animals or include the application of recombinant DNA technology. Another significant hazard is manifested in the increasing development of animal-related allergies or occupation-related asthma.[7] Repetitive motion injuries,[8] compassion fatigue[9] and increased activities by individuals or organizations opposed to the use of animals in research are additional hazards relevant to the animal research setting.

20:8 What sources of information can the IACUC access relative to hazards found in the animal research setting?

Opin. The common hazards include chemical, physical, biological, mechanical, and environmental hazards and vary with the type of research, geographic location, research species, facility design, and duration of exposure. The IACUC should rely on institutional health and safety officers and the AV to collect, collate, and share material from local, state, and federal agencies relevant to common hazards. (See 20:30.) Health and safety information and regulations may also be accessed via the internet. This often can be achieved by performing a keyword search on agency names mentioned in this chapter. The *Guide* (pp. 170–172) also recommends a comprehensive listing of references.

20:9 What are the relevant institutional committees and departments that the IACUC should interface with relative to personnel safety?

Opin. The IACUC has an oversight responsibility (see 20:1) related to personnel safety issues associated with animal protocols. To assure that personnel safety issues are effectively addressed as part of the review process, the IACUC must seek input and communicate with the committees and departments specifically charged with regulatory compliance, worker safety, and environmental health. These groups and individuals interact directly with agencies such as OSHA, the Nuclear Regulatory Commission (NRC), and the Centers for Disease Control and Prevention (CDC). These committees and departments usually include the following:

- Environmental Health & Safety/Occupational Safety provides assistance related to common physical hazards, personal protective equipment assessments, and chemical use, including anesthetics and disinfectants.
- Biological Safety provides assistance related to biological containment practices, human-derived material use, safe handling of sharps, biological safety cabinet use, and biohazardous waste management.
- Radiation Safety provides assistance with radiation safety training, signage, personnel exposure monitoring, and area surveys for radioactive contamination. The scope of controls for radiological hazards is determined by the Institutional Radiation Safety Committee. When radiological hazards are used in conjunction with animals, the IACUC must maintain open communication with the Radiation Safety Committee, or a radiation safety officer, to assure that radiation safety compliance items have been addressed before IACUC approval is granted.

- Occupational Health provides medical screening of personnel who are enrolled in the occupational health program, as well as information on allergies and hygiene.
- A Chemical Hygiene Committee may, or may not, be established at an institution (see 20:18). If such a committee is not established, a chemical hygiene officer with oversight for animal research facilities should be included in the IACUC review for protocols involving hazardous chemicals, including anesthetics.
- An Institutional Biosafety Committee (IBC) is required for an institution if it receives any NIH funds for recombinant DNA research. This committee is charged with review of research involving recombinant DNA molecules (see 20:22), although many IBCs also review the use of biological hazards, including potentially infectious agents and toxins. When recombinant DNA molecules are used in conjunction with animals, the IACUC and the IBC should work jointly to assure that all regulatory compliance items have been addressed before either committee approves a project.

Physical and Chemical Hazards

20:10 What is meant by a *physical hazard*?

Reg. According to *Occupational Health and Safety in the Care and Use of Research Animals*[1] (pp. 32–43) physical hazards are those generally associated with

- Flammable materials
- Pressurized vessels
- Machinery
- Noise
- Temperature
- Vibration
- Electricity
- Nonionizing and ionizing radiation (also considered a radiation hazard)
- Ultraviolet radiation
- Lasers
- Illumination
- Sharp objects

Occupational exposure limits and protective measures for some of these hazards are covered under specific OSHA standards in the US:

- Occupational Noise Exposure[3]
- Nonionizing Radiation[10]
- Ionizing Radiation[11]
- Electrical[12,13]

In addition, a hazardous chemical may have present physical hazard properties that will pose a danger to personnel and the environment such as being flammable, explosive, reactive with water and other incompatible materials, etc. These properties are fully described in the OSHA Hazard Communication Standard.[14]

Opin. Excessive noise can be produced by machinery and animals, especially pigs and dogs. Exposure to intense noise over time will result in hearing loss. If normal talking or phone conversation is not possible because of excessive noise, the noise level should be assessed by a safety professional. Electricity is of concern if electrical equipment is out of date, poorly maintained, or improperly grounded. Compliance with electrical code requirements is necessary, as is equipment maintenance.

(See 20:34–20:38 for specific information on the management of radiological hazards.)

The IACUC should seek the assistance of occupational health and safety professionals to assess the potential physical hazards associated with animal protocols and animal holding facility operations.

20:11 Should physical methods of euthanasia be considered a physical risk?

Reg. Physical methods of euthanasia as defined by the American Veterinary Medical Association (AVMA) *Guidelines for the Euthanasia of Animals*[15] include captive bolt, gunshot, cervical dislocation, decapitation, electrocution, kill traps, thoracic compression, exsanguination, maceration, stunning, microwave irradiation, and pithing. According to the 2013 AVMA *Guidelines*, "Since most physical methods involve trauma, there is inherent risk for animals and humans. Extreme care and caution should be used." PHS Policy (IV,C,1,g) requires euthanasia methods to be consistent with the AVMA recommendations unless a deviation for scientific reasons is justified, in writing, by the investigator. APHIS/AC Policy 3 states, "The method of euthanasia should be consistent with the current AVMA *Guidelines for the Euthanasia of Animals*, the American Association of Zoo Veterinarians (AAZV) Guidelines for Euthanasia of Nondomestic Animals or the European Commission Working Party documents."[16]

Opin. As with all physical procedures and practices, the skill and expertise of the person performing the procedure determine the degree of risk. Prior to approval of such methods, the IACUC should evaluate the proficiency and level of training of individuals performing these procedures and be assured that the equipment is in good working order.

20:12 How can the IACUC evaluate physical hazard risk?

Opin. Potential physical hazards for animals and research personnel working with them should be considered as part of the protocol review process. The IACUC may also use required animal facility inspections as opportunities to observe practices and procedures involving potential physical hazards, including those that that fall outside of the scope of protocol activities, such as cage wash, feed handling, facility cleaning, etc. However, the IACUC should involve institutional occupational health and safety professionals in the evaluation process because control of such hazards is within the purview of these individuals' responsibilities.

TABLE 20.1

Examples of Chemical Hazards

Physical Hazards	Health Hazards
Chemicals that pose one of the following hazardous effects: • Explosive • Flammable (gases, aerosols, liquids or solids) • Oxidizer (liquid, solid or gas) • Self-reactive • Pyrophoric (liquid or solid) • Self-heating • Organic peroxide • Corrosive to metal • Gas under pressure • In contact with water emits flammable gas	Chemicals that pose one of the following hazardous effects: • Acute toxicity (any route of exposure) • Skin corrosion or irritation • Serious eye damage or eye irritation • Respiratory or skin sensitization • Germ cell mutagenicity • Carcinogenicity • Reproductive toxicity • Specific target organ toxicity (single or repeated exposure • Aspiration hazard

20:13 What is meant by a *chemical hazard*?

Reg. A *hazardous chemical* is defined by OSHA[14] as any chemical which is classified as a physical hazard or a health hazard, a simple asphyxiant, combustible dust, pyrophoric gas, or hazard not otherwise classified. See Table 20.1.

Opin. The IACUC and AV should seek the assistance of the institution's industrial hygiene professionals or chemical safety officer when questions arise related to the hazard status and assessment of an experimental compound or previously unused chemical on an animal protocol.

20:14 What are some typical chemical hazards that can be found in a research environment where animals are used?

Opin. Researchers commonly use a wide variety of chemicals. Toxic agents, carcinogens, euthanasia agents, chemotherapeutic and other test agents can be used to induce disease, study metabolic processes, euthanize animals or reduce and reverse the effects of disease. Other chemicals are used in analytic techniques. Most laboratories have flammable solvents, corrosive liquids, and a variety of toxic chemicals. Common animal facility cleaning supplies include solvents, acids, disinfectants, and sanitizers such as bleach or quaternary ammonium compounds. Some of these are used for manual cleaning tasks and others in mechanical devices such as cage washers or foamers. These cleaning agents can be an exposure hazard in a number of scenarios including improper mixing or dispensing technique, mechanical equipment failure, or improper storage of cleaning compounds resulting in container failure.

20:15 What is a Material Safety Data Sheet (MSDS)?

Reg. A material safety data sheet (or "safety data sheet" under forthcoming regulatory changes) is a document containing chemical hazard identification and safe handling information. It is prepared in accordance with the OSHA Hazard Communication Standard[14] also known as the "Right-to-Know" law. Chemical manufacturers and distributors must provide the purchasers of hazardous chemicals with an appropriate MSDS for each hazardous chemical or product purchased.

Beginning June 1, 2015, all chemical manufacturers and distributors must use a uniform 16-section format for safety data sheets to enhance the effectiveness of these safety resource documents.

Opin. Copies of the MSDS for hazardous chemicals at a given worksite must be readily accessible to employees in that area. As a source of detailed information on hazards, they must be located close to workers and readily available during working hours. MSDSs for hazardous chemicals used in the animal facility are required by law. Accessibility, however, does not necessarily require having copies available in each room or attached to each chemical. Central repositories (e.g., main office) can store all available MSDSs. Each area housing hazardous chemicals should post information identifying the location of MSDSs for that area. Many facilities use computer-based retrieval systems, loaded with software containing large numbers of MSDSs.

20:16 Should all approved anesthetic and euthanasia agents be considered hazardous?

Reg. Approved anesthetic and euthanasia agents are considered hazardous under OSHA regulations if they meet one or more of the criteria identified for hazardous chemicals (see 20:13).

Opin. The degree of hazard varies from agent to agent. Factors that determine the degree of hazard include the specific agent and its properties, exposure concentrations, work practices, engineering controls, personal protective equipment used, and route of administration. An industrial hygienist (usually a member of the institution's environmental health and safety group) should be consulted regarding the hazard status of such agents. This individual should also assess the administrative practices for these agents to determine the need for specific practices and safety equipment to reduce personnel exposure. These assessments should be used to establish procedures and training related to use, storage, and disposal of anesthetic and euthanasia agents. Training in proper techniques involving inhalation and non-inhalation agents should be provided.

20:17 How can an IACUC evaluate a chemical risk?

Opin. An IACUC could partner with an industrial hygienist within its organization (or hired as a consultant) or any other health and safety professional with specific knowledge and experience (e.g., toxicologist) to review and evaluate chemical risks. Areas for review should include the following

- Chemical agents used
- Engineering controls
- Special personal protective equipment required
- Route of delivery to the animal (feed, gavage, aerosol, injection)
- Route of excretion
- Precautions for handling animals exposed to specific chemicals
- Animal disposal
- Bedding disposal
- Cage decontamination

- Special precautions
- Storage and delivery procedures
- Level of training and experience of personnel

20:18 Is a separate chemical hazard committee necessary?

Reg. The use of an institutional chemical hygiene committee is an option suggested by the OSHA Laboratory Safety Standard[17] to be established by the institution if appropriate. Currently, OSHA does not mandate a chemical hazard committee. However, if hazardous chemicals are used in the laboratory or animal facility, a chemical hygiene plan is required by OSHA. This plan not only outlines all relevant safety procedures and practices as required by law, but also must designate personnel who are responsible for implementing the plan including the assignment of a chemical hygiene officer.

Opin. Establishment of a separate chemical hygiene committee is a recommended option in overseeing institutional compliance with all aspects of chemical safety. In addition, review of animal research projects involving hazardous chemicals should be done in cooperation with the chemical hygiene officer or industrial hygiene professional with these designated duties.

20:19 What institutional resources can the IACUC access to determine whether there is a need for personnel respiratory protection for a specific animal protocol?

Reg. The OSHA Respiratory Protection Standard[4] requires the institution to develop and implement a written respiratory protection program with worksite-specific procedures and elements for required respirator use. This program, and the use of respirators, are necessary if employees are exposed to air contaminated with harmful dusts, fogs, fumes, mists, gases, smokes, sprays, or vapors, and these contaminants cannot be controlled or eliminated by other means (e.g., engineering controls). Animal care–related examples of contaminants and their applications may include

- Disinfection of animal rooms with aerosolized chemicals
- Project-specific feed preparations containing carcinogens
- Working with or near animals *and* being identified as having animal allergies

If an employer is required to have a respiratory protection program, the employer or institution must designate a program administrator who is qualified by appropriate training or experience to administer or oversee the respiratory protection program and conduct the required evaluations of program effectiveness. This individual is the key resource for the IACUC for matters associated with respiratory protection.

Opin. Effective respiratory protection programs require a knowledgeable program administrator, an occupational health professional who can provide medical evaluations, and an understanding, at the institutional level, of the restrictions and requirements for use of respirators by employees. The IACUC should confer with the respiratory protection program administrator (usually a member of environmental health and safety or the occupational health department) whenever the need for respiratory protection is identified or suggested, as part of an animal protocol.

20:20 Can outer garments that are worn in animal rooms (e.g., scrub suits) also be worn outside of the animal room?

Reg. According to the *Guide* (p. 20), outer garments worn in animal rooms should not be worn outside the animal facility unless they are covered by other outerwear. However, if scrub suits are routinely covered by other outerwear, such as a disposable gown or jump suit, while working in animal areas, the scrub suit can be worn outside of the animal facility once the disposable outerwear is removed.

Opin. Outerwear serves to limit contact of outer garments with allergens and other contaminants in the facility environment, thus reducing the probability of transferring those contaminants out of the facility. The same outerwear may also be prescribed as personal protective equipment to limit the wearer's exposure to a hazard (such as in an animal biosafety level 2 room). When outerwear is worn as PPE, it is important to assure that removal and replacement of the outerwear is performed upon exit from the area where the hazardous agent is present to avoid spreading associated contaminants to other areas of the facility.

Biological and Radiological Hazards

20:21 What is the definition of a *biological hazard*?

Reg. Currently there are no regulatory-based definitions for biological hazards. A regulatory definition for bloodborne pathogens has been published in the OSHA standards.[5] Recombinant DNA molecules are defined in the *NIH Guidelines for Research Involving Recombinant or Synthetic Nucleic Acid Molecules*.[18] Certain infectious materials are defined by the Department of Health and Human Services (HHS),[19] USDA,[20,21] and Department of Transportation regulations[22] related to their transfer, acquisition, and transportation.

Opin. The term, *biological hazards* (or *biohazards*), commonly refers to agents or materials of biological origin that are potentially hazardous to humans, other animals, or plants. Although there is no standardized definition for biohazards, domestic institutions often include infectious agents, materials containing recombinant DNA molecules, toxins of biological origin, and human—or nonhuman primate—derived materials in the definition.

20:22 What is meant by *recombinant and synthetic nucleic acid molecules* and what are some examples?

Reg. *Recombinant and synthetic nucleic acid molecules* are defined by the *NIH Guidelines*[18] as

1. Molecules that are constructed by joining synthetic nucleic acid molecules and that can replicate in a living cell, i.e., recombinant nucleic acids.
2. Nucleic acid molecules that are chemically or by other means synthesized or amplified, including those that are chemically or otherwise modified but can base pair with naturally occurring nucleic acid molecules, i.e., synthetic nucleic acids.
3. Molecules that result from the replication of those previously described.

Under the *NIH Guidelines,*[18] common examples of recombinant and synthetic nucleic acid molecule use that require registration with, and approval by, the IBC would include

- Creation of transgenic animals (including rodents)
- Acquisition, breeding or use of transgenic animals other than rodents
- Introduction of microorganisms or viruses (including viral vectors) containing recombinant and synthetic nucleic acid molecules in an animal model
- Introduction of cells or tissues containing recombinant and synthetic nucleic acid molecules in an animal model

Opin. The ability to isolate, manipulate, and express genetic material has resulted in a new field of basic and applied research called *genetic engineering.* By using recombinant DNA techniques, genes can be isolated, expressed, and transferred, resulting in a multitude of possible applications. The development of transgenic and knockout animals (e.g., rodents) has proved valuable for studying a variety of diseases. Other applications involve the use of genetically modified infectious agents for vaccine development. Transgenic technology has also been used to alter certain aspects of the biochemistry, hormonal balance, and protein products in livestock, develop recombinant human insulin, growth hormone and blood clotting factors.

20:23 Must an institution working with any biological hazards have a Biological Safety Officer (BSO)?

Reg. The only agency currently requiring the position of a BSO is the NIH as part of the requirements for recombinant DNA research. As outlined in the *NIH Guidelines,* "the institution shall appoint a Biological Safety Officer if it engages in large-scale research or production activities involving viable organisms containing recombinant DNA molecules."[18] In addition, "the institution shall appoint a Biological Safety Officer if it engages in recombinant DNA research at BL3 or BL4."[18]

Opin. Most institutions have gone beyond this basic requirement and have appointed a Biological Safety Officer to oversee all aspects of biological safety. Biological Safety Officers commonly manage institutional programs addressing the OSHA requirements for bloodborne pathogens and compliance with the U.S. Department of Health and Human Services and USDA programs (i.e., Select Agents), develop biological safety operating procedures, evaluate the potential for zoonotic disease transmission, and provide training and education related to biosafety topics. To maintain a high standard of proficiency and education, certification and testing programs for biological safety professionals have been developed by the American Biological Safety Association (ABSA)[23] and the National Registry of Microbiologists (NRM). The IACUC should consult the Biological Safety Officer in all areas related to biosafety.

20:24 Do transgenic or knockout rodents that have been commercially purchased or received from another institution fall under the *NIH Guidelines?*[18]

Reg. As outlined in Appendix C-VII of the *NIH Guidelines*[18] the purchase or transfer of transgenic rodents for experiments that require Biosafety Level 1 (BL1) containment are exempt from the *NIH Guidelines*. The 2011 revision of the *NIH Guidelines* added an exemption for crossing transgenic rodents under certain conditions. In short, the conditions which prohibit exemption are based upon the use of certain viruses used in the creation of the transgene. The details of the exemption application can be found at Appendix C-VIII of the *NIH Guidelines*.

Opin. Transgenic rodents generated at the institution are still covered by the *NIH Guidelines* and require, at a minimum, Institutional Biosafety Committee (IBC) notification simultaneous with the initiation of the work, even if the rodents require only BL1 containment. BL1 is the lowest level of physical containment and appropriate for well-characterized agents not known to cause disease, and of minimal potential hazard to laboratory personnel and the environment. Generally, combinations of laboratory practices, containment equipment, and facility design can be made to achieve different levels of physical containment, also referred to as *biosafety levels*. Four levels of physical containment (BL1 to BL4) are commonly used: from BL4, the highest level of physical containment, to BL1, the lowest. For a comprehensive overview of biosafety levels as defined by the *NIH Guidelines*, refer to Appendix G of that document.

20:25 Should an IACUC member be a voting member of the institutional biosafety committee (IBC)?

Reg. There is currently no requirement that an IACUC member be a voting member of the IBC. However, the *NIH Guidelines*[18] require that one member of the IBC shall have expertise in animal containment principles if the institution is involved in recombinant DNA molecule research with animals, and those projects require containment practices that would fall under Appendix Q of the *NIH Guidelines*. In short, these containment practices would apply to larger species of animals that may not be maintained in enclosed caging systems that place an environmental barrier between the animals and the room environment because of size or growth requirements.

Opin. At smaller institutions, it may be advantageous to use an individual's expertise in a variety of functions. Institutions, therefore, might select an IACUC member with the appropriate expertise, or an institutional veterinarian to be part of the local IBC. Alternately, the Biosafety Officer may be a valuable ad hoc member of the IACUC, especially for screening of protocols that may be subject to the *NIH Guidelines*.[18]

20:26 Assume an agent transmissible to humans, such as Macacine herpes virus 1 (B-virus), is known or suspected to be harbored by an animal, but the virus is not part of the proposed research. Should this protocol be reviewed and approved by the Institutional Biosafety Committee (IBC) before receiving final IACUC approval?

Opin. In lieu of specific regulatory requirements, guidelines and recommendations for implementing a safety and health program specific for B-virus and nonhuman primates have been established by the CDC[6,24,25] and others.[26,27] None of these guidelines or recommendations requires a separate IBC review or approval of research involving animals considered to be infected with herpes B-virus prior to final IACUC approval. However, an institution that is working with or housing animals of the *Macaca* genus should have an established program to control this zoonotic hazard including training, specific SOPs, and exposure response procedures as recommended by the references cited. Development and oversight of the program should be verified as part of an IACUC review of protocols involving macaques or other Old World primates. The BSO or the Institutional Biosafety Committee Chair (if there is no BSO) and the AV should be consulted as part of this verification.

Herpes B-virus is one of the examples for which specific guidelines are recommended to control a zoonotic disease hazard associated with a given species. When a protocol under review involves animals that are known, or strongly suspected, to be infected with agents transmissible to humans, the IACUC must consult the BSO or the IBC regarding applicable institutional programs or reviews related to biological safety. Research animals should be housed and handled in accordance with the animal biosafety level practices outlined in the *BMBL*.[6]

20:27 What general guidelines can the IACUC follow to assure that biohazardous materials are being used safely in the animal facility and in individual laboratories?

Reg. The use of biohazardous materials, including recombinant DNA, is guided by a combination of laboratory practices and techniques, safety equipment, and special facilities commonly referred to as *biosafety levels*. The CDC/NIH *BMBL*[6] is one reference that contains information on four different biosafety levels specific for laboratories and vertebrate animals. The procedures and practices outlined in this publication have been widely accepted as a standard and are used as a basis for a comprehensive biosafety program. Similar information is outlined in Appendices G and Q of the *NIH Guidelines*[18] Nevertheless, the application of Appendix Q is intended for recombinant DNA research involving animals that may not be maintained in enclosed caging systems that place an environmental barrier between the animals and the room environment because of size or growth requirements. When applicable, compliance is mandatory for institutions receiving NIH funding as outlined in Section 1-D of the *NIH Guidelines*.[18] Local and state agencies should be consulted for all waste-related issues pertaining to biohazardous materials because specific regulations may exist for the treatment and disposal of biohazardous waste.

Opin. Because of the increasing complexity associated with the use and disposal of biohazardous materials, the IACUC needs to utilize the IBC and biosafety professionals for advice, oversight, and compliance. One publication (*BMBL*)[6] can be used as a general guidance document and it should be available to all IACUC members. Additional biological safety informational resources are available through the American Biological Safety Association web site, www.absa.org.[23] It is strongly recommended that the BSO serve as a liaison between the IACUC and the IBC. This will assure that animal protocols that require both IBC and IACUC approval are reviewed in the most effective and timely manner possible. Projects may require specific protocols, safety procedures, safety equipment, inspection, waste disposal, and other considerations. By engaging the BSO throughout the protocol development and approval process the IACUC can best support the biosafety needs of their animal researchers.

20:28 What generally accepted procedure should an IACUC require relative to recapping of needles in the animal facility?

Reg. Sharp devices, regardless of contamination status, constitute a recognizable hazard when these devices are handled in any occupational setting.

Unnecessary recapping of needles in an occupational setting is a hazardous practice and could be cited under Section 5(a) of the Occupational Safety and Health Act, the OSHA General Duty Clause, which states, "Each employer shall furnish to each of his employees employment and a place of employment which are free from recognized hazards that are causing or are likely to cause death or serious physical harm to his employees."[28] In short, if a hazard that is present in the workplace can feasibly be eliminated and no actions are taken to do so, the employer may be cited for violation with the OSHA General Duty Clause.

For institutions that need to comply with CDC and NIH standards, recapping is addressed in the biosafety level criteria sections of these documents. Under Section V of the BMBL,[6] the need for the development and implementation of policies for the safe handling of sharps is specifically cited.

Opin. Accidental sharp injuries involving animal fluids are a significant occupational hazard and have resulted in human infection.[29,30] Prudent practice should require that needles are not recapped and are disposed of immediately after use in approved sharps containers. Conversely, capped needles to be discarded should not be uncapped before disposal in the sharps container. In certain animal procedure circumstances, recapping will be necessary. However, this should only be done when all non-recapping alternatives, including the use of devices with guards or automated needle retraction function, have been explored and determined infeasible. If recapping must be performed, personnel should be trained to use a "one-handed scoop" technique. (i.e., a technique that does not permit the non-dominant hand to be placed in front of the needle).

If safety needles will be adopted, personnel must be trained in the use of these devices before use in a "real" application to minimize potential exposure risk for personnel and injury for the animals involved in the procedures. The BSO should be consulted for assistance with establishing safe sharps handling and disposal practices in both laboratory and animal research environments.

20:29 What common personal protective devices should be required by the IACUC for personnel who enter a room housing macaque monkeys?

Reg. As a result of exposures with a fatal outcome to Macacine herpesvirus 1 (B-virus), the CDC and the National Institute of Occupational Safety and Health (NIOSH) have issued specific guidance documents on personal protective equipment.[24,25] The *Guide* (pp. 21, 22) specifically states that personnel exposed to nonhuman primates should be provided with such protective items such as gloves, arm protectors, masks, face shields and goggles. According to the *BMBL*,[6] macaques should be handled under strict barrier precaution protocols. Gloves, masks, and protective gowns are recommended for personnel working with non-human primates, especially macaques. Protective face shields minimize the risk of droplet splashes to mucous membranes. The use and type of personal protective equipment should be determined by on-site risk assessment and the work being performed.

Opin. The selection of appropriate personal protective equipment (PPE) should be based on the following process:

- Identification of the most likely hazards to be encountered
- Assessment of the risk and any adverse effects caused by unprotected exposure
- Identification of all other control measures available and feasible to be used instead of personal protective equipment
- Performance characteristics for the required protection (e.g., splash or impact protection)
- Need for decontamination (e.g., reuse or disposal)
- Assessment of any constraints that might negatively influence the use of PPE (e.g., vision, dexterity)

In general, PPE, such as gowns, aprons, gloves, and eye or face protection, is considered the last line of defense. It should only be relied on when the specific hazard cannot be removed or contained with any other control measures. The primary hazard posed by macaques is a potential herpes-B virus exposure through infectious body fluids, bites, or scratches caused by the animal or contaminated objects. The PPE selected should provide protection against the physical hazards as well as fluid exposure. Primary routes of entry are the mucous membranes, which require appropriate splash protection in the form of splash goggles and masks (for the nose and mouth). The use of face shields provides additional face protection. Other routes of entry include parenteral exposure through cuts, breaks in the skin, and needle sticks. PPE selection should address these risks and be based on the task performed. Heavy-duty reinforced leather gloves are appropriate for restraining animals, but inappropriate for handling syringes. In the latter case, double gloving is recommended, since it facilitates the safe changing of gloves (due to contamination) without compromising skin protection. Depending on the task performed, and in accordance with local and state regulations, PPE may have to be disposed of as biohazardous waste. It is highly recommended to use CDC[24] and NIOSH[25] guidelines as the basis for a comprehensive B-virus protection program.

20:30 What role does the AV have in evaluating the impact of biological hazards in the animal facility?

Opin. Biological hazards, in general, or specific infectious agents can have a significant effect on the health of animals. Disease prevention needs to be an essential component of a program of comprehensive veterinary care and biosecurity as identified in the *Guide* (p. 109). The program of veterinary care is the responsibility of the AV, who is certified or has training or experience in laboratory animal science and medicine involving the species being used. It is in the institution's best interest to support the AV in all aspects of a disease prevention program. There is a special need for close collaboration among the IACUC, the AV, and the BSO when a research project involves the use of naturally occurring pathogens, experimentally induced infectious disease, or wild caught domestic and exotic species. The AV and the BSO can assess the potential impact and highlight any concerns for the IACUC.

20:31 Is it necessary or advisable for an IACUC to request documentation that cell lines, tumors, or non-sterile biologic fluids are free of adventitious agents before a protocol receives final IACUC approval?

Opin. The *Guide* (p. 113) states that "transplantable tumors, hybridomas, cell lines, blood products, and other biologic materials can be sources of both murine and human viruses that can contaminate rodents or pose risks to laboratory personnel," and that appropriate testing should be considered to detect these agents in biological materials. Introduction of adventitious agents into an animal colony can alter research results, cause disease and death of animals, and waste animal life and time. Some agents have zoonotic potential under certain conditions. Therefore, it is strongly advised that the IACUC request documentation to assure such agents are not introduced into the animal facility via cell lines, tumors or other biological materials. Responsibility for oversight of this effort is frequently assigned to the AV as part of the veterinary care program.

20:32 What is a Select Agent?

Reg. Biological agents and toxins that are considered by the Department of Health and Human Services (HHS), or the USDA, as having the potential to pose substantial harm or a severe threat to human, animal, or plant health or plant products are regulated as Select Agents. Under the HHS[19] and USDA[20] regulations, anyone who wishes to transfer or possess any quantity of a Select Agent pathogen must successfully complete an intensive registration and approval process through the appropriate regulatory agency. This process includes a security clearance component for all who will have access to the Select Agent. Possession of Select Agent–listed toxins does not require this intensive registration process with the federal agencies if specified quantities of the Select Agent or toxin are not exceeded by the individual PI.

Opin. All researchers (and research-related committees such as the IACUC) should be aware of which agents and toxins are currently regulated as Select Agents. Current lists, including specific exemptions, are available through the applicable federal agency web sites. Institutions should be vigilant in implementing policies to assure that Select Agents are not unknowingly introduced to, and possessed by, the institution. The BSO and Chemical Hygiene Officer can contribute significantly in this regard.

20:33 How does the use of a Select Agent in an animal study impact the IACUC review process?

Opin. The review process should not change for protocols involving animals. However, this process should include direct communication with the PI and the Responsible Official for the Select Agents. Unique access restrictions and emergency response requirements for Select Agent work can be best clarified by these individuals. The Responsible Official is the individual who is charged by the institution with the authority and responsibility of ensuring compliance with HHS and/or USDA Select Agent regulations. In some instances, this individual will be the BSO. In all instances, the BSO will be affiliated with Select Agent regulatory compliance activities. Animal studies involving the use of a Select Agent should also be reviewed by the IBC.

Radiological Hazards

20:34 What is meant by a radiological hazard?

Reg. According to Title 10, Code of Federal Regulations, Part 20.1003 the definition is as follows: "Radiation (ionizing radiation) means alpha particles, beta particles, gamma rays, x-rays, neutrons, high-speed electrons, high-speed protons, and other particles capable of producing ions."[31] Radiation may be machine produced (x-ray machines, accelerators), by-product materials (e.g., radioisotopes produced by a reactor) or naturally occurring atoms that emit radiation, such as uranium, radium, or other naturally radioactive atoms.

Opin. The level of hazard is determined by the amount of radiation, type of radiation, chemical form, method of use (procedures and protocols), and other factors. Radiological hazards must be assessed carefully by a qualified radiation protection professional. Many radiation safety issues are more public relations and regulatory compliance than real risk problems.

20:35 What are typical sources of radiological hazards in the research animal environment?

Opin. Typical sources of ionizing radiation in the animal research environment are radioactive materials administered as part of a research protocol or machine-produced radiation, all of which can be used for treatment, as diagnostic, or research tools. Some commonly used radionuclides include H-3, C-14, In-111, Cr-51, I-131, and Tc-99_m. These radioisotopes may be administered to animals as radiolabeled antibiotics, chemical toxins, and blood flow tracers for trauma and injury studies. X- or neutron radiation may be administered to animals to study brain or other physiological characteristics, to treat cancers, or to pursue other areas of research or treatment. Sophisticated imaging technologies now in common use, such as positron emission tomography and computed tomography, present unique radiation hazards. The use of such radiologic technology tools in animal studies can present exposure hazards for personnel handling these animals during and after exposure. Careful review and oversight by the radiation safety staff are needed.

20:36 Is external beam radiation, such as X-irradiation, considered a radiation hazard of concern to an IACUC?

Reg. External beam radiation is regulated as a radiation hazard and therefore should be recognized as such by the IACUC. Sources of external radiation beams could include radioactive material (e.g., irradiators), X-ray devices, or accelerators. Radioactive material must be obtained, stored, used, and processed for disposal according to the constraints of both the applicable regulations and the site's radioactive material license, but the regulatory agency, and hence the applicable regulations, varies by location. Most states are agreement states, meaning those states have an agreement with the Nuclear Regulatory Commission (NRC) to set forth and enforce their own state regulations for radiation protection and issue state radioactive material licenses. Radioactive material users in non-agreement states, nuclear reactor facilities, and all federal facilities in any state (e.g., military bases,

Veterans Administration medical centers, etc.) are always regulated directly by the NRC, except for Department of Energy (DOE) sites, which have their own regulations. X-ray devices and accelerators are regulated by the states. The regulatory controls imposed by this regulatory framework are generally commensurate with the associated hazard, and the IACUC must coordinate with the site's Radiation Safety Officer to ensure that the applicable regulatory requirements have been addressed prior to approving such radiation use.

Opin. External beam radiation effects should be considered a radiation hazard. Therapeutic or research uses involving high doses of X-radiation that potentially can cause somatic effects should be considered similar to other procedures that cause pain and suffering for animals. The IACUC should communicate with the radiation safety department to assure that protocols involving X-irradiation, as well other forms of radiologic technologies are properly planned, documented and monitored.

The use of external beam radiation in the animal research environment is normally limited to therapeutic purposes (diagnostic and treatment). Nevertheless, it should be included in the overall animal care program oversight of the IACUC, as it can be a hazard. If a research protocol requires the use of X-radiation, that use requires IACUC review and approval. Therapeutic or research uses involving high doses of X-radiation that potentially can cause somatic effects should be considered similar to other procedures causing pain and suffering for animals.

All human health aspects related to external beam radiation must be addressed by institutional safety professionals in the radiation safety and occupational health areas. This area is usually beyond the IACUC's expertise and scope.

20:37 How can an IACUC evaluate radiological hazard risk?

Reg. Radiological risk potential is reflected in the rigor of the applicable federal, state, and local regulations for radiation. In general the regulations require that each use of radiation must be authorized by the Radiation Safety Officer and, for larger programs, the Radiation Safety Committee, providing a built in mechanism for knowledgeable review of the radiological hazards involved in any proposals that come before the IACUC.

Opin. The IACUC should evaluate only the effects of animal experiments involving radiation. The radiation safety officer and the institutional radiation safety committee must assess radiation risk, determine precautions and management practices, and document the radiation protection evaluation. They also must inspect and assure safety and compliance with the regulations. In certain cases, precautions must be taken by animal care staff. The instructions for precautions are often posted on the door of the room or on the cages of the animals for which the precautions are necessary. Animal care staff who will handle or assist with the management of animals that have been administered radioactive materials, or who will handle or manage the radioactive waste or bedding, must be trained in radiation safety at their institution.

20:38 What kind of documentation relative to an investigator's approved radioisotope use may be provided to the IACUC from a radiation safety office?

Reg. Regulations require the radiation safety office to maintain sufficient records to document compliance with the license conditions and applicable regulations.

This will include records of the program content and implementation, audits and radiation surveys, individual radiation monitoring results, radioactive material inventory, dose to members of the public, waste records, etc. The authorization documentation (i.e., which animal users are authorized to use radioactive material or radiation emitting devices) will generally be of most interest to the IACUC. Coordination with the Radiation Safety Officer is needed to ensure that all required documentation of radiation use associated with animals is maintained.

Opin. The safe use of radiation and radioactive materials entails a very comprehensive and strict program of safety and compliance performed by qualified radiation safety professionals. IACUCs should initiate and maintain a close and friendly working relationship with the radiation safety managers at their institutions and defer to their findings. The idea of combining radiation and animal use authorization documents may appear to offer an appealing opportunity to reduce duplication of effort, but generally the requirements and business rules of the IACUC and radiation safety programs are too different to make this approach practicable.

References

1. National Research Council. 1997. *Occupational health and safety in the care and use of research animals.* Washington, D.C.: National Academy Press.
2. National Institutes of Health, Office of Laboratory Animal Welfare. Frequently asked questions G.2, What is required for an occupational health and safety program? http://grants2.nih.gov/grants/olaw/faqs.htm#instresp_2.
3. Office of the Federal Register, Code of Federal Regulations, Title 29, Part 1910.95. Occupational Noise Exposure, Washington, D.C.
4. Office of the Federal Register, Code of Federal Regulations, Title 29, Part 1910.134, Respiratory Protection. Washington, D.C.
5. Office of the Federal Register, Code of Federal Regulations, Title 29, Part 1910.1030, Bloodborne Pathogens. Washington, D.C.
6. Centers for Disease Control and Prevention and National Institutes of Health. 2009. Biosafety in Microbiological and Biomedical Laboratories, 5th ed. Washington, D.C.: Government Printing Office.
7. National Institute of Occupational Health Alert. 1998. Preventing Asthma in Animal Handlers, U.S. Department of Health and Human Services, Publication No. 97–116.
8. Institute for Laboratory Animal Research. 2000. Strategies That Influence Cost Containment in Animal Research Facilities, p. 28. Washington, D.C.: National Academy Press.
9. American Association for Laboratory Animal Science. 2009. Cost of Caring—Human emotions in the care of laboratory animals. www.aalas.org/pdf/06-00006.pdf
10. Office of the Federal Register, Code of Federal Regulations, Title 29, Part 1910.97. Nonionizing Radiation, Washington, D.C.
11. Office of the Federal Register, Code of Federal Regulations, Title 29, Part 1910.1096, Ionizing Radiation. Washington, D.C.
12. Office of the Federal Register, Code of Federal Regulations, Title 29, Part 1910 Subpart S, Electrical. Washington, D.C.
13. Office of the Federal Register, Code of Federal Regulations, Title 29, Part 1910.301–1910.399, Electrical. Washington, D.C.
14. Office of the Federal Register, Code of Federal Regulations, Title 29, Part 1910.1200, Hazard Communication, Washington, D.C.

15. American Veterinary Medical Association. 2013. *AVMA Guidelines for the euthanasia of animals: 2013 edition.* https://www.avma.org/KB/Policies/Documents/euthanasia.pdf.
16. U.S. Department of Agriculture, Animal Plant and Health Inspection Service. Animal Care Policy Manual. 2011. Policy 3, Veterinary Care. http://www.aphis.usda.gov/animal_welfare/policy.php?policy=3.
17. Office of the Federal Register, Code of Federal Regulations, Title 29, Part 1910.1450, Occupational Exposure to Hazardous Chemicals in Laboratories. Washington, D.C.
18. National Institutes of Health. 2012. Notice pertinent to the March 2013 revisions of the *NIH guidelines for research involving recombinant or synthetic nucleic acid molecules (NIH guidelines).* http://oba.od.nih.gov/oba/rac/Guidelines/NIH_Guidelines_prnnew.pdf.
19. Office of the Federal Register, Code of Federal Regulations, Title 42, Parts 72 & 73, Possession, Use, and Transfer of Select Agents and Toxins. Washington, D.C.
20. Office of the Federal Register, Code of Federal Regulations, Title 7, Part 331, and Title 9, Part 121, Agricultural Bioterrorism Protection Act of 2002, Possession, Use and Transfer of Biological Agents and Toxins, Final Rule, Washington, D.C.
21. Office of the Federal Register, Code of Federal Regulations, Title 9, Animal and Animal Products, Parts 92–95, 122, Washington, D.C.
22. Office of the Federal Register, Code of Federal Regulations, Title 49, Parts 171–180, Hazardous Materials Regulations. Washington, D.C.
23. American Biological Safety Association, Mundelein, IL 60060. http://www.absa.org Accessed Feb. 5, 2013.
24. Centers for Disease Control and Prevention. 1987. Guidelines for the prevention of herpes virus simiae (B-virus) infection in monkey handlers. *MMWR* 36:680–682, 687–689.
25. National Institute for Occupational Safety and Health. April 1998. Health Hazard Evaluation Report: 98-0061-2687, Yerkes Primate Research Center, Lawrenceville, GA.
26. Holmesd, G.P., L.E. Chapman, J.A. Stewart et al. 1995. Guidelines for the prevention and treatment of B-virus infections in exposed persons. The B virus working group. *Clin. Infect. Dis.* 20:412–439.
27. Institute for Laboratory Animal Research. 2003. Occupational Health and Safety in the Care and Use of Nonhuman Primates. Washington, D.C.: National Academy Press.
28. Office of the Federal Register, 29 U.S.C 654, Section 5, Occupational Health and Safety Act of 1970, Washington, D.C.
29. Miller, C.D., J.R. Songer, and J.F. Sullivan. 1987. A twenty-five year review of laboratory acquired human infections at the National Animal Disease Center, *AIHA J.*, 48:271–275.
30. Biosafety in the Laboratory: Prudent Practices for Handling and Disposal of Infectious Materials. 1989. Washington, D.C.: National Academy Press.
31. Office of the Federal Register, Code of Federal Regulations, Title 10, Part 20.1003, Standards for Protection against Radiation, Washington, D.C.

21

Personnel Training

Howard G. Rush and Melissa C. Dyson

Introduction

The AWAR, the PHS Policy, and the *Guide* require institutions to provide training for personnel engaged in animal research. Since the second edition of this book, additional information on the recommended content of training courses has become available in the eighth edition of the *Guide*. Many symposia and seminars have been sponsored by national professional organizations to address these issues and a great deal of resource material has become available. Numerous methods have arisen to provide training, which range from one-on-one sessions to online tutorials. In response to the demand for increased training, many institutions have created specific positions for trainers as part of the IACUC staff. IACUCs increasingly utilize training and retraining as a means to achieve compliance, especially in response to episodes of noncompliance encountered with research personnel.

A survey was conducted by the editors and authors of this book. IACUC chairs from 297 institutions responded to questions on personnel training. In this chapter, the survey results are expressed as the number of responses per number of respondents and percent of respondents. Thus, if more than one answer was provided, the sum of all responses will be greater than the number of respondents. If no answer was provided by some respondents, the number of respondents will be less than 297.

21:1 Is there any requirement in either the AWAR or the PHS Policy for general training in laboratory animal care and use?

Reg. The AWAR require that specific topics be included in institutional training programs. The *Guide* provides recommendations on the content of personnel training programs. The PHS Policy (IV,A,1) does not specify training content, but it does require that assured institutions use the *Guide* as a basis for their animal care and use program.

The relevant sections of the AWAR are §2.32,c,1–5 as shown below.

c. Training and instruction of personnel must include guidance in at least the following areas:
   1. Humane methods of animal maintenance and experimentation, including:
      i. The basic needs of each species of animal;
      ii. Proper handling and care for the various species of animals used by the facility;

iii. Proper pre-procedural and post-procedural care of animals; and
iv. Aseptic surgical methods and procedures;
2. The concept, availability, and use of research or testing methods that limit the use of animals or minimize animal distress;
3. Proper use of anesthetics, analgesics, and tranquilizers for any species of animals used by the facility;
4. Methods whereby deficiencies in animal care and treatment are reported, including deficiencies in animal care and treatment reported by any employee of the facility. No facility employee, committee member, or laboratory personnel shall be discriminated against or be subject to any reprisal for reporting violations of any regulation or standards under the Act;
5. Utilization of services (e.g., National Agricultural Library, National Library of Medicine) available to provide information:
   i. On appropriate methods of animal care and use;
   ii. On alternatives to the use of live animals in research;
   iii. That could prevent unintended and unnecessary duplication of research involving animals; and
   iv. Regarding the intent and requirements of the Act.

In addition to the training requirements in §2.32,c,1–5, the AWAR (§2.38,I,1–3) set requirements for development and implementation of institutional contingency (emergency) plans. A component of this requires employee training on such plans. Specifically, "the facility must provide and document participation in and successful completion of training for its personnel regarding their roles and responsibilities as outlined in the plan."

The PHS Policy does not specify training program content to the same degree as the AWAR; nevertheless, PHS Policy (IV,C,1,f) requires the IACUC, in its review of protocols, to determine that personnel conducting procedures are appropriately qualified and trained. In addition, the PHS Policy (IV,A,1,g) requires the institution to include in its NIH/OLAW Assurance a "synopsis of [the] training or instruction in the humane practice of animal care and use, as well as training or instruction in research or testing methods that minimize the number of animals required to obtain valid results and minimize animal distress, offered to scientists, animal technicians, and other personnel involved in animal care, treatment, or use."

The *Guide* (pp. 15–17) specifies that "all research groups should receive training in animal care and use legislation, IACUC function, ethics of animal use and the concepts of the Three Rs, methods for reporting concerns about animal use, occupational health and safety issues pertaining to animal use, animal handling, aseptic surgical technique, anesthesia and analgesia, euthanasia, and other subjects, as required by statute." In contrast, the *Guide* does not specify the content of training programs for animal care personnel. Instead, it states only that "[animal care] staff should receive training and/or have the experience to complete the tasks for which they are responsible."

Opin. The U.S. Government Principles and the AWA emphasize that animals should not be subjected to unnecessary pain or distress. Animal care and research personnel who are well trained can perform animal research techniques with greater skill and fewer adverse outcomes for the animals used, leading to less pain and distress. Consequently, it is important for institutions to provide training to animal care and research personnel to ensure humane care for animals used in research.

21:2 Is there any requirement in either the AWAR or the PHS Policy for additional training in the care and use of a particular species?

Reg. The AWAR (§2.32,c,1,i; §2.32,c,1,ii; §2.32,c,3) require training on the basic needs of the relevant species, proper handling and care for various species used by the facility, and the proper use of anesthetics, analgesics, and tranquilizers for any animals used by the facility.

The PHS Policy (IV,A,1) does not specify any type of species-specific training, but it does require that Assured institutions use the *Guide* as a basis for their animal care and use program.

The *Guide* (p. 16) states that research personnel should have the necessary knowledge and expertise for the specific animal procedures proposed and the species used.

Opin. There is a vast diversity of species used in biomedical research. Although it is estimated that rats and mice account for more than 90% of the animals used in research, numerous other species have been utilized and will continue to be utilized, depending on the suitability of particular models for the research being conducted. It is impossible for any one person to be familiar with all species that are or might be used in research. Therefore, institutions must provide animal care and research personnel with access to training when experiments are planned using new species.

21:3 Do the AWAR or the PHS Policy state whether or not training is required for specific research procedures (e.g., performing an arterial cut-down)?

Reg. Procedure-specific training specified in the AWAR (§2.32,c,1,iii; §2.32,c,1,iv) includes training on pre- and post-procedural care of animals and aseptic surgical methods and procedures. The PHS Policy does not require procedure-specific training, and, as noted previously, the *Guide* (p. 16) indicates that procedure-specific training may be necessary to ensure that research personnel who perform experimental manipulations on animals can conduct their research humanely.

Opin. Innumerable animal research techniques are described in the scientific literature for use in the many disciplines in biomedical research. It is impossible for any one person to be skilled in all techniques commonly used in animal research. Training personnel should be prepared to teach common animal research techniques to individuals who conduct animal research. In addition, they should identify individuals at their institution who work in research laboratories and have unique skills in performing specific research procedures with animals. These individuals should be cultivated as ancillary training staff that can be called upon to help train personnel when training requests for their particular skills are received. Engaging research personnel to participate in the training effort can be a fruitful and rewarding means to expand the training provided at an institution while conserving fiscal resources.

21:4 Who is responsible for assuring that research and animal care personnel working with animals are adequately trained?

Reg. According to the AWAR (§2.32,a), the research facility is responsible for ensuring that research personnel are qualified to perform their duties. To accomplish this, the research facility must make training available to personnel engaged in animal care and use. The HREA (§495,c,1,b) and the PHS Policy (IV,A,1,g) require the awardee institution to assure that all personnel involved in animal care, treatment, or use have training available to them. Both the AWAR (§2.31,d,1,viii) and the PHS Policy (IV,C,1,f) specify that the IACUC, in its review of protocols, should determine that personnel are appropriately qualified and trained. The *Guide* (p. 15) states that the institution is responsible for providing the resources to support training of animal care and use personnel and the IACUC is responsible for oversight of training and evaluation of the training program.

Opin. The responsibility for assuring that personnel are adequately trained must be shared by the institutional officers, the IACUC, the animal care and veterinary staff, and the PI. Overall, it is an institutional responsibility to ensure that animal care and research personnel are adequately trained to care for research animals and conduct animal research procedures. The institution must be willing to commit the personnel and financial resources to accomplish this objective. Nevertheless, in practical terms, the responsibility for assuring that training is available lies with the IACUC, as oversight of the animal care and use program is an IACUC responsibility. The IACUC is able to identify the personnel who should be trained and coordinate the resources and activities necessary to provide the training. The responsibility for actually providing the training usually rests upon the veterinary staff or the staff that supports the IACUC. These are the individuals who possess the scientific, clinical, and technical skills that the animal care and research staff need to perform their duties. Certainly, some specialized training may have to be provided by research personnel. Finally, the PI has a responsibility to convey to her or his staff the importance of receiving proper training in order to conduct humanely the animal studies in which they will be participating. The PI's attitude toward training sets the tone for laboratory personnel with regard to their participation in the institution's training program.

21:5 Should personnel be adequately trained before a research project begins, or can they be trained during the course of a project?

Opin. The AWAR and PHS Policy do not provide guidance on when or how training is provided. However, the *Guide* (p. 17) does state that all animal users should have adequate training before beginning animal work. Some types of training might be so specialized that it can only be provided by other members of the research staff in a mentor–trainee setting. It would be wise to provide training in a stepwise fashion, starting with the most basic techniques and progressing to the most complex and specialized.

Surv. At your institution, must personnel be adequately trained before a research project begins, or can they be trained during the course of a project via observations, assisting a fully trained person, etc.?

- In most instances, the person must be fully trained before the project begins — 98/294 (33%)
- In most instances, the person must have the basic skills needed before the project begins, but more detailed skills can be gained during the course of the project — 147/294 (50%)
- In most instances, all hands-on training can be gained during the course of the project — 44/294 (15%)
- I don't know — 0/294 (0%)
- Other — 5/294 (2%)

Most institutions represented by survey respondents require that personnel be given either full or basic training before working with animals.

21:6 Is training in animal care and use mandatory for personnel who will be working with animals?

Opin. Training for animal care and research personnel is certainly desirable in order to ensure the highest level of humane care. However, from a regulatory standpoint, the AWAR does not actually stipulate that training be mandatory for personnel working with animals. The AWAR (§2.32,a) state that it is the responsibility of the research facility to ensure that personnel are qualified to perform their duties and that "this responsibility shall be fulfilled in part through the provision of training and instruction to those personnel." The *Guide* (p. 15) emphasizes that all personnel must be adequately educated, trained, and/or qualified in basic principles of laboratory animal science. This implies that training, *per se*, is not necessarily mandatory, but that personnel have a particular knowledge base if they are involved in animal care and use. For example, an institution could choose not to require training for a new employee who had been trained at another institution or had acquired the necessary skills to perform his duties through experience.

21:7 Is it practical for the IACUC to rigidly enforce attendance for mandatory training courses?

Opin. Enforcement of attendance at mandatory training sessions must be viewed in the context of the institutional culture. If the IACUC is touted as a regulatory enforcement body, then noncompliance will likely not be tolerated. On the other hand, if the IACUC is the facilitator of research, then investigator cooperation will likely be high and compliance will not be a problem.

21:8 What actions can an IACUC take if members of the research team do not complete their training?

Opin. Enforcement of institutional policies that require training is essential to maintaining a compliant animal care program. Methods to ensure compliance with training policies must be achievable and should limit the access of users to animals. Common methods of ensuring completion of required animal use training include limiting protocol approval until training completion or restriction from animal research until training is completed.

Surv. What are the consequences of failure to complete required training by the Principal Investigator or members of the research team? Check as many boxes as appropriate.

- Not applicable as we have no formal training program — 15/295 (5%)
- There are no consequences — 4/295 (1%)
- Protocol or amendment approval is contingent upon completion of training — 191/295 (65%)
- Employment is contingent upon completion of training — 11/295 (4%)
- We report non-compliance with training requirements to the IACUC — 90/295 (31%)
- Personnel are restricted from performing animal research procedures until training is completed — 210/295 (71%)
- I don't know — 1/295 (<1%)
- Other — 7/295 (2%)

21:9 Should an IACUC approve a protocol with novel procedures based on the surgical or other procedural experience of the investigator or other personnel involved?

Reg. Although the PHS Policy and the AWAR are mute on this matter, the *Guide* (p. 28) provides some insight into this issue. If novel procedures are encountered in protocols under review, the *Guide* suggests that the IACUC consider the use of pilot studies to identify and define humane endpoints.

Opin. In addition to pilot studies, the IACUC can require that novel procedures be observed by a veterinarian or other qualified staff member to ensure that they are performed in a humane and painless manner.

21:10 Person A wants to learn a new technique for Protocol A but will use animals on Protocol B under the supervision of a person from Protocol B. Should Person A be listed only on Protocol A or on both protocols?

Opin. The IACUC should consider the degree of involvement and the invasiveness of the procedure in determining whether to require Person A to be listed on Protocol B. For example, if someone only needs to observe a minimally invasive technique used in another laboratory, the IACUC would be justified in not requiring Person A to be listed on Protocol B. On the other hand, if someone must learn a complicated surgical procedure that would take extensive training by a skilled mentor, inclusion on the protocol should be required. Additionally, if training is regularly provided under Protocol B, the training should also be incorporated into the protocol and approved by the IACUC.

21:11 Would principal investigators who do not themselves perform procedures on animals have to receive animal care and use training?

Opin. Investigators who will not be conducting procedures on animals themselves should still be required to take basic introductory training in order to become familiar with federal, state, and institutional policies and procedures. As supervisors of research personnel who will be performing animal research procedures, they have a responsibility to convey the importance of utilizing the most humane methods in animal research and of regulatory compliance.

Surv. At your institution, do Principal Investigators, who will not be performing any hands-on animal work, have to partake in the institutional animal care and use training program?

• We have no training program	5/291 (2%)
• Yes, but only basic training in animal care and use. More detailed training is not required.	103/291 (35%)
• Yes, they are required to complete the same training as staff who perform animal research procedures	108/291 (37%)
• No, no additional training of any kind is required	64/291 (22%)
• I don't know	1/291 (<1%)
• Other	10/291 (3%)

21:12 If a person has successfully completed an animal care and use training course at a different institution, should the IACUC at a new institution require that she retake a similar course at that institution or otherwise demonstrate his or her capabilities?

Reg. See 21:1–21:3.

Opin. As the IACUC examines the credentials of animal use personnel in the course of conducting its review of proposed activities, any training that has occurred at another institution should be taken into account. In order to do this, the IACUC may request the individual to provide detailed information on the nature of training at the previous institution. The amount of "credit" awarded by the IACUC will likely vary with the institution's policies and the content of the training at the previous institution. Regardless of the nature of the training at the previous institution, the individual should take any orientation training offered in order to acquire institution-specific information that would not be acquired in any other way. (See 21:13.)

21:13 Can training be waived if a person has received training at another institution?

Opin. In the course of conducting its review of proposed activities, the IACUC examines the credentials of animal use personnel. If an individual has received training at another institution, it is appropriate for the IACUC to take this training into account. In order to do this, the IACUC may request the individual to provide detailed information on the nature of training at the previous institution. The amount of "credit" awarded by the IACUC will likely vary with the institution's policies and the content of the training at the previous institution. Regardless of the nature of the training at the previous institution, the individual should take any introductory training that is specific to the institution and is policies and practices. (See 21:12.)

Surv. Does your IACUC waive training requirements for persons having documented training at another institution?

- Not applicable. We have no training program. 6/295 (2%)
- It varies with circumstances but we can waive some training requirements 194/295 (66%)
- We never waive training requirements 87/295 (29%)
- I don't know 6/295 (2%)
- Other 2/295 (>1%)

21:14 Can someone with experience in surgery or anesthesia in human patients perform these procedures on animals without demonstrating their competence to do so?

Reg. The *Guide* (p. 16) and the PHS Policy (IV,C,1,f) state that research personnel should have the necessary knowledge and expertise for the specific animal procedures proposed and the species used. In addition, the *Guide* (p. 115) indicates that researchers trained in human surgery may need additional training before performing surgery on animals. The AWAR require the IACUC to ensure the PI is qualified for what s/he is doing (see 21:1,21:2). Experience with humans does not qualify a person with regard to animals. Also, the AWAR (§2.32,c,3) require training specifically for the proper use of anesthetics for the species of animals being used. Although similarities exist and can make training easier, it is dangerous to assume that techniques for humans will be the same for animals.

Surv. Does your IACUC allow persons experienced with human patients (such as physicians who are anesthesiologists or surgeons) to perform anesthesia or surgery on laboratory animals without providing evidence of having acquired similar skills with laboratory animals?

- Not applicable 96/293 (33%)
- Yes, we usually do 6/293 (2%)
- Yes, we might do so, but we evaluate each request on a case-by-case basis 66/293 (23%)
- No, we do not 119/293 (41%)
- I don't know 4/293 (1%)
- Other 2/293 (>1%)

21:15 Should all personnel performing animal care and use procedures receive the same training?

Surv. Does your IACUC require that short-term personnel (e.g., summer students, visiting professors, etc.) fulfill the same training requirements as full time investigators/educators and their staff?

- Not applicable 20/294 (7%)
- Yes, short-term employees have to fulfill the same training requirements 232/294 (79%)
- No, short-term employees have somewhat modified training requirements 35/294 (12%)
- No, short-term employees have no training requirements 2/294 (<1%)
- I don't know 3/294 (1%)
- Other 2/294 (<1%)

21:16 If a person from another institution or a commercial organization is to demonstrate how to perform a procedure on a live animal, how can the IACUC assure that a person is qualified to perform the procedure?

Opin. This circumstance is most often encountered at large academic medical centers with well-developed continuing medical education programs. However, visiting experts or trainers may be asked to demonstrate procedures on animals in many types and sizes of institutions. Such programs are important for providing training to faculty and local practitioners on new techniques and equipment. When institutions encounter such requests, it is important to develop mechanisms to manage them effectively rather than forbid them outright so as not to inhibit education and training.

Surv. At your institution, if a representative of a commercial company is to demonstrate the proper use of a new surgical product (e.g., for intestinal anastomosis) on a live animal, how does your IACUC assure that the representative is properly trained in the use of the product on the species to be used? Check all that apply

- Not applicable 90/293 (31%)
- We accept a written statement from the Principal Investigator that the person is qualified 69/293 (24%)
- We review the written qualifications of the person 129/293 (44%)
- We have a veterinarian watch the person perform the procedure 42/293 (14%)
- We have a veterinarian watch the person perform the procedure and the veterinarian has the authority to stop the procedure 84/293 (29%)
- The person must first demonstrate the procedure to our IACUC representative before training others 11/293 (4%)
- I don't know 21/293 (7%)
- Other 8/293 (3%)

21:17 Should students working with live animals during a class be required to participate in an animal use training program?

Opin. Students working in research laboratories should receive the same training as all other research personnel. Students who utilize animals in classroom settings could receive abbreviated training specific to the procedures being conducted. This may be provided by the IACUC staff or by the course instructor.

Surv. If your institution has courses in which students use live animals for any reason or length of time, are they required to participate in an animal use training program? More than one response is possible.

- Not applicable 145/293 (49%)
- Students in courses do not receive training 16/293 (5%)
- Course instructors provide training to students in the course 99/293 (34%)
- Students in courses receive the same or similar training as that provided to research personnel 57/293 (19%)
- Students in courses receive their training from the same individuals who train research personnel 24/293 (8%)
- Students in a single training lab get less (or no) training compared to students who will be using animals throughout the course 7/293 (2%)
- Other 5/293 (2%)

21:18 Can a high school student perform animal research if he or she works under the guidance of an investigator with an IACUC-approved protocol?

Opin. It has become increasingly common for high school students to work as volunteers, students, or employees in animal research settings at colleges and universities. There is no inherent reason that a high school student cannot perform research procedures using animals if he or she is under the guidance of an investigator with approval to conduct the research. High school students should receive whatever training is necessary for them to perform their duties. Their training should be the same as that provided to college-level students and regular employees of the institution.

21:19 Is there any lower age limit below which the IACUC should disallow student participation in animal care and research?

Opin. There is no specific reason to place an arbitrary age limit on student participation in animal care and use. However, it is certainly uncommon for students in elementary and middle school to volunteer or seek employment in research settings. IACUCs are advised to treat any such request on a case-by-case basis. Any student, regardless of age, must be adequately supervised by an investigator with approval to conduct the animal studies. Other factors such as labor laws and institutional policies on employment and volunteers potentially may limit the participation of younger students in animal research.

21:20 In terms of animal care and use, who is ultimately responsible for the activities of a student?

Opin. The PI who has approval for animal use bears the responsibility for any and all personnel, including students working in his or her laboratory. In the course of review and approval of her or his protocol, the PI should provide assurance to the IACUC that all personnel working with animals will be adequately trained to perform their duties. The investigator must understand that this responsibility extends to students working in the laboratory. Ultimately the IACUC (*Guide* p. 15) and the research facility (AWAR §2.32,a; §2.33,b,1) are responsible for compliance and ensuring that staff are appropriately qualified to work with animals in research.

21:21 Does the unaffiliated member of an IACUC need additional or supplemental training beyond that which is provided to other committee members?

Reg. According to the *Guide* (p. 17) IACUC members should be provided formal orientation training covering the animal care program, relevant legislation, regulations, guidelines and policies, animal facilities and laboratories, processes or protocols, and program review. Similarly, the AWAR (§2.32,a) states that the institution ensure that all scientists, research technicians, animal technicians, and other personnel involved in animal care, treatment, and use are qualified to perform their duties. This would include unaffiliated members.

Opin. It is advisable to provide the same training to all IACUC members, not just the unaffiliated member(s).

Surv. Does your IACUC provide any additional training to the unaffiliated member of your IACUC that is not provided to other committee members?

- Yes, routinely 59/294 (20%)
- Yes, but only upon request of the unaffiliated member 56/294 (19%)
- No we do not 169/294 (57%)
- I don't know 8/294 (3%)
- Other 2/294 (<1%)

21:22 At what interval is periodic retraining of animal users required?

Surv. If your institution requires repeated training of animal users, how often does this retraining occur?

- Not applicable 91/291 (31%)
- Every other year 33/291 (11%)
- Every third year 65/291 (22%)
- Every fourth year 4/291 (1%)
- Every fifth year 6/291 (2%)
- Less frequent than every fifth year 0/291 (0%)
- It varies with the species or procedures to be performed 65/291 (22%)
- Other 27/291 (9%)

21:23 Should retraining be utilized as remediation for instances of protocol noncompliance?

Opin. Retraining can be an effective tool in addressing gaps in knowledge, understanding, or skills. However, protocol noncompliance can be caused by a combination of many factors including poor or insufficient training, inadequate oversight and staff management or other factors. It is important to assess each noncompliance situation to determine the potential causes and match remediation efforts to the needs of each unique situation.

Surv. Does your IACUC require research personnel to undergo retraining as remediation for instances of protocol noncompliance?

- Not applicable 48/293 (16%)
- Yes, occasionally 120/293 (41%)
- Yes, frequently 104/293 (35%)
- No 12/293 (4%)
- I don't know 5/293 (2%)
- Other 4/293 (1%)

21:24 Which categories of personnel should provide animal care and use training?

Opin. Institutions utilize a variety of personnel to provide training including veterinarians, veterinary technicians, laboratory animal technicians and technologists, and research personnel. No doubt the choice of individuals to fill these roles is affected by the size of the institution, its mission (academic, industrial, etc.), the organization of the IACUC and animal care services, the physical resources, the animal population and species maintained, and the training and skill of personnel (*Guide* p. 15).

Surv. At your institution, who actually trains people in the care and use of laboratory animals? Check as many boxes as appropriate.

- Not applicable. We have no training program 1/294 (<1%)
- An outside private company 18/294 (6%)
- The veterinary staff, including veterinarians and/or veterinary technicians 219/294 (74%)
- The animal care staff, including animal care technicians, supervisors, and/or managers 218/294 (74%)
- The IACUC staff 51/294 (17%)
- Research personnel 203/294 (69%)
- Designated training program personnel 87/294 (30%)
- Other institutional personnel 30/294 (10%)
- Other 24/294 (8%)

In the survey, veterinary personnel (74%), animal care personnel (74%), and research personnel (69%) were most often enlisted to provide training. A significant number of institutions (30%) have designated training program personnel. This is the case at the authors' institution where a Training Core has been established that is responsible for training of all animal care and animal research personnel.[1] The category "Other" included a wide variety of responses such as online resources (e.g., AALAS Learning Library and the Collaborative Institutional Training Initiative [CITI]), PIs, and outside consultants.

21:25 What departments or individuals should be responsible for providing training of personnel that are potentially exposed to hazards as part of their job duties?

Opin. Many individuals from existing groups at an institution, such as the IACUC staff, animal care program staff, and members of the research team, may participate in providing training to personnel exposed to hazards. However, personnel in these areas are not experts or authorities in occupational health issues. As a result, development and guidance of the training program should be the responsibility of the institution's Occupational Safety and Risk Management Office (or its equivalent). However, monitoring and assuring the effectiveness of the program is the IACUC's responsibility.

Surv. Who is responsible for providing training of personnel potentially exposed to hazards? Check as many boxes as appropriate.

- Not applicable 3/295 (1%)
- The IACUC 31/295 (11%)
- The Principal Investigator 173/295 (59%)
- Animal facility personnel 83/295 (28%)
- The Occupational Safety and Risk Management Office (or its equivalent) 244/295 (83%)
- Other 7/295 (2%)

21:26 What subject matter should be included in animal care and use training programs?

Reg. See 21:1.

Surv. Which of the following topics are included in your training and education programs for animal users? Check as many boxes as appropriate.

- Not applicable, we have no training and education program 4/295 (1%)
- Purpose and general functions of the IACUC 278/295 (94%)
- Organizational structure of the institutional animal care and use program 251/295 (85%)
- The politics and ethics of animal experimentation 223/295 (76%)
- The 3 Rs and means of implementing them 258/295 (87%)
- Overview of the occupational health and safety program 242/295 (82%)
- Experimental endpoints and humane endpoints 227/295 (77%)
- Overview of institutional policies and procedures 245/295 (83%)
- Additional information on anesthesia and analgesia (if appropriate to your program) 209/295 (71%)
- Acceptable euthanasia methods 251/295 (85%)
- Record keeping requirements 237/295 (80%)
- How to report concerns about animal care and use 276/295 (94%)
- How and when to request veterinary help 254/295 (86%)
- Basic care and use of common laboratory animals 259/295 (88%)
- Basic surgery information (if applicable to your program) 181/295 (61%)

- Hands-on training in basic research techniques (e.g., animal handling, injections) 215/295 (73%)
- Hands-on training surgical techniques 162/295 (55%)
- I don't know 2/295 (<1%)
- Other 20/295 (<1%)

The survey results indicate several common themes that should be included in institutional training programs, namely institution-specific information on the IACUC, the animal care and use program, and policies; the 3R's and humane endpoints; the occupational health and safety program; experimental procedures (anesthesia and analgesia, euthanasia; methods); the provision of veterinary care; and reporting concerns. Surgical information was provided at less than two thirds of institutions.

21:27 What type of information might the IACUC consider as essential to basic required training?

Opin. Basic training will vary somewhat from one institution to another, but certain common elements will most assuredly be incorporated. These include historical perspectives on animal use, the current animal research climate, animal care and use regulations and policies, institutional policies and procedures, the institutional animal care and use application, the institutional occupational health and safety program, and the procedures for registering concerns about animal use. Some institutions may also include species-specific information on the basics of normal and abnormal behavior, signs of illness, and research techniques. As noted in 21:1, the AWAR (§2.32,c,1–5) define specific training topics that should be included in the training program. Similarly, the *Guide* (pp. 15–17) includes topics that should be included in training of animal care and use personnel.

21:28 What information is useful for training of IACUC members?

Opin. The AWAR do not explicitly identify IACUC members as needing training, but it can be inferred from §2.32,a, which states that all individuals involved in animal care, treatment, and use be qualified to perform their duties. The *Guide* (p. 17) states that IACUC members should be provided formal orientation training covering the animal care program, relevant legislation, regulations, guidelines and policies, animal facilities and laboratories, processes or protocol and program review and be provided opportunities for ongoing education as needed. Training on routine and minor procedures (handling, restraint, injections, etc.) might be valuable for some members who have limited experience with animals, but training on complex procedures such as surgery is probably not warranted. In addition to material provided by the institution, a variety of online resources are available such as the American Association for Laboratory Animal Science Learning Library, IACUC.org, the Collaborative Institutional Training Initiative, and NIH/OLAW. NIH/OLAW offers online seminars on various topics. Attendance at meetings sponsored by national organizations may be helpful. Public Responsibility in Medicine and Research (PRIM&R) is a national organization that promotes ethical standards in research. They sponsor training for professionals in human and animal research protection. IACUC 101 is an educational workshop sponsored by

NIH/OLAW that provides training to IACUC members, administrators, veterinarians, animal care staff, researchers, regulatory personnel, and compliance officers. Scientists Center for Animal Welfare offers IACUC Training Workshops similar to IACUC 101. The IACUC Administrators Association sponsors Best Practice Meetings where topics are determined by the participants and varied institutional practices are discussed.

Surv. When training new IACUC members, what topics are covered? Check all appropriate responses. Check as many boxes as appropriate.

- We have no training program for new IACUC members 15/295 (5%)
- Training is via on-the-job observations only. There is no formal program 67/295 (23%)
- Animal use justification 195/295 (66%)
- What constitutes an appropriate literature search 162/295 (55%)
- Animal number justification 182/295 (62%)
- Means of alleviating pain and distress 185/295 (63%)
- Euthanasia techniques 177/295 (60%)
- Means of conducting semiannual inspections 205/295 (69%)
- General policies of the IACUC 243/295 (82%)
- Training and skill qualification standards for animal users 130/295 (44%)
- Means of reporting and responding to concerns about animal welfare at our institution 205/295 (69%)
- Expectations for housing and general care of animals 174/295 (59%)
- Other 38/295 (13%)

21:29 What strategies can be used to deliver training to animal care and use personnel?

Opin. One of the most effective means of instruction for animal care and use training is the individual or small group hands-on training session, sometimes termed a "wet lab." This setting provides the optimal conditions for student–teacher interaction and is excellent for fostering skill development. This type of training session is ideal for teaching animal research techniques such as handling, restraint, anesthesia, and surgery. On the other hand, the wet lab is not an efficient use of time or personnel when it is necessary to present introductory material such as federal laws and national standards, institutional policies and procedures, the institutional animal care and use program, and institutional occupational health programs. For this type of material, slide presentations, videos, written material, and computer-based training are more appropriate and efficient. This assertion is borne out by the results of the following survey, which demonstrate a range of training strategies at the surveyed institutions.

Surv. At your institution, what approaches are used to provide education and training to animal care and use personnel? Check as many boxes as appropriate.

- Not applicable. We have no training program. 3/294 (1%)
- We use live lectures (with or without slides) 170/294 (58%)
- We use live individual or small group hands-on training 230/294 (78%)

• We subscribe to a commercial training program (e.g., CITI)	133/294 (45%)
• We use video presentations	106/294 (36%)
• We used computer-based training (e.g., autotutorials)	152/294 (52%)
• We use written materials (e.g., handouts, institutional manuals, newsletters)	210/294 (71%)
• I don't know	3/294 (1%)
• Other	6/294 (2%)

21:30 How should the IACUC inform the research community of changes in regulatory guidance, institutional policies, and standards of practice?

Opin. While changes in regulations happen slowly, changes in IACUC or veterinary policies and guidelines may occur frequently. Communication of these changes in policies, practices or expectations of research animal users is essential for animal care program compliance. It can be difficult to communicate changes to an "information saturated" community. Practices employed often include email, newsletters, websites or town hall meetings, protocol renewal dates or post-approval monitoring visits.

Surv. How does your IACUC provide updates to animal care and use personnel on changes in regulatory guidance, institutional policies, and standards of practice? Check as many boxes as appropriate.

• We do not provide any such updates	18/295 (6%)
• We have required retraining or continuing education classes	98/295 (33%)
• We use e-mail, newsletters, website, and town hall meetings	192/295 (65%)
• We provide this information at post approval monitoring visits	77/295 (26%)
• We provide this information when protocol are being renewed	154/295 (52%)
• I don't know	4/295 (1%)
• Other	18/295 (6%)

21:31 How should training be documented and what data should be maintained?

Reg. While the PHS Policy does not require that training records be maintained, the *Guide* (p. 15) states that all program personnel training should be documented. In addition, the AWAR (§3.8,I,3) require documentation of training on the institution's contingency plan.

Opin. It is important to maintain training records to demonstrate compliance with the responsibility of the IACUC to ensure that personnel are trained and qualified. Logically, it is difficult to meet these obligations without a mechanism to track the training and experience of individuals engaged in animal research. The most common methods of documentation are filing the sign-in sheets from training classes and maintaining an electronic database of personnel training. The most common types of records maintained include the personnel identification and organizational affiliation, class dates, class name or type, and the trainer's identity.

Surv. 1 How, and to what extent, does your institution document training and education efforts? Select all that apply.

- We do not document our training and education efforts 14/295 (5%)
- We maintain the sign-in sheets from training classes 174/295 (59%)
- We maintain electronic records from online training 191/295 (65%)
- We maintain an electronic database of personnel training records 161/295 (55%)
- We provide trainees with certificates of training 99/295 (34%)
- Other 15/295 (5%)

Surv. 2 What records does your institution keep relative to training and education of animal use personnel? Select all that apply.

- Not applicable as we do not keep such records 11/294 (4%)
- We maintain the person's name and organizational affiliation 181/294 (62%)
- We maintain the names of classes or training sessions 219/294 (74%)
- We maintain the dates of training sessions 230/294 (78%)
- We maintain the names of the trainers 126/294 (43%)
- We maintain a description of each person's training and experience in the IACUC approved protocol 156/294 (53%)
- Other 8/294 (3%)

21:32 What resources are available for provision and documentation of training for animal users?

Opin. The most common methods for documenting and/or providing training include learning management system software, general database software, protocol management software, and web-based training management programs from an external vendor or organizations.

Surv. Does your institution utilize any of the following to document or provide training to animal users? Check as many boxes as appropriate.

- Learning management system software that is supported by your institution and provides class registration, online web modules, transcripts, and training documentation 74/292 (25%)
- General database software such as Microsoft Excel, Access, etc. 78/292 (27%)
- Protocol management software that can track and document a person's training 44/292 (15%)
- Web-based training management programs from an external vendor or organization (e.g., AALAS learning Library, CITI) 183/292 (63%)
- We do not use any of the above methods 66/292 (23%)

21:33 Should the IACUC be provided with regular reports on training activities?

Opin. The PHS Policy and the AWAR do not specifically require reports on training activities. However, the *Guide* (p. 15) places the responsibility for providing oversight and for evaluating the effectiveness of the training program with the IACUC. Therefore, it would be important to provide training reports to the IACUC at regular intervals, perhaps as part of the semiannual program evaluation. Even if training information is incorporated into the protocol, summary reports to the IACUC could be very useful in evaluating training effort and staffing levels. According to the survey, most institutions provide some type of reports to their IACUC's.

Surv. What documentation of training activities must be provided to the IACUC? Select all that apply. Check as many boxes as appropriate.

- We do not provide reports of training activities to our IACUC — 70/291 (24%)
- We provide the IACUC with a summary report on training activities as part of the semiannual report to the Institutional Official — 86/291 (30%)
- We provide reports on training activities only upon request from the IACUC — 78/291 (27%)
- We update the IACUC at its regular meetings — 78/291 (27%)
- Other — 33/291 (11%)

21:34 What are some effective ways to provide and document continuing education?

Reg. The *Guide* (pp. 16–17) recommends that institutions provide opportunities and support for continuing education for animal care and use personnel.

Opin. Continuing education is a wise investment for institutions because it improves personnel engagement in the institution's success and improves employee morale. Continuing education opportunities can be provided by existing staff (trainers, supervisors, veterinarians, etc.), by institutional human resource and development personnel, by outside trainers, or by off-site training sessions at local, regional, or national meetings.

21:35 What forms of assessment can be used to evaluate personnel that have received training?

Opin. Some form of assessment is necessary to ensure that personnel have sufficient mastery of the procedures that they will be performing such that the welfare of animals being used will not be compromised.

Surv. As part of the training for your animal care and use program, is any form of evaluation, assessment, or testing utilized? Check as many boxes as appropriate.

- Not applicable. We have no training program. — 4/294 (1%)
- No, testing is not required as part of our training program — 47/294 (16%)
- Yes, written examinations — 111/294 (38%)
- Yes, demonstration of competency during training sessions — 170/294 (58%)

- Yes, worksite evaluation of competency by training or IACUC personnel — 98/294 (33%)
- Yes, worksite evaluation of competency by supervisory personnel — 129/294 (44%)
- Other — 15/294 (5%)

21:36 How can training versus qualifications be effectively evaluated by the IACUC?

Reg. The *Guide* (p. 16) and the PHS Policy (IV,C,1,f) state that research personnel should have the necessary knowledge and expertise for the specific animal procedures proposed and the species used. The AWAR (§2.32,a) require that personnel be qualified, and that training can be used, in part, to provide the necessary qualifications.

Opin. In order for the IACUC to evaluate personnel training and qualifications, investigators should provide the IACUC with specific information on their prior experience and training with the species and procedures proposed in their protocol as well as the experience and training of their staff members. The IACUC then should determine, in the course of reviewing the investigator's protocol, whether the research personnel have sufficient training or experience to conduct the proposed procedures. For example, an investigator may be trained as a human surgeon but, without specific training or experience in animal surgery, may not be qualified to perform surgical procedures in animals. Such an individual may need additional training on species-specific anatomy, physiology, behavior, anesthesia, and analgesia. Prior training or experience of the research staff can also be taken into consideration.

When semiannual inspections of animal care and use facilities are performed, the IACUC also can gain some insight into the adequacy of training by interviewing research personnel during these inspections. This approach not only identifies inadequacies in training after protocol approval has taken place, but also can be useful as an audit of the adequacy of the review process in identifying personnel in need of training. (See 21:37.)

21:37 How can the IACUC determine when an individual requires additional training?

Opin. Ideally, additional training needs would be identified during basic or primary training or workplace assessments. However, in practice it can be difficult to predict the competency of personnel in the work place when they leave the training environment. Often inappropriate work practices are noted during laboratory inspections, post-approval monitoring visits and these observations can result in the assignment of additional training requirements. Likewise, additional training requirements may be assigned as remediation for protocol noncompliance or animal use concern reports to the IACUC.

Surv. What criteria are used by your institution to indicate to the IACUC that an individual requires more training? Check as many boxes as appropriate.

- Not applicable — 16/295 (5%)
- Poor performance on peri- or post-training evaluation process (e.g., tests, competency evaluation, worksite assessment) — 149/295 (51%)
- Supervisor or PI recommendations — 193/295 (65%)

- Noncompliance reports to the IACUC 222/295 (75%)
- I don't know 7/295 (2%)
- Other 16/295 (5%)

21:38 What metric(s) can be used to evaluate the effectiveness of a training program?

Reg. The *Guide* (p. 15) places the responsibility for providing oversight and for evaluating the effectiveness of the training program with the IACUC. Further, because the training program is part of the facility's animal care and use program, it must be evaluated semiannually (§AWAR 2.31,c,1; AWAR §2.31,c,3; PHS Policy IV,B,1).

Opin. Evaluation of training should occur at several levels extending from evaluations of the course by attendees, evaluating trainee knowledge and skills during or immediately after classes and evaluating the effects of long term application and change in staff behaviors and skills. Effective long term behavior change is difficult to instill and difficult to evaluate. This requires routine post-approval monitoring and/or quality assurance programs and staff dedicated to perform this oversight. Additionally, appropriate workplace practices are affected by education and training, staff management and oversight, and institutional culture. It is difficult to separate the effects of all of these issues to gain a final conclusion on the effectiveness of training.

Surv. What metric(s) are used to evaluate the effectiveness of your training program? Check as many boxes as appropriate.

- Not applicable 8/293 (3%)
- We do not use any metrics 140/293 (48%)
- Data or reports from our post-approval monitoring assessments 84/293 (29%)
- Data or reports from worksite assessments by trainers, IACUC, animal care and veterinary staff 104/293 (35%)
- Data or reports from supervisors and Principal Investigators' assessments of trainees 68/293 (23%)
- Data from compliance reports to the IACUC 81/293 (28%)
- Other 11/293 (4%)

Reference

1. Dyson, M.C. and H.G. Rush. 2012. Institutional training programs for research personnel conducted by laboratory-animal veterinarians. *J. Vet. Med. Educ.* 39:160–168.

22

Confidential and Proprietary Information

Marilyn J. Chimes and Priya Sankar*

Introduction

The purpose of this chapter is to provide a basic framework for understanding how the concepts of confidential information and proprietary information relate to IACUC activities. In providing institutional review and approval for research activities using animals, IACUCs might review information considered confidential or proprietary by the researchers. Even information that may not seem confidential or proprietary, such as a researcher's name, may necessitate protective treatment under certain circumstances.

IACUCs at government-supported institutions, such as public universities, are subject to state laws for open meetings and Freedom of Information acts for "public records"; however, even among such IACUCs, the application of these access-to-information laws varies. Some IACUCs have open meetings, and, therefore, whatever information is discussed by the committee becomes public information. Others have closed meetings and may not have the automatic release of information into the public domain. Similarly, state laws on access to public records may lead to radically different results, as some jurisdictions require release. Some jurisdictions may require release but may also provide exemptions that preserve confidentiality under certain circumstances. It is important to seek legal advice in determining jurisdictional requirements.

In reading the court decisions cited in the following sections, it is important to keep in mind that a court's decision is based on the set of facts before it. Thus, a different factual situation might lead to a different legal decision. This is one reason why courts within the same jurisdiction may appear to reach contradictory decisions. It also is important to remember that a published decision provides binding precedent only within the court's jurisdiction and only as to the issue addressed. Of course, a broad and well-written decision might provide persuasive authority in other jurisdictions and to other related issues. Thus, for example, a decision by the Vermont Supreme Court, the state's highest court, on the applicability of that state's public records act to an IACUC is binding only within Vermont. The Vermont decision, however, might provide persuasive reasoning that would guide judges in other states when determining the applicability of their state public records laws to an IACUC in their state. Readers are encouraged to check with their institution's legal counsel or state attorney general for specific laws and interpretations in their particular state.

* The authors thank Katheryne Lawrence for researching and updating the laws and case law cited in select sections of this chapter as well as drafting and editing select sections in this chapter. The authors are grateful also to Laure Bachich Ergin and Kathryn A. Donohue for their contributions to this chapter in the second edition of *The IACUC Handbook*.

22:1 What are the sources of law that can apply to an IACUC's activities?

Reg. IACUCs at institutions using any live or dead warm blooded animals (other than birds, rats of the genus *Rattus*, and mice of the genus *Mus*, bred for use in research) for research, testing, teaching, or experimentation purposes must comply with the federal AWA, found at 7 U.S.C. §§2131 through 2159, particularly §2143,b (setting forth the establishment and duties of the IACUC) and §2157 (prohibiting release and wrongful use of confidential information by members of the IACUC). IACUCs at such institutions must also comply with the federal regulations promulgated under the AWA (the AWAR, found at 9 C.F.R. §1.1–4.11), particularly AWAR §2.31 (detailing IACUC functions). The requirements specified in AWAR §2.32 (personnel qualifications), AWAR §2.33 (AV and adequate veterinary care), AWAR §2.35 (recordkeeping requirements), and AWAR §2.36 (annual report) also are relevant to IACUC activities. The AWA and AWAR have been in effect since 1966, with multiple amendments over the years.

IACUCs at federal or state institutions or institutions receiving federal or state funding are additionally subject to the laws, regulations, and policies applicable to the funding source or the institution. For example, 42 U.S.C. §289,d, created by the Health Research Extension Act of 1985 (HREA), sets forth requirements of an IACUC at institutions receiving funds from the PHS. In addition to the AWA and AWAR, institutions receiving PHS funding for research must conform to the PHS Policy, which is applicable to any PHS-conducted or PHS-supported use of a live vertebrate animal (whether warm blooded or cold blooded) for research, research training, experimentation, or biological testing, wherever that activity is performed. PHS Policy II requires every individual researcher receiving PHS support for an activity involving animals to be affiliated with or sponsored by an institution which can and does assume responsibility for compliance with this Policy, unless other arrangements are made with the PHS. U.S. Department of Veterans Affairs facilities must follow VHA Handbook 1200.07, *Use of Animals in Research,*[1] which requires compliance with the PHS Policy even if PHS funds are not received. Institutions receiving funding from certain sources and all institutions accredited by AAALAC must comply with the *Guide*[2] and the Federation of Animal Science Societies' *Guide for the Care and Use of Agricultural Animals in Research and Teaching.*[3] Although AAALAC accreditation is voluntary and not mandated by law, the governing body of an institution may require accreditation or conformance with the standards described in these guides. The *U.S. Government Principles*, for example, which forms the basis for the PHS Policy, refers to the *Guide* for guidance in following its principles.

IACUCs at institutions with facilities in other countries or that review protocols to be performed in other countries must ensure that those protocols comply with all applicable laws of those countries. For example, one such set of standards that may apply is that of the *European Convention for the Protection of Vertebrate Animals used for Experimental and Other Scientific Purposes* (CETS No. 123).[4]

State laws and regulations also may apply to IACUCs at institutions required or allowed by their state to be licensed or inspected by state authorities and may be more stringent than federal laws. For example, state or local laws or ordinances may impose restrictions on obtaining animals from certain sources such as municipal animal control facilities or shelters. Additionally, IACUCs should be cognizant of state anti-cruelty laws which could apply to activities at their institutions, especially in the minority of states that do not exempt research activities from animal cruelty laws. All states have laws prohibiting animal cruelty and many states have significantly modified those statutes in recent years to increase penalties for violations. The AWA (§2145,b) expressly authorizes the USDA to cooperate with state and local authorities in animal welfare matters.

Statutes, regulations, and policies governing other activities of an institution may apply to activities involving animals used for research, teaching, or testing, and thus apply to an IACUC's activities as well. These could include laws of the Environmental Protection Agency (EPA), the Fish and Wildlife Service, or the Food and Drug Administration (FDA), such as the Good Laboratory Practices regulations,[5] among others. Certain protocols, such as studies involving biohazardous agents or radioactive materials, may implicate other federal, state, and local laws, regulations, and policies. An IACUC's activities also may be impacted by regulations of the Centers for Disease Control and Prevention (CDC) and Occupational Safety and Health Administration as they affect the safety and health of persons working with or near animals. The Endangered Species Act;[6] the Chimpanzee Health Improvement, Maintenance, and Protection Act;[7] the Migratory Bird Treaty Act;[8] or similar laws protecting certain animals may also occasionally factor into decisions by an IACUC.

IACUCs also should know that under the Animal Enterprise Terrorism Act,[9] it is a federal crime to damage or interfere with the operations of an enterprise using animals for research, education, or testing. This law prohibits conduct including threats, vandalism, property damage, criminal trespass, harassment, and intimidation. Additionally, many states have laws protecting animal research facilities from criminal acts such as arson, economic sabotage, and similar acts of intimidation.

Finally, federal, state, local, and institutional laws not specific to animal-related activities can apply to an IACUC's activities, such as laws pertaining to patents or trade secrets, laws requiring public disclosure of records or meetings open to the public, and laws protecting personal privacy. The federal Freedom of Information Act[10] (FOIA) requires the federal government to provide access to records in its possession, which may include institutions' annual reports and inspection reports filed with the APHIS/AC pursuant to the AWA. All states have laws providing for the public disclosure of state agency activities and records, which may apply to state universities. IACUCs at universities, other nonprofit organizations, and small businesses receiving federal government funding for research should be aware of the Bayh-Dole Act (Patent and Trademark Act Amendments of 1980),[11] which allows these institutions to retain title to and file for patents on inventions developed under that funding as well as promote those inventions commercially.

Opin. Many sources of law can bear on an IACUC's activities. They include, in order of hierarchy, federal and state constitutions, federal and state statutes, federal and state regulations, federal and state case law, and institutional policies.

Constitutions set out the fundamental, broad legal principles that govern the nation or the state that enacted them. The principle of academic freedom, for example, is often considered to be grounded in the First Amendment to the U.S. Constitution. State constitutions are sometimes found to be more protective of certain rights (such as the right of privacy) than is the federal Constitution, even where the wording in the two documents is similar.

Statutes are laws enacted by the federal Congress, a state legislature, or, in some states, by a vote of the people. In a particular state, both federal laws and the laws of that state apply. Laws are codified, or collected and organized systematically, by dividing them into titles or chapters by subject matter. The AWA, for example, is codified in the U.S. Code at Title 7, which includes laws pertaining to agriculture. The HREA is codified in various places within the U.S. Code, most particularly with respect to animal research at Title 42, which collects laws protecting the public health and welfare, and was implemented by the PHS Policy. Most federal and state codes and statutes are available for viewing online.

Regulations are rules enacted by a federal or state agency pursuant to requirements or authority granted in a statute. As with statutes, these rules are collected and typically organized by topic. Pursuant to the AWA, for example, the USDA has enacted specific rules (the AWAR) that are found in the Code of Federal Regulations (C.F.R.) at Title 9, which sets forth the regulations governing activities of the APHIS and other rules pertaining to animals and animal products. Strict administrative procedures govern the adoption of regulations. Typically, the agency is required to publish a notice of proposed rulemaking and to provide an opportunity for comment. Federal notices of this sort are published in the Federal Register; states have similar publications for the rulemaking activities of their agencies. Regulations tend to be much more detailed than statutes.

Case law is the set of written opinions by judges interpreting constitutions, statutes, regulations, and other sources of law when deciding a particular case. Not all lawsuits result in written case law. Only published decisions of appellate-level judges (those who review trial-level decisions) can be cited as binding precedent in other lawsuits. Comparatively few lawsuits result in this kind of decision. The vast majority of filed lawsuits settle prior to trial; only a few that go to trial are appealed, and only some of those that are appealed result in a decision written for publication. The citation to a case indicates where to find the published decision, which court made the ruling, and in what year.

22:2 What sources of authority other than those noted in 22:1 might apply to an IACUC's activities?

Reg. The *U.S. Government Principles for the Utilization and Care of Vertebrate Animals Used in Testing, Research, and Training* address compliance with the AWA and other applicable federal laws, guidelines, and policies and generally provide a set of overarching principles for ensuring that the use of research animals is justified and humane. These principles are incorporated in the PHS Policy. The PHS Policy applies to the use of live vertebrate animals in any activity supported or conducted by any PHS agency, including the NIH, FDA, and CDC. The PHS Policy mandates compliance with the AWA and AWAR, as applicable, the *Guide,* and the *AVMA Guidelines for the Euthanasia of Animals.* VHA Handbook 1200.07, *Use of Animals in Research,*[12] applies to research, testing, and teaching activities involving laboratory animals in the Department of Veterans Affairs and mandates compliance with the PHS Policy, the *Guide,* and the *AVMA Guidelines for the Euthanasia of Animals.* Institutions that are accredited by AAALAC or receive funding from certain sources must comply with the *Guide* and the Federation of Animal Science Societies' *Guide for the Care and Use of Agricultural Animals in Research and Teaching*.[13] The *AVMA Guidelines for the Euthanasia of Animals*,[14] prepared by the American Veterinary Medical Association, is an essential reference for IACUCs; deviations from these recommendations should be extremely rare and only when scientifically justified, with any such waiver documented in writing.

Other sources of authority will apply to activities conducted in countries other than the United States.

Opin. In addition to the types of legal authority described, other sources of authority may apply, with varying degrees of force, to IACUC activities, even in a legal setting such as a lawsuit. The PHS Policy, the *Guide*, and the *AVMA Guidelines for the Euthanasia of Animals* carry a great deal of authority as the standards for research animal care and use. In addition, a court is likely to consider as presumptively appropriate other standards or references set out or condoned by AAALAC or similarly influential professional organizations. Several government agencies have guidelines or requirements applicable to animal activities that should be followed in IACUC-approved protocols when appropriate and possible. Also, some professional scientific associations publish authoritative guidelines relevant to animal activities relating to their areas of interest. Courts often look for evidence, such as these references, of a "community standard" among similar facilities if there is a dispute as to whether an institution acted properly.

Other important sources of authority are any internally adopted policies that apply to the IACUC. These can be policies adopted by the animal facility, university, department, or any entity with authority to set policy for the IACUC, including the IACUC itself. If an IACUC is to be subject to an internal policy that transcends legal requirements, however, it is important that the IACUC be able to comply with that policy. Having a policy with standards that are not met may place the IACUC in a more vulnerable legal position than having no policy at all.

Contractual obligations may apply to an IACUC's activities. Research involving third parties (such as collaborations or sponsored research) and research that involves transfers of third-party information or materials will likely be governed by contracts that restrict the use and disclosure of information. Such contracts may be identified as confidentiality agreements, memoranda of understanding, collaborative research agreements, sponsored research agreements, or material transfer agreements. Agreement names vary widely and although agreements may share the same name, they may contain differing obligations, restrictions, and rights. The terms and conditions of each agreement must be carefully read and understood. Research-related agreements usually contain language allowing the use and disclosure of information within an institution and to regulatory bodies in compliance with law; this permits an IACUC to perform its necessary functions. However, any disclosures beyond those strictly necessary to comply with law may violate such third-party agreements. Violations of contractual confidentiality restrictions can give rise to claims of breach and civil damages.

Finally, IACUC activities also may need to comply with an institution's policies on conflicts of interest or other policies relating generally to committee activities.

22:3 What is proprietary or confidential information?

Reg. Federal law defines certain types of proprietary information, including copyrights, patents, trademarks, and trade secrets. Under federal law, *copyright* protection can attach to "original works of authorship fixed in any tangible medium of expression."[15] A *patent* can be obtained for a "new and useful process, machine, manufacture, or composition of matter, or any new and useful improvement thereof."[16] A *trademark* is something that distinguishes the goods or services of the owner from those of others.[17] A *trade secret* is information that has economic value by virtue of its not being generally known to or readily ascertainable by the public.[18] *Confidential information* may be defined by federal or state law or by private agreements or policies. Confidential information can include trade secrets, personal information, individually identifiable health information, personnel records, information protected by a statutorily defined privilege such as the attorney–client or doctor–patient privilege, or other types of information.

Several federal laws protect individuals' privacy. For example, the federal Privacy Act limits the type of information that federal agencies, the military and other government institutions may collect, maintain, use, and disclose on individuals.[19] Most states have additional laws protecting individuals from disclosure of their confidential information by others, through constitutional provisions, statutes, or regulations.

Opin. *Proprietary information* is information that has the legal status of personal property, that is, information that someone exclusively owns, exercises control over, or uses. The term is usually used for trade secrets, confidential business or financial information, or information about an invention that has not yet been disclosed in a filed patent application. IACUCs must be very careful not to disclose such information inappropriately or inadvertently. Proprietary information also includes any intellectual property, including patents, trademarks, copyrights, and proprietary software. Patents, trademarks, and copyrights are publicly known, but they are considered personal property because unauthorized use or copying by someone other than their owner constitutes infringement. Other types of proprietary information are not known by the public and their public disclosure can cause irreparable harm to their owner.

A *trademark* or *service mark* is a word, name, symbol, device, or any combination, used or intended to be used to identify and distinguish the goods or services of one seller or provider from those of others, and to indicate the source of the goods or services. Although federal registration of a mark is not mandatory, it has several advantages, including notice to the public of the registrant's claim of ownership of the mark,[20] legal presumption of ownership nationwide, and exclusive right to use the mark on or in connection with the goods or services listed in the registration.[21] Marks may also be registered with states, but a trademark need not be registered at all to be valid and subject to infringement. An unregistered trademark may be identified by a ™ symbol, while a registered mark can be identified by a ® symbol. The Nike swoosh design and the Kleenex® name for facial tissues are examples of trademarks. Rights in a mark cease if the mark is not actively used for a period of time and registrations must be periodically renewed.[22]

Copyright is an author's exclusive right to copy or modify his or her original expression.[23] Copyright protection can attach to original works of authorship fixed in any tangible medium of expression, including literary works, pictorial and graphic works, and sound recordings.[24] Journal articles (including drafts), laboratory notebooks, drawings, and photographs are examples of works with copyright protection. Ideas, procedures, processes, principles, and the like, cannot be protected by copyright. The term of a copyright typically runs until 70 years after the author's death.[25]

A *trade secret* may be any secret, commercially valuable information or compilation of information used in a business that gives the owner an advantage over competitors who do not know or use it and is the end product of either innovation or substantial effort. It may include any form or type of financial, business, scientific, technical, economic, or engineering information, including a compilation, plan, formula, pattern, design, prototype, device, method, technique, process, procedure, program, or code, whether tangible or intangible, and whether or however stored, compiled, or memorialized.[26] A person who asserts that information qualifies as a trade secret must show that reasonable efforts were made to keep the information confidential, the information has economic value, and the information is not generally known to or readily ascertainable by the public. When properly safeguarded, a trade secret may remain valuable for an indefinite period of time. One of the best-known examples of a well-kept trade secret is the formula for Coca-Cola®.

A U.S. *patent* gives its owner the right to exclude others from making, using, selling, or offering to sell in the U.S., or importing into the U.S.,[27] the "process, machine, manufacture, or composition of matter"[28] claimed in the patent. The term of a U.S. patent generally extends from the date of issuance until 20 years from the date of filing of the patent application.[29] A patent document includes a description of the invention, drawings (if necessary), and one or more claims. The claims particularly point out and distinctly claim the subject matter the applicant regards as the invention and define the scope of the protection afforded by the patent. A patentable invention must be useful, novel, and nonobvious.[30] Under the novelty requirement, no U.S. patent will be issued if the invention was in public use or described in any printed publication more than one year before the application for the patent was filed;[31] thus, disclosure to the public of information describing an invention can forever preclude the inventor from receiving a patent. In most other countries, this grace period is shorter or does not exist at all. In the early stages of a research project, investigators may not be able to identify reliably all the information that may be important to a future patent application and know which information must be protected to prevent compromising the value of the research, so all potentially patentable information must be kept confidential. Laboratory notebooks, drawings, and dated laboratory photographs are common evidence of inventions and thus should generally be treated as confidential. Even live animals may be evidence of a patentable invention.[32]

Under the Patent and Trademark Act Amendments of 1980, popularly known as the Bayh–Dole Act,[33] ownership of patentable inventions arising from federally funded research can be vested in the entities performing the research, including universities, other nonprofit organizations, and small businesses.[34] The holder of the patent rights can license those rights to private companies for further development and commercialization and collect royalties if the invention is sold or used. Thus, inappropriate disclosure by an IACUC could result in the foreclosure of an opportunity for a significant income stream to the IACUC's institution. Funding agreements require that if the grantee elects not to file an application for patent, the grant recipient must disclose inventions to the government in time for the government to file a patent application, and that the government be licensed to practice the invention if the grantee obtains a patent.[35] Federal agencies are authorized to maintain the confidentiality of information disclosing an invention the federal government funded for a reasonable period to permit a patent application to be filed.[36] If the grantee patents the invention but does not appropriately utilize it, the Bayh–Dole Act permits the government to step in and grant licenses to other companies to make the products or use the methods it helped finance for the benefit of the public.[37]

Confidential information is any information that is intended to be held in confidence or kept secret pursuant either to state or federal law or by private agreement or policy. Confidential information can include trade secrets and patentable information or educational, health, and personnel records. It can also include research information subject to a confidentiality agreement or nondisclosure agreement. Some information is considered confidential as a result of a statutorily defined privilege such as the attorney–client or doctor–patient privilege; an attorney should be consulted regarding the limits and requirements of these types of privileges because disclosure to another person or to the IACUC can destroy the confidentiality of such privileged information.

Marking something "confidential" is sometimes a condition of protecting information from disclosure under state and federal laws, as well as private agreements. But that action alone has no legal significance and does not protect information from disclosure. Legal protection is provided either through the law or under a contractual agreement between parties.

22:4 Under what circumstances might proprietary or confidential information disclosed to an IACUC become available to the public?

Reg. IACUCs of public institutions may be required to open their meetings to the public under state open meetings laws. Agencies of the federal government must disclose records in compliance with the federal FOIA.[38] Agencies of a state or local government must disclose records in accordance with the public records laws of their particular state or municipality. Documents and records in the possession or control of a government agency must be made available to the public for inspection and copying, unless some compelling justification exists for keeping them confidential.

Under AWA §2157 it is unlawful for any member of an IACUC to use or release to the public or reveal to another person any confidential information of the research facility, including "any information that concerns or relates to (1) the trade secrets, processes, operation, style of work, or apparatus; or (2) the identity, confidential statistical data, amount or source of any income, profits, losses, or expenditures, of the research facility." Such wrongful use or disclosure may result in removal from the IACUC, fines, and/or imprisonment. In addition, any institution or individual who is injured by such disclosure or use may bring a lawsuit against the IACUC member and recover damages and the costs of the suit from the IACUC member.

§2.35,f of the AWAR requires APHIS/AC inspectors to maintain the confidentiality of information reviewed for inspection purposes and prohibits them from removing materials from research facilities' premises "unless there has been an alleged violation, they are needed to investigate a possible violation, or for other enforcement purposes." The public release of any information or materials copied or removed from a research facility by an APHIS/AC inspector is governed by FOIA.

The federal FOIA provides nine exemptions to disclosure of records. One of the nine exemptions from disclosure under the federal FOIA is for "trade secrets and commercial or financial information that is obtained from a person and is privileged or confidential."[39] Privileged information is information that would ordinarily be protected from disclosure in civil litigation by a recognized evidentiary privilege, such as the attorney–client privilege. Confidential information is defined by FOIA as information that, if disclosed by the government,

1. May impair the government's ability to obtain necessary information in the future
2. Would substantially harm the competitive position of the person who submitted the information
3. Would impair other government interests, such as program effectiveness and compliance; or
4. Would impair other private interests.[40]

Another FOIA exemption prohibits release of "personnel and medical files and similar files the disclosure of which would constitute a clearly unwarranted invasion of personal privacy."[41] Internal government communications are also exempted from disclosure under FOIA.[42]

IACUCs should give special attention in circumstances involving research integrity,[43] particularly if the circumstances may lead to law enforcement actions against those involved. If research misconduct is committed in connection with research subject to IACUC oversight, IACUC records can ultimately become evidence in a law enforcement action. Under certain circumstances, FOIA provides an exemption for records or information compiled for law enforcement purposes.[44] State or local public records laws may have similar exemptions. Legal advice should be obtained prior to responding to FOIA requests relating to research involving potential research misconduct.

Opin. Information disclosed to an IACUC might become available to the public if it is discussed in an open meeting, if the information is included in a document or report submitted by the IACUC to a government agency, or if the information is disclosed in response to a public request because the IACUC is considered an agency of the local, state, or federal government. Whether a document or report was required by law or was submitted or maintained voluntarily is irrelevant to whether it will be disclosed. In litigation, confidential or proprietary information may be required to be disclosed to an adverse party when the information is relevant and necessary to the action, although it may be withheld from public disclosure under a protective order.

The confidentiality of proprietary or confidential information must be scrupulously maintained by an IACUC and each individual member of an IACUC. Inappropriate disclosure of confidential or proprietary information can destroy the value of years of research efforts and have devastating effects on the careers of researchers or the profits of a company or institution. Additionally, if investigators lack confidence that their information will be handled appropriately, they may be reluctant to provide information that the IACUC needs to evaluate proposals properly and meet its legal obligations. Moreover, inappropriate disclosures may lead to researchers becoming the object of harassment or terrorism by persons opposed to the use of animals in research, potentially endangering their lives or laboratories. Unauthorized disclosure or use of confidential or proprietary information by an IACUC member may also lead to removal from the IACUC, payment of a fine, imprisonment, and/or a civil lawsuit brought by the individual or institution injured by the use or disclosure.

If confidential or proprietary information is released to a government agency, even when that submission is required by law or involuntary, the information can become available to the public under FOIA or state or local public records laws. Only information created or obtained by a government agency or government contractor and actually in the agency's or contractor's possession or control can be disclosed; information the government merely has a right to obtain or may have relied on but did not obtain is not available to the public. Receiving a grant from a government agency does not make a private organization a government agency subject to FOIA, unless the government extensively supervises the private organization on a day-to-day basis; similarly, receiving a federal grant does not make a state university subject to the federal FOIA.[45] A private organization is not necessarily an agency of the government when designated by the government to perform certain functions, even if it has decision-making authority.[46] Any such arrangement must be examined individually, considering several factors including the type and degree of control exercised by the government over the private entity as well as the portion of the entity's business which is performed by contract with the government.[47]

Each state has its own version of a FOIA or public records act that applies to its own government entities.[48] Additionally, each federal and state government agency operates under a specific set of rules or regulations governing its disclosures of records to the public.[49] Many states and local governments also have laws requiring that meetings of government boards, commissions, councils, committees, and the like, be open to the public.[50] The FOIA, public records laws, and open meetings laws are based on the principle that public officials and institutions are accountable to the people. Under this principle, the public is entitled to know how its government makes decisions and to review the documents and data behind those decisions.

According to FOIA and these other laws, "any person" regardless of motive, purpose, or need may seek government documents. Pursuant to these laws, every government agency must make available to the public, upon request, any information in its possession or control unless the information falls within the scope of one of several specified exemptions. Courts interpret these exemptions very narrowly. To prevent information from being disclosed under these laws, sufficiently specific arguments must be made regarding the exemption under which the information falls and the substantial harm that disclosure would cause. The institution wanting to withhold the information has the burden of proving that an exemption protects the information from disclosure.

Open meetings or public meetings laws, sometimes called "open door" or "sunshine" laws, require that meetings of "public bodies" be open to the public. These laws usually include a provision that the committee or other entity may adjourn into an executive session closed to the public when discussing privileged or confidential information.

The few courts that have considered any of these laws in relation to IACUCs have reached differing conclusions, although it must be understood that they were interpreting different laws. Under the federal FOIA, grant applications submitted to a government agency have been required to be disclosed by that agency, under the reasoning that the research designs of scientists not engaged in profit-oriented research are not commercial information or trade secrets.[51] Courts have also required disclosure of IACUC records, reports, and minutes not actually in the government's possession when a government-owned animal research facility is operated by a contractor.[52]

Some courts have ruled that an IACUC at a state university is a public body subject to the state's open meeting law and a public agency subject to the state's public records act.[53] These courts have required the release, upon request under the state's public records law, of applications for IACUC approval,[54] including unfunded grant proposals.[55] However, other courts have reached the opposite conclusion, reasoning that even an IACUC at a state university was created by order of the federal government and thus is accountable only to the USDA and federal funding agencies, not the state, even though the IACUC may be monitoring a state institution. Those courts conclude that disclosure of IACUC records is governed by the federal FOIA and the state open meetings laws have no application to IACUC meetings.[56]

Institutions have generally, but not always, been successful in withholding from disclosure the names and contact information of researchers and staff under exemptions in the public records laws relating to privacy or safety. One court held that the names and work addresses of the animal researchers "serve to document the organization, functions, and operations" of a state university's activities and must be disclosed as public records,[57] but another court held that its state law allowed any information "concerning research" to be withheld from disclosure.[58]

Confidential and proprietary information such as trade secrets is always exempt from public disclosure, provided it is adequately proven to actually be confidential and proprietary, because exemptions are construed narrowly. For example, an Oregon court held that the names of experimental drugs and the names of companies for which research was performed constituted proprietary information exempt from disclosure under the state's public records law, but required the release of "daily logs" pertaining to the care of individual animals housed at a regional primate research center.[59]

Another possible circumstance under which proprietary or confidential information disclosed to an IACUC might become available to the public is illustrated by a case in which documents received in response to a FOIA request, including IACUC minutes and correspondence between the IACUC and an investigator, were used to support a *qui tam* ("whistleblower") action under the federal False Claims Act.[60] The False Claims Act imposes liability on those who defraud the government.[61] It encourages the uncovering of such fraud by permitting private persons (i.e., "whistleblowers") to bring *qui tam* actions on behalf of the government and allowing them to share in any monetary recovery.[62] The action alleged that a research scientist submitted a fraudulent grant application to the NIH.

In light of the varying court opinions as to what information is confidential or proprietary and what information the public has a right to see, IACUCs should seek guidance from their institutional legal counsel regarding preparation of reports, meeting minutes, and other documents, as well as how to appropriately conduct meetings open to the public. The general guideline should be that reports include everything requested in sufficient detail to be responsive to the purpose, but nothing more. If provided, guidance from the agency requesting a report should be followed.

22:5 How specific should IACUC protocols be when proprietary information related to techniques, compounds, or devices is at stake?

Reg. The AWA (§2143,a,6,B) states that neither the AWA nor the AWAR should be "construed to require a research facility to disclose publicly or to the Institutional Animal Committee during its inspection, trade secrets or commercial or financial information which is privileged or confidential." AWA §2157 prohibits any member of an IACUC from using, releasing to the public, or revealing to another person any confidential information of the research facility. The AWAR (§2.35,f) require APHIS/AC inspectors to maintain the confidentiality of any information they obtain in connection with an inspection of a research facility and to disclose it only when necessary in accordance with FOIA, which exempts the release of privileged and confidential information. Similarly, the HREA (§495,e) states that none of its guidelines or regulations "may require a research entity to disclose publicly trade secrets or commercial or financial information which is privileged or confidential."[63]

However, under the AWAR (§2.31,d; §2.31,e) and PHS Policy (IV,C,1), the IACUC is obligated to review proposed activities related to the care and use of animals and to determine that the proposed activities are in accordance with the AWA and, when applicable, the PHS Policy. Therefore, IACUC protocols must be sufficiently detailed to permit this required review.

Opin. Protocols submitted to an IACUC must be sufficiently detailed to permit the IACUC to evaluate adequately the proposed procedures involving animals, including understanding any potential physical, physiological, and behavioral discomfort, pain, and/or distress the animals may experience as a result of the procedures and related husbandry methods. Protocols must include enough detail to assure the IACUC that no alternatives to the use of animals are available and that the procedures will not unnecessarily duplicate previous experiments. IACUC reviewers must be provided sufficient information to understand whether proposed procedures will be performed with appropriate sedatives, analgesics, or anesthetics, or that the withholding of such agents is scientifically justified. Reviewers must also be able to evaluate whether surgery will be performed on animals and, if so, whether appropriate preoperative, operative, and postoperative care will be provided and whether any animal will be used in more than one surgical procedure.

If a proposed activity related to the care or use of animals utilizes a proprietary method, compound, or device, disclosure in the protocol of that proprietary information may be necessary for the IACUC to adequately conduct its review of the proposed activity.

Protocols also must include sufficient specificity to permit the IACUC to understand whether proposed procedures may cause conditions for which animals may need medical care, including the provision of anesthetics, analgesics, or tranquilizers, and whether that medical care will be provided. The method of euthanasia of each animal associated with or involved in the proposed study must also be adequately explained in a protocol submitted to an IACUC.

The IACUC must be provided with sufficient information to be able to evaluate the adequacy of the training and other qualifications of all personnel who will conduct the proposed procedures involving animals.

Including descriptions and explanations of proposed procedures in a protocol submitted to an IACUC that are sufficiently detailed to permit the IACUC to evaluate the protocol adequately in accordance with the IACUC's legal responsibilities as just outlined may require the disclosure of confidential or proprietary information. Techniques, compounds, devices, and animal models may be patentable or may have significant commercial value if not publicly known, but the IACUC may need information about them to be able to understand properly any potential discomfort, distress, or pain the animals involved may experience as a result of the procedures or related husbandry methods.

Protection of confidential or proprietary information is no excuse for an inadequate review by an IACUC of proposed procedures involving animals. The IACUC must have sufficient facts to be able to perform its legal and ethical responsibilities and make informed decisions about the welfare of the research animals. Accordingly, the IACUC, and each individual IACUC member, must follow appropriate procedures to provide assurance to investigators that their secrets will not be revealed in ways that could cause harm to the researchers or loss of the value of their intellectual property. Additionally, IACUCs must diligently protect information revealed and entrusted to them and not permit disclosure to anyone not needing to know that information. See 22:8 below for procedural mechanisms an IACUC may adopt to protect confidential and proprietary information of researchers. Also see 8:19 for survey responses to who has access to IACUC protocol forms.

22:6 Must a sponsor of research be identified on the IACUC protocol forms?

Reg. Institutions may require their employees to disclose any outside financial support they receive or anticipate receiving in connection with research or other activities so they can identify any potential conflicts of interest or other ethical issues. Disclosure of sponsorships to the IACUC on protocol forms may be included in these institutional policies.

Government employees are required to comply with applicable laws on ethical conduct and financial disclosure and the implementing guidance and policies of their particular employer. Employees of the Department of Health and Human Services (HHS) and uniformed service officers in the Public Health Service Commissioned Corps on active duty, for example, are bound by the Standards of Ethical Conduct for Employees of the Executive Branch contained in 5 C.F.R. Part 2635. These employees are also subject to the executive branch-wide financial disclosure regulations at 5 C.F.R. Part 2634, the Employee Responsibilities and Conduct regulations at 5 C.F.R. Part 735, and the HHS regulations regarding conduct at 45 C.F.R. Part 73.[64]

Institutions and individual investigators that apply for or receive research funding from PHS granting agencies, including the NIH, must comply with Financial Conflict of Interest (FCOI) regulations in 42 C.F.R. Part 50 Subpart F. Similar regulatory requirements apply to institutions and investigators performing research under PHS contracts and are found at 45 C.F.R. Part 94. Note that intellectual property rights and interests, unless assigned to the institution, are included within the definition of financial interest in these regulations. Under these regulations, significant financial interests must be disclosed to the institutional designated official(s) when an application is submitted to the NIH for funding, upon discovering or acquiring a new significant financial interest, and on an annual basis. If a significant financial interest is determined to constitute a financial conflict of interest, the institution must make the investigator's name, the sponsor's name, and information regarding the significant financial interest publicly accessible through a website or written response to a request for the information.[65]

Opin. There is no *per se* legal requirement to identify a sponsor on protocols submitted to an IACUC. However, protocols for research involving funding from a government agency, a government contract, or a cooperative agreement with a government agency require disclosure of financial sponsors to ensure that the design, conduct, and reporting of funded research is not biased by any conflicting financial interest of the investigators. Therefore, institutions having investigators with such government involvement may adopt IACUC protocol forms for the convenience of the investigators that comply with applicable government regulations, so investigators need not prepare multiple forms.

Conflict of interest policies require researchers to disclose the identity of sponsors and the nature and extent of any financial relationship between the researcher and sponsor, in order to preserve the integrity of research and prevent potential researcher bias. Institutional policies may require disclosure of a financial sponsor. Scientific journals, where researchers may wish to publish the results of their IACUC-approved animal studies, also frequently require authors to disclose financial support from entities other than their employers, as well as financial relationships with entities that could be perceived to influence the submitted work.

Although not required, IACUCs may choose to request identification of financial sponsors on IACUC protocol forms, as well as any management plan for a financial conflict of interest, to aid their review of activities involving animals. AWAR §2.31,d requires IACUCs to review proposed activities to ensure that all procedures involving animals will avoid or minimize discomfort, distress, and pain to the animals, as well as ensure that the activities do not unnecessarily duplicate previous experiments and that the PI considered alternatives to procedures that may cause more than momentary or slight pain or distress to the animals. A financial sponsor may, intentionally or inadvertently, negatively affect animal welfare through its influence on how a proposed activity is conducted or its hope for certain experimental outcomes. Investigators may believe, perhaps naively, that their actions are not biased by their desire to please their sponsors. Fully-informed examinations of protocols, animal activities, and animal living conditions by IACUC members who are concerned primarily with the animals' welfare and not beholden to sponsors would seem to better conform to the purpose of IACUC review.

Even if a research sponsor's name is not listed on IACUC documents, sponsor information may be obtainable by a public records or FOIA request to a public research institution or about a taxpayer funded activity.

Some agreements concerning the performance of research may require a sponsor's identity to be maintained as confidential to prevent the sponsor's competitors from learning about the research. This situation weighs against disclosure of sponsor information on protocol forms to the extent that it is not necessary for regulatory compliance or compliance with institutional policy. When the name of a sponsor, in conjunction with other information in the protocol, could reveal confidential or proprietary information of value to the sponsor's competitors, the sponsor's name should be deleted before the protocol form is released to anyone outside the IACUC.[66]

Surv. Does your IACUC protocol form ask for the identification of the financial sponsor of the study (if any)?

- Not applicable 17/294 (6%)
- Yes 211/294 (72%)
- No 63/294 (21%)
- I don't know 2/294 (1%)
- Other 1/294 (0.3%)

22:7 Can researchers protect their own identities on IACUC protocol forms?

Reg. AWAR §2.31,c,6 and PHS Policy IV,B,6 give the IACUC the authority to accept, require modifications of, or withhold approval of animal activities. To be able to approve a proposed animal activity, an IACUC must know the identities of the personnel who will be performing each part of a protocol to determine whether they are adequately and appropriately qualified and trained.

Scientists, technicians, and other personnel involved with animal care, treatment, or research must be appropriately qualified and trained in humane practices and research methods; this is required by the AWA (§2143,d); AWAR (§2.31,d,1,viii), and the PHS Policy (IV,C,1,f). The HREA mandates that applicants for grants, contracts, or cooperative agreements involving animal research which are administered by the NIH or any national research institute provide assurances that "scientists, animal technicians, and other personnel involved with animal care, treatment, and use by the applicant have available to them instruction or training in the humane practice of animal maintenance and experimentation, and the concept, availability, and use of research or testing methods that limit the use of animals or limit animal distress."[67]

In addition to the identities of those persons performing procedures, the identities of other personnel that may be involved with a protocol may need to be disclosed for purposes of identifying conflicts of interest. Institutions and individual investigators that apply for or receive research funding from PHS granting agencies, including the NIH, must comply with FCOI regulations in 42 C.F.R. Part 50 Subpart F. Similar regulatory requirements apply to institutions and investigators performing research under PHS contracts and are found at 45 C.F.R. Part 94. Note that intellectual property rights and interests, unless assigned to the institution, are included within the definition of financial interest in these regulations. Under these regulations, significant financial interests must be disclosed to the institutional designated official(s) when an application is submitted to the NIH for funding, upon discovering or acquiring a new significant financial interest, and on an annual basis. If a significant financial interest is determined to constitute a financial conflict of interest, the institution must make the investigator's name, the sponsor's name, and information regarding the significant financial interest publicly accessible through a website or written response to a request for the information.[68]

Opin. The IACUC is required to determine that the personnel conducting animal research are properly qualified by their education and training. The identity of all personnel conducting procedures on animals must be disclosed in order for the IACUC to make this determination. Additionally, the identity of the PI overseeing the work is important to the evaluation of the protocol and should be disclosed to the IACUC. Identities may also be relevant for purposes of evaluating conflicts of interest in accordance with HHS regulations and institutional policies. If a researcher has a significant financial interest that is determined to be a financial conflict of interest, the institution must make the researcher's name, title, and role in the research publicly accessible through a website or written response to a request for such information.[69]

If IACUC records are requested by the public under FOIA or a state's public records law, or if IACUC meetings are open to the public, the identities of the researchers may or may not be able to be protected from disclosure, depending on the applicable law. Names and contact information of researchers are usually held to be exempt from disclosure, but biographies, credentials, and other information that could identify specific individuals may be required to be disclosed if such information is publicly available from other sources.[70] (See 7:12.)

22:8 What procedural mechanisms may an IACUC adopt to identify and protect information considered proprietary by the researchers?

Reg. Under §2157 of the AWA, it is unlawful for any member of an IACUC to release to the public any confidential information of the research facility. Similarly, subsection (e) of the HREA states that none of its guidelines or regulations "may require a research entity to disclose publicly trade secrets or commercial or financial information which is privileged or confidential."[71] Confidential information includes trade secrets, processes, operations, style of work, apparatus, the identity, confidential statistical data, amount or source of any income, profits, losses, or expenditures of a research institution. The AWA (§2157) also prohibits any member of an IACUC from using or attempting to use to his or her advantage, or revealing to another person, any information entitled to protection as confidential information of the research facility.

Opin. While serving on an IACUC, members receive information from researchers. Some of the information will be confidential or proprietary. AWA §2157 imposes a duty on IACUC members not to disclose to anyone outside the IACUC any confidential information they learn through service on the IACUC and not to use any such information in their own research without the express permission of the owner of the information. Thus, it is important for an IACUC to implement mechanisms to first identify potentially confidential and proprietary information and then protect it from inappropriate disclosure.

Flagging. Although all information submitted to and created by an IACUC should be handled securely, researchers' confidential and proprietary information should be especially protected. Investigators could be requested to flag or mark confidential or proprietary information specially, including a designation on the title page that such information is included. Such information could be included as an attachment to a standard protocol form. In an electronic document, confidential or proprietary information could be included in a separate document that is encrypted or protected with an additional password. Under FOIA and state public records laws, documents containing both confidential and non-confidential information are subject to disclosure. When such documents are requested, the confidential information should be redacted prior to disclosure. Flagging or marking information considered confidential or proprietary, or including such information in an attachment, would facilitate identification and segregation of the information by the IACUC or institution in the event of a request by the public for inspection of IACUC records or the preparation of an IACUC report or submission to a government agency. However, unless information falls within the scope of a defined exemption in the applicable public records law, the information will nevertheless have to be disclosed upon request.

Investigators should be advised that although all information submitted to an IACUC will be treated confidentially to the extent permitted by law, some information may become available to the public through state public records requests or discussion in open meetings. Only information within the scope of a legal exemption to disclosure can be assured of protection; flagging or labeling portions of a submission as confidential or proprietary cannot guarantee nondisclosure but can assist the IACUC or the institution in reviewing a record in the event of a public records request or open meeting. Investigators should be instructed to identify and flag selectively only information that is truly confidential or proprietary.

Confidentiality or Nondisclosure Agreements. As a general rule, any information provided by researchers to an IACUC should be distributed only to those persons who have a need for the information. No information received by an IACUC member from a researcher should be forwarded, discussed, or otherwise shared with anyone not essential to the review of the procedures or related husbandry, whether or not the information seems confidential or proprietary. To emphasize IACUC members' duty not to disclose information, requiring that they sign confidentiality or nondisclosure agreements is a good practice.

Security. Basic security measures should always be taken to protect information about the use of animals in research or persons conducting such research because of the controversial nature of such activities and the ease with which they can be misinterpreted. These security measures include, at a minimum, building access controls, escorting visitors, and restrictions on photography.

Additional security measures should be taken to protect documents, data, and other written or recorded information provided by researchers to an IACUC, as well as IACUC meeting minutes in which such information is discussed. These measures should include storage in a locked cabinet in an area with restricted access and shredding or similar destruction of duplicative or otherwise unnecessary records. IACUC members should be encouraged to destroy their personal copies of IACUC-related documents or return them to the IACUC secretary when they are no longer needed. A clear and concise policy for the retention and destruction of records should be followed, and materials regularly destroyed when the retention period expires.

Electronic Security. Electronic materials are readily copied and distributed, so electronic protocol submissions, distribution, and reviews must be tightly controlled. The computers used for electronic storage, transmission, or discussion of researchers' information should be access-controlled and password-protected, including a requirement to enter a password after a period of nonuse, and encryption of IACUC-related documents should be considered. In establishing security policies and procedures for the protection of researchers' electronic confidential or proprietary information, institutions may find useful the specifications for the protection of electronic health information in the security regulations promulgated under the Health Insurance Portability and Accountability Act of 1996 (HIPAA).[72] These regulations describe administrative, physical, and technical safeguards to prevent access to sensitive information by unauthorized persons. Institutions that include or are affiliated with a medical center may have personnel and technical resources experienced in working under the HIPAA regulations available to assist in developing appropriate procedures and policies for the protection of proprietary information.

22:9 What responsibilities do IACUCs and IACUC members have to protect records? What guidelines should be followed to protect confidential and proprietary information?

Reg. IACUCs must, under AWAR §2.35 and PHS Policy IV,E, maintain certain records and reports for at least three years. AWAR §2.35,f and PHS Policy IV,E,2 require records that relate directly to proposed activities involving animals and proposed significant changes in ongoing activities reviewed and approved by the IACUC to be maintained for the duration of the activity and for an additional three years after the completion of the activity. These records must be available for inspection and copying by authorized USDA or funding federal agency representatives. In addition, an institution has a legal duty to identify, preserve, and produce any records relating to a potential or actual legal claim.

Under AWA §2157, it is unlawful for any member of an IACUC to use or release to the public or reveal to another person any confidential information of the research facility, including "any information that concerns or relates to (1) the trade secrets, processes, operations, style of work, or apparatus; or (2) the identity, confidential statistical data, amount or source of any income, profits, losses, or expenditures of the research facility." Such wrongful use or disclosure may result in removal from the IACUC, fines, and/or imprisonment. In addition, any institution or individual who is harmed by such disclosure or use may bring a lawsuit against the IACUC member and recover damages and the costs of the suit from the IACUC member.

Opin. The IACUC is required by law, under AWAR §2.35 and PHS Policy IV,E, to retain a copy of every approved protocol and to record and maintain minutes of all committee meetings and deliberations for at least three years. These laws further require that IACUC records that relate directly to proposed activities and proposed significant changes in ongoing activities involving animals must be retained for three years after the completion of the activity. These records must be available for inspection and copying by authorized representatives of the USDA or the funding federal agency (if applicable). The IACUC should serve as the custodian for such records and should maintain those records in a secure location, whether physically, electronically, or both. There is no legal requirement for individual members to maintain separate copies of protocols or other similar records. Maintenance of separate records increases the risk of disclosure of confidential and proprietary information they may contain. In addition, in the event the institution is required to produce such records in response to a legal claim, it will be required to produce all existing copies of such records. If an institution has implemented policies and procedures that (1) require original records to be maintained by a designated custodian and (2) require destruction of all other copies of such records once no longer needed for their intended use, it will only be required to produce the original record. However, the institution must be able to show that the policies and procedures are actually followed. Thus, if an IACUC does not require individual members to destroy such records at the end of meetings or after final action, or if an IACUC has not implemented policies and procedures to retain official records and destroy unnecessary copies, it may be required by law to identify, preserve, and produce both original records and all copies.

In today's paper and electronic environment, the task of identifying and producing all existing copies can require extensive resources and financial cost. Thus, IACUCs should have policies and procedures that include collecting copies of protocols and other documents reviewed by committee members, either at the end of the relevant meeting or when the committee's review is completed. Such practices not only protect confidential and proprietary information contained in the records but also reduce legal and financial risk to the IACUC and institution. These practices should be governed by institutional record management policy. The IACUC practices should not conflict with the institution's other record retention requirements, including preserving records related to a potential or pending legal claim or retaining such records in compliance with other federal laws and regulations pertaining to animal-based research with which the IACUC must comply.

Surv. Does your IACUC require that protocols or other sensitive documents be turned in, destroyed, or electronically deleted at the end of an IACUC meeting or once a final action has been taken by the IACUC?

- Not applicable 10/295 (3%)
- Yes 97/295 (33%)
- No 181/295 (61%)
- I don't know 2/295 (1%)
- Other 5/295 (2%)

22:10 May an IACUC be sued for breach of proprietary confidence?

Reg. AWA §2157 addresses disclosures or wrongful uses of proprietary information by a member of an IACUC. This statute prohibits the release by an IACUC member of any confidential information that "concerns or relates to (1) the trade secrets, processes, operations, style of work, or apparatus; or (2) the identity, confidential statistical data, amount or source of any income, profits, losses, or expenditures, of the research facility." This statute also makes it illegal for any member of an IACUC "(1) to use or attempt to use to his advantage; or (2) to reveal to any other person, any information which is entitled to protection as confidential information" under the statute. Penalties for violation of this law include a fine of up to $10,000 and imprisonment of up to three years, as well as removal from the IACUC; the violator also will be required to pay the damages, costs, and attorneys' fees of the injured investigator and/or the research institution. Thus, IACUC members have a duty not to disclose to anyone outside the IACUC any confidential information they learn through service on the IACUC and not to use or attempt to use any such information to their own advantage without the express permission of the owner of the information. In addition, any institution or individual who is injured by such disclosure or use may bring a lawsuit against the IACUC member and recover damages and the costs of the suit from the IACUC member.

Opin. An IACUC can be sued for breach of proprietary confidence. IACUCs are created under and obtain their authority from the AWA, which explicitly requires that IACUC members maintain the confidentiality of proprietary information they receive in the course of performing their responsibilities, specifies penalties for violations of confidentiality, and provides remedies for injured persons. Violation of this law could be the basis for a lawsuit against the IACUC, the research institution, or an individual member of the IACUC. The HREA also provides authority for IACUCs at institutions receiving funding from certain sources; it states that it should not be interpreted to require the IACUC or research entity to disclose trade secrets or commercial or financial information which is privileged or confidential. The HREA thus could potentially provide another basis for a lawsuit against an IACUC that releases such information, although the HREA does not specifically provide for rights and remedies of persons harmed by disclosures.

Additionally, a legal claim could be based on the improper disclosure of confidential proprietary information by an IACUC during an open meeting or in response to a request under a state public records act, or in a submission by the IACUC to a government agency. Further, a claim for breach of proprietary confidence also could be supported by inadvertent disclosure of confidential proprietary information. Inadvertent disclosure might occur if IACUC records or an IACUC member's notes are not stored securely or are disposed of in a way that permits viewing of confidential information by unauthorized persons.

A state university successfully avoided breaching proprietary confidence in *Mississippi State University v. People for the Ethical Treatment of Animals, Inc.*[73] The university had performed research for a company under contracts that included non-disclosure agreements requiring secrecy of information and no disclosure of confidential information. PETA sought IACUC records under the state public records law for projects, tests, and experiments funded by the company. The court held that the university was a public body as defined by state law and therefore the IACUC protocol forms were public records, but the data and information recorded on IACUC protocol forms was not public property because it was owned, furnished, or required by the company. The completed protocol forms in this case contained confidential commercial, proprietary, and trade secrets information about the company's research, so the university was able to convincingly argue that they fit within a statutory exemption permitting nondisclosure.

22:11 Must an IACUC at a state institution open its meetings to the public?

Reg. To promote transparency in government, every state has some version of law requiring all government business to be conducted in open meetings to which the public has access. The court's opinion in *Oklahoma Association of Municipal Attorneys* states why open meetings laws exist: "If an informed citizenry is to meaningfully participate in government or at least understand why government acts affecting their daily lives are taken, the process of decision making as well as the end results must be conducted in full view of the governed."[74]

However, AWA §2157 prohibits IACUCs from releasing to the public any confidential information of their research facilities.

Opin. Whether an IACUC at a state institution is required to open its meetings to the public depends on the wording and interpretation of the state's open meetings law. Open meetings laws, also referred to as "open door" or "sunshine" laws, require that meetings of public bodies be open to the public. If either the IACUC or the state institution is considered a public agency or public body as defined under the open meetings law of its state, then the IACUC may be required to hold its meetings open to the public. Nevertheless, these laws usually include a provision that the committee or other entity may adjourn into an executive session closed to the public when discussing confidential or proprietary information or information defined by law as privileged, such as communications between an attorney and client or doctor and patient.

The few state courts that have considered open meetings laws in relation to IACUCs have reached different conclusions, although it must be understood that they were interpreting different state laws (see also 22:4).

In *Animal Legal Defense Fund, Inc. v. Institutional Animal Care & Use Committee of University of Vermont*,[75] the Supreme Court of Vermont ruled that the IACUC at a state university was a committee of the University and thus was subject to the state's open meetings law.

In *Dorson v. State of Louisiana*,[76] a court in a different state held that the IACUC at a state university was created by the order of the federal government and thus was accountable only to the USDA and NIH, not the state, even though the entity monitored by the IACUC was a state institution. That court concluded that the state open meetings law had no application to IACUC meetings. The highest court in New York also held, in *In re American Society for the Prevention of Cruelty to Animals v. Board of Trustees of the State University of New York*,[77] that meetings of the IACUC of a state university did not need to be open to the public because the IACUC's powers and functions derived solely from federal law.

Surv. Are your IACUC meetings open either in full or in part to the general public?

- Not applicable 9/294 (3%)
- Yes, and we are a public institution 52/294 (18%)
- Yes, and we are a private institution 7/294 (2%)
- No, and we are a public institution 68/294 (23%)
- No, and we are a private institution 151/294 (51%)
- Other 7/294 (2%)

22:12 Is an IACUC at a state institution required to make its records available to the public?

Reg. The federal FOIA,[78] state public records acts, and state open meetings laws require public bodies, such as government agencies and state universities, to make their information and records available to the public upon request.

Opin. Whether an IACUC at a state institution is required to make its records, including records not submitted to another entity, available to the public varies by state. This issue depends on the wording and interpretation of the state's public records law, also referred to as the state's FOIA. Public records laws generally require that records in the possession or control of a public body be accessible for inspection and duplication by the public except for specific exceptions enumerated in the law. Courts generally construe state public records acts strongly in favor of disclosure.

The determination of whether an IACUC must disclose records to the public is a multi-step analysis. The first piece of the analysis is whether the IACUC is a public body. If either the IACUC or the state institution is considered a public agency or public body, as defined in the public records law, then the IACUC may be required to make its records available to the public. Secondly, it must be determined whether any of the specific exemptions found in the public records law apply. Finally, disclosure depends on whether the records have been already disclosed to a federal government agency, which may make the records available to the public under the federal FOIA.

Few courts have considered whether IACUCs are public bodies subject to records disclosure. Those courts that have considered public records laws in relation to IACUCs have reached different conclusions, although it must be understood that they were interpreting different state laws (see also 22:4). Some courts have ruled that an IACUC at a state university is a public agency subject to the state's public records act.[79] These courts have required the release, upon request under the state's public records law, of applications for IACUC approval,[80] including unfunded grant proposals.[81] However, other courts have reached the opposite conclusion, reasoning that even an IACUC at a state university was created by order of the federal government and thus is accountable only to the USDA and federal funding agencies, not the state, even though the IACUC may be monitoring a state institution. Those courts conclude that disclosure of IACUC records is governed only by the federal FOIA[82] and the IACUC is not an entity "performing a governmental or proprietary function for the state."[83]

State public records laws, like the federal FOIA, typically include a list of exemptions. If one or more of the enumerated exemptions applies, the otherwise presumptively public document, or portions of it, may be withheld from disclosure. The institution seeking to withhold the information has the burden of proving that the information fits within an exemption. Exemptions are interpreted very narrowly. See 22:14 for more information on exemptions.

Institutions have generally, but not always, been successful in withholding from disclosure the names and contact information of researchers and staff under exemptions in the public records laws relating to privacy or safety. One court held that its state law allowed any information "concerning research" to be withheld from disclosure,[84] but another court held that the names and work addresses of the animal researchers "serve to document the organization, functions, and operations" of a state university's activities and must be disclosed as public records.[85] The court in *In re Physicians Committee for Responsible Medicine v. Hogan* stated that public employees have a diminished expectation of privacy with regard to details of their public employment.[86] That court found that the public has a legitimate interest in being granted access to records detailing the research activities of state employees in order to assess the scientific value and social utility of the research being performed. The court permitted contact information of individual researchers to be withheld, but required disclosure of information identifying researchers and describing details of their work in response to a request under the state's freedom of information law.

Confidential and proprietary information such as trade secrets is always exempt from public disclosure, provided it is adequately proven to actually be confidential and proprietary. For example, the court in *In Defense of Animals v. Oregon Health Sciences University*[87] found the names of drug company sponsors and experimental drugs to be exempt from disclosure because they constituted sensitive business records that would not ordinarily be disclosed to the companies' competitors. That court also held that staff member names were exempt from disclosure because the public interest in disclosure was not outweighed by the university's interest in nondisclosure.

Although not directly involving IACUC records, the opinion of the Supreme Court of Ohio in *State ex rel. Physicians Committee for Responsible Medicine v. Board of Trustees of Ohio State University*[88] is instructive. An advocacy organization had filed a writ of mandamus to compel a state university to release certain photographic and videotaped records created by the college of medicine in connection with its research on spinal cord injuries using laboratory mice and rats. The applicable public records statute exempted from public disclosure "intellectual property records" created during research, but only if those records had not been "publicly released, published, or patented."[89] In this case, the state university had lent some of the requested records to scientists and research trainees and shown a few of them to scientists at medical conferences. Acknowledging its responsibility to construe the public records act strictly in favor of disclosure, the court nevertheless held that the intellectual property records exception to disclosure was available in this case because of the efforts the state university took to protect the confidentiality of the records. The state university kept the records in secure cabinets within a locked office accessible only to members of the research laboratory, lent them only to persons who signed nondisclosure agreements, and showed them only to small groups of scientists and researchers under controlled circumstances. The court found that this limited sharing, for purposes related to the research, did not constitute prior public release, so the advocacy organization was not entitled to any of the records under the state public records act.

In the event an IACUC or state institution is not considered a public body under its state public records laws or it is considered a public agency but exemptions apply, the public could still access IACUC records if the records were ever disclosed to the federal government. Once IACUC records are disclosed to the federal government or an agency of the federal government, such as semiannual inspection or annual reports, such records, which were safe under state public records laws, may be considered public records as defined under FOIA and may then be available to the public, unless they fall within a FOIA exemption. Information the federal government merely has a right to obtain or may have relied on but never actually obtained is usually not available to the public.

Responses to survey question 6:22 show the impact of FOIA and similar state laws on IACUC minutes.

22:13 Is an IACUC at a private institution subject to FOIA or a state's public records or open meeting laws and therefore required to make its records and meetings open to the public?

Reg. The federal Freedom of Information Act,[90] state public records acts, and state open meetings laws require public bodies, such as government agencies and state universities, to make their information and records available to the public upon request. These laws do not apply to private nongovernmental institutions.

Opin. As a general rule, private institutions are not required to make their records or meetings available to the public. However, if a private institution is supported by, affiliated with, or considered a state-aided, state-owned, or public institution, the state open meeting and public records laws may apply. Simply receiving a grant from a government agency does not make a private organization a public agency subject to these laws; nor does a private organization necessarily become a public body by performing specific functions for the government. But a government cannot delegate away its statutory responsibility to govern transparently by "outsourcing" its governmental functions. IACUC records from a government-owned animal research facility operated by a private contractor must be disclosed under FOIA because the government is considered to own those records.[91] This may be true also under state public records acts.[92] Arrangements between private entities and governments must be examined individually, considering several factors including the type and degree of control exercised by the government over the private entity as well as the portion of the entity's business which is done by contract with the government.[93]

When a private institution or IACUC at a private institution discloses any of its records, such as semiannual inspection or annual reports, to the federal government, such records become available to the public under FOIA, because they now constitute public records. Similarly, reports prepared by a APHIS/AC representative following inspection of a private research facility are public records. Information the government merely has a right to obtain or may have relied on but never actually obtained is usually not available to the public.

Even involuntary disclosure of IACUC records to the government may make those records available to the public. In *In Defense of Animals v. U.S. Department of Agriculture*, records relating to the USDA's investigation of a contract research organization, including IACUC records, were required to be disclosed under FOIA.[94] The research organization argued that disclosure would be perceived as a breach of its confidentiality agreements with its customers and cause it to lose study sponsors, but the court held that the only relevant harm of disclosure was whether a competitor could reap a commercial windfall from the use of disclosed proprietary information and the organization had failed to carry its burden of proving that harm.

22:14 How should IACUCs respond to FOIA or state public records act requests?

Reg. The federal FOIA[95] requires a federal agency to provide access to records in its possession or control to requesters. This may include providing access to an institution's annual reports and inspection reports filed with the APHIS/AC pursuant to the AWA, as well as an institution's Assurance statement and other reports made to NIH/OLAW. All states have their own public records laws, also referred to as state freedom of information acts. These state laws provide for the public disclosure of state agency activities and records, which may apply to state universities.

AWA §2157 states that it is unlawful for any member of an IACUC to release to the public any confidential information of the research facility. Confidential information includes the trade secrets, processes, operations, style of work, and apparatus of the facility, as well as the identity, confidential statistical data, amount or source of any income, profits, losses, or expenditures of the facility. It is also unlawful for any member of an IACUC to use or attempt to use to his or her advantage, or to reveal to another person, any information entitled to protection as confidential information of the research facility.

Opin. Many freedom of information requests for IACUC documents originate from animal rights activist groups and the numbers of requests from such groups are increasing.[96] Requests for information should be made directly to the federal or state agency, but an institution and its IACUC should take certain steps to protect confidential, protected, or proprietary information in anticipation of such requests.

Responding to FOIA Requests: Facts and Resources[97] is a helpful reference guide for researchers about FOIA and state open records laws developed jointly by the National Association for Biomedical Research (NABR), Society for Neuroscience (SfN), and Federation of American Societies for Experimental Biology (FASEB).

Most institutions' legal counsels maintain procedures for responding to governmental requests such as subpoenas. Institutions and their IACUCs should process freedom of information requests under a similar procedure. Individuals knowledgeable in FOIA and related state laws with an understanding of what information is subject to disclosure should process such requests. IACUC members and administrative staff should be trained on and become knowledgeable of FOIA and similar state laws. Guidelines or best practices should be established for developing records giving due consideration to the records most likely to be requested. Records used for governmental submissions should only provide necessary information and should not include extraneous information. If responding in connection with an alleged noncompliance situation, references to particular individuals by name should be avoided. Rather, titles or descriptions of individuals should be used. Finally, disclosure of confidential or proprietary information should be strictly avoided.

Some freedom of information requests made under state public records laws can be denied due to an exemption. State public records laws typically include a list of exemptions. If one or more of the enumerated exemptions applies, the otherwise presumptively public document, or portions of it, may be withheld from disclosure. Courts have generally construed state public records acts strongly in favor of disclosure. The entity wishing to withhold information has the burden of proving the applicability of an exemption. In addition to the enumerated exceptions, federal law or constitutional principles might preclude disclosure in a particular situation.

This section discusses some of the more common enumerated exceptions, as well as the other sources of legal authority that might be relevant. Note that more than one exception or other law might apply and that state laws differ significantly, making it difficult to generalize.

Some public records acts include an exception for research-related documents. They differ widely in their wording and interpretation. In *Robinson v. Indiana University*, an Indiana court held that university records on animal use in research were "information concerning research" and thus exempt under that state's Public Records Act.[98] By contrast, in *Progressive Animal Welfare Society v. University of Washington*, the Washington Supreme Court held that unfunded grant proposals, as a whole, were not "valuable formulae" or "research data" as defined in Washington's public records act, although much of the material in the grant proposals did fall within that exemption.[99]

Public records acts typically include an exception aimed at protecting the privacy of employees. Whether an IACUC can delete (redact) names and addresses of researchers or others when responding to a public records request has been the subject of litigation, with differing results. In *State ex rel. Thomas v. Ohio State University,* the court required a state university to release the names and work addresses of animal research scientists. The university had, for security reasons, redacted that information from research-related documents provided in response to a public records request. The court acknowledged the likelihood of increased harassment but noted that there did not appear to be the "high potential ... for victimization" required to establish a constitutionally based exemption and if criminal conduct resulted from the disclosure it "should be punished by criminal sanctions."[100] The court in *In re Physicians Committee for Responsible Medicine v. Hogan* similarly refused to protect the identity of researchers and the details of their research "merely because the passions of unknown terrorists or criminals might be inflamed" and stated that "public employees have a diminished expectation of privacy with regard to details of their public employment."[101] That court permitted redaction of phone numbers, email addresses, building addresses, and similar information under a personal privacy exemption.

In contrast, in *S.E.T.A. UNC-CH, Inc. v. Huffines,* the court ruled that under that state's public records act, the identity of researchers and staff members could be withheld. The court reasoned that "public policy does require that any information contained in the [research] applications relating to the names of the researcher and staff members, their telephone numbers, addresses, their experience, and the department name be redacted from the IACUC applications."[102] That court also ruled that applications that are not approved need not be made public.

Many states have exceptions for "preliminary" or "draft" information, and these exceptions might preclude release of unapproved or unfunded proposals. In addition, intellectual property and public policy arguments often have great force when the research is at this early stage.

It is important to keep in mind that other state statutes outside the state public records laws may provide exemptions from disclosure for research records. Statutes protecting trade secrets as well as any anti-harassment statutes specifically relevant to animal researchers may not protect research records in their entirety but may protect various portions of the records from disclosure.

Freedom of information requests made under FOIA can also be denied due to an exemption. Some state courts have determined that the federal FOIA rather than the state public records act applies to the records of an IACUC.[103] The federal FOIA provides nine exemptions to disclosure of records. One of the nine exemptions from disclosure under the federal FOIA is for "trade secrets and commercial or financial information that is obtained from a person and is privileged or confidential." Privileged information is information that would ordinarily be protected from disclosure in civil litigation by a recognized evidentiary privilege, such as the attorney–client privilege. Confidential information is defined as information that, if disclosed by the government, (1) may impair the government's ability to obtain necessary information in the future; (2) would substantially harm the competitive position of the person who submitted the information; (3) would impair other government interests, such as program effectiveness and compliance; or (4) would impair other private interests.[104]

Another exemption under FOIA is for internal government communications[105] and yet another important exemption is for "records about individuals if disclosure would constitute a clearly unwarranted invasion of their personal privacy."[106]

In deciding whether to release records that contain personal or private information about a person, the foreseeable harm of invading that person's privacy must be weighed against the public benefit that would result from the release.

The federal FOIA includes no exemption from disclosure for information qualified under the so-called privilege of academic freedom, which proponents trace to the First Amendment of the United States Constitution. The few cases that have considered the protection of material under an academic freedom claim in circumstances such as those an IACUC might encounter have rejected claims that this "privilege" prohibits disclosure.[107]

22:15 How does the principle of academic freedom apply to IACUC activities?

Reg. The Supreme Court of the United States has recognized the principle of academic freedom.[108] It said: "The essentiality of freedom in the community of American universities is almost self-evident. No one should underestimate the vital role in a democracy that is played by those who guide and train our youth. To impose any straitjacket upon the intellectual leaders in our colleges and universities would imperil the future of our Nation. No field of education is so thoroughly comprehended by man that new discoveries cannot yet be made. ... Teachers and students must always remain free to inquire, to study and to evaluate, to gain new maturity and understanding; otherwise our civilization will stagnate and die."[109]

Opin. Researchers are often aware of the principle of academic freedom and are interested in whether it protects from public disclosure information related to their research. Academic freedom is a multifaceted principle, invoked in support of allowing scholars to study, discuss, and publish ideas free of inappropriate restraints. This principle is invoked in a variety of contexts, from debates over tenure to challenges over publication rights. This principle is explicitly reflected in the policies of numerous universities and academic associations. In addition, it is generally acknowledged to be firmly grounded in the free speech protections of the federal Constitution and state constitutions.[110]

One facet of academic freedom is often referred to as research scholars' privilege or academic privilege. Some have argued that this privilege is rooted in the First Amendment[111] and that it protects documents that would otherwise be available under public records statutes, because releasing the information would chill research, the unfettered discussion of ideas, and other activities that underlie academic freedom.[112] This is the context in which IACUCs are most likely to face the potential applicability of the principle of academic freedom.

Unfortunately, in this context, the principle of academic privilege has generally not fared well in the courts. In *University of Pennsylvania v. Equal Employment Opportunity Commission* (EEOC), the Supreme Court of the United States firmly rejected the university's attempts to expand academic freedom to protect confidential peer review materials from disclosure.[113] The university argued that peer review materials from tenure review files deserved protection under First Amendment academic freedom principles, claiming that the disclosure of such documents to the EEOC and ultimately the public, pursuant to a subpoena, would have a "chilling effect" on the tenure evaluation process and affect the "quality of instruction and scholarship" at the university. The Court disagreed, finding the university's reliance on past academic freedom cases misplaced since under the facts of the case, there were no attempts by the government to control university speech.

A year later, in *S.E.T.A. UNC-CH, Inc. v. Huffines,* the North Carolina Court of Appeals reviewed the same issue, but this time in the context of IACUC disclosure of records pursuant to state public record laws. That court, following *University of Pennsylvania v. EEOC,* rejected the academic privilege argument and ordered the IACUC records to be disclosed.[114] Similar decisions were made by the Supreme Courts of Ohio and Washington, which both refused to expand academic freedom to protect IACUC records from disclosure to the public.[115]

So it appears that the role academic freedom plays in IACUC activities is rather small, at least as far as protecting research records is concerned. Until the Supreme Court rules differently, IACUCs would be better off understanding their state public record laws well in addition to any statutory exemptions that their legislature has seen fit to provide.

22:16 What mechanisms are available to protect against harassment and vandalism?

Reg. To intentionally damage or cause the loss of any property, including animals or records, used by an animal research facility, or to conspire to do so, is a federal crime under 18 U.S.C. §43.[116] Many states have enacted similar laws making it a crime to enter an animal facility with the intent to commit prohibited acts. The prohibited acts vary from state to state and may include: (1) interfering with facility operations; (2) defacing or damaging the facility; (3) releasing, stealing or killing animals; (4) taking photographs; and/or (5) committing acts to defame the facility. Punishment for violation of these federal and state laws can include a fine, imprisonment for a term of years or life, and a requirement for restitution, depending on the extent of the damage or loss caused.

Opin. The use of animals in research is a privilege granted to those investigators and programs that commit to meeting the highest ethical and regulatory standards. However, some opponents of animal research have engaged in criminal activity such as vandalism and harassment. Although comparatively infrequent, notable incidents have involved bombings, arson, threats of bodily harm, and publication of private information on the Internet.

An IACUC cannot prevent criminal activity by third parties, but it can adopt practices that minimize risk to researchers. For example, researchers' home addresses, telephone numbers, and social security numbers should never be requested or used in IACUC submissions. Files, offices, computers, and animal housing areas and laboratories should be locked and/or password protected. Documents containing confidential information should be shredded before disposal or recycling. Computer records should be backed up in case of theft or destruction of computers, but all computer disks and tapes should be erased or destroyed before disposal or recycling. Academic institutions in particular abhor secrecy in favor of the free exchange of information; however, they might be able to learn from their counterparts in industry about appropriate security measures. On a personal level, researchers and IACUC members would be well advised to adopt practices designed to thwart identity theft, such as shredding personal financial information, maintaining unlisted home telephone numbers, and regularly changing passwords.

Another way to protect against crimes such as harassment and vandalism is to educate the public about the nature and purpose of animal research, the care and effort devoted to protocol review, animal care, the prevention or minimization of pain, the fact that animal studies are usually necessary to satisfy FDA requirements for safety and efficacy data as a prerequisite to human trials, and the benefit that animal research has given to humankind through the introduction of new drugs, devices, and therapies. (Although FDA regulations and guidance suggest that animal studies should be performed in most cases to demonstrate safety and efficacy prior to human clinical trials of new drugs and medical devices, validated in vitro tests can substitute for certain whole animal tests.[117] In limited situations, when definitive human efficacy studies cannot ethically be conducted because doing so would expose healthy human volunteers to lethal or permanently disabling toxic substances, studies in animals alone can serve as the basis for FDA approval of a new drug for human use.[118])

The institution's legal counsel should be educated on the risks to the institution and its employees so that counsel can be prepared if an incident occurs. Counsel should also be encouraged to join the "animal law" committees of the American Bar Association and state and local bar associations, which are deliberating proposed legislation that would impair or eliminate animal research. These associations tend to be overly represented by opponents of animal research, with little balance to any debate. Participation by more institutions' legal counsel could help ensure that legislation considers all sides of the issues.

Research facilities should evaluate the applicable state criminal and civil laws regarding trespassing, harassment, and property destruction and establish institutional policies to prevent and respond to such activities. If a public facility is involved, the constitutional right to free speech also must be considered, for example, in establishing boundaries for a protest. In some states, an individual may obtain an injunction based on a reasonable belief that he or she may be injured by the commission of animal research–related vandalism or harassed because of animal research activities. If an injunction is issued to prevent anticipated vandalism or harassment, the violation of the injunction itself gives rise to legal action. Although this type of statutory provision might be valuable in some circumstances, its practical benefits are limited by the fact that it requires researchers to file court documents identifying themselves, the harassers, and the anticipated unlawful action sufficiently in advance of that action. A number of states have enacted laws that classify some of the tactics employed by animal rights groups as terrorism. The benefit of these statutes is that they impose greater criminal penalties and might therefore deter opponents from engaging in illegal behavior. Another benefit is that terroristic threats, a tactic employed by some opposition groups, are actionable. Institutions can protect researchers by lobbying state lawmakers in support of such legislation.

Research associations are often excellent resources for practical steps that can be taken to reduce the level of confrontation over animal research. The National Association for Biomedical Research and its state affiliates, for example, have helpful resources regarding community outreach, relations with animal rights groups, and response to protests.

22:17 What mechanisms should be used by an IACUC and institution to ensure research collaborations between institutions are conducted in a compliant manner?

Reg. The *Guide* (p. 15) states that institutions collaborating in research should enter into a formal written understanding (e.g., a contract, memorandum of understanding, or agreement) that addresses each party's responsibilities regarding offsite animal care and use, animal ownership, and IACUC review and oversight. Research collaborations are not addressed in the AWAR.

Opin. Generally, any research collaborations between institutions should be memorialized in a written agreement to not only address responsibilities with regard to research but also to allocate legal rights and obligations among the parties. Any research collaboration agreement (such as a contract or memorandum of understanding) should

1. Designate which institution's IACUC will be responsible for reviewing and monitoring the protocol(s)
2. Designate which institution's policies will apply to the researcher and the research
3. Establish guidelines for animal care and use activities that require cooperation and coordination among the institutions, such as animal transport
4. Establish guidelines for the sharing of information regarding the research, program(s), as well as issues arising from the research (such as lapses, terminations, suspensions, or noncompliance)
5. Establish guidelines for handling noncompliance and, specifically, whether the institutions will investigate separately or jointly

Depending on the nature of the research, collaborating institutions may choose to address other legal rights and obligations such as treatment of confidential and proprietary information, intellectual property, indemnification, insurance, publications, and use of names and trademarks. Institutions may also want to establish guidelines on how to manage certain media or public requests and/or inquiries.

By taking the time to establish a written agreement that clearly outlines each party's rights and responsibilities, collaborating institutions will create a productive relationship and environment that facilitates their research. In addition, the institutions are better situated to identify and address any noncompliance as well as to effectively communicate with governmental agencies and the public.

22:18 How should an IACUC protect confidential information and the rights of a person against whom an allegation of protocol noncompliance or animal mistreatment has been made?

Reg. Under the PHS Policy (IV,C,6) and the AWAR (§2.31,d,6) an IACUC may suspend a previously approved activity involving animals for reasons of protocol noncompliance. If the IACUC suspends an activity, the IO must, in consultation with the IACUC (AWAR §2.31,d,7; PHS Policy IV,C,7) review the reasons for suspension, take appropriate corrective action, and if APHIS/AC regulated animals are involved, report that action with a full explanation to APHIS/AC (and any federal agency funding that activity). If only PHS Policy regulated animals are involved, it is reported to NIH/OLAW with a full explanation. The PHS Policy and AWAR are silent on whether an investigator can appeal the decision by the IO. (See 29:37; 29:38.)

Defamation is a false statement about a person which causes that person to suffer harm such as mental anguish or loss of reputation. Slander is a spoken defamatory statement. Libel is a written or otherwise recorded defamatory statement. Most states recognize *per se* defamation in which the alleged defamatory statement itself is presumed to have caused damage, eliminating any need to prove that the statement caused harm to the person. Statements against a person's professional character or standing are considered *per se* defamation in most jurisdictions.

Opin. Allegations of protocol noncompliance or animal mistreatment can have a significant impact on the accused person. Such allegations can negatively affect the person's current employment as well as future professional opportunities. If allegations are not kept confidential, the person may suffer reputational harm. Reputational harm may impact the manner in which the accused person is treated by colleagues, the professional community, and even the general public if the allegations are particularly egregious. Even worse, the person may face retaliation from animal extremist groups.

IACUCs should have a policy that clearly outlines the process of investigating allegations and issuing determinations resulting from such investigations. The policy should clearly define the rights and responsibilities of the IACUC, the IO, and the accused person. The accused person should be afforded an opportunity to defend against the allegations. For example, the accused person should be allowed to provide information in the form of records, testimony, or statements. In addition, the process should allow the accused person an opportunity to appeal any corrective actions. An appeals process will give an investigator the opportunity to introduce or re-introduce information that may not have been adequately considered in determining the corrective actions.

IACUC policy should require that allegations and investigations of protocol noncompliance and animal mistreatment be treated in confidence. Allegations that are not kept in confidence can lead to reputational harm of the accused person within the institutional community or professional community. If the allegations are particularly egregious, the accused person can suffer reputational harm with the general public. If the accused person suffers significant reputational harm that impacts his or her professional future, the IACUC runs the risk of a *per se* defamation claim. In today's electronic environment, information, without any consideration for its veracity, can be distributed world-wide within seconds and retraction of such information is nearly impossible. Therefore, IACUCs should maintain and distribute documentation and communications regarding allegations and investigations in a secure, confidential manner. See 29:23.

Surv. 1 Does your IACUC allow for appeals of any sanctions imposed on an animal user? Assume that any such appeals are made to the IACUC. See survey 29:37.

Surv. 2 During an investigation of an allegation, does your IACUC have a formal policy about protecting the confidentiality of research or other records? See survey 29:14.

Surv. 3 Does your IACUC have a formal policy about protecting the rights and confidentiality of the person against whom an allegation has been made? See survey 29:14.

22:19 What role should legal counsel play in IACUC activities?

Reg. The AWAR do not specifically require an institution's legal counsel or any attorney to serve as a member of the IACUC. Under PHS Policy (IV,A,3,b), the IACUC must comprise at least five members including, at a minimum, a veterinarian, a practicing scientist experienced in animal research, an individual "whose primary concerns are in a nonscientific area (for example, ethicist, lawyer, member of the clergy)," and a member of the community who is not affiliated with the institution or a family member of a person affiliated with the institution. Thus, it is not necessary to have an institution's legal counsel serve on the IACUC.

Opin. An IACUC should maintain a good working relationship with the institution's legal counsel. Whether or not legal counsel is a voting member of the IACUC, the IACUC should seek advice from counsel regarding its legal obligations. Legal counsel can provide training to new IACUC members regarding the legal responsibilities discussed in this Chapter. Also, it is particularly important for an IACUC to seek legal advice when research poses potential compliance issues. Attorneys are trained to read and interpret the law and can serve as a valuable resource in understanding regulations, particularly when application of the regulations and guidance do not provide a clear resolution. An IACUC should seek legal advice in conducting investigations. Legal counsel can advise on how to properly conduct an investigation in accordance with the IACUC's policies and procedures, frame questions for interviews to obtain the necessary information to assess the noncompliance, and assist in determining whether acts under investigation constitute noncompliance based on the findings. In addition, legal counsel should be consulted whenever responding to third party or governmental inquiries such as FOIA requests and subpoenas.

References

1. Department of Veterans Affairs. (2011). *Use of Animals in Research*. http://www1.va.gov/vhapublications/ViewPublication.asp?pub_ID=2464.
2. Institute for Laboratory Animal Research, National Research Council of the National Academy of Sciences. (2011). *Guide for the Care and Use of Laboratory Animals*, 8th edition. Washington, D.C.: The National Academies Press. http://www.nap.edu/catalog.php?record_id=12910#toc.
3. Committees to revise the Guide for the Care and Use of Agricultural Animals in Research and Teaching. (2010). *Guide for the Care and Use of Agricultural Animals in Research and Teaching* (3rd edition). Champaign, IL: Federation of Animal Science Societies. http://www.fass.org/docs/agguide3rd/Ag_Guide_3rd_ed.pdf.
4. Council of Europe. (1991). http://conventions.coe.int/treaty/Commun/QueVoulezVous.asp?CL=ENG&NT=123.
5. 21 C.F.R. §§ 58.1–58.219.
6. 16 U.S.C. §§ 1531–1544.
7. 42 U.S.C. § 283m.
8. 16 U.S.C. §§ 703–712.
9. 18 U.S.C. § 43.
10. 5 U.S.C. § 552.
11. 35 U.S.C. §§ 200–212, with implementing regulations at 37 C.F.R. Part 401.
12. Department of Veterans Affairs. (2011). *Use of Animals in Research*. http://www1.va.gov/vhapublications/ViewPublication.asp?pub_ID=2464.
13. Committees to Revise the Guide for the Care and Use of Agricultural Animals in Research and Teaching. (2010). *Guide for the Care and Use of Agricultural Animals in Research and Teaching* (3rd edition). Champaign, IL: Federation of Animal Science Societies. http://www.fass.org/docs/agguide3rd/Ag_Guide_3rd_ed.pdf.

14. American Veterinary Medical Association. (2013). *AVMA Guidelines for the Euthanasia of Animals: 2013 Edition*. https://www.avma.org/KB/Policies/Documents/euthanasia.pdf.
15. 17 U.S.C. § 102.
16. 35 U.S.C. § 101 et seq.
17. 15 U.S.C. § 1051 et seq.
18. 18 U.S.C. § 1839.
19. 5 U.S.C. § 552a.
20. 15 U.S.C. § 1072.
21. 15 U.S.C. § 1065.
22. 15 U.S.C. § 1059.
23. 17 U.S.C. § 106.
24. 17 U.S.C. § 102.
25. 17 U.S.C. § 302.
26. 18 U.S.C. § 1839(3).
27. 35 U.S.C. § 154.
28. 35 U.S.C. § 101.
29. 35 U.S.C. § 154.
30. 35 U.S.C. § 103.
31. 35 U.S.C. § 102.
32. See, e.g., Transgenic Animal Expressing Alzheimer's Tau Protein, U.S. Patent No. 8,288,608 (filed July 9, 2003).
33. 35 U.S.C. § 200 et seq.
34. 35 U.S.C. § 202.
35. 35 U.S.C. § 202(c).
36. 35 U.S.C. § 205.
37. 35 U.S.C. § 203.
38. 5 U.S.C. § 552.
39. 45 C.F.R. § 5.65; 5 U.S.C. § 552,b,4.
40. 45 C.F.R. § 5.65.
41. 5 U.S.C. § 552(b)(6); 45 C.F.R. § 5.67.
42. 5 U.S.C. § 552(b)(5); 45 C.F.R. § 5.66.
43. See, e.g., 42 C.F.R. Part 93 (applicable to misconduct in research supported by PHS).
44. 5 U.S.C. § 552(b)(7).
45. See *Forsham v. Harris*, 445 U.S. 169 (1980).
46. *Public Citizen Health Research Group v. Dep't of Health, Educ. & Welfare*, 668 F.2d 537, 543 (D.C. Cir. 1981).
47. *Public Citizen Health Research Group v. Dep't of Health, Educ. & Welfare*, 668 F.2d 537, 545 (D.C. Cir. 1981) (Tamm, J., concurring).
48. See, e.g., FOIAdvocates, State Public Record Laws, http://www.foiadvocates.com/records.html (accessed July 9, 2013).
49. See, e.g., 7 C.F.R. Part 1 (USDA); 21 C.F.R. Part 20 (FDA); 45 C.F.R. Part 5 (HHS).
50. See, e.g., http://sunshinereview.org/index.php/Portal:WikiFOIA.
51. *Washington Research Project, Inc. v. Dep't of Health, Educ. & Welfare*, 504 F.2d 238 (D.C. Cir. 1974) (summary statements and site visit reports of initial review groups consisting of nongovernmental consultants with no legal decision-making authority held to be exempt from public disclosure as intra-agency memoranda).
52. *In Defense of Animals v. Nat'l Inst. of Health*, 543 F. Supp. 2d 83 (D.D.C. 2008); *Sangre de Cristo Animal Protection, Inc. v. U.S. Dep't of Energy*, 1998 U.S. Dist. LEXIS 23505 (D.N.M. Mar. 10, 1998) (names and identities of contractor employees permitted to be redacted).
53. E.g., *Animal Legal Defense Fund, Inc. v. Institutional Animal Care & Use Comm. of Univ. of Vermont*, 616 A.2d 224 (Vt. 1992).

54. *S.E.T.A. UNC-CH, Inc. v. Huffines*, 399 S.E.2d 340 (N.C. Ct. App. 1991) (permitting redaction of names and contact information, as well as training and experience, of researcher and staff; applications not approved could be withheld); *In re Am. Soc'y for the Prevention of Cruelty to Animals v. Bd. of Trs. of the State Univ. of N.Y.*, 556 N.Y.S.2d 447 (N.Y. Sup. Ct. 1990) (permitting redaction of names, contact information, and grant number or application number of funding source).
55. *Progressive Animal Welfare Soc'y v. Univ. of Washington*, 884 P.2d 592 (Wash. 1994) (certain information held exempt from disclosure, including personal information of researchers; raw research data and hypotheses; budget breakdowns; and peer reviewers' comments).
56. *Dorson v. State of Louisiana*, 657 So.2d 755 (La. Ct. App. 1995); *In re Am. Soc'y for the Prevention of Cruelty to Animals v. Bd. of Trs. of the State Univ. of N.Y.*, 591 N.E.2d 1169 (N.Y. 1992).
57. *State ex rel. Thomas v. Ohio State Univ.*, 643 N.E.2d 126 (Ohio 1994).
58. *Robinson v. Indiana Univ.*, 659 N.E.2d 153 (Ind. Ct. App. 1995) (withholding animal care and use applications and other information requested about individual research projects).
59. *In Defense of Animals v. Oregon Health Sci. Univ.*, 112 P.3d 336 (Or. Ct. App. 2005).
60. *United States v. Catholic Healthcare West*, 445 F.3d 1147 (9th Cir. 2006).
61. 31 U.S.C. § 3729.
62. 31 U.S.C. § 3730.
63. Codified at 42 U.S.C. § 289d(e).
64. 5 C.F.R. § 5501.101.
65. 42 C.F.R. § 50.605.
66. See *Mississippi State Univ. v. People for the Ethical Treatment of Animals, Inc.*, 992 So.2d 595 (Miss. 2008); *In Defense of Animals v. Oregon Health Sci. Univ.*, 112 P.3d 336 (Or. Ct. App. 2005).
67. 42 U.S.C. § 289d(c).
68. 42 C.F.R. § 50.605.
69. 42 C.F.R. § 50.605.
70. See *Sangre de Cristo Animal Protection, Inc. v. U.S. Dep't of Energy*, 1998 U.S. Dist. LEXIS 23505 (D.N.M. 1998)(names and identities of contractor employees were allowed to be redacted); *In re Physicians Comm. for Responsible Med. v. Hogan*, 918 N.Y.S.2d 400 (N.Y. Sup. Ct. 2010) (allowing redaction of contact information); *In re Am. Soc'y for the Prevention of Cruelty to Animals v. Bd. of Trs. of the State Univ. of N.Y.*, 556 N.Y.S.2d 447 (N.Y. Sup. Ct. 1990)(allowing redaction of names and contact information); *S.E.T.A. UNC-CH, Inc. v. Huffines*, 399 S.E.2d 340 (N.C. Ct. App. 1991) (allowing redaction of names and contact information, as well as training and experience, of researcher and staff); *In Defense of Animals v. Oregon Health Sciences University*, 112 P.3d 336 (Or. Ct. App. 2005) (holding staff member names exempt from disclosure); *Progressive Animal Welfare Soc'y v. Univ. of Washington*, 884 P.2d 592 (Wash. 1994) (holding exempt from disclosure the personal information of researchers). *Contra State ex rel. Thomas v. Ohio State Univ.*, 643 N.E.2d 126 (Ohio 1994).
71. Codified at 42 U.S.C. § 289d(e).
72. See 45 C.F.R. § 164.302 et seq.
73. *Mississippi State Univ. v. People for the Ethical Treatment of Animals, Inc.*, 992 So.2d 595 (Miss. 2008).
74. *Oklahoma Ass'n of Municipal Attorneys*, 577 P.2d 1310 (Okla. 1978).
75. *Animal Legal Defense Fund, Inc. v. Institutional Animal Care & Use Comm. of Univ. of Vermont*, 616 A.2d 224 (Vt. 1992).
76. *Dorson v. State of Louisiana*, 657 So.2d 755 (La. Ct. App. 1995).
77. *In re Am. Soc'y for the Prevention of Cruelty to Animals v. Bd. of Trs. of the State Univ. of N.Y.*, 591 N.E.2d 1169 (N.Y. 1992).
78. 5 U.S.C. § 552.
79. *Animal Legal Defense Fund, Inc. v. Institutional Animal Care & Use Comm. of Univ. of Vermont*, 616 A.2d 224 (Vt. 1992).
80. *In re Am. Soc'y for the Prevention of Cruelty to Animals v. Bd. of Trs. of the State Univ. of N.Y.*, 556 N.Y.S.2d 447 (N.Y. Sup. Ct. 1990) (permitting redaction of names, contact information, and

grant number or application number of funding source); *S.E.T.A. UNC-CH, Inc. v. Huffines*, 399 S.E.2d 340 (N.C. Ct. App. 1991) (permitting redaction of names and contact information, as well as training and experience, of researcher and staff; applications not approved could be withheld).

81. *Progressive Animal Welfare Soc'y v. Univ. of Washington*, 884 P.2d 592 (Wash. 1994) (certain information held exempt from disclosure, including personal information of researchers; raw research data and hypotheses; budget breakdowns; and peer reviewers' comments).
82. *Dorson v. State of Louisiana*, 657 So.2d 755 (La. Ct. App. 1995).
83. *In re Am. Soc'y for the Prevention of Cruelty to Animals v. Bd. of Trs. of the State Univ. of N.Y.*, 591 N.E.2d 1169 (N.Y. 1992).
84. *Robinson v. Indiana Univ.*, 659 N.E.2d 153 (Ind. Ct. App. 1995) (withholding animal care and use applications and other information requested about individual research projects).
85. *State ex rel. Thomas v. Ohio State Univ.*, 643 N.E.2d 126 (Ohio 1994).
86. *In re Physicians Comm. for Responsible Med. v. Hogan*, 918 N.Y.S.2d 400 (N.Y. Sup. Ct. 2010).
87. *In Defense of Animals v. Oregon Health Sci. Univ.*, 112 P.3d 336 (Or. Ct. App. 2005).
88. *State ex rel. Physicians Comm. for Responsible Med. v. Bd. of Trs. of Ohio State Univ.*, No. 2005-0612, 2006 WL 508325 (Ohio Mar. 15, 2006).
89. Ohio Rev. Code § 149.43.
90. 5 U.S.C. § 552.
91. *In Defense of Animals v. Nat'l Inst. of Health*, 543 F. Supp. 2d 83 (D.D.C. 2008); *Sangre de Cristo Animal Protection, Inc. v. U.S. Dep't of Energy*, 1998 U.S. Dist. LEXIS 23505 (D.N.M. Mar. 10, 1998).
92. See *Clarke v. Tri-Cities Animal Care & Control Shelter*, 181 P.3d 881 (Wash. Ct. App. 2008) (privately-run corporation providing animal control services under contract with municipal agency subject to public disclosure act).
93. *Public Citizen Health Research Group v. Dept of Health, Educ. & Welfare*, 668 F.2d 537, 545 (D.C. Cir. 1981) (Tamm, J., concurring).
94. *In Defense of Animals v. U.S. Dep't of Agric.*, 656 F. Supp. 2d 68 (D.D.C. 2009).
95. 5 U.S.C. § 552.
96. See, e.g., Margaret Snyder, *Freedom of Information and Openness*, http://www.iom.edu/~/media/Files/Activity%20Files/Research/NeuroForum/2011-JUL-26/Snyder.pdf.
97. NABR, SfN, & FASEB, *Responding to FOIA Requests: Facts and Resources*, available at http://www.nabr.org/uploadedFiles/nabrorg/Content/Animal_Law/Responding_to_FOIA_Requests.pdf.
98. *Robinson v. Indiana Univ.*, 659 N.E.2d 153 (Ind. Ct. App. 1995).
99. *Progressive Animal Welfare Soc'y v. Univ. of Washington*, 884 P.2d 592 (Wash. 1994).
100. *State ex rel. Thomas v. Ohio State Univ.*, 643 N.E.2d 126 (Ohio 1994).
101. *In re Physicians Comm. for Responsible Med. v. Hogan*, 918 N.Y.S.2d 400 (N.Y. Sup. Ct. 2010).
102. *S.E.T.A. UNC-CH, Inc. v. Huffines*, 399 S.E.2d 340 (N.C. Ct. App. 1991).
103. E.g., *Dorson v. State of Louisiana*, 657 So.2d 755 (La. Ct. App. 1995); *In re Am. Soc'y for the Prevention of Cruelty to Animals v. Bd. of Trs. of the State Univ. of N.Y.*, 584 N.Y.S.2d 198 (N.Y. App. Div. 1992).
104. 45 C.F.R. § 5.65.
105. 45 C.F.R. 5.66; see, e.g., *Washington Research Project, Inc. v. Dep't of Health, Educ. & Welfare*, 504 F.2d 238 (D.C. Cir. 1974).
106. 45 C.F.R. § 5.67.
107. See *Univ. of Penn. v. Equal Employment Opportunity Comm'n*, 493 U.S. 182 (1990); *S.E.T.A. UNC-CH, Inc. v. Huffines*, 399 S.E.2d 340 (N.C. Ct. App. 1991).
108. See *Sweezy v. New Hampshire*, 354 U.S. 234 (1957); *Keyishian v. Bd. of Regents*, 385 U.S. 589 (1967).
109. *Sweezy v. New Hampshire*, 354 U.S. 234, 250 (1957).
110. See *generally* Chang, A.W., *Resuscitating the Constitutional "Theory" of Academic Freedom: A Search for a Standard beyond Pickering and Connick*. 2001. Stan. Law Rev. 53:915; Byrne, J.P. 1989. Academic Freedom: A "Special Concern of the First Amendment." *Yale Law J.* 99:251.
111. *Keyishian v. Bd. of Regents*, 385 U.S. 589, 603-04 (1967).

112. See, e.g., Levinson-Waldman, R. (2011). *Academic Freedom and the Public's Right to Know: How to Counter the Chilling Effect of FOIA Requests on Scholarship.* http://www.acslaw.org/sites/default/files/Levinson_-_ACS_FOIA_First_Amdmt_Issue_Brief.pdf (suggesting possible ways to change laws to protect academic documents from disclosure requests).
113. *Univ. of Penn. v. Equal Employment Opportunity Comm'n*, 493 U.S. 182 (1990).
114. *S.E.T.A. UNC-CH, Inc. v. Huffines*, 399 S.E.2d 340 (N.C. Ct. App. 1991).
115. *State ex rel. Thomas v. Ohio State Univ.*, 643 N.E.2d 126 (Ohio 1994); *Progressive Animal Welfare Soc'y v. Univ. of Washington*, 884 P.2d 592 (Wash. 1994).
116. See, e.g., *United States v. Fullmer*, 584 F.3d 132 (3rd Cir. 2009).
117. See, e.g., 21 C.F.R. § 312.23(a), § 812.27(a).
118. 21 C.F.R. § 314.610.

23

General Concepts of the Program Review and Facility Inspection

Joseph D. Thulin and Kenneth P. Allen

Introduction

The AWAR, PHS Policy, and *Guide* all require the IACUC to periodically review the animal care and use program and inspect the animal facilities. While the *Guide* (p. 25), which was not written solely for a U.S. audience, recommends that these activities occur at least annually, the AWAR and PHS Policy both require them to be conducted at least every 6 months. Although the program review and facilities inspection are listed as separate IACUC responsibilities in the regulatory documents, they are integrally related. Indeed at many institutions, and perhaps at the vast majority, the two are orchestrated as a set of coordinated, semiannual events. Hence, the single term, "semiannual," is used frequently to express the collective activities of both of these IACUC functions.

It can be argued that the semiannual review of the program and inspection of facilities are the most important responsibilities of the IACUC. Admittedly, the fact that these are listed first among the several requisite functions of the IACUC in both the AWAR and PHS Policy does not necessarily indicate that the framers of these documents considered them to be the most important. Nevertheless, it is in a properly conducted "semiannual" that the IACUC most comprehensively fulfills its purpose of evaluating the care, treatment, housing, and use of animals, and for assessing the institution's conformance to regulatory requirements and adherence to adopted standards.

The material presented in this chapter builds on that of the authors of the previous edition (Stephen K. Curtis and Karen James), and much of their text is retained. Among other things, the revisions for this edition reflect the adoption of the 8th edition of the *Guide* by NIH/OLAW and AAALAC and the experiences and opinions of the current authors. While we address implementation of the program review and facilities inspection as required by the AWAR and the PHS Policy and recommended by the *Guide*, readers are cautioned that state and local laws may add other requirements.

23:1 What is meant by reviewing the program of animal care and use?

Reg. Both the PHS Policy (IV,B,1) and the AWAR (§2.31,c,1) require the IACUC to review the institution's program of animal care and use every 6 months. Neither of these documents provides a succinct definition of "program," although the PHS Policy (IV,A,1) does list the elements that must be fully described in an institution's Animal Welfare Assurance with NIH/OLAW. The *Guide* (p. 6) defines the animal care and use program as "the policies, procedures, standards, organizational structure, staffing, facilities, and practices put into place by an institution to achieve the humane care and use of animals in the laboratory and throughout the institution." The PHS Policy (IV,A,1) specifies that the institution is to use the *Guide* as the basis for evaluation, whereas the AWAR (§2.31,c,1) requires the evaluation to be based on the standards of the AWAR. None of these documents prescribe a specific methodology for conducting the program review.

Opin. The definition of the animal care and use program in the *Guide* expresses well the scope of elements to include in the program review—regardless of whether the institution is required to use the *Guide* as the basis of its evaluation for regulatory or accreditation reasons. The review should consist of a comprehensive evaluation of each of those elements listed in the above definition. In so doing and while referring to the standards for the program set forth in the PHS Policy, AWAR, and the *Guide*, as applicable, the IACUC should determine the extent to which their program meets those standards. Institutions and their IACUCs have a great deal of flexibility in determining the specific methods used to conduct the program review; however, to ensure all the facets of the program are adequately considered, it is advisable to use as guidance for the review, an outline derived from the content of PHS Policy, AWAR, and the *Guide*. The outline should list the various program facets and standards derived from the applicable standards. Institutions may develop their own outlines or use one of the many that are readily accessible through the web. NIH/OLAW provides an excellent version of a program review checklist on its website.[1] This document provides a window into the scope of review that NIH/OLAW considers appropriate for the reviews.

23:2 What are the similarities and differences between the program review and the facility inspection?

Reg. In addition to the program review (see 23:1), the PHS Policy (IV,B,2) and the AWAR (§2.31,c,2) also require the IACUC to inspect all of the institution's animal facilities, including satellite facilities/animal study areas, at least once every 6 months. The *Guide* (p. 25) recommends inspection of facilities and animal use areas at least annually. In addition to the frequency with which they must occur, the PHS Policy and AWAR have other requirements that are the same for both of these IACUC functions. For example, the AWAR require both the reviews and inspections be conducted by at least two voting members of the IACUC (see 23:8) and the PHS Policy and the AWAR have the same requirements for both the program review and facilities inspection in terms of categorizing identified deficiencies and reporting results to the IO. The PHS Policy (IV, B,3, footnote 8) states that the IACUC may determine the best means of conducting an evaluation of the institution's programs and facilities.

Opin. The program review and facility inspection are two discrete yet interrelated duties of the IACUC. The program review encompasses all of the policies, plans, standard procedures, and systems under which the institution fulfills its obligation to care for and use animals in a lawful and ethical manner. On the other hand, the facility inspection is a physical and visual assessment of the animals, the environments in which the animals are housed and used, buildings, and related equipment. An important aspect of the facility inspection is to look at records/documents associated with the care and use of animals such as animal medical records, laboratory records of animal use, animal room service logs, cage washer maintenance records, etc. During the program review the IACUC evaluates the expectations for animal care and use, while in the facility inspection the committee validates the degree to which those expectations are effectively implemented. The outcomes or conditions observed during the facility inspection can indicate deficiencies that should be addressed in the program review. For example, if all of the rodent cages are found to be poorly cleaned during a facility inspection, that condition may be due to a poor program element such as an inappropriate SOP or a poor training program. Together, facility inspection and program review provide a good internal audit of the institution's animal care and use program. See Table 23.1.

TABLE 23.1

Similarities and Differences in the Program Review and Facility Inspection

Feature	Program Review	Facility Inspection
Directed by	As assigned by IACUC Chair	As assigned by IACUC Chair
Primary focus of review	Policies, plans, procedures, organization, staffing, etc.	"Bricks and mortar," equipment, animals, and outcomes
General references in the *Guide*	Chapter 2–Animal Care and Use Program Chapter 3–Environment, Housing and Management Chapter 4–Veterinary Care	Chapter 3–Environment, Housing and Management Chapter 4–Veterinary Care Chapter 5–Physical Plant
General reference in the AWAR	Part 2, Subpart C–Research Facilities	Part 3–Standards
How review is conducted	• Talking around a table • Asking questions of management; persons who know and supervise the program • Reviewing written policies, plans, and procedures • Reviewing documentation • Receiving oral/written reports • Reporting on what is read or spoken	• Walking around the animal and research facilities • Asking questions of the animal care and research staff • Examining records typically kept in the animal facility or lab, e.g., animal health/use records • Visual observation and physical assessment • Generally reporting on what you see
Minimal recommended participation	• At least two members of the IACUC Generally need more people to conduct the review; representative persons familiar with the policies, plans, and procedures and how they are used: Veterinarians IACUC administrators Maintenance Animal care management Occupational health Research administrators	• At least two members of the IACUC • Fewer people needed; will talk with persons encountered during the review: • Animal care • Maintenance personnel • Persons doing the work in research laboratories
Inspection of animals	Generally not	Always
Physical review of laboratories	No	Yes
Ratification of findings and reports	Majority of IACUC members' signatures	Majority of IACUC members' signatures

23:3 What is the purpose of the program review and the facility inspection?

Reg. The semiannual evaluation of the animal care and use program and facilities is a primary means by which the IACUC fulfills its oversight responsibility, particularly in terms of assessing the extent to which the institution adheres to the provisions of the PHS Policy, AWAR, and *Guide*, as applicable, and developing plans to address identified deficiencies.

Opin. Most animal research facilities in the U.S. (depending on the source of funding and the species of animals used) fall under the jurisdiction of the PHS Policy or the AWAR. Both documents mandate semiannual program and facility evaluations (see 23:1 and 23:2). Therefore, an institution could take the view that these activities are necessary merely to meet regulatory expectations; however, such a narrow view would not meet the spirit and intent of the requirements.

The program review and facility inspection should be considered a central element of post-approval monitoring (PAM, see Chapter 30). The review helps ensure that animal health and well-being are optimized and pain and suffering are minimized. The inspections give the committee members time to take a firsthand look at the animals, procedures, holding facilities, and laboratories that they have been discussing during their meetings. The inspections allow investigators to evaluate their own labs continually and ideally maintain or improve the standards within their own area. Monitoring also helps ensure that all personnel are trained in the care and use of animals and protected from occupational hazards associated with animals. In addition, a well-managed and documented animal care and use program helps improve public confidence in the research conducted at the facility and engenders increased support for government-sponsored research in general. Commercial entities may find that clients will seek out laboratories that adhere to the highest standards of animal care, and clients may be willing to pay a premium for high-quality work. In other words, a well-managed animal care and use program is a vital component of good scientific research.

Surv. How useful is the semiannual program review to your IACUC?

- Of little value; we do it for compliance purposes — 29/294 (10%)
- Of moderate value; it sometimes helps us set policies, correct problems, and assure the quality of our program — 147/294 (50%)
- Very valuable; it often helps us develop new policies, correct problems, and is a key part of our quality assurance program — 117/294 (40%)
- I don't know — 1/294 (0.3%)

23:4 What laws, guidelines, and policies delineate what to identify when an IACUC inspects an animal facility and reviews the program of animal care and use?

Reg. Two basic documents mandate and govern the administrative aspects of the facility inspection and the program review: PHS Policy (IV,B) and the AWAR (§2.31,c).

(See 23:1 and 23:2.) It should be noted that PHS Policy (IV,B) requires the IACUC to use "the *Guide* as a basis for evaluation" when conducting the program review and facility inspection. In addition, if agricultural research is conducted, the *Guide for the Care and Use of Agricultural Animals in Agricultural Research and Teaching* (Ag Guide)[2] should be consulted.

Opin. Other references also should be used to evaluate an animal care and use program. The primary standard for conducting an animal care and use program is the *Guide.* It is customary for research facilities to follow the *Guide* even if they have no legal obligation to do so. Both the *Guide* and Ag Guide contain extensive references; a few of the more frequently used references are the following:

- Semiannual program review and facility inspection checklist[1]
- U.S. Government Principles for the Utilization and Care of Vertebrate Animals Used in Testing, Research, and Training (part of the PHS Policy)
- Occupational Health and Safety in the Care and Use of Research Animals[3]
- Biosafety in Microbiological and Biomedical Laboratories[4]
- Guide to the Care and Use of Experimental Animals[5]
- Institutional Animal Care and Use Committee Guidebook[6]
- Tutorial: PHS Policy on Humane Care and Use of Laboratory Animals[7]

23:5 How frequently should program reviews and facility inspections be conducted?

Reg. PHS Policy (IV,B,1; IV,B,2) and the AWAR (§2.31,c,1; §2.31,c,2) require inspection of the animal facilities and review of the program of animal care and use at least once every 6 months. The *Guide* (p. 25) states that these activities "... should occur at least annually or more often as required (e.g., by the Animal Welfare Act and PHS Policy)."

Opin. It is generally accepted that the interval between reviews and inspections should not exceed 6 months. However, that is not usually interpreted to mean that the interval cannot exceed 182.5 (365 ÷ 2) days! Nor is it usually interpreted so strictly that if a review is completed on May 15, the institution is out of compliance if the next review is not completed by November 15. Anecdotally, NIH/OLAW and USDA/APHIS/AC will accept a leeway of up to a month, i.e., evaluations conducted more than seven months apart would be considered non-compliant by these agencies. Nevertheless, conventional wisdom dictates that major departures from the 6-month rule should be prevented. (See 24:6.)

23:6 Can the twice-yearly inspections occur in November and again during the following January?

Reg. (See 23:5.)

Opin. No. (See 23:5.) The regulatory language dictates an inspection every 6 months, not twice a year. The IACUC may choose to conduct inspections more frequently, but not less frequently. Consequently, if the committee followed a November inspection with another the following January, the next inspection would be due 6 months later, i.e., in July. It is advisable, therefore, to plan on conducting inspections at 6-month intervals, in this case either May and November or January and July.

23:7 Must the program review and facility inspection be conducted at the same time?

Opin. No. At small institutions, the IACUC often will conduct both the program review and the facility inspection on the same day. At larger facilities, however, time constraints typically make it necessary to conduct the reviews over several days or even weeks during the reporting period. At very large facilities, inspections may be accomplished by inspecting one sixth of the facility each month during a 6-month period. It can be advantageous to conduct the facility inspection prior to the program review in that some program deficiencies might only come to light during the inspection. (See 23:13.) Alternatively, the IACUC might elect to discuss and compare at a convened meeting the results of both the program review and inspection prior to preparation of the final report to the Institutional Official. A sample semiannual schedule is shown in Table 23.2.

23:8 Does the entire IACUC need to participate in the program review and facility inspection?

Reg. To comply with the AWAR (§2.31,c,3), a subcommittee participating in the facility inspection and program review must be composed of at least two committee members and "no Committee member wishing to participate in any evaluations... may be excluded." Under PHS Policy (IV,B,3, footnote 8, 2002 reprint) the IACUC "may, at its discretion, determine the best means of conducting an evaluation of the institution's programs and facilities." The AWAR (§2.31,c,3) require that the majority of the members of the IACUC review and sign the report.

Opin. Commonly, the entire IACUC conducts the program review and facility inspection; however, this is not necessary. At many institutions, one or more IACUC subcommittees are charged to conduct the reviews and facility inspections. This is particularly apt for large institutions in that spreading the workload among multiple subcommittees can reduce the overall burden borne by individual committee members. It should be noted that regardless of the method chosen, the IACUC retains responsibility for the animal care and use program. As such, the full IACUC should review the final results of the review and inspection and approve the final report before submission to the IO.

TABLE 23.2

Sample IACUC Inspection and Program Review Schedule

Event/reporting period[a]	April 1–September 30	October 1–March 31
Program review	1st week of October	1st week of April
Facility inspection	2nd week of October	2nd week of April
Semiannual report	November	May

[a] Do not confuse this reporting period with that for the annual APHIS/AC or AAALAC report. This reporting period is merely a 6-month period for the semiannual report to the IO. Any 6-month period could be used.

23:9 Minimally, who should be included on each program review and inspection team?

Reg. (See 23:8.)

Opin. There are large variations in how institutions form teams/subcommittees to conduct the program review and inspection. Ideally, a typical team would consist of several IACUC members with at least one scientist, one veterinarian, and, where possible, one unaffiliated member. Other useful members (not necessarily from the IACUC) are the animal care supervisor, a person with occupational health experience, and a member of the building operations/maintenance department. These internal consultants can round out the committee's subject matter expertise in their evaluation of the program and facility. No IACUC member who wishes to participate should be excluded from participation (see 23:8). A useful practice at large facilities is to have as one member of the team with prior knowledge of the facility, someone who participated in the last evaluation, and a new member to provide a fresh or unbiased review (see 23:21). It is important to keep the inspection and review process vibrant so that the process does not become routine and just another walk around the facility or a "rubber stamping" of the program.

Surv. Who participates in the semiannual program review at your institution? More than one response is possible.

• A previously selected IACUC subcommittee (which reports back to the full IACUC)	92/292 (32%)
• A previously selected IACUC subcommittee (which does not report back to the full IACUC)	6/292 (2%)
• The entire IACUC at all times	189/292 (65%)
• Selected external persons as needed (e.g., maintenance department representative)	51/292 (17%)
• Other	22/292 (8%)

23:10 Can outside consultants be used by the IACUC during inspection and review?

Reg. The PHS Policy (IV,B,3, footnote 8, 2002 reprint) states that the "... IACUC may invite ad hoc consultants to assist in conducting the evaluation. However, the IACUC remains responsible for the evaluation and report." The AWAR (§2.31,c,3) have similar language.

Opin. Yes; however, two members of the IACUC must participate in the inspection and review to satisfy the AWAR (see 23:8). Most institutions possess ample expertise within their own organization and ad hoc consultants typically are not necessary for facility inspection and review purposes. In practice, outside consultants (other than veterinary consultants retained by some institutions) are not commonly used. Yet in situations where internal expertise is limited or additional validation is sought, the use of an outside consultant can be valuable. For example, an ad hoc consultant might be employed when evaluating ABSL-3/4 laboratory space if the IACUC does not possess the necessary expertise to evaluate the facility. The use of internal consultants may also be desirable (see 23:9).

23:11 Can a recent AAALAC site visit be used in lieu of a semiannual program review and facility inspection?

Reg. According to NIH/OLAW Notice OD-00-007,[8] a site visit by AAALAC can be used as the semiannual program review and inspection of the animal facilities, provided the requisite elements of PHS Policy (IV,B,3) are met. (See 23:8.) According to a clarification provided by APHIS/AC during the review of this chapter, an AAALAC site visit can be used as a semiannual program review and inspection of animal facilities as long as the requirements under AWAR §2.31c and §2.35a3 are fulfilled such as

- The IACUC adheres to the requisite frequency of program reviews and facility inspections. (See 23:5; 23:6.)
- At least two members of the IACUC must participate in both the program review and facility evaluation components of the site visit.
- All members of the IACUC must have the opportunity to participate with the AAALAC site visitors in the program review and facility evaluation.
- A properly convened quorum of the IACUC should review a summary of the findings from the site visit, determine whether any identified deficiencies are minor or significant, decide on suitable corrective action, and set a schedule for corrections to be completed.
- Committee actions relative to the review of the site visit findings need to be properly recorded in the official minutes of the IACUC.
- Minority opinions must be included if submitted.
- Relevant sections of the AAALAC site visit report are available to the USDA inspector.

Opin. Yes. An AAALAC site visit entails a comprehensive review of the animal care and use program and facility, and the IACUC may conduct a semiannual program review and facility inspection in conjunction with the site visit. In some cases, particularly at larger institutions, the AAALAC site visitors might not evaluate all the facilities that require a semiannual inspection by the IACUC, e.g., all laboratories where survival surgery is conducted. In such cases, the committee will need to ensure those areas are inspected subsequent to the site visit. It should be noted that the IACUC may add to the list of deficiencies and might decide to exclude other items identified by AAALAC. Action taken by the IACUC to formalize the AAALAC site visit should be completed shortly after AAALAC presents their initial verbal recommendations. Since the AAALAC visit is going to be a type of IACUC subcommittee, it would be best to convene a regular IACUC meeting soon after the AAALAC site visit. The purpose of the meeting would be to discuss the AAALAC findings and any other issues/deficiencies identified by the IACUC and to endorse the report as a formal IACUC review. If too much time is allowed to elapse, appropriate action may be delayed including the timely correction of deficiencies. Timely committee action will also aid in writing the post–AAALAC visit response, which usually is due to AAALAC within 10 business days after the site visit.

23:12 What is an effective method for conducting the program review?

Reg. Footnote 8 of the PHS Policy (August 2002 reprint) states that the IACUC may, at its discretion, determine the best means of conducting an evaluation of the institution's programs and facilities.

Opin. The AWAR and PHS Policy do not specify the method to be used for conducting the semiannual program review. As such, there are large variations in the methods employed among institutions. However, there are two accepted formats in common use. One involves a discrete review at 6-month intervals of the entire program, typically at a single meeting or group of meetings scheduled within a relatively short window of time. The other involves an ongoing or rolling review process in which a portion of the program, e.g., program management (*Guide* pp. 13–24) is reviewed as part of normal (generally monthly) IACUC meetings, such that the entire program is evaluated in the meetings held over the course of 6 months. The authors prefer the former in that a discrete review conducted over a short period of time can simplify planning and documentation processes. However, we acknowledge that the rolling review method can help with time management for some committees. In all instances, either the full committee or a subcommittee may conduct the review.

Regardless of the format used, the committee's review process should address all programmatic aspects described in the *Guide* (p. 11). As mentioned previously, checklists based on the outline of the *Guide* can be very helpful in this regard (see 23:1). Often, all of the information required to determine whether a particular standard is being met is not at hand when merely reviewing a checklist. As such, many IACUC's have implemented procedures to ensure relevant information is presented to the committee for evaluation. Commonly, these methods include interviews of or presentations/reports by key program personnel, and review of program documents such as IACUC policies/procedures and animal facility standard operating procedures, and research records.

Surv. What methods does your IACUC use to conduct its semiannual program review of humane care and use of animals? More than one response is possible.

- We use the NIH/OLAW checklist that is on the OLAW website 232/295 (79%)
- We use a checklist prepared by our own institution 102/295 (35%)
- We interview key people (e.g., veterinarian, IACUC Chair) 88/295 (30%)
- Our IACUC chairperson or IACUC office staff prepare a list of items to discuss 131/295 (44%)
- We review selected Standard Operating Procedures of the animal facility 135/295 (46%)
- We review selected research records 83/295 (28%)
- We have special presentations regarding various aspects of the animal care and use program 38/295 (13%)
- We review current IACUC policies and procedures 186/295 (63%)
- We do part of our program review at every full committee meeting 58/295 (20%)
- Other 10/295 (3%)

23:13 Can the review of animal care and use programs be based on observations made during the inspection of the facilities? (See 23:2; 23:7)

Opin. While observations made during the facility inspection certainly do provide information relevant to the program, facility inspection findings should not be used as the sole basis for the program review. Conditions in the animal facility and laboratories where animals are used can reflect the overall quality of the program, and the IACUC should determine whether deficiencies identified during the inspection have programmatic implications. However, assessment of the facility inspection results would not likely result in the evaluation of all aspects of the program. More than a visual inspection of the facilities is required to determine whether the animal care and use program is operating in accordance with federal requirements. For example, it is not likely that problems with the composition of the IACUC or documentation of the occupational health program will be identified during a facility inspection. Therefore, the IACUC should have discrete processes for both the program review and facility inspection.

23:14 Can the review of the animal care and use program be conducted while inspecting the facilities?

Opin. Although conceivable, especially at institutions with very small programs, it is not advisable for a committee/subcommittee to carry out the program review and inspection simultaneously. As suggested in 23:13, it would be highly unlikely that a physical inspection of the facility would provide information sufficiently representative of the state of all aspects of the program. In addition, it generally would be impractical to attempt to go through a list of program elements and determine the extent to which the standards for those elements are met while simultaneously viewing and assessing the facility. For these reasons, separate (either temporally or by different subcommittees) sessions for the program review and facility inspections are recommended.

23:15 Should the IACUC evaluate the animal facility disaster plan as part of the program review? If so, what are the requirements for a disaster plan?

Reg. The *Guide* (pp. 35, 74–75) requires facilities to have a disaster plan that takes into account both animals and humans. The animal facility plan should be part of the overall institutional disaster plan. Some key elements of the plan include: the actions necessary to prevent animal pain, distress, and deaths due to loss of critical systems/infrastructure; how the facility will preserve animals that are necessary for critical research activities or are irreplaceable; and preparation and training of key personnel, e.g., the AV and colony manager, having responsibilities in the event of a disaster; and how significant personnel absences will be handled. In December 2012, APHIS/AC published in the Federal Register (Vol. 77, No. 250, p. 76815) final rules amending the AWAR to add requirements for contingency planning. The AWAR requirements (§2.38,i,4; §2.38,l) are generally similar to the *Guide* recommendations; however, there is the explicit requirement to provide and document the training of personnel regarding their roles and responsibilities outlined in the facility's plan. NIH/OLAW has published guidance on this topic.[9,10]

Opin. Since it is an important component of the overall animal care and use program, the IACUC should evaluate the animal facility disaster plan as part of the semiannual program review. It is a relatively straightforward task to compare the disaster plan content against the guidance listed above. However, it is advisable for IACUCs to assess the currency and readiness of the plan. To this end, the committee should ask questions of the animal facility director/manager and perhaps institutional safety personnel about training staff concerning the plan. Questions might include: What training is given to the animal care staff regarding the disaster plan? Does management ever conduct drills or tabletop exercises to simulate disaster situations and practice responses? What processes are in place to ensure that all aspects of the plan are kept current, e.g., lists of responsible personnel, emergency notification contact lists/calling trees, etc.? Questions such as these will help the IACUC determine whether the disaster plan is merely a document or a set of processes that are ready to be implemented should the need arise. An NIH/OLAW webinar on disaster planning can be viewed at http://grants.nih.gov/grants/olaw/educational_resources.htm.

23:16 How much time should the IACUC expect to devote to conducting the program review?

Opin. There are many factors that influence the amount of time the IACUC devotes to the program review. Most important are the specific methods and level of detail chosen to perform the review. If an IACUC decides to have a single meeting in which they merely run down a checklist of program components, that committee might complete its work in as little as one hour for each review. However, such a quick and probably superficial review is not likely to yield very valuable results and could result in an inadequate assessment of the program. An appropriate and complete program review requires the input of individuals and groups having the proper expertise to conduct an evaluation of the program as it exists and determine the extent to which it meets (or deviates from) the regulatory requirements and other standards. In well-functioning programs, the IACUC can rely on program area experts, e.g., the AV, IACUC support staff, and occupation health and safety personnel, to provide the committee information and preliminary assessments of the various aspects of the program. This preliminary work can save time for the committee as a whole. Assuming there is adequate preparation and depending on the size and complexity of the program, it is reasonable to expect that a formal subcommittee meeting to review the entire program can usually be completed within a period of about 2 hours. However, the state of the program and identification of deficiencies can lengthen that time. Further full committee review will take additional time—up to 2 hours.

It is worth noting that the amount of time required to complete a thorough program review is not necessarily proportional to the size of the institution. Small institutions might lack certain program elements, such as biocontainment facilities, that add complexities to the review; however, all institutions regardless of size subject to the same general requirements for program organization, implementation, and oversight. Table 23.3 illustrates a typical review process.

TABLE 23.3

Typical Steps for the Program Review Process

Time Frame	Activity	Who	What	Time Required
2–4 weeks prior to a meeting	Individual review: remind area experts about pending review; suggest they review appropriate checklist and evaluate their specific part of the program	Area expert: veterinarians, occupational health, facilities maintenance, animal care, IACUC coordinator, training coordinator, other key persons Note: Some area experts will also be committee members	Using a checklist or the *Guide*, individual will review his/her aspect of the program along with that of the regulatory requirement	Varies with program; a large dynamic program will take more time Problems identified at this level should be corrected prior to any further action; this part of the review is essentially a self-evaluation
Subcommittee meeting day	Subcommittee review	Should include at least two members of the IACUC as well as area experts as indicated	Collectively reviews the program in terms of the combined and accumulated knowledge of the area experts Makes recommendations to the full committee	Up to 2 hours; longer meeting may be unproductive
Within 30 days of subcommittee review	Full IACUC review	Members of the IACUC and invited area experts as desired	Review findings of the subcommittee; establish corrective action and schedules of completion Significant deficiencies may need more expeditious handling	Up to 2 hours; longer meeting may be unproductive; with good preparation at previous levels meeting may be much shorter
Within 30–45 days of full committee review; some significant deficiencies may need more timely action	Post any administrative paperwork	IACUC Administrator Note: At small facilities a committee member may serve as the Administrator	Complete minutes, send out notice of required corrective action, and complete other required documentation Significant deficiencies may need more expeditious handling	As long as needed; should generally complete process in 30–45 days
See above	Final report	IACUC Chair and IACUC Administrator	Signed by committee and forwarded to the IO	See above

Surv. Approximately how much time is devoted by the IACUC membership to the semiannual program review? Estimate the total time as an entity, not the cumulative time of each individual member.

- Not applicable, we do not do semiannual reviews 1/295 (0%)
- Approximately 1–2 hours 97/295 (33%)
- Approximately 2–4 hours 115/295 (39%)
- Approximately 4–6 hours 39/295 (13%)
- Over 6 hours 40/295 (14%)
- Don't know 3/295 (1%)

23:17 Can alternate members of the IACUC perform semiannual inspections that involve PHS Policy and AWA regulated species?

Reg. The AWAR and PHS Policy do not address this issue specifically. As stated previously, the AWAR (§2.31,c,3) requires the inspection team to consist of at least two IACUC members. NIH/OLAW guidance on this topic states, "An alternate may only contribute to a quorum and function as an IACUC member if the regular member for whom they serve as alternate is unavailable to participate in IACUC business, whether it is because that member is unable to attend the meeting, has to leave the meeting early or arrive late, or is recused from participating due to a real or potential conflict of interest."[11] As such, if the alternate functions as an IACUC member during semiannual program review and facility inspections, they may do so only if the regular member for whom they serve is unavailable. Using alternates to conduct inspections solely to reduce the workload of regular members of the IACUC (for example, using an alternate to serve on a different inspection group from the regular member) would not be consistent with the cited guidance. However, the guidance does not preclude alternate members from participating as *ad hoc* consultants (see 23:10) or in addition to the two regular members required by the AWAR.

Opin. Alternates should participate in the semiannual program reviews and facility inspections. Clearly, an alternate member may participate instead of the regular IACUC member when the regular member is unable to take part in the inspection. Beyond that, however, the participation of alternate members should be viewed as an important element in their development. Alternate members should be appropriately trained and experienced to fulfill their role, i.e., to substitute for an unavailable regular member. Given the centrality of the semiannual program review and inspection to the IACUC's function, it follows that alternate members should be thoroughly familiarized with this process. In instances where alternate members are inexperienced, this is best achieved by their participation as *ad hoc* consultants or "extras" with teams of at least two regular voting members. But even for experienced alternates, e.g., former regular members, this form of participation is a good way for them to maintain familiarity with the program and process. Alternates developed in this way will be better prepared to step in for a regular member when required. (See 5:24.)

Surv. Does your IACUC allow an alternate member of the committee to perform inspections if the IACUC position for which he/she is the alternate is performing inspections at the same time but in a different location? Assume this question is for USDA covered species only.

- Not applicable 140/292 (48%)
- Yes 69/292 (24%)
- No 65/292 (22%)
- No response 18/292 (6%)

23:18 What areas should the IACUC include in the semiannual facility inspection?

Reg. Both the PHS Policy (IV,B,1; IV,B,2) and the AWAR (§2.31,c,1; §2.31,c,2) require the IACUC to inspect the institution's animal facilities once every 6 months. (See 15:1.) The PHS Policy requires the inspection to include "satellite facilities," which are defined in the Policy (III,B) as those areas that are "...outside of a core facility or centrally designated or managed area in which animals are housed for more than 24 hours." However, NIH/OLAW has further expanded the definition to include "...areas where any form of surgical manipulations (minor, major, survival, non-survival) are performed..."[12] The AWAR (§1.1) requires the inclusion of "animal study areas," which are defined as areas "...outside of a core facility or centrally designated or managed area in which animals are housed for more than 12 hours." The *Guide* (p. 25) recommends that in addition to the animal facility the inspection include "animal use areas"; however, the term is not further qualified.

Opin. The semiannual facility inspection should include at a minimum all areas, whether they be for the purpose of animal housing, conducting animal procedures, or for support functions, e.g., cage wash, that are within the "envelope" of the central animal housing facility(ies), those areas that meet the definitions of satellite facility and/or animal study area, and performance areas such as imaging, behavioral testing units, specialized housing, transport vehicles. In our opinion, however, best practice would be for the committee to include not only those areas specified above, but also all areas where animals are used—even areas where animals are taken for non-invasive procedures or euthanasia and tissue collection. Yet, it could be exceedingly difficult for some IACUCs, particularly at large institutions, to visit every six months every single laboratory/area where animals are used. In such cases, it is essential that the IACUC implement alternative means of assessing those activities and spaces. NIH/OLAW[13] has stated that when considering IACUC responsibilities for semiannual review, it is important to keep in mind that each Assured institution, acting through its IACUC or facility veterinarian, is responsible for all animal-related activities at the institution regardless of where the animals are maintained or the duration of their stay. NIH/OLAW FAQ E.1 states that "Institutions have discretion with regard to how they oversee areas used for routine weighing, dosing, immunization, or imaging, but should monitor such areas on a random or fixed schedule to effectively oversee activities at the institution." Therefore, IACUCs have broad discretion in the way they oversee animal work that is conducted in laboratories. The degree, frequency, and method of IACUC oversight of these other animal use areas may depend on the nature of the activity or the species involved. The methods may involve, for example, random or scheduled visits (which may be less frequent than semiannual) by IACUC members, the veterinary staff, or PAM staff. Whatever method is chosen, the authors recommend that documentation/reports of the oversight activities be presented to the full committee for its consideration by the committee and inclusion in the semiannual report to the IO. (See 24:12.)

Surv. What areas does your IACUC include in your semiannual facility inspection? More than one answer is possible.

- Central animal holding areas 289/295 (98%)
- Small departmental animal holding areas 229/295 (78%)
- Laboratories holding animals for over 12 or 24 hours, depending on the species 200/295 (68%)
- Laboratories holding animals for any length of time 224/295 (76%)
- Service areas such as cage wash, food storage, cage storage 287/295 (97%)
- Survival surgery areas 264/295 (89%)
- Nonsurvival surgical areas 241/295 (82%)

23:19 Can an IACUC inspection team inspect areas more often than twice a year?

Reg. (See 23:5.)

Opin. Yes, facilities may be inspected as often as the IACUC deems necessary. The PHS Policy and the AWAR do not preclude IACUC inspections more frequently than once every 6 months. An IACUC may "reset the clock" if it wishes to perform an inspection early and then perform the next inspection 6 months later. More frequent inspections may be required to verify that previously identified concerns have been corrected. Some IACUCs conduct inspections on an ongoing basis. For example, an IACUC may set aside a time each month to visit laboratories where animal research is conducted. Under this system, an individual laboratory may be inspected more frequently. It is also a common practice to evaluate new areas before housing animals.

Members of the animal care staff, such as directors and managers, are often members of the IACUC. In addition, at least one veterinarian must be a member. These individuals are usually in the animal facility on a regular basis between formal IACUC inspections and, therefore, have the opportunity to evaluate and report to the IACUC more frequently than every 6 months. Individual(s) performing PAM visits on behalf of the committee also should report findings to the IACUC on a frequent basis and many animal holding/use areas will be evaluated more than twice per year as a result of such inspections combined with the semiannual inspections.

23:20 Should animal facilities be inspected only if animals are present at the time of inspection?

Reg. (See animal facility definitions in 15:1.)

Opin. Not necessarily. Within the central animal facility there may be holding rooms that are empty or procedure/surgery rooms that are not in use at the time of the inspection. These areas still should be evaluated for acceptability for the intended use. Similarly, animals might not be present in satellite facilities or other animal use areas outside of the central facility at the time the IACUC inspection. In these cases, the committee should still evaluate the spaces and review records and interview staff about the activities in order to assess their adequacy. If it is likely that animals will never again be housed or used in a particular area, the area can be removed from the list of those locations to be inspected.

23:21 What information should be provided to the facility inspection team, either before or during the inspection, about the animal activities that are approved for each area they inspect?

Opin. In order to conduct a proper evaluation, the team needs to have some knowledge about the animal activities that are approved for the areas they inspect. At the outset of the inspection, the team should have knowledge of the approved protocols, species used, and general categories of procedures performed, e.g., survival surgery, in each location that is to be inspected. This information usually can be included in a list provided to the inspection team. During the inspection, it is advisable for the committee to ask questions of the investigators and research staff concerning the details of how animal procedures are conducted. In addition, the team should have the capability of referencing in real-time the currently approved protocols. This permits the comparison of what is observed with what has been approved by the IACUC. Methods to accomplish this may include bringing hard-copies of the approved protocols along on the inspection or ensuring access to electronic documents at computer stations, laptops, or mobile computing devices.

Surv. During the semiannual facility inspection, how do the inspectors know what procedures are being performed (or were performed) on protocols that were approved for an investigator or educator? More than one answer is possible.

- Not applicable — 14/294 (5%)
- We provide the inspectors with broad-based information on the investigator's approved protocols (e.g., survival surgery on mice) — 116/294 (39%)
- We provide the inspectors with fairly detailed information on the investigator's approved protocols (e.g., survival splenectomy on mice) — 66/294 (22%)
- We give the inspectors the protocol number and expect them to ask the appropriate questions — 38/294 (13%)
- We do not give the inspectors any information but expect them to ask the appropriate questions — 67/294 (23%)
- We provide hard-copies of the investigator's protocols to the inspectors — 36/294 (12%)
- We provide electronic access to the investigator's protocols via a mobile device (e.g., laptop, tablet, etc.) — 45/294 (15%)
- Don't know — 2/294 (1%)
- Other — 34/294 (12%)

23:22 Should the facility inspection team evaluate animal medical records and animal experimentation records?

Opin. Yes. When conducting the semiannual inspection, the IACUC is only getting a "snap shot" view. They are evaluating the conditions and practices as they are at the time of the inspection. Reviewing animal records is an effective way of expanding the temporal dimension of the inspection. Animal health and experimentation records are important in substantiating that over time animal care and use is provided consistent with applicable standards and IACUC approvals. Aspects of animal care and use that are particularly important to evaluate are those having to do with prevention and amelioration of pain and distress. These would include the administration of anesthetics/analgesics and monitoring parameters associated with humane endpoints. When inspecting the central animal facility, the team should review a sampling of animal medical records kept by the veterinary staff. The records selected for review might be based on prior observations of the animals. For example, if the team observes nonhuman primates that have had surgical procedures, they might choose the records of those particular animals for review. Similarly, if the team observes post-surgical rats in the animal facility, they might choose to review the investigator's animal use records for those particular animals. By reviewing these records the team gains useful insight about the care actually provided to animals. (See also 18:38; 18:39; 26:10–26:15; 27:21; 27:26; 27:35.)

Surv. During semiannual inspections, do your inspectors usually look at medical records, including the use of anesthetics and analgesics (if applicable)?

- Not applicable 19/293 (6%)
- We ask them to do so but I don't know if they actually do look at records 20/293 (7%)
- They routinely look at medical records during semiannual inspections 175/293 (60%)
- They are not obligated to look at medical records during semiannual inspections 73/293 (25%)
- Don't know 2/293 (1%)
- Other 4/293 (1%)

23:23 When doing semiannual inspections, is it a good idea to have some IACUC members "specialize" in certain areas year after year, or is it a better idea to have members rotate among areas?

Opin. There is benefit to having inspection teams consisting of a combination of people who have inspected an area, or areas, previously in addition to new members. Retaining members on a team that have inspected area(s) previously is particularly valuable as it facilitates an historical perspective. Such individuals are available to offer their previous experience regarding the areas and to answer questions the new members may pose regarding the previous inspection. Conversely, changing or adding new members to an inspection team will add a "new set of eyes." The new members may offer fresh perspectives, or different expertise to the inspection team. This scenario is beneficial to all members involved and should result in a more thorough evaluation than scenarios where members repeatedly inspect the same areas or rotate every inspection.

Surv. For semiannual inspections, do you always send the same people to the same places or do your rotate the responsibilities?

• We are a small institution and everybody goes to the same places during every inspection	122/293 (42%)
• We are large enough to rotate assignment areas but we usually do not do that	12/293 (4%)
• We usually try to rotate assignment areas	158/293 (54%)
• I don't know	0/293 (0%)
• Other	1/293 (0.3%)

23:24 Should the IACUC allow a previously unused area to be used for live animal–related work without a formal IACUC inspection?

Reg. Both the PHS Policy (IV,B,6) and AWAR (§2.31,c,6) indicate that the IACUC should "review and approve, require modifications in (to secure approval) or withhold approval of those activities related to the care and use of animals." According to NIH/OLAW, "Institutions are responsible for oversight of all animal-related activities regardless of how long or where the activity occurs."[12] The *Guide* (p. 25) indicates that IACUC oversight functions include "regular inspection of facilities and animal use areas."

Opin. No research project involving live animals should be conducted prior to a review and approval by the IACUC. As a component of that review the IACUC should assess the suitability of the proposed space for planned animal-related work. The methods for inspecting the area should be similar to those used for conducting the semiannual inspection, for example, forming an inspection team of at least two members of the IACUC to evaluate the appropriateness of new animal housing or study area. A common approach is to have the AV (or other program veterinarian) and one additional IACUC member conduct the inspection and report back to the full committee. In any case, work on the protocol should not commence until the area has been evaluated and approved by the IACUC. (See 25:4.)

Surv. Does your IACUC allow a previously unused area to be used for live animal related work (e.g., survival rodent surgery) without a formal IACUC inspection?

• Yes, routinely without any additional conditions	6/295 (2%)
• Yes, but the PI must notify the IACUC prior to use	6/295 (12%)
• Veterinary review and approval is required by our IACUC	101/295 (34%)
• One or more IACUC members must review and approve the area	107/295 (36%)
• I don't know	11/295 (4%)
• Other	34/295 (12%)

23:25 Is there a conflict of interest when the director of the animal facility (an IACUC member) participates on a team that inspects the core animal facility?

Reg. The PHS Policy (IV,C,2), AWAR (§2.31,d,2), and *Guide* (p. 26) all contain language indicating the need for recusal of IACUC members from deliberations on animal use proposals for which they have a conflict of interest, i.e., are personally involved. However, none of these documents have a similar statement with regards to the facility inspection. To the contrary, with respect to the semiannual program review and facility inspection the AWAR (§2.31,c,3) state that "no committee member wishing to participate in any evaluation conducted...may be excluded."

Opin. There might be a perception of a conflict of interest because the director, who often is the AV, would be in a position of having to criticize his or her own area of responsibility. However, in practice this is seldom a problem. Animal program directors desire a well-managed and regulation-compliant animal care facility/program since they are often faced with annual unannounced inspections by APHIS/AC and periodic review by AAALAC. Using this same logic, animal program directors often seek input from others regarding the animal facility since programmatic improvements are in their best interest. Directors seldom hide problems; rather, they are the ones to point out problems and potential problems to newer members of the team. Due to his or her background, the director is often the most knowledgeable individual on the IACUC regarding animal care standards. Thus, the director, as well as others involved in management of the animal facility, are valuable resources for other committee members and should be available to answer questions during the facility inspection of the animal facility.

Nevertheless, if there remains concern about the perception of a conflict of interest, the IACUC may consider assigning members other than the facility director to formally conduct the review and permit the director to participate as an observer and to provide information. If the IACUC believes that the director, particularly if the director also is the AV, must be excluded for fear of interfering or otherwise corrupting the inspection process, this is a sign of more serious programmatic problems that are beyond the scope of this text.

Surv. To avoid potential conflicts of interest, does your IACUC attempt to assure that during semiannual inspections the attending veterinarian or other member(s) of the IACUC who are responsible for the daily operations of your central animal facility(s) do not inspect the central animal facility(s)?

- Not applicable 29/294 (10%)
- Yes, we try to prevent this from happening 90/294 (31%)
- No, we make no particular effort to prevent this from happening 163/294 (55%)
- I don't know 1/294 (0%)
- Other 11/294 (4%)

23:26 Is there a conflict of interest when an investigator who also is an IACUC member participates on a team that inspects their own laboratory?

Reg. (See 23:25.)

Opin. As with the scenario in 23:23 regarding the animal facility director, this situation can be, and typically is, considered a conflict of interest by the IACUC. In this case, the individual should not perform the primary evaluation of their own research laboratory; rather, other committee members should serve in that capacity. But just as the facility director should not be precluded from participating in the evaluation of the animal facility, neither should an investigator be excluded from the inspection of their laboratory because they are an IACUC member. Just as the facility director is a valuable resource during the inspection of the animal facility, the researcher/investigator, whether or not they are an IACUC member, should be present during the laboratory inspection so that they may answer questions and provide information to the inspection team as necessary.

Surv. To avoid potential conflict of interest, does your IACUC attempt to assure that during semiannual inspections investigators do not inspect their own laboratories?

• Not applicable	24/291 (8%)
• Yes, we try to prevent this from happening	201/291 (69%)
• No, we make no particular effort to prevent this from happening	63/291 (22%)
• I don't know	0/291 (0%)
• Other	3/291 (1%)

23:27 Should facility inspections be announced or unannounced?

Opin. Opinions differ, but most institutions favor announced inspections. There are no requirements in the AWAR, *Guide*, or PHS Policy stipulating announced or unannounced inspections. The advantages of both have been briefly discussed.[6] An unannounced inspection gives the inspection team a clear picture of what is routinely occurring in the animal facility and laboratories where animals are used. On the other hand, announced inspections allow essential persons to be available to answer questions. Announced visits also give investigators the incentive to review their areas and procedures with their lab personnel; they may motivate some PIs to elevate the standards within their labs, especially if they are provided with a listing by the IACUC of what are some common types of infractions. These higher standards will ideally prevail through the entire year. As there are benefits from both announced and unannounced inspections, an IACUC may want to try both approaches to determine what best fits their particular institution. In any case, the choice is that of the individual IACUC. (See 25:6.)

In addition to semiannual IACUC inspections, IACUC oversight of animal activities may be enhanced by a PAM program and these visits may be announced and/or unannounced. PAM is discussed in the *Guide* (pp. 33–34). PAM visits may supplement semiannual inspections so such inspections do not need to be unannounced. In addition, PAM visits are generally targeted for one specific area or laboratory and a more thorough evaluation of the animal care/procedures can be performed than during a semiannual inspection. PAM visits offer additional oversight that supplements semiannual facility inspections. (See Chapter 30.)

Surv. Are your facility inspections announced or unannounced?

- Not applicable; we do not do semiannual facility inspections 0/294 (0%)
- Usually announced 214/294 (73%)
- Usually unannounced 66/294 (22%)
- We leave it to the discretion of the inspectors 12/294 (4%)
- I don't know 0/294 (0%)
- Other 2/294 (1%)

23:28 Should the IACUC inspectors speak with individuals participating in research using animals?

Opin. Yes, this is a generally accepted practice and is highly desirable. There are two main groups of people with whom the committee should interact during the inspection: the research staff and the animal resource staff. Conversations with individuals from these groups often yield valuable information about the program and facilities. For example, through informal interviews at the time of the inspection, it is possible to verify that SOPs are being followed, that persons are appropriately trained, and that anesthesia and euthanasia methods are appropriate and in accordance with approved protocols. This allows the inspection team to be more efficient and obtain the information they seek in a timely manner.

Some institutions require that each laboratory have a representative present to answer questions from the inspection team. Other IACUCs schedule inspections at times when persons are normally working and question those persons available on a less formal basis. The inspection process can and should be a learning experience for both the IACUC and the staff. Free and open communication facilitates this learning process.

The IACUC should take time at the end of a visit to give feedback to the persons involved. When warranted, positive as well as negative comments should be made. IACUCs may consider having an "exit interview" with interested persons as a means of exchanging information.

23:29 What is a deficiency, and what criteria are used to categorize a deficiency as minor or major (significant)?

Reg. Neither the PHS Policy nor the AWAR explicitly define the term "deficiency." However, the implicit definition is any unapproved departure from the provisions of the *Guide*, PHS Policy, or AWAR. (See PHS Policy IV,B,3; AWAR §2.31,c,3). Regarding the criteria that distinguish a significant deficiency from a minor, the AWAR and PHS Policy are in concordance: "A significant deficiency is one which... in the judgment of the IACUC and the Institutional Official, is or may be a threat to the health or safety of the animals" (AWAR §2.31,c,3; PHS Policy IV,B,3).

Note: While many IACUCs use "major" and "significant" interchangeably, the term "major" is not present in the regulatory lexicon as a qualifier or category of deficiencies. NIH/OLAW offers guidance regarding minor deficiencies and notes, "Generally, a minor deficiency refers to a problem for which an immediate solution is not necessary to protect life or prevent distress (e.g., peeling or chipped paint)."[14]

Opin. Given that a deficiency is any unapproved departure from the provisions of the regulatory standards,[15] it follows that a thorough knowledge of those standards is necessary in order to identify a deficiency. One could argue that for any deficiency the IACUC should be able to specifically identify by chapter and verse the regulatory provisions from which there has been an unapproved departure. Once the committee has determined that a finding represents an unapproved departure from the regulatory standards, i.e., that it is a deficiency, the IACUC can then consider whether the finding represents an actual or potential threat to the health and safety of the animals. If the IACUC judges that it does, then the finding will be categorized as significant. If not, the finding by default is considered minor.

While it often is straightforward to determine if a finding meets the threshold of being a deficiency and if the deficiency meets the criteria to be classified as significant, there are many cases when such discrimination is not easy. The IACUC actually has broad discretion in this regard. It is critical that the IACUC ensures that its membership includes individuals having in-depth knowledge of the regulatory standards, as well as an understanding of the nuances of their application. The authors recommend that all deficiencies be presented at a regularly convened meeting of the IACUC, so that the full committee can make the final determination as to the nature of deficiencies identified in the program review or facility inspection.

Surv. What criteria does your IACUC use to determine if a deficiency found on a semiannual inspection is minor or significant?

- Not applicable 0/293 (0%)
- We have a list of commonly found deficiencies that our IACUC has classified as minor or significant 31/293 (11%)
- We let the inspectors decide if the deficiency would be harmful to the health or safety of the animal 94/293 (32%)
- We let the IACUC decide if the deficiency would be harmful to the health or safety of the animal 148/293 (51%)
- We let either the IACUC chairperson or the IACUC administrative office decide if the deficiency would be harmful to the health or safety of the animal 12/293 (4%)
- I don't know 1/293 (0.3%)
- Other 7/293 (2%)

23:30 During semiannual program review and inspection, who decides if a deficiency is a significant deficiency or a minor deficiency?

Reg. (See 23:29.)

Opin. It should be noted that according to the regulatory guidance provided (AWAR §2.31,c,3; PHS Policy IV,B,3) both the IACUC and IO have joint responsibility for deciding if a deficiency is significant. That being said, having the inspection team/subcommittee preliminarily classify findings as minor or significant at the time of the inspection can be beneficial. For example, if a finding is preliminarily called significant, a discussion with appropriate individuals in the animal facility or laboratory, as appropriate, can be initiated so that the deficiency is corrected as soon as possible. In addition, this initial determination by the inspection team will facilitate a subsequent review by the full committee. As noted above, the authors' advice is that the full committee should make the final determination as to whether a deficiency is registered as minor or significant in the report to the IO. (See 23:34.)

Surv. During semiannual inspections, who decides if a deficiency is a significant deficiency or a minor deficiency?

- Not applicable — 0/295 (0%)
- We ask the inspectors to make that determination — 42/295 (14%)
- The inspectors make that determination and then the full IACUC either agrees to or modifies that decision — 101/295 (34%)
- The full IACUC makes that determination — 79/295 (27%)
- We have a subcommittee make the initial determination then the full IACUC either agrees to or modifies that decision — 22/295 (7%)
- The IACUC chairperson makes the decision — 10/295 (3%)
- The IACUC administrative office makes the decision — 2/295 (1%)
- We use an inspection checklist that defines categories of minor/significant deficiencies — 34/295 (12%)
- I don't know — 0/295 (0%)
- Other — 5/295 (2%)

23:31 What are some examples of minor and significant deficiencies?

Reg. (See 23:29.) NIH/OLAW offers guidance.[14]

Opin. Any deficiency that could cause injury, death, or severe distress to animals or people would be considered significant. Examples could include failures in heating, ventilating, and air-conditioning systems and the related electrical systems and power failures of sufficient duration to affect important areas such as surgical suites. Not only can program or facility deficiencies be included as significant (if they meet the definition of the term) but accidents and natural disasters can result in a reporting of a significant deficiency. For example, if a hurricane tore off a roof of an area where animals were kept, and injury, death or severe distress to the animals resulted, it can be reported as a significant deficiency, although Acts of God are not generally considered deficiencies. Additional examples of significant deficiencies would include deviations from an approved protocol such as withholding analgesics without IACUC approval, or performing unapproved surgery, particularly if such procedures result in clinical signs of illness in research animals. If significant deficiencies are identified at an NIH/OLAW Assured institution, and are either programmatic or PHS supported, they must be reported to NIH/OLAW.

Although the PHS Policy and AWAR distinguish between significant and minor deficiencies only in terms of the actual or potential threat to the health or safety of the animals and people, some institutions elect to classify as significant other serious deficiencies that may not directly affect animals. One example of this might be a deficiency in the occupational health and safety program that is judged to threaten the health or safety of personnel. The authors concur that there are some specific situations, such as the example noted, when it would be reasonable for the IACUC to classify as significant a serious deficiency that does not impact animal health or welfare; however, we caution against a broad expansion of the classification so, for example, there is a complete blurring between a significant deficiency and an event that requires reporting to a regulatory agency. (See 23:39.) NIH/OLAW offers guidance on reportable incidents.[16]

Examples of minor deficiencies would include expired food, unsealed wood in an animal surgery area, missing ceiling tiles, a dripping faucet, and small cracks in the wall. These deficiencies do not directly impact the health of research animals, or pose a direct threat to their safety. They are, however, problems that should be addressed.

23:32 Should deficiencies and possible corrective actions be discussed at the time of the inspection?

Opin. In general, yes, although opinions vary. If the inspection team identifies a deficiency they consider to be significant, the responsible individual (investigator, facility manager, or veterinarian) should be informed immediately of the problem so they can correct it as soon as possible. Delaying notification about a potentially significant deficiency until the entire committee is informed only prolongs the problem. The IACUC can always request additional action be taken to correct a deficiency at a later date. By discussing the deficiency with the appropriate persons at the time of the inspection, the inspecting subcommittee can make sure that their recommended course of action and date for correction are workable before presenting them to the entire IACUC for review.

There are times when it might be prudent to withhold discussion of findings at the time of the inspection, pending a full IACUC discussion. If the reviewing subcommittee is unsure of the nature of the deficiency, it is advisable to withhold judgment. On the other hand, if a problem is serious but corrective action is not immediately required, then consideration by full committee review is warranted. For example, while reviewing animal records the inspecting subcommittee discovers that an investigator had performed major survival surgical procedures on rodents but did not administer postoperative analgesics as specified in the IACUC-approved protocol; however, all the animals had been euthanized prior to the inspection. Although this is a serious noncompliance issue, and the health and the well-being of the animal was affected at the time of the surgery and during recovery, no immediate intervention may be required by the reviewers. Nevertheless, the reviewers should immediately report their findings so that the IACUC can assess the situation and, if warranted, take immediate action. In accordance with NIH/OLAW's guidelines, the IACUC would also determine whether the violation warranted reporting to NIH/OLAW, depending on the funding source for the activity, its impact on PHS-supported work, or if it was found to be a program-wide problem.

23:33 An IACUC has a policy of requiring investigators to euthanize animals by two methods (e.g., carbon dioxide followed by a physical method) to assure the animals are dead. During a semiannual inspection it is determined that a PI is using only one of the two approved methods. Is this considered a minor or a significant deficiency?

Reg. (See 23:29.) NIH/OLAW offers guidance on this issue.[14,16]

Opin. Minor deficiencies are problems for which immediate resolutions are generally not necessary to protect life or prevent distress. If the IACUC considered it necessary to require two methods of euthanasia to ensure death, it would follow that if only one method were used, there would be a possibility that the animal would be in distress if a quick and painless death did not occur. Thinking in this mode would lead one to list it as a significant deficiency. There would be no question about this if during the inspection animals were found in a disposal refrigerator/freezer and conscious. If animals were euthanized using only one method and no distress was found by the inspection team, either by direct observation or from speaking to personnel working in the lab, the committee might decide to classify it as a minor deficiency. The infraction would be the investigator's lack of adherence to the methodology described in the protocol only. That being said, NIH/OLAW guidance on reporting includes, "Failure to adhere to IACUC-approved protocols" and this would be considered a reportable event based on NIH/OLAW guidance.[16]

23:34 How are results of inspections and reviews recorded and processed and do semiannual inspection reports need to be signed by a majority of IACUC members?

Reg. The PHS Policy (IV,B,3; IV,E,1,d), AWAR (§2.31,c,3), and *Guide* (p. 25) all require that upon completion of the review and inspection a report is to be forwarded to the IO. However, only the AWAR (§2.31,c,3) requires the report to be signed by a majority of the IACUC members. The HREA of 1985 (§495,b,3,C) references a requirement for certification that the review has been conducted, but does not specify how this is to be done. (See 6:1; 23:37–23:40). NIH/OLAW guidance specifies that "Final reports of the semiannual evaluations and inspections are considered full committee actions and should be reviewed and endorsed by a majority of the IACUC."[17]

Opin. Three methods are often used to document the facility inspection and program review:

- Completed review/inspection forms or checklists
- IACUC minutes
- Summary reports

The method or format used depends on the complexity of the institution and the preference of the IACUC. The results/findings of reviews and inspections should be recorded at the time the evaluations are conducted. Checklists such as those provided by NIH/OLAW (see 23:1) or customized forms are valuable tools for recording outcomes. Raw data from the inspections and reviews should be compiled as needed to facilitate IACUC review of the results, tracking of follow up actions, and inclusion in the semiannual report to the IO. The use of spreadsheet or database software is highly useful in this regard and is strongly recommended. In cases where hard copy raw data subsequently are transcribed into an electronic format, e.g., spreadsheet or database, the original hard copy documents should be retained. The final report to the IO may include an executive summary of the IACUC's findings. However, the report should in all cases include the details of any programmatic or facility deficiency identified by the committee. Often such details, particularly for minor facility deficiencies, are provided in the form of an appendix to the report. The written report must be certified, and the most effective way to document certification is via signatures of the IACUC members at a convened meeting that involves a discussion regarding the facility inspection and program review.

23:35 How should the IACUC decide whether a deficiency item identified during the facility or program review is actually included on the final report?

Reg. The PHS Policy (IV,B,3) and the AWAR (§2.31,c,3) both state that the report to the IO "...must identify specifically any departures from the provisions..." of the respective regulatory standards, i.e., the PHS Policy, the *Guide*, or the AWAR. NIH/OLAW guidance provides numerous examples of significant deficiencies.[17]

Opin. All items that the IACUC determines to be deficiencies, whether minor or significant, should be included in the report. It is important for the IACUC to have a robust process for determining whether program review and inspection findings are judged to be deficiencies and whether deficiencies are classified as minor or significant. (See 23:32.) IACUCs should avoid the temptation to exclude items because they think they will reflect poorly on the institution. The semiannual report to the IO must include identification of any deficiencies in the program or facility, including classification as either significant or minor and a reasonable and specific plan and schedule for correction. Reports with minor items listed demonstrate that the inspection process is working.

23:36 How about findings that may be insignificant, for example, a single crowded cage in a room of 1000 mouse cages or a burned-out light fixture in an animal room? Should findings such as these be included in the report, and should the IACUC have a mechanism to track them?

Reg. (See 23:29 and 23:35.)

Opin. The threshold here should be whether the IACUC considers a finding to be a deficiency or not. If so, it should be included in the report and its correction tracked. If the IACUC determines that the finding does not represent a departure from the provisions of the regulatory standards, i.e., is not a deficiency, then it need not be included in the report nor tracked.

IACUCs frequently encounter items that are of some import but do not meet the criteria for a programmatic deficiency. Such could be the case in the discovery of one overcrowded mouse cage in a room of 1000 cages. Depending on the specifics, an IACUC might or might not deem this to be a deficiency. But even if it is not, it would be prudent to point out the overcrowded cage to the animal care and/or investigative staff so they can attend to it. Similarly, a single burned out light fixture might not have sufficient impact on the room environment to rise to the level of a deficiency, but it still is worthwhile to take note of it so that it can be addressed. While isolated items like these might not end up as deficiencies and accounted in the report, there may be other non-deficiency items that do warrant a mention in the report. Examples could include concern about aging equipment that has not yet failed or outmoded procedures or drugs for which better options are available. The IACUC may wish to have these kinds of findings included in the report under headings such as, "Other Comments," "Notes," or "Recommendations."

23:37 How does your IACUC follow up on significant deficiencies to assure they have been corrected within the time frame specified by the IACUC?

Reg. In accordance with the PHS Policy (IV,B,3) and AWAR (§2.31,c,3), "a reasonable and specific plan and schedule" must be given for correction of each deficiency. It does not matter whether the deficiency is designated as minor or significant. Furthermore, "any failure to adhere to the plan and schedule that results in a significant deficiency remaining uncorrected shall be reported in writing within 15 business days by the IACUC, through the IO, to APHIS/AC and any Federal agency funding that activity" (AWAR §2.31,c,3). In accordance with the PHS Policy (IV,F,3), "The IACUC, through the Institutional Official, shall promptly provide OLAW with a full explanation of the circumstances and actions taken with respect to... any serious or continuing noncompliance with this Policy [or] any serious deviation from the provisions of the *Guide*." Furthermore, a recent policy notice from NIH/OLAW states that prompt reporting is required if there is a "failure to correct deficiencies identified during the semiannual evaluation in a timely manner."[16]

Opin. IACUCs should devise plans that are attainable within the timetable established. Significant deficiencies must be handled immediately and plans formulated to ensure the health and well-being of the animals. The criteria for determining appropriate deadlines for correction are set in concert with a reasonable timetable for the type of correction required. Thus, for significant deficiencies, there may be two timetables—one to ensure the animal's health and well-being immediately and a second to correct the deficiency. The type of deficiency discovered will determine the appropriate timetable for correction. For example, a facility issue (bricks and mortar) may take longer to correct than a programmatic issue in which a correction can be put into motion just by changing a policy. An issue that can be corrected by the IACUC, animal care staff, or investigative staff may be addressed more rapidly than a problem that involves outside contractors and facility issues, which can involve a long procurement process or allocation of funding. Temporary "quick fix" solutions may be used while planning and implementing the final solution.

Surv. How does your IACUC follow up on significant deficiencies to assure they have been corrected within the time frame specified by the IACUC? More than one response is possible.

- Not applicable 1/292 (0.3%)
- We have no follow up procedure 8/292 (3%)
- The original inspectors perform the follow up and report back to the IACUC 50/292 (17%)
- The IACUC administrative office performs the follow up 122/292 (42%)

- The IACUC chairperson designates the person(s) who will perform the follow up — 136/292 (47%)
- I don't know — 2/292 (1%)
- Other — 36/292 (12%)

23:38 Who decides the appropriate resolution of specific deficiencies? How is this done?

Reg. Both the PHS Policy (IV,B,3) and AWAR (§2.31,c,3) require the IACUC to report on the deficiencies found during the facility and program review and to categorize them as either minor or significant (major). The IACUC is required to develop a "reasonable and specific plan and schedule" for correcting each deficiency. A deficiency is resolved when the IACUC determines that it has been corrected.

Opin. The IACUC has the ultimate responsibility for determining what actions are needed to correct a deficiency. However, it is best to involve the stakeholders who will be affected by the committee's decision. As appropriate, the IACUC should consult the AV, the investigator, and the maintenance department. If major funding is needed to correct an item, it will also be necessary to consult the IO or other administrative persons to set a reasonable plan for correction. Such involvement makes for a more collegial atmosphere and ultimately a better solution.

23:39 What action is taken if a significant deficiency is not corrected in accordance with the established schedule?

Reg. In some cases institutions are required to report uncorrected significant deficiencies. The AWAR (§2.31,c,3) are explicit in this regard, requiring that "Any failure to adhere to the plan and schedule that results in a significant deficiency remaining uncorrected shall by reported in writing within 15 business days by the IACUC, through the Institutional Official, to APHIS and any Federal agency funding that activity." While not specifically citing significant deficiencies, the PHS Policy (IV,F,3) requires prompt reporting of any serious or continuing noncompliance with the PHS Policy, any serious deviation from the provisions of the *Guide*, or any suspension of an activity by the IACUC. If a significant deficiency meets any of these criteria, NIH/OLAW must be notified.[16]

In addition to the reporting requirement specified above, the IACUC has the option of suspending the activity associated with the significant deficiency. The AWAR (§2.31,c,8; §2.31,d,6) authorize the IACUC to suspend an activity involving animals if it is not being conducted in accordance with the description of that activity provided by the PI and approved by the committee. This may only occur at a convened meeting of a quorum of the IACUC with the suspension vote of a majority of the quorum present. The PHS Policy (IV,C,6) allows the IACUC to "suspend an activity that it previously approved if it determines that the activity is not being conducted in accordance with applicable provisions of the Animal Welfare Act, the *Guide*, the institution's Assurance...the IACUC may suspend an activity only after review of the matter at a convened meeting of a quorum of the IACUC and with the suspension vote of a majority of the quorum present." As applicable, suspensions must be reported to APHIS/AC and any federal funding agency as per AWAR §2.31,d,7.

Opin. Because a significant deficiency has, or may have, an adverse effect on animal health and safety it must be corrected quickly. In some instances the initial action need not be the final solution, but the action must protect the animals. For example, if a dog run is found with a broken fence with potential to cause an injury, the immediate action would be to remove the dog to a suitable enclosure. The action will correct the significant deficiency. The dog run can then be fixed as time permits.

Another option that may be taken by the IACUC in response to a lack of attention regarding correction of a significant deficiency is the suspension of an animal activity. Typically, this is reserved only for situations involving an immediate threat to animal health or well-being that is not resolved or a continued lack of compliance in regard to a significant deficiency or deficiencies, and most IACUCs use this only in situations with continued noncompliance or egregious violations. IACUCs should use due diligence to assure that corrective action is taken in a timely manner.

23:40 For how long should facility inspection and program review records be maintained?

Reg. PHS Policy (IV,E,2) states "All records shall be maintained for at least three years; records that relate directly to applications, proposals, and proposed significant changes in ongoing activities reviewed and approved by the IACUC shall be maintained for the duration of the activity and for an additional three years after completion of the activity." The AWAR (§2.35,f) uses similar wording.

Opin. In practice, there are two opinions regarding record keeping. Many institutions maintain records indefinitely because of the difficulty in segregating records into activities that are either complete or ongoing. At these institutions, records are kept as long as storage is available. However, some institutions have concerns about storing unnecessary records because of freedom of information laws. These institutions generally store minutes and other general records for only 3 years and specific protocols for the duration of the project plus 3 years. It is unusual, in the authors' opinion, for regulatory bodies to request records older than 3 years. It should be noted that many institutions currently use, or are currently considering, electronic protocol management systems. With such systems, records that date beyond 3 years may be identified with minimal effort and deleted from the system. Conversely, such records can be archived and maintained indefinitely depending on institutional preference. (See 22:12–22:14.)

23:41 How should the IACUC address minority opinions relative to the semiannual inspection and program review?

Reg. Both PHS Policy (IV,F,4) and AWAR (§2.31,c,3) mandate that minority reports, if any, be included with the semiannual report and PHS Policy requires minority views to be included in the annual report to NIH/OLAW.

Opin. Minority reports are infrequent because most IACUCs are able to devise reasonable compromises to difficult problems. Dissenting members generally do so by voting "no" on an issue and requesting that a notation be made in the regular committee minutes. However, a member(s) may write a full minority opinion if so desired. This opinion must be included as part of the official report.

23:42 Who receives the findings from the program review and facility inspection?

Reg. In accordance with PHS Policy (IV,B,3) and the AWAR (§2.31,c,3), the IACUC must submit a report of the inspection and program review to the IO.

Opin. It also is useful to submit a copy of the report to those who are responsible for correcting deficiencies, such as the facility management, the AV, and the facilities maintenance department. Members of the IACUC should receive a copy of the report as well.

23:43 Other than the IO and the IACUC, who has legal access to the IACUC's semiannual reports to the IO?

Reg. According to NIH/OLAW, semiannual reports must be submitted to NIH/OLAW if requested.[17] PHS Policy (IV,E,2) requires that "All records shall be accessible for inspection and copying by authorized OLAW or other PHS representatives at reasonable times and in a reasonable manner." In addition, PHS Policy (IV,A,2) requires institutions lacking AAALAC accreditation must submit its most recent semiannual report from the IACUC to NIH/OLAW. According to the AWAR (§2.38,a) "each research facility shall furnish to any APHIS official any information concerning the business of the research facility which the APHIS official may request in connection with the enforcement of the provisions of the Act and standards."

Opin. The regulatory verbiage used may be interpreted to mean that NIH/OLAW and APHIS/AC may request semiannual reports, but they should only receive semiannual reports for cause or as part of the Assurance negotiation noted above for institutions lacking AAALAC accreditation. Legal access depends on whether the facility is private or public and depends on applicable federal regulations and state laws. Legal counsel should be sought to address this issue. (See 22:4.)

23:44 Is the IACUC responsible for compliance oversight for the use and proper storage of federally controlled drugs used in animal-based research?

Reg. PHS Policy (IV,B,2) indicates that the IACUC must inspect all of the institution's animal facilities once every six months using the *Guide* as a basis for evaluation. The *Guide* (p.34) suggests that the "handling and use" of controlled substances be addressed during semiannual inspections. The sample semiannual program review checklist from NIH/OLAW includes a section on drug storage and control.[1] Even though some guidance is given, there is no regulatory requirement that the IACUC serve as the regulatory body for DEA–regulated drugs that are used in research animals.

Opin. No. The IACUC shares with the AV the responsibility for ensuring the use of all drugs (particularly anesthetics, analgesics, and anti-infective agents) in animal subjects is consistent with current accepted laboratory animal veterinary practice and congruent with the activities approved by the IACUC. However, this responsibility does not imply that the IACUC's regulatory compliance role should include enforcement of federal, state or local controlled substances laws and regulations. The institution should provide resources/support for a separate program to monitor storage and use of controlled substances. For example, a controlled drug officer could provide oversight for the ordering, use, and destruction of controlled substances. During the semiannual inspection, an IACUC indeed should look at controlled and non-controlled drugs and how they are stored. However, the committee's primary focus should be on making sure that the drugs are used in accordance with the approved research protocols and within established expiration dates. That said, if the committee observes deviations from controlled substances storage or recordkeeping requirement, it may still be considered a minor deficiency, and the appropriate institutional personnel should be notified.

Surv. During semiannual inspections, do the inspectors check for expired drugs (including controlled substances, if applicable)?

- Not applicable 10/295 (3%)
- Yes, they do for all drugs 232/295 (79%)
- Yes, they do for all drugs except for controlled substances 16/295 (5%)
- No, they do not 33/295 (11%)
- I don't know 0/295 (0%)
- Other 4/295 (1%)

23:45 Can the IO overrule a significant deficiency finding of the IACUC on a semiannual report?

Reg. The PHS Policy (IV,B,3) and AWAR (§2.31,c,3) both state that "a significant deficiency is one which...in the judgment of the IACUC *and* [emphasis added] Institutional Official, is or may be a threat to the health or safety of the animals."

Opin. The regulatory language could be interpreted to mean that a deficiency is significant only if both the IACUC and IO judge it to be so. In practice, however, the IO typically does not participate in the IACUC's process for determining whether a deficiency is significant and the IO is merely informed of the committee's designation, e.g., via the semiannual report, after the fact. Rarely, a situation may arise in which the IO does indeed disagree with the IACUC with respect to the designation of a significant deficiency. In such a case, it would not mean that the IO is "overruling" the committee. It may just mean that the deficiency might not technically be considered significant because both parties did not designate it so. If this situation were to occur, it would be in the best interest of the institution if the IACUC and IO would jointly consider the matter and come to an agreement in terms of the severity of the deficiency. If the IO and the IACUC could not come to an agreement, consultation with the regulatory agencies (i.e. NIH/OLAW and/or APHIS/AC, as applicable) may be warranted.

23:46 Do all conditions that jeopardize the health or well-being of animals have to be reported to NIH/OLAW?

Reg. PHS Policy (IV,F,3) indicates that "the IACUC, through the IO, shall promptly provide OLAW with a full explanation of the circumstances and actions taken with respect to (a) any serious or continuing noncompliance with this Policy; (b) any serious deviation from the provisions of the *Guide*; or (c) any suspension of any activity by the IACUC." NIH/OLAW offers guidance regarding reporting.[16,18] A number of situations that would require reporting are included such as any "conditions that jeopardize the health or well-being of animals, including natural disasters, accidents, and mechanical failures, resulting in actual harm or death to animals." There are situations where animal morbidity and mortality are noted but reporting is not warranted such as, "death of animals that have reached the end of their natural life spans... death or failures of neonates to thrive when husbandry and veterinary medical oversight of dams and litters was appropriate... infrequent incidents of drowning or near-drowning of rodents in cages when it is determined that the cause was water valves jammed with bedding (frequent problems of this nature, however, *must* be reported promptly along with corrective plans and schedules)."

Opin. No. The IACUC should utilize professional judgment with respect to reporting while adhering to the guidance provided by NIH/OLAW. If a committee cannot come to a consensus regarding a decision on reporting, NIH/OLAW may be consulted for guidance in such situations.

23:47 In the event of a naturally occurring disaster such as a hurricane, burst water pipes, etc. in which animals are not hurt or distressed, is it necessary to report the same to NIH/OLAW or AAALAC (for accredited institutions)?

Reg. NIH/OLAW offers specific guidance on this issue and states, "inoperable HVAC, electrical or watering systems, failure of such systems sufficient to affect critical housing and operational areas, and situations such as natural disasters that cause injury, death, or severe distress to animals"[14] and "conditions that jeopardize the health or well-being of animals, including natural disasters, accidents, and mechanical failures, resulting in actual harm or death to animals"[16] are reportable. The *Guide* (pp. 74–75) offers some verbiage regarding requirements for an emergency plan and required response in the event of an emergency. See 23:15.

Opin. Natural disasters that cause severe distress, injury, or death are considered significant deficiencies. As a significant deviation from the *Guide*, and based on the guidance from NIH/OLAW, such deficiencies must be reported. This issue has been addressed in the literature and as a significant deviation from the *Guide*, it must be reported to NIH/OLAW.[16] AAALAC requires accredited institutions to report adverse events as a result of "Natural disasters" and the results of "OLAW/USDA investigations."[19]

Acknowledgments

The foundational work of Stephen K. Curtis and Karen James, authors of the previous edition of this chapter, is gratefully acknowledged.

References

1. National Institutes of Health, Office of Laboratory Animal Welfare. 2012. Semiannual Program Review and Facility Inspection Checklist. http://grants.nih.gov/grants/olaw/sampledoc/cheklist.htm.

2. Committees to Revise the Guide for the Care and Use of Agricultural Animals in Agricultural Research and Teaching. 2010. *Guide for the Care and Use of Agricultural Animals in Agricultural Research and Teaching*, 3rd ed. Savoy: Federation of Animal Science Societies.
3. National Research Council. 1997. *Occupational Health and Safety in the Care and Use of Research Animals*. Washington, D.C.: National Academy Press.
4. U.S. Public Health Service, Centers for Disease Control and Prevention. 2007. *Biosafety in Microbiological and Biomedical Laboratories*, 5th edition, Washington, D.C.: U.S. Government Printing Office.
5. Canadian Council on Animal Care. 1993/1984. *Guide to the Care and Use of Experimental Animals*. Vols. I and II. Ottawa: Canadian Council on Animal Care.
6. Applied Research Ethics National Association/Office of Laboratory Animal Welfare. 2002. *Institutional Animal Care and Use Committee Guidebook*, 2nd edition. Bethesda: National Institutes of Health.
7. National Institutes of Health, Office of Laboratory Animal Welfare. 2011. Tutorial: *PHS Policy on Humane Care and Use of Laboratory Animals*. http://grants.nih.gov/grants/olaw/tutorial/index.htm.
8. National Institutes of Health, Office of Extramural Research. 1999. Notice NOT-OD-00-007, Utilization of AAALAC activities as semiannual program evaluation. http://grants.nih.gov/grants/guide/notice-files/not-od-00-007.html.
9. National Institutes of Health, Office of Laboratory Animal Welfare. 2012. Frequently Asked Questions, G.3, Do awardee institutions need animal facility disaster plans? http://grants.nih.gov/grants/olaw/faqs.htm#instresp_3.
10. National Institutes of Health, Office of Laboratory Animal Welfare. 2012. Disaster planning and response resources. http://grants.nih.gov/grants/olaw/disaster_planning.htm.
11. National Institutes of Health, Office of Laboratory Animal Welfare. 2011. Notice NOT-OD-11-053, Guidance to reduce regulatory burden for IACUC administration regarding alternate members and approval dates. http://grants.nih.gov/grants/guide/notice-files/NOT-OD-11-053.html.
12. National Institutes of Health, Office of Laboratory Animal Welfare. 2012. Frequently Asked Questions, E.1. Should the IACUC inspect laboratories or other sites where investigators use animals? http://grants.nih.gov/grants/olaw/faqs.htm#prorev_1.
13. National Institutes of Health, Office of Laboratory Animal Welfare. 2008. The Public Health Service responds to commonly asked questions. http://grants.nih.gov/grants/olaw/references/ilar91.htm.
14. National Institutes of Health, Office of Laboratory Animal Welfare. 2012. Frequently Asked Question, E.2, How does the IACUC distinguish between significant and minor deficiencies? http://grants.nih.gov/grants/olaw/faqs.htm#prorev_2.
15. National Institutes of Health, Office of Laboratory Animal Welfare. 2012. Departures from the *Guide*. http://grants.nih.gov/grants/olaw/departures.htm.
16. National Institutes of Health, Office of Laboratory Animal Welfare. 2005. Notice NOT-OD-05-034, Guidance on prompt reporting to OLAW under the PHS Policy on Humane Care and Use of Laboratory Animals. http://grants.nih.gov/grants/*Guide*/notice-files/not-od-05-034.html.
17. National Institutes of Health, Office of Laboratory Animal Welfare. 2011. Semiannual report to the institutional official. http://grants.nih.gov/grants/olaw/sampledoc/ioreport.htm.
18. Office of Laboratory Animal Welfare. 2013. Reporting noncompliance. http://grants.nih.gov/grants/olaw/reporting_noncompliance.htm.
19. Association for Assessment and Accreditation of Laboratory Animal Care International. 2012. Frequently Asked Questions, I.2. Maintaining accreditation. Reporting requirements. http://www.aaalac.org/accreditation/faq_landing.cfm#I2.

24

Inspection of Animal Housing Areas

Christine A. Boehm and Debra L. Hickman*

Introduction

Animal housing facilities must be inspected by the IACUC at least once every 6 months, according to the AWAR (§2.31,c,2) and PHS Policy (IV,B,2). Nevertheless, aside from this simple directive, a few subsequent policy clarifications, and the detailed "standards" of animal housing and care provided in the AWAR and the *Guide*, these regulatory authorities allow institutions the flexibility to develop programs that function effectively and efficiently for their specific research program. An IACUC may approach implementation of the inspection requirement in a variety of ways. The information provided in this chapter is intended to assist IACUCs in developing an effective animal housing facility inspection program tailored to the needs of their individual institutions.

Information, ideas, and opinions expressed in this chapter were compiled from the results of surveys of IACUCs at institutions located throughout the United States. The first survey, conducted in 1997, gathered information from 93 IACUCs on how their inspections were managed and administered. The most recent survey included information from over 290 IACUCs from institutions of varying size, type, and culture.

While the information provided in this chapter may represent the way in which many, or even most, IACUCs conduct animal housing and care inspections, the common practices described should not be interpreted as the best practices for every individual institution. The methods that an institution's IACUC employs depend on the size, type, and culture of the institution. The "best" inspection methods are likely to be very different for large versus small facilities, centralized versus decentralized facilities, corporate organizations versus educational institutions, and so forth. IACUCs are encouraged to exercise their professional judgment and creativity in developing an effective animal housing facility inspection program tailored to the particular needs of their individual institutions.

24:1 What is the definition of an animal housing facility?

Reg. (See 15:1; 24:12.)

* This chapter builds upon the contributions of Patricia A. Ward, the author of this chapter in the first and second editions of *The IACUC Handbook*.

24:2 What is the purpose of conducting IACUC inspections of animal housing facilities?

Reg. The IACUC must fulfill its regulatory responsibility to inspect animal housing areas at least once every 6 months to evaluate compliance with applicable guidelines (AWAR §2.31,c,2; PHS Policy IV,B,2; HREA §495,b,3,A). The recommendations of the *Guide* (p. 25) specify that the IACUC should inspect the animal facilities at least annually.

Opin. The IACUC must represent its institution both by advancing the institutional mission and by serving as the institutional conscience. By inspecting animal housing areas at least twice a year, the IACUC can ensure that the animal housing and care program is furthering the institutional mission by providing high-quality animals (through proper attention to animal welfare), maintaining a suitable environment for research activities, and safeguarding the health and safety of its personnel. In addition to ensuring the institution's compliance with applicable regulations, the IACUC must be confident that the animal housing and care program is accomplishing these goals in a scientific, humane, and ethical manner.

24:3 Should an institutional guideline, SOP, or policy be developed for IACUC inspection of animal housing facilities?

Reg. There is no regulatory requirement for IACUCs to establish guidelines, SOPs or policies regarding inspection of animal housing facilities. The AWAR (§2.31,c,3) states that the IACUC may determine the best means of conducting evaluations.

Opin. Many IACUCs have found the creation of guidelines, SOPs or policies regarding inspection of animal housing facilities to be helpful in the administration and management of the inspection process. Clarifying expectations for IACUC members conducting inspections, veterinary and administrative staff supporting inspections, and managers of facilities undergoing inspection can promote consistent and thorough inspections and enhance rapport among all parties. Issues addressed in the guidelines, SOPs, or policies can include the following:

- Who conducts the IACUC inspections
- The orientation or training inspectors receive
- How inspections are scheduled
- Whether inspections are to be announced or unannounced
- How inspectors should prepare for each inspection
- How inspections will be facilitated
- Which sites are to be inspected
- Which general issues are considered during inspections
- Findings of previous inspections as examples
- How findings will be communicated, documented, and distributed
- How the IACUC will follow up deficiencies
- How long inspection documents will be retained
- Expectations of timing to correct deficiencies

24:4 Who conducts the IACUC inspections of animal housing areas?

Reg. While the PHS Policy (IV,B,2), the *Guide* (pp. 25, 34), and the AWAR (§2.31,c,3) specify animal facility inspection as an IACUC function, the AWAR offer the most explicit "minimal" answer to this question: "No Committee member wishing to participate... may be excluded. The IACUC may use subcommittees composed of at least two Committee members and may invite ad hoc consultants to assist."

Opin. At many institutions, particularly those with smaller or centralized animal facilities, the entire IACUC participates in the animal housing facility inspection. However, in larger or decentralized facilities, the IACUC may divide into subcommittees or appoint a single subcommittee to tackle the job. Such subcommittees must include at least two IACUC members if housing areas for AWA-regulated species will be inspected (AWAR §2.31,c,3). Duly appointed and qualified alternate members of the IACUC may conduct the inspections in place of corresponding regular members.[1] A site visit from the AAALAC may be utilized for the semiannual animal facility inspection, as long as the participation and endorsement of the IACUC meet the requirements of the AWAR and the PHS Policy.[2] Members of the veterinary or IACUC staff also may participate in the inspection to provide expertise, continuity, and support.

24:5 Should IACUC members receive orientation or training in preparation for conducting facility inspections?

Reg. The *Guide* (p. 24) states that "it is the institution's responsibility to provide suitable orientation, background materials, access to appropriate resources, and, if necessary, specific training to assist IACUC members in understanding and evaluating issues brought before the committee."

Opin. Most IACUCs arrange for some form of orientation or training for incoming members that includes information about inspection of animal housing facilities. This training is especially beneficial to the nonscientist and nonaffiliated members of the IACUC. Information presented in the inspection training session or packet can include the following:

- An overview of the size and scope of the institution's animal housing facilities, including maps or floor plans
- Institutional guidelines, SOPs, or policies regarding IACUC inspections
- Copies of regulatory, professional, and institutional animal housing and care standards
- Guidelines for evaluating animal well-being
- The institutional inspection checklist, if used
- Sources of additional information (e.g., web sites or information about contacting key laboratory animal organizations)

Most IACUCs find it worthwhile to spend a significant portion of the training effort reviewing the various established standards that apply to the areas to be inspected. These include the animal housing and care standards provided in the AWAR, the *Guide*, and the *Guide for the Care and Use of Agricultural Animals in*

Agricultural Research and Teaching (Ag Guide).[3] As applicable, IACUCs may also consider reviewing the various animal management documents produced by, among others, the Institute of Laboratory Animal Resources and the National Research Council.[4] A working understanding of these standards not only enables IACUC members to evaluate regulatory compliance issues in the facilities inspected but also helps to develop their professional judgment with regard to the more important issues of animal welfare, research integrity, and personnel health and safety.

24:6 What periods are appropriate for IACUC inspections of animal facilities?

Reg. Both the AWAR (§2.31,c,2) and PHS Policy (IV,B,2) specify that animal facilities be inspected at least once every 6 months. (See 23:5–23:7.) The *Guide* (p. 25) states that facilities should be inspected at least annually.

Opin. For the most part, meeting this requirement is not difficult for institutions where the animal facilities can be inspected in a relatively short time. For larger or decentralized facilities, however, meeting this requirement can be more difficult. If completing the inspection will take several days, the IACUC is faced with management of a large time commitment. Under these circumstances, most IACUCs will establish regular "inspection windows" at 6-month intervals and break the inspection up over several days within each window. Alternatively, an IACUC can opt to distribute inspection trips throughout the entire 6-month period. In this case, or if the inspection window spans more than a few weeks, care must be taken to inspect the various sites in approximately the same order for each 6-month period. Otherwise, the inspection interval for some sites might exceed the 6-month maximum. This could occur if a particular site is inspected early in one 6-month period and late in the next.

24:7 Should inspections be announced or unannounced?

Opin. (See 23:27.)

24:8 How should IACUC inspectors prepare for each inspection?

Opin. Before embarking on each inspection, most IACUC inspectors want to review the sites to be visited with particular respect to entry and exit restrictions, previous inspection findings or deficiency history and experimental activities being conducted. Entry and exit restrictions may affect the way inspectors dress for the inspection, the order in which various areas are visited, and the extent to which exposure of inspectors to certain other animals is prohibited for a specified period before or after the inspection. Reviewing the previous inspection findings or deficiency history of the sites to be inspected may help IACUC inspectors focus attention on problem areas or remind them to express appreciation for improvements made. Some IACUC inspectors may wish to review experimental protocols approved for use in the areas to be inspected. Doing so can alert them to animal use procedures of particular concern, such as those requiring special consideration during protocol review (*Guide* p. 25), and allow IACUC inspectors to make pertinent observations or inquiries during the inspection visit. Alternatively, some IACUC inspectors choose to wait until after the inspection to review the experimental protocols of animals noted during the inspection.

24:9 What strategies can be used to facilitate inspections?

Opin. Every IACUC looks for ways to facilitate the inspection process. Including administrative or veterinary staff on the inspection team is one popular strategy used to provide expertise, continuity, and support. Another is the use of the previous inspection report or an inspection checklist, to use as a reference or to record findings during the inspection. NIH/OLAW has made a sample checklist available on its website[5] and institutions may download and modify it to suit their own programs. Some IACUC inspection teams carry copies of pertinent regulations and standards, or equipment for recording findings (photo, audio, or video) or validating conditions (temperature or humidity reader, light meter, smoke bottle, etc.). (See 24:11.)

24:10 What electronic technology might the IACUC use to aid in the conduct of the semiannual program review and facility inspection?

Opin. While the advent of newer technologies such as iPads®, other tablet computers, and data management systems may streamline the IACUC inspection process, most institutions have not progressed to the point where it is feasible to incorporate these technologies into their semiannual inspection processes. As the technologies become further integrated into the animal facility and protocol management systems, we may see more adaptation of these systems in the future.

Surv. Which of the following technologies does your IACUC employ to facilitate the conduct and documentation of the semiannual program review and facility inspection? Check as many boxes as appropriate.

• Mobile devices (e.g., laptop, tablet device, etc.) for recording findings	30/292 (10.3%)
• Commercial IACUC/animal resource data management product(s)	10/292 (3.4%)
• Internally-developed data management system	86/292 (29.5%)
• None of the above	160/292 (54.8%)
• Other	38/292 (13.0%)

"Other" responses included written inspection forms or checklists that are compiled into electronic versions after the inspection.

24:11 Should the program review necessarily precede the facility inspection during the semiannual review?

Opin. While there is no right or wrong way with respect to the sequence of events for the program review and facility inspection, it is advantageous for members of the inspection team to review all of the SOPs and policies associated with the animal program before inspecting the animal facility. This allows the inspection team to be confident in their evaluation of the animal facility.

Surv. Which of the following best describes the sequencing of your semiannual program review and facility inspection?

• Program review prior to the facility inspection	83/293 (28.3%)
• The facility inspection prior to the program review	111/293 (37.9%)
• Program review while conducting the facility inspection	18/293 (6.1%)
• Program review at the same time as the facility inspection, but by different pairs/groups of IACUC members	22/293 (7.5%)
• It varies; we have no set format	56/293 (19.1%)
• I don't know	0/293 (0%)
• Other	3/293 (1.0%)

24:12 What sites should be included during the inspection of animal housing facilities?

Reg. The AWAR (§1.1, Animal) define an animal as a live or dead warm-blooded animal, and does not include all vertebrates. The AWAR also specifically excludes birds bred for research, rats of the genus *Rattus*, mice of the genus *Mus*, horses not used for research purposes, and other farm animals such as livestock or poultry used or intended for use as food or fiber or the improvement of animal nutrition, breeding, management, or production efficiency. The PHS Policy (III,A) applies to any live vertebrate animal used in research, research training, experimentation, or biological testing where PHS funding is involved. Most IACUCs resolve this discrepancy simply by including the housing areas of all vertebrate animals, regardless of species or type of use, in their inspections. Some IACUCs elect to inspect the housing areas of invertebrate species as well, especially when the inclusion of such areas benefits the program as a whole.[8]

What constitutes an animal housing area is also a matter of regulatory definition (see 15:1). The AWAR (§2.31,c,2) specify that animal facilities, excluding wild habitats but including study areas, must be inspected. While a definition of an animal facility is not provided in the AWAR, a housing facility (AWAR §1.1, Housing Facility) is defined as any land or shelter intended to house animals, divided into indoor housing facility (structure or building with environmental controls), sheltered housing facility (facility that provides protection from the elements), and outdoor housing facility (housing that does not meet any other definition of housing). The AWAR (§1.1, Study Area) further defines a study area as any area where animals are housed for more than 12 hours.

The PHS Policy (III,B) defines an animal facility as "any and all buildings, rooms, areas, enclosures, or vehicles, including satellite facilities, used for animal confinement, transport, maintenance, breeding, or experiments inclusive of surgical manipulation. A satellite facility is any containment outside of a core facility or centrally designated or managed area in which animals are housed for more than 24 hours." The PHS Policy definition is quite comprehensive, including support and experimental surgery areas along with housing areas. (See 23:18; 25:1.)

Opin. Many IACUCs, particularly those wishing to obtain or maintain accreditation by AAALAC, want to comply with the animal housing inspection requirements of the AWAR, PHS Policy, and the *Guide*. To accomplish this, virtually all IACUCs visit all areas where animals are housed or kept (including study areas, satellite facilities, and laboratories) for more than 12 hours (AWAR-regulated species) or 24 hours (all other vertebrate species) during their semiannual inspections. Typically, areas where rats and mice are kept are not inspected if those animals are held there for less than 24 hours, unless survival or non-survival surgery is conducted at that location (see NIH/OLAW FAQ E.1). Temporarily unoccupied animal housing areas also are visited by most IACUCs. Corridors and anterooms contiguous with these areas are included in most inspections. Field research locations are not usually inspected by IACUCs unless animals are held in captivity for more than 12 hours or the IACUC is particularly interested in observing the field research procedures or setting. Cage washing and other sanitation facilities are included in the semiannual inspections, as are diet preparation and food and bedding storage areas. Most IACUCs also inspect areas where caging, equipment, and supplies are stored. Loading docks and transport vehicles, particularly those dedicated to traffic of animals and animal-related supplies and equipment, are usually inspected, as are carcass storage and disposal areas. Animal procedure areas contiguous with the animal housing facilities, including surgical suites, procedure rooms, and euthanasia stations, are also usually included in the semiannual inspection of the animal housing and support areas. (See 23:18.)

24:13 What general issues should be considered when inspecting animal housing facilities?

Reg. The AWAR (§2.31,c,1) require the IACUC to use the AWAR as a guide for semiannual inspections. PHS Policy (IV,B,1) requires the IACUC to use the *Guide* as a basis for evaluation.

Opin. Paramount among inspection issues are animal well-being, research integrity, and personnel health and safety. Proper management of animal care and use promotes all three, and IACUC inspectors seek to evaluate the success of the program in this context. To do this during an inspection, most IACUCs try to organize the effort and attention they give to the various aspects of the animal care and use program. Many IACUCs will utilize currently available tabulations, such as the table of contents of the *Guide* or the Facility Inspection Checklist[5] available from NIH/OLAW, to provide such structure. Some IACUCs prefer to focus their attention on issues of particular concern in the type of animal facilities at their institutions. (See also 24:9.)

No matter how the inspection is structured, the many issues that should be considered by IACUC inspectors can be overwhelming once they actually enter an animal housing area to conduct an inspection. One popular approach is for IACUC inspectors to begin with the condition of the animals and expand their focus to the following general categories in any convenient order. Note that contingency plans for emergency situations (AWAR §2.38; *Guide* p. 35) invariably have a physical component (e.g., animal rack movements, emergency supply storage, evacuation routes) and these areas should be included during the inspection of animal facilities. This is particularly true if there is an intent to shelter animals in place during an emergency situation (AWAR §2.38,l,1).

- Animal health and well-being
- Primary enclosures, environmental conditions, and physical plant
- Housekeeping, disinfection, and sanitation
- Water, food, bedding, and other supplies (including expired drugs and appropriate marking of secondary containers)
- Animal euthanasia, carcass storage, and carcass disposal
- Labels, signage, and records
- Personnel, training, and occupational safety
- Transportation vehicles
- Documentation of the above as appropriate, including contingency plans (AWAR §2.38,l,2)

24:14 How should inspection findings be communicated, documented, and distributed?

Reg. Both the AWAR (§2.31,c,3) and PHS Policy (IV,B,3) require that reports of inspection findings be forwarded to the IO at the close of every 6-month inspection period. This is usually done in conjunction with the IACUC's semiannual evaluation of its institutional program for humane use and care of animals. Inspection findings may be summarized in the semiannual report to the IO, or copies of detailed site-by-site inspection records may simply be attached to the report. Sample reports are available on NIH/OLAW's website.[9] In either case, the inspection findings must distinguish significant deficiencies from minor deficiencies (significant deficiencies represent a threat to the health or safety of animals; see 23:31) and include a plan and schedule for correction (AWAR §2.31,c,3; PHS Policy IV,B,3). The records and reports of the inspections must be maintained for a period of three years (AWAR §2.35,a,3).

Opin. In addition to reporting inspection findings to the IO, IACUCs generally communicate their findings to animal facility managers in order to assist them in correcting deficiencies and to obtain full compliance. This communication can occur through many means, depending on the culture of the institution. If inspections are arranged with animal facility managers in advance, most IACUC inspectors ask questions and discuss concerns with those managers throughout the inspection visit. Alternatively, and particularly if facility managers are not present during the inspection, IACUC inspectors may arrange for a verbal interview with the managers either at the conclusion of the inspection visit or at a later time. In addition to verbal communication, most IACUCs will present facility managers with a written report of the inspection findings. This may be a simple list or table of deficiencies or completed checklist presented at the conclusion of the inspection or a more formal inspection report sent at a later date. In either case, these written inspection reports generally contain the following information:

- Date and site location
- Names of IACUC inspectors
- Deficiencies observed and whether each deficiency was significant or minor
- A directive and date for correction

In addition, many IACUCs also will include one or more of the following in their inspection reports:

- The administrative unit responsible for the facility inspected
- Names of facility representatives present at the inspection
- A list, including description and contents, of all rooms inspected
- Whether deficiencies observed were first-time or previously cited
- The specific standard or regulation violated by each deficiency
- A general summary, comment, or commendation

24:15 How should the IACUC follow up on deficiencies?

Reg. Both the AWAR (§2.31,c,3) and PHS Policy (IV,B,3) require that program and facility deficiencies be included in the semiannual report to the IO, that a plan and date for correction be indicated for each deficiency, and that minor deficiencies be distinguished from significant deficiencies (deficiencies that may threaten the health or safety of animals). Additionally, the AWAR require that failure to adhere to the correction schedule for significant deficiencies be reported within 15 days by the IACUC through the IO to APHIS/AC and any federal agency funding the cited activity (AWAR §2.31,c,3). PHS Policy (IV,F,3,a) requires prompt reporting to NIH/OLAW of serious or continuing noncompliance with the Policy. (See 25:12.)

Opin. When an IACUC encounters a situation that represents a threat to the health or safety of animals, prompt corrective action should be taken and a thorough follow-up conducted in order to ensure that the situation does not recur. However, the vast majority of deficiencies cited by the IACUC on inspection are minor and the nature of follow-up action depends on the culture of the institution. Many IACUCs re-inspect the cited facility at the correction deadline to verify resolution of the deficiency. Most IACUCs specifically assess the correction of previously cited deficiencies at subsequent inspections. Occasionally, an IACUC encounters previously cited minor deficiencies that either have not been corrected or have recurred. These deficiencies usually do not become "elevated" to major deficiencies unless the uncorrected minor deficiency has become a threat to the health or safety of animals. When there are recurrent or uncorrected minor deficiencies, the IACUC, using authority granted to it by its institution and federal regulations can employ a variety of strategies to obtain compliance. Examples include the following:

- Increased frequency of IACUC inspections
- IACUC interview of responsible persons
- Notification of superiors in the chain of command
- IACUC-issued verbal or written warnings
- Withdrawal of IACUC approval to house or use animals
- Discipline of personnel in accordance with institutional policy

These actions are generally taken after discussion and approval by the IACUC and with the engagement of the IO.

24:16 For how long should inspection documents be retained?

Reg. Both the AWAR (§2.35,a,3) and PHS Policy (IV,E,2) require that semiannual reports to the IO, which include the IACUC's animal facility inspection findings, be retained by the institution and be available for inspection for at least 3 years.

Opin. If additional inspection records are created, the length of time they are retained is at the discretion of the IACUC. Many IACUCs retain these records for the same 3 years as the semiannual reports to the IO. Most, however, find both the semiannual reports to the IO and additional inspection records to be a valuable documentation of animal facility management history. These IACUCs are likely to retain such records for an extended period. IACUCs at institutions subject to federal or state freedom of information acts or open records laws should consult appropriate persons at their institution for advice on their records retention program. (See 22:12–22:14.)

24:17 How should the IACUC assess the well-being of animals during inspection?

Reg. IACUC inspectors should endeavor to assess the physical and psychological well-being of all animals. According to the *Guide* (p. 41), "An appropriate Program … provides environments, housing, and management that are well suited for the species or strains of animals maintained and takes into account their physical, physiologic, and behavioral needs, allowing them to grow, mature, and reproduce normally while providing for their health and well-being." To accomplish this, the structural environment, social environment, and activity pattern of the animals must be considered (*Guide* p. 64). The standards of the AWAR (§3.8; §3.81) include specific provisions for the exercise of dogs and environmental enhancement of nonhuman primates.

The *Guide* (p. 63) notes that "High levels of repetitive, unvarying behaviors (stereotypies, compulsive behaviors) may reflect disruptions of normal behavioral control mechanisms due to housing conditions or management practices." Identification of these types of behaviors can indicate to IACUC inspectors that the psychological well-being of the animals demonstrating the behavior, and perhaps the colony in general, has not been adequately addressed.

Opin. During animal housing facility inspections, the first thing to capture the attention of IACUC inspectors is likely to be the condition of the animals housed. By virtue of training or experience, most IACUC inspectors will readily distinguish normal healthy animals from those exhibiting signs of illness, injury, pain, or distress. Discovery of animals in the latter group should prompt further inquiry into the cause of the condition and the way it is being managed. If the condition is known to be experimentally induced, IACUC inspectors may wish to verify that the condition and its management are consistent with information provided in the IACUC-approved protocol. If the condition is not related to the experimental procedure, inquiry into the colony health status and other recent illnesses or injuries may be in order, and management practices questioned if undesirable patterns are identified. In either instance, IACUC inspectors should be assured that all animals are checked daily, that veterinary care is available around the clock if needed, contact information is current, the staff is aware of how to contact the veterinarian when needed, and that the veterinary staff is promptly notified of animal health problems and closely monitors the effectiveness of the implemented treatment and management plan (*Guide* pp. 112–113; AWAR §2.33,b).

If the animal housing and care program is adequately providing for the psychological well-being of animals, IACUC inspectors should expect to see primary enclosures with features that accommodate the animals' species-specific postures and motor activities, management practices that cater to species-specific needs for cognitive stimulation and social interaction, and compatible social groupings where appropriate. (See 27:1.)

24:18 How should the IACUC evaluate animal enclosures during inspection?

Reg. Both the *Guide* (pp. 57–63) and the AWAR (§3.6; §3.15 ;§3.28; §3.53; §3.80; §3.101; §3.104; §3.128) provide specific standards for primary enclosures.

Opin. To evaluate compliance with these standards, IACUC inspectors generally ask themselves the following questions when examining an animal enclosure:

- Is the enclosure of appropriate design for the comfort, security, and observation of the animals housed?
- Are animals able to engage in normal postural positions?
- Are animals able to engage in species-specific behaviors?
- Is it constructed of durable and sanitizable materials?
- Is space adequate for the size and number of animals contained?
- Is the enclosure clean and in good repair?
- Are food and water readily accessible?
- Does the enclosure in any way jeopardize the well-being of the animals housed therein?

The *Guide* (p. 41) has clear expectations of the performance outcomes that are expected for animal housing environments. It is understood that the laboratory environment will not fully approximate the natural environment but the institution should take steps to ensure that animals are able to engage in species specific behaviors, move through normal postural adjustments, and have social interactions that are appropriate for the species. A basic understanding of the natural history of the species in use (e.g., social or isolated, burrowing, foraging, etc.) will assist the inspectors in their evaluation of the appropriateness of the animal enclosures for meeting the well-being of animals. The inspectors should note the condition of the animals, balance that with the institutional policies and guidelines, and evaluate them in the context of current literature regarding natural history and husbandry recommendations.

24:19 What environmental parameters should the IACUC evaluate during inspection of an animal facility?

Reg. Specific standards for proper animal housing environmental conditions are provided in the *Guide* (pp. 43–45) and AWAR (Part 3). The PHS Policy (IV,A,1) uses the *Guide* as a basis for the development and implementation of an animal care and use program.

Opin. IACUC inspectors generally find some environmental parameters, such as extremes of temperature and humidity, inadequate ventilation, excessive odor, and chronically loud noise, to be readily assessable the moment they enter an animal housing room. Other parameters require a little more investigation. For example, adequate assessment of the lighting cycle and magnitude and frequency of temperature fluctuation require more than momentary sensory perception by the IACUC inspectors. Additionally, depending on the type of primary enclosure used, the environmental conditions experienced by the animals within the enclosure may be very different from those perceived by the IACUC inspectors in the larger room environment. Records of HVAC function at the room and facility level should be reviewed as part of the semiannual inspection process to ensure that the macroenvironment consistently meets the intended standards. Most IACUC inspectors interview animal facility managers about the monitoring methods used for such situations and review the monitoring documentation to ensure that the appropriate environmental conditions are being provided. Included in this discussion and record review should be the results of tests of or actual experience with emergency alarms and response systems.

24:20 How should the IACUC evaluate the animal facility's physical plant during inspection?

Reg. As indicated in 24:17, IACUC inspectors should consult the *Guide* and AWAR for specific standards regarding the animal housing facility physical plant.

Opin. In addition to ensuring compliance with these standards, IACUC inspectors generally evaluate the facilities in terms of design, construction, building systems, fixtures and equipment, security, maintenance, and housekeeping. Room and facility level records may have to be evaluated to ensure that these expectations are being met appropriately. To satisfy most IACUC inspectors, the facility's physical plant should meet the following criteria:

- The design of the structure should be conducive to achieving the goals of animal well-being, research integrity, and personnel health and safety.
- Construction should be structurally sound, surface finishes should be sanitizable, and physical barriers should be in place to prevent entry of vermin.
- Building systems (electrical, plumbing, heating, ventilation, air-conditioning, etc.) should be reliable and sufficient to support facility demands.
- The facility should be outfitted with fixtures and equipment appropriate to the housing, care, and use of the species housed. The surfaces of these items should be sanitizable.
- Equipment requiring periodic service or certification should be properly maintained.
- Appropriate security should include provisions for preventing both escape of animals and entry of unauthorized personnel.
- The facility structure and building systems should be maintained in good repair.
- In general, the facilities should be clean and free of unnecessary clutter.

24:21 How should the IACUC assess the adequacy of housekeeping, disinfection, and sanitation practices during inspection?

Reg. The *Guide* (pp. 69–73) recommends frequent bedding changes and cleaning and disinfection of primary enclosures and animal housing rooms. The AWAR (§3.11;

§3.31; §3.56; §3.84; §3.106; §3.107; §3.131) provide specific minimal requirements for cleaning, sanitization, housekeeping, and pest control in animal housing facilities.

Opin. During inspections, most IACUC members evaluate these issues through direct observation and examination of records. Terrestrial animals are examined to ensure that they are clean and dry. Bedding should not be excessively soiled. Primary enclosures should not exhibit an accumulation of soil. Likewise, animal housing rooms should neither exhibit an accumulation of soil or waste, nor be excessively cluttered (most IACUC inspectors discourage storage of equipment and supplies that are not used in the routine care and use of the animals). All surfaces should be sealed, intact and sanitizable, and agents used for sanitation and disinfection should be appropriate for the purpose, be properly stored, and be used according to the manufacturers' directions. IACUC inspectors should see evidence of an effective vermin control program. Waste material should be removed from the animal rooms in a timely fashion and handled in a manner that minimizes the potential to introduce contaminants or attract vermin to the animal facility. Many IACUC inspectors also evaluate the effectiveness of an animal housing facility's housekeeping, disinfection, and sanitation program in terms of colony health. They inquire about recent incidences and spread of disease and identify questionable housekeeping, disinfection, or sanitation practices that may contribute to colony health problems. (See 24:19; 24:24.)

Inspection of facilities for washing and sanitizing caging and equipment also are included in the IACUC's evaluation of sanitation practices. Most IACUC inspectors want to see that these facilities do not exhibit an accumulation of soil and that caging and equipment are processed in such a way as to minimize the recontamination of clean items. Chemical agents and automatic washing equipment used for sanitation and disinfection of caging and equipment should be appropriate for the purpose used, be properly stored, and be used according to manufacturers' directions. Most IACUC inspectors want to verify the effectiveness of the sanitation process. This can be done by inspecting records of validation test results (e.g., cage washer temperature indicator strips and cultures of sanitized surfaces). If the process calls for sterilization of caging or equipment, the performance of sterilizers is similarly evaluated by most IACUC inspectors.

24:22 How should the IACUC evaluate the adequacy of the water, food, bedding, medications, and other supplies provided to or used for animals?

Reg. Prior to inspection, IACUC inspectors are advised to consult the *Guide* (pp. 65–69) and AWAR (Part 3) for specific standards regarding the quality, storage, and provision of water, food, bedding, medications, and other supplies.

Opin. IACUC inspectors should be assured that all supplies that have contact with animals are appropriate, safe, and effective. Food, bedding, medications, and so on, can usually be evaluated by direct observation during inspection. Acceptable animal food must be appropriate for the species, within its recommended shelf life, and stored under conditions that minimize the potential for premature deterioration, spoilage, contamination, or infestation. Food products should be labeled and dated with the expiration date. It should be available to animals ad libitum or provided in an appropriate ration and available to all individuals in a group unless veterinary or IACUC-approved research protocols dictate otherwise. Bedding should be absorbent, nonnutritive, nontoxic, and stored in a manner that minimizes the potential for spoilage, contamination, or infestation. Food and bedding should be free of foreign objects and normal in appearance, texture, and odor. Medications, treatments, and other veterinary or experimental supplies should be appropriate for the type of animal and research intended, medical grade, within designated expiration dates, and stored according to package directions (IACUC inspectors should be aware that special storage and record keeping requirements are imposed by the regulations of the Drug Enforcement Agency for controlled substances).[6] Unless veterinary or IACUC-approved research protocols dictate otherwise, drinking water should be available to animals ad libitum, potable, and free of contaminants. To assess the latter, most IACUC inspectors interview facility managers about the source of the water provided (municipal, well, purified, etc.) and the integrity of the system that delivers the water to the animals (plumbing systems, manifold flushing practices, backflow prevention, etc.). Inspectors will also usually examine the results of any water analysis conducted and inquire about colony health problems in which animal drinking water may have played a role.

24:23 What aspects of animal euthanasia, carcass storage, and carcass disposal should concern the IACUC during inspection?

Opin. Most IACUC inspectors visit areas for animal euthanasia, carcass storage, and carcass disposal in conjunction with their inspection of animal housing facilities. Euthanasia areas should be clean and the methods employed should be consistent with the recommendations of the American Veterinary Medical Association's *AVMA Guidelines for the Euthanasia for Animals: 2013 Edition.*[7] (See Chapter 17.)

Carcasses are generally stored in dedicated refrigerators or freezers, with special provisions for the labeling and storage of carcasses contaminated with infectious, radioactive, or chemical hazards, including ether. Whether carcass disposal is handled by the institution internally or contracted to a professional waste management company, IACUC inspectors should be apprised of the disposal methods and be assured that these methods comply with all applicable regulations, including local requirements and restrictions. IACUC inspectors also may wish to review any facility records associated with animal euthanasia and carcass disposal, particularly those required by the AWAR (§2.35,b) for documentation of the final disposition of dogs and cats. Standards on animal euthanasia, carcass storage, and carcass disposal also can be found in the *Guide* (pp. 73–74, 123–124, 142).

24:24 What labels, signage, and records should be evaluated by the IACUC during inspection?

Opin. In an animal housing and care facility managed in accordance with the AWAR, PHS Policy, and the recommendations of the *Guide*, most IACUC inspectors can expect to see labels, signage, and records associated with nearly every aspect of the program. Animal enclosures should be labeled with pertinent information (see the *Guide* p. 75) about the animals and their intended use (e.g., PI and IACUC-approved protocol number). Information for reporting concerns regarding animal welfare should be clearly posted (*Guide* p. 24). Animal food should be labeled with the species formulation and the milling or expiration date. Various cleaning, husbandry, veterinary, and experimental substances found in the animal facility should be labeled with the name and opening or expiration date of each item. Usage records should accompany controlled substances as required by the regulations of the Drug Enforcement Administration.[6] Hazardous substances and contaminated animals or materials should be clearly identified when the hazard is present and labeled with precautions. Special instructions should be prominently posted to assist personnel to comply with non-routine or critical procedures. Animal housing areas should be posted with current emergency contact information. Husbandry activities should be recorded on log sheets. Environmental readings should also be logged or recorded on automatic monitoring system printouts. Animal monitoring records (post-procedural, food and water consumption, enrichment, exercise, etc.) are usually present at the animal housing location and should be reviewed. Animal health, clinical and veterinary treatment records, as well as animal receiving and disposition records, should be maintained in an appropriate location and inspected by the IACUC. Records of sanitation and disinfection, including cage washer and sterilizer performance, also should be available for review by IACUC inspectors.

While not all of these labels, signs, and records are specifically required by the AWAR, PHS Policy, and recommendations of the *Guide*—and certainly not every activity in the animal housing facility need be recorded—many of these records and documents will prove to be useful animal facility management and inspection tools. Some IACUCs choose to require that extensive records be kept of animal housing activities and parameters, particularly to document to outside authorities that there is compliance with standards. Others rely on the state of their facilities and animal care programs to demonstrate compliance and forgo record keeping not specifically required by applicable regulations and standards.

24:25 What personnel issues should concern the IACUC during inspection?

Opin. One of the highlights of the IACUC animal housing facility inspection process is the opportunity for dialog between IACUC members and personnel working in the animal facility. Such personnel can include members of the research staff, as well as facility managers and animal care and support personnel. In their conversations, most IACUC inspectors try to evaluate the extent to which personnel are qualified, either through experience or training, to perform their duties, and whether or not experimental procedures and work practices are consistent with IACUC-approved animal use protocols, institutional guidelines, policies, and SOPs. As APHIS/AC mandates training of all staff regarding the implementation of the contingency plan and their role (AWAR §2.38,l,3), the IACUC inspectors should discuss this issue with key personnel.

Some of the most important personnel issues IACUC members consider during their inspections of animal housing facilities concern the occupational health and safety of animal care, use, and support personnel. IACUC inspectors should look for evidence that personnel are informed about and protected from the various health and safety risks associated with exposure to animals and the hazardous experimental agents and procedures used with animals. Examples of such evidence include elements of facility design and management (such as space delineation and air pressure differentials) that separate personnel areas and activities from those of animals, especially where hazardous agents are present. Compliance of personnel with appropriate personal hygiene and protective equipment recommendations also can indicate an effective occupational health and safety program. (See Chapter 20.)

Additionally, it is important that personnel who are working with animals understand that they are able to anonymously report concerns about animal welfare to the IACUC for investigation. Because personnel who are reporting animal welfare concerns have an expectation of whistle-blowing protection, ensuring that the reports can be made anonymously and that all personnel know how to make those anonymous reports is of the utmost importance. (See 29:8; 29:9.)

24:26 What methods are useful for assessing remote animal holding locations?

Opin. Off-site animal holding study locations can be very challenging to inspect as they may require a significant time commitment from IACUC members for travel to and from the facility. Some facilities may have multiple facilities spread out over multiple geographic locations, thus making the logistics of arranging the IACUC inspections difficult as well. Off-site locations also may take more time to inspect as the holding areas are very different from those IACUC inspectors are accustomed to, and thus there is a large deal of education that must also happen before and during the site visit. While video recordings may be acceptable, it is up to each individual IACUC to develop a policy regarding off-site animal housing areas that is acceptable for their institution and still ensures the health and safety of the animals.

Surv. As part of the semiannual facility inspection, how does your IACUC evaluate remote off-site animal holding or study locations (e.g., agricultural/field stations)?

- We do not have any remote animal holding/study locations — 169/290 (58.3%)
- The committee or subcommittee visits the locations — 87/290 (30.0%)
- We evaluate records from the site, but there is no on-site inspection by the committee — 7/290 (2.4%)
- We evaluate video recordings from the remote site — 5/290 (1.7%)
- We do not inspect or evaluate our remote animal holding/study locations — 15/290 (5.2%)
- Other — 7/290 (2.4%)

24:27 Should the IACUC include high containment areas in the semiannual inspection?

Opin. Animal Biosafety Level (ABSL) 3 or 4 animal laboratories present special challenges to IACUCs because of the advanced level of training that may be required to enter these spaces. However, with careful consideration, IACUCs can select members of a subcommittee to enter these areas, such as employee health and safety personnel who are familiar with these practices.

Surv. As part of the semiannual facility inspection, how does your institution's IACUC evaluate high containment animal holding/study locations (i.e., ABSL-3/4 animal laboratories)?

- We do not have any high containment facilities — 215/294 (73.1%)
- The committee or subcommittee enters the facility(s) to conduct the inspection — 67/294 (22.8%)
- Evaluation of records from the facility(s), but the committee does not enter the high containment facility — 5/294 (1.7%)
- We use video recordings from the facility — 2/294 (0.7%)
- We do not inspect or evaluate our high containment facility(s) — 1/294 (0.3%)
- Other — 4/294 (1.4%)

References

1. National Institutes of Health, Office of Extramural Research. 2011. Notice NOT-OD-011-053, Guidance to reduce regulatory burden for IACUC administration regarding alternate members and approval dates. http://grants.nih.gov/grants/guide/notice-files/NOT-OD-11-053.html.
2. National Institutes of Health, Office of Extramural Research. 1999. Guidance regarding reduction of regulatory burden in laboratory animal welfare. Notice OD-00-007. http://grants.nih.gov/grants/guide/notice-files/not-od-00-007.html.
3. Committees to revise the Guide for the Care and Use of Agricultural Animals in Research and Teaching. 2010. *Guide for the Care and Use of Agricultural Animals in Agricultural Research and Teaching*, 3rd edition. Savoy: Federation of Animal Science Societies. http://www.fass.org/docs/agguide3rd/Ag_Guide_3rd_ed.pdf.
4. Institute for Laboratory Animal Research. 2013. http://dels.nas.edu/ilar/.
5. National Institutes of Health, Office of Laboratory Animal Welfare. 2012. Sample Semiannual facility inspection checklist animal housing and support areas. http://grants.nih.gov/grants/olaw/sampledoc/cheklist.htm.
6. Office of the Federal Register. 1988. Controlled Substances Act, amended May 1, 1987, Title 21, Food and Drugs, Chapter 13, Drug Abuse Prevention and Control, 21 CFR §1301.11–§1308.15, Schedule of Controlled Substances, Washington, D.C.
7. American Veterinary Medical Association. 2013. *AVMA guidelines for the euthanasia for animals: 2013 edition*. Schaumburg: American Veterinary Medical Association.
8. Association for the Assessment and Accreditation of Laboratory Animal Care International. 2012. Frequently Asked Questions B.2. http://www.aaalac.org/accreditation/faq_landing.cfm#A2.
9. National Institutes of Health, Office of Laboratory Animal Welfare. 2013. Semiannual report to the institutional official. http://grants.nih.gov/grants/olaw/sampledoc/ioreport.htm.

25

Inspection of Individual Laboratories

Neil S. Lipman and Scott E. Perkins

Introduction

One of the IACUC's principal roles is to ensure that animals used in research, testing, and teaching are used humanely by trained staff. Although IACUC review of proposals describing planned animal activity is an important mechanism for meeting this role, observation of animal use provides the most direct assurance that these goals are attained. At some institutions, animals are never removed from the animal facility as all activities are performed within animal holding or procedure rooms. Nevertheless, it is common for animals to be transported from the animal facility to an investigator's laboratory for experimental use. While animal resource personnel can readily observe the appropriateness of activities conducted within the animal facility, this task becomes significantly more difficult, especially at large institutions, when animal research is conducted in an investigator's laboratory.

Although there is no specific regulatory directive requiring IACUCs to inspect investigators' laboratories and observe ongoing activities (unless animals are held in these areas for more than 12 hours; See 25:1), the IACUC is charged with ensuring that animal use is conducted humanely by appropriately trained personnel and that these activities have been approved, in advance, by the IACUC. It is this chapter authors' opinion that laboratories in which animals are used should be visited by the IACUC periodically in order to meet the spirit of applicable regulations, policies, and the *Guide*.

25:1 Is it necessary to inspect investigators' laboratories where research with animals is conducted?

Reg. The AWAR (§1.1, Study Area; §2.31,c,2) require inspections if animals are maintained in the laboratory for more than 12 hours, whereas the PHS Policy (III,B; IV,B,2) requires inspections if animals are maintained in the laboratory for more than 24 hours. (See 25:2.)

Opin. Strictly speaking, neither the AWAR nor PHS Policy mandates inspection of investigators' laboratories where research with animals is conducted unless animals are maintained in the laboratory longer than a specified period, as noted (see below). However, the *Guide* (p. 25) does specify that the IACUC, as part of its oversight functions, should conduct regular inspections of animal activity areas at least annually or more often if required by the AWAR or PHS Policy. The *Guide* also suggests in this same chapter (p. 34) that laboratory inspections be considered as one component of a post-approval monitoring program. It is also noteworthy that AAALAC requires institutions seeking or renewing accreditation to identify protocols in which procedures are conducted on animals outside animal resource program managed areas as a component of the program description. These areas may be visited during the site visit based upon the nature of the procedures conducted at the site and the site visitors' discretion. Therefore, many institutional IACUCs assume this responsibility in order to meet their regulatory and institutional charge of ensuring that animals are used humanely by trained personnel. Laboratory visitation frequently provides the added benefit of enhancing dialog between the committee and investigative staff, frequently improving the IACUC's understanding of proposals. (See 23:18; 23:20; 24:12.)

NIH/OLAW, in its FAQ E.1[1] has addressed the IACUC inspections of laboratories where investigators use animals as follows:

> Institutions are responsible for oversight of all animal-related activities regardless of how long or where the activity occurs. Satellite facilities (defined by PHS Policy as a containment outside a core or centrally managed area in which animals are housed for more than 24 hours) and areas where any form of surgical manipulations (minor, major, survival, non-survival) are performed must be inspected at least once every six months by the IACUC as part of the semiannual evaluation. Institutions have discretion with regard to how they oversee areas used for routine weighing, dosing, immunization, or imaging, but should monitor such areas on a random or fixed schedule to effectively oversee activities at the institution. USDA requires semi-annual inspection of "animal study areas" defined as areas where USDA covered animals are housed for more than 12 hours.

APHIS/AC supports the position that although laboratories do not have to be inspected if animals are held there less than 12 hours, IACUC inspection is encouraged to the extent that it helps the IACUC and the institution fulfill its responsibilities under the AWA to oversee all animal care and use activities at the institution.[2]

Surv. During your semiannual inspections, which laboratories that hold or house animals are inspected by your IACUC? More than one response is possible.

- Not applicable — 17/295 (5.8%)
- Inspect all laboratories that hold or house animals for any length of time — 235/295 (79.7%)
- Only those holding or housing animals for more than 12 hours (USDA covered species) — 31/295 (10.5%)
- Only those holding or housing animals for more than 24 hours — 17/295 (5.8%)
- We inspect a random selection of laboratories — 10/295 (3.4%)
- We inspect only those laboratories deemed by the IACUC to be of increased concern — 12/295 (4.1%)
- Other — 10/295 (3.4%)

25:2 Is it necessary to inspect and evaluate every laboratory where animals are used at the time of each IACUC inspection?

Reg. AWAR (§1.1; §2.31,c,2) require that any *study area* (defined as any building room, area, enclosure, or other containment outside a core facility or centrally designated or managed area in which animals are housed for more than 12 hours) be inspected at least once every 6 months by a subcommittee of at least two IACUC members.

The PHS Policy (III,B) defines an *animal facility* as "any or all buildings, rooms, areas, enclosures, or vehicles, including satellite facilities, used for animal confinement, transport, maintenance, breeding, or experiments inclusive of surgical manipulation. A satellite facility is any containment outside of a core facility or centrally designated or managed area in which animals are housed for more than 24 hours." All facilities, including satellite facilities, must be inspected at least once every 6 months by the IACUC (PHS Policy IV,B,2). (See 15:11; 25:1.) The *Guide* (p. 25) does specify that the IACUC, as part of its oversight functions, should conduct regular inspections of animal activity areas at least annually or more often if required by the AWAR or PHS Policy. (See 25:1.)

Opin. As addressed in 25:1, many IACUCs conduct routine laboratory inspections in order to meet their oversight responsibility of the institution's animal care and use program. It should be noted that the AWAR (§2.31,c,2) require that the IACUC inspect all animal facilities where animals are kept for more than 12 hours, including animal study areas, at least once every 6 months. Other animal facilities include surgical rooms, food storage, and cage wash areas.[2] Some of these areas may not house animals for more than 12 hours or may not house animals at all. Nevertheless, APHIS/AC will inspect these areas and cite any noncompliances; therefore, it is appropriate to correct any noncompliances found during IACUC inspections before they are cited by APHIS/AC inspectors.

It is also the authors' experience that AAALAC site visit teams will often request to visit laboratories outside of the centrally managed animal resource program where survival surgical procedures on any species are performed due to the invasive nature of the procedures and the potential for post-procedural complications. As we noted in the introduction to this chapter and in 25:1, it is our opinion that such inspections are appropriate.

25:3 How frequently should each laboratory where animals are used be inspected?

Reg. Research laboratories in which APHIS/AC-covered species are maintained for more than 12 hours must be inspected by a subcommittee of at least two IACUC members every 6 months (AWAR §1.1, Animal; §1.1, Study Area; §2.31,c,2–3). IACUCs serving institutions required to comply with the PHS Policy (III,B; IV,B,2) must inspect laboratories in which animals are maintained for more than 24 hours at least once every 6 months. As stated in 25:1, the *Guide* (p. 25) does specify as part of its oversight functions should conduct regular inspections of animal activity areas at least annually or more often if required by the AWAR or PHS Policy.

Opin. As there are no other regulatory requirements mandating laboratory inspection other than what is recommended in the AWAR and PHS Policy, it is at the individual IACUC's discretion whether or not and how often they should inspect animal research laboratories at their institution that do not meet these time requirements. However, as excerpted in 25:1, NIH/OLAW states in their response to FAQ E.1

that areas where any form of surgical manipulations (minor, major, survival, non-survival) are performed must be inspected at least once every six months by the IACUC as part of the semiannual evaluation.[1] It is the authors' opinion that at least select animal research laboratory inspections should be conducted by the IACUC at least annually in order to ensure compliance with both the content and the spirit of applicable regulations and guidelines. It may be prudent to inspect select areas more frequently, for example every 6 months, as part of the semiannual inspection, based on the nature of the activities conducted at the site. (See 25:1.)

25:4 Should a laboratory be approved by the IACUC before permitting an investigator to use the laboratory to conduct procedures on animals?

Reg. The AWAR (§2.31,d,1) require the IACUC, as part of the protocol review process, to ensure that all components of animal care and use are in compliance with the AWAR prior to approval of the protocol. PHS Policy (IV,C,1) requires that the project will be conducted in accordance with the AWA and the *Guide*. (See 25:1.)

Opin. Other than as noted for the AWAR and the PHS Policy, there are no regulatory requirements mandating the approval of laboratories by the IACUC. Nevertheless, it is the IACUC's responsibility to ensure that the proposed activity is conducted at a suitable location. It is the authors' opinion that there are significant advantages of having the IACUC inspect laboratories before providing approval for work with animals. Inspection of the laboratory provides the IACUC the opportunity to determine whether the facilities and equipment are suitable for the proposed activity. In addition, it provides an opportunity for the investigative staff to gain a better understanding of IACUC functions and provides IACUC members a more thorough understanding of the proposed project. At institutions where IACUC approval of sites is not feasible because of time constraints, the IACUC can delegate this responsibility to a member of the institution's animal resource program's staff. (See 23:24.)

25:5 How does the IACUC determine the location of laboratories used for research with animals?

Opin. The most direct method for determining which laboratories are used to conduct research on animals is to request that information in the animal use proposal reviewed by the IACUC. This information is frequently used to generate a database. Additionally, laboratories may be identified on the basis of information provided by the animal resources staff or the institution's administration.

Surv. How does your IACUC determine the location of laboratories used for research with animals? More than one response is possible.

• Not applicable	25/295 (8.5%)
• From information on the IACUC protocol form	221/295 (74.9%)
• From a separate database that we periodically update	55/295 (18.6%)
• From requests to investigators to identify their laboratory's location	51/295 (17.3%)
• From the knowledge of animal care personnel	115/295 (39.0%)
• I don't know	1/295 (0.3%)
• Other	16/295 (5.4%)

25:6 Should the IACUC inspection be announced or unannounced? If an IACUC inspection team misses a laboratory during an inspection or if a member of the investigative staff is not present in the laboratory at the time of the inspection, should they go back at an alternative time to complete the inspection?

Opin. There are advantages and disadvantages of both methods. An announced site visit generally ensures that a knowledgeable staff member is available to provide laboratory access, discuss animal research activities, and answer questions. Staff availability permits dialog, which is advantageous to both the IACUC members and research staff, who frequently gain a better understanding of the IACUC's functions and activities. However, when conducting announced site visits, IACUC members may be exposed to an artificial environment not reflective of the normal state of the laboratory, personnel conduct, or animal use. Further, it is not uncommon for laboratories to cancel animal research activities when expecting a site visit. (See 23:27.)

Unannounced inspections permit the IACUC members to view the laboratory and its activities under normal operating conditions. However, research staff may not be present, potentially limiting laboratory access, or if they are present, ongoing research responsibilities may limit them in interfacing with IACUC members.

Institutions conducting announced inspections may also conduct unannounced inspections if the IACUC has specific concerns that may not be uncovered if inspections are announced.

The determination as to whether a laboratory should be revisited if it is missed is dependent on federal and state regulatory requirements and the regulatory and institutional charge of the IACUC. The AWAR (§2.31,c,2) and the PHS Policy (III,B; IV,B,2) require that study areas be inspected at least once every 6 months by the IACUC to maintain compliance. (See 25:1; 25:2.) Therefore, if the area is mandated for inspection and access cannot be obtained, the authors recommend rescheduling a visit within the required inspection timetable.

Surv. During a semiannual laboratory inspection, if a laboratory cannot be inspected (for any reasons) does your IACUC make an attempt to revisit the lab during the same inspection cycle or is the inspection deferred until the next semiannual inspection?

• Not applicable	66/292 (22.6%)
• We typically attempt to revisit the lab during the current semiannual inspection time frame	200/292 (68.5%)
• We typically wait for next semiannual inspection	10/292 (3.4%)
• We leave the decision to the inspection team	5/292 (1.7%)
• We leave the decision to the IACUC office or IACUC chairperson	8/292 (2.7%)
• Other	3/292 (1.0%)

25:7 Is it necessary for a PI or his or her designee to be present during the inspection?

Opin. Although not mandated by the AWAR or PHS Policy, there are clear advantages in having a knowledgeable research staff member present during the inspection. (See 25:6.) If the inspections are announced, the IACUC may stipulate that the PI or designee be present during the inspection. With unannounced visits, the IACUC cannot ensure that access can be gained to the area.

25:8 Prior to the semiannual inspection, should the IACUC members who are visiting the laboratory be briefed on procedures conducted in the laboratory?

Opin. It is advantageous for IACUC members to be briefed on procedures occurring in the laboratory prior to the inspection, by the AV, a knowledgeable IACUC member, or by consulting the protocol. Alternatively, the PI or his or her designee can provide this information at the time of the inspection.

25:9 What areas and issues should the IACUC focus on during inspection of the laboratory?

Opin. IACUC members should

- Ensure appropriate facilities, personnel, and equipment are present to provide adequate veterinary care
- Observe the laboratory and equipment used for cleanliness and adequate sanitation
- Assess safety issues such as the following:
  - Food or drink consumption within the laboratory
  - Anesthetic gas scavenging
  - Appropriate use and handling of hazardous substances
  - Use of appropriate animal handling procedures
  - Sharps as well as syringes and needles are handled appropriately
- Observe the condition of animals, if they are present
- Confirm compliance with approved IACUC protocols including ascertaining that procedures and staff have been described and included in the approved protocol
- Ensure staff understand the requirement for gaining IACUC approval in advance of utilizing a different species and/or changing/adding an experimental procedure(s) in accordance with institutional policy
- Evaluate the adequacy of training
- If surgery is performed, determine adherence to aseptic technique, suitability and understanding of anesthesia monitoring, and postoperative care including pain relief
- Review methods of euthanasia
- Verify that all pharmaceuticals are handled and stored appropriately and are not expired
- If using controlled substances ensure their storage, use, and record keeping is in compliance with federal, state, and/or institutional policies
- Review records, especially when conducting survival surgical procedures (See 25:10)
- Confirm that equipment requiring regular maintenance or certification, e.g., anesthetic vaporizers, biological safety cabinets, and guillotines are maintained/certified as appropriate

- If the laboratory is approved for animal housing, determine that the animals' husbandry needs are met, environmental conditions are both stable and suitable and appropriate records are maintained
- Consider using NIH/OLAW's Semiannual Program Review and Checklist[3]

Surv. If your IACUC inspects areas (other than a central laboratory animal facility) in which animals are permitted to be housed for more than one day, which of the following are typically evaluated by the inspectors? More than one response is possible.

- Not applicable 108/294 (36.7%)
- Light levels 156/294 (53.1%)
- Temperature 176/294 (59.9%)
- Humidity 159/294 (54.1%)
- Air changes/hour 107/294 (36.4%)
- Daily veterinary rounds (if daily rounds are made) 73/294 (24.8%)
- Daily husbandry rounds 147/294 (50%)
- Other 22/294 (7.5%)

25:10 What records should the IACUC inspection team review when visiting the laboratory?

Opin. Record evaluation is influenced by the activities that are conducted in the laboratory. If the laboratory is approved for the housing and care of animals, the PI's staff should maintain, and the IACUC members should periodically examine, those records indicating the frequency of cage changing, environmental monitoring (e.g., temperature, humidity, and light cycles), laboratory sanitation, and daily animal observations (including weekends). Records should be similar to those maintained by animal resource personnel in the animal facility. The following records, if applicable to the use of the laboratory, should be maintained and made available to the IACUC members during an inspection:

- Numbers of animals used per unit time (period in which the IACUC can compare actual animal usage to the number of animals requested in the protocol)
- Procedures conducted
- Treatment or drug administration
- Anesthetic, surgical, and postoperative care
- Controlled substances log
- Dosimetry for radiation exposure if using radionuclides or exposed to ionizing radiation
- Any institutionally required documentation when using hazardous substances
- Copies of institutionally required policies, handbooks, or guidelines

25:11 What criteria can the inspection team use to make sure that acceptable standards are being met in the laboratory?

Opin. There are many criteria that can and should be utilized to determine that acceptable standards are used in the research laboratory. IACUC members should utilize the *Guide* and the AWAR, as well as IACUC and institutional policies, as the basis for determining the suitability of the laboratory for conducting research with animals. Depending on the nature of the research conducted and the institution, the Good Laboratory Practices Act,[4] the Occupational Safety and Health Act,[5] the Bloodborne Pathogen Standard,[6] Standards for Protection against Radiation,[7] Lab Safety Standard,[8] Controlled Substances Act,[9] NIH Recombinant DNA Guidelines,[10] and Centers for Disease Control and Prevention's *Biosafety in Microbiologic and Biomedical Laboratories*[11] also may be used as criteria.

25:12 How should the IACUC proceed after identifying an item of deficiency in a laboratory?

Reg. The AWAR (§2.31,c,3) require the IACUC to identify deficiency items as significant or minor and develop a plan and timetable for correction. The AWAR (§2.31,c,3) define a *significant deficiency* as "one which, with reference to Subchapter A, and, in the judgment of the IACUC and the Institutional Official, is or may be a threat to the health or safety of the animals." If the established timetable is not adhered to and a significant deficiency remains uncorrected, the deficiency must be reported to APHIS/AC and any federal agency funding the activity within 15 days (AWAR §2.31,c,3). Similarly, the PHS Policy (IV,B,3) requires the IACUC to draft reports that are submitted to the IO of its inspection findings. As with the AWAR, identified deficiencies must be distinguished as significant or minor on the basis of whether they are or may be a threat to the health or safety of the animals. If program or facility deficiencies are noted, the report to the IO must contain a reasonable and specific plan and schedule for correction of each deficiency. Noncompliance must be promptly reported to NIH/OLAW (PHS Policy IV,F,3,a). Serious deviations from the provisions of the *Guide* must be promptly reported to OLAW (PHS Policy IV,F,3,b). See also NIH/OLAW guidance on reporting departures from the *Guide*, Eighth Edition at http://grants.nih.gov/grants/olaw/departures.htm. (See 24:15; 26:16.)

Opin. Although not specifically addressed in the regulations, the authors recommend expanding the criteria for a significant deficiency to include those that may be, or are, a threat to the health or safety of personnel in addition to animals. The authors find it most effective for deficiencies to be discussed with the responsible individuals or their designee, if available, at the time of the site visit. The ensuing dialog may clarify misinterpretations. Deficiencies should be reviewed by the IACUC and categorized as described. A plan and timetable for correction should be established. A written notice should be sent from the IACUC to the responsible individual, describing the stated deficiency and a plan and timetable provided for correction. The authors advise that the IACUC develop a mechanism by which the responsible individual can contest the deficiency if there is disagreement. Further, the responsible individual should be requested to provide written documentation that the deficiency is corrected. The IACUC may wish to re-inspect to ensure the correction is effective.

Alternatively, the authors are aware of institutions that effectively use a multicopy form that is completed by the site visitors at the time of the site visit. The form contains a checklist for the inspectors to use to identify compliance or lack thereof. If deficiencies are noted, a timetable for correction based on preapproved criteria established by the IACUC is provided at the time of the site visit. The form is signed by the responsible individual or designee at the time of the site visit, and a copy is retained by the laboratory. The form stipulates that a formal response is required from the responsible individual for all noncompliant items.

Surv. As part of your semiannual laboratory inspection, how does your IACUC inform an investigator of a significant deficiency? More than one response is possible.

- Not applicable 12/294 (4.1%)
- The inspection team notifies the laboratory during the inspection 160/294 (54.4%)
- The IACUC sends a written notification after the inspection 251/294 (85.4%)
- The IACUC calls the laboratory after the inspection 46/294 (15.7%)
- I don't know 4/294 (1.4%)
- Other 4/294 (1.4%)

All communications should convey the date of expected resolution. The authors recommend that the IACUC either re-inspect the area to determine whether the item was corrected by the specified date or confirm correction during the next scheduled inspection. Depending on the severity and repetition of the item of noncompliance, the IACUC may impose sanctions on the investigator (see Chapter 29).

25:13 What criteria should be used for evaluation of biological safety cabinets (BSCs) and laboratory hoods (also referred to as chemical or fume hoods)?

Reg. The CDC's *Biosafety in Microbiological and Biomedical Laboratories, 5th Edition,*[11] *Appendix A, Section VII-Certification of BSCs,* indicates that a BSC be certified, preferably by an accredited field certifier, before it is placed in service, after it is relocated or annually, using the National Sanitation Foundation (NSF) Standard 49 (NSF/ANSI Standard 49-2007) specifications. NSF 49 states that it is customary for the person conducting the designated tests to affix to the cabinet a certificate of adequate performance when the cabinet meets all field test criteria. Laboratory hood performance is addressed in OSHA's Laboratories Standard[7] (OSHA CFR Title 29 Part 1910.1450) requiring each institution to establish methods for evaluating hood performance in their required Chemical Hygiene Plan.

Opin. Current certification of each BSC should be confirmed during a laboratory inspection by examining the certification label on the cabinet. The IACUC inspection team should be acquainted with their institution's laboratory hood certification process and verify the hood is appropriately certified. If the cabinet or hood is in use, the authors recommend observing its use to ensure safe operational procedures are followed.

25:14 What standards should be applied to animals being housed in laboratories?

Reg. The AWAR (§2.31,c,2) require the same standards to be used (Title 9, Chapter I, subchapter A-Animal Welfare) for evaluating the animal study area as are used for the animal facility if the former meets the criteria (>12 hour housing) for inspection. The PHS Policy (IV,B,2), as with the AWAR, makes no distinction in evaluation criteria between the animal facility and laboratory if the 24 hour housing requirement is met; therefore, the *Guide* is used as the basis for laboratory evaluation.

Opin. In the authors' experience, laboratory housing for extended periods is frequently necessitated by several factors: the need for the animals to be maintained in close proximity to unique equipment that is not found in or would be difficult or impossible to relocate to the animal facility; the need to house the animals under unique environmental conditions which cannot be replicated in the animal facility; and/or the transportation of animals to and from the animal facility would confound experimental results. When there is a need for frequent or prolonged laboratory housing, the expectation, in the authors' opinion, is that the laboratory would be dedicated to this purpose and would be constructed similarly to an animal holding room or procedure laboratory in the animal facility. There may be situations in which the essential need for laboratory housing is for a limited time and infrequent. In this circumstance, the IACUC may consider giving greater latitude with respect to laboratory conditions; however, the IACUC must ensure that the animal's welfare and safety are not at risk, there is no risk to personnel that will access the laboratory and the environmental conditions and stability are conducive with obtaining repeatable and reliable results. Routine monitoring of animal welfare by appropriately trained staff is essential for all laboratory housing.

References

1. National Institutes of Health, Office of Laboratory Animal Welfare. 2012. Frequently Asked Questions. http://grants.nih.gov/grants/olaw/faqs.htm.
2. U.S. Department of Agriculture, Animal and Plant Health Inspection Service. 2013. *Animal Welfare Inspection Guide*. http://www.aphis.usda.gov/animal_welfare/downloads/Inspection%20Guide%20-%20November%202013.pdf.
3. National Institutes of Health, Office of Laboratory Animal Welfare. 2012. Semiannual program review and checklist. http://grants.nih.gov/grants/olaw/sampledoc/cheklist.htm.
4. Office of the Federal Register. 1978. Code of Federal Regulations, Title 43, Good Laboratory Practice Regulations. Washington, D.C. http://www.gpo.gov/fdsys/pkg/CFR-2011-title21-vol1/pdf/CFR-2011-title21-vol1-part58.pdf.
5. Office of the Federal Register. 1984. Code of Federal Regulations, Title 29, Part 1910, Occupational Safety and Health Standards, Subpart G, Occupational Health and Environmental Control. Washington, D.C. http://www.osha.gov/pls/oshaweb/owastand.display_standard_group?p_toc_level=1&p_part_number=1910.
6. Office of the Federal Register. 1991. Code of Federal Regulations, Title 29, Part 1910.1030, Bloodborne Pathogens, Occupational Health and Safety Act, Washington, D.C. http://www.osha.gov/pls/oshaweb/owadisp.show_document?p_table=STANDARDS&p_id=10051.
7. Office of the Federal Register. 1992. Code of Federal Regulations, Title 10, Chapter 1 Part 20, Standards for protecting against radiation nuclear regulatory commission regulations, Washington, D.C. http://www.nrc.gov/reading-rm/doc-collections/cfr/.

8. Office of the Federal Register. 1990. Code of Federal Regulations, Title 29, Part 1910, Department of Labor, Occupational Safety and Health Administration, Occupational exposures to hazardous chemicals in laboratories: Final rule. Washington, D.C. http://www.osha.gov/pls/oshaweb/owadisp.show_document?p_table=standards&p_id=10106.
9. U.S. Department of Justice. 2010. Pharmacist's Manual: An informational outline of the Controlled Substances Act. Springfield: Office of Diversional Control. http://www.deadiversion.usdoj.gov/pubs/manuals/pharm2/pharm_manual.pdf.
10. U.S. Department of Health and Human Services. 2011. Guidelines for research involving recombinant DNA molecules (NIH guidelines). http://oba.od.nih.gov/oba/rac/Guidelines/NIH_Guidelines.pdf.
11. U.S. Department of Health and Human Services. 2009. Biosafety in microbiological and biomedical laboratories, 5th edition. HHS Publication No. (CDC) 21-1112. Washington, D.C.: U.S. Government Printing Office. http://www.cdc.gov/biosafety/publications/bmbl5/BMBL.pdf.

26

Inspection of Surgery Areas

Scott E. Perkins and Neil S. Lipman

Introduction

The AWAR (§2.31,c,2) require IACUCs to inspect "all of the research facility's animal facilities, including animal study areas" for species regulated by the AWAR. The PHS Policy (IV,B,2) requires any and all animal facilities and areas to be inspected at least every 6 months including satellite facilities and surgical areas. The PHS Policy (IV,A,1) also requires adherence to the *Guide* (p. 25), which recommends inspection of the animal facilities and animal care and use areas at least annually or more often as required (e.g., by the AWA and the PHS Policy). The *Guide* and AWAR specify requirements for surgical facilities and procedures for conducting survival surgery on rodents and AWAR-regulated species. The IACUC must approve all surgical procedures involving animals to ensure that appropriately trained individuals conduct the research in a humane manner. This chapter is intended to help the reader to comply with these regulations and recommendations.

26:1 Should all surgical sites be visited and evaluated at the time of each IACUC inspection?

Reg. There are federal requirements for IACUCs to inspect areas used to conduct survival surgical procedures on AWAR-regulated species[1] regardless of whether or not the site is located within an animal facility (*Guide* p. 25). The AWAR (§2.31,c,2) require that animal facilities, including study areas, be inspected at least once every 6 months.

The PHS Policy (IV,B,2) also requires any and all animal facilities to be inspected at least every 6 months. Specifically, the PHS Policy (III,B) includes "any and all buildings, rooms, areas, enclosures, or vehicles, including satellite facilities, used for animal containment, transport, maintenance, breeding, or experiments inclusive of surgical manipulation." This includes areas used for non-survival surgery.[1] PHS Policy (IV,A,1) also requires adherence to the provisions of the *Guide* (p. 25), which states "Program review and facilities inspections should occur at least annually or more often as required (e.g., the Animal Welfare Act and PHS Policy)."

Opin. The AWAR (§2.31,c,2) and PHS Policy (IV,B,2) require semiannual inspection of surgical areas.[1] Therefore, all surgical areas, including surgical facilities outside a core animal facility, should be inspected semiannually by the IACUC to maintain compliance with and meet the spirit of the AWAR, PHS Policy, and *Guide*. Both the AWAR (§2.31,d,1,ix) and the *Guide* (pp. 115–120, 144–145) detail facility and procedural requirements for conducting survival surgical procedures on rodents or USDA-regulated species.

Surv. 1 Which surgical sites does your IACUC visit as part of its semiannual inspection? More than one response is possible.

- Not applicable 36/295 (5.3%)
- Survival surgical sites for USDA covered species 91/295 (13.5%)
- Survival surgical sites for all species 246/295 (83.4%)
- Non-survival surgical sites for USDA covered species 88/295 (13.0%)
- Non-survival surgical sites for all species 212/295 (71.9%)
- I don't know 0/295 (0%)
- Other 3/295 (0.4%)

Surv. 2 During the semiannual inspection of surgical facilities, are your inspectors given a checklist of items to evaluate?

- Not applicable 37/292 (12.7%)
- Yes, we use the NIH/OLAW aseptic surgery checklist 97/292 (33.2%)
- Yes, we use an in-house developed checklist 82/292 (28.1%)
- No 72/292 (24.7%)
- I don't know 2/292 (0.7%)
- Other 2/292 (0.7%)

26:2 What is the minimal frequency for evaluating surgical sites?

Reg. (See 26:1.)
Opin. (See 26:1.)

26:3 What facilities are required for conducting survival surgery on non-rodent and rodent species?

Reg. The IACUC must differentiate between major and non-major surgical procedures when evaluating facilities for performing survival surgical procedures, as facility requirements depend on the procedure conducted (see 26:5 for definitions).

Both the AWAR (§2.31,d,1,ix; §2.31,d,1,x) and the *Guide* (pp. 144–145) specify facility requirements for conducting major survival surgery on non-rodent species. The AWAR (§2.31,d,1,ix) require that "major operative procedures on non-rodents will be conducted only in facilities intended for that purpose which shall be operated and maintained under aseptic conditions." The AWAR do not require dedicated facilities when conducting major operative procedures at field sites; however, aseptic technique must be followed (§2.31,d,1,ix).

The *Guide* (pp. 144–145) provides significantly more detail pertaining to facility requirements, recommending that non-rodent surgical facilities include a variety of functional components, usually but not always separated by physical barriers. These include areas for surgical support, animal preparation, surgeon's scrub, operating room, and postoperative recovery. The need to provide all of these areas depends on the size and scope of the surgical program. With respect to ventilation, the *Guide* (p. 139) states that "areas of surgery should be kept under relative positive pressure with clean air."

Opin. It is the chapter authors' opinion that the minimal requirements for most operative suites are

- Area for animal preparation
- Area for surgical scrub
- Operating room

The surgical suite should be designed and operated in a manner to control microbial contamination. This can be accomplished by

- Minimizing traffic in the area
- Adhering to rigorous sanitation and disinfection programs
- Limiting equipment present within the operating room to only essential movable equipment (such as surgical and support tables and anesthesia and monitoring equipment)
- Outfitting the surgical suite with a ventilation system that provides directional airflow maintaining the operating room at positive pressure with respect to the animal preparation and surrounding areas (under most circumstances)

Facility requirements for conducting non-major surgery on non-rodents or major survival surgery on rodents are less stringent. The AWAR (§2.31,d,1,ix) do not require dedicated facilities for conducting these procedures but do specify the use of aseptic technique. An area used for such procedures can be a room or portion of a room, which is easily sanitized and is dedicated for that purpose at the time of surgery. The laboratory bench or surgical table should be disinfected prior to and after use, be free of ancillary equipment, be located in an area free of drafts, and have a suitable temperature.

26:4 What facilities are required for conducting non-survival surgery on animals?

Opin. There are no facility-specific requirements when conducting non-survival surgical procedures. The area used for non-survival surgery should be a room or portion of a room, which is easily sanitized and at the time of surgery, dedicated for that purpose. The laboratory bench or surgical table should be disinfected prior to and after use, be free of ancillary equipment, and be located in an area free of drafts with a suitable temperature. In general, adherence to aseptic technique is unnecessary. The surgical site should be clipped and the instruments clean, and the surgeon should wear gloves. The IACUC should evaluate the duration of non-survival surgery as procedures can occasionally last for sufficient periods for sepsis to develop. In these cases, adherence to aseptic technique may be advisable. Additionally, the IACUC should evaluate the need for aseptic technique or "aseptic" facilities, if certain procedures are being performed (e.g., harvesting an organ for transplantation). (See 18:21.)

26:5 How does one differentiate a major from a minor surgical procedure?

Reg. The AWAR (§1.1) defines a *major operative procedure* as "any surgical intervention that penetrates and exposes a body cavity or any procedure which produces permanent impairment of physical or physiological functions." (See 18:4.) The PHS Policy itself provides no definition of major or minor surgical procedures, but it does require adherence to the provisions of the AWA for regulated species (PHS Policy II) and of the *Guide* (PHS Policy IV,A,1).

The *Guide* (pp. 117–118) defines a major survival surgical procedure as one that "penetrates and exposes a body cavity, produces substantial impairment of physical or physiological functions, or involves extensive tissue dissection or transection." It defines *minor survival surgery* as one in which the procedure "does not expose a body cavity and causes little or no physical impairment."

NIH/OLAW provides the following guidance in FAQ F.13: "OLAW recognizes the authority of the IACUC to determine whether specific manipulations used in research are major operative procedures. The IACUC's determination must be based on a detailed description of the procedure and the anticipated or actual consequences, as characterized by the investigator. In some cases, the classification by the IACUC of a procedure as major or minor may be readjusted post-procedurally depending on clinical outcome. If the IACUC, after thorough review, determines that the surgical procedure only penetrates but does not expose a body cavity and that the procedure does not produce substantial impairment, the IACUC may conclude that it is not a major operative procedure. Any laparoscopic surgery that produces substantial impairment of physical or physiological function must be considered a major operative procedure."

Opin. Although both the AWAR and the *Guide* provide a definition of a major operative procedure, the classification of some surgical procedures requires IACUC interpretation, which is based on the procedure's extent, effects, and potential for development of complications. Major operative procedures include laparotomy, thoracotomy, and limb amputation. However, APHIS/AC acknowledges that the IACUC has latitude to determine whether a laparoscopic procedure is a major or minor operative procedure.[2] Minor operative procedures include wound suturing, peripheral vessel cannulations, and certain routine farm animal procedures. The *Guide* (p. 117) indicates that "laparoscopic surgeries and some procedures associated with neuroscience research, (e.g., craniotomy, neurectomy) may be classified as major or minor depending on their impact on the animal." (See 18:8.)

26:6 Assume an animal is regulated by the AWAR and the PHS Policy. Under what circumstances can a major survival surgical procedure be performed in an area other than a dedicated surgical suite approved by the IACUC?

Reg. The AWAR (§2.31,d,1,ix) require that "major operative procedures on non-rodents will be conducted only in facilities intended for that purpose which shall be operated and maintained under aseptic conditions." The AWAR (§2.31,d,1,ix) do not require dedicated facilities when conducting major operative procedures at field sites, although aseptic procedures are required.

PHS Policy (IV,A,1) requires adherence to the *Guide* (p. 62), which states, "Unless an exception is specifically justified as an essential component of the research protocol and approved by the IACUC, aseptic surgery should be conducted in dedicated facilities or spaces." Nevertheless, the *Guide* (pp. 117–118) does provide exceptions for emergencies and field sites: "Emergency situations sometimes require immediate surgical attention under less than ideal conditions," and "Generally, agricultural animals maintained for biomedical research should undergo surgery with techniques and in facilities compatible with the guidelines set forth in this section. However, some minor and emergency procedures commonly performed in clinical veterinary practice and in commercial agricultural settings may take place under field conditions. Even when conducted in an agricultural setting, however, these procedures require the use of appropriate aseptic technique, sedatives, analgesics, anesthetics, and conditions commensurate with the risk to the animal's health and well-being."

Opin. It is the authors' opinion, in accordance with the *Guide*, that major operative procedures can be performed in a non-dedicated surgical suite when scientifically justified (e.g., unique instrumentation in the laboratory) and approved by the IACUC. In the event of an emergency when the animal cannot be transported to the dedicated operating room, this should be considered a clinical medical requirement, determined by veterinary medical judgment, and does not require IACUC approval.

26:7 Can non-survival surgery on non-rodent species, or survival surgery or non-survival surgeries on rodents, be performed in survival surgical facilities used for non-rodents? If yes, are there any special procedures that should be followed?

Opin. IACUCs may permit the conduct of survival rodent surgery or non-survival surgical procedures in survival surgical facilities if either of the following conditions is met:

- Non-survival procedures that are conducted utilize the same standards that apply to survival procedures in accordance with the standards of the *Guide* (p. 118).
- The operating room is thoroughly sanitized and disinfected prior to conduct of a survival surgical procedure in non-rodents (*Guide* p. 118).

26:8 What areas and issues should the IACUC focus on during inspection and evaluation of surgical areas?

Reg. PHS Policy (IV,B,2) requires any and all animal facilities (as defined in PHS Policy III,B) to be inspected at least every 6 months, including "any and all buildings, rooms, areas, enclosures, or vehicles, including satellite facilities, used for animal confinement, transport, maintenance, breeding, or experiments inclusive of surgical manipulation." PHS Policy (IV,A,1) requires compliance with the *Guide* (pp. 115–120), which includes IACUC evaluation of aseptic procedures, postoperative evaluation and care, anesthetic monitoring, and other elements. Other sections of the *Guide* provide further information on OHS (*Guide* pp. 17–23), design of facilities for aseptic surgery (*Guide* pp. 144–145), and other considerations. The AWAR (§2.32), PHS Policy (IV,C,1,f), U.S. Government Principle VIII, and the *Guide* (p. 115) all address the need for proper qualifications and experience of personnel working with animals (see 26:1).

Opin. The AWAR (§2.31,d,1,ix) and the *Guide* (pp. 116–120, 144–145) describe facility and procedural requirements for conducting survival surgical procedures in non-rodents. In the authors' opinion, minimal evaluation of standards for surgical areas should include the following:

- Assessment of personnel training
- Adherence to aseptic technique
- Scavenging of waste anesthetic gases
- Correct use of unexpired pharmaceuticals
- Scientifically justified and IACUC approved use of non-pharmaceutical-grade chemicals and other substances
- Proper use of equipment and procedures for peri-operative monitoring and care
- Maintenance of peri-operative records
- Availability of appropriate equipment and pharmaceuticals for emergencies
- Proper disposal of sharps

Surv. During the semiannual inspection of surgical areas, which of the following items are routinely evaluated by the inspectors? More than one response is possible.

• Not applicable	33/294 (1.3%)
• We have no policy on this issue	26/294 (1.0%)
• The general conditions of the physical plant	218/294 (74.1%)
• Personnel training and surgeons' competence	105/294 (35.7%)
• Scavenging of anesthetic waste gases, if used	188/294 (63.9%)
• Maintenance of gas anesthesia machines, if used	202/294 (68.7%)
• Availability of appropriate surgical instruments	145/294 (49.3%)
• Availability of appropriate surgical equipment	160/294 (54.4%)
• Drug storage and drug inventory	220/294 (74.8%)
• General cleanliness	241/294 (82.0%)
• Records of disinfecting the area	104/294 (35.4%)
• Procedures used for instrument and supply sterilization	160/294 (54.4%)
• Methods of testing autoclaves and appropriate test records	132/294 (44.9%)
• Proper preparation of the animal (i.e., skin prep)	102/294 (34.7%)
• Adherence to aseptic technique during surgery	125/294 (42.5%)
• Use of appropriate surgical garb	123/294 (41.8%)
• Posting of emergency contact information	168/294 (57.1%)
• Peri-operative monitoring and records	156/294 (53.1%)
• I don't know	2/294 (0.7%)

26:9 What reference sources are available to help evaluate the issues in question 26:8?

Opin. The IACUC should utilize the AWAR, the *Guide* and its associated references, applicable state regulations, and institutional policies to develop acceptable standards for surgical facilities and their operation. The IACUC can also consult references, as well as expert consultants, to assist in the development of standards for conducting surgery. Additional references include veterinary and human surgery and anesthesia texts, texts and references on experimental surgery, the Good Laboratory Practices Act,[3] the Occupational Safety and Health Act,[4] the Blood Borne Pathogen Standard,[5] Standards for Protection against Radiation,[6] Lab Safety Standard,[7] information on the Controlled Substances Act,[8] Recombinant DNA Guidelines,[9] the Centers for Disease Control and Prevention (CDC) publication *Biosafety in Microbiologic and Biomedical Laboratories*,[10] NIH/OLAW Semiannual Program Review and Facilities Inspection Checklist,[11] and references on disinfection and sanitation control.[12]

26:10 What records should the inspection team review when visiting non-rodent and rodent surgical areas? (See 18:38.)

Opin. The IACUC should confirm that procedures and staff training, experience, and qualifications have been accurately described and included in the approved animal care and use proposal. The IACUC should review pertinent animal health and experimental records including the following:

- Peri-operative records describing the surgical procedure performed
- Experimental agents administered
- Dose and route of anesthetics and analgesics administered
- Required records for controlled substances
- Frequency and type of intra-operative monitoring procedures performed
- Postoperative monitoring and observations
- Quality assurance records for sterilization of materials (periodically reviewed)
- All records should be made available for IACUC review whether they are maintained with the animal, in the surgical facility, or by the investigator

26:11 How detailed should surgical and postoperative records be? What are the essential elements of an adequate record?

Reg. The *Guide* (pp. 75–76) states, "Medical records for individual animals can also be valuable, especially for dogs, cats, nonhuman primates, and agricultural animals (Suckow and Doerning 2007). They should include pertinent clinical and diagnostic information, date of inoculations, history of surgical procedures and postoperative care, information on experimental use, and necropsy findings where applicable." The AWAR (§2.33,b,5) require "adequate pre-procedural and post-procedural care in accordance with current established veterinary medical and nursing procedures." APHIS/AC Policy 3 states, "Health records are meant to convey necessary information to all people involved in an animal's care. Every facility should have a system of health records sufficiently comprehensive to demonstrate the delivery of adequate health care."[13]

The HREA (§495,a,2,B) requires "appropriate pre-surgical and post-surgical veterinary medical and nursing care for animals" used in biomedical and behavioral research. (See 18:38; 18:39.)

Opin. It is the opinion of the authors that the AWAR and HREA regulations could be interpreted as surgical records being a required component of adequate veterinary care. Records should be detailed and contain all relevant information including, but not limited to

- Animal identification
- Project identification
- Detailed description of the surgical procedure
- Experimental agents administered
- Dose and route of anesthetics and analgesics administered
- Frequency (e.g., time interval) and type (e.g., blood pressure, heart rate) of intraoperative monitoring procedures performed
- Peri-operative administrations and treatments (e.g., fluids, other medications)
- Postoperative monitoring and observations recorded
- Postoperative treatments administered

Surv. Does your IACUC or animal facility provide investigators with specific forms to document the surgical information typically required by your IACUC (such as drugs used, vital signs monitoring, and observations)?

- Not applicable 40/295 (13.6%)
- Yes, and the form must be used 48/295 (16.3%)
- Yes, either the provided form or an investigator designed facsimile is required to be used 66/295 (22.4%)
- The use of a form is suggested but not required 34/295 (11.5%)
- No 98/295 (33.2%)
- I don't know 4/295 (1.4%)
- Other 5/295 (1.7%)

26:12 Should surgical records be kept for all species or only certain ones?

Opin. The *Guide* (pp. 75–76) states that records are especially valuable for dogs, cats, non-human primates, and farm animals. Although there are no specific record requirements in the AWAR or PHS Policy, APHIS/AC inspectors expect records, in accordance with established veterinary medical practice and as interpreted from the AWAR (§2.33,b,5) to be maintained. PHS Policy also requires adherence to the requirements of the AWAR for regulated species (PHS Policy II) and the *Guide* (PHS Policy IV,A,1).

It is the opinion of the authors that surgical records should be maintained for all species to provide documentation that appropriately trained individuals conducted the research in a humane manner. (See 18:38–18:39.)

Surv. For which species does your IACUC require that adequate surgical records be maintained?

- Not applicable 33/294 (11.2%)
- We have no policy 28/294 (9.5%)
- For USDA-covered species only (and we perform surgery on more than just USDA covered species) 45/294 (15.3%)

- For USDA-covered species (and we perform surgery only on USDA covered species) 10/294 (3.4%)
- For all mammalian species 161/294 (54.8%)
- I don't know 5/294 (1.7%)
- Other 12/294 (4.1%)

26:13 Should surgical records be kept for each individual animal or can they be maintained by groups?

Opin. There are no regulatory requirements for having individual surgical records for each individual animal. The *Guide* (pp. 75–76) notes that clinical records for individual animals can be useful and states that they should include the history of surgical procedures and postoperative care information. It is the opinion of the authors that records can be maintained by groups for rodents, but non-rodents should have surgery records maintained for individual animals. (See 26:10; 27:16.)

26:14 Do records for survival surgery differ from those maintained for animals undergoing non-survival surgical procedures?

Opin. Neither the AWAR nor the *Guide* stipulates that record keeping should be less stringent or different for animals undergoing non-survival procedures. Records should be maintained in accordance with established veterinary medical standards as noted in 26:10 and 26:12. (See 18:38–18:39.)

Surv. Does your IACUC have different intraoperative record keeping requirements for survival vs. non-survival major surgical procedures (assume the same species would be used)?

- Not applicable 41/294 (13.9%)
- We have no policy 53/294 (18.0%)
- Our intraoperative record keeping requirements are the same 126/294 (42.9%)
- Our intraoperative record keeping requirements are less rigid for non-survival procedures 56/294 (19.1%)
- I don't know 18/294 (6.1%)

26:15 What are the record keeping requirements when performing surgery on rodents?

Reg. The *Guide* (pp. 75–76) states that clinical records for animals are valuable and describes the kind of information to be maintained. Rodents are not excluded from these statements.

Opin. At the authors' institutions, records are required for rodents undergoing surgical procedures, although the frequency of monitoring procedures and details for intraoperative monitoring procedures are not required. Additionally, the records may be maintained for groups. Rodent records should include the following:

- A description of the surgical procedure conducted
- Experimental agents administered

- Dose and route of anesthetics and analgesics administered
- Postoperative monitoring and observations
- Treatments administered during the perioperative period

26:16 How should the inspection team proceed after identifying an item of noncompliance in a surgical area?

Reg. The IACUC should identify the item as significant or minor and develop a timetable for correction as specified in the AWAR (§2.31,c,3) and the PHS Policy (IV,B,3). (See 25:12.)

Opin. The policy for contacting the appropriate individual varies depending on the magnitude of the item. IACUC options for notification include the following:

- Briefing the responsible individual from the laboratory at the time of inspection, or later by telephone or e-mail
- Sending a copy of the inspection report to the responsible individual
- Sending a formal letter documenting the noncompliant item

All communications should convey the date of expected resolution. Depending on the severity and repetition of the item of noncompliance, the IACUC may impose sanctions on the investigator such as a letter of reprimand from the IO, cessation of research for a specified time or permanently, mandatory training, restriction of specific staff from conducting procedures, or increased inspection frequencies. The IACUC may revisit the area prior to the next site visit to confirm correction. (See 25:12 and Chapter 29.)

26:17 Can unrelated major survival surgeries be conducted simultaneously in a single operating room on multiple non-rodent animals of the same or different species? Does the response differ for rodent species?

Opin. Neither the AWAR nor the *Guide* has restrictions on the performance of more than one surgery, on the same or different species, in the same operating room. There are unique situations (e.g., inter- and intraspecies transplantation) when conducting surgery on more than one animal in the same operating room is justifiable.

Because of the potential for infectious agent transmission and behavioral incompatibility, it is not advisable to conduct survival surgical procedures on more than one species simultaneously unless the species are of similar microbiological status; for example, specific pathogen-free rodents. Conducting simultaneous multiple survival surgeries on the same species is acceptable as long as the nature of each surgical procedure conducted does not place the other animals undergoing surgery at increased risk (e.g., different microbiological status of animals or an infectious disease study).

26:18 Can a major survival and a non-survival operative procedure be performed simultaneously on non-rodent species in the same operating room?

Opin. When justifiable, as described in 26:17, survival and non-survival surgical procedures can be conducted in the same operating room. However, both surgeries should be conducted with aseptic technique, as if they were survival procedures, because strict aseptic technique should be followed in accordance with the regulations and with the purpose of providing the highest level of sanitation for the animal undergoing the survival procedure.

26:19 What level of training should the IACUC require for individuals performing survival surgical procedures? Should the IACUC verify whether individuals have adequate training and skills to perform major operative procedures?

Reg. The IACUC has the responsibility for ensuring that individuals conducting surgery, irrespective of species used, have the necessary skills to execute the surgical procedure successfully. Both the AWAR (§2.32) and PHS Policy (IV,c,1,f) require that all personnel involved in animal care, treatment, and use are qualified and trained to perform their duties. U.S. Government Principle VIII stipulates that "investigators and other personnel shall be appropriately qualified and experienced for conducting procedures on living animals." In addition to the conduct of the surgery, individuals associated with the activity must be skilled at administering and monitoring anesthesia and have the capability of providing suitable postoperative care.

Opin. Although the IACUC's responsibility is clear, neither the PHS Policy nor the AWAR describe implementation methods. There are a variety of mechanisms that the committee can use to meet this responsibility. (See Chapter 21.) IACUCs frequently garner information pertaining to an investigator's training and skills via required descriptions of them in the initial proposal submitted to the IACUC. Although not all individuals on a specific project may have suitable skills to carry out the project, the IACUC must determine that there is sufficient skilled staff available to meet the project's goals. Clearly, the variety and scope of surgical procedures conducted in a research or training setting vary considerably, requiring the IACUC to make this determination on a case-by-case basis. The IACUC may request additional information or invite the investigative staff to meet with them to review their experience and training.

At many institutions the animal resource program personnel are responsible for the oversight of a centralized surgical facility. The committee frequently relies on the animal resource program's veterinary or paraprofessional staff for information pertaining to an individual's skill or training. Alternatively, the committee may request that one or several of its members observe procedures on a select number of animals before giving the project full approval. Adequacy of training should be ascertained prior to the start of the project; however, it is important for continuing assessment to be conducted. An infrequently used option is that the committee may request the aid of a consultant to make an assessment for projects of a sensitive nature or requiring highly specialized staff.

26:20 What constitutes aseptic technique for major operative procedures? (See 18:18.)

Reg. The AWAR (§2.31,d,1,ix) and the *Guide* (pp. 115–120) require that major survival surgical procedures on all animals be conducted utilizing aseptic technique.

Opin. Aseptic technique is a body of practices employed to prevent microbial contamination of living tissues or sterile materials by excluding, removing, or killing microbial organisms. Aseptic technique requires strict adherence to practices in several areas. The laboratory bench or surgery table should be clean, disinfected, and free of ancillary equipment. Devices or equipment (e.g., animal restraining devices, monitoring equipment, stereotaxic devices) that will be required in the surgical field also should be disinfected. These practices reduce or eliminate potentially infectious organisms and the substrates on which they grow.[14,15] All surgical instruments, implantable devices, and materials that will contact the surgical site or are implanted in the animal must be sterilized. The sterilization method selected is dependent on time considerations, specialized equipment available, and the composition of the material to be sterilized. Sterilization monitoring devices should be routinely utilized to validate sterilization techniques. Surgical instruments and supplies should be prepared prior to sterilization so that they can be opened without being touched.[16]

The regulatory requirements for preparation of the surgeon for rodent surgery are less rigorous than those required for non-rodents. Either a clean or disposable laboratory coat or a surgical scrub shirt should be worn by the surgeon and assistants. A surgical mask (which serves to prevent contamination of the wound by droplets of saliva from the surgical team) and a surgical cap, for individuals who have long hair, which can potentially fall into the surgical field, should be worn. The surgeon should wash his or her hands with a disinfectant and don sterile surgical gloves by a method that maintains sterility (*Guide* p. 118).[17]

Preparation of the surgical site involves removal of fur by clipping or using a depilatory, and disinfection of the skin by scrubbing with suitable agents. The surgical site should be covered with a sterile drape. The goal of aseptic technique is to prevent the surgeon, the surgical site, and all instruments and equipment utilized from causing infection and subsequent disease.[16]

26:21 If performing major operative procedures on multiple rodents, can a single set of surgical instruments be used on multiple animals? (See also 18:20.)

Opin. Groups of non-USDA regulated rodents are frequently surgically manipulated during a single session. Care must be taken to prevent contaminating one animal from another. Sterilized instruments should always be used when initiating a surgical session. New sterile gloves should be donned any time they become contaminated; preferably, separate sterile surgical instruments should be provided for each animal subject. When a limited number of surgical instruments are available and separate sterile surgical packs are not feasible, the following options may be considered:

- Use two sets of instruments. One set is used for incising and manipulating the skin, which is considered a potentially contaminated site because of microbial flora. Once the initial incision and skin manipulation are complete, these instruments are set aside and not used again without prior sterilization. A second set of instruments is used to manipulate deeper, sterile tissues. The second set of instruments (which is shared) should be rinsed with sterile saline solution between animals. Care must be exercised to maintain the sterility of the surgical gloves by avoiding contacting tissues with fingers when utilizing this option.
- Utilize a glass bead sterilizer to sterilize instrument tips between animals. Recognize that only the instrument tips are sterilized. Care must be taken to prevent touching tissues with instrument handles or other non-sterile

instrument parts. Because non-sterile instrument handles are held with gloved fingers, contact of tissues with the fingers is prevented.

- Have two or more sets of sterile instruments available so that the contaminated set can soak in cold sterilant solution for the minimal time required by the solution manufacturer, typically 15 minutes, to kill vegetative bacteria prior to reuse. Instruments must be rinsed thoroughly with sterile saline or water prior to reuse.

References

1. National Institutes of Health, Office of Laboratory Animal Welfare. 2012. Frequently Asked Questions E.1, Should the IACUC inspect laboratories or other sites where investigators use animals? http://grants.nih.gov/grants/olaw/faqs.htm#prorev_1.
2. Petervary, N. 2013. Best practices for IACUCs in the evaluation of multiple major operative procedures. *Lab Anim. (NY)* 42:330–333.
3. Office of the Federal Register. 2011. Code of Federal Regulations, Title 43, Good Laboratory Practice Regulations, Washington, D.C. http://www.gpo.gov/fdsys/pkg/CFR-2011-title21-vol1/pdf/CFR-2011-title21-vol1-part58.pdf.
4. Office of the Federal Register. 1984. Code of Federal Regulations, Title 29, Part 1910, Occupational Safety and Health Standards, Subpart G, Occupational Health and Environmental Control, Washington, D.C. http://www.osha.gov/pls/oshaweb/owastand.display_standard_group?p_toc_level=1&p_part_number=1910.
5. Office of the Federal Register. 1991. Code of Federal Regulations, Title 29, Part 1910.1030, Bloodborne Pathogens, Occupational Health and Safety Act, Washington, D.C. http://www.osha.gov/pls/oshaweb/owadisp.show_document?p_table=STANDARDS&p_id=10051.
6. Office of the Federal Register. 1992. Code of Federal Regulations, Title 10, Chapter 1 Part 20, Standards for protecting against radiation. Nuclear Regulatory Commission Regulations, Washington, D.C. http://www.nrc.gov/reading-rm/doc-collections/cfr/.
7. Office of the Federal Register. 1984. Code of Federal Regulations, Title 29, Part 1910, Occupational Safety and Health Standards, Subpart G, Occupational Health and Environmental Control, Washington, D.C. http://www.osha.gov/pls/oshaweb/owastand.display_standard_group?p_toc_level=1&p_part_number=1910.
8. Drug Enforcement Administration. 2010. *Pharmacist's Manual: An informational outline of the Controlled Substances Act of 1970.* Washington, D.C. http://www.deadiversion.usdoj.gov/pubs/manuals/pharm2/pharm_manual.pdf.
9. U.S. Department of Health and Human Services. 2011. Guidelines for research Involving recombinant DNA molecules (NIH Guidelines). http://oba.od.nih.gov/oba/rac/Guidelines/NIH_Guidelines.pdf.
10. U.S. Department of Health and Human Services. 2009. *Biosafety in microbiological and biomedical laboratories*, 5th edition. HHS Publication No. (CDC) 21-1112. Washington, D.C.: U.S. Government Printing Office. http://www.cdc.gov/biosafety/publications/bmbl5/BMBL.pdf.
11. National Institutes of Health, Office of Laboratory Animal Welfare. 2012. Semiannual program review and checklist. http://grants.nih.gov/grants/olaw/sampledoc/cheklist.htm.
12. Block, S. (ed.). 2001. *Disinfection, sterilization, and preservation,* 5th edition. Philadelphia: Lippincott Williams & Wilkins.
13. U.S. Department of Agriculture, Animal and Plant Health Inspection Service. 2011. *Animal Care Resources Guide,* Policy 3, Veterinary Care. http://www.aphis.usda.gov/animal_welfare/downloads/policy/Policy%203%20Final.pdf.

14. Cockshutt, J. 2003. Principles of surgical asepsis. In *Textbook of Small Animal Surgery*, 3rd edition, ed. D. Slatter, Chapter 10. Philadelphia: W.B. Saunders.
15. Hobson, H.P. 2003. Surgical facilities and equipment. In *Textbook of Small Animal Surgery*, 3rd edition, ed. D. Slatter, Chapter 13. Philadelphia: W.B. Saunders.
16. Mitchell, S.L. and J. Berg. 2003. Sterilization. In *Textbook of Small Animal Surgery*, 3rd edition, ed. D. Slatter, Chapter 11. Philadelphia: W.B. Saunders.
17. Shmon, C. 2003. Assessment and preparation of the surgical patient and operating team. In *Textbook of Small Animal Surgery*, 3rd edition, ed. D. Slatter. Philadelphia: W.B. Saunders.

27

Assessment of Veterinary Care

Mark A. Suckow and Bernard J. Doerning

Introduction

While any animal may experience spontaneous illness, animals used in research or educational programs are at risk for possible husbandry related problems or potential adverse events due to specific procedures they have undergone. For these reasons, the availability of proper veterinary care is essential to an institutions animal care and use program. Specific care and treatment of all animals should always be guided by the professional judgment of a veterinarian.

The overall goal of the program of veterinary care is to assure that the animals remain healthy and are used in a humane and judicious manner. In this regard, the AV has a responsibility to the animals to ensure their wellbeing, to the institution to promote regulatory compliance, and to the investigator to facilitate research. Of these, the wellbeing of the animals is paramount.

A challenge for the IACUC is to identify means by which to evaluate the scope and adequacy of the program of veterinary care. The IACUC should recognize that the implementation of a sound program of veterinary care relies on the professional guidance and judgment of an AV who has training or experience in laboratory animal medicine. Useful parameters for the IACUC to consider include

1. The provisions for preventive, clinical and emergency veterinary care
2. The completeness and accuracy of medical records that allow for retrospective evaluation of the program
3. The training and number of staff available to assist the AV or investigators in delivery of acceptable care
4. The mechanism for reporting problems and the response of the AV and his or her staff to such information

27:1 What is meant by adequate veterinary care?

Reg. Adequate veterinary care is an essential part of every animal care program. The AWAR (§2.33,a) state, "Each [registered] research facility shall have an attending veterinarian who should provide adequate veterinary care to its animals." The PHS Policy (IV,C,1,e) requires the IACUC to assure that "medical care for animals will be available and provided as necessary by a qualified veterinarian." Specific descriptions of proper veterinary care are not contained in the PHS Policy;

however, the *Guide* (p. 105), which is used as a basis for evaluation by the PHS Policy, lists the following as components of adequate veterinary care:

- Animal procurement and transportation
- Preventive medicine
- Clinical care and management
- Surgery
- Pain and distress
- Anesthesia and analgesia
- Euthanasia

The *Guide* (p. 14), in addition to indicating that the program of veterinary care is the responsibility of the AV, endorses the American College of Laboratory Animal Medicine's (ACLAM) guidelines for *Adequate Veterinary Care.*[1] These guidelines specify that adequate veterinary care requires veterinary access to all animals and their medical record, in order to provide competent clinical, preventive and emergency veterinary care.

Opin. *Adequate veterinary care* implies a minimal acceptable standard; however, an AV by virtue of specialty training and/or experience should strive to develop their institution's program of veterinary care to a level that exceeds this minimum. General aspects of adequate veterinary care encompass three overlapping areas: animal husbandry, animal health, and study coordination (see 27:6).

Animal husbandry is much more than cleaning cages and feeding the animals. Prevention of disease and injury is a key objective of the animal husbandry program. The following aspects of animal husbandry are areas of concern as they relate to veterinary care:

Animal procurement (see also 27:32): Potential vendors should be carefully evaluated because newly acquired animals can introduce disease into established colonies. A health surveillance program to screen or monitor incoming animals or potential vendors can be used to select appropriate sources of animals. All animals should be observed upon arrival at the facility for signs of illness.

Acclimation and quarantine (see also 27:32): In general, animals should be allowed a period of acclimation and physiologic stabilization after arrival at the facility to allow them to recover from shipping stress and permit them to adapt to their new surroundings. Animals known or suspected to be from contaminated sources should be quarantined in such a way that risk to other animals is minimized. Quarantine often involves special precautions, including restriction of personnel access, maintenance of air pressure differentials between quarantine and other facility areas, and strict disinfection procedures.

Separation of species and source: Different species and animals from different sources are best maintained in separate housing units. Separating animals by species is useful to reduce anxiety due to interspecies conflict and to minimize the possibility of interspecies disease transmission (*Guide* pp. 11–12). Similarly, animals from different sources should be separated to control possible transmission of infectious disease between groups. Separation can be achieved by housing animals in separate rooms, in various forms of isolators or in specifically designed individually ventilated cage (IVC) systems.

Daily animal care: All animals should be housed in cages of appropriate size and style for the species and cleaned and maintained in accordance with the standards of the AWAR and the *Guide*. In addition, it is imperative that all animals are observed every day to assure that their health and living conditions remain acceptable. An SOP describing methods and equipment to be used in animal care is often useful to standardize the expectations of care and procedures to be followed. It is further useful to utilize a written or electronic check-off sheet to document when animals or room conditions have been evaluated and when procedures associated with animal care have occurred. This provides a permanent record that can be used to evaluate the consistency of animal care over time.

Animal health is intricately related to veterinary care. Ill or injured animals must receive prompt medical attention when needed. Important issues to address include the following:

Observation: All animals must be observed daily (*Guide* p. 74; AWAR §2.33,b,3) and any unexpected deaths and deviations from normal behavior or appearance must be reported to the veterinarian. Sick or injured animals require a plan for alerting the veterinary staff to the need for prompt attention.

Reporting of animal health problems (also see 27:21): Many different mechanisms exist to report animals in need of veterinary attention. Whether it is electronic record-keeping, notebooks, cage cards or "animal health forms" a process must be defined to capture and record progress of care. (See 27:28–27:29.) Documentation by the veterinarian in writing of his or her findings, recommendations, and the progression of the animal condition is often a legal requirement and an effective way to assist the IACUC in evaluation of the adequacy of veterinary care.

Anesthesia and analgesia: According to the AWAR (§2.31,d,1,iv,A), "Procedures that may cause more than momentary or slight pain or distress to the animals will be performed with appropriate sedatives, analgesics or anesthetics, unless withholding such agents is justified for scientific reasons." The PHS Policy (IV,C,1,b) is essentially the same as the AWAR. Similarly, the *Guide* (p. 120) states, "An integral component of veterinary medical care is prevention or alleviation of pain associated with procedural or surgical protocols." It is the duty of the veterinarian to advise the investigator on proper use of these drugs and to assure that animals are being adequately medicated. Use of such agents should be carefully documented in writing. The IACUC may find evaluation of such records to be useful. (See Chapter 16.)

Survival surgery: A major responsibility of the AV is to assure that survival surgical procedures are conducted in proper facilities by trained individuals using acceptable methods. Again, records of procedures performed and documentation of qualifications of personnel can be used by the IACUC to evaluate this aspect of veterinary care. (See Chapter 18.)

Intraoperative, postoperative and post-procedural care: Additional care and attention are often critical to the successful outcome of surgical or other invasive procedures that involve anesthesia. For example, intraoperative hemodynamic and body temperature maintenance; care immediately after surgery including administration of supportive fluids, analgesics, and other drugs; monitoring of vital signs such as temperature, pulse, and respiration; provision of supplemental heat if the animal is hypothermic; and observation of the animal until it is awake and ambulatory. Longer-term care might include administration of antibiotics or other drugs, evaluation of the surgical site for signs of dehiscence or infection, and overall continued attention to the general medical needs of the animal. For all intraoperative, postoperative and post-procedural care, it is important that accurate records be maintained for retrospective evaluation if problems are found. This record also allows the IACUC to evaluate this aspect of veterinary care. Pre-op care is also important, and a requirement for adequate pre-op and pre-procedural care is in the AWAR (§2.31,d,1,ix; §2.33,b,5).

Study coordination: *Pre-study veterinary consultation.* Investigators' discussions of research proposals with the veterinarian are a useful component of adequate veterinary care and required when the research may result in more than momentary pain or distress to the animal (AWAR §2.31,d,1,iv,B). The AWAR (§2.33,b,4) state that the program of veterinary care includes "guidance to principal investigators and other personnel involved in the care and use of animals regarding handling, immobilization, anesthesia, analgesia, tranquilization, and euthanasia."

Training and assessment of skills: Adequate veterinary care is greatly augmented by training and evaluation of skills for anyone planning to conduct animal research. As part of the program of veterinary care, the AV should ensure that a program is in place to accomplish this.

27:2 Must the AV be a laboratory animal specialist?

Reg. The AWAR (§2.31,b,3,i) and the PHS Policy (IV,3,b,1) indicate that the IACUC must include at least one member who is a doctor of veterinary medicine with training or experience in laboratory animal science or medicine who has direct or delegated responsibility for activities involving animals at the research facility. By definition, this individual is usually the AV, but the AWAR indicate that another veterinarian with delegated program responsibility for activities involving animals can serve as the veterinary representative to the IACUC (§2.33,a,3). The AWAR (§1.1, Attending Veterinarian) also mandate that the AV be a person who has graduated from a veterinary school accredited by the American Veterinary Medical Association's Council on Education, or has a certificate issued by the American Veterinary Medical Association's Education Commission for Foreign Veterinary Graduates, or has received equivalent formal education as determined by the USDA. The PHS Policy (IV,C,1,e) states that the IACUC must determine that "medical care for animals will be available and provided as necessary by a qualified veterinarian," although the Policy does not specifically define *qualified*.

The *Guide* (p. 106) indicates that the AV must be an individual who has a veterinary degree and has training or experience in laboratory animal science and medicine. However, the *Guide* does not specifically require the AV to serve as the veterinary member of the IACUC.

Opin. The answer to this question is yes, if a *specialist* is defined as someone who has specialized training or experience. Individuals may have several areas of specialization, including laboratory animal science and medicine. Thus, the use of a consultant as the AV is acceptable even if he or she devotes major effort to other areas of veterinary medicine. The American Veterinary Medical Association recognizes laboratory animal medicine as a specialty practice area of veterinary medicine. (See 5:27.)

Completion of an internship at a laboratory animal facility as part of the veterinary professional curriculum is one means by which individuals might gain minimal experience in laboratory animal medicine. Sometimes, individuals become specialists as a result of on-the-job experience under the supervision of an experienced AV. Some choose to complete residencies in laboratory animal medicine at academic, military, or industrial institutions. Such residencies allow for concentrated study and experience in the specialty. An individual who has completed an approved residency or has approved practical experience, has an approved publication in the field of laboratory animal medicine, and has successfully passed the appropriate examination is eligible for certification as a Diplomate by the American College of Laboratory Animal Medicine (ACLAM). Diplomate status within ACLAM is widely viewed as evidence of significant training and experience in laboratory animal medicine.

Membership and involvement in relevant professional organizations can be interpreted as evidence of specialized interest in laboratory animal science and medicine. For example, activity in the American Association for Laboratory Animal Science (AALAS) or the American Society of Laboratory Animal Practitioners (ASLAP) would likely improve the professional knowledge of individuals active in the profession.

A veterinary practitioner who lacks specific training in laboratory animal medicine might still ably fill the role of AV as a consultant. For example, if the species maintained at the facility are those with which the practitioner routinely works, his or her experience would be appropriate. In instances when a veterinary practitioner serves as a consulting veterinarian for a research facility, it is important for that individual to supplement his or her knowledge base where needed. This might be done by enrollment in professional associations such as AALAS or ASLAP; attendance at professional meetings that focus on laboratory animal medicine, completion of appropriate online coursework, or provision of appropriate texts or other material for the veterinarian to expand his or her knowledge base.

Surv. Did your Attending Veterinarian have any formal training, coursework, or experience with all or most of the laboratory animal species in your animal facility prior to becoming your Attending Veterinarian?

- Yes, to the best of my knowledge 269/293 (91.8%)
- No, to the best of my knowledge 16/293 (5.5%)
- I don't know 8/293 (2.7%)
- Other 0/293 (0%)

27:3 Is it necessary for all organizations housing or using laboratory animals to have an AV?

Reg. Any institution that is under the purview of AAALAC, the AWAR, or the PHS Policy is required to have an AV (see 27:1). Other organizations that do not fall under the purview of those standards may still choose to employ an AV on a full-time or consulting basis.

Opin. The AV typically has several functions within the institution[1] (AWAR §2.33,a; AWAR §2.33,b; *Guide* pp. 14, 27, 30, 34). For example, it is the duty of the AV to assure that the husbandry of the animals is appropriate for the species. The AV is responsible for developing and implementing programs for preventive medicine to minimize the likelihood of animals' developing clinical illness or injury. Furthermore, the AV directs the diagnosis and treatment of clinical disease affecting the animals. In addition, the AV typically advises investigators and the IACUC on methods to relieve or alleviate pain and distress, perioperative care, and surgical procedures. The AV must also be familiar with applicable state veterinary acts, inclusive of licensure requirements, particularly in the discharging of certain official duties such as signing interstate health certificates or verifying rabies vaccination or tuberculosis status of animals. In short, the role of the AV is to assure that animals remain healthy and are used in a judicious and humane manner (see 27:7).

Several types of facilities are not required under the existing regulations to have an AV. For example, institutions that use only rats of the genus *Rattus* and mice of the genus *Mus,* bred specifically for use in research or teaching, are not required to have an AV if they lack AAALAC accreditation and do not receive PHS or other federal funds for research or teaching using animals. Such facilities should still develop a written plan for animal health in the event of unexpected illness or injury.[2] In addition, if animals are to be used in surgical or other invasive procedures, it is advisable to identify an individual who can serve as the AV.

27:4 Under what circumstances would a full-time versus a part-time versus a consultant veterinarian be adequate?

Reg. The AWAR (§2.33,a,1) require that institutions using a consultant veterinarian have a written program of veterinary care and that the AV make regularly scheduled visits to the facility. The PHS Policy (IV,C,1,e) does not specify a requirement for a full-time versus part-time AV. Nevertheless, the policy requires that the IACUC assure that medical care be "available and provided as necessary." PHS Policy (IV,A,3,b,1) and the AWAR (§2.33,a,1) also require that there be a veterinarian with direct or delegated program authority and responsibility who will serve on the IACUC. §2.33,b of the AWAR requires availability of appropriate care, including emergency care. All this suggests that the veterinary staff needs to be adequate to the extent that competent veterinary care can be provided whenever the need arises. In this regard, the IACUC is charged with determining the level of veterinary service that must be procured.

Opin. PHS Policy (IV,C,1,e) suggests that the veterinary staff must be sufficient to provide the appropriate medical care whenever the need arises. In this regard, the IACUC is charged with determining the level of veterinary service that must be procured.

The need for the services of the AV or other ancillary staff varies extensively with the size and scope of animal research related operations of the institution. In this sense, it is not possible to outline specific standards for the need of a full-time or other veterinarian.

The American College of Laboratory Animal Medicine (ACLAM) recommends that a formal arrangement for veterinary care exist in all instances, while recognizing that the specific arrangement may vary with the number and species of animals involved and the nature of the experimentation at the institution.[1]

Institutions where research involving survival surgery or other invasive procedures is performed on a routine basis should seek greater veterinary involvement than institutions where this does not occur. Similarly, other conditions that merit increased veterinary involvement are shown in Table 27.1.

The IACUC should consider the factors listed in Table 27.1 and determine whether the available veterinary staff can provide the necessary service and oversight. Each facility has its own unique characteristics and needs. Some small facilities may need a full-time veterinarian, while some larger ones may require only a part-time or consulting veterinarian. Experience of the authors has shown that institutions characterized by three or more of the factors listed can generally justify the need for a full-time veterinarian. Facilities that are characterized by few or none of the listed factors may find a part-time veterinarian (who is a regular employee of the institution) or even a consultant veterinarian (one who provides veterinary care on a fee-for-service basis to the institution) to be sufficient.

Veterinary support in addition to the AV may be needed as the number of animals in the facility and the number of relevant factors increase. Such support can be in the form of additional veterinarians or, if appropriate, ancillary veterinary staff such as veterinary technicians or trained animal care technicians (see 27:9). Veterinary activities performed by technicians should be conducted under the direction of a qualified veterinarian. Communication between the veterinarian and technicians should occur regularly to ensure that the veterinarian is aware of any animal health or care issues and can advise the technical staff accordingly (AWAR §2.33,b,3; *Guide* pp. 106, 114). Under any arrangement, veterinary services must be available at all times, including evenings, weekends, and holidays (AWAR §2.33,b,2; and see 27:6). It is important that the information for contacting the veterinarian is communicated to all research and animal care personnel and kept updated.

TABLE 27.1

Factors That Increase Need for Veterinary Services

Animal-Related Factors	Research-Related Factors
• Large number of animals in the colony • Use of phylogenetically higher mammals such as dogs, cats, and nonhuman primates • Use of immunodeficient animals • Use of animal models with special medical or biologic needs	• Use of biohazards • Use of chemical hazards • Use of radiologic hazards • Procedures requiring the use of anesthetics or analgesics, particularly survival surgery • Survival surgical procedures involving extensive perioperative care • Anticipated pain or distress

27:5 If a veterinary consultant is used when the AV is unavailable, what criteria might the IACUC use to assess the adequacy of his or her credentials?

Reg. As discussed in 27:2, the AWAR, the PHS Policy, and the *Guide* describe qualifications for the AV; however, none of these documents describe qualifications for other veterinary staff who may provide care to the animals. The AWAR (§2.31,d,1,vii) and the PHS Policy (IV,C,1,e) state that medical care for animals will be available and provided by a qualified veterinarian; however, neither document describes what criteria might be used to establish a veterinarian other than the AV as "qualified."

Opin. Absent any specific guidance from the AWAR, PHS Policy, or the *Guide*, it would be reasonable for IACUCs to use those criteria which define a qualified AV as a reference point. Often, a consultant will have less direct experience with laboratory animal species than the AV (or an expert consultant might have more). Nonetheless, he should have at least some familiarity with those species in terms of handling and common medical issues, and the IACUC should seek to assess the consulting veterinarian's background for any evidence of this experience and more importantly, he should understand what resources are available for the prevention, diagnosis, and treatment of disease in laboratory animal species. In this regard, the IACUC should ensure that such resources are recognized by the consulting veterinarian and are readily available. This might mean that printed materials are available; that there is ready access to internet resources; or that continuing education opportunities for the consulting veterinarian are offered. Importantly, the IACUC should evaluate the consulting veterinarian's availability and accessibility. In other words, does the consultant have sufficient time away from other commitments to be able to respond to concerns or emergencies involving animals at the facility? Is he or she located a reasonable distance from the facility so that he might quickly respond to the needs of ill animals? Can he or she be easily reached by phone or other means, such that consultation might be readily offered in that way? Though there are not specific criteria available for answering these questions, the IACUC should assure themselves that sound veterinary care can be quickly provided when needed.

27:6 If a consultant is used, how frequent should visits be and how can emergency care be provided if the consultant is not available?

Reg. The frequency of visits to a facility by a consultant can vary significantly with the size and scope of animal use activities (*Guide* p. 14). The AWAR state that "regularly scheduled" visits should be made to the facility (§2.33,a,1).

Opin. Visits should be frequent enough to provide the same degree of oversight to each study as would be provided to studies in a larger facility employing full-time veterinary staff. The experience of the authors is that onsite visits that occur less frequently than once a month diminish the effectiveness of the AV if animals are housed continuously at the facility. As the scope and complexity of the facility increase, so should the frequency of onsite visits by the consultant. Visits by the consulting veterinarian should be regularly scheduled and supplemented with additional visits as needed.

The availability of consultants to respond to emergency situations may be limited by time or distance constraints. Nonetheless, emergency veterinary care must be provided (AWAR §2.33,b,2; *Guide* p. 114). Trained veterinary or animal care technicians can sometimes provide such care under the direction of the veterinarian, through instructions communicated by such means as telephone, fax, email, or prior written instructions and guidelines. In contrast, some facilities make prior arrangements with a local practicing veterinarian to provide occasional emergency care onsite. Any veterinarian employed in this capacity should be provided with initial training regarding the species of animals and the type of research at the facility. Both the veterinarian and the facility should be comfortable with each other's expectations before entering this relationship.

27:7 What are the AV's responsibilities?

Reg. The principal responsibility of the AV is to develop and implement an effective program of veterinary care (see 5:28; 27:3; 27:32). Such a program has multiple components, including preventive medicine; proper use of anesthetics, analgesics, and tranquilizers; perioperative care; euthanasia; diagnostic procedures; and provision of veterinary care, emergency care, and aspects of animal husbandry[1] (AWAR §2.33,b,1–§2.33,b,5; *Guide* p. 105). These responsibilities and associated aspects are summarized in Table 27.2.

Opin. Although the AV has a principal responsibility to assure the wellbeing of the animals being used at the institution, he or she also has responsibilities to the institution and to the investigators. Ultimately, the AV should facilitate the ability of

TABLE 27.2
Responsibilities of the Attending Veterinarian

Type of Responsibility	Specific Aspects of Responsibility
Diagnostic procedures	Direct or perform procedures to identify disease etiologies
Provision of medical care	Prescribe treatments to treat ill or injured animals
Preventive medicine	Develop programs for quarantine and isolation, monitoring of vendors, monitoring of colony health, and routine vaccinations and parasite control
Use of anesthetics, analgesics, sedatives, and tranquilizers	Advise investigators on and assure proper use of methods to relieve or reduce pain and distress
Perioperative care	Oversee pre-surgical preparation, surgical procedures, and postsurgical care of animals
Euthanasia	Assure that animals are euthanized when appropriate and in a humane manner
Animal husbandry	Assure that programs for disinfection, housing, nutrition, breeding, and environmental enrichment are appropriate
Use of hazards	Work with hazards oversight professionals to assure safe use of biologic, radiologic, and chemical hazards
Use of animals	Advise investigators and the IACUC on the appropriateness of specific techniques and methods in animals, and the availability of alternative animal and non-animal models
Occupational health	Advise IACUC and health professionals on aspects of occupational health program

the investigator to perform animal research in a humane manner and consistent with regulatory principles. Typically, the IACUC and the AV have a close working relationship. The IACUC relies on the professional judgment of the AV in animal health issues, and the AV turns to the IACUC for support if resistance to proper animal care and use is encountered.

27:8 What is an appropriate chain of command for decision making with respect to veterinary care, and what is an appropriate chain of communication with respect to animal health issues?

Reg. The AV is responsible for the health and well-being of all animals used at the institution according to the *Guide* (p. 14). AWAR (§2.33,a,2) state with respect to the AV and adequate veterinary care that the facility should have an AV who provides adequate veterinary care and that the AV have appropriate authority to ensure the provision of adequate veterinary care and to oversee the adequacy of other aspects of animal care and use. Further, both this section of the AWAR (§2.33,b,3) and the *Guide* (p. 14) indicate that the AV must be provided with sufficient authority, access to animals and resources to manage the program of veterinary care. A mechanism of direct and frequent communication is required to provide timely and accurate information concerning animal health to the AV.

Opin. Clearly, the AWAR and the *Guide* designate the AV as the appropriate authority to ensure the provision of adequate veterinary care. Most facilities interpret this to mean that in an emergency or life-threatening situation, or one in which an animal is clearly suffering and study personnel cannot be contacted, the designated veterinary staff has the authority to initiate treatment that may or may not include euthanasia (also see 27:29–27:30). It should be stressed, however, that every effort should be made to contact study personnel prior to initiating treatment or euthanasia. There may be a simple explanation for clinical signs being observed–such as a lethargic animal's recovery from anesthesia. Further, communication with the study personnel is crucial if vital blood or tissue samples need to be collected from the animal.

A good system of communication between the facilities' veterinary staff and research personnel is critical. The veterinarian needs to be aware of study details to diagnose and develop a treatment plan effectively. The study director must be aware of veterinary care diagnostics and treatment plans in order to consider potential study interactions.

The ideal situation would develop as follows: problem identified, veterinary observation, study director contact and discussion, diagnostic and treatment plan discussed, mutually acceptable concurrence, implementation and completion of plan, detailed notes to the medical records. This works well for most nonemergency situations, but when something must be done immediately and either the veterinarian or the research personnel cannot be reached it is best to have a policy to delegate authority and responsibilities.

In some cases, Good Laboratory Practice (GLP) regulations are interpreted that, since a study director has overall responsibility for the technical conduct of the study, that the involvement of veterinary personnel is cited as being disallowed. GLP regulations in no way restrict or prohibit veterinarians from fulfilling their responsibilities as defined in the AWAR. It is important to communicate and work with the user groups to develop a mutually acceptable pathway for providing animal health services.

Surv. In general, who makes the decisions regarding the veterinary care provided to animals at your institution?

• The AV or another qualified veterinarian	259/293 (88.4%)
• Our veterinary technician	3/293 (1.0%)
• The principal investigator or study director	19/293 (6.5%)
• One or more persons in our animal care group	8/293 (2.7%)
• I don't know	0/293 (0%)
• Other	4/293 (1.4%)

"Other" responses included

- Veterinarian together with PI
- Both the AV and clinical veterinarians
- AV or alternate veterinarian in consultation with the PI
- Animal colony supervisor in consultation with the AV

27:9 What staff, other than veterinarians, are useful in providing adequate veterinary care? What level of support staff is considered adequate?

Opin. Ancillary personnel such as veterinary technicians, research technicians, PIs, and trained animal care technicians often provide valuable support to the program of veterinary care (see 27:4). Such individuals usually interact closely with the animals on a daily basis and can serve as a primary point for observation of animal health. The IACUC through oversight, program review and semiannual inspections contribute to the integrity of the institutions standards of veterinary care.

The size of the staff needed to provide adequate veterinary care is completely dependent upon the size and complexity of the animal facility. Unreported or untreated illness or injury, insufficient feed or water, or signs of insufficient husbandry can all indicate inadequate observation of (or response to) the animals and may signal a need for additional staff.

Any animal housed in a research facility should be checked at a minimum of once per day, every day (AWAR §2.33,b,3; *Guide* p. 74). This need can increase with the invasiveness or potential for untoward outcomes of the study. It is within the scope of the IACUC's animal welfare oversight to suggest that more frequent observations are integrated into a research protocol.

27:10 What type of training should be provided to animal care and research staff with respect to veterinary care issues?

Opin. Section 2.32 of the AWAR requires training of personnel in pre- and post-procedural care, aseptic surgery, and pain-relieving drugs. There are numerous resources available to augment in-house training programs. Conferences, webinars and online courses can help personnel, as described in 27:9, that might potentially examine animals exhibiting signs of illness. It is furthermore critical for the IACUC to assure that these individuals are familiar with both normal and abnormal appearances and behaviors of animals (Table 27.3). Training provided or evaluated by the veterinary staff should ensure that personnel are able to recognize

TABLE 27.3

Some Indications of Illness in Animals

Body System	Abnormal Clinical Signs
Respiratory system	Nasal discharge, labored breathing, blue tint to mucosal membranes
Gastrointestinal system	Vomiting, diarrhea, thin appearance
Integumentary system	Bleeding or seeping wounds, alopecia, observed persistent scratching, closed wounds
Nervous system	Inability to move normally, overly aggressive behavior, lethargy
Circulatory system	Edema in extremities, ascites, labored breathing

that an animal may require veterinary attention. While useful, it is not critical that ancillary personnel be able to diagnose illness; instead, they should understand the procedures for notifying the veterinary staff if an animal is even suspected of being ill. Though a clear mechanism to report sick or injured animals to the AV must exist, the rigor of reporting may vary with the nature of the specific study; thus, a dermal abrasion in an otherwise normal animal should be called to the attention of the veterinary staff but does not likely constitute an emergency. For certain animal models, it is important that personnel also learn to recognize medical nuances specific to that model. For example, in an animal model of hemophilia it would be important for the staff to understand the urgent nature of even a seemingly minor wound. Similarly, in an animal model of Parkinson's disease, it would be important to instruct personnel that tremor of limbs is an expected outcome.

Some staff may also benefit from training related to provision of veterinary care. In this regard, it may be useful to train personnel with respect to administration of drugs by various routes, bandaging of surgical wounds, or other common veterinary care procedures. The IACUC should ensure that proficiency has been documented for ancillary personnel expected to provide veterinary care.

27:11 To what extent and under what circumstances can the AV delegate clinical responsibilities to other veterinarians or non-veterinarians?

Opin. The AV may delegate the responsibility for observation of animals and "hands-on" care to other veterinarians or technicians provided that it can be documented that such individuals are adequately trained in the care of the species and in the procedures involved. All animals should be observed every day by someone trained to evaluate the health and wellbeing of that species (AWAR §2.33,b,3; *Guide* p. 74; also see 27:1). The ultimate responsibility for assuring proper clinical care remains with the AV, and that person should establish a mechanism for receiving animal health information that will ensure fulfillment of this responsibility (AWAR §2.33,b,3; *Guide* p. 14).

27:12 Are PIs responsible for any aspect of adequate veterinary care?

Opin. Investigators play a significant role in veterinary care. It is critical to recognize that veterinary care is not exclusively the treatment of sick animals. It includes the proper care and use of animals in such a way that animals remain healthy, are treated humanely, and recover normally from experimental manipulations. Beginning with protocol initiation, an appropriate animal model must be chosen to ensure a project is successful. Potential adverse reactions or outcomes should be identified by the investigator and plans made to address them. Indeed, the IACUC should establish a mechanism which allows investigators to easily report unanticipated clinical outcomes related to novel genotypes or procedures performed on animals (see 27:27; 27:42). Investigators and their technicians will observe animals frequently during the procedure and are, therefore, in an excellent position to report potential problems to the veterinary staff. For example, weight loss in a cage of group-housed animals may not be apparent to a caretaker whereas review of the investigator's study notes with individual body weights of animals could detect the problem.

Depending on the staffing of the facility, PIs or their technicians may be responsible for administering treatments prescribed by the veterinarian. As part of this, they should be instructed on procedures to contact the veterinary staff when animals are found in poor clinical condition.

For these reasons, it is critical to view the role of the PI and his or her staff, along with the AV, as providers or facilitators of veterinary care. Although the AV is ultimately responsible for provision of proper veterinary care, the IACUC should recognize that there are others who have an ethical, if not a regulatory, responsibility to impact the health and welfare of research animals positively.

27:13 What aspects of a veterinary care program can and should be addressed under SOPs?

Reg. The AWA and AWAR do not describe SOPs and are therefore silent on this issue. The PHS Policy (IV,A,1) requires that the Assurance statement fully describe the institution's program for animal care and use, using the *Guide* as a basis for the development of that program. The *Guide* (p. 12) encourages the establishment and periodic review of written procedures that ensure consistent application of *Guide* standards, one of which is the provision for adequate veterinary care.

Opin. It is worthwhile for institutions to have a written description of veterinary care program which can be reviewed by the IACUC during the semiannual program review. While the contents of this document should be customized for the facility, it should address areas such as daily assessment of animal health; prevention, control, diagnosis, and treatment of animal disease and injury; guidance to researchers on proper handling, restraint, anesthesia, analgesia, and euthanasia methods; training of research personnel in appropriate surgical techniques and procedures; emergency preparedness; and monitoring of surgical procedures and postsurgical care. Additionally a facility may utilize or augment their program of veterinary care with documents typically referred to as *standard operating procedures* (SOPs). In general terms, the function of an SOP is to provide written guidance in sufficient detail so that a person with the appropriate education, training, and experience can perform the procedure described in the SOP. Research facilities that wish to conduct animal studies under Good Laboratory Practices (GLP) conditions are required by the Food and Drug Administration to maintain SOPs (21 CFR Part 58). Specifically, Sect. 58.81 requires the following:

(a) A testing facility shall have standard operating procedures in writing setting forth nonclinical laboratory study methods that management is satisfied are adequate to ensure the quality and integrity of the data generated in the course of a study.

(b) Standard operating procedures shall be established for, but not limited to, the following:

1. Animal room preparation
2. Animal care
3. Receipt, identification, storage, handling, mixing, and method of sampling of the test and control articles
4. Test system observations
5. Laboratory tests
6. Handling of animals found moribund or dead during study
7. Necropsy of animals or postmortem examination of animals
8. Collection and identification of specimens
9. Histopathology
10. Data handling, storage, and retrieval
11. Maintenance and calibration of equipment
12. Transfer, proper placement, and identification of animals

Non-GLP facilities often operate under SOPs as well. Alternatively, and sometimes additionally, both GLP and non-GLP facilities may develop guidelines or policies. These are documents that help define the conduct of procedures or situations that may have been a source of controversy or earlier deliberation for which the IACUC has reached a consensus. For example, the IACUC might approve a policy to allow with certain restrictions the controversial practice of chronic housing of rodents in wire bottom cages. (See 14:41.)

Surv. Which of the following items are reviewed and approved via the standard protocol review process of your institution? Check as many boxes as appropriate.

- Sentinel animal program 186/277 (67.2%)
- Incoming animal quarantine program 150/277 (54.2%)
- Surgical rederivation of rodents 95/277 (34.3%)
- Veterinary surgical interventions for clinical reasons 68/277 (24.6%)
- Veterinary medical interventions for clinical reasons 85/277 (30.7%)
- Animals used for training animal care staff 180/277 (65.0%)
- Animals used for training research or teaching staff 202/277 (72.9%)
- I don't know 14/277 (5.1%)
- Other 14/277 (5.1%)

"Other" responses included

- We don't quarantine or derive as most of our animals are in house less than 1 month.
- We do not have any of the other programs.
- Question is confusing so nothing checked.

- Project specific procedures.
- None of these apply.
- No policy at the current time.
- Any general animal procedure.
- Animals can't be brought to the facility without an IACUC approved protocol.
- Animal holding protocol for housing animals not assigned to active protocols.

27:14 What equipment and pharmaceuticals might be considered essential items for an adequate program of veterinary care?

Reg. The AWAR (§2.33,b,1–§2.33,b,2) state that, among other considerations, appropriate facilities and equipment must be available to prevent, control, diagnose, and treat diseases and injuries. The *Guide* (p. 105) describes components of an effective program of veterinary care to include preventive medicine; surveillance, diagnosis, treatment, and control of disease; management of protocol-associated disease or disability; anesthesia and analgesia; surgery and postsurgical care; assessment of well-being; and euthanasia. Thus, these documents do not specify the essential items required for adequate veterinary care and allow the AV and the institution to decide which items are essential to achieve the programmatic objectives described.

Opin. The specific items needed to provide adequate veterinary care will vary widely with the species housed within the facility and the type of research being conducted. For example, a program in which specific pathogen-free rodents are used and euthanized for collection of tissues within several days of arrival will have a much different need for specialized equipment and supplies than a program in which nonhuman primates undergo survival surgery and are maintained for several years afterward. In general, the facility should rely upon the advice of the AV in deciding which equipment and supplies are necessary to the needs of the animals and the research in a way that meets the objectives stated in the AWAR and the *Guide*. Most veterinary care programs find that routine supplies include items such as topical or injectable antibiotics, bandage materials, materials for euthanasia and necropsy, material and equipment for surgical repair of wounds in an aseptic manner, supplies for microscopic examination of samples, sterile fluids, antiseptics, syringes and needles, and means to provide supplemental warmth, such as a circulating water heating pad. These items are only examples of the kinds of supplies that would be useful to address the typical clinical needs of animals in a research facility. The list is by no means complete and each veterinary care program should have on hand supplies to meet its specific needs. For example, radiographic and other imaging equipment is essential to some programs, but not for others. Though availability of needed supplies is essential and suggests that the veterinary care program need not necessarily have all equipment and supplies on hand but rather only have ready access to such resources, adequate veterinary care is facilitated to a far greater extent when the veterinary care program has direct ownership of these resources.

27:15 If a drug not approved in the protocol is used in an animal for experimental purposes, should the IACUC regard this as an item of non-compliance? Should it be reported to OLAW?

Reg. The AWAR (§2.31,e,3) requires a complete description of the use of the animal. The description is incomplete if it does not include all experimental drugs. The AWAR

(§2.31,d,4; §2.31,e,1–5) require IACUC approval of significant changes in the use of animals. Adding a drug is a significant change and would be considered a non-compliance if done without IACUC approval.

Opin. Both the AWAR (§2.31,c,6) and the PHS Policy (IV,B,6) indicate that the IACUC is expected to review and approve activities related to animal care and use. The *Guide* (pp. 25–26) indicates that an animal use protocol should provide a clear and concise sequential description of procedures involving the use of animals; information on the impact of the proposed procedures on the animals' well-being; and consideration of the use of hazardous materials and provision of a safe working environment. The AWAR (§2.31,d) similarly describe the need for consideration for procedures that might result in pain or distress to animals. Failure to specify a drug in the protocol limits the ability of the IACUC to determine if these provisions of the *Guide*, the PHS Policy, and the AWAR have been met, thus this situation would be considered an item of non-compliance.

NIH/OLAW states[3] that failure to adhere to an IACUC-approved protocol is one example of a reportable situation. If a drug was used to provide veterinary clinical care to an individual animal, such use would not be reportable. In contrast, it would be regarded as a significant, reportable change if a drug not indicated in the approved protocol is used for experimental purposes. If changes in drugs are made by a veterinarian for humane reasons during the course of a study, such use can be initiated but should be followed by IACUC approval of a protocol amendment which describes and justifies the change before any subsequent use of the drug.

Surv. If a drug that is not approved in the IACUC protocol is used for experimental purposes, does your IACUC regard this as an item of non-compliance?

- Yes, it is considered non-compliant 257/292 (88.0%)
- No, it is not considered to be non-compliant 7/292 (2.4%)
- I don't know 5/292 (1.7%)
- Not applicable 20/292 (6.8%)
- Other 3/292 (1.1%)

Surv. If a drug that is not approved in the IACUC protocol is used for experimental purposes, would your IACUC report this finding to OLAW?

- Yes, we would report this 94/295 (31.9%)
- We may or may not report this; it is situation dependent 159/295 (53.9%)
- No, we would probably not report this finding 17/295 (5.8%)
- I don't know 6/295 (2.0%)
- Not applicable 18/295 (6.1%)
- Other 1/295 (0.3%)

27:16 Is it necessary to use only pharmaceutical quality drugs when they are available for animal use?

Reg. The *Guide* (p. 31) states that pharmaceutical-grade chemicals or substances should be used, when available, for all animal-related procedures, and that use of non-pharmaceutical grade compounds should be approved by the IACUC as part of a protocol. NIH/OLAW is in agreement on this point and emphasizes that the use of pharmaceutical-grade compounds is expected and will help avoid toxicity or side-effects that might result from use of non-pharmaceutical grade materials.[4] APHIS/AC concurs as described in Policy 3 (Veterinary Care)[2] which states that investigators are expected to use pharmaceutical-grade medications whenever they are available.

Opin. NIH/OLAW defines the use of a pharmaceutical-grade compound as a drug that is approved by the U.S. Food and Drug Administration for which a chemical purity standard has been established by the United States Pharmacopeia-National Formulary or British Pharmacopeia. For many test compounds, pharmaceutical-grade material is not available, thereby requiring, by default, the use of non-pharmaceutical grade compound. The *Guide* (p. 31), NIH/OLAW[4] and APHIS/AC[2] all recognize that there may be circumstances in which pharmaceutical-grade materials may not be available or must otherwise be used for scientific purposes. In such cases, use must be approved by the IACUC, with cost-savings not regarded as appropriate justification. When a specific compound is not available as a pharmaceutical-grade material, the PI might also consider using other, similar compounds available as pharmaceutical-grade material as long as such use would not be expected to alter the scientific integrity of the study. NIH/OLAW states that when non-pharmaceutical grade materials are used the IACUC is expected to evaluate the possibility of adverse events by giving consideration to such factors as grade; purity; sterility; acid-base balance; pyrogenicity; osmolality; stability; site and route of administration; compatibility of components; side effects and adverse reactions; storage; and pharmacokinetics.[4] In this regard, it would seem reasonable that the IACUC might develop a plan for monitoring animals for adverse events or side-effects when non-pharmaceutical grade materials are used. (See 16:18.)

Surv. Under what conditions would your IACUC approve non-pharmaceutical quality drugs to be used in animals? Check all appropriate responses. Figures in parentheses indicate percent of all respondents who selected the response.

- Under no circumstances 28/295 (9.5%)
- When pharmaceutical quality drugs are not available 184/295 (62.4%)
- When scientific need requires the use of non-pharmaceutical quality drugs 204/295 (69.2%)
- I don't know 12/295 (4.1%)
- Not applicable 14/295 (4.8%)
- Other 6/295 (2.0%)

27:17 Can expired drugs, biologics, or supplies be used for terminal procedures?

Reg. NIH/OLAW states that use of expired materials is not consistent with adequate veterinary care, and that euthanasia agents, anesthetics and analgesics should not be used for terminal procedures if expired.[5] Allowance for use of expired materials is given for circumstances in which the manufacturer verifies efficacy of the material beyond the expiration date or if the investigator can convincingly demonstrate to the IACUC that use of expired materials would neither harm animal welfare nor compromise the scientific integrity of the study. APHIS/AC policy is in agreement with NIH/OLAW; and, although APHIS/AC indicates that some expired materials can be used in acute terminal procedures if such materials do not adversely affect animal well-being or compromise the validity of the scientific study, all euthanasia agents, anesthetics, emergency drugs, and analgesics used must be within their dates of expiration.[2] Further, facilities that allow use of expired materials under these limited circumstances are expected by APHIS/AC to have a policy covering the use of such materials and/or require investigators to describe in their animal activity proposals the intended use of expired materials. The *Guide* does not specifically address the use of expired materials in animals; however, it is stated (p. 105) that institutions must provide a veterinary program that offers a high quality of care.

Opin. It is widely accepted that use of expired materials in survival procedures is inconsistent with proper veterinary care. Though some institutions have allowed use of expired materials only in non-survival procedures, it is clear that both APHIS/AC and NIH/OLAW consider this to be acceptable only when such materials do not include those that would be used for pain relief (anesthetics or analgesics), euthanasia, or physiologic stability (emergency drugs). Although these agencies allow the possibility to use such expired materials, it would be only under the very unusual circumstance in which an investigator could convince the IACUC that the materials could not possibly pose any threat to animal well-being or to the scientific integrity of the study. Doing so would represent a substantial task, as the investigator might be expected to demonstrate that no physiologic stress would occur even in the anesthetized animal. The *Guide* supports the positions of NIH/OLAW and APHIS/AC, though in less specific terms.

Apart from the sole issue of expiration, it seems reasonable that, unless proved otherwise, expired materials might have compromised purity or sterility. As described by NIH/OLAW,[4] the IACUC is responsible for considering potential adverse consequences of using non-pharmaceutical grade materials and should evaluate purity and sterility as two of several factors. This suggests that expired materials could be considered as non-pharmaceutical grade if sterility or purity is potentially compromised and use would be discouraged as described in 27:16.

Surv. Does your IACUC allow the use of expired drugs, biologics, or supplies for terminal procedures?

- Yes, we do 49/293 (16.7%)
- No, we do not 224/293 (76.5%)
- Not applicable 12/293 (4.1%)
- I don't know 8/293 (2.7%)

27:18 Can the AV give approval to change an anesthetic or analgesic without the approval of the IACUC?

Reg. The AWAR state that the AV has direct or delegated authority for activities involving animals (§1.1) and that medical care for animals will be available and provided as necessary by a qualified veterinarian (§2.31,d,1,vii). Further, the AWAR state that the AV shall have authority to ensure the provision of adequate veterinary care (§2.33,a,2) and that research facilities shall have programs of adequate veterinary care that include the use of appropriate methods to prevent, control, diagnose, and treat diseases and injuries (§2.33,b,2). The *Guide* (p. 105) states that an adequate program of veterinary care must be provided and that a component of such a program includes effective management of clinical disease, disability, or related health issues. The *Guide* (p. 106) also states that the veterinary care program is the responsibility of the AV. Further, the *Guide* (p. 114) indicates that provisions for emergency care of animals must be in place and that the veterinarian must have ultimate authority for the course of action with respect to a sick animal. The PHS Policy (IV,C,1,e) states that medical care for animals will be available and provided as necessary by a qualified veterinarian.

Opin. It is apparent from the AWAR, PHS Policy, and the *Guide* that the AV is granted latitude to exercise professional judgment in the delivery of veterinary care when there is an immediate clinical need. However, if a change in anesthetic or analgesic was made for reasons other than a pressing and immediate need, approval of the IACUC would be required. For example, if an animal had undergone a surgical procedure and the approved analgesic was not providing sufficient pain alleviation, the AV could properly decide to change the analgesic regimen for that individual animal. In contrast, if the investigator decided that a change in analgesic regimen might provide better pain relief in a particular study, the AV would be extending beyond his or her authority in granting approval; instead, approval of the IACUC would be needed for this change.

Surv. Does your Attending Veterinarian have the authority to approve a change in an anesthetic or analgesic drug without the approval of the IACUC?

- Yes, routinely 54/293 (18.4%)
- Yes, but only in a clinical emergency 171/293 (58.4%)
- No 51/293 (17.4%)
- Not applicable 11/293 (3.8%)
- I don't know 6/293 (2.0%)

27:19 Are there any circumstances under which veterinary care can be provided at a facility not affiliated with the institution, such as a private veterinary clinic? If so, what sort of criteria might the IACUC use to determine the acceptability of such sites and what sort of oversight should the IACUC provide?

Reg. The *Guide* (p. 112) states that methods of disease prevention, diagnosis, and therapy should be those currently accepted in veterinary practice. The AWAR (§2.33,b,2; §2.33,b,3) state that a program of veterinary care includes the availability of appropriate facilities, personnel, and equipment to prevent, control, diagnose, and treat diseases and injuries. Though no specific mention is made of utilizing off-site resources, these passages imply that the importance to animal health offered by certain equipment, methods, and expertise is substantial. One might infer, then, that if the necessary resources cannot be practically located at the institution, the primacy of animal health would lend merit to the use of off-site resources.

Opin. At some institutions, particularly small ones, the expertise or equipment needed to make accurate diagnoses or to provide proper treatment may not exist. For example, a small institution that maintains rabbits producing a valuable polyclonal antibody may not have radiographic equipment or orthopedic expertise to diagnose and treat a rabbit suspected of fracturing its leg. Though euthanasia is an option in such cases, that approach may not represent judicious use of animals if the condition is one that could be reasonably treated so that the animal regains health as well as research usefulness. In such cases, diagnostic and treatment options at other facilities, such as nearby academic institutions, zoos, or private veterinary clinics, are an acceptable means for providing veterinary care if the animal can be easily and safely transported to such a site. Further, it could be useful for veterinary professionals from such sites to visit the location of the ill animal if needed equipment is mobile.

The IACUC should focus on several oversight aspects when it is determined that the medical needs of an animal are best served by transport to an off-site location. First, it should be clear that there is benefit to the animal in instances when off-site locations may be utilized. For example, the need for specialized imaging equipment needed for the diagnostic effort of a nonhuman primate may be evident, but if the animal is moribund and has a poor prognosis, there may be limited advantage to such an effort. Second, the IACUC should be certain that transport of the animals does not place personnel at risk. Transport of animals infected with a biohazardous agent, for example, would be unwise. Third, the specific methods for transport, the location and business entity of the off-site facilities, and the overall infrastructure, including expertise, being made available should be understood by the IACUC. Though specific instances of need for use of off-site facilities cannot always be anticipated, the IACUC should expect the veterinary staff to generate a written plan if the possibility exists for use of such facilities. While it may not be necessary for the IACUC to conduct an inspection of seldom used off-site facilities, it is advisable, and an inspection is certainly warranted, if use of such sites is likely to occur. If off-site facilities are to be used as part of a contingency plan for veterinary care, the IACUC could, instead, evaluate such sites by written description from the AV after his or her visit to the site. If use of an off-site facility is a routine part of the provision of health care, the facility should be included in the semiannual inspection of facilities.

27:20 Is it acceptable for private pets to receive veterinary care at the veterinary facility of a research institution? Is the IACUC responsible for oversight of such activity? (See 15:8.)

Opin. There is nothing in the AWAR, PHS Policy, or the *Guide* that specifically prohibits an animal facility from providing care to private pets. Because these animals are not research animals we can only assume that IACUC involvement should center on the program of adequate veterinary care under which the facility would be operating. Even though pet animals may not be covered under regulatory requirements set forth by APHIS/AC, there are a number of other issues that must be addressed, including the following:

- The appropriate state and federal licenses of those conducting the procedures.
- The liability to the individual and institution in case of problems.
- The potential for cross-contamination of the in-house research animals.
- A fair and equitable program describing whose pet animals will be treated.
- The time required for the veterinary staff to provide such care and the possibility that other duties of the veterinary staff might not be completed as a result.

The IACUC may be asked to monitor issues such as these in order to provide assurances and safeguards for the research facility.

27:21 Which animal health and veterinary care records should be kept? How detailed should these records be? (See also 18:38; 18:39; 23:22; 26:10–26:15; 27:26; 27:35.)

Reg. The APHIS/AC Policy 3[2] specifies that "health records are needed to convey necessary information to all people involved in an animal's care. Every facility should have a system of health records sufficiently comprehensive to demonstrate the delivery of adequate health care." The *Guide* (pp. 75–76) provides recommendations on clinical records for laboratory animals.

Opin. The purpose of any record keeping system is to document the occurrence of a procedure, result, or event. All records must be detailed enough to explain adequately to a person knowledgeable in the subject what occurred to the animal. Records associated with adequate veterinary care serve two main purposes. They assist in facility management and document an adequate veterinary care program to regulatory officials. All records should be maintained as long as they may be useful to the facility. It should be noted that veterinary medical records only document animal health; many research animals can be maintained in a state of good health without maintenance of a medical record.[6]

Some regulatory documents have sections that pertain to the retention of records. The majority of such requirements relate to the product submission and approval process. In many instances, particularly for records related to studies conducted under Good Laboratory Practice (GLP) standards, permanent archiving of records is done (21 CFR Part 58.195). In the case of the Federal Insecticide, Fungicide, and Rodenticide Act (FIFRA), records need to be retained for the life of the product (40 CFR Part 160.195). It stands to reason that these retained records should include those related to animal health. By law, certain records, such as acquisition records of dogs and cats, are required to be retained for a minimum of 3 years after completion of the study (AWAR §2.35,f).

Records having importance with respect to veterinary care include the following:

Standard operating procedures and animal health: Activities such as observation of animals, provision of feed and water, exercise, provision of environmental enrichment, and evaluation of environmental parameters are usually checked daily and should be documented. Many facilities have found it useful to develop an SOP detailing what is to be included in routine animal husbandry. A daily check-off form can then be used to document that a procedure has occurred. Many facilities employ a monthly room care sheet to document activities occurring less frequently such as cage changing or disinfection.

Veterinary care records (see 27:1; 27:26; 27:28): Any animal exhibiting abnormal behavior or signs suggestive of illness or injury should be reported to the veterinary staff. Written documentation should describe the veterinarian's findings, diagnostic procedures, treatments, diagnosis and prognosis, follow-up care, and progress of the animal. It is also important for the records to note the provision of preventive medical care to the animal;[6] this would include items such as physical examination, vaccination, dosing with anthelmintics, hoof or nail trimming, grooming, and other procedures typical of health maintenance for the species. The records should also indicate the study protocol number and include notation for experimental procedures performed on the animal. Euthanasia and the euthanasia method should be noted and the results of any necropsy included.

Colony animals (see 27:26; 27:28): Maintenance of long-term animal colonies requires an additional level of record keeping. For example, production records in a breeding colony may offer clues on disease transmission or genetic abnormalities. It is typical to maintain individual animal records on dogs, cats, nonhuman primates, and occasionally farm animals. These individual records serve as a reference for growth, development, health-related observations, and preventive medical care.

Surgery and postsurgical and post-procedural care: Records should detail preoperative preparation of the animal; use of anesthetics, analgesics, and other medications; and monitoring of the animal during surgery and during recovery. Records should be sufficiently detailed and organized for the IACUC to determine readily that procedures are being conducted in a manner that minimizes the risk of pain or distress to the animals. In addition, it should be clear that the veterinary staff has responded expeditiously to any significant problems.

Surv. Which of the following would be expected to be found on an animal health record of any species at your institution? Check all that apply. Figures in parentheses indicate percent of all respondents who selected the response

- The identification of the animal 250/294 (85.0%)
- Medical or surgical treatments that are not research related 201/294 (68.4%)
- General clinical observations 217/294 (73.8%)
- Surgical records 217/294 (73.8%)
- Anesthesia/analgesia records 222/294 (75.5%)
- Dates of treatments or observations 243/294 (82.7%)
- Times of treatments or observations 191/294 (65.0%)
- Initials or signature of person making observations 228/294 (77.6%)
- Enrichment activities 91/294 (30.9%)
- Exercise times 49/294 (16.7%)
- I don't know 19/294 (6.5%)
- Not applicable 22/294 (7.5%)
- Other 6/294 (2.0%)

27:22 Are veterinary medical records best maintained in hard copy or an electronic format?

Reg. APHIS/AC Policy 3[2] requires that information contained in the veterinary record must be readily available, but it can be maintained in whatever format is convenient for the licensee or registrant. Similarly, the *Guide* (pp. 75–76) states that clinical history information should be readily available to investigators, veterinary staff, and animal care staff. The *Guide* does not prescribe a specific format for such records but recognizes that information can be appropriately recorded in formats ranging from notation on cage or identification cards to detailed computerized records.

Opin. Several factors influence the format in which veterinary medical records are maintained. For large, complex animal care programs with a broad scope involving a multitude of species, invasive procedures, chronic studies, and procedures likely to result in substantial pain or distress to the animals, an electronic record system would have clear advantage in that such a format can allow rapid and efficient data synthesis and access. Such systems can be used, for example, to determine quickly how many animals undergoing a certain procedure have had postoperative infections within a given time frame. For programs of smaller scope, hard copy records are usually sufficient as long as the information that has been entered can be easily accessed and allows retrospective evaluation of veterinary care. Hard copy records can be maintained effectively as detailed written records in individual files for each animal, as written records in laboratory notebooks, or even as notations made on an identification or cage card. The specific degree to which hard copy records are maintained will depend upon the scope of the program and the typical severity of clinical problems that might be expected.

Surv. What format does your veterinary care unit use for its medical records?

- Paper medical records 121/292 (41.4%)
- Electronic medical records 11/292 (3.8%)
- A combination of paper and electronic medical records 104/292 (35.6%)
- I don't know 32/292 (11.0%)
- Not applicable 24/292 (8.2%)
- Other 0/292 (0%)

27:23 Are cage cards an acceptable location for recording veterinary medical information?

Reg. There is no specific regulatory passage in either the AWAR or the PHS Policy that specifically forbids notation of veterinary medical information on cage cards. The *Guide* (p. 75) states only that "identification cards should include the source of the animal, the strain or stock, names and locations of the responsible investigators, pertinent dates (e.g., arrival date, birth date, etc.), and protocol number, where applicable."

Opin. Though use of cage cards for recording of veterinary medical information is not specifically mentioned in the AWAR, PHS Policy, or the *Guide*, the amount and scope of information that can be noted on a cage card are substantially limited by the size of the card. As described in 27:21, the expectations of APHIS/AC in terms of information included in a medical record is broad enough to rule out use of a cage card generally for animals under AWAR purview and maintained for extended periods; still, notes made on a cage card in those cases might serve as a useful summary of relevant clinical events that are recorded in greater detail in a more formal clinical record. In contrast, for animals such as rodents or others that will not be maintained for long periods and are generally free of illness, notations made directly on the cage card may be an efficient way to record veterinary medical information; with large colonies of such animals, it can be efficient to have the limited clinical information available with the animal.

27:24 Who should be allowed access to veterinary care records and where should the records be maintained?

Reg. APHIS/AC Policy 3[2] states that animal health records must be "readily available," and the presumption is that this policy refers to availability of the records to representatives of APHIS/AC acting under the auspices of the AWAR. The *Guide* (p. 76) indicates that medical records for animals should be readily accessible to investigators, veterinary staff, and animal care staff.

Opin. Veterinary care records exist to guide the clinical care of an individual or group of animals. For this reason, such records should be readily accessible to members of the veterinary staff, including veterinary technicians. When animal health records are maintained by the PI, in a research notebook, for example, some mechanism should be established so that the veterinary staff can easily access the clinical information. It is the difficulty with providing for this sort of seamless flow of information that argues against maintaining health records in research notebooks or research labs. For obvious reasons, it is critical that both the IACUC and representatives from regulatory or accrediting agencies be allowed easy and complete access to veterinary care records; without such access it is impossible to determine accurately whether veterinary care is being provided at an acceptable level. Further, it is important for animal care and research personnel to be provided access to veterinary care records relevant to animals under their charge. Such personnel often are the initial point at which ill or injured animals are identified, and these same personnel typically both are curious about and have some degree of responsibility to assure that sick or injured animals are receiving appropriate care and veterinary attention. For all of these reasons, it is logical for most veterinary records to be physically maintained within the veterinary services unit.

Surv. At your institution, who has access to animal health records? Check as many boxes as appropriate. Figures in parentheses indicate percent of all respondents who selected the response

- Not applicable 15/294 (0.1%)
- The attending veterinarian 259/294 (88.1%)
- The veterinary staff, in general 183/294 (62.2%)
- Animal care staff (if separate from the veterinary staff) 189/294 (64.3%)
- The investigator/educator and that person's staff 210/294 (71.4%)
- The IACUC Chairperson 198/294 (67.3%)
- The IACUC (all committee members) 189/294 (64.3%)
- The Institutional Official 194/294 (66.0%)
- IACUC administrators 127/294 (43.2%)
- Our legal counsel 69/294 (23.5%)
- Any person who requests to see the records 36/294 (12.2%)
- I don't know 7/294 (2.4%)
- Other 7/294 (2.4%)

27:25 For how long should veterinary records be maintained?

Reg. There exists no regulatory requirement for how long veterinary records should be maintained, though other records for individual animals are required to be maintained for a minimum of 3 years after completion of the activity (AWAR §2.35,f). The *Guide* does not suggest a specific period for which veterinary records should be maintained. (See 16:29.) For PHS funded activities, NIH Grants Policy requires grantees to retain financial and programmatic records, supporting documents, statistical records, and all other records that are required by the terms of a grant, or may reasonably be considered pertinent to a grant, for a period of 3 years from the date the annual Federal Financial Report is filed.[7]

Opin. The purpose of veterinary records is to document the health of animals on a specific study and of animals under the purview of the animal care program in general. In the former instance, it stands to reason that veterinary records should be maintained for a period that will allow the scientific community to review the results of the research. In the case of records for animals being used to generate data to meet the needs of a regulatory directive (e.g., Food and Drug Administration), the records should be permanently archived. Beyond the regulatory aspects, an archiving system for veterinary records is useful in that it allows retrospective review and analysis of animal health trends. Such information can help guide and direct veterinary approaches to health care for the broader population of animals within the program. From this latter perspective, it may not be necessary to archive the entire veterinary record for every animal, but a database of clinical information distilled from the records could still prove very useful.

Surv. For what length of time does your institution maintain animal health records?

- We never discard animal health records 34/294 (11.6%)
- Not applicable 19/294 (6.5%)
- For three years after the last entry into the record 34/294 (11.6%)
- For three years after the study is completed 97/294 (33.0%)
- It varies by species 19/294 (6.5%)
- I don't know 81/294 (27.9%)
- Other 8/294 (2.7%)

27:26 Should every animal in the facility have a health record or should every colony have one? (See also 27:11; 27:12.)

Reg. APHIS/AC Policy 3[2] essentially states that health records sufficiently comprehensive to demonstrate the delivery of adequate health care are required for species regulated under the AWAR. Though the policy stops short of requiring entirely separate records for each animal, it is clearly expected that notations for individual treatment of an animal must be made as entries specific to that animal.

Opin. In general, all observed illness or abnormalities should be entered into a clinical record; however, an alternate view holds that many animals, such as rodents, can be successfully maintained without formal medical records.[6] Records are frequently more detailed for some species than for others. For example, individual health records for dogs, cats, nonhuman primates, and agricultural animals are useful (*Guide* p. 75).

A useful rule of thumb is that individual records should be maintained for those animals that receive routine periodic individual health evaluations. For example, individual records should be maintained for dogs, detailing vaccinations, parasite evaluations, any treatments administered and periodic physical examinations.

Separate records are not needed for animals that may have periodic evaluation by examination of several representative individuals from the colony. Rather, the health information of such animals can be recorded as a colony health record. An alternative rule of thumb is that individual records should be maintained for large long-lived animals (*Guide* p. 75). In either case, the interpretation is that individual records are useful for species such as dogs, cats, nonhuman primates, rabbits, and farm animals, and of less value for rodents or common non-mammalian species. Alternatively, individual surgical records can be very useful for animals that have had invasive procedures, regardless of species. If an animal in a colony receives treatment as an individual (e.g., for an injury), a record of that treatment should be maintained.

Surv. At your institution, for which species are individual animal health records prepared, even if a health problem has not arisen? Check as many boxes as appropriate. Figures in parentheses indicate percent of all respondents who selected the response.

- We do not maintain any individual health records 19/289 (6.6%)
- Only species covered by the AWAR 51/289 (17.6%)
- All mammalian species 104/289 (36.0%)
- All vertebrate species 29/289 (10.0%)
- Common laboratory rats and mice 24/289 (8.3%)
- Not applicable 19/289 (6.6%)
- Other 17/289 (5.9%)

27:27 What records and information should the IACUC review if a PI proposes to create an animal having a novel genotype with an unknown phenotype?

Reg. The *Guide* (p. 28) recognizes that whether genetic manipulation is targeted or random, the resulting phenotype may be unpredictable and can affect the animal's well-being. As a result, the *Guide* offers that the IACUC should be informed when characterization of a novel phenotype reveals a condition that negatively affects an animal's well-being. The PHS Policy includes no specific reference to this issue, though it does state (IV,A,1) that institutions having an Assurance are required to use the *Guide* as a basis for developing and implementing an institutional program for activities involving animals. The AWAR do not directly address this issue, but general guidance might be taken from a statement (§2.31,d,1,v) which emphasizes that animals experiencing severe, chronic pain or distress that cannot be alleviated will be euthanized at the end of the procedure or earlier.

Opin. For the IACUC to properly engage its charge as described in the *Guide* and AWAR, it must understand the impact on clinical health and well-being of phenotypes which result from genetic manipulation. In this regard, the IACUC should identify a mechanism by which unexpected outcomes and phenotypes counter to well-being are reported to the IACUC. Such a mechanism might best encumber the investigator to inform the IACUC, perhaps by a standard form, by email notification, or other documented means. This idea might be extended to any unexpected outcome of experimental manipulation of an animal, even those not involving genetic manipulation. Indeed, the *Guide* (p. 28) indicates that genetically modified animals are an example of a model for which increased monitoring for unexpected outcomes could be implemented, suggesting an expectation that unexpected adverse outcomes of any experimental model should be reported to the IACUC. (See 27:42.)

When an investigator proposes to create an animal having a novel genotype, the IACUC would likely need to review certain information in order to adequately understand the scope of the phenotype effect on animal well-being. For instance, if the genetic alteration is expected to result in a specific disease phenotype, the investigator should indicate so as part of the protocol approval process. Similarly, any steps that will be taken to mitigate distress to affected animals, including euthanasia at defined endpoints, should be described. If the genetic alteration is intended to mimic a naturally occurring aberration of humans or other animals, the IACUC might choose to review relevant literature to assess the likelihood and degree of clinical disease and distress. Further, the effect of alteration on similar genes might be considered by the IACUC in assessment of likely effect. After approval and once the work has begun, the IACUC might ask the veterinary staff to make a clinical assessment of animals with the novel genotype to assess the degree of clinical disability, if any, and effect on animal well-being. Such an assessment might include review of any reports of animal illness or deaths of the animals, or review of colony production records to assess if there is any evidence of fetal or neonatal mortality. Similarly, the IACUC might ask the investigator to provide a summary, written or verbal, of the phenotypic characterization of the animals, with emphasis on traits that could impact animal health and well-being.

27:28 How sick or abnormal should an animal be to merit veterinary attention?

Opin. Every animal that is found to be ill or injured, is behaving abnormally, or is suspected of being ill or injured should be reported to the AV or to the veterinary staff as directed by the institution's program of veterinary care. In some cases, the veterinary staff may determine that no specific treatment is merited, while in other cases treatment may be needed. It is the veterinary staff, under the direction of the AV, which should make a judgment regarding the severity of illness or injury and subsequent course of action.

27:29 Are there any circumstances under which an investigator may rightfully refuse veterinary care for his or her research animals?

Reg. The AWAR state that the AV must have appropriate authority to ensure adequate veterinary care (§2.33,a,2). Similarly, the *Guide* (p. 105) states that a veterinary program that offers a high quality of care and ethical standards must be provided regardless of the number of animals or species maintained. Because the PHS Policy (IV,A,1) requires adherence to the AWAR and principles in the *Guide*, clearly the expectations with respect to veterinary care are the same.

Opin. It is unequivocal that the AV has been granted authority by APHIS/AC and the PHS Policy to make decisions regarding the veterinary care provided to animals. The *Guide* (p. 114) states that for animals on research protocols, the veterinarian or their designee should make every effort to discuss any problems with the PI or project director to jointly determine the most appropriate course of treatment or action. Essentially, then, the AV has the final say in matters related to veterinary care, but it is preferable that agreement between the veterinarian and the investigator be reached. This principle must not be interpreted as endowing the AV with authority to make veterinary decisions that contradict veterinary parameters outlined in an approved protocol. For example, if the protocol has approved a procedure involving unrelieved pain and distress (support with scientific justification), the AV would not have the authority to administer treatments to mitigate that pain and distress. In instances in which the animal is in immediate peril, the AV may need to make clinical decisions without consulting the investigator.

It is conceivable that an investigator may choose to euthanize animals rather than pursue the option of treatment recommended by the veterinarian. Depending upon the clinical circumstances and the nature of the research, such a decision may be practical and reasonable. Similarly, it is within the authority of the veterinarian to euthanize an animal for clinical reasons (*Guide* p. 114).

27:30 If an animal is clearly suffering, can the IACUC allow a veterinarian to euthanize the animal if the research laboratory personnel cannot be contacted?

Reg. The AWAR state that the AV must have the authority to ensure adequate veterinary care (§2.33,a,2). The *Guide* (p. 14) states that adequate veterinary care must be provided to animals and that the veterinary care program is the responsibility of the AV. The *Guide* further states (p. 114) that the veterinarian must have the authority to treat the animal, remove it from the experiment, institute appropriate measures to relieve severe pain or distress, or euthanize the animal if necessary. Because the PHS Policy (IV,A,1) requires adherence to the AWAR for AWAR-regulated species and the principles in the *Guide*, it is clear that the expectations with respect to veterinary care are the same.

Opin. *Veterinary care* is defined in the AWAR (§2.33,b,4) as including a provision of guidance by the AV to PIs regarding euthanasia. While commonly interpreted to mean that the AV provides guidance in terms of methods for euthanasia rather than the decision to euthanize a suffering animal, another interpretation is supported by an earlier section in the AWAR (§2.31,d,v) that states that animals experiencing severe or chronic pain that cannot be relieved will be euthanized. Likewise, the *Guide* (p. 105) considers euthanasia to be a component of the veterinary care program for which the AV has responsibility. The *Guide* (p. 14) also endorses the ACLAM *Guidelines for Adequate Veterinary Care*[1] and its position that the veterinarian must have the responsibility and authority to assure that handling, restraint, anesthesia, analgesia and euthanasia are administered as required to relieve pain and such suffering in research animals, provided such intervention is not specifically precluded in protocols reviewed and approved by the IACUC. It is apparent that the intent of these documents is to assign to the AV authority to euthanize animals that are suffering and for which the IACUC has not approved procedures that involve suffering or death as an end point.

Surv. If an animal is clearly suffering, does your institution allow a veterinarian to euthanize the animal if the research laboratory personnel cannot be contacted?

• We have no policy on this matter	13/294 (4.4%)
• Yes, the veterinarian has that authority	273/294 (92.9%)
• No, the veterinarian does not have that authority	1/294 (0.3%)
• I don't know	5/294 (1.7%)
• Other	2/294 (0.7%)

27:31 How often should notations be made in the animal health care records for an animal reported to be ill?

Reg. The AWAR require that all regulated animals be observed daily to assess their health and well-being (§2.33,b,3).

Opin. Clinically ill or abnormal animals should be followed until resolution, either through return to clinical normalcy or euthanasia or determination that the relevant condition is clinically minor with little likelihood of increased severity. Records should document that follow-up care and evaluation have occurred.

The exact frequency at which follow-up evaluations should be made varies considerably with the clinical condition and depends upon the sound professional judgment of the veterinarian. Animal suffering, pain or distress indicates severe conditions that might necessitate evaluation several times daily, while other conditions might be appropriately evaluated once every several weeks. In general, follow-up evaluations should occur and be documented in the records at a frequency such that significant increases in the severity of the condition could reasonably be expected to be detected. For most conditions, this usually involves at least daily evaluation with more frequent evaluations if necessary. For less threatening or improving conditions, weekly or biweekly evaluations are often sufficient.

Repeated instances of ill animals' progressing to a more severe disease status in the absence of corresponding notations in the record may indicate that animals are not being examined frequently enough. Conversely, if notations exist for such animals without adjustments to treatment, that could suggest a decline in the animal's clinical condition is being duly noted but ignored by the veterinary staff. In either circumstance, the IACUC must further investigate the situation.

27:32 How can the IACUC evaluate programs for preventive medicine and health monitoring? What type of programs should be expected?

Opin. Preventive medicine and health monitoring programs are designed to assure that healthy animals are acquired initially and that these animals remain healthy. To accomplish this, regularly scheduled evaluations of animals and vendors should be performed. The IACUC should inspect the veterinary care records (see 27:35) for accordance with a written description of the preventive medicine program, as well as visually examine animals for general appearance as an indicator of health status. Further, the IACUC should ascertain that animal health evaluations indicate the animals remain healthy and a timely and appropriate response is made when evaluations indicate a problem (see 27:1).

The program of preventive medicine should begin with evaluation of the vendor or source of the animals. Regular reports of the health status of the vendor's animals should be reviewed by the AV or his or her designate to determine that healthy animals are being acquired by the institution. If this review determines that health problems exist at the vendor's facilities, then the AV should coordinate arrangements to identify alternate vendors or other means that ensure that healthy animals are being used.

At the time of arrival at the facility, animals should be inspected by the veterinary staff and then separated from other more established groups of animals (see 27:1) until, in the professional judgment of the veterinary staff, the newly arrived animals do not pose a threat to the established animals. For example, animals such as dogs, cats, and nonhuman primates often receive a physical examination upon arrival as part of the preventive medicine program and are then separated from established animals.

The program of preventive medicine also should include procedures such as appropriate vaccinations, parasite evaluations, and testing for tuberculosis for nonhuman primates. Livestock, such as sheep and cattle, should have routine deworming or fecal evaluations (or both). Other procedures are often included depending on the facility and the species involved. These procedures should be repeated periodically, per veterinary medical standards. Periodic, often annual, physical examinations should be performed for animals such as cats, dogs, and nonhuman primates. Chronic colony health issues such as dental care and weight management should be included in a prevention program for animals maintained over many years. Geriatric colonies may also necessitate more frequent examinations. Routine physical examinations on other animals such as rabbits, guinea pigs, and other small animals can be useful for animals maintained for extended periods. Often, such examinations are brief and may be conducted by trained personnel other than the AV.

Biosecurity (*Guide* p. 109) is an integral part of any preventive medicine program. Rodent colonies are often evaluated for microbial contamination (*Guide* pp. 112–113). In general, a small number of animals (often sentinel animals placed in the room specifically for the purpose of health evaluation) are euthanized, blood is collected for serologic detection of microbial contaminants, tissues are evaluated histologically, and the gastrointestinal tract and pelt are evaluated for parasites. Recommendations for the exact procedures and number of animals evaluated in this way are presented in detail elsewhere.[8]

Examples of activities that can be associated with a preventive medicine program are shown in Table 27.4.

The exact procedures and scope of activities vary significantly with the species and type of animals and the type of research. Activities listed are examples only.

TABLE 27.4

Typical Components of a Preventive Medicine Program

Aspect of Preventive Medicine	Typical Activity
Animal vendor surveillance	Review of vendor health status reports, onsite visits to vendor when possible
Animal transportation	Evaluate method of transportation for minimization of animal stress and risk of disease. Dedicated delivery, climate control
Processing of newly received animals	Inspection and examination of animals and crates, initial vaccinations, medications, parasite evaluations
Periodic reevaluation (usually every 6–12 months)	Physical examination, vaccination, tuberculosis testing (nonhuman primates), dental prophylaxis
Rodent health surveillance	Serologic and histopathologic evaluation for microbial contaminants, pelt examination for ectoparasites, examination of gastrointestinal tract for endoparasites

27:33 Should the IACUC attempt to evaluate animal health? If so, how?

Opin. Yes, the IACUC should periodically evaluate the health of the animals, since the IACUC is responsible for evaluating the institutions animal care and use program which includes veterinary care (AWAR §2.31,c,1; AWAR §2.31,d,1,C,vii; AWAR §2.33,a; PHS Policy IV,B,1; PHS Policy IV,C,1,e; *Guide* pp. 26, 30). The IACUC typically relies on the AV for advice with respect to animal health concerns and this has the potential to create a conflict of interest since the AV is a member of the IACUC and evaluator of his or her own program of veterinary care. This conflict is avoidable when all resources are actively pursuing the best animal care program.

The most basic way for the IACUC to evaluate animal health is to observe the animals during the semiannual inspections and to speak with those responsible for animal care. Table 27.3 lists some clinical observations that might be evidence of an animal health problem.

The number of animals with indications of illness is a valuable measure of animal health the IACUC can use. Of course, animals found moribund or dead can also signal an animal health problem.

The IACUC should also examine animal health, anesthesia and analgesia, and surgical records for an indication of the robustness of the veterinary care program (see 27:16). It is useful to spot-check records detailing animal health over the preceding 6 months, examine records of animal health monitoring activities such as routine fecal examinations for endoparasites or routine serologic monitoring of rodent colonies. It is worthwhile to query technical staff involved with experimental procedures to determine whether they are aware of any animal health problems.

If the preventive care aspect of the program of veterinary care (including careful vendor selection) is effective, the IACUC should find few problems. However, it is not unusual for the IACUC to identify occasional problems, since spontaneous disease and consequences of experimental procedures can sometimes lead to clinical illness even in well-managed animal facilities. The key for the IACUC is to evaluate the *adequacy of the mechanism for reporting problems and the response to any problems* (see 27:35). For example, timeliness of efforts to respond to and diagnose a problem, medications administered, supportive care provided, advice to investigators on technique refinement, and efforts to isolate sick animals to prevent disease spread could all be considered evidence of an appropriate response to a clinical problem. The precise response will vary with the specific problem.

27:34 How can the IACUC determine that the veterinary response to health monitoring findings is reasonable and adequate?

Opin. Having determined that the program for preventive medicine and health monitoring is properly designed and implemented, the IACUC must further ensure that results of these efforts are interpreted correctly and that appropriate action is taken when potential problems are identified. To accomplish this, the IACUC must evaluate the response of the veterinary staff, in particular the AV. The semiannual program reviews are a good place to start. How has the program changed in the last 6 months? Are there more or fewer clinical problems? Are the same issues coming up repeatedly and if so is there a plan proposed to address the issue? Is the staff proactive or reactive?

In general, the IACUC should see evidence of a veterinary staff that responds expeditiously to identified problems related to animal health, monitoring and surveillance. The precise response can vary and can include continued monitoring of animals, isolation of suspect animals, specific treatment of animals, or euthanasia of all affected or exposed animals. Again, the appropriateness of the response depends on the type and species of animals, the nature of the research, and the professional judgment of the AV and the veterinary staff.

Additionally, the IACUC should further examine health records to determine whether the response of the veterinary staff has been successful. If not, the IACUC should determine that the AV is developing other strategies to resolve the problem. The IACUC must be assured that problems are not being ignored, and that the veterinary staff does not abandon the care of an animal.

27:35 How can the IACUC assure that veterinary care records are adequate?

Opin. Records for veterinary care should be evaluated for completeness and for evidence of an adequate response to reported problems (see 27:21; 27:26). For example, records should clearly identify the specific animals involved, the clinical history, the source of the animal, the investigator, the age and gender of the animal, the date of animal receipt, and the types and dates of experimental or therapeutic procedures that have been performed on the animals (*Guide* pp. 75–76).

Appropriate diagnostic efforts and prescribed treatments should be documented. In addition, the record should indicate each time the animal is medicated or otherwise treated. The clinical progress of the animal should be noted until the clinical problem has resolved or the animal has been euthanized. Evidence that an animal languished until it died or was finally euthanized without reasonable attempts to diagnose and treat the problem can be considered as indicative of inadequate veterinary care (see 27:33). In addition, proper records should reflect the substance of any discussions with investigators or other staff on ways to minimize problems in the future. Individuals examining animals or providing treatment or other actions should sign or initial their notations.

Records can be considered to be inadequate if the IACUC cannot readily understand the steps taken to identify, report, diagnose, discuss, treat, and continue to evaluate an animal until the humane resolution of the case.

27:36 For what type of studies does the IACUC require direct veterinary involvement (e.g., requiring a veterinarian to check Category E animals or new surgical procedures)? For these studies, what is an appropriate period for this oversight?

Reg. The IACUC is charged with ensuring that "personnel conducting procedures on the species being maintained or studied will be appropriately qualified and trained in those procedures" (AWAR §2.31,d,1,viii; PHS Policy IV,C,1,f). The *Guide* (p. 114) states that the veterinarian or his or her designee should discuss with the investigator any problems so that a course of action may be determined.

Opin. A commonly used system, employed by many IACUCs, is to require direct veterinary involvement in evaluating pain and in training and documenting the surgical skills of the researchers. It is important to remember that this requirement for training as specified in the AWAR does not only apply to APHIS/AC annual report Category E protocols (i.e., those that involve significant pain and distress that are not relieved by anesthetics, analgesics, or euthanasia because they would adversely affect the animal activity; AWAR §2.37,b,7) and new surgical procedures, but to all procedures being conducted on animals. Because of expected pain, potential pain, and the skill required for surgical procedures, an IACUC may impose additional means of oversight for Category E protocols and those requiring new surgical procedures and in this way guarantee that those conducting the procedures are adequately trained and competent. Although direct veterinary involvement is effective, other methods of performing this task should not be overlooked. Experienced in-house investigators or external resources may be more familiar with the specific procedures and better suited for training or qualifying the competency of the person performing the technique. They may also be more attuned to the subtle signs of pain that may occur. The duration of this oversight should be directly related to the skill and competence of the individuals performing the procedures. Some leeway must be given during the training period to those learning the procedures, but as a rule, individuals learning techniques should have a documented experienced trainer nearby, preferably in the same room, who can provide guidance, suggestions, and assistance as necessary. The role of the veterinarian or his or her designee should conclude the process by assessing and documenting the appropriate skill levels and reporting this to the IACUC.

Surv. For which of the circumstances described below does your IACUC require direct veterinary involvement? Check as many boxes as appropriate. Figures in parentheses indicate percent of all respondents who selected the response.

• Usually for observing major survival surgical procedures that have not been performed previously at our institution, but on Animal Welfare Act covered species only	47/293 (16.0%)
• Usually for observing major survival surgical procedures that have not been performed previously at our institution on any vertebrate species	136/293 (46.4%)
• Usually for observing procedures performed on Animal Welfare Act covered species in pain/distress category E	32/293 (10.9%)
• Usually for observing procedures performed on any species in pain/distress category E	85/293 (29.0%)
• Only if specifically requested by the IACUC	120/293 (41.0%)
• None of the above	22/293 (7.5%)
• I don't know	12/293 (4.1%)

27:37 What is an appropriate organizational relationship among the IACUC office, the veterinary care staff, and the IO?

Reg. The AWAR (§2.31,a) state that the Chief Executive Officer (CEO) of the research facility is the official who appoints the IACUC, qualified through the experience and expertise of its members to assess the research facility's animal program, facilities, and procedures. The IACUC reports its evaluations to the IO, as required by AWAR §2.31,c,3. The *Guide* (p. 25) likewise states that the IACUC reports to "the IO about the status of the program." The *Guide* (p. 13) further indicates that a shared responsibility relationship should exist between the IO, the AV, and the IACUC for the overall direction of the animal care and use program. NIH/OLAW states[9] that there should be direct and clean lines of responsibility and authority so that the organization is able to respond quickly to situations that may arise; and that key elements include the IO, the IACUC, and the AV. Further, the IO should have authority to allocate resources to maintain the program in accordance with recommendations from the IACUC and the AV; and the IO should clearly define and assign responsibilities for program aspects such as training and occupational health.

Opin. Often it is difficult to distinguish among the roles of the IACUC, veterinary staff, facility operations, and even the IO. Some smaller facilities with limited staffing may require one individual to fill multiple roles. To guard against possible conflict of interest—where those inspecting and assessing the operations are also those in charge of them—the trend in larger facilities has been to create an obvious separation of responsibilities. This system provides for the checks and balances similar to those we have seen in government. The executive branch appoints (CEO) and supports (IO) the IACUC and facility. The judicial branch (IACUC) interprets and enforces regulations, and the legislative branch (veterinary care and facilities operations) oversees the daily operational needs. The trick then becomes one of effectively maintaining good communications and close working relations. A key point to feedback to each group and to the research community is that the overall goal of the process is to provide for optimal research conditions.

27:38 How might a veterinary services unit work for the IACUC, with the IACUC, or against the IACUC?

Opin. Often, the veterinary services unit works closely with the IACUC and although this scenario may seem ideal in some ways, it raises the issue of how separate the two groups are. Do the groups work well together because they are made up of the same players? If the groups tend to be the same, then they are really working with themselves. It could be argued that an efficient IACUC and veterinary services unit will at times find that they are working for and against each other.

Expanding on the checks and balances analogy from the previous question, each group (IACUC office, the veterinary care staff, facility operations, and the IO) has the responsibility of to ensure that a single component of the program does not gain power to the point at which they can control all functions. There are times an IACUC may call on the veterinary staff to work for them, utilizing the expertise and training of the laboratory animal veterinarian to help develop policies or detail programs of care and use. Likewise, the groups could work against each other if either is becoming too complacent or authoritarian in their activities. A diverse IACUC through their interactions and resulting discussions will perform a valuable service to the institution.

27:39 What types of veterinary issues might the IACUC commonly find during the semiannual programmatic review and facility inspection?

Opin. The IACUC is required to review the program of veterinary care to assure that it is designed and implemented in such a way as to provide an adequate level of veterinary care. This means that programs, personnel, and other resources are available for the diagnosis, treatment, control, and prevention of animal disease and injury. Veterinary records detail the delivery of this program and should therefore be reviewed to assess adequacy of implementation by the veterinary and research staffs. While planned programs are sometimes lacking in scope and breadth, underfunding of the veterinary effort often results in lack of needed equipment. Records, particularly those maintained by research staff, are frequent sources of concern as insufficient detail or data entry indicates a lack of attention to provision of animal health support and/or a lack of attention to proper record keeping. If the IACUC believes that veterinary care is inadequate, such a finding should be cited as a deficiency which may be either minor or major depending on the potential impact on the animals. Under such circumstances, it would be reasonable for the IACUC to discuss with the veterinarian such concerns and. As well, the IACUC might consider enlisting the service of a consultant veterinarian with appropriate expertise to offer an additional opinion with respect to the rigor of veterinary care.

Surv. During the semiannual program review, what types of program-wide problems have you ever identified from the list below? Check as many boxes as applicable. Figures in parentheses indicate percent of all respondents who selected the response.

- We do not review the program of veterinary care during the semiannual program review — 10/290 (3.4%)
- Inadequate veterinary care as provided by our laboratory animal veterinarians — 13/290 (4.5%)
- Inadequate veterinary care as provided by veterinary technicians — 11/290 (3.8%)
- Inadequate veterinary care provided by research teams — 48/290 (16.6%)
- Inadequate medical records kept by veterinary services — 32/290 (11.0%)
- Inadequate medical records kept by research groups — 115/290 (39.7%)
- None of the above — 147/290 (50.7%)
- I don't know — 6/290 (2.1%)

27:40 What considerations are required when an experimental protocol includes the restriction of food or water?

Reg. The AWAR (§2.38,k,1) requires research facilities to comply with the standards of care as set forth in Part 3-Standards. The *Guide* (p. 31) states that animals should be closely monitored to ensure that food and fluid resources meet their nutritional needs. Body weights should be recorded at least weekly, more frequently if the restrictions are greater. Written records should be maintained for each animal to document daily food and fluid consumption as well as clinical or behavioral changes.

Opin. Food or fluid restriction may be required to achieve desired results of various research protocols. Ideally these protocols will be planned to require the least restriction necessary to achieve the desired scientific results. While a research protocol must indicate deviations from an institutions standard husbandry practice, most IACUCs do not expect significant monitoring for restrictions of 24 hours or less. For restrictions greater than 24 hours or procedures that limit access to food or water to specific time periods the *Guide* (p. 31) suggests special attention be paid to the level of restriction, potential adverse consequences of regulation and methods for assessing animal health and well-being. Depending on the severity of the restrictions, monitoring of food and fluid intake as well as body weights and condition should be documented.

It is further suggested that when food is used as a training or conditioning reward that a highly desired food treat is used as a positive reinforcement as opposed to restricting feed so that the animal will work in order to receive basic nutritional needs.

Surv. In a study in which an animal's food or water is restricted (other than for pre-surgical fasting), which of the following parameters are typically monitored at your institution? Check as many boxes as appropriate. Figures in parentheses indicate percent of all respondents who selected the response.

- The animal's body weight 227/295 (77.0%)
- The animal's hydration status 182/295 (61.7%)
- The animal's behavior 219/295 (74.2%)
- We do not monitor any parameters 5/295 (1.7%)
- Not applicable 55/295 (18.6%)
- Other 8/295 (2.7%)

"Other" responses included

- Urine, feces
- Hibernating animals are not disturbed
- Food intake for water restriction
- Fecal output
- Depends on length of restriction
- Comparison to weight matched controls
- Body condition score, general blood chemistry
- Animal's coat condition, posture

27:41 What provisions for veterinary care should the IACUC expect in the case of field studies?

Reg. The AWAR (§1.1) defines field study as a study conducted on free-living wild animals in their natural habitat. However, this term excludes any study that involves an invasive procedure, harms, or materially alters the behavior of an animal under study. This distinction must not be overlooked. Studies conducted in the field that are invasive, harm, or materially alter the behavior of an animal under study fall under IACUC purview to ensure they comply with the institutions animal care and use program. The *Guide* (p. 32) advocates that many issues associated with field studies are similar to those for species maintained and used in the laboratory. This suggests that the provisions for veterinary care during field studies should be included in the overall program of veterinary care. The AWAR (§2.31,d,1,ix) does provide for one exception: that operative procedures conducted at field sites need not be performed in dedicated facilities, but must be performed using aseptic procedures. NIH/OLAW[10] has provided the following guidance on IACUC oversight of field studies: "If the activities are PHS-supported and involve vertebrate animals, the IACUC is responsible for oversight in accord with PHS Policy. IACUCs must know where field studies will be located, what procedures will be involved, and be sufficiently familiar with the nature of the habitat to assess the potential impact on the animal subjects. If the activity alters or influences the activities of the animal(s) that are being studied, the activity must be reviewed and approved by the IACUC (e.g., capture and release, banding). If the activity does not alter or influence the activity of the animal(s), IACUC review and approval is not required (observational, photographs, collection of feces). ...When capture, handling, confinement, transportation, anesthesia, euthanasia, or invasive procedures are involved, the IACUC must ensure that proposed studies are in accord with the *Guide* (page 32). The IACUC must also ensure compliance with the regulations and permit requirements of pertinent local, state, national, and international wildlife regulations."

Opin. Investigators preparing for field studies should consult with veterinarians and other subject matter experts in the design of their protocols. While the goal of the veterinary care program should be similar to that of laboratory animal operations, differences in capture, restraint, identification, surgery drug dosages and other aspects of the work may be significantly different in their implementation. The IACUC should also ensure that the occupational health and safety issues are addressed so that the health and safety of the animals and the persons in the field are not compromised.

27:42 Should there be a simple procedure for investigators or others to report to the IACUC unanticipated adverse events that happen during the course of research?

Reg. The IACUC must conduct continuing reviews of approved activities at intervals determined by the IACUC (PHS Policy, IV,C,5) and AWAR (§2.31,d,5). Adequate medical care for animals must be available (AWAR, §2.33,a; PHS Policy, IV,C,1,e; *Guide* p. 14).

Opin. An unanticipated adverse event is an unplanned outcome outside of the parameters described in an IACUC-approved protocol that did or potentially can result in physical harm, health deterioration, or distress to an animal subject. It can affect one animal, a subgroup of animals, or all animals in a study. Unanticipated adverse events are not considered protocol violations unless they emanate from protocol noncompliance or animal abuse. (See 27:27.) Some examples of unanticipated adverse events are

- Mortality greater than 10% of that predicted in the IACUC protocol or greater than 10% of the animal subjects in a study group if no mortality is predicted.
- Post-operative complications that are objectively or subjectively judged by the investigative team or a qualified veterinarian to be significantly greater than anticipated.

- Pain or distress or declining animal subject health that is objectively or subjectively judged by the investigative team or a qualified veterinarian to be significantly greater than anticipated.
- Accidental physical injury to an animal subject that results in more than momentary and mild pain or distress.
- Unanticipated injection reaction near or distant to an injection site that results in more than momentary and mild pain or distress.

Many IACUCs request an annual report of adverse events arising from animal activities during the past year. While some such events may be restatements of incidents already known to the IACUC, other incidents, such as unanticipated health problems arising from unique phenotypes (*Guide* p. 28) or excessive anesthetic deaths may or may not have been known to the IACUC or AV. The prevention of recurrences may have benefitted from timely veterinary consultation and additional observations. Therefore, it is to the benefit of animals, the investigator, and the institution to have an easy method for the prompt voluntary reporting of unanticipated adverse events. An Unanticipated Adverse Event Form that is readily available on the IACUC web site or from the IACUC or animal facility offices, can be a significant aid to animal welfare when its stated goal is to help the animal using community prevent recurrences of a problem.

Unanticipated adverse events also can occur during animal transport and NIH/OLAW expects shipping institutions to report adverse events that happen during transport.[11]

References

1. American College of Laboratory Animal Medicine. Position Statements: Adequate Veterinary Care. http://www.aclam.org/Content/files/files/Public/Active/position_adeqvetcare.pdf. Accessed February 5, 2013.
2. United States Department of Agriculture, Animal and Plant Health Inspection Service. 2011. *Animal Care Resource Guide*, Policy 3, Veterinary Care. http://www.aphis.usda.gov/animal_welfare/policy.php?policy=3.
3. National Institutes of Health, Office of Laboratory Animal Welfare. 2005. Notice NOT-OD-05-034. Guidance on prompt reporting to OLAW under the PHS Policy on Humane Care and Use of Laboratory Animals. http://grants.nih.gov/grants/guide/notice-files/NOT-OD-05-034.html.
4. National Institutes of Health, Office of Laboratory Animal Welfare. 2012. Frequently Asked Questions F.4, May investigators use non-pharmaceutical-grade compounds in animals? http://grants.nih.gov/grants/olaw/faqs.htm#useandmgmt_4.
5. National Institutes of Health, Office of Laboratory Animal Welfare. 2012. Frequently Asked Question F.5, May investigators use expired pharmaceuticals, biologics, and supplies in animals? http://grants.nih.gov/grants/olaw/faqs.htm#useandmgmt_5.
6. Medical Records Committee. 2007. Medical records for animals used in research, teaching, and testing. Public statement from the American College of Laboratory Animal Medicine. *ILAR J.* 48: 37–41. http://www.aclam.org/Content/files/files/Public/Active/position_medrecords.pdf.
7. National Institutes of Health, Office of Extramural Research. 2012. NIH grants policy statement. Part II: Terms and Conditions of NIH Grant Awards Subpart A: General—File 6 of 6 http://grants.nih.gov/grants/policy/nihgps_2012/nihgps_ch8.htm#_Toc271264975).

8. Committee on Infectious Diseases of Mice and Rats. 1991. *Health surveillance programs, in infectious diseases of mice and rats. A report of the Institute of Laboratory Animal Resources*, 21. Washington, D.C.: National Academy Press.
9. National Institutes of Health, Office of Laboratory Animal Welfare. 2012. Frequently Asked Questions G.4, What kind of administrative organization works best for ensuring compliance? http://grants.nih.gov/grants/olaw/faqs.htm#instresp_4.
10. National Institutes of Health, Office of Laboratory Animal Welfare. 2012. Frequently Asked Questions A.6., Does the PHS Policy apply to animal research that is conducted in the field? http://grants.nih.gov/grants/olaw/faqs.htm#instresp_4.
11. National Institutes of Health, Office of Laboratory Animal Welfare. 2012. Frequently Asked Questions F.12, What are the institution's responsibilities in ensuring that animals are shipped safely and in reporting adverse events that occur in shipment of animals to or from the institution? http://grants.nih.gov/grants/olaw/faqs.htm#useandmgmt_12.

28

Laboratory Animal Enrichment

Kathryn Bayne and Jennifer N. Camacho

Introduction

In the United States, the concept of addressing the *behavioral needs* of research animals first came to the forefront through the 1985 amendments to the Animal Welfare Act (PL 99–198, "The Improved Standards for Laboratory Animals Act") which directed the Secretary of Agriculture to develop regulations to ensure an adequate physical environment to promote the psychological well-being of nonhuman primates and to provide for the exercise of dogs. These stipulations were considered progressive at the time. The contemporaneous *Guide for the Care and Use of Laboratory Animals*[1] did not yet recognize this important aspect of the welfare of laboratory animals, stating that meeting the biologic needs of the animals involved "maintenance of body temperature, urination, defecation, and, if appropriate, reproduction…." However, the resulting regulations pertaining to exercise for dogs and psychological well-being for nonhuman primates prompted a significant amount of research into laboratory animal behavior, and more specifically, environmental enrichment. The output of this research effort included significant changes in the way in which nonhuman primates were housed in laboratories and consideration of promoting physical activity in dogs.

This evolution in the consideration of the behavioral needs (i.e., not just the biological needs) of laboratory animals was reflected in the seventh edition of the *Guide*,[2] which devoted an entire section to the "behavioral management" of laboratory animals. While the principal target species of this section were nonhuman primates and dogs, the recommendations were sufficiently broad to be applicable to a variety of research animal species. A behavioral management program was described as including three elements: (1) the structural environment; (2) the social environment; and (3) activity (both physical and cognitive). Methods to enhance each of these areas were described to improve the welfare of the animal. As the momentum increased for providing enrichment to nonhuman primates, additional guidance was developed and offered by the federal government. For example, the USDA and NIH/OLAW collaborated on publishing a series of six booklets that provide an introduction to "Enrichment for Nonhuman Primates"[3] and more recently have produced a seminar recording on the subject, "AC and OLAW Perspective on Nonhuman Primate Enrichment and Social Housing".[4] In addition, USDA has published a bibliography, "Environmental Enrichment for Nonhuman Primates"[5] that includes general references pertaining to environmental enrichment, primate behavior and abnormal behavior, as well as species specific references.

In the last decade or so, the scope of environmental enrichment has increased and a large number of research studies have evaluated the impact of enrichment on rodents, aquatic species, agricultural animals, and many other laboratory animals used in biomedical research. The newest edition of the *Guide*[6] echoes this further maturation of the field in a section entitled "Environmental Enrichment" that more broadly addresses enrichment techniques for both large and small laboratory animals, as well as aquatic animal species. NIH/OLAW has developed a Frequently Asked Questions (FAQ) website[7] to further clarify their position on some aspects of the provision of enrichment devices. The provision of enrichment is generally considered a key element of addressing the principle of Refinement in the Three Rs and thus continues to gain momentum in implementation across species when assurance is provided that neither the heath of the animal nor scientific goal of the study is compromised by its provision.

The survey distributed to IACUC Chairs included 17 questions related to environmental enrichment. The comments of the survey respondents were reviewed, and the results are presented here.

28:1 How does your IACUC interpret the term "laboratory animal enrichment"?

Reg. Although it does not define the term, the *Guide* (p. 52) states that the "The primary aim of environmental enrichment is to enhance animal well-being by providing animals with sensory and motor stimulation, through structures and resources that facilitate the expression of species-typical behaviors and promote psychological well-being through physical exercise, manipulative activities, and cognitive challenges according to species-specific characteristics." The *Guide* (p. 53) also notes that "Well-conceived enrichment provides animals with choices and a degree of control over their environment, which allows them to better cope with environmental stressors…." The AWAR (§3.81) address enrichment only for nonhuman primates and, like the *Guide,* do not define the term. However, the term "environment enhancement" is used in the regulations to refer to the social, structural, and activity-related aspects of behavioral management. Requirements include social grouping and providing a means of expressing non-injurious species-typical activities, such as cage complexities, manipulanda, foraging, etc.

Opin. Environmental enrichment is an approach to animal housing that includes both social and non-social stimuli that are provided to foster the expression of species-typical behaviors and reduce the occurrence of abnormal behaviors.[8] It is a component of animal husbandry that is designed to enhance the quality of captive animal care and improve animal welfare by identifying and providing the environmental stimuli necessary for optimal behavioral, physiological and physical well-being. In practice, this covers a multitude of innovative and imaginative techniques, devices, and practices aimed at offering the animals the opportunity to express species-appropriate behaviors in a safe and controlled manner.[9]

Care should be taken to enrich the physical environment in the primary enclosure by providing means of expressing only non-injurious species-typical activities and ensuring that any enrichment provided does not pose the potential for harm to the animal.[10] Species differences should be considered when determining the types or methods of enrichment, and thus a basic understanding of the normal behavior of the species in question is requisite to a successful enrichment program. Examples of environmental enrichments include perches, swings, mirrors, nesting material, hiding places, and other cage complexities; providing objects to manipulate; varied food items; social opportunities; using foraging or task-oriented feeding methods; providing control over some aspect of the environment, such as visual access to other animals, music, or video; and providing interaction with the caregiver or other familiar and knowledgeable person consistent with personnel safety precautions. Of course, provision of the presumed enrichment must be well thought out and with a sound knowledge of the species normal behavior, constraints imposed by the research, safety considerations, and the caregiver's level of experience.[11]

Surv. How does your IACUC interpret the term "laboratory animal enrichment"? Multiple answers were acceptable for this question.

Many respondents defined enrichment as more than one of these possible answers, thereby indicating good familiarity with the multiple dimensions of enrichment. Respondents placed relatively equal emphasis on defining laboratory animal enrichment as care practices that had an end goal of improving the animal's welfare by promoting the physical and psychological well-being of the animals or facilitating the expression of species-typical behaviors and those that had a more proximate definition of the term as the inclusion of items or structures in the cage. Further, many respondents agreed that enrichment should aid in reducing abnormal behavior. The following data reflect the distribution of interpretations of the term "laboratory animal enrichment" from the 294 unique respondents to the question.

- Not applicable 13/294 (4.4%)
- Care practices that promote physical and psychological well-being 245/294 (88.3%)
- Care practices that promote species-typical behavior 235/294 (79.9%)
- Care practices that provide caging complexity 113/294 (38.4%)
- Care practices that provide objects to the animals (toys, nesting material, perch, etc.) 230/294 (78.2%)
- Care practices that are aimed at decreasing aberrant behavior 166/294 (56.5%)
- I don't know 5/294 (1.7%)
- Other 0/294 (0%)

28:2 Which species does your IACUC typically include in its laboratory animal enrichment program?

Reg. The AWAR (§3.8) require that research institutions "develop, document, and follow an appropriate plan to provide dogs with the opportunity for exercise" through group housing in cages, pens or runs that provide at least 100% of the required space for each dog, by maintaining singly-housed dogs in twice the minimum floor space, or by providing access to a run or open area at a specified frequency and duration. The AWARs (§3.81) also specify that a research institution must "develop, document, and follow an appropriate plan for environment enhancement adequate to promote the psychological well-being of nonhuman primates." The regulations stipulate that the plan must address social housing, environmental enrichment (such as cage complexities, food items, human interaction), special considerations (e.g., infant and young juvenile animals, animals showing signs of psychological distress, animals whose activity is restricted, primates unable to see and hear other animals of their own or compatible species, and great apes over 110 pounds), restraint devices and exemptions.

The *Guide*, which applies to all vertebrate animals (p. 2), describes possible enrichment techniques for a variety of animals, ranging from rodents to nonhuman primates. The *Guide* (p. 4) establishes the Three Rs as the foundation of the recommendations contained in the document and notes that Refinement includes modifications to husbandry procedures (p. 5). These two recommendations, taken together, suggest that the *Guide* endorses providing all animals with species-relevant enrichment.

Opin. The 1996 version of the *Guide*[2] very clearly articulated different types of enrichment that could be provided to the diverse range of laboratory animals in use across the country, and thus by implication promoted the use of enrichment for species in addition to those that were highlighted in the AWARs. In the intervening 15 years between the publication of the 7th and 8th editions of the *Guide*, numerous publications have described different methods of providing enrichment to various species of laboratory animals,[12–15] and thus enrichment has become more "mainstream" in its application beyond nonhuman primates, such that it is now generally considered an integral component of animal husbandry. Although possible negative, or at the least unanticipated, consequences of providing enrichment have long been noted in the literature,[10,16–18] more recently the wholesale provision of enrichment to rodents has engendered concern among some scientists.[19] Yet, the wealth of literature offering evidence of the beneficial effects of some enrichment techniques provides a strong counterpoint to the argument against rodent enrichment.[20–23] Therefore, it is likely that the trend toward encompassing an ever-expanding number of species in the enrichment program, supported by scientific investigation into the effects of that enrichment, will continue.

Surv. Which species does your IACUC typically include in its laboratory animal enrichment program? Multiple answers were acceptable for this question.

The survey results support that animals (nonhuman primates) for which enrichment is required are, for the most part, receiving it. Due to the limitations of the survey questions, the explanation for the approximately 25% of respondents who indicated they did not provide enrichment to nonhuman primates is not available. However, primates may be exempted from the enrichment program for reasons of health or condition, or in consideration of their well-being (see 28:4). There are also potential scientific reasons for such an exemption, and indeed some comments provided suggest this was the case, such as concern regarding interference with toxicology studies and exemptions based on requests to conform with National Toxicology Program (NTP) requirements. Therefore, the survey data likely reflect a portion of the respondents' animals that have been legitimately exempted from the enrichment program. Of particular note is the high percentage of respondents that provide enrichment to rabbits, rats and mice. The provision of enrichment to rodents has elicited considerable discussion in the laboratory animal medicine and science community, and it appears that most institutions provide some form of enrichment to these animals. Interestingly, just over half of the respondents with agricultural animals provide enrichment to those species and less than half of respondents provide enrichment to aquatic animals. The following data reflect the distribution of answers to the question from the unique respondents for each species of animal:

- Nonhuman primates 93/124 (75.0%)
- Dogs 100/129 (77.5%)
- Rabbits 140/166 (84.3%)
- Rats 193/216 (89.3%)
- Mice 198/222 (89.2%)
- Agricultural animals 62/113 (54.9%)
- Fish 45/125 (36.0%)
- Amphibians 49/120 (40.8%)
- Other 49/79 (62.0%)

28:3 How is your enrichment program funded?

Opin. Since the provision of an enrichment program is mandated by federal law (see 28:4) for some species, it is critical that adequate resources be identified to ensure compliance with this institutional responsibility. An ongoing budgetary commitment to the enrichment program will allow for the program to evolve as new information becomes available and to stay dynamic. Inadequate funding can cause a program to be ineffective and not meet regulatory requirements or recommendations of the *Guide*. The NIH states in the manual, *Institutional Administrator's Manual for Laboratory Animal Care and Use*[24] (p. 13) that "sustained and visible support from institutional officials is absolutely essential to establishing and maintaining a high quality animal care and use program.... They can assure sufficient monetary and personnel resources are allocated to the institution's program." The enrichment program should not be viewed as separate or an "extra" program but rather as an integral component of the entire animal care and use program. In addition to funding enrichment through per diem charges, institutional personnel often implement low cost and innovative approaches to providing environmental enrichment.

Surv. How is your enrichment program funded? Multiple answers were acceptable for this question.

Survey responses indicated that the enrichment program is primarily funded out of the animal facility budget. Slightly more than one-third of the responses indicated that the enrichment program was funded out of the animal facility's per diem charge, while just under one-quarter of the responses indicated that enrichment was paid for out of the animal facility budget, but not as a part of the per diem charge. Other sources of funding for the enrichment program were quite minor. Figures in parentheses indicate percent of all respondents who selected the response.

- Not applicable; we do not have an enrichment program 62/292 (21.2%)
- From the animal facility budget, but independent of a per diem charge 72/292 (24.6%)
- From the animal facility budget, as part of the per diem charge 115/292 (39.4%)
- From the IACUC's budget 2/292 (0.7%)
- From an independent institutional budget 13/292 (4.4%)

- No specific budget; enrichments are donated or collected by the animal care staff — 20/292 (6.8%)
- I don't know — 14/292 (4.7%)
- Other — 13/292 (4.4%)

28:4 If animals are exempt from all or part of your enrichment program, what are the criteria used to justify the exemption?

Reg. The AWARs stipulate that nonhuman primates may be exempted from the enrichment program and dogs may be exempted from the exercise program for scientific, health or behavioral reasons (§3.81,e and §3.8,d respectively). Specific examples offered as acceptable justification for exempting a primate from social housing include overly aggressive behavior/incompatibility, a contagious disease, debilitation because of the age of the animal, or other conditions such as arthritis. Records of exemptions must be maintained by the research facility and must be made available to USDA officials or officials of any pertinent funding Federal agency upon request (AWAR §3.81,e,3; §3.8,d,3).

PHS Policy (II; IV,A,1) requires institutions to comply with the AWAR and the *Guide*. Additional guidance for the social housing requirement of nonhuman primates is offered by the NIH/OLAW in the context of NIH/OLAW FAQ F.14,[25] which states "Exemptions to the social housing requirement must be based on strong scientific justification approved by the IACUC or for a specific veterinary or behavioral reason. Lack of appropriate caging does not constitute an acceptable justification for exemption."

Opin. The management of an enrichment program includes, of necessity, provisions for allowing exemptions for select animals from participating in aspects of the program. Exemptions may be based on concerns related to the health of the animal (either related to the experiment or due to natural causes) or the scientific goals of the study. Exemptions from an enrichment program should not be considered as "all or nothing" participation by the animal. An animal may be exempted from only a portion of the program, while still enjoying the benefits of other elements of the program (e.g., the animal may be precluded from social housing but offered a variety of nonsocial enrichments, such as food treats or toys). For example, an animal's role in research involving infectious diseases, atypical rearing conditions, physical restraint, surgery, pain, substance abuse, or aggression should not be excluded *a priori* from the enrichment program.[26] Rather, protocols which may pose restrictive environments for animals should be re-evaluated periodically to determine if new technologies may be available that could reduce the restrictions placed on the animal.

Four criteria are proposed for assessing animal well-being in addition to general physical health:[21]

- The animal's ability to cope effectively with environmental changes.
- The animal's ability to engage in beneficial species-typical behavior.
- The absence of maladaptive or pathological behavior.
- The presence of a balanced temperament.

When evaluating an animal for an exemption based on health or behavior, it is important to understand fully what is "normal" for an animal in that condition. For example, an aged animal frequently has reduced perceptual and locomotor capabilities[27] and arthritis has been shown to alter behavior.[28] Thus, the enrichment program may need to be modified to accommodate limitations imposed by the health or condition of the animal.

Surv. If animals are exempt from all or part of your enrichment program, what are the criteria used to justify the exemption? Multiple answers were acceptable for this question.

Although more than 15% of the responses indicated that the institution did not have an enrichment program in place. The connotation of there being a "program" in place appeared to have been interpreted differently from an unstructured, less formal manner of providing enrichment, as exemplified by a comment that enrichment was provided on a case-by-case basis. In addition, a commenter indicated a lack of understanding as to what type of enrichment should be provided to certain species (e.g., zebrafish and amphibians). Some responders indicated that enrichment was not mandatory or it was managed at the discretion of the investigator. However, the majority of responses confirmed that a research, animal health or behavioral contraindication was the reason for an exemption to be granted, with the majority of exemptions based on a scientific justification. Figures in parentheses indicate percent of all respondents who selected the response.

- We do not have an enrichment program 45/292 (15.4%)
- There are no exemptions from our enrichment program 56/292 (19.2%)
- Scientific justification based on research needs 172/292 (58.9%)
- Medical exemption based on the determination of a veterinarian 111/292 (38.0%)
- Behavioral exemption based on incompatibility with group housing 119/292 (40.7%)
- I don't know 10/292 (3.4%)
- Other 10/292 (3.4%)

28:5 If exemptions for enrichment activities are allowed, who makes that decision?

Reg. The AWAR (§3.81,e,1) state that "the attending veterinarian may exempt an individual nonhuman primate from participation in the environment enhancement plan because of its health or condition, or in consideration of its well-being. The basis of the exemption must be recorded by the attending veterinarian for each exempted nonhuman primate. Unless the basis for the exemption is a permanent condition, the exemption must be reviewed at least every 30 days by the attending veterinarian." The IACUC also may exempt an individual nonhuman primate for scientific reasons (§3.81,e,2). The AWARs (§3.8,d,1; §3.8,d, 2) apply the same animal health criteria for exemptions to the canine exercise program made by the AV as well as "scientific reasons set forth in the research proposal that it is inappropriate for certain dogs to exercise" which must be IACUC approved and reviewed periodically.

In addition, the *Guide* strongly emphasizes social housing of social animals, but recognizes that an animal may need to be housed individually. The *Guide* (p. 64) states that in such cases the "need for single housing should be reviewed on a regular basis by the IACUC and veterinarian."

Opin. It is beneficial to use input from a team of knowledgeable personnel at the institution to make determinations of exemptions for health or behavioral reasons. Behavioral staff, the technicians that work daily with the animal(s) of concern, and the research staff can offer valuable perspective to the veterinarian or IACUC regarding changes observed in the animal's activities and the potential significance of those changes (e.g., compromised access to food or water due to other animals in the social group, the animal is withdrawn/depressed, increased observations of aggression, etc.). Different insights to the situation may shed light on nuances in management practices that may be contributing to the issue, possibly yield information regarding a previously undetected pattern of concerns, and yield innovative solutions to address the issue.

Surv. If exemptions for enrichment activities are allowed, who makes that decision? Multiple answers were acceptable for this question.

In the majority of cases, the IACUC or the AV was responsible for deciding that an animal(s) would be exempted from the enrichment program. Fifteen percent of the respondents attributed this decision to animal care technicians. A small number of commenters noted that the PI made the determination to exempt an animal from the enrichment program (presumably for scientific reasons), and a few others indicated that an enrichment specialist or behaviorist made the determination for those cases that had a behavioral basis of concern. Figures in parentheses indicate the percent of all respondents who selected the response.

- Not applicable; we do not have an enrichment program 46/280 (16.4%)
- Usually the IACUC 163/280 (58.2%)
- Usually a subcommittee of the IACUC, with or without a veterinarian's input, as needed 9/280 (3.2%)
- The Attending Veterinarian 102/280 (36.4%)
- Usually animal care technicians with veterinary input, if needed 42/280 (15.0%)
- I don't know 5/280 (1.8%)
- Other 12/280 (4.3%)

28:6 Should an institution have dedicated enrichment personnel?

Opin. For the enrichment program to be implemented in a coordinated and efficient manner, there should be a responsible party. If an individual is made responsible for the program, that person should seek input from the scientific, veterinary, and animal care staffs. In the broad sense, dedicated enrichment personnel may be those individuals at the cage/pen level providing and monitoring each animal's enrichment. But, personnel may also have a portion of their duties dedicated to enrichment vis-à-vis their role on an enrichment oversight committee (which may be a subcommittee of the IACUC or a stand-alone committee that reports to the IACUC). A committee that is comprised of various categories of personnel benefits from the diversity of perspective requisite for the program to be successful.[29] A committee may gain the added benefit of greater involvement and vesting in the enrichment program by committee members if they not only assist in developing the enrichment program but participate in keeping it current and vital.

In the absence of a dedicated person or committee, the monitoring of the enrichment program should be conducted by different constituents of the animal care and use program. For example, implementation of the enrichment program should be assessed by the IACUC during the semiannual facility inspections and the scope of the program should (must, if using nonhuman primates regulated under the AWA) be reviewed by the IACUC during semiannual program reviews. The animal care staff and AV should be assessing the success of the program periodically. This assessment should include whether animals use the enrichments, whether the enrichments are safe for the animals, and whether the enrichments result in improved well-being of the animals. Outside consultants are often retained to assist in either establishing this system of monitoring or periodically conducting the assessment.

Surv. Does your institution have dedicated enrichment personnel? Multiple answers were acceptable for this question.

A majority of responses indicated that the institution did not employ dedicated enrichment staff, with less than 10% indicating that dedicated enrichment staff implemented the enrichment and only about 6% indicating that dedicated enrichment personnel were involved in programmatic decisions. Rather, this program component was more generally implemented by the husbandry staff and embedded in the daily scope of work in providing essential animal care services. Figures in parentheses indicate percent of all respondents who selected the response.

- We do not have dedicated enrichment personnel 205/294 (69.7%)
- Our animal facility has dedicated enrichment personnel that make programmatic decisions 18/294 (6.1%)
- Our animal facility has dedicated enrichment personnel that carry out enrichment initiatives 28/294 (9.5%)
- Providing enrichment is part of the husbandry staff's job description 85/294 (28.9%)
- I don't know 5/294 (1.7%)
- Other 2/294 (0.7%)

28:7 What human resources are allocated for your enrichment program?

Reg. The *Guide* (p. 53) states that "Enrichment programs should be reviewed by the IACUC, researchers, and veterinarian on a regular basis to ensure that they are beneficial to animal well-being and consistent with the goals of animal use. They should be updated as needed to ensure that they reflect current knowledge."

Opin. The way in which institutions allocate human resources for the implementation of the enrichment program often varies with the size of the institution, the species used at the institution, and whether it is an academic institution. For example, large programs may have the funds to employ a staff member dedicated to the behavior and enrichment program (e.g., primate research centers), whereas smaller programs may exclusively rely on the animal care staff to create, implement, monitor and update the enrichment program. Institutions that use only rodents are more likely to rely on husbandry staff to manage the enrichment program. Academic institutions have the opportunity to involve students in the support of aspects of the program. Regardless of the type of individual who manages the program, the IACUC should (must, if using nonhuman primates regulated under the AWA) ensure that he or she is adequately qualified through training or experience. In addition, if multiple individuals are involved in implementing the program, care should be taken that standard operating procedures or other institutional guidance documents are followed to maintain a high level of consistency in the delivery of the program across the institution.

Surv. What human resources are allocated for your enrichment program? Multiple answers were acceptable for this question.

The greatest number of responses indicated that the specific technician caring for the animals requiring enrichment provides the enrichment for those animals. Less than 10% of the responses indicated that a specific person was employed to implement the enrichment program; a slightly higher percent of responses (approximately 19%) indicated that the responsibility was rotated among the animal care or veterinary staff. Very few institutions use students or outside volunteers to assist with the program. Figures in parentheses indicate percent of all respondents who selected the response.

- Not applicable; we do not have an enrichment program — 61/293 (20.8%)
- We have one or more persons specifically hired to implement the animal enrichment program — 25/293 (8.5%)
- We rotate responsibilities among our animal care or veterinary staff — 55/293 (18.8%)
- The person caring for the particular animals provides the enrichment for those animals — 166/293 (56.6%)
- We use students whenever we can — 14/293 (4.8%)
- We use outside volunteers whenever we can — 1/293 (0.3%)
- I don't know — 15/293 (5.1%)
- Other — 11/293 (3.7%)

28:8 Who receives training regarding environmental enrichment at your institution?

Reg. The AWAR (§2.32,b; §2.32,c) require training and instruction of personnel to include "humane methods of animal maintenance." The PHS Policy (IV,A,1,g) stipulates inclusion of a "synopsis of training or instruction in the humane practice of animal care and use" in the NIH/OLAW Animal Welfare Assurance document. Personnel qualifications and training also are emphasized in the *Guide* (p. 16), which states that "Personnel caring for animals should be appropriately trained ... and the institution should provide for formal and/or on-the-job training to facilitate effective implementation of the Program and the humane care and use of animals." The *Guide* (p. 53) also notes that, "Personnel responsible for animal care and husbandry should receive training in the behavioral biology of the species they work with to appropriately monitor the effects of enrichment as well as identify the development of adverse or abnormal behaviors." NIH/OLAW emphasizes in its FAQ F.14[25] pertaining to social housing that "Staff performing nonhuman primate socialization should be trained and competent in the procedure and knowledgeable about the animals."

Opin. The value of providing training on enrichment and the recognition of normal and abnormal behavior may vary with the species used at the institution because of the inherent complexities of observing behavioral changes and of providing enrichment to nonhuman primates versus rats and mice. Regarding nonhuman primate enrichment, the National Research Council[26] notes (p. 44) that the success of an enrichment program is dependent on personnel having knowledge of and experience with nonhuman primate behavior, and that "periodic training of staff to acquaint them with advances in the field is essential." Safety training also may be considered an essential topic in a behavioral management training program (pp. 19–20).[30] Hazards to staff associated with forming or separating social pairs or groups of nonhuman primates due to possible resulting aggression, or associated with animals reaching out of cages to grab (and possibly scratch) personnel as they attach or remove enrichment items affixed to the cage front should be addressed with relevant personnel. Training should be provided to the variety of personnel at the institution involved in implementing and monitoring the enrichments. This could include research staff, veterinary medical personnel, husbandry staff and members of the IACUC. Because the IACUC is responsible for oversight of the enrichment program and is evaluating it during the semiannual program reviews and facility inspections, it is important that Committee members receive training so that they can adequately assess the success of the program during these activities.

Surv. Who receives training regarding environmental enrichment at your institution? Multiple answers were acceptable for this question.

The majority of responses clustered around two possible answers: (1) no specific training pertaining to enrichment is provided, or (2) training is provided to the husbandry staff. As frequently the enrichment training is provided by the veterinarian and other senior program personnel, it is not surprising that approximately one-third of respondents indicated that the veterinary staff received specific training on enrichment. Of note, only one-fifth of respondents indicated that IACUC members received training in enrichment. Training at conferences, seminars and online (such as those available through the AALAS Learning Library[31] or certification preparatory material) should all be considered. Figures in parentheses indicate percent of all respondents who selected the response.

- We do not provide any specific training on environmental enrichment — 115/292 (39.4%)
- Training is provided to IACUC members — 61/292 (20.9%)
- Training is provided to Principle Investigators and their research staff — 73/292 (25.0%)
- Training is provided to animal care staff — 140/292 (47.9%)
- Training is provided to the veterinary medical staff — 90/292 (30.8%)
- Specific enrichment personnel are hired with prior enrichment experience — 11/292 (3.8%)
- I don't know — 19/292 (6.5%)
- Other — 4/292 (1.4%)

28:9 What is the most commonly used form of training regarding environmental enrichment offered to your staff?

Opin. Seminars at local, regional, or national meetings; scientific publications, autotutorial DVDs, webinars; consultants who provide training at the institution; and on-the-job training are good means for providing staff training.

Surv. What is the most commonly used form of training regarding environmental enrichment offered to your staff? Multiple answers were acceptable for this question.

Although the greatest number of responses indicated that training was embedded in the SOPs associated with implementing the enrichment program, respondents were also divided among the use of online resources or that training resources regarding enrichment were not applicable. A slightly lower percent indicated that personnel were sent to off-site training. Fewer than 20% of the respondents indicated that formal training sessions were provided in-house or that print resources were made available on the topic. Very few institutions turned to outside experts to provide training. Figures in parentheses indicate percent of all respondents who selected the response.

- Not applicable 85/292 (29.1%)
- We provide formal training sessions in-house 53/292 (18.1%)
- We send people to various conferences, workshops, etc. that entirely or in part address enrichment topics 64/292 (21.9%)
- We use on-line resources (e.g., webinars, electronic forums, etc.) 86/292 (29.4%)
- We use outside experts to provide training 16/292 (5.5%)
- Training is intrinsic in our Standard Operating Procedures that pertain to enrichment 91/292 (31.2%)
- We provide books and journals that include enrichment topics 51/292 (17.5%)
- I don't know 35/292 (12.0%)
- Other 11/292 (3.8%)

28:10 Are your institution's enrichment program and associated records reviewed by the IACUC as part of the semiannual program review?

Reg. The AWARs (§2.31,c,1) state that the IACUC shall "review at least once every six months the program for humane care and use of animals using Title 9, Chapter I, subchapter A-Animal Welfare, as the basis for evaluation." This reference includes the sections pertaining to dog exercise and environmental enhancement to promote psychological well-being. The *Guide* (p. 25) ascribes to the IACUC the responsibility for "oversight and evaluation of the entire Program and its components as described in other sections of the *Guide*," which includes the environmental enrichment and social housing sections. Specifically, the *Guide* (p. 53) states that "Enrichment programs should be reviewed by the IACUC, researchers, and veterinarian on a regular basis to ensure that they are beneficial to animal well-being and consistent with the goals of animal use." The PHS Policy states (IV,B,1) that the

IACUC must "review at least once every six months the institution's program for humane care and use of animals, using the *Guide* as a basis for evaluation."

Opin. In accordance with the AWAR, institutions that have an enrichment plan for nonhuman primates and/or a canine exercise program must review those aspects of the animal care and use program semiannually. Similarly, if the institution holds a PHS Assurance or is accredited by AAALAC, the behavioral management program described in the *Guide* should be reviewed as part of the semiannual program review. Some elements of these programs may also be evaluated during the facility inspections, for example assessing the degree of implementation of the plan, how consistently it is implemented across the institution, whether the implementation conforms to institutional guidance, and gauging its effectiveness in terms of the animals' behavior or other measures of well-being. The IACUC has a responsibility to review the animal care and use program semiannually and to ensure its completeness, appropriateness, and currency.

Surv. Are your institution's enrichment program and associated records reviewed by the IACUC as part of the semiannual program review?

Approximately one-third of the respondents indicated that the IACUC reviews the enrichment program as part of the semiannual program review, and approximately one-quarter of respondents stated they don't maintain enrichment records that the IACUC could review and approximately another 10% stated the IACUC doesn't review the enrichment program or enrichment records during the semiannual program review.

- Not applicable; we do not have an enrichment program — 62/290 (21.4%)
- No, because we do not keep formal records for the IACUC to review — 75/290 (25.9%)
- No, the IACUC does not review the enrichment program or records during its semiannual program review — 38/290 (13.1%)
- Yes, the IACUC does review the enrichment program and records during its semiannual program review — 99/290 (34.1%)
- I don't know — 9/290 (3.1%)
- Other — 7/290 (2.4%)

28:11 How is the efficacy of your enrichment program assessed (e.g., animal usage, rate or occurrence of aberrant behavior, etc.)?

Reg. The AWAR (§2.31,c,1) and PHS Policy (IV,B,1) require that the IACUC review at least once every 6 months the program for humane care and use of animals. The *Guide* (p. 9) states that the IACUC has oversight and evaluation responsibility for the animal care and use program and the components of the program described in the *Guide*. The *Guide* (p. 25) also recommends program review and facilities inspections occur at least annually, or more often as required.

Opin. Morgan et al.[32] reviewed various methods of assessing enrichment programs. Methods included reliance on others to inform the IACUC, common sense, and empiricism. They noted that a significant disadvantage of reliance on others is that acceptance is often based on the authority of the other person (the tendency to accept that a point of information is true simply because it is that of an authority figure). They suggested that common sense is a good starting point for evaluating enrichment, although its main disadvantage is that common sense is grounded on individual interpretations and biases. However, they strongly argued that empirical testing of enrichment strategies is necessary to validate their use both from the standpoint of the value to the animal and for judicious expenditure of limited resources.

An assessment of the enrichment program has the greatest relevancy when it is considered in the context of the welfare of the laboratory animal. Specifically, the question should be asked whether the enrichment used (within any constraints imposed by the research goals) is fostering laboratory animal welfare or is it of null value to the animal. This question broadens the scope of the evaluation of the enrichment program to more properly focus on the animal rather than the enrichment itself, such as use of the toy, consumption of a food treat, etc. For many laboratory animal species, a reduction in abnormal behaviors (e.g., obsessive-compulsive behaviors) or an increase in the breadth of expression of normal behaviors (e.g., nest building, foraging, etc.) are indicators of improved well-being and the literature is rich with reports of ways to achieve these for numerous species. Also, on a pragmatic level, expenditure of limited fiscal resources for enrichment methods that are not achieving the goal of improving laboratory animal welfare is an unsound management practice.

Surv. How is the efficacy of your enrichment program assessed (e.g., animal usage, rate or occurrence of aberrant behavior, etc.)? Multiple answers were acceptable for this question.

For those institutions that had an enrichment program, the highest number reported that they depend on the veterinary or husbandry staff to report problems to the IACUC. Reliance on the veterinary and animal care staff presumes that there are established metrics that are known and consistently applied by personnel to validate the enrichment practice. Thirteen percent relied on AAALAC to evaluate efficacy of the program during the on-site assessment by the AAALAC representatives. More than a third of responders noted that there is no true assessment of the efficacy of the enrichment program. Slightly more than a quarter of respondents indicated that the IACUC is involved to some degree in assessing the efficacy of the enrichment program. Figures in parentheses indicate percent of all respondents who selected the response.

- Not applicable; we do not have an enrichment program 62/294 (21.1%)
- We have no true assessment of the efficacy of our program 99/294 (33.7%)
- The IACUC does their best to assess the program during the semiannual program review 80/294 (27.2%)
- We consider the program successful if AAALAC is pleased with it during its site visit 38/294 (12.9%)
- We depend on the veterinary medical or animal care staff to advise the IACUC of any problems with the enrichment 120/294 (40.8%)
- I don't know 12/294 (4.1%)
- Other 4/294 (1.4%)

28:12 What information is maintained in your enrichment records?

Reg. The AWAR (§3.81) require that dealers, exhibitors, and research facilities develop, document, and follow an environment enhancement plan that is adequate to promote the psychological well-being of nonhuman primates. The plan must include specific provisions to address the social needs of nonhuman primates that are social in nature, a description of the environmental enrichment provided, and special considerations (i.e., those primates requiring special attention). They further require that records of exemptions of animals from the plan be maintained (AWAR §3.81,e,3) and be made available to USDA officials or representatives of federal funding agencies upon request. A similar plan must be developed to provide dogs with the opportunity for exercise and must include written standard procedures to be followed in providing the opportunity for exercise. It must address dogs housed individually, dogs housed in groups, methods and period of providing the opportunity to exercise and exemptions (AWAR §3.8).

PHS Policy (II; IV,A,1) requires institutions to comply with the AWAR as applicable and to follow the *Guide*. The *Guide* (p. 75) notes that records of rearing and housing histories are useful for the management of many species. Institutions must keep records of departures from the *Guide* and IACUC approval of any departures (PHS Policy IV,B,3).

Opin. Documentation not only serves as a verifiable record of enrichment and socialization activities, but also validates the effectiveness of ongoing enrichments or provides evidence of the need to modify the enrichment and socialization program. The National Research Council[26] suggests (p. 26) that nonhuman primate enrichment programs should include a mechanism whereby there are "protocols for diagnosing the cause of physical impairments and abnormal behavior, determining when remediation is necessary, developing remediation plans, assessing the effectiveness of remediation, and maintaining appropriate records."

Surv. What information is maintained in your enrichment records? Multiple answers were acceptable for this question.

More than half of respondents indicated that enrichment records were not kept. This was the largest percentage for any of the possible answers to the question. A range of from approximately 20%–26% of respondents replied that records of rotation schedules of enrichments, enrichment use, exemptions, abnormal behavior, or information regarding social housing were maintained. Figures in parentheses indicate percent of all respondents who selected the response.

- We do not keep enrichment records 159/292 (54.4%)
- Rotation schedule of manipulative objects used for enrichment 76/292 (26.0%)
- Food treat usage 71/292 (24.3%)
- Observed usage and/or effects of the enrichment 48/292 (16.4%)
- Exemptions from the enrichment program 67/292 (22.9%)
- Information on animal behavior (normal or if aberrant behaviors are observed) 64/292 (21.9%)
- Details on social (pair or group) housing 66/292 (22.6%)
- I don't know 28/292 (9.6%)
- Other 7/292 (2.4%)

28:13 Are your investigators concerned that enrichment may introduce a scientific variable into their work?

Reg. The *Guide* (p. 54) acknowledges that "Some scientists have raised concerns that environmental enrichment may compromise experimental standardization by introducing variability, adding not only diversity to the animals' behavioral repertoire but also variation to their responses to experimental treatments...," but that "A systematic study in mice did not find evidence to support this viewpoint (Wolfer et al. 2004), indicating that housing conditions can be enriched without compromising the precision or reproducibility of experimental results." The *Guide* (p. 54) notes that "Further research in other species may be needed to confirm this conclusion." But to strengthen the case for providing enrichment, the *Guide* (p. 54) concludes the discussion by stating, "However, it has been shown that conditions resulting in higher-stress reactivity increase variation in experimental data (e.g., Macrì et al. 2007). Because adequate environmental enrichment may reduce anxiety and stress reactivity (Chapillon et al. 1999), it may also contribute to higher test sensitivity and reduced animal use (Baumans 1997)."

Opin. Proponents of enrichment and scientists who work with animals living in enriched environments recognize that the presence of enrichment can influence the data generated.[10] In some species, the effects on research of providing enrichment may not be as profound as they are in other species (e.g., dogs or nonhuman primates vs. rodents). Regardless, a prudent course of action is to evaluate the type of enrichment being provided to the animal on a study-by-study basis and assess with the investigator whether it may confound research results. If so, then perhaps another type of enrichment could be considered, or the animal may need to be exempted from the enrichment program, with approval by the IACUC. A recent review of mouse enrichment[33] illustrates the point that enrichment may influence the scientific data by introducing an independent variable, or it may facilitate scientific enquiry by stimulating new research directions. Toth et al.[19] argue in favor of standardizing the laboratory rodent's environment to maximize the sensitivity of the test and to minimize the number of animals used that will achieve appropriate power of the statistical test. They also make a strong case that the type of enrichment available to the animals in the study, and when it was made available, should be reported by the researcher in the published study. Others have argued equally persuasively that standardization can lead to inaccurate results and poor reproducibility of the data.[34] Standardization as a necessary component of behavioral phenotyping of mutant mice has been adjured;[35] but, others suggest that systematic variation of genetic and environmental backgrounds is a sound approach to augmenting the validity of study results, including behavioral phenotyping,[36] and have gone so far as to suggest that environmental standardization is a cause of poor reproducibility of research results.[37] However, even basic procedures in the animal facility, such as cage change and transporting rodents in a cage from the animal holding room to a procedure room, can affect the physiology of the animals[38] and there are numerous environmental variables such as noise, temperature, light level and others that contribute to a lack of standardization across studies.[36] Thus, it appears that as researchers examine more closely various housing and husbandry aspects of their research subjects, there are multiple entry points for variability in a study and researchers should be aware of these and assess them in the context of the specifics of the study.

Surv. Are your investigators concerned that enrichment may introduce a scientific variable into their work?

If only responses from institutions that had an enrichment program are examined (and excluding those that replied they did not know if there was a concern or "other"), then the responses (n = 232) are almost evenly divided between those who stated investigators had not expressed a concern and those who stated their investigators had expressed a concern (47.8% vs. 52.2%), with a slightly higher percentage indicating that a concern had been expressed (ranging from infrequently to high frequency) that enrichment was introducing a variable into their research.

- Not applicable; we do not have an enrichment program 51/294 (17.4%)
- No, this has not been a problem 111/294 (37.8%)
- Yes, but infrequently 101/294 (34.4%)
- Yes, with moderate to high frequency 20/294 (6.8%)
- I don't know 9/294 (3.1%)
- Other 2/294 (0.7%)

28:14 If your investigators have concerns about using animal enrichment, what are their objections?

Reg. The *Guide* (p. 53) states "Not every item added to the animals' environment benefits their well-being." Enrichment can serve as a vector for disease transmission, pose a risk of injury if ingested, or in the case of food enrichment (treats) can lead to obesity in animal models.[39] In addition, some specific enrichment items have been reported to increase stress in certain species. For example, marbles are used as stressors in mouse anxiety studies.[40] The *Guide* (p. 53) elaborates on the potential problems of providing enrichment in some cases, "In some strains of mice, cage dividers and shelters have induced overt aggression in groups of males, resulting in social stress and injury (Bergmann et al. 1994; Haemisch et al. 1997)." Emphasis is placed on the IACUC, researchers and veterinarians to ensure that the environmental enrichment provided are safe and beneficial to the animals' well-being and consistent with the animal-use goals.

Opin. Baumans[41] has suggested that with the provision of enrichment, the mouse can exhibit more species-appropriate behavior, and thus may be able to better cope with unexpected changes to its environment and respond more uniformly to different challenges. Abnormal behavior, stress, fear and anxiety, and impaired thermoregulation are potentially confounding variables that may adversely affect the outcome of animal-based studies, and consequently increase variation in the data. However, neurological, physiological and immunological effects have been described in rodents and other laboratory animal species offered different types of enrichment (for example, see a brief review of the effects on mice by Bayne and Würbel).[33] Therefore, it behooves investigators and IACUCs to be aware of the potential influence enrichment, or lack thereof, may have on the study animals and thus the resulting data.

Concern about possible harm to the animal has merit as reports of injury to animals[42–45] or death[16,46] resulting from various social or non-social types of enrichment appear in the literature despite the clear intent of offering these enrichments as a means to enhance animal well-being. In addition, Bayne et al.[17] observed that

Kong® toys presented to nonhuman primates were not completely sanitized after going through a mechanical cage washer; therefore, some attention should be given to how toys are distributed across a primate colony, and further evaluation of the efficacy of sanitation of enrichment devices should be undertaken.

Despite these cautions, there is a large body of evidence to support the many positive effects of enrichment on promoting the welfare of numerous species of laboratory animals, when the enrichment technique is chosen carefully and the entire enrichment program is monitored appropriately. The introduction of new enrichment methodologies should be done carefully, and possibly incrementally, to assess any possible negative consequences associated with its use. Additionally, if one form of enrichment is precluded due to the nature of the study, alternative enrichment devices or techniques should be considered before removing the animal(s) from the enrichment program completely. Publishing information regarding the types of enrichment used successfully in various experimental paradigms is encouraged to further refine enrichment programs across institutions.

Surv. If your investigators have concerns about using animal enrichment, what are their objections? Multiple answers were acceptable for this question.

One-half of the respondents indicated that investigators did not have concerns about using enrichment. The next most common response (37.0%) was related to a potential unknown effect on the research data, followed by the concern that *any* change in husbandry will introduce an experimental variable. Concern that there would be a negative impact on the animal through injury or other undesirable consequence collectively accounted for approximately 28% of concerns. Issues related to cost, hygiene and ease of access or visibility collectively accounted for 14.8% of expressed concerns. Figures in parentheses indicate percent of all respondents who selected the response.

- Not applicable 146/292 (50.0%)
- Concern that there will be an unforeseen effect on the scientific data 108/292 (37.0%)
- Concern that any change in husbandry will introduce an experimental variable 87/292 (29.8%)
- Cost 16/292 (5.5%)
- Hygienic concerns 11/292 (3.8%)
- Concern that social housing may result in injury to the animals 57/292 (19.5%)
- Concern that other harm may occur to the animal 25/292 (8.6%)
- Ease of access, visibility or handling of the animals 16/292 (5.5%)
- I don't know 11/292 (3.8%)
- Other 6/292 (2.0%)

28:15 Which of the following species do you typically house individually (when there is no scientific, veterinary medical or behavioral reason not to socially house the animals)?

Reg. The AWAR (§3.81,a) state "The environment enhancement plan must include specific provisions to address the social needs of nonhuman primates of species known to exist as social groups in nature." The AWAR allow exemption from social housing if the animal has an infectious disease, shows aggressive behavior, or is incompatible with other animals. The AWAR (§3.81,a,2) state that individually housed nonhuman primates must be able to see and hear nonhuman primates of their own or compatible species, unless the veterinarian determines that this contact would compromise the well-being of the animal.

The *Guide* (p. 64) states, "Single housing of social species should be the exception and justified based on experimental requirements or veterinary-related concerns about animal well-being. In these cases, it should be limited to the minimum period necessary, and where possible, visual, auditory, olfactory, and tactile contact with compatible conspecifics should be provided." The authors of the *Guide* recognized that social housing is not always possible for experimental, health, and behavioral reasons.

Regarding the social housing of nonhuman primates, NIH/OLAW states in FAQ F.14[25] that, "There is universal agreement among oversight agencies that nonhuman primates should be socially housed.... Single housing of social primates is considered an exception.... Exemptions to the social housing requirement must be based on strong scientific justification approved by the IACUC or for a specific veterinary or behavioral reason. Lack of appropriate caging does not constitute an acceptable justification for exemption. When necessary, single housing of social animals should be limited to the minimum period necessary."

AAALAC has issued a Position Statement[47] that "Social housing will be considered by AAALAC International as the default method of housing unless otherwise justified based on social incompatibility resulting from inappropriate behavior, veterinary concerns regarding animal well-being, or scientific necessity approved by the IACUC... The institution's policy and exceptions for single housing should be reviewed on a regular basis and approved by the IACUC... and/or veterinarian."

Opin. The provision of opportunities for social housing of compatible animals is perhaps the most effective means of environmental enrichment for nonhuman primates and other social species of laboratory animals. A compatible social partner is an ever-changing stimulus that can also provide comfort through mutual grooming and physical proximity, aid in thermoregulation, serve as a playmate, etc. Because social relationships among some species of animals, such as primates, are frequently based on hierarchical relationships that require caution when forming social pairs or groups of animals and can change with time, ongoing monitoring of the animals' compatibility is essential to prevent harm resulting from aggressive encounters. Some strains of rodents engage in barbering behavior or more overt aggression that results in wounding of cage mates, possibly due to territorial behavior expressed over an enrichment item.[48] Other strains of mice provided only nesting material as an enrichment can be housed compatibly.[23] Social housing of rabbits has recently received greater attention. Efforts to evaluate the effects of social housing on rabbits have supported the challenge associated with this type of housing. Nevalainen et al.[49] confirmed that female rabbits were compatible in a social housing setting only until sexual maturity, and Love[50] provided a thorough review of the behavioral biology of the rabbit in the wild with a well-reasoned extrapolation of approaches and obstacles to social housing these animals in laboratories. Therefore, personnel should be familiar with the behavior of the animal (at the species as well as individual level), the research use of the animals, and the possible consequences of social housing on the research objectives before commencing social housing.

Surv. Which of the following species do you typically house individually (when there is no scientific, veterinary medical or behavioral reason not to socially house the animals)? Multiple answers were acceptable for this question.

The majority of responses indicated that animals were not singly housed for any reason other than those offered in the question. Rabbits appeared to be the animal housed individually most often for an alternate reason, with a small percentage of responses noting single housing of animals occurred for a variety of other species for other reasons. Figures in parentheses indicate percent of all respondents who selected the response.

- Not applicable 154/272 (56.6%)
- Nonhuman primates 10/272 (3.7%)
- Dogs 22/272 (8.1%)
- Cats 12/272 (4.4%)
- Rabbits 55/272 (20.2%)
- Swine 24/272 (8.8%)
- Sheep or goats 9/272 (3.3%)
- Rats 27/272 (9.9%)
- Mice 20/272 (7.3%)
- Frogs 7/272 (2.6%)
- I don't know 18/272 (6.6%)
- Other 11/272 (4.0%)

28:16 If your institution usually houses nonhuman primates individually, which points below describe the rationale for doing this?

Reg. (See 28:15.)

Opin. The emphasis on social housing of nonhuman primates over the last two decades has resulted in greater acceptance of this as a routine manner of housing, with any exemption necessitating strong justification. Alternative patterns of social housing, such as part-time social housing or restricted physical contact (e.g., through the use of grooming-contact bars), allow more flexibility in meeting research goals, while still affording the animals social opportunities.

Surv. If your institution usually houses nonhuman primates individually, which points below describe the rationale for doing this? Multiple answers were acceptable for this question.

A majority of responses indicated that individual housing of nonhuman primates was not applicable. The next most frequent response was concern that social housing would result in harm to the primates through bouts of aggression. For those that specified a scientific reason for individual housing of primates, toxicology studies, use of telemetry systems, surgical manipulations, behavioral testing, and infectious disease studies were given as examples of this type of justification. Cost of social housing cages, inadequate room for social housing, personnel safety with regard to handling the animals and ease of their accessibility all received a negligible percent response rate. Figures in parentheses indicate percent of all respondents who selected the response.

- Not applicable 234/291 (80.4%)
- Social housing would likely affect study results 20/291 (6.9%)
- We are concerned about aggression and possible injury to animals 40/291 (13.7%)
- Social housing cages cost too much 0/291 (0%)
- We don't have room for social housing cages 5/291 (1.7%)
- Single housing provides easier access to animals 5/291 (1.7%)
- There is a scientific or veterinary medical reason to keep animals separated 26/291 (8.9%)
- We think single housing is safer for people having to handle the primates 3/291 (1.0%)
- I don't know 4/291 (1.4%)
- Other 4/291 (1.4%)

28:17 Does your institution supplement the animals' standard diet by providing food treats as part of an enrichment program for the following species?

Reg. With regard to nonhuman primates, the AWAR (§3.81.b) state "Examples of environmental enrichments include… varied food items; using foraging or task-oriented feeding methods…." The *Guide* (p. 67) notes that "In some species (e.g., nonhuman primates) and on some occasions, varying nutritionally balanced diets and providing "treats," including fresh fruit and vegetables, can be appropriate and improve well-being. Scattering food in the bedding or presenting part of the diet in ways that require the animals to work for it (e.g., puzzle feeders for nonhuman primates) gives the animals the opportunity to forage, which, in nature, normally accounts for a large proportion of their daily activity." The *Guide* (p. 53) offers the caution that "foraging devices can lead to increased body weight (Brent 1995)."

Opin. Creating opportunities for nonhuman primates to express foraging behavior has been proposed as a beneficial enrichment technique for many years, though the degree of benefit is variable in terms of the length of time stereotypic behavior is mitigated[51–53] or based on the type of foraging device used.[54] Food enrichment with rabbits has also been evaluated,[55] with the finding that rabbits spent more time engaged with the food enrichment than non-food enrichment items and that a specific type of food enrichment was preferred. Scattering food in a pen containing socially housed rabbits has been recommended to reduce boredom and decrease incidents of aggression.[56] Chemically defined food items are commercially available, which can be especially useful for rabbits used in toxicology studies.[57] Food enrichment has been recommended in cats[58] and dogs[59] to reduce boredom and stimulate activity that simulates natural behaviors of searching for and obtaining food. Cautions regarding monitoring these animals for weight gain and for destruction of toys and other devices in which the food items are placed are offered. Food enrichment for rodents has been suggested as a means of stimulating natural foraging behaviors and reducing boredom.[60] Novel food enrichments are also proposed for rodents,[61,62] though potential contamination of feed provided on the cage floor should be a consideration.

Surv. Does your institution supplement the animals' standard diet by providing food treats as part of an enrichment program for the following species?

Species of Animal	Supplemented	Not Supplemented	Not Housed	Housed, But I Don't Know
Nonhuman primate	75/248 (30.2%)	4/248 (1.6%)	159/248 (64.1%)	10/248 (4.0%)
Dog	66/245 (26.9%)	12/245 (4.9%)	148/245 (60.4%)	19/245 (7.8%)
Cat	42/237 (17.7%)	10/237 (4.2%)	167/237 (70.5%)	18/237 (7.6%)
Rabbit	94/249 (37.8%)	31/249 (12.5%)	95/249 (38.2%)	29/249 (11.7%)
Rat	53/257 (20.6%)	142/257 (55.3%)	31/257 (12.1%)	31/257 (12.1%)
Mouse	47/259 (18.2%)	158/259 (61.0%)	20/259 (7.7%)	34/259 (13.1%)
Agricultural animal	40/246 (16.3%)	31/246 (12.6%)	148/246 (60.2%)	27/246 (11.0%)

When the responses indicated the species was not housed at the institution, or the responder did not know the answer to the question, those data were removed from the tally and the following results regarding supplementation of food items were obtained:

Species of Animal	Supplemented
Nonhuman primate	75/79 (94.9%)
Dog	66/78 (84.6%)
Cat	42/52 (80.8%)
Rabbit	94/125 (75.2%)
Rat	53/195 (27.1%)
Mouse	47/205 (22.9%)
Agricultural animal	40/71 (56.3%)

These data suggest that, with the exception of rodents, most laboratory animals receive food enrichment of some type, with nonhuman primates most often noted as being offered food enrichment (nearing 100%), perhaps because of the rich literature on this topic for these animals.

28:18 For species that have food treats offered as part of the enrichment program, what type of food enrichment is provided?

Opin. Because many studies require a well-defined diet for the subject animals, commercially available food treats whose ingredients are well-characterized are often used. There is also a degree of convenience in purchasing food treats designed for specific laboratory animals, though cost can be a consideration.

Surv. For species that have food treats offered as part of the enrichment program, what type of food enrichment is provided? Multiple answers were acceptable for this question. Figures in parentheses indicate percent of all respondents who selected the response.

- Not applicable 117/292 (40.1%)
- Commercially available treats manufactured specifically for laboratory animals 119/292 (40.7%)
- Human food treats (e.g., popcorn, marshmallows, candy, etc.) 56/292 (19.2%)
- Human foods (cereals, fresh produce, dried fruit & nuts, meat or fish, etc.) 104/292 (35.6%)
- I don't know 27/292 (9.2%)
- Other 13/292 (4.4%)

The data suggest that responders replying that they provide food enrichment supply either a commercially available product or food used for human consumption that is nutritious rather than "junk" food.

28:19 Does your institution provide manipulanda (e.g., toys/puzzles/chains/chewing material) as part of your enrichment program?

Reg. The AWAR (§3.81.b) state "Examples of environmental enrichments include providing perches, swings, mirrors, and other increased cage complexities; providing objects to manipulate…." The *Guide* (p. 53) recommends offering manipulative activities and cognitive challenges as part of the enrichment program, specifically "manipulable resources such as novel objects and foraging devices for nonhuman primates; manipulable toys for nonhuman primates, dogs, cats, and swine; wooden chew sticks for some rodent species; and nesting material for mice."

Opin. Species for which the provision of enrichment has a broader literature base (e.g., nonhuman primates, agricultural animals, rodents) or for which the behavior of the animals is well understood (e.g., dogs and cats) appear to receive enrichment more commonly than those species for which consideration of enriching the environment is a relatively new husbandry practice (e.g., fish, amphibians). But even this trend may be changing, as the increased use of fish in research has led to more reports in the literature regarding the influence of enrichment on fish behavior. For example, studies have demonstrated more flexible behavior in fish reared in structurally complex environments, the promotion of social learning, and the expression of dominance displays that match those expressed in wild counterparts.[63–65] Of note, as has been observed with rodents, the effects (or lack of effect) of enrichment varies with the species of fish.[66] As more scientific studies are published regarding the effects of enrichment on aquatic species, husbandry practices may more routinely incorporate suitable enrichment methods for these animals as well.

The National Agricultural Library's Animal Welfare Information Center has links to several databases pertaining to enrichment techniques for a variety of laboratory animals.[67] However, it is not all-inclusive and thus literature searches for current information on enrichment for a specific species of animal (e.g., through PubMed) is an excellent adjunctive approach to designing a meaningful enrichment program that is based on the successful experience of others, with a reduced risk of harm to the animals.

Surv. Does your institution provide manipulanda (e.g., toys/puzzles/chains/chewing material) as part of your enrichment program?

Species of Animal	Provided	Not Provided	Not Housed	Housed, But I Don't Know
Nonhuman primate	79/249 (31.7%)	3/249 (1.2%)	161/249 (64.7%)	6/249 (2.4%)
Dog	79/247 (32.0%)	7/247 (2.8%)	152/247 (61.5%)	9/247 (3.6%)
Cat	53/241 (22.0%)	8/241 (3.3%)	169/241 (70.1%)	11/241 (4.6%)
Rabbit	102/253 (40.3%)	27/253 (10.7%)	96/253 (37.9%)	28/253 (11.1%)
Rat	141/267 (52.8%)	81/267 (30.3%)	26/267 (9.7%)	19/267 (7.1%)
Mouse	146/268 (54.5%)	84/268 (31.3%)	18/268 (6.7%)	20/268 (7.5%)
Other rodents	76/241 (31.5%)	42/241 (17.4%)	103/241 (42.7%	20/241 (8.3%)
Fish	8/241 (3.3%)	94/241 (39.0%)	109/241 (45.2%)	30/241 (12.5%)
Amphibians	9/236 (3.8%)	78/236 (33.1%)	124/236 (52.5%)	25/236 (10.6%)
Agricultural animal	46/244 (18.9%)	25/244 (10.3%)	148/244 (60.7%)	25/244 (10.3%)

When the responses indicated the species was not housed at the institution, or the responder did not know the answer to the question, those data were removed from the tally and the following results regarding the provision of manipulanda were obtained:

Species of Animal	Manipulanda Provided
Nonhuman primate	79/82 (96.3%)
Dog	79/86 (91.9%)
Cat	53/61 (86.9%)
Rabbit	102/129 (79.1%)
Rat	141/222 (63.5%)
Mouse	146/230 (63.5%)
Other rodents	76/118 (64.4%)
Fish	8/102 (7.8%)
Amphibians	9/87 (10.3%)
Agricultural animal	46/71 (64.8%)

The results clearly demonstrate a significant difference in the number of institutions that enrich mammalian species versus aquatic species.

28:20 Does your institution provide structural enrichment to the animals' cage/enclosure (e.g., huts, tubes, boxes, shelves, perches, beds, nesting material) as part of an enrichment program?

Reg. The AWAR (§3.81.b) recommend that primate enrichment include "providing perches, swings, mirrors, and other increased cage complexities…." The *Guide* (p. 53) states that "Examples of enrichment include structural additions such as perches and visual barriers for nonhuman primates (Novak et al. 2007); elevated shelves for cats (Overall and Dyer 2005; van den Bos and de Cock Buning 1994) and rabbits (Stauffacher 1992); and shelters for guinea pigs (Baumans 2005)." The *Guide* (p. 53) further notes that, "Well-conceived enrichment provides animals with choices and a degree of control over their environment, which allows them to better cope with environmental stressors (Newberry 1995). For example, visual barriers allow nonhuman primates to avoid social conflict; elevated shelves for rabbits and shelters for rodents allow them to retreat in case of disturbances (Baumans 1997; Chmiel and Noonan 1996; Stauffacher 1992); and nesting material and deep bedding allow mice to control their temperature and avoid cold stress during resting and sleeping (Gaskill et al. 2009; Gordon 1993, 2004)."

Opin. The commercial production of various enrichment devices for several types of research animals has increased significantly over the last several years. Many of these items increase the structural complexity of the animal's cage environment. Examples for rodents include round, square, arched or tented tunnels; huts/shacks (which are either destructible or indestructible); balls; igloos/domes (with or without a running wheel); and a variety of types of nesting material such as wood, paper and cotton. Dogs housed in runs or larger caging are often provided with raised beds. Felines housed in rooms or large caging are provided with climbing structures and shelves. Similarly, nonhuman primates are provided opportunities for perching, climbing, hiding, swinging, and numerous other activities through the use of added cage "furniture," the specific type depending on the species of primate and quantity of space available.

Surv. Does your institution provide structural enrichment to the animals' cage/enclosure (e.g., huts, tubes, boxes, shelves, perches, beds, nesting material) as part of an enrichment program?

Species of Animal	Provided	Not Provided	Not Housed	Housed, But I Don't Know
Nonhuman primate	66/250 (26.4%)	11/250 (4.4%)	166/250 (66.4%)	7/250 (2.8%)
Cat	52/240 (21.7%)	7/240 (2.9%)	173/240 (72.1%)	8/240 (3.3%)
Rabbit	79/249 (31.7%)	50/249 (20.1%)	99/249 (39.8%)	7/249 (2.8%)
Rat	186/268 (69.4%)	43/268 (16.0%)	26/268 (9.7%)	13/268 (4.9%)
Mouse	204/271 (75.3%)	39/271 (14.4%)	16/271 (5.9%)	12/271 (4.4%)
Other rodents	110/248 (44.4%)	17/248 (6.9%)	105/248 (42.3%)	16/248 (6.4%)
Fish	46/241 (19.1%)	53/241 (22.0%)	115/241 (47.7%)	27/241 (11.2%)
Amphibians	51/239 (21.3%)	34/239 (14.2%)	131/239 (54.8%)	23/239 (9.6%)
Agricultural animal	38/241 (15.8%)	29/241 (12.0%)	151/241 (62.7%)	23/241 (9.5%)

When the responses indicated the species was not housed at the institution, or the responder did not know the answer to the question, those data were removed from the tally and the following results regarding the provision of structural enrichment were obtained:

Species of Animal	Structural Enrichment Provided
Nonhuman primate	66/77 (85.7%)
Cat	52/59 (88.1%)
Rabbit	79/129 (61.2%)
Rat	186/229 (81.2%)
Mouse	204/243 (83.9%)
Other rodents	110/127 (86.6%)
Fish	46/99 (46.5%)
Amphibians	51/85 (60.0%)
Agricultural animal	38/67 (56.7%)

The results demonstrate that the use of structural complexities is quite common for many laboratory animals. Not surprisingly, these were frequently used with rodents, felines and nonhuman primates. In contrast to the responses to the question regarding the provision of manipulanda, many institutions (almost half of the responses) provide structural enrichment to aquatic animals. Overall, this appears to be a well-accepted enrichment method.

References

1. National Research Council. 1985. *Guide for the Care and Use of Laboratory Animals.* U.S. Department of Health and Human Services, Public Health Service, National Institutes of Health.

2. National Research Council. 1996. *Guide for the Care and Use of Laboratory Animals*. Washington, D.C.: National Academies Press.
3. U.S. Department of Agriculture and Office of Laboratory Animal Welfare/U.S. Public Health Service. 2005. *Enrichment for Nonhuman Primates*. http://grants.nih.gov/grants/olaw/request_publications.htm.
4. National Institutes of Health, Office of Laboratory Animal Welfare. 2012. AC and OLAW perspective on nonhuman primate enrichment and social housing. [Online seminar] http://grants.nih.gov/grants/olaw/educational_resources.htm#special-seminars.
5. U.S. Department of Agriculture, National Agricultural Library, Animal Welfare Information Center. 2012. Environmental enrichment for nonhuman primates. http://www.nal.usda.gov/awic/pubs/Primates2009/primates.shtml.
6. Committee for the Update of the Guide for the Care and Use of Laboratory Animals. 2011. *Guide for the care and use of laboratory animals*. Washington, D.C.: National Academies Press.
7. National Institutes of Health, Office of Laboratory Animal Welfare. 2012. Frequently Asked Questions: PHS Policy on Humane Care and Use of Laboratory Animals. http://grants.nih.gov/grants/olaw/faqs.htm.
8. Bayne, K., S. Dexter, H. Mainzer et al. 1992. The use of artificial turf as a foraging substrate for individually housed rhesus monkeys (*Macaca mulatta*). *Anim. Welf.* 1:39–53.
9. Fillman-Holliday, D. and M. Landi. 2002. Animal care best practices for regulatory testing. *ILAR J.* 43:49–58.
10. Bayne, K. 2005. Potential for unintended consequences of environmental enrichment for laboratory animals and research results. *ILAR J.* 46:129–139.
11. Wolfle, T.L. 1991. Psychological well-being: The billion dollar solution. In *Through the looking glass: Issues of psychological well-being in captive nonhuman primates*, ed. M.A. Novak, and A.J. Petto, 119–128. Washington, D.C.: American Psychological Association.
12. Van de Weerd, H.A. 1997. Preferences for nest material as environmental enrichment for laboratory mice. *Lab. Anim.* 31:133–143.
13. Olsson I.A. and K. Dahlborn. 2002. Improving housing conditions for laboratory mice: A review of "environmental enrichment." *Lab. Anim* 36:243–270
14. Moons, C.P., P. Van Wiele, and F.O. Odberg. 2004. To enrich or not to enrich: Providing shelter does not complicate handling in laboratory mice. *Contemp. Top. Lab. Anim. Sci.* 43:18–21.
15. Wolfer, D.P., O. Litvin, S. Morf et al. 2004. Laboratory animal welfare: Cage enrichment and mouse behaviour. *Nature* 432(7019):821–822.
16. Line, S.W., K.N. Morgan, H. Markowitz et al. 1990. Behavioral responses of female long-tailed macaques (*Macaca fascicularis*) to pair formation. *Lab. Prim. News.* 29:1–5.
17. Bayne, K, S.L. Dexter, J.L. Hurst et al. 1993. Kong® toys for laboratory primates: Are they really an enrichment or just fomites? *Lab. Anim. Sci.* 43:78–85.
18. Van Loo, P.L.P, C.L.J.J. Kruitwagen, L.F.M. Van Zutphen et al. 2000. Modulation of aggression in male mice: Influence of cage cleaning regime and scent marks. *Anim. Welfare.* 9: 281–295.
19. Toth L.A., K. Kregel, L. Leon et al. 2011. Environmental enrichment of laboratory rodents: The answer depends on the question. *Comp. Med.* 61:314–321.
20. Armstrong, K.R., T.R. Clark, and M.R. Peterson. 1998. Use of corn-husk nesting material to reduce aggression in caged mice. *Contemp. Top. Lab. Anim. Sci.* 37(4):64–66.
21. Würbel, H. 2001. Ideal homes? Housing effects on rodent brain and behaviour. *Trends Neurosci.* 24:207–211.
22. Kimura, T., M. Kubota, and H. Watanabe. 2009. Significant improvement in survival of tabby jimpy mutant mice by providing folded-paper nest boxes. *Scan. J. Lab. Anim. Sci.* 36: 243–249.
23. Gaskill, B.N., C.J. Gordon, E.A. Pajor et al. 2012. Heat or insulation: Behavioral titration of mouse preference for warmth or access to a nest. *PLoS One* 7(3):e32799.
24. National Institutes of Health. 1988. *Institutional administrator's manual for laboratory animal care and use*. Bethesda: Office for Protection from Research Risks, NIH Publication No. 88-2959.

25. National Institutes of Health, Office of Laboratory Animal Welfare. 2012. Frequently Asked Question F.14, "Is social housing required for nonhuman primates when housed in a research setting?" http://grants.nih.gov/grants/olaw/faqs.htm#useandmgmt_14.
26. National Research Council. 1998. *The Psychological Well-Being of Nonhuman Primates.* Washington, D.C.: National Academies Press.
27. Bayne, K. 1985. Qualitative observations of idiosyncratic behavior in old monkeys. In *Behavior and pathology of aging in rhesus monkeys*, eds. R.T. Davis, and C.W. Leathers, 201–222. New York: Alan R. Liss.
28. Rothschild, B. 2012. Interpretation of primate behavior, ambulation and biomechanics: Caveat arthritis. *J. Primatol.* 1(3):1000e112.
29. Stewart, K.L., and S.S. Raje. 2001. Environmental enrichment committee: Its role in program development. *Lab. Anim. (N.Y.).* 30(8):50.
30. National Research Council. 2003. *Occupational health and safety in the care and use of nonhuman primates.* Washington, D.C.: National Academies Press.
31. American Association of Laboratory Animal Science. AALAS Learning Library. http://www.aalaslearninglibrary.org/. Accessed December 9, 2012.
32. Morgan, K.N., S.W. Line and H. Markowitz. 1998. Zoos, enrichment, and the skeptical observer: The practical value of assessment. In *Second nature: Environmental enrichment for captive animals*, eds. D.J. Shepherdson, J.D. Mellen and M. Hutchins, 153–171. Washington, D.C.: Smithsonian Institution Press.
33. Bayne, K. and H. Wurbel. 2012. Mouse enrichment. In *The laboratory mouse*, 2nd edition, ed. H. Hedrich, 545–564. New York: Elsevier.
34. Richter S.H., J.P. Garner, B. Zipser et al. 2011. Effect of population heterogenization on the reproducibility of mouse behavior: A multi-laboratory study. *PLoS One* 6(1):e16461.
35. Van der Stay, F.J. and T. Steckler. 2002. The fallacy of behavioral phenotyping without standardization. *Genes Brain Behav.* 1:9–13.
36. Würbel, H. 2002. Behavioral phenotyping enhanced—Beyond (environmental) standardization. *Genes Brain Behav.* 1:3–8.
37. Richter, S.H., J.P. Garner, and H. Würbel. 2009. Environmental standardization: Cure or cause of poor reproducibility in animal experiments? *Nat. Methods* 6:257–261.
38. Gerdin, A.-K., N. Igosheva, L.-A. Roberson et al. 2012. Experimental and husbandry procedures as potential modifiers of the results of phenotyping tests. *Physiol. Behav.* 106:602–611.
39. Brent, L. 1995. Feeding enrichment and body weight in captive chimpanzees. *J. Med. Primatol.* 24:12–16.
40. De Boer, S.F, and J.M. Koolhaas. 2003. Defensive burying in rodents: Ethology, neurobiology and psychopharmacology. *Eur. J. Pharmacol.* 463:145–161.
41. Baumans, V. 2010. The laboratory mouse. In *The UFAW handbook on the care and management of laboratory and other research animals*, 8th edition, eds. R. Hubrecht, and J. Kirkwood, 276–310. Oxford: Wiley-Blackwell.
42. Shomer, N.H., S. Peikert and G. Terwilliger. 2001. Enrichment-toy trauma in a New Zealand White rabbit. *Contemp. Top. Lab. Anim. Sci.* 40(1):31–32.
43. Bazille, P.G., S.D. Walden, B.L. Koniar et al. 2001. Commercial cotton nesting material as a predisposing factor for conjunctivitis in athymic nude mice. *Lab Anim. (N.Y.)* 30(5):40–42.
44. Rowan, K.E.K., and L. Michaels. 1980. Injury to young mice caused by cottonwool used as nesting material. *Lab. Anim.* 14:187.
45. Hahn, N.E., D. Lau, K. Eckert et al. 2000. Environmental enrichment-related injury in a macaque (*Macaca fascicularis*): Intestinal linear foreign body. *Comp. Med.* 50:556–558.
46. Wolfle, T.L. 2005. Environmental enrichment. *ILAR J.* 46(2):79–82.
47. Association for Assessment and Accreditation of Laboratory Animal Care International. 2012. Social housing. http://www.aaalac.org/accreditation/positionstatements.cfm#social.
48. Howerton, C.L., J.P. Garner, and J.A. Mench. 2008. Effects of a running wheel-igloo enrichment on aggression, hierarchy linearity, and stereotypy in group-housed male CD-1 (ICR) mice. *Appl. Anim. Behav. Sci.* 115(1):90–103.

49. Nevalainen, T.O., J.I. Nevalainen, F.A. Guhad et al. 2007. Pair housing of rabbits reduces variances in growth rates and serum alkaline phosphatase levels. *Lab. Anim.* 41:432–440.
50. Love, J.A. 1994. Group housing: Meeting the physical and social needs of the laboratory rabbit. *Lab Anim Sci.* 44:5–11.
51. Byrne, G.D., and S.J. Suomi. 1991. Effects of woodchips and buried food on behavior patterns and psychological wellbeing of captive rhesus monkeys. *Am. J. Primatol.* 23:141–151.
52. Bayne, K., H. Mainzer, S. Dexter et al. 1991. The reduction of abnormal behaviors in individually housed rhesus monkeys (*Macaca mulatta*) with a foraging/grooming board. *Am. J. Primatol.* 23:23–35.
53. Novak, M.A., J.H. Kinsey, M.J. Jorgensen et al. 1998. Effects of puzzle feeders on pathological behavior in individually housed rhesus monkeys. *Am. J. Primatol.* 46:213–227.
54. Gottlieb, D.H., S. Ghirardo, D.E. Minier et al. 2011. Efficacy of 3 types of foraging enrichment for rhesus macaques (*Macaca mulatta*). *J. Am. Assoc. Lab. Anim. Sci.* 50:888–894.
55. Harris, L.D, L.B. Custer, E.T. Soranaka et al. 2001. Evaluation of objects and food for environmental enrichment of NZW rabbits. *Contemp. Top. Lab. Anim. Sci.* 40(1):27–30.
56. Morton, D.B., M. Jennings, G.R. Batchelor et al. 1993. Refinements in rabbit husbandry. Second report of the BVAAWF/FRAME/RSPCA/UFAW Joint Working Group on Refinement. *Lab. Anim.* 27:301–329.
57. Bayne, K. 2003. Environmental enrichment of nonhuman primates, dogs and rabbits used in toxicology studies. *Toxicol. Pathol.* 31(Suppl):132–137.
58. McCune, S. 2010. The domestic cat. In *The UFAW Handbook on the Care and Management of Laboratory and Other Research Animals*, 8th edition, eds. R. Hubrecht, and J. Kirkwood, 453–472. Oxford: Wiley-Blackwell.
59. MacArthur Clark, J., and C.J. Pomeroy. 2010. The laboratory dog. In *The UFAW Handbook on the Care and Management of Laboratory and Other Research Animals*, 8th edition, eds. R. Hubrecht, and J. Kirkwood, 432–452. Oxford: Wiley-Blackwell.
60. Baumans, V. 2005. Environmental enrichment for laboratory rodents and rabbits: Requirements of rodents, rabbits and research. *ILAR J.* 46:162–170.
61. Brown, C. 2009. Novel food items as environmental enrichment for rodents and rabbits. *Lab Anim. (N.Y.)* 38:119–120.
62. Brown, C. 2010. Organic wheatgrass as environmental enrichment. *Lab Anim. (N.Y.)* 39:74–75.
63. Moberg, O., V.A. Braithwaite, K.H. Jensen et al. 2011. Effects of habitat enrichment and food availability on the foraging behaviour of juvenile Atlantic Cod (*Gadus morhua L*). *Environ. Biol. Fish* 91:449–457.
64. Strand, D.A., A.C. Utne-Palm, P.J. Jakobsen et al. 2010. Enrichment promotes learning in fish. *Mar. Ecol.-Prog. Ser.* 412:273–282.
65. Berejikian, B.A., E.P. Tezak, S.C. Riley et al. 2001. Competitive ability and social behaviour of juvenile steelhead reared in enriched and conventional hatchery tanks and a stream environment. *J. Fish Biol.* 59:1600–1613.
66. Brydges, N.M. and V.A. Braithwaite. 2009. Does environmental enrichment affect the behaviour of fish commonly used in laboratory work? *Appl. Anim. Behav. Sci.* 118(3):137–143.
67. U.S. Department of Agriculture, National Agricultural Library, Animal Welfare Information Center. 2012. http://awic.nal.usda.gov/research-animals/environmental-enrichment-and-exercise.

29

Animal Mistreatment and Protocol Noncompliance

Jerald Silverman

Introduction

One of the most contentious roles of the IACUC is its mandate to review allegations of animal mistreatment and protocol noncompliance. This must be done carefully and with the utmost concern for due process and confidentiality as careers may be severely hurt by unfounded accusations and unsubstantiated rumors. Nevertheless, the welfare of animals cannot be compromised. The result is the thin line on which the IACUC walks. The intent of this chapter is to provide the reader with guidance to many of the problems that IACUCs face when confronted with allegations of animal mistreatment and protocol noncompliance. It is important to remember that allegations remain nothing more than allegations until proven otherwise.

Because of the seriousness and sensitivity of this IACUC role, it is very important that the procedures it will use to review and investigate complaints be formalized ahead of time. The time to hunt for the most appropriate operating procedure is not when a complaint reaches the IACUC. Although IACUC policies can never account for all contingencies, it is wise to have a reasonable number available. It is important to know what authority your IACUC has under the AWAR and PHS Policy. This is because the AWAR and PHS Policy do not provide all the answers to the myriad problems the IACUC faces when complaints are presented to it. Lastly, it also is important to know the extent of additional authority, if any, that your institution has granted to your IACUC. It is this additional authority that often allows an IACUC to expeditiously address administrative and regulatory issues without having to call for an IACUC meeting every time such issues arise.

No IACUC works in a vacuum. Rules, regulations, policies, and the like, require a conscientious IACUC and strong institutional support. Without the latter, the most conscientious IACUC is severely handicapped when reviewing and investigating allegations of protocol noncompliance and animal mistreatment.

29:1 What is meant by animal mistreatment and protocol noncompliance?

Reg. Neither the PHS Policy nor the AWAR directly define animal mistreatment; however, U.S. Government Principle IV states that the proper use of animals includes the avoidance or minimization of discomfort, distress, and pain when consistent with sound scientific practices. The AWA (§2131,1) states that a purpose of the Act is "to insure that animals intended for use in research facilities or for exhibition purposes or for use as pets are provided humane care and treatment." AWA §2143,a,3,A states that animals used in experimental procedures will be covered by practices causing minimal pain and distress. The *Guide* (p. 2) notes that its basic goal is to promote the humane care and use of laboratory animals.

Likewise, protocol noncompliance is not directly defined. However, both the PHS Policy (IV,B,6) and the AWAR (§2.31,c,6) require IACUC approval for animal use activities. Therefore, for the limited scope of this chapter (i.e., protocol noncompliance and animal mistreatment) it follows that any animal use activity not approved by the IACUC constitutes protocol noncompliance.

Opin. Animal mistreatment has been defined as physical or psychological wrongful or abusive treatment of an animal.[1] Examples include hitting animals, taunting animals, animal neglect, or not providing food for punitive reasons.

Protocol noncompliance indicates that procedures or policies approved by the IACUC are not being followed.[1] Examples include performing of unauthorized surgery, participation of unauthorized persons in a research project, or injection of drugs that the IACUC has not approved. It is not unusual to have some overlap between animal mistreatment and protocol noncompliance. For example, the unauthorized restraint of a nonhuman primate for an unusually long time can potentially entail both animal mistreatment and protocol noncompliance. (See 29:46.)

When faced with protocol noncompliance the IACUC's first step, if possible, should be to determine if a revision to the protocol would make the activity compliant without compromising animal welfare. On the other hand, when faced with animal mistreatment, the IACUC must take immediate action to stop the mistreatment and provide for the needs of the animal.

There may be instances when issues not directly affecting animals become part of an IACUC approved protocol and can lead to compliance concerns. For example, as part of the protocol review process questions may arise about aspects of occupational health and safety (OHS). The PHS Policy (IV,A,1), which uses the *Guide* as the basis for implementing an animal care and use program, requires the implementation of an OHS program (*Guide* p. 17). The establishment and functioning of this program is the responsibility of the research institution and may encompass multiple departments; nevertheless, the IACUC is responsible for the animal care and use program and often acts as a gatekeeper for animal-related issues, not approving an animal activity until the proper OHS or other assurances are established. Discussions in this chapter concerning protocol noncompliance are limited to issues directly affecting animals.

NIH/OLAW has commented on departures from the *Guide* in terms of the standards that must be met, should be met, and may be met.[2] In many of the examples provided by NIH/OLAW, these departures involve the IACUC or vivarium at the level of the animal care and use program and may not involve noncompliant activities in an individual IACUC-approved protocol.

29:2 What role does an institution's administration have in the overall concept of animal mistreatment or protocol noncompliance?

Reg. Under the PHS Policy, the IO, by signing an Animal Welfare Assurance with NIH/OLAW, commits the institution to compliance with the PHS Policy. Noncompliance (such as animal mistreatment or protocol noncompliance) is ultimately the responsibility of the IO who represents the institution's administration (PHS Policy III,G; IV,A). The IO has a similar responsibility for assuring compliance with the AWAR (AWAR §1.1, Institutional Official). The IO has the right to suspend (halt) a protocol (independent of the IACUC) but cannot allow an animal activity to occur without IACUC approval (AWAR §2.31,d,8). Although the IO bears ultimate responsibility for compliance, the IO and the IACUC share the responsibility for the review and investigation of animal welfare concerns (*Guide* p. 23).

Opin. Note: To avoid confusion the words suspend or suspension as used in this chapter indicate a temporary or permanent cessation of an animal activity that is imposed by the IACUC in accordance with federal Regulations (AWAR §2.31,d,6; PHS Policy IV,C,6). Words such as stop and halt are used to indicate a temporary or permanent cessation of an animal activity by an individual such as the AV or the IO that is imposed using authority granted by the research institution.

The IACUC is a regulatory committee that functions as an agent of the institution (PHS Policy IV,B; AWAR §2.31,c). The institution must make it known that the IACUC not only helps to assure animal welfare but is an indispensable link in helping protect the researcher against unwarranted accusations and assuring the integrity of the institution.

Because regulatory committees are often disparaged by investigators, it must be understood that an IACUC cannot properly function without general institutional ethical and administrative support. The institution, by actions and words, must stand behind the IACUC. The IO can be a powerful ally and should have a clear understanding of his or her responsibilities and authority under the AWAR and PHS Policy and communicate this support to the IACUC and investigators.

29:3 What is the initial responsibility of the IACUC relative to allegations of animal mistreatment or protocol noncompliance according to the AWAR and the PHS Policy?

Reg. Under the AWAR (§2.31,c,4), the IACUC must "review and, if warranted, investigate concerns involving the care and use of animals at the research facility." This includes complaints from the public and from laboratory or research facility personnel or employees. The PHS Policy (IV,B,4) is slightly broader and states that the IACUC must "review concerns involving the care and use of animals at the institution." The *Guide* (p. 23) reiterates this and the NIH/OLAW website Frequently Asked Questions (FAQ) G.8 states that "the IACUC must consider allegations of noncompliance or animal welfare issues as concerns that must be addressed in accordance with relevant PHS Policy provisions and Animal Welfare Regulations (AWRs)."[3]

Opin. A key phrase in the AWAR is *if warranted*, as not all complaints need to be investigated in depth. For example, the complaint "The dogs prefer Brand A dog food to Brand B" requires far less action than "Dr. X is performing unapproved splenectomies on dogs." The IACUC Chair or appropriate designees may make an initial decision about the depth of review required based on the nature of the complaint but the IACUC itself should make the final decision. (See 29:4.) In both the AWAR and PHS Policy, it is prudent for the IACUC to interpret *institution* and *research facility* as any site where the IACUC has jurisdiction.

Surv. 1 At your institution, how is the decision made to inform the entire IACUC of an allegation of protocol noncompliance or animal mistreatment?

• We have no policy in place	27/293 (9.2%)
• The IACUC administrative staff makes the decision	5/293 (1.7%)
• The IACUC chairperson makes the decision	69/293 (23.6%)
• Another IACUC-affiliated person makes the decision (e.g., veterinarian, Institutional Official)	5/293 (1.7%)
• All allegations are brought to the attention of the IACUC	178/293 (60.8%)
• I don't know	3/293 (1.0%)
• Other	6/293 (2.0%)

Surv. 2 At your institution, how quickly is an allegation of protocol noncompliance or animal mistreatment brought to the attention of the entire IACUC?

• We have never had any allegations	63/294 (21.4%)
• It varies with the allegation	44/294 (15.0%)
• We typically notify the full IACUC as soon as possible	97/294 (33.0%)
• We typically wait until the next full committee meeting	60/294 (20.4%)
• We typically form a subcommittee to investigate the complaint and only bring it to the attention of the IACUC after we receive the report from the subcommittee	26/294 (8.8%)
• A subcommittee typically does all the work and it is rarely brought to the attention of the full IACUC	2/294 (0.7%)
• I don't know	1/294 (0.3%)
• Other	1/294 (0.3%)

29:4 Should the IACUC take investigative action on all allegations of animal mistreatment or protocol noncompliance?

Reg. The IACUC is responsible for determining whether an activity is in accordance with the AWA, the *Guide*, the institution's NIH/OLAW Assurance, the PHS Policy (PHS Policy IV,C,6), and the AWAR (AWAR §2.31,d,6). Thus, the IACUC must consider all allegations of noncompliance and determine whether there is sufficient reason to investigate further. (See 29:2; 29:44.)

Opin. Although the IACUC must review all allegations, some may not require a thorough investigation while others may require a full investigation. All concerns must be brought to a quorum of the full IACUC and it is not the prerogative of IACUC chairperson or any other person to unilaterally decide on how to resolve a complaint. It is the expectation of NIH/OLAW and APHIS/AC that "concerns about animal activities at an institution must be reviewed at a convened meeting of the IACUC."[4] There are at least five circumstances when IACUCs should consider becoming actively involved.

- When an allegation should be, but is not, satisfactorily resolved at the local level.[1]
- When any reasonable person would consider that gross animal mistreatment or protocol noncompliance has occurred. This is based on the allegation itself, although it may subsequently be found to be untrue.[1]

- When there are repeated minor instances of noncompliance or mistreatment involving the same person or research group.[1] See 29:44 for one suggested method for IACUCs to evaluate continuing minor instances of non-compliance.
- When the IACUC cannot clearly decide whether an allegation is worth investigating, it is probably better to investigate.[1]
- When there are repeated problems of significant noncompliance that were previously brought to the attention of the IACUC.

The PHS Policy (IV,F,3,a) mentions but does not define specifically "continuing noncompliance." Examples of what might be considered continuing noncompliance under the PHS Policy are failure to correct deficiencies identified during the semiannual evaluations in a timely manner and "chronic failure to provide space for animals in accordance with recommendations of the *Guide*..."[5,6] The APHIS/AC *Animal Welfare Inspection Guide*[7] provides additional examples such as

- A non-compliant item cited on the last APHIS/AC inspection
- The same problem previously found in one location and now found in another location
- The same problem previously found for one species and now found in another species

Any investigation of an allegation should be done in a timely manner. Memories can quickly fade and the desire to move forward can ebb as time passes. (See 29:59.)

29:5 Who can bring allegations of animal mistreatment or protocol noncompliance to the IACUC?

Reg. The AWAR (§2.31,c,4) state that the general public and institutional employees may make complaints. The PHS Policy is silent on this issue; however, it does state that "all institutions are required to comply, as applicable, with the Animal Welfare Act and other Federal statutes and regulations relating to animals" (PHS Policy II). Further, there is nothing in the PHS Policy that precludes the IACUC from acting upon concerns raised from any source. In addition, it is not unusual for APHIS/AC veterinary medical officers to identify such problems to the IACUC Chair or the AV. (See 29:24; 29:50.)

Opin. Free and open communication is the first step in building the trust that is necessary for people to voice their concerns. Any limitations on who can present allegations to the IACUC sends a message to employees and the public that the IACUC or the institution has something to hide or does not have a full commitment to animal welfare.

Surv. At your institution, who can bring allegations of animal mistreatment or protocol noncompliance to the IACUC? Check as many boxes as appropriate.

- Any employee 289/294 (98.3%)
- The public in general 158/294 (53.7%)
- Students (if applicable) 194/294 (66.0%)
- I don't know 0/294 (0%)
- Other 11/294 (3.7%)

29:6 What communication pathways should be followed in order to bring allegations of animal mistreatment or protocol noncompliance to the attention of the IACUC?

Reg. The AWAR (§2.32,c,4) state that research facility personnel must receive training on how to report alleged deficiencies in animal care and treatment but there are no specific statements in the AWAR as to how to comply with this requirement. PHS Policy via the *Guide* (pp. 23–24) states that the institution must develop methods for reporting animal welfare concerns and mechanisms for reporting concerns should be posted in prominent locations in the animal facility and on applicable institutional websites. The *Guide* (p. 24) also notes that multiple points of contact (e.g., IO, IACUC Chair, and AV) are recommended.

Opin. Institutions have developed various procedures of their own and most are focused on providing open and easy communication. Any allegation must eventually reach the Chair of the IACUC. If the allegation is against the Chair, a written policy should state who will help adjudicate the complaint. As an example, this might be the Vice-Chair of the IACUC, the AV, or the IO.

Typical communication methods include direct verbal conversations and written letters to the AV, IACUC Chair, IO, IACUC members, college deans, and others. This information must be efficiently passed on to the committee Chair. It is helpful to have a written IACUC-initiated policy on how to do this.

Surv. At your institution, how can people bring concerns about animal mistreatment or protocol noncompliance to the attention of your IACUC? Check as many boxes as appropriate.

• Via the Attending Veterinarian	271/295 (91.9%)
• Via the IACUC chairperson	288/295 (97.6%)
• Via any IACUC member	277/295 (93.9%)
• Via the Institutional Official	271/295 (91.9%)
• Via an institutional administrator (e.g., dean, IACUC administrator)	221/295 (74.9%)
• Via a telephone, computer, drop box, or similar hotline	230/295 (78.0%)
• I don't know	0/295 (0%)
• Other (included training/compliance specialists, other veterinarians, IACUC office)	12/295 (4.1%)

29:7 How are employees, students, researchers, and others trained and informed that they have the right to bring complaints to the IACUC about animal care and use?

Reg. (See 29:6.)

Opin. It is suggested that training be provided for all institutional personnel. Many methods can be used as is shown in the following survey.

Surv. At your institution, how are people notified of the means available to them to bring their concerns to the attention of the IACUC? Check as many boxes as appropriate.

• We have no formal mechanism in place	33/294 (11.2%)
• During a formal training session	173/294 (58.8%)
• From handouts to investigators and their research teams	92/294 (31.3%)
• Through posted signs	200/294 (68.0%)

• Through a periodic newsletter	26/294 (8.8%)
• Through an IACUC or animal facility web site	139/294 (47.3%)
• I don't know	1/294 (0.3%)
• Other	7/294 (2.4%)

29:8 Must a complainant be identified to the IACUC or the IACUC Chair?

Reg. There are no federal animal welfare laws, regulations, or policies requiring the identification of a complainant. The *Guide* (p. 24) states that "the process [of reporting a concern] should include a mechanism for anonymity."

Opin. Required identification might potentially deter certain people from making complaints. Some academic or other institutions have policies that require the identification of a complainant before action can be taken. Although this requirement can lead to a conflict between the IACUC and the institution, in practice it need not. The IACUC itself can act as the complainant in order to maintain the confidentiality of the true complainant (whether known or not). It is prudent for the IACUC and the institution to develop policies for handling such procedural conflicts well before a problem arises. Confidentiality should be maintained when requested by the complainant. See Chapter 22 for additional confidentiality discussions.

Surv. At your institution, must a complainant identify himself or herself before the IACUC will consider a complaint?

• We have no policy	35/294 (11.9%)
• The complainant must openly identify himself/herself to the IACUC	5/294 (1.7%)
• The complainant must be identified but only to the IACUC Chairperson	17/294 (5.8%)
• The complainant does not have to be identified	229/294 (77.9%)
• I don't know	6/294 (2.0%)
• Other [included identification to the IO]	2/294 (0.7%)

29:9 Is there protection against repercussions if a person makes a complaint to the IACUC; that is, are "whistle blowers" protected?

Reg. The AWAR (§2.32,c,4) provide specific protection for employees, IACUC members, or laboratory personnel against discrimination or other reprisals for reporting violations of the AWAR or the AWA itself. As a result, legal action can be taken against the facility in the event a reprisal can be proven. Such action may lead to a monetary penalty (AWAR §4.11,a,2). The employee of the facility, however, has no private right of action under the AWA or AWAR.

The PHS Policy states that all institutions are required to comply, as applicable, with the AWA and other federal statutes and regulations (PHS Policy II; U.S. Government Principle I). The *Guide* (p. 24) states that there should be compliance with applicable whistle blower policies, nondiscrimination against the concerned or reporting party, and protection from reprisals.

Opin. Although the PHS Policy does not offer the specific protection against whistle blowers that is written into the AWAR, the statement in the *Guide* (p. 24) is a standard required for PHS Assured institutions (see PHS Policy IV,A,1). This provides a greater degree of protection to whistle blowers than was previously offered. Nevertheless, NIH/OLAW has written that "as the PHS Policy does not contain explicit whistle blower protections, whistle blower anonymity will be honored only to the extent allowed under the FOIA [Freedom of Information Act], or other applicable laws, regulations or policies."[8]

29:10 How common are IACUC investigations of allegations of animal mistreatment or protocol noncompliance?

Opin. In this author's experience, allegations are not common although this undoubtedly varies with the culture of individual institutions. Whereas it is appropriate to handle problems as expeditiously as possible, there are times when problems should be called to the attention of the IACUC (e.g., unauthorized surgery). Not doing so makes the IACUC somewhat of a paper tiger and defeats its self-regulatory role.

Surv. At your institution, has the IACUC ever been involved with investigating allegations of animal mistreatment or protocol noncompliance?

• Not that I know of	91/292 (31.2%)
• Yes, but rarely	43/292 (49.0%)
• Yes, occasionally (e.g., two or three times a year)	42/292 (14.4%)
• Yes, fairly often (e.g., about every other month)	16/292 (5.5%)
• I don't know	0/292 (0%)
• Other	0/292 (0%)

29:11 To what extent should allegations of animal mistreatment or protocol noncompliance be documented?

Reg. The PHS Policy (IV,E,1,b) requires the institution to maintain minutes of IACUC meetings, activities of the IACUC, and IACUC deliberations. The *Guide* (p. 24) states that reported concerns and corrective actions taken should be documented. Reported allegations that are discussed at IACUC meetings, or acted upon by the IACUC, must be recorded. The AWAR (§2.35,a) also have a requirement for maintaining records of IACUC meetings.

Opin. It is unlikely that all complaints taken to the IACUC will be fully documented in writing and signed. Thus, the person who receives the complaint (often the AV or IACUC Chair) must obtain and record as much information as possible. If the complaint is verbal, questions such as, Did you see this yourself? When did this happen? and Where did this happen? are appropriate. If possible, one should obtain the name of a contact person (such as the complainant) and a means of contacting that person.[1] (See 29:13.)

29:12 Must allegations of noncompliance identified by external agencies (e.g., during an APHIS/AC inspection or AAALAC site visit) be investigated?

Reg. Yes. Such allegations must be investigated and reported to NIH/OLAW if they meet the reporting requirements of the PHS Policy at IV,F,3[6] and IV,B,4 and the AWAR (§2.31,d,6). (See 29:50.)

Opin. One of the functions of the IACUC, as noted above, is to investigate concerns involving the care and use of animals at its institution. The Federal regulation and policies noted above are not limited to concerns raised only from within the home institution. (See 29:50 for the APHIS/AC and NIH/OLAW Memorandum of Understanding.)

29:13 What types of documentation might an IACUC need to investigate a complaint fully?

Opin. Necessary documentation varies with the specific situation. Assuming the IACUC as a whole or the designated person (e.g., the Chair) determines that further investigation is warranted, the investigating persons can determine what information they must collect to investigate an allegation properly. It may be necessary for the IACUC to interview people, examine animals, or obtain surgical records, housing records, and even an investigator's research records (in the last example, confidentiality must be assured). (See 29:23.) The IACUC's own communications or policies may have to be examined. Any records obtained should be directly related to the allegation made. Because the IACUC does not have any authority to subpoena people or records, institutional support is crucial to the investigative process. A previous survey of IACUC chairpersons indicated that a majority of IACUCs only initiated an investigation if there was, at the minimum, enough documentation to suggest a significant problem.[9] (See 29:2; 29:11.)

29:14 Should the IACUC let all concerned persons know the basis of a complaint and the procedures to be followed? Who might be "concerned persons"?

Opin. Assuming the IACUC has determined that a complaint should be fully investigated, it is prudent to inform all involved persons about the nature of the complaint and procedures to be followed. This method will help assure due process (the protection of the legal and ethical rights of all concerned individuals).[1] Concerned persons can include the alleged violator and that person's immediate superior (if the alleged violator is not the PI). The IO is often informed of the problem at or before this time. Clearly, individual circumstances will dictate final decisions on whom to notify.

Surv. 1 At your institution, when is a person notified of an allegation of animal mistreatment or protocol noncompliance against a person or his/her staff?

• We have no policy in place	37/294 (12.6%)
• We provide notification as soon as an allegation reaches the IACUC	95/294 (32.3%)
• We provide notification after an initial investigation by the IACUC	107/294 (36.4%)
• We provide notification only if the IACUC decides to proceed with an investigation	32/294 (10.9%)
• I don't know	11/294 (3.7%)
• Other (included never had an allegation, legal office makes the determination, varies with the situation)	12/294 (4.1%)

Surv. 2 Does your IACUC have a formal policy about protecting the rights and confidentiality of the person against whom an allegation has been made?

- We have no such formal policy 112/293 (38.2%)
- We do have a formal policy 154/293 (52.6%)
- I don't know 22/293 (7.5%)
- Other 5/293 (0.7%)

29:15 Once a complaint has reached the IACUC, what is the responsibility of the IACUC Chair?

Opin. The Chair oversees the efforts described in 29:3. The Chair should assure that the investigation moves forward and due process is followed. If an allegation is placed against the Chair, then the Vice-Chair or another member designated by the IO would oversee any activities surrounding the investigations of the allegation.

29:16 Should the IACUC approach each complaint on an individual basis or is a standard operating procedure for handling complaints more appropriate?

Reg. Although each complaint is likely to have its own unique needs, the AWAR and PHS Policy require the IACUC to review all complaints. (See 29:3; 29:4.)

Opin. It is suggested that the IACUC have an SOP that states the general procedures to be used for most allegations reaching the committee. This helps prevent careless errors in due process or confidentiality. It is important that the persons formulating the policy for handling complaints and investigations be representative of the institution's research community as a whole in order to achieve broad support for its recommendations.

29:17 Can the IACUC Chair appoint an investigative subcommittee?

Reg. Responsibility for the review and investigation of animal welfare concerns rests with the IO and the IACUC (*Guide* p. 23; AWAR §1.1, Institutional Official).

Opin. Yes. There is nothing in the AWAR or PHS Policy that details how an IACUC should proceed with an investigation if the IACUC determines one is needed. The choice of procedures is entirely at the discretion of the IACUC, with input, as needed, from the IO. In addition to an investigative subcommittee the Chair can do his or her own investigation, the Chair can ask the AV to perform an initial investigation, and so forth. It is strongly recommended that an IACUC policy on this topic be established well before a potential problem arises. (See 29:16; 29:18.)

29:18 What might be the composition of an appropriate investigative subcommittee?

Opin. There are no federal regulations or policies defining the composition of an investigative subcommittee of an IACUC. Investigative subcommittees can be composed of one or more persons, including non-IACUC members. Some IACUCs assign one person (often a veterinarian) to do all (or at least preliminary) investigations, while others establish a larger subcommittee. If the IACUC is small, the entire IACUC may act to investigate an allegation. Some IACUCs attempt to assure that a scientist be on an investigative subcommittee if the allegation is against a scientist. It is suggested that in a large IACUC there be broad enough representation on a subcommittee to help assure due process and acceptance of its recommendations.

29:19 Can the IACUC use non-IACUC members as part of an investigative subcommittee?

Opin. Yes (see 29:18). There is no federal regulation or policy prohibiting this. However, since the IACUC is charged with reviewing all claims of animal mistreatment or protocol noncompliance (see 29:4), it is suggested that only the IACUC or IO appoint such member(s) and that they report to the IACUC directly or through the IACUC subcommittee of which they are a part. The IACUC and IO retain the responsibility for investigative subcommittees under their auspices and any subsequent IACUC actions or determinations.

29:20 Should the IACUC keep a complainant informed of the *progress* of an IACUC investigation?

Opin. Assuming the complainant is known, this becomes an IACUC and institutional choice. It is this author's opinion that limited feedback should be provided to a complainant during the course of an investigation. That feedback should be little more than updates (if requested) on where the IACUC is in the investigation and a possible date for the conclusion of the investigation. Full feedback during the course of the investigation, in this author's opinion, can potentially lead to unwarranted pressure on the investigative committee and intentional or unintentional leaks of confidential information.

Surv. At your institution, is a known complainant informed of the progress of any ongoing investigation?

- There has never been an investigation 77/293 (26.3%)
- We do not keep the complainant informed 30/293 (10.2%)
- We usually provide the complainant with limited information 110/293 (37.5%)
- We usually provide the complainant with detailed information 52/293 (17.7%)
- I don't know 14/293 (4.8%)
- Other 10/293 (3.4%)

29:21 Should the IACUC keep a complainant informed of the *results* of an IACUC investigation?

Reg. The *Guide* (p. 23) states that there should be communication of findings to the concerned employee unless the allegation was made anonymously. The AWAR has no comparable statement.

Opin. In this author's opinion, providing limited results to a complainant once an investigation is completed is appropriate. It reinforces the credibility of the IACUC to both the institution and the public. There is a downside to full disclosure; the complainant may not be satisfied with the IACUC's determination and he or she may initiate actions against the institution or an individual. Likewise, the alleged violator may do the same and also claim defamation of character. The results of the investigation should be made known to the full committee in a timely manner and documented in the IACUC's records (*Guide* p. 24), although the specific details to be included become a matter of institutional policy. Institutional policy must be established well in advance. An earlier survey indicated that most IACUCs provide at least some of the results of an investigation to the complainant.[9]

29:22 Is legal or other representation on behalf of the accused person or the institution allowed during the course of an IACUC investigation?

Opin. PHS Policy and the AWAR are silent on this question. As shown by the following survey, most IACUCs have not addressed this issue and there is no AWAR or PHS Policy guidance on the issue. One suggestion is to allow legal or other representation as advisory only to the accused and/or the IACUC, not as an active participant in an investigation or hearing.

Surv. 1 Is legal or other representation on behalf of the accused person or on behalf of your institution allowed during the course of an IACUC investigation?

- We have no policy in place 168/294 (57.1%)
- We allow legal (or other) representation for the accused only 1/294 (0.3%)
- We allow legal (or other) representation for the institution only 4/294 (1.4%)
- We allow legal (or other) representation for all parties 52/294 (17.7%)
- We do not allow legal (or other) representation 10/294 (5.8%)
- I don't know 54/294 (18.4%)
- Other 5/294 (1.7%)

Surv. 2 If legal or other representation is allowed at your institution, which of the follow statements apply? Check as many boxes as appropriate.

- Not applicable 138/277 (49.8%)
- The representative of the institution may ask questions 48/277 (17.3%)
- The representative of the accused may ask questions 44/297 (14.8%)
- I don't know 83/297 (27.9%)
- Other 8/297 (2.7%)

29:23 What can the IACUC do to ensure confidentiality of research or other records that are reviewed during the course of an investigation?

Opin. Various procedures are in use. Some IACUCs keep no records of investigative proceedings, some keep limited records, and many instruct people involved with the proceedings to maintain confidentiality. Formal confidentiality agreements can be considered. It must be remembered that allegations remain allegations until proven otherwise, and unintentional leaks of information can adversely affect people's lives. It is suggested that records be kept as they may be crucial for further reference, particularly if APHIS/AC or NIH/OLAW needs to conduct their own investigation. The availability of records should be restricted to those persons having a legitimate need to have access to them. Use of locked cabinets or secured computers may be necessary. (See 29:11.)

Surv. During an investigation of an allegation, does your IACUC have a formal policy about protecting the confidentiality of research or other records?

- We have no such formal policy 112/293 (38.2%)
- We do have a formal policy 154/293 (52.6%)
- I don't know 22/293 (7.5%)
- Other 5/293 (1.7%)

29:24 How might the AV report to the IACUC or PI animal activities that occurred but were not included in an approved IACUC protocol? (See 29:6.)

Reg. The *Guide* (p. 24) states that there should be multiple methods for reporting concerns to the IACUC.

Opin. The nature of the activity often dictates actions. An extra toy placed in a cage by a researcher's technician might be welcomed rather than vilified and likely will lead to no report. If a protocol does not include walking a dog up and down a hallway in the animal facility, but it is being done for the benefit of the animal, the AV might suggest to the PI that a protocol addendum (amendment) be made. For significant activities not included in an IACUC protocol (e.g., a major surgical procedure) the most direct route is personal contact with the IACUC Chair followed by a written description of the alleged violation. The AV also may contact the IO or request another IACUC member to contact the Chair. If the AV is the IACUC Chair, he or she should call (if necessary) an emergency meeting of the committee. It is not necessary for the AV to be identified as the source of the complaint. (See 29:5.)

In order to correct the problem rapidly, immediate communication between the AV and PI is strongly advised, if at all possible.

29:25 What is meant by *sanctions*?

Reg. There is no definition of the word sanction in the AWAR, and the word does not appear in the PHS Policy, including the *Guide.* The PHS Policy (IV,C,6; IV,C,7) addresses suspension of an activity by the IACUC. NIH/OLAW has defined an IACUC suspension as any IACUC intervention that results in the temporary or permanent interruption of an animal activity.[6] If the IACUC places an ongoing project on hold or requires a temporary cessation, these actions are synonymous with suspension. Furthermore, PHS Policy, the AWAR, and the *Guide* presume that all ongoing animal activities have received prospective review and approval. Accordingly, the IACUC's authority to suspend unauthorized activities is always implied, if not explicit.[10]

Opin. Sanctions are often mentioned on the NIH/OLAW FAQ website[3] but without a specific definition of the word. For the purposes of the questions that follow, sanctions refer to any coercive action taken by the IACUC to help ensure compliance with the AWAR or PHS Policy. Sanctions are only one of the many tools the IACUC can use to help assure the welfare of laboratory animals and rarely should have to be the first used. The IACUC's suspension of a previously approved activity can be considered a type of sanction. (See 29:26.) In the discussions that follow, we often segregate suspensions from other forms of sanctions.

The IACUC always has the authority to make recommendations to the IO regarding any aspect of the institution's animal care and use program (PHS Policy IV,B,5; AWAR §2.31,d,5). If the IACUC believes that the institution should impose institutional sanctions (e.g., a reprimand for repeated violations or revocation of a PI's privilege to conduct research with animals), the IACUC should make those recommendations to the IO.

Surv. Has your IACUC ever applied sanctions (e.g., retraining, suspended protocol, curtailment of animal facility use) to persons as a consequence of a verified finding of protocol noncompliance or animal mistreatment?

- Not applicable 27/291 (9.3%)
- There have been no sanctions that I am aware of 84/291 (28.9%)
- We have on rare occasions applied sanctions of some sort 115/291 (39.5%)
- We often apply sanctions as a consequence of a verified finding of protocol noncompliance or animal mistreatment 60/291 (20.6%)
- I don't know 3/291 (1.0%)
- Other 2/291 (0.7%)

29:26 What sanctions may the IACUC impose if allegations of animal mistreatment or protocol noncompliance are verified?

Reg. The AWAR (§2.31,c,8) and PHS Policy (IV,B,8) only allow for suspension of an activity previously approved by the IACUC. Thus, any other sanctions imposed by the IACUC, the IO, or the institution are imposed through institutional policy. (See 29:25; 29:40; 29:52–54.) PHS Policy (IV,F,3,a) states "The IACUC, thorough the Institutional Official, shall promptly provide OLAW with a full explanation of the circumstances and actions taken with respect to: a. any serious or continuing noncompliance." Under the PHS Policy (IV,B,3) institutions are required to remediate instances of noncompliance and are empowered to take appropriate steps to prevent recurrence.

Opin. Other than a suspension of an animal activity, sanctions used via institutional authority can include denied access to the animal facility, written reprimands, personnel retraining, monetary fines, employment termination, etc.

29:27 Can the IACUC suspend a previously approved animal activity or a person?

Reg. An activity, as defined in the AWAR (§1.1, Activity), is any element of research, testing, or teaching procedures that involve the care and use of animals. The PHS Policy (I) defines activities as research, research training, and biological testing. The AWAR (§2.31,d,6) and PHS Policy (IV,B,8; IV,C,6) specifically allow the IACUC to suspend a previously approved activity.

Opin. The IACUC can suspend a previously approved animal activity in whole or in part after review of the matter at a convened quorum of the IACUC, and with a suspension vote of a majority of the quorum present. See 29:29 for a discussion of partial versus complete suspension of an animal activity. NIH/OLAW has also clarified that the IACUC can suspend a person, not just an animal activity, and such a suspension is reportable to NIH/OLAW.[11]

29:28 Is there any regulatory requirement to have a "face-to-face" meeting of a quorum of the IACUC in order to vote on the suspension of an animal activity?

Reg. PHS Policy (IV,C,6) and the AWAR (§2.31,d,6) state that a vote to suspend an animal activity can occur "only after review of the matter at a convened meeting of a quorum of the IACUC and with the suspension vote of a majority of the quorum present."

Opin. The regulatory requirement noted previously suggests, but does not explicitly state, that a face-to-face committee meeting is required. NIH/OLAW has provided guidance that telephone or video conferencing are acceptable for the conduct of official IACUC business requiring a quorum, provided that certain criteria are met.[12] APHIS/AC has endorsed the NIH/OLAW guidance.[13] Given the seriousness of suspending an animal activity, it is this author's opinion that in all but the most exceptional of circumstances there should be a face-to-face meeting of the IACUC when a vote to suspend an animal activity is needed.

There may be unique circumstances in which there is an immediate need to halt an animal activity without a meeting of the full IACUC because of a significant danger to animal welfare. This is discussed in 29:33–36.

29:29 Must the IACUC suspend an entire previously approved activity or can it suspend parts of a previously approved activity?

Reg. An activity, as defined in the AWAR (§1.1, Activity), is any element of research, testing, or teaching procedures that involve the care and use of animals. As defined by PHS Policy I, an activity includes research, research training, and biological testing.

Opin. The AWAR and PHS Policy are silent as to whether the IACUC has the authority to suspend only portions of a previously approved activity. The AWAR (§2.31,d,6) and the PHS Policy (IV,B,8) both speak of an "activity involving animals." Neither states that an entire IACUC protocol must be suspended. It, therefore, may be interpreted to allow the IACUC the authority to suspend either a full activity or a portion of such an activity. This opinion has been supported by NIH/OLAW and APHIS/AC[14] and NIH/OLAW also has written that "a portion of a protocol may be suspended and the rest of the activity may continue to retain to be approved."[11] As an example, an IACUC can determine that only a surgical component of an ongoing research protocol need be suspended. (See 29:53.)

29:30 Does the IACUC, the IO, or both decide on the sanctions to be taken?

Reg. The AWAR (§2.31,d,7) and the PHS Policy (IV,C,7) state that the IO, in consultation with the IACUC, must take appropriate corrective action after a suspension of a previously approved activity. (See 29:25.)

Opin. The wording of the AWAR and PHS Policy as stated above suggests that the IO has somewhat more authority than the IACUC in the development of corrective actions after suspensions. In practice, most IACUCs decide what sanctions are to be taken.[9]

29:31 If sanctions are imposed by the IACUC, do they occur via a formal full committee vote or can the IACUC Chair take action on behalf of the committee?

Reg. A formal committee vote at a convened meeting of a quorum of the IACUC is required for the suspension of a previously approved activity (AWAR §2.31,d,6; PHS Policy IV,C,6).

Opin. Neither the AWAR nor PHS Policy addresses the question of sanctions (as defined in 29:25) other than suspension of a previously approved activity. The method for approval of the imposition of sanctions (other than a suspension) to help prevent recurrence should become a policy of the IACUC in consultation with the IO. It is suggested that any sanctions taken (whether a suspension of a previously approved activity or any other form of sanction) have an affirmative vote by a majority of a quorum of the full committee at a convened meeting.

29:32 If an allegation of animal mistreatment or protocol noncompliance is verified, must the IACUC apply sanctions?

Reg. PHS Policy (IV,C,6) states that the IACUC may suspend an activity if it determines that the activity is not being conducted in accordance with applicable provisions of the AWA, the *Guide*, the institution's Assurance, or IV,C,1,a–IV,C,1,g of the PHS Policy. The AWAR (§2.31,d,7) and the PHS Policy (IV,C,7) state that the IO, in consultation with the IACUC, must take appropriate corrective action in the event of a suspension of a previously approved activity.

Opin. If, in the opinion of the IACUC, sanctions are not appropriate, they need not be applied. A clearly minor and unintentional misinterpretation of an IACUC policy that has created no problem for an animal is an example of a verified allegation of protocol noncompliance that might lead to an explanation, not a sanction.

For sanctions other than a suspension of a previously approved activity, institutional and IACUC policy should be established and followed. It is strongly suggested that any authority the IACUC, IO, or any other institutional representative has to impose sanctions (other than suspensions) be unambiguous.

29:33 Can the IACUC Chair apply sanctions (other than suspension of a previously approved protocol) without full IACUC approval if the IO is in agreement with the sanctions?

Opin. The only sanction the IACUC is empowered to employ under the AWAR and PHS Policy is the suspension of a previously approved activity, and a vote for suspension must occur at a meeting of a quorum of the full committee (see 29:26). Thus, the application of any other sanctions by the Chair, with or without full IACUC approval, is a matter of institutional policy. (See 29:37.) There may be unique circumstances under which a Chair might unilaterally impose a sanction, pending rapid review by the IACUC (see 29:36; 29:58).

29:34 Can the AV suspend (halt) an ongoing IACUC-approved research project or related activity?

Reg. There is nothing in the AWAR or PHS Policy that specifically authorizes the AV to suspend a research project. However, the AWAR (§1.1 Attending Veterinarian; §2.33,a,2) state that the AV must have direct or delegated authority for activities of animal care and use and must be able to provide adequate veterinary care. The PHS Policy has a similar statement (IV,A,3,b,1). The *Guide* (p. 114) requires that the veterinarian have the authority to remove an animal from an experiment in the case of a pressing health problem.

Opin. (See 29:2 for the usage of the words suspend and halt.) These regulations have been interpreted by some IACUCs as giving the AV authority to halt an ongoing project, subject to IACUC review. Such an in-house policy can become part of an adequate veterinary care statement for APHIS/AC (for consulting AVs) or the PHS Assurance statement if accepted by those agencies. It should be disseminated to investigators. The IACUC should be immediately notified if a halt occurs. In this author's opinion, the AV should have the authority from his or her institution to halt an ongoing study, subject to rapid review by the IACUC. Any such action by the AV, even if not sustained by the IACUC, must be reported to the federal government.[6,15] (See 29:58.)

Surv. Does your institution give your Attending Veterinarian the authority to suspend an animal use project, subject to subsequent timely review by the IACUC?

• Not applicable	5/294 (1.7%)
• Yes, my institution gives our Attending Veterinarian that authority	232/294 (78.9%)
• No, my institution does not give our Attending Veterinarian that authority	43/294 (14.6%)
• I don't know	12/294 (4.1%)
• Other	2/294 (0.7%)

29:35 Can a veterinarian other than the AV stop a previously approved research or teaching project?

Opin. There is nothing in the AWAR or PHS Policy that specifically authorizes the AV or any other veterinarian to halt a previously approved activity. Any such authorization is a matter of institutional policy. (See 29:34.) In this author's opinion, the veterinarian with primary responsibility for laboratory animal science issues (typically, the AV) should have the authority to halt an ongoing project, subject to subsequent rapid review by the IACUC. It also should become part of an adequate veterinary care statement for the APHIS/AC (if a consulting veterinarian is used). Any such action by the veterinarian, even if not sustained by the IACUC, must be reported to the federal government.[6,15] (See 29:58.)

29:36 Can the IACUC Chair halt an ongoing IACUC-approved research project prior to a meeting of the full IACUC? (See 29:33.)

Opin. There is nothing in the AWAR or PHS Policy that gives such authority to the IACUC Chair. Nevertheless, an institution may delegate such authority to the

IACUC Chair pending review by the full committee. Such an in-house policy should become part of the PHS Assurance statement and should be disseminated to investigators. In the opinion of this author, the IACUC Chair should have the authority to halt an ongoing study, subject to subsequent rapid review by the IACUC. Any such action, even if not sustained by the IACUC, must be reported to the federal government.[6,15] (See 29:58.)

Surv. At your institution, does the IACUC chairperson have the authority to suspend [halt] a research or teaching project?

- Not applicable — 2/293 (0.7%)
- Yes, our chairperson has institutional authority to suspend an animal use protocol — 145/293 (49.5%)
- No, our chairperson does not have institutional authority to suspend an animal use protocol — 125/293 (42.7%)
- I don't know — 15/293 (5.1%)
- Other — 6/293 (2.1%)

29:37 Can an appeal be made to the IACUC or other institutional authority if sanctions are imposed?

Reg. The AWAR and PHS Policy are silent on this issue. Nevertheless, officials of an institution cannot approve an animal activity that has not been approved by the IACUC (AWAR §2.31,d,8; PHS Policy IV,C,8).

Opin. An institution, as part of its own policy, may allow appeals on sanctions to be made to the IACUC, the IO, or other persons, although only the IACUC can reverse one specific sanction, that being the suspension of an animal activity (AWAR §2.31,d,8; PHS Policy IV,C,8; and see 29:49). Appeals may be appropriate as the IACUC is not infallible or (for example) may not have followed the correct procedures for suspending a protocol. Another pair of eyes or the presentation of a different perspective on a problem might suggest alternative solutions to the IACUC. There is somewhat of a precedence for an IACUC to consider appeals to sanctions as the AWAR (§2.31,d,4) and PHS Policy (IV,C,4) allow for appeals of an IACUC decision to withhold approval of an animal activity (i.e., withheld approval of a protocol or a protocol amendment). Thus, an appeal process can be considered for two types of sanctions: those to suspend an animal activity and those for sanctions other than suspension of an animal activity. If appeals to the IACUC are permitted for sanctions *other than* the suspension of an animal activity then the IACUC should carefully consider whether or not this type of an appeal can also be made to the IO or another institutional authority. If the IACUC decides to allow such appeals to the IO or another institutional authority then the IACUC should be prepared to have its sanction overridden by the IO or another institutional authority. (See 9:56; 9:57; 29:38; 29:39.)

For the IACUC to run smoothly and accomplish its task, it is strongly recommended that all policies concerning sanctions be clearly understood by the IACUC, IO, and all animal users.

Surv. Does your IACUC allow appeals of any sanctions imposed on an animal user? Assume that any such appeals are made to the IACUC.

• Not applicable	24/293 (8.2%)
• We do allow for appeals	140/293 (47.8%)
• We do not allow for appeals	16/293 (5.5%)
• I don't know	23/293 (7.8%)
• We have no policy	89/293 (30.4%)
• Other	1/293 (0.3%)

29:38 Can an appeal of sanctions imposed by the IACUC be made to an APHIS/AC administrator or to the NIH/OLAW?

Opin. No. There are no provisions for such appeals. The PHS Policy (V,A,5) allows the NIH/OLAW to waive provisions of its policy, but that section refers to general programmatic decisions, not decisions on a given sanction.

29:39 Can an institution or IO override the IACUC and impose lesser sanctions on a person?

Reg. An IO may not override an IACUC suspension (IACUC withdrawal of approval). In the event of an IACUC suspension, the IO is required to take appropriate corrective action and report that action with a full explanation to NIH/OLAW (PHS Policy IV,C,7) or APHIS/AC (AWAR §2.31,d,7).

Opin. Although the IO may not override an IACUC suspension of an animal activity, other sanctions that are imposed via institutional policy do potentially open the door to overriding. (See 29:37.)

29:40 Can an institution or IO impose sanctions that are more severe than those imposed by the IACUC?

Reg. The PHS Policy and the AWAR do not preclude any institutional or IO sanctions. As noted in NIH/OLAW FAQ B.10,[3] these sanctions may be even more severe than those imposed by the IACUC.

Opin. Yes. There is nothing to prohibit the institution or its authorized representative from imposing a sanction more severe than that recommended by the IACUC. The AWAR (§2.31,d,7; §2.31,d,8) imply that the research institution may do this as does PHS Policy (IV,C,7; IV,C,8).

29:41 Can an IACUC impose sanctions if the vertebrate animals used are not currently covered by the AWAR (e.g., common laboratory rats, *Rattus norvegicus*) and funding for the study is from nonfederal sources?

Opin. The AWAR (§1.1, Animal) and PHS Policy (III,A) clearly state which nonhuman animals are regulated by their respective agencies. (See 12:1.) Most IACUCs, as an institutional policy, oversee the care and use of all vertebrate animals in biomedical research, teaching, and product testing, whether or not those animals are under the auspices of pertinent federal regulations or policies. Indeed, many institutional Animal Welfare Assurances to the NIH/OLAW are worded in a manner that includes all vertebrate animals used in research, teaching, or testing, not just those of studies funded by the PHS. If, as part of an institutional policy, an IACUC is empowered to apply sanctions (in addition to the suspension of a previously approved activity) then in many instances those sanctions can be applied to all vertebrate animals.

29:42 An investigator performs an animal activity that does not have IACUC approval. Must the IACUC take a formal vote in order to suspend an activity that it never approved?

Reg. The purpose of the IACUC is "to assess the research facility's animal program, facilities and procedures." (AWAR §2.31,a; PHS Policy IV,A,3,a uses similar language.) The AWAR (§2.31,c,6) and the PHS Policy (IV,B,6) authorize the IACUC to approve activities involving the care and use of animals. Suspension of an ongoing animal activity requires a formal vote of a quorum of the full committee. (See 29:26.) NIH/OLAW in its FAQ B.9 addresses this by stating: "The PHS Policy, *Guide*, and the USDA Animal Welfare Regulations presume that all ongoing animal activities have received the required prospective review and approval. An activity that has been undertaken without prior approval should be halted and subsequently reported to OLAW because it constitutes serious noncompliance."[3]

Opin. In addition to FAQ B.9 noted above, NIH/OLAW and APHIS/AC have written,"... PHS Policy, USDA Regulation, and the [seventh edition of the] *Guide* language presume that all ongoing animal activities have received prospective review and approval. Accordingly, the IACUC's authority to suspend unauthorized activities is always implied, if not explicit."[10] On the basis of this rationale, the NIH/OLAW guidance in FAQ B.9, and the IACUC's responsibility to oversee animal-based research, the IACUC can suspend an activity it never formally approved. Further, it would require a vote of a quorum of the full committee to suspend that activity. This opinion is based on the premise, noted previously, that IACUC approval is presumed to be necessary to begin an animal activity. It therefore follows that the suspension of that activity should have a properly taken suspension vote from the IACUC. Another option would be for the institution itself to give an individual, such as the IACUC Chair or IO, the authority to suspend an animal activity. (See 29:36; 29:61.) Such authority should have strict limitations as it can override an important function of the IACUC. It would still require notification of the suspension to the appropriate federal agencies.[6,15]

29:43 An IACUC voted to suspend a researcher's animal use activity if any instance of protocol noncompliance were to be confirmed during a certain period. Another instance of noncompliance was confirmed during that period. Is it necessary for the IACUC to take another vote to suspend the protocol?

Reg. A formal vote of a majority of the quorum present at a convened meeting of the IACUC is required for the suspension of a previously approved activity (AWAR §2.31,d,6; PHS Policy IV,C,6).

Opin. This scenario approaches an unclear area of compliance, as the IACUC had already voted to suspend the researcher's animal use activity if a confirmed instance of protocol noncompliance occurred. Nevertheless, it would be appropriate to take a revote since noncompliance can range from a very minor incident to a significant incident. Further, the circumstances surrounding a particular incident can influence the committee's decision making. Because the IO, in consultation with the IACUC, has to review the reasons for the suspension and take appropriate corrective action (AWAR §2.31,d,7; PHS Policy IV,C,7) it might be inappropriate to put the IO in this position if the incident was quite minor. There is nothing in the AWAR or PHS Policy that prohibits the IACUC from reversing a decision if it is correctly done and documented. Indeed, the AWAR (§2.31,d,4) and PHS Policy (IV,C,4), referring specifically to withholding approval of an animal activity, allow for the IACUC to reconsider its decision. One can extrapolate from this that reconsidering a previous action is a reasonable and proper function of the IACUC; however, if there is to be a reversal or other change to a previous action, the IACUC must do so by voting. (See 29:37; 29:49.)

29:44 For helping to determine when instances of continuing protocol noncompliance or animal abuse require direct IACUC involvement and possible sanctions, a point system can be considered. What is meant by a point system and how might it be implemented?

Reg. PHS Policy (IV,F,3,a) requires prompt reporting to NIH/OLAW of any serious or continuing noncompliance with the PHS Policy. The PHS Policy is silent on the types of continuing noncompliance that might prompt such a report. The AWAR (§2.31,d,7) requires notification of APHIS/AC and any federal funding agency of a suspension of a previously approved animal activity and the reason for the suspension. There is no mention of "continuing noncompliance." (See 29:4.)

Opin. Neither the PHS Policy, the *Guide*, nor the AWAR provide specific examples of what constitutes serious enough incidents of continued protocol noncompliance to report to APHIS/AC or NIH/OLAW. The *Guide* (p. 13) notes that the general evaluation of the animal care and use program is left to the judgment of the IACUC, AV, and IO. Examples of continuing deficiencies have also been provided by APHIS/AC (see 29:11). Other examples of reportable situations have been identified by NIH/OLAW[6] (see 29:52). Whereas certain actions, either individually or repeated, are clearly significant enough to consider evoking sanctions (e.g., performing major survival surgery without IACUC approval), others are not quite as clear (e.g., forgetting to replace a watering device promptly after an animal has fully recovered from anesthesia). Repeated problems of the latter sort in the same research group, even if they do not cause an animal well-being problem, can present a problem for an IACUC. Assigning a point value for certain types of noncompliance actions, with a predetermined total "score" that triggers the IACUC to consider sanctions or a report to a federal agency, can help an IACUC decide when enough is enough. Thus, repetitive problems are summed up over a predetermined period.

For example, an IACUC might use a total score of 100 points within 6 months to trigger committee deliberations on a possible sanction. This can be the suspension of an animal activity or a lesser sanction if the committee has its organization's authority to impose lesser sanctions. Not replacing a water bottle immediately after full recovery from anesthesia might be assigned 5 points whereas not observing an animal as often as stated in the protocol might be assigned 20 points. Both of these examples assume that animal well-being was not compromised. If animal well-being is negatively impacted as part of a PHS funded research, teaching, or testing project the activity is reportable to NIH/OLAW.

It is suggested that if a point system is to be considered, general categories of noncompliance activities be defined ahead of time to prevent the IACUC from becoming bogged down with determining the relative importance of every conceivable noncompliance scenario that reaches the committee. Further, a point system, if used, should not be considered so inflexible as to preclude the IACUC from taking an alternative course of action.

29:45 How long should a suspension of a previously approved activity last?

Opin. The duration is at the discretion of the IACUC. It is unusual for an activity to be permanently suspended. Most activities are suspended for the length of time that the IACUC believes will prevent recurrence of the problem and allows any remedial actions to occur. Any suspension must include the enactment of corrective actions to help prevent future problems (AWAR §2.31,d,7; PHS Policy IV,C,7).

29:46 The PHS Policy (IV,F,3) speaks of providing NIH/OLAW with a full explanation of the circumstances and actions taken with respect to any serious or continuing noncompliance with PHS policy or any serious deviation from the provisions of the *Guide*. What are some examples of serious IACUC protocol noncompliance?

Opin. *Noncompliance* in the context of PHS Policy IV,F,3 refers to the entirety of the PHS Policy. Protocol noncompliance is a subcategory. NIH/OLAW has provided many examples of noncompliance with the overall PHS Policy and some specific examples of protocol noncompliance.[6] They include (but are not limited to) beginning projects without IACUC approval, inadequate postprocedural care, failure to maintain appropriate animal-related records, and failure to ensure death of animals after euthanasia procedures.

Although *serious protocol noncompliance* is not synonymous with significant deficiencies found during semiannual IACUC inspections, the concept that a significant deficiency is any action that "is or may be a threat to the health or safety of animals" (PHS Policy IV,B,3; AWAR §2.31,c,3) can be applied to certain instances of serious noncompliance with an IACUC protocol. Therefore, additional examples of serious protocol noncompliance are not providing anesthetics or analgesics as required in the protocol, performing unapproved multiple survival surgical procedures, using an unapproved species, and using an unapproved euthanasia method. Further examples that constitute serious protocol noncompliance include having improperly trained persons working with animals, changing the PI without IACUC approval, and using more animals than approved. See 29:4; 29:44 for a discussion of continuing noncompliance.

29:47 If an animal activity is suspended can an investigator continue to breed his or her animals but not use them for research?

Reg. See 29:25 for a definition of a suspension. An activity is any element of research, testing, or teaching procedures that involve the care and use of animals (AWAR §1.1, Activity; PHS Policy I).

Opin. A properly voted suspension (AWAR §2.31,d,5; PHS Policy IV,C,6) terminates a previously approved animal activity and breeding is an animal activity. There is nothing in the AWAR or PHS Policy that excludes animal breeding as an activity. If animal breeding was not part of the original protocol it cannot be added to a fully suspended protocol until after the IACUC reapproves the protocol. If only a portion of a protocol was suspended, animal breeding can continue if breeding was not the suspended portion but if breeding was not part of the original protocol it can occur only if an amendment to the protocol is approved by the IACUC. (See 29:29.)

29:48 If an animal activity is suspended, who is financially responsible for the upkeep of animals that are affected by the suspension?

Reg. (See 29:25 for a definition of suspension.)

Opin. USDA/APHIS is silent on this issue but the PHS has stated that no costs with live vertebrate animals may be charged to the NIH if there is not a valid Animal Welfare Assurance and IACUC approval of the activity. This includes paying *per diem* costs and funds used to support salaries for persons doing technical work associated with the animals or anyone who is associated with the animal activities. Additionally, equipment and supplies associated with animal activities cannot be charged to NIH funds.[11] NIH expects recipients of grants (i.e., the institution receiving the grant) to continue to maintain and care for animals during periods when animal activities are conducted in the absence of a valid Assurance or IACUC approval. However, on a case by case basis, the granting Institute or Center of the NIH may allow expenditures of NIH grant funds for the care and maintenance of animals.[5,16,17] (See 29:50.)

29:49 If a protocol is suspended, how is it reactivated?

Reg. The AWAR and the PHS Policy are silent on this issue. NIH/OLAW and APHIS/AC have provided guidance on reactivation stating: "APHIS and OLAW expect that the suspended protocol, along with any amendments added as a result of the suspension, will be reinstated only by review and approval according to the IACUC protocol approval process."[18]

Opin. The IACUC should lift a suspension only after it has determined that the activity will be in full compliance with the PHS Policy, the NIH/OLAW Animal Welfare Assurance, AWAR, and the *Guide*. Conceptually, reactivating a suspended protocol (in whole or in part) is analogous to approving a new protocol or approving a major amendment to a protocol. Either requires approval via full committee or designated member review. The IACUC should establish a policy to provide for reactivation because a suspension can be for a finite time period or have an undefined endpoint. In some instances a person or a procedure, rather than a protocol, is suspended. Reactivation of the person or procedure can be accomplished by having the PI submit an amendment to the protocol and having the IACUC handle the amendment as described in Chapter 10.

29:50 Apart from sanctions imposed by an IACUC, what sanctions can NIH/OLAW and APHIS/AC impose on a research or teaching institution for protocol noncompliance?

Opin. Subsequent to a citation of a violation of the AWA by an APHIS/AC veterinary medical officer, APHIS/AC Investigative and Enforcement Services personnel may investigate an alleged violation if an institution does not take corrective measures to come into compliance with the AWA. Investigations that find AWA violations are acted on in a variety of ways, depending on their severity. Many infractions are settled with an official notice of warning or a stipulation offer. Official letters of warning notify your institution that further infractions can result in more stringent enforcement actions. Stipulations allow alleged violators to pay a penalty in lieu of formal administrative proceedings. In cases of serious or chronic violations, consequences become more substantial.[19] Any research facility that violates any regulation may be assessed a civil penalty of not more than $10,000 for each such violation. Each violation and each day during which a violation continues is considered a separate offense.[20]

NIH/OLAW evaluates all allegations or indications of noncompliance with the PHS Policy or an institution's Assurance. NIH/OLAW will typically allow an institution to clarify an allegation of noncompliance or remedy a confirmed problem. If after an appropriate investigation NIH/OLAW determines that noncompliance has occurred, it may restrict its approval of an institution's Assurance. This may include (for example) suspending an institution's Assurance's applicability to some or all projects until specified corrections have been implemented; requiring prior NIH/OLAW review of some or all research conducted under the Assurance; requiring additional training for some or all investigators; or requiring special reporting to OLAW. If deemed appropriate, NIH/OLAW may withdraw its approval of an institution's Assurance and if that should happen no research can move forward using PHS financial support until an appropriate Assurance is approved by NIH/OLAW. Details of this process have been published.[8] Office of Management and Budget circular A-21 informs educational institutions receiving Federal grants and contracts that any costs attributable to any federal sponsored agreement (such as a grant or contract) cannot be shifted to any other federally sponsored agreement in order to avoid restrictions on their use caused by circumstances such as an expired IACUC protocol or suspension of an animal activity.[21] On a practical level this means that most educational institutions will have to use non-federal funds to cover the cost of maintaining animals on an expired or suspended IACUC protocol or if certain restrictions on its Assurance are imposed by NIH/OLAW. (See 29:48.)

It should be recognized that NIH/OLAW and APHIS/AC work together under a Memorandum of Understanding which allows each agency to provide the other with information that is pertinent to the enforcement of the AWAR and PHS Policy.[22]

29:51 If sanctions are imposed, which persons or agencies must be informed and who does the informing?

Reg. PHS Policy (IV,C,7; IV,F,3) and the AWAR (§2.31,d,7) refer only to the suspension of previously approved activities. Under that circumstance, the IO, in consultation with the IACUC, must promptly report the suspension along with a full explanation to NIH/OLAW. PHS Policy (IV,F,3) also requires the IACUC, through the IO, to report to NIH/OLAW any serious or continuing noncompliance with the PHS Policy or serious deviations from the provisions of the *Guide*. The conduct of animal activities when a protocol has been suspended, expired, or does not have IACUC approval must be reported to the NIH Institute or Center supporting the grant award.[5] If the species used is under the auspices of the AWAR, the suspension and a full explanation must be sent to APHIS/AC and any federal agency funding the suspended activity AWAR (§2.31,d,7). The APHIS/AC regional office is the appropriate place to report the suspension. Any sanctions imposed by the IACUC should be recorded in the minutes of the IACUC meeting (AWAR §2.35,a,1; §2.35,a,2) and are subject to inspection by APHIS/AC (AWAR §2.35,f).

Opin. The IACUC, through the IO, makes an appropriate report to NIH/OLAW. Similar to the federal funding agency notification requirement for APHIS/AC noted above, the suspension and a full explanation must be sent to the PHS funding agency. It is the responsibility of the PI or the Authorized Organization Representative (also known as the Signing Official) to report the suspension to the appropriate NIH Grants Management Officer.[11]

Reports to NIH/OLAW about noncompliance findings that did not incur a suspension of a previously approved activity will typically include details of any sanctions that were applied by the IACUC via institutionally granted authority. (See 29:52.)

29:52 Other than suspending an animal activity, what types of IACUC-related problems are to be reported to the NIH/OLAW, APHIS/AC, and federal funding agencies?

Reg. PHS Policy (IV,F,3) requires the IACUC, through the IO, to report to NIH/OLAW any serious or continuing noncompliance with the PHS Policy or serious deviations from the provisions of the *Guide*. Details can be found in NIH Notice NOT-OD-05-034.[6] The requirements to report suspensions of previously approved activities (PHS Policy IV,C,7; IV,F,3; AWAR §2.31,d,7) are discussed in 29:51 and 29:53. The NIH Grants Policy Statement requires grantee institutions to report to the Institute or Center supporting the grant award instances of serious noncompliance with section IV,F,3 of the PHS Policy.[5]

Opin. NIH/OLAW: NIH/OLAW offers examples of actions that require reporting to NIH/OLAW.[6] They include (but are not limited to) the following (and see 30:16):

- IACUC suspension or other institutional intervention that results in the temporary or permanent interruption of an activity due to noncompliance with the PHS Policy, AWA, the *Guide*, or the institution's Animal Welfare Assurance. Note that a suspension of an activity includes a suspension of a person (see 29:27).
- Failure to adhere to IACUC-approved protocols.
- Serious or continuing noncompliance with the PHS Policy.
- Serious deviations from the provisions of the *Guide*, such as shortcomings in programs of veterinary care, occupational health, or training.
- Failure to maintain appropriate animal-related records.
- Failure to ensure the death of an animal after euthanasia.
- Failure of animal care and use personnel to carry out veterinary orders (e.g., treatments).

APHIS/AC: In addition to reporting suspensions the AWAR (§2.31,c,3) also require reporting of any uncorrected significant deficiency (found on the semiannual review) that is uncorrected within the time period determined by the IACUC.

The report is made to APHIS/AC and any federal funding agency within 15 business days after the time period has expired. The report is made by the IACUC, through the IO.

Federal Funding Agencies: Suspensions of a protocol using animals covered by the AWAR must be reported to the appropriate federal funding agency (AWAR §2.31,d,7). For activities supported by the PHS, NIH Notices NOT-OD-07-044[16] and NOT-OD-10-081[17] state that performing an animal activity without an IACUC approval (including performing activities on a suspended protocol), performing an animal activity without a valid Assurance on file with NIH/OLAW, or performing an animal activity *after* the expiration of an IACUC approval must be reported to the Institute or Center supporting the award.[5] Any such report to an Institute or Center is an institutional requirement; it is not a mandated responsibility of the IACUC (see 29:51).

Surv. Has your IACUC ever informed NIH/OLAW or USDA/AC (as applicable) about an activity that led to a protocol suspension (in whole or in part), a serious or continuing noncompliance with the PHS Policy, or a serious deviation from the provisions of the *Guide*?

- We have never informed NIH/OLAW or USDA/AC of any such actions — 93/293 (31.7%)
- We have informed either NIH/OLAW or USDA/AC, but very rarely — 129/293 (44.0%)
- We have informed either NIH/OLAW or USDA/AC, with moderate frequency — 46/293 (15.7%)
- I don't know — 12/293 (4.1%)
- Other — 13/293 (4.4%)

29:53 If only parts of a previously approved activity are suspended by the IACUC (e.g., the surgical portion of a large project), must the IACUC notify the NIH/OLAW or APHIS/AC?

Reg. PHS Policy (IV,F,3) requires the IACUC, through the IO, to report promptly any serious or continuing noncompliance with PHS Policy or serious deviation from the provisions of the *Guide* or any suspension of an activity by the IACUC, along with a full explanation of the circumstances and actions taken by the IACUC. An activity is any element of research, testing, or teaching procedures that involve the care and use of animals (AWAR §1.1, Activity; PHS Policy I).

The AWAR (§2.31,d,6) allow the IACUC to suspend an activity if it is not being conducted in accordance with the description that was provided by the PI and approved by the IACUC. The AWAR (§2.31,d,7) require a full explanation as part of the reporting.

Opin. The wording in the AWAR, PHS Policy, and NIH Notice NOT-OD-05-034[6] refers to an animal activity, which need not be the entire IACUC protocol. Many suspensions of a part of an ongoing approved activity would meet one or more of the concerns mentioned in the regulatory section of this question and, therefore, would be reportable even if the entire protocol were not suspended. In some instances (e.g., excessive mortality or morbidity rate), the IACUC might suspend part of an ongoing activity until more information can be gathered. If all or part of an animal activity is suspended, even if temporarily, NIH/OLAW or APHIS/AC and the appropriate federal funding agency (for animals covered by the AWAR) should be notified.[6,15] (See 29:29; 29:52; 29:58.)

29:54 Must a federal funding agency be informed if allegations of animal mistreatment or protocol noncompliance are verified, but sanctions are not imposed?

Reg. A federal funding agency must be informed if a suspension of a previously approved activity occurs and the species involved are under the auspices of the AWAR (§2.31,d,7). The NIH Grants Policy Statement requires instances of serious noncompliance to be reported to NIH/OLAW and the Institute or Center supporting the grant award.[5]

Opin. Questions 29:51 and 29:52 describe conditions under which an institution receiving PHS funds must inform certain federal funding agencies of problems with animal use activities. This notification is an institutional requirement, not specific to the IACUC. The requirement under the AWAR to inform federal agencies about the suspension of an animal activity should not be confused with the need to notify appropriate federal funding agencies if major deficiencies were noted during the semiannual program review and inspection of animal care and use areas that were not corrected in the appropriate period (AWAR §2.31,c,3).

29:55 Must *nonfederal* funding agencies be informed of full or partial suspensions of a previously approved activity?

Opin. Nonfederal funding agencies may establish whatever criteria are appropriate to their needs. These criteria may be binding on an institution. The IACUC may be requested by the nonfederal agency to provide information to it about suspensions or other sanctions.

29:56 If an IACUC suspension of a previously approved activity occurs, what information should be provided to NIH/OLAW, APHIS/AC, and federal funding agencies?

Reg. (See 29:46; 29:51–29:53.)

Opin. NIH/OLAW has provided guidance on this issue.[6] It is appropriate to include the federal grant or contract number, the Animal Welfare Assurance number (when applicable), and the category of people involved (e.g., research technicians, post-doctoral fellow). It is also suggested that a full and clear statement of the problem be provided along with the planned (or completed) corrective action. Any other details of the suspension (e.g., length of time) should be included. See NOT-OD-05-034 for additional information.[6]

29:57 An institution has an IACUC but it does not house or use animals that fall under the auspices of the AWAR or PHS Policy. Must the institution inform APHIS/AC and NIH/OLAW of any suspensions of a previously approved activity?

Reg. No. If no activity involving PHS-supported animals is conducted at the institution, the institution does not have reporting obligations to NIH/OLAW. Likewise, if the suspended animal use does not include species regulated by the AWAR (e.g., invertebrates), there is no reporting obligation to APHIS/AC.

Opin. Under the circumstances described, the applicable federal regulations and policies do not pertain to the institution. The words institutional animal care and use committee do not guarantee that a facility has indirect federal oversight of its animal care and use activities. Nevertheless, the NIH/OLAW Assurance may include vertebrate animals in studies that are not PHS funded. Under those circumstances, NIH/OLAW must be notified. (See 29:41.)

29:58 Should NIH/OLAW or the APHIS/AC be informed that a project has been suspended pending the completion of an IACUC investigation?

Reg. (See 29:33–29:36.)

Opin. There is no AWAR or PHS Policy that specifically requires APHIS/AC or NIH/OLAW notification about an activity that is under investigation. However, a gray area arises if the IACUC has properly discussed and voted to suspend an activity until an investigation is completed. Any suspension, even a temporary suspension, must be reported to APHIS/AC or NIH/OLAW,[6] but because full details are not yet known or verified, the IACUC cannot properly comply with the AWAR (§2.31,d,7) or PHS Policy (IV,C,7) on reporting the suspension along with a *full explanation* of the problem and corrective actions. The NIH/OLAW and APHIS/AC have recommended that they be informally advised of any suspension that occurs under these unique circumstances and formally notified under the previously noted regulation and policy if the IACUC verifies the complaint leading to the suspension.[6,23] (See 29:59.)

29:59 How soon after a previously approved activity is suspended must that suspension be reported to APHIS/AC or NIH/OLAW?

Reg. The PHS Policy (IV,F,3) requires institutions to report "promptly" any serious or continuing noncompliance with the PHS Policy or serious deviation from the provisions of the *Guide*. This includes the temporary or permanent interruption of an activity involving animals by the IACUC.[6,24] The AWAR (§2.31,d,7) does not provide any general or specific time frame for notifying APHIS/AC of a suspension of an activity involving animals.

Opin. Although the AWAR do not specify any time frame for reporting a suspension to APHIS/AC, this should be done promptly, along with the reasons for the suspension and the specific activities or plans for corrective action (see 29:56). NIH/OLAW requests similar information[24] and has emphasized that promptly means without delay.[6] NIH/OLAW understands that it may take some time to develop a full report, as required by PHS Policy (IV,F,3) and therefore recommends that a preliminary report be transmitted to NIH/OLAW as soon as possible and followed up by a thorough report once final action has been taken.[6] (See 29:58.)

29:60 Does the IACUC have to report to either NIH/OLAW or APHIS/AC that an investigator has voluntarily halted part or all of an animal activity in response to an IACUC investigation?

Reg. PHS Policy (IV,F,3) requires that the IACUC inform NIH/OLAW of any suspension of any activity by the IACUC, any serious deviation from the *Guide*, or any serious or continuing noncompliance with the PHS Policy. The AWAR (§2.31,d,7) state that any animal activity that is suspended by the IACUC must be reported to APHIS/AC and any Federal agency funding that activity. An activity is any element of research, testing, or teaching procedures that involve the care and use of animals (AWAR, §1.1, Activity). Additionally, under the AWAR any uncorrected significant deficiency must also be reported (§2.31,c,3).

Opin. There may be times when an investigator tries to avoid having the IACUC report to NIH/OLAW or APHIS/AC a suspension of an animal activity on his or her protocol by "voluntarily" stopping all or part of that activity for a period of time that is apt to be satisfactory to their IACUC. See 29:2 for the usage in this Opinion of the words suspend and halt.

If a protocol violation is significant enough to warrant a suspension, whether voluntarily halted or formally suspended by the IACUC, APHIS/AC expects the IACUC to notify the IO and develop a reasonable plan and time frame for correcting the violation. This information must be part of the IACUC's records and available to the APHIS/AC inspector. A failure to comply with the plan developed by the IACUC must be reported to APHIS/AC and any Federal agency funding the research.[25] In response to a hypothetical scenario in which an investigator voluntarily halted a research activity, APHIS/AC did not make a specific statement as to whether or not the investigator's action was reportable to APHIS/AC and any Federal funding agency. Indeed, the explanation given implied to this author that such a report was not required.[25] Nevertheless, as noted in 29:61, not all suspensions that are reportable to APHIS/AC must originate with a vote taken by the IACUC. An activity that is halted by the IO is also reportable[15] and it would seem to follow that a voluntary halt in an animal activity that is accepted by a vote of the IACUC taken at a properly convened meeting of the committee would also be reportable to APHIS/AC.

Included among examples of activities that are reportable to NIH/OLAW is an "IACUC suspension or other institutional intervention that results in the temporary or permanent interruption of an activity due to noncompliance with the [PHS] Policy, Animal Welfare Act, the *Guide*, or the institution's Animal Welfare Assurance."[6] A voluntary halt of an animal activity that is accepted by the IACUC at a properly convened meeting of the committee appears to be an "other institutional intervention" that falls under this example.

Surv. Does your IACUC report an animal user's voluntary suspension [halt] of an animal activity to either NIH/OLAW or USDA/AC (as appropriate)?

- Not applicable 69/293 (23.6%)
- It depends on the circumstances 114/293 (38.9%)
- Yes, we usually report a voluntary suspension 39/293 (13.3%)
- No, we usually do not report a voluntary suspension 40/293 (13.7%)
- I don't know 28/293 (9.6%)
- Other 3/293 (1.0%)

29:61 An IO stops an animal activity even though the IACUC voted not to suspend the same animal activity. Is this action reportable to the NIH/OLAW or APHIS/AC since it did not arise from an IACUC vote?

Reg. The PHS Policy and the AWAR do not preclude any IO sanctions. (See 29:40.) A formal full committee vote is required for the suspension of a previously approved activity (AWAR §2.31,d,6; PHS Policy IV,C,6).

Opin. Although the IO cannot approve an animal activity that has not been approved by the IACUC (AWAR §2.31,e,8; PHS Policy IV,C,8) there is nothing in the PHS Policy or AWAR that prohibits the IO from halting an animal activity if the IO has the institutional authority to do so (see 29:40). In a scenario similar to that described in this question, APHIS/AC responded that although a suspension did not occur strictly according to regulatory requirements, it was done with the institution's decision to allow the IO to halt the activity and therefore would have to be promptly reported to APHIS/AC and any federal agency funding that activity.[15] The same rationale is maintained by NIH/OLAW.[6,14] This is further supported by the NIH/OLAW position that judgments of the degree of seriousness associated with noncompliance with the PHS Policy or deviation from the provisions of the *Guide* are not the exclusive prerogative of the IACUC.[26] An institution or IO may decide not to support a project and no longer accept funding for reasons other than noncompliance. This is not reportable to NIH/OLAW but should be discussed with the PHS funding component.

29:62 Assume a study is initiated without IACUC approval, the problem is discovered, and when the PI is told to halt all animal work the response is that halting the work will mean that all the data collected to date will be meaningless and the lives of the animals will have been wasted. What should the IACUC do?

Reg. The AWAR (§2.31,c,6) and the PHS Policy (IV,B,6) authorize the IACUC to approve activities involving the care and use of animals. Beginning an animal activity without IACUC approval constitutes serious noncompliance with the AWAR and PHS Policy (see 29:42). Suspension of an ongoing animal activity requires a formal vote of a quorum of the full committee (see 29:26; 29:31).

Opin. The lives of the animals should not be wasted yet the IACUC should not accept the PI's explanation without investigating the validity of the claim in a timely manner. In the interim, the committee should institute detailed IACUC oversight of the research, define any limitations on further animal use within the study, and require the immediate submission of a protocol for IACUC review. If it is determined that the PI's claim is not valid, the project must be stopped until (and if) it is approved by the IACUC. The only sanction allowable to the IACUC under the PHS Policy and AWAR is the suspension of an ongoing animal activity (see 29:26). However, the committee can impose any sanctions it deems appropriate via authority it may have received from the research institution. An additional consideration is that arguments not to waste animals "may be moot when it is understood that data from activities not conducted in compliance with applicable federal regulations may not be publishable. From that perspective the animals... may already have been wasted."[14]

Whether or not all or part of the project is formally suspended, informally halted, or whether the IACUC uses institutional authority to impose additional sanctions on the PI, the fact remains that this action constitutes serious noncompliance with the PHS Policy and AWAR and must be reported to those agencies (PHS Policy IV,F,3; AWAR §2.31,d,7).[6,15] (See 29:52; 29:53; 29:65.)

29:63 If Institution A subcontracts part of a PHS grant to Institution B and there is a verified occurrence of significant protocol noncompliance or animal mistreatment at Institution B, is it the responsibility of Institution A or B to report this issue to NIH/OLAW?

Reg. PHS Policy (IV,F,3) requires prompt notification to NIH/OLAW of serious or continuing noncompliance with the PHS Policy, serious deviations from the *Guide*, or the suspension of an activity by the IACUC.

Opin. NIH/OLAW has written that the grantee (the holder of the main grant) always retains responsibility for assurance with the PHS Policy when both the grantee and the subcontractor have NIH/OLAW Assurances (see NOT-OD-07-044).[16] However, a Memorandum of Understanding (MOU) should be in place between the two institutions which defines responsibilities when a situation such as described above occurs (*Guide* p. 15). It is particularly important for the MOU to clarify who will report the occurrence to NIH/OLAW.

29:64 Does a veterinary school have to report to NIH/OLAW the suspected abuse by the owner of a privately owned animal that is part of a clinical trial if the trial was approved by both the IACUC and a separate Clinical Trials Committee? If the NIH/OLAW Assurance includes all vertebrate species used in research, teaching, or testing, does it matter if the clinical trial is sponsored by a Federal or by a private agency?

Reg. The PHS Policy covers live vertebrate animals used in research that is supported by the PHS (PHS Policy 1). The AWAR cover research with regulated species (AWAR §1.1 Activity; AWAR §1.1 Animal).

Opin. Animals treated at a veterinary school as routine clinical patients are not subject to the PHS Policy or AWAR. However, if the animals are treated in response to the requirements of a federal grant or contract designed to systematically collect and evaluate biomedical data (i.e., research), the study is subject to IACUC review under the PHS Policy and AWAR.[27] This is reiterated in NIH/OLAW FAQ[3] A-8. If a clinical trial is not federally funded but the institution's PHS Assurance covers all vertebrate animals used in research, regardless of the funding source, then IACUC review of the clinical trial protocol is still required under the PHS Policy but not under the AWAR. This has been affirmed by NIH/OLAW and APHIS/AC, writing in response to a hypothetical clinical trial scenario: "Since the institution's Assurance covered all species used for research, OLAW would expect the IACUC to review the [clinical] trial. USDA would expect likewise, [but only] if the clinical trial was conducted under funding from a Federal agency." [28]

Neither the PHS Policy nor the AWAR differentiate between animals owned by an institution and privately owned animals.[27] Nevertheless, to clarify the relationship between NIH/OLAW and APHIS/AC with a private owner of an animal, both agencies have stated that "Abuse of a privately owned animal by the owner is outside the jurisdiction of USDA and OLAW, even if the animal is on a PHS-supported study."[28]

29:65 A PI gently takes a small amount of blood from a dog's cephalic vein but does so without having an IACUC-approved protocol. Is this considered serious noncompliance with PHS Policy and the AWAR and should it be reported to those agencies and the appropriate funding agencies?

Reg. The AWAR (§2.31,c,6) and the PHS Policy (IV,B,6) authorize the IACUC to approve activities involving the care and use of animals. The PHS Policy (IV,F,3,a) requires the prompt reporting of any serious noncompliance with the PHS Policy.

Opin. The fact that the procedure was a common one and was performed gently does not negate the fact that there was no IACUC oversight of the research. Both NIH/OLAW and APHIS/AC have reaffirmed that in a situation such as this, lack of harm done to an animal or the intent of an investigator *cannot* negate the fact that an animal activity was performed without animal welfare oversight. This can lead to significant consequences including possible disallowance of charges against a grant.[8] Additionally, previous guidance from NIH/OLAW makes it clear that performing research without IACUC approval is serious noncompliance and subject to reporting of the action to NIH/OLAW.[6,26] The same rationale can be extrapolated for the purposes of reporting such an activity to APHIS/AC and any sponsoring federal agencies, although the AWAR do not specifically mandate reporting unless the IACUC also suspends all or part of the approved protocol (see 29:52). Nevertheless, as noted in 29:1, this is considered protocol noncompliance under the PHS Policy and AWAR. Further, the AWA (§2143,a,7,B,iii) requires a research facility to provide an explanation for any deviation from the standards of the provisions of the AWA during an APHIS/AC inspection and provide a similar explanation on the APHIS/AC annual report. The latter requirement is reiterated in the AWAR (§2.36,b,3). The AWA (§2143,b,4,A,ii) requires that during the IACUC's semiannual inspections of research facilities the committee is to note any violation of the standards or deviations of research practices from the originally approved proposals that might adversely affect animal welfare.

29:66 Has protocol noncompliance occurred if an investigator changes an anesthetic from one drug to another, without notifying the IACUC, if the AV states (after the change was made) that new anesthetic was appropriate for the species, the procedure, and was used in an appropriate manner?

Reg. The AWAR (§2.31,c,6) and the PHS Policy (IV,B,6) authorize the IACUC to approve activities involving the care and use of animals. The AWAR (§2.31,c,7) and PHS Policy (IV,b,7) require the IACUC to review and approve significant changes regarding the use of animals in ongoing activities.

Opin. The immediate issue is whether the change of an anesthetic drug is a significant change. NIH/OLAW notice NOT-OD-05-034 states that the implementation of any significant change to IACUC-approved protocols without prior IACUC approval is reportable to NIH/OLAW.[6] NIH/OLAW FAQ D.9 specifically lists a change in an anesthetic agent as being a significant change requiring IACUC approval.[3] In the author's opinion, only under very unusual conditions involving the welfare of an animal, such as one requiring an immediate intraoperative decision about an anesthetic drug—and without a qualified veterinarian being available for consultation—would such a change be considered as being made under emergency conditions and required for the welfare of the animal rather than noncompliance with an approved protocol. Nevertheless, the IACUC would still have to thoroughly review and document all aspects of the incident and decide whether or not the anesthetic change was appropriate under the unique circumstances described.

29:67 Is the failure to perform all procedures listed on an IACUC protocol considered protocol noncompliance?

Reg. PHS Policy IV,B,7 and the AWAR (§2.31,c,7) require the IACUC to review proposed significant changes regarding the use of animals in ongoing activities.

Opin. Although NIH/OLAW has provided examples of significant changes to ongoing activities,[24] NIH/OLAW FAQ D.9 notes that the IACUC still maintains "discretion to define what it considers a significant change, or to establish a mechanism for determining significance on a case-by-case basis."[3] The failure to administer analgesia as indicated on an approved protocol would likely be of greater concern to an IACUC than would be the failure to take one final scheduled blood sample if analysis of the earlier samples provided the needed research information. The burden is twofold. First, the IACUC should have clear guidelines for researchers as to what it considers to be significant and non-significant changes to ongoing activities or how that determination is to be made. In turn, the researcher must take those guidelines seriously. NIH/OLAW has commented that when a researcher has doubts about a change in a research activity it is appropriate to contact their IACUC.[29] This is useful advice if the person is unsure if a proposed change might be considered significant by his or her IACUC.

29:68 Is continued or serious noncompliance with the PHS Policy to be construed as research misconduct under the auspices of the NIH Office of Research Integrity?

Opin. No. The Office of Research Integrity (ORI), working under the regulations of 42 CFR Part 50, Subpart A, directs the PHS research integrity activities on behalf of the Secretary of Health and Human Services (with the exception of the regulatory research integrity activities of the Food and Drug Administration). The ORI oversees research misconduct defined as fabrication, falsification, plagiarism, or other practices that seriously deviate from those that are commonly accepted within the scientific community for proposing, conducting, or reporting research. Its charge does not include overseeing compliance with regulations or policies governing the proper use of animals in biomedical research.

References

1. Silverman, J. 1994. IACUC handling of mistreatment or noncompliance. *Lab Anim. (NY)* 23(8):30–32.
2. Office of Laboratory Animal Welfare. 2012. Departures from the *Guide*. http://grants.nih.gov/grants/olaw/departures.htm.
3. Office of Laboratory Animal Welfare. Frequently Asked Questions. PHS Policy on Humane Care and Use of Laboratory Animals. http://grants2.nih.gov/grants/olaw/faqs.htm.
4. Brown, P. and C.A. Gipson. 2010. A word from OLAW and USDA. *Lab Anim.* (NY) 39:167.

5. Office of Extramural Research. 2012. NIH Grants Policy Statement. Part II. Subpart A. Section 4.1.1.5. http://grants.nih.gov/grants/policy/nihgps_2012/nihgps_ch4.htm#olaw_reporting.
6. Office of Laboratory Animal Welfare. 2005. NOT-OD-05-034. Guidance on prompt reporting to OLAW under the PHS Policy on Humane Care and Use of Laboratory Animals. http://grants.nih.gov/grants/guide/notice-files/not-od-05-034.html.
7. U.S. Department of Agriculture, Animal and Plant Health Inspection Service. 2013. *Animal Welfare Inspection Guide*. http://www.aphis.usda.gov/animal_welfare/downloads/Inspection%20Guide%20-%20November%202013.pdf.
8. Brown, P.A. Aug. 8, 2007. Compliance Oversight Procedures. National Institutes of Health, Office of Laboratory Animal Welfare. http://grants.nih.gov/grants/olaw/ComplianceOversightProc.pdf.
9. Silverman, J. 2007. Animal mistreatment and protocol noncompliance. In *The IACUC Handbook*, second edition, ed. J. Silverman, M.A. Suckow, and S. Murthy, 543–569. Boca Raton: CRC Press.
10. Garnett, N., and W.R. DeHaven. 1998. The view from USDA and OPRR. *Lab Anim. (NY)* 27(9):17.
11. Office of Laboratory Animal Welfare. 2008. Transcript of OLAW online IACUC staff seminar: December 4, 2008. http://grants.nih.gov/grants/olaw/081204_seminar_transcript.pdf.
12. Office of Laboratory Animal Welfare. 2006. Notice NOT-OD-06-052. Guidance on Use of Telecommunications for IACUC Meetings under the PHS Policy on Humane Care and Use of Laboratory Animals. http://grants.nih.gov/grants/guide/notice-files/not-od-06-052.html.
13. Garnett, N. and C. Gipson. 2001. Suggestions to bring electronic protocol system into compliance. *Contemp. Topics Lab. Anim. Sci.* [Letter to Editor]. 40(6):8.
14. Brown, P., and C. Gipson. 2006. A word from OLAW and USDA. *Lab Anim. (NY)* 35 (10):18.
15. Garnett, N.L. and C.A. Gipson. 2003. A word from OLAW and USDA, *Lab Anim. (NY)* 32(9):19.
16. National Institutes of Health. 2007. NOT-OD-07-044. NIH policy on allowable costs for grant activities involving animals when terms and conditions are not upheld. http://grants.nih.gov/grants/guide/notice-files/not-od-07-044.html.
17. Office of Laboratory Animal Welfare. 2010. Guidance on confirming appropriate charges to NIH awards during periods of noncompliance for activities involving animals. NOT-OD-10-081. http://grants.nih.gov/grants/guide/notice-files/NOT-OD-10-081.html.
18. Brown, P and C. Gipson. 2012. A word from OLAW and USDA. *Lab Anim. (NY)* 41:251.
19. U.S. Department of Agriculture. AWA Enforcement Information. http://www.aphis.usda.gov/animal_welfare/enforcement.shtml (accessed May 23, 2012).
20. 7 U.S.C. 54 §2149. http://corpuslegalis.com/us/code/title7/violations-by-licensees (accessed May 23, 2012).
21. Office of Management and Budget. 2004. Circular A-21. Cost principles for educational institutions. Sect. C,4,b. Revised May 10, 2004. http://www.whitehouse.gov/omb/circulars_a021_2004/.
22. Office of Laboratory Animal Welfare. 2011. Memorandum of Understanding Among the Animal and Plant Health Inspection Service U.S. Department of Agriculture and The Food And Drug Administration, U.S. Department of Health and Human Services, and The National Institutes Of Health, U.S. Department of Health and Human Services, Concerning Laboratory Animal Welfare. MOU 225-06-4000. http://grants.nih.gov/grants/olaw/references/finalmou.htm.
23. DeHaven, W.R. and N. Garnett. 1998. Personal communication.
24. Potkay, S., N.E. Garnett, J.G. Miller, et al. 1995. Frequently asked questions about the Public Health Service Policy on Humane Care and Use of Laboratory Animals. *Lab Anim. (NY)* 24(9):24–26.
25. Brown, P. and C. Gipson. 2009. A word from OLAW and USDA. *Lab Anim. (NY)* 38:47.
26. Garnett, N.L. A word from OLAW. 2002. *Lab Anim. (NY)* 31(5):21.
27. Brown, P. and C. Gipson. 2012. A word from OLAW and USDA. *Lab Anim. (NY)* 39:68.
28. Gipson, C. and C. Wigglesworth. 2006. A word from USDA and OLAW. *Lab Anim. (NY)* 31 (6):17.
29. Brown, P. Protocol deviations revisited [Correspondence]. 2009. *Lab Anim. (NY)* 38:45.

30

Postapproval Monitoring

Ron E. Banks

Introduction

Regulatory agencies have provided the framework for the operation of institutional animal care and use programs. Institutions have an obligation to oversee and manage animal research with a sense of integrity and focus. Toward this end, institutions establish and charge the IACUC, sometimes called the Animal Ethics Review Board or Animal Subjects Committee, with the performance of these institutional tasks. Therefore, on behalf of the institution, the IACUC has the overarching responsibility to ensure animal welfare and well-being.

Researchers work to move the boundaries of science, and most researchers do so with a focused intent on humane care and ethical use. This chapter recognizes that, on occasion, the drive for scientific advancement and the need for regulatory processes may not sufficiently connect. Unintended missteps have the potential of causing serious programmatic and research consequences. It is the goal of postapproval monitoring (PAM) to mitigate noncompliance on the part of the institution and the individual.

Consider one example where PAM might serve to prevent unintended missteps. NIH/OLAW policy provides for a maximum IACUC approval period of three years; and researchers are typically required to submit a proposal for IACUC review/approval that covers a three year period. In the author's opinion, this is one of the most difficult aspects of being a researcher. It can be difficult to accurately forecast experimental directions 3–6 months down the road; to specifically define the precise research direction of a project over a period of three years is even more challenging a task. Many researchers follow their curiosity, based upon experimental outcomes, undertaking studies which may be different from those in the approved protocol or amendment. An engaged PAM program can assist researchers with identification of minor missteps before they become a major regulatory problem for the institution and the individual.

This chapter begins with the premise that the AWA and its amendments, APHIS/AC Policies, the PHS Policy, and the various funding agency requirements remain valid as a central expectation that the institution will provide sufficient integrity of process to ensure that animals are being used as approved and as expected. While the time-tested activity of the IACUC semiannual review continues to be a necessary function of the animal care and use program, many animal research programs have greater diversity and complexity than in the past. Earlier methods of animal use oversight may no longer be as effective or sufficient to assure granting agencies, the public, regulatory agencies, and legislative bodies of the appropriate quality of care and use for animals. An institution with an effective PAM process has a simple and effective means of ensuring the institution's good name, extending the oversight of the IACUC in a practical and reasonable manner, enhancing the effective utilization of the AV and the clinical care staff, and providing collegial facilitation and communication for the researcher.

Readers of this chapter should note the author's use of the term Compliance Liaison refers to the individual, while PAM refers to the process. Institutions are advised that the selection of a term which espouses collegiality and support for the research team is a critically important aspect of building a PAM program which will work. As much as possible, the institution should discourage a vision of "auditing," "inspecting," or "monitoring." An effective Compliance Liaison engaging a collegial PAM process should be viewed as an institutional partnership with the researcher rather than as the "keeper of the kingdom," or "guardian of the animal's well-being."

30:1 What are the goals of a PAM program?

Reg. According to the AWAR (§2.31,c,1; §2.31,c,2) the IACUC should review, at least once every 6 months, the research facility's program for humane care and use of animals and inspect, at least once every 6 months, all of the research facility's animal facilities, including animal study areas. The AWAR (§2.31,d,5) also requires continuous review of animal activities at least annually. The PHS Policy (IV,B,1) similarly notes that the IACUC should, with respect to PHS-conducted or supported activities, review at least once every 6 months the institution's program for humane care and use of animals, using the *Guide* as a basis for evaluation. Neither the AWAR nor the PHS Policy use the term PAM, while the *Guide* (pp. 33–34) does suggest that PAM may serve as one tool of the IACUC to assure it is fulfilling its obligations of on-going oversight.

The PHS Policy (IV,C,1) also indicates that the IACUC should review proposed projects or significant changes in ongoing research projects for conformity with the Policy and the *Guide*. It further states that the IACUC shall conduct continuing review of activities covered by the Policy at appropriate intervals as determined by the IACUC, but not less than once every 3 years (PHS Policy IV,C,5). Further, the IACUC may suspend an activity that it previously approved if it determines that the activity is not being conducted in accordance with applicable provisions of the AWAR(§2.31,d,6), the *Guide*, the institution's Assurance, or the PHS Policy (PHS Policy IV,C,6).

The *Guide* (p. 33) states that PAM helps to ensure animal well-being and to provide opportunities to refine research procedures. Its 8th edition significantly enhanced the discussion and validity of PAM as a practical and effective method that may enhance the performance of the IACUC. The *Guide* notes that PAM may have many faces, including all types of protocol monitoring after the IACUC initial approval. The *Guide* also notes an important attribute of a functioning PAM—that of identifying opportunities to refine research procedures.

The *U.S. Government Principles for the Utilization and Care of Vertebrate Animals Used in Testing, Research, and Training* state in Principle VIII, "Investigators and other personnel shall be appropriately qualified and experienced for conducting procedures on living animals. Adequate arrangements shall be made for their in-service training, including the proper and humane care and use of laboratory animals."

Opin. While not enunciated in the *Guide*, refinement of a research procedure has the mutual benefits of improving animal well-being *and* improving research outcomes, and perhaps the reliability of research data. The point is that PAM is recognized by the *Guide* as more than an audit function or a game of "gotcha." PAM is a supportive and valuable animal research program tool.

The ability of many IACUCs to assure all aspects of the regulatory requirements is challenging, especially considering that many IACUC members are volunteers. Some institutions have an IACUC Office to assist with programmatic issues but IACUC service is largely a voluntary activity serviced by busy professional and non-affiliated members. As such, institutions must often seek adjunctive measures which extend the capabilities of the IACUC without engaging additional time commitment by the volunteer members. One such measure is the development of a PAM program. The goals of a PAM program include

- Ensuring animal well-being. Ensuring animal well-being is placed first in the list to remind us of a core necessity when using animals in research.
- Keeping the IACUC informed about program status and process. Compliance Liaisons serve a valuable role to communicate IACUC positions on matters of animal care, education, or training needs.
- Protecting the institution's reputation. PAM can provide evidence for the argument that animal care and use is performed properly.
- Assuring husbandry consistency across the institution. Consistency is important for animal well-being and reliable research outcomes.
- Serving as a resource to the research community. Compliance Liaisons assist with collegial facilitation of reports, rewrites, and new submissions; making the task of application submission, review, and approval smoother and more efficient.
- Supporting the advancement of strong science through a coordinated and sound basis for animal use.
- Assist with training research staff in animal care or animal use. Serving as an on-site educator may eliminate an unintended adverse event.
- Facilitating regulatory compliance. Since the author believes most people will follow the rules if they know what the rules are, having a Compliance Liaison who can work alongside researchers is helpful in sharing information about a particular rule or requirement.

The attention and focus of the Compliance Liaison should be in the order as outlined—at least with respect to the first item (ensuring animal well-being) and the last item (providing regulatory compliance). Programs that focus on compliance as the first goal often become adversarial in nature and not well supported by the research community. In contrast, programs that focus on ensuring animal well-being achieve regulatory compliance as a by-product of more collegial interactions. Programs of PAM must resist the natural development of characterizations such as "animal police" or "compliance watchdogs." A concerted effort of the institution, the IACUC, and the PAM participants should be to facilitate good behavior rather than trying to find and punish poor behavior.

30:2 "PAM Officer," "Animal police," or "Compliance Liaison"—what is the difference?

Reg. The AWAR (§2.31,c,3) allow the IACUC to determine the best way to conduct evaluations of the research facility's program. There are no regulatory directives or other federal guidelines that suggest how the individual charged with implementing the PAM program should function.

Opin. Words have consequences. Titles set the tone for a process. Though the title of the individual charged with PAM may seem trivial, it is not. Perception is often reality, and if the research community believes that they are not trusted, or that they are being accused of poor animal care practices by having the "animal police" intrude into their laboratory uninvited, the ability to develop a "culture of compliance" or a strong and mutually supportive program is compromised. On the other hand, if the research community views the Compliance Liaison as a member of the institution's research support team who by his or her actions assists the researchers in scientific inquiry while staying compliant with regulations and institutional expectations, the chances of success are much greater and the outcome of a collegial and productive environment—a culture of compliance—is more likely.

A critical aspect for institutional consideration in the early phases of PAM development is the selection of the proper title for the person responsible for PAM. Institutions which have chosen terms such as liaison rather than officer have set the foundation for a more collegial atmosphere and greater researcher support. No one likes to live fearful of receiving a ticket for speeding; almost everyone will be nervous when being watched by the "animal cops." Most are willing to accept encouragement to slow down for the potholes ahead or gracious assistance from a research colleague (a liaison) who helps keep the research on track.

30:3 Is PAM the same as the IACUC semiannual review?

Reg. According to the AWAR (§2.31,c) and PHS Policy (IV,B,1–IV,B,5) with respect to activities involving animals, the IACUC, as an agent of the research facility, shall

1. Review, at least once every 6 months, the research facility's program for humane care and use of animals.
2. Inspect, at least once every 6 months, all of the research facility's animal facilities, including animal study areas.
3. Prepare reports of its evaluations and submit the reports to the IO.
4. Review, and, if warranted, investigate concerns involving the care and use of animals at the research facility resulting from public complaints received and from reports of noncompliance received from laboratory or research facility personnel or employees.
5. Make recommendations to the IO regarding any aspect of the research facility's animal program, facilities, or personnel training.

The *Guide* (p. 33) notes that PAM may help "ensure the well-being of the animals and may also provide opportunities to refine research procedures. Methods include continuing protocol review; laboratory inspections (conducted either during regular facilities inspections or separately); veterinary or IACUC observation of selected procedures; observation of animals by animal care, veterinary, and IACUC staff and members; and external regulatory inspections and assessments."

Opin. No, PAM is not the same as an IACUC semiannual review. The most important goals of a PAM program are: (1) finding evidence of good performance and documenting the adherence to an IACUC approved activity; and (2) developing cultural change which results in improved animal care and more consistent animal use practices. IACUC semiannual review and PAM are parallel processes, both functions of the IACUC oversight methodology, which may be used to buttress the process of on-going oversight. It is important to note that while the semiannual inspection and PAM are parallel and mutually supportive, the activities are not the same, nor do they have the same specific outcomes.

Most IACUC semiannual inspections focus assessment on facilities, environment, storage, refrigeration, controlled drugs, and record keeping, especially if the semiannual review involves laboratory visits by appointment. It is less common for IACUC semiannual inspections to observe specific procedures of the approved protocol and usually even less opportunity to view ongoing research. For example, the IACUC may happen upon a rodent surgical procedure in progress and it will be even less likely that the IACUC will have the time to observe the preparatory steps, the intraoperative measures taken, and the postsurgical recovery procedure during the inspection. A Compliance Liaison performing PAM has the necessary time and required skill set to review those same laboratories and focus on procedures and activities, comparing the approved protocol or SOP with the observed practice. One might consider this in terms of the similarities and differences between the semiannual review and PAM by comparing the process to the manufacture of a car. The IACUC semiannual review is the quality assurance of the finished product sitting on the factory floor; the PAM is the road test—the real test of the performance of the product. PAM will identify problems not anticipated during manufacture; PAM will identify maintenance issues or areas where supplemental instruction to the driver would be helpful.

30:4 How is the PAM process different from IACUC semiannual inspections and program review?

Opin. The processes differ by purpose and practical process. The IACUC review process generally attends to facilities and documentation; rarely are procedures observed by IACUC semiannual inspection teams, and if so only a snippet of the entire procedure can be observed due to time constraints. In contrast, the Compliance Liaison focuses his or her assessment on ongoing care and use practices and experimental congruence with IACUC approvals (or SOPs). The IACUC semiannual inspection and review and PAM are complementary activities, each contributing an important bit of knowledge to the IACUC's information set regarding compliance with regulations and institutional expectations. Together, the two processes provide a more complete picture regarding outcome of compliance efforts than either could provide alone.

The processes differ also by persona and provision. The IACUC semiannual process is required, focused, routine, and performed by volunteer members of the committee. It does not routinely capture animal use activities in progress; nor does the semiannual evaluation likely assess the preparation, performance, or post-procedure care practices. The IACUC semiannual process often rotates the duty of inspection to a specific facility among its members. There are several advantages of such a practice, but there is also the disadvantage of inconsistency in a review system that is uncomfortable for the researchers. The PAM process, on the other hand, provides a resource that can dedicate the time to observe all phases of an experiment, giving a top-to-bottom assessment of protocol adherence. Compliance Liaisons provide a desired level of consistent review, can communicate with the researchers, develop a sense of collegiality with the process, can assist research staff with an understanding of the expectations of the PAM process, and can assure that the application of oversight is institution-wide. PAM facilitates communication with the research technical staff (below the level of the PI) that will encourage enhanced communication with the animal program, because research staff will know exactly who to call with questions—they have already developed a rapport with the Compliance Liaison during prior PAMs.

30:5 Now that I know what PAM is (see 30:2), what is PAM not?

Reg. There is no regulation that directly addresses the need for PAM (see 30:2) but the *Guide* (pp. 33–34) supports the concepts behind PAM. Additional information on postapproval monitoring is provided in NIH/OLAW FAQ G.6. The AWAR (§2.31,d,5) notes that the IACUC shall conduct continuing reviews of activities covered by this subchapter at appropriate intervals as determined by the IACUC, but not less than annually. It also states (§2.31,c,3):

- The IACUC may determine the best means of conducting evaluations of the research facility's programs and facilities; no committee member wishing to participate in any evaluation conducted under this subpart may be excluded. The IACUC may use subcommittees composed of at least two committee members and may invite ad hoc consultants to assist in conducting the evaluations; however, the IACUC remains responsible for the evaluations and reports as required by the act and regulations.
- The IACUC must review, and, if warranted, investigate concerns involving the care and use of animals at the research facility resulting from public complaints received and from reports of noncompliance received from laboratory or research facility personnel or employees.

Opin. It is sometimes easier to define an activity in terms of what it is not, and this may be the case with PAM:

- Compliance Liaisons are not "animal police" nor is a PAM an "animal police action."
- Compliance Liaisons should never be the "heavy" for the IACUC (it is the job of the IACUC to effectively oversee and manage the program for animal care and use).
- Compliance Liaisons are not the hatchet arm for the IO or the AV. Attitudes of this nature will severely disrupt and even destroy the ability of the Compliance Liaison to perform PAM in a manner which serves the institution by extending the oversight and educational functions of the IACUC.
- PAM, as a management tool of the IACUC, is not specifically required by APHIS/AC, NIH/OLAW, or AAALAC. Even so, all three agencies have recognized the value and benefit of a well-engaged PAM process.
- PAM is also not a mandate of any funding agency. Even so, most if not all funding agencies recognize the value of using PAM. All agencies leave the ultimate expression of program oversight to the institution and the IACUC.

- PAM is not any easier at large institution than it is at small institutions. While the nature and extent of PAM may differ among institutions, all have the same obligations to ensure competent and appropriate animal use oversight, and all have essentially the same tools to accomplish this task. At a smaller institution, the PAM staff may be part-time or may be engaged in other aspects of the program or even in non-IACUC activities. In this case, institutions should be sensitive to perceptions of conflict of interest. (Note: Generally there are fewer perceptions of conflict when the Compliance Liaison resides entirely within the "IACUC office.") At larger institutions, multiple PAM staff may be required to oversee the animal use across campus, and the expense of permanent Compliance Liaisons will be greater and the conflicts may be more political than at smaller institutions.
- PAM is not new ground or new regulation. The federal regulations have remained essentially unchanged since 1985, though the 8th edition of the *Guide* (pp. 33–34) has a concise but thorough narrative regarding PAM. Over the years, most programs have grown in scope and complexity, and as the clothes of a child no longer fit a mature adult (while the need for protection from the elements remains the same) so the institution must consider new methods to meet the same need for covering the institution's vulnerable natures. The operational processes may be new to some, but the requirement for program oversight is a long-standing expectations. What is new is that the public now expects an increased level of confidence that animal care and use activities are being properly conducted. The development of a PAM process is a simple and logical extension of the IACUC's obligation with respect to these concerns, while supporting the continued growth and maturation of the research enterprise.

30:6 Are quality assurance (QA) and PAM the same thing?

Opin. Maybe. The question is really one more of process than nomenclature. If the process is strictly regulatory in nature, then the research community may regard it as a QA activity or as an inspection. If the process is more collegial, providing clear commendation of positive observations, offering constructive suggestions toward resolution of concerns, providing (or scheduling) education as necessary, and assisting the research laboratory with open IACUC communication, then the program will assume a greater degree of respect, consistency and integrity (for the purpose of this discussion, this is referred to as PAM).

30:7 Is a PAM program a requirement?

Reg. While there is no specific regulation that directly addresses the need for PAM (see 30:5), according to the AWAR (§2.31,c) and PHS Policy (IV,B,7; IV,B,8) the IACUC (1) is empowered to review and approve, require modifications in (to secure approval), or withhold approval of proposed significant changes regarding the care and use of animals in ongoing activities; and (2) authorized to suspend activities involving animals. In addition, the AWAR (§2.31,d,1–6) and PHS Policy (IV,B,2; IV,B,3) state that the IACUC is responsible for (1) the review of care and treatment of animals in all animal study areas and facilities of the institution at least semiannually and (2) maintenance of appropriate records of reviews conducted. These regulatory passages suggest the need for ongoing, active oversight of animal research activities. Additionally, PHS-Assured institutions are required to follow the *Guide,* and the *Guide* (pp. 33–34) gives several advantages of engaging a PAM process at the institution. NIH/OLAW FAQ G.6 states: "Continuing IACUC oversight of animal activities is required and can be accomplished through a variety of mechanisms. Monitoring animal care and use is required by the PHS Policy, but the Policy does not explicitly require specific post approval monitoring (PAM) procedures to compare the practices described in approved protocols and SOPs against the manner in which they are actually conducted. IACUCs are charged, however, with program oversight and as such are responsible for program evaluations, reviews of protocols, reporting noncompliance, ensuring that individuals who work with animals are appropriately trained and qualified, and addressing concerns involving the care and use of animals at the institution."

Opin. There is a clear expectation that the IACUC will employ a mechanism to assure that animal care and use are in accordance with regulations and guidelines. All institutions using animals under the purview of the AWAR and the PHS Policy are required to assure that animals are being used in a manner that is approved by the IACUC. Funding agencies require that animals be used in a manner in which the grant funds were provided. The AWAR (§2.31,c,8) and the PHS Policy (IV,C,6) state that IACUCs must be authorized to suspend activities involving animals in accordance with certain specifications. The PHS Policy (IV,B,1–IV,B,3) and the AWAR (§2.31,c,8) state that the IACUC must evaluate and prepare reports on all of the institution's programs and facilities (including satellite facilities) for activities involving animals at least twice each year. The *Guide* (p. 25) states that the animal care program should be reviewed and the animal facilities inspected annually or more often as required. (See 30:2.) The institution is granted the freedom to choose the degree, depth, or complexity of its oversight processes, including PAM. Any oversight process should be responsive to the size of the program, the numbers of species, the activities using animals, the degree of invasiveness of approved activities, the level of care required, and many other variables.

While none of the agencies requires a PAM program, all require oversight of the process at an appropriately detailed level to ensure compliance and humane care of animals. In certain instances, the IACUC may be able to oversee the needs of the institution's program fully and without the use of Compliance Liaisons; in other cases the IACUC may not have enough members or time to accomplish the task adequately. Recognizing a common requirement for an oversight process, the question becomes, By what method does the institution assure what was approved by the IACUC is/was being performed as approved? Is your methodology for oversight working? What is an appropriate tool for assessing the effectiveness of your training program? For most programs, use of the semiannual review alone is not sufficient. Compliance Liaisons implementing a thorough PAM process will significantly improve institutional assurance that legal and ethical obligations are met.

30:8 What methods are useful for implementation of PAM?

Reg. The AWAR (§2.31,d,5) state that the IACUC shall conduct continuing reviews of activities covered by this subchapter at appropriate intervals as determined by the IACUC, but not less than annually. The AWAR (§2.31,c,3) make two statements related to the methods used for PAM in helping the IACUC fulfill its obligation to perform ongoing and annual reviews:

- The IACUC may determine the best means of conducting evaluations of the research facility's programs and facilities; no committee member wishing to participate in any evaluation conducted under this subpart may be excluded. The IACUC may use subcommittees composed of at least two committee members and may invite ad hoc consultants to assist in conducting the evaluations; however, the IACUC remains responsible for the evaluations and reports as required by the act and regulations.
- The IACUC must review, and, if warranted, investigate concerns involving the care and use of animals at the research facility resulting from public complaints received and from reports of noncompliance received from laboratory or research facility personnel or employees.

The PHS Policy (IV,C,3) also indicates that the IACUC may use consultants to assist in the review of complex issues. None of these references prohibit the use of PAM, and all provide sufficient grounds for employment of an "extra-IACUC" process for assuring the integrity of program oversight. The AWAR do not, however, prescribe a specific method for the IACUC to undertake PAM should it choose to do so. The *Guide* (pp. 33–34) notes that PAM should be considered in its broadest sense. NIH/OLAW FAQ G.6 indicates that PAM may be a useful tool of the IACUC to gather programmatic information and assure overall programmatic compliance.[1]

Opin. There are three general methods for conducting PAM:

1. IACUC members conduct monitoring
2. Veterinary staff conducts monitoring
3. Dedicated staff (either full or part-time) to perform PAM

Sometimes, institutions may choose a hybrid of these basic models, although a simple approach is usually best. The key is not so much the specific method, but choice of the method that best ensures and supports the institutional objectives. As a general rule, the more complex or intensive research enterprises benefit by employing dedicated Compliance Liaison staff. Smaller, more limited programs might be able to accomplish the task successfully by having IACUC members conduct PAM. Contrasting with options one and three above, in the author's experience, using the veterinary staff to conduct PAM is the least desirable choice regardless of program size. If researchers view the veterinary staff as monitors, the veterinary staff may be viewed as roadblocks or police. Since it is desirable for there to be a partnership between the veterinary staff and the research staff, the general use of the veterinary staff for PAM should be avoided. Of course, veterinary staff should be encouraged to alert the Compliance Liaison of any concerns related to animal use; and such concerns can then be reviewed by the Compliance Liaison, who might subsequently recommend the researcher contact the veterinary staff for assistance in solving the problem. It is always better for the veterinary staff to be viewed as part of the solution rather than as part of the "enforcement." (See 30:10.)

30:9 What are the advantages and disadvantages of having IACUC members perform PAM?

Opin. In most situations, the disadvantages outweigh the advantages of IACUC members conducting the PAM. While IACUC members can serve effectively in this role, they frequently have problems with availability (members are generally very busy volunteers with full-time occupations); consistency (different members performing reviews differently) can be a problem; and, conflict of interest (the reviewer and reviewed laboratories may have collaborated or may have competed for the same research space or other resources) can also be a concern. IACUC members already dedicate significant effort to the review of animal use proposals and their modifications; participate in semiannual program reviews and facility inspections, and many other special assignments. While critically important activities, committee membership and participation in committee activities are not income-generating activities for the researcher or the institution; and therefore not a preferred place for the IACUC volunteers to focus their time and energies. The interests of most institutions are best served by having researchers focus on their own research, rather than their spending time observing the research techniques of others. Further, while IACUC members are usually well intended, they generally do not have the depth of familiarity with regulatory requirements and generally do not possess sufficiently broad background in animal research and care to provide the institution with a consistent PAM process. Having a large number of IACUC members performing PAM will introduce a degree of inconsistency that may be discouraging to the researcher and marginally productive for the institution. Lastly, for many IACUCs, maintaining reviewer anonymity is an important protection for the IACUC membership to discourage institutional territorial battles; therefore, sending an IACUC member to review a colleague's laboratory may present conflicts that interfere with consistent and accurate assessment, effective communication, and teamwork.

30:10 Is it advisable for the veterinary staff to serve as the eyes and ears of the IACUC?

Reg. The AWAR (§2.31,d,1,vi; §2.31,d,1,vii) note that the AV or other scientist trained and experienced in the proper care, handling, and use of the species being maintained or studied must direct the housing, feeding, and nonmedical care of the animals, and that medical care will be provided by a qualified veterinarian. The *Guide* (p. 106) states that the AV is responsible for the overall program of veterinary care and further indicates (p. 34) that the veterinary staff might be used to observe increased risk procedures for adverse events. The PHS Policy (IV,C,1,e) indicates that medical care for animals will be provided by a qualified veterinarian. None of these documents prescribe a role for the veterinary staff in performing PAM for the IACUC.

Opin. Though the veterinary staff often has the most extensive background in animal health, the demands on them are often such that there is little time during the course of the day to trade hats, stop providing veterinary care or husbandry, and start PAM for animal use. Secondly, although the veterinary staff may be in a position to provide assessment of the use of animals, a significant conflict may arise if the PI perceives the veterinary staff as choosing to perform PAM versus animal care or investigator support. Using the veterinary staff as the sole (or primary) mechanism for assuring adherence to an IACUC-approved protocol can become self-defeating. The institutional goal should be for the veterinary staff to assist researchers, and to encourage, improve, and assure high-quality animal care. A veterinary staff that regularly reports deviations may discourage openness and communication between the research and veterinary staffs. Additionally, certain research activities do not occur at times when the veterinary staff is present, thus presenting a significant complication for PAM that is solely performed by the veterinary staff. Further, because PAM should review the entire animal care program—including veterinary care—the veterinary staff would have a conflict of interest with respect to PAM. It is generally preferable to prevent any conflicts of interest and maintain PAM and veterinary care as separate and distinct operations.

30:11 What is meant by a dedicated PAM staff? If the institution is not large enough or has the fiscal resources to employ someone full-time, what are the options?

Opin. The term dedicated does not refer to a personnel allotment but rather an assignment of function. The importance of having dedicated staff is focused more on the assignment of responsibility and prevention of conflicts than on the length of business engagement. A dedicated staff member may spend only a percentage of his/her time performing the activities of PAM. Other appropriate duties of the PAM employee may include coordinating IACUC business, providing training for new faculty or staff, serving as IACUC recorder, or assisting with report generation to the regulatory agencies.

30:12 What are the advantages of using the dedicated staff model of PAM?

Opin. Having a dedicated staff to perform PAM allows other groups, such as the IACUC and veterinary staff, to focus on the aspects of animal care and use for which they are responsible. For example, by using dedicated PAM staff, the veterinary staff can focus on animal health and facility management. Having dedicated PAM staff allows the veterinary staff to serve more directly as facilitators to enhance animal care and assist the investigators with proper animal use.

A dedicated PAM staff can also serve as a focused continuing education and training service, addressing specific issues of noncompliance as opportunities for training. For example, when problems are identified in a laboratory, the Compliance Liaison might provide information on required, improved, or enhanced alternate techniques or processes. In this way, the Compliance Liaison can serve as a user-friendly resource and at the same time help to ensure that activities are performed in conformance with the requirements and expectations of the IACUC. In most circumstances, PAM staff members will interact with the research associates and technicians more than PIs, thus establishing communications networks that are not routinely employed by the IACUC. Over time, the PAM staff gains an institutional memory and can provide a sense of stability to the IACUC.

The most valuable role of PAM staff is providing the IACUC with firsthand reports of ongoing activities in the laboratory on a regular and recurring basis. The amount of PAM time given to a particular activity should be determined by the needs of the program. For example, in institutions where rodent breeding represents the majority of the animal activity, the time needed to perform a reasonable level of PAM will likely be less than at an institution where the principal activity involves survival surgery.

30:13 What types of individuals are best suited to serve as Compliance Liaison staff?

Opin. The selection of individuals who will perform PAM should be less concerned with credentials than with perspective and personality. The IACUC requires clear and accurate information; therefore, individuals who have excellent observation and good communication skills are ideal. The best Compliance Liaisons observe those small aspects of laboratory activities that can impact the care or use of animals. For example, assume PAM was performed to review post-procedural care of surgery animals. During the assessment, the Compliance Liaison also observed whether animals were being checked with sufficient frequency (as IACUC approved) while in the laboratory, reviewed each instance of drug administration noting how is was recorded, whether there were thermometers located near the animals (versus in the room but distant to the animals), and whether proper aseptic technique is being used for all survival surgeries. Each of these observations are indirectly related to the primary issue of assessing post-procedural care, but collectively may have impact upon animal well-being. A great Compliance Liaison observes things that are not required and builds an assessment based upon expected and unexpected observations.

The PAM staff also requires good communication skills. The researchers may at times be less than pleased with perceived intrusion into their laboratories. Compliance Liaisons who can communicate in a firm but non-adversarial style are worth their weight in institutional gold. These staff provides quiet confidence and assures professionalism in all interactions.

The author interviews prospective Compliance Liaisons by asking that they describe what they observed from the time they entered the building until they arrived at the IACUC Office. The goal of such an exercise is to identify those individuals who have good situational awareness and are able to observe and note conditions and situations without being told to do so and without being tunnel-minded. The author believes the best Compliance Liaisons have the ability to make "lemonade out of lemons turning adverse interactions into positive teachable moments while describing the size of the lemon tree and noting the number of lemons on each branch observing the work around them in detail."

Beyond these skills, individual programs might have specific issues that would benefit from individuals with certain skills and backgrounds. For example, an institution having a surgical program would be well served to find someone with a good understanding of surgical technique and related skills (e.g., veterinarian, technician with surgery experience, or M.D. surgeon). An institution conducting a significant amount of antibody production would be best served by having a PAM staff member who has relevant experience and knowledge regarding expected standards and procedures in that area.

Generally speaking, "nice people" who have good observation skills can be trained to assess the accuracy of IACUC approvals and expectations with the performance of the procedures in the laboratory or animal care facility. It is far more difficult to teach people to speak well and present a pleasing style, no matter what their degree of technical expertise.

30:14 What might be the job description for Compliance Liaison staff?

Opin. While specific job tasks may vary from institution to institution (and be based on specific needs of the institution's program for animal care and use) a general template for performing protocol assurance and other IACUC-related duties might include general categories of responsibility that are highly parallel to the obligations of the IACUC, such as the following:

- Audits for consistency with approved activities
- Communicate effectively with PIs, lab managers, veterinary staff, and IACUC and animal care program leadership concerning observations made
- Conduct observations and audits of approved animal use activities by monitoring and auditing approved animal use activities of the types noted (e.g., compliance review, procedure observation, surgical facility assessment, laboratory assessment) (also see 30:17)
- Report activities in a clear, concise, and accurate manner to the IACUC, laboratory managers, PIs, and veterinary staff
- Communication and program enhancement ideas for the IACUC
- Develop (or assist in development) of systems and relationships to support communication and information exchange between the IACUC and the research community
- Assist PIs in developing and implementing corrective action for relevant findings in laboratory assessment reports
- As instructed by the IACUC, prepare and assist in accomplishing corrective actions of IACUC-defined noncompliant items
- Identify and report other program concerns that may arise through the work process
- Draft revised SOPs
- Recommend development of other activities or training events that enhance the institution's program for animal care and use
- Benchmark procedures against other institutions and determine best practices for the local campus construct, making recommendations as appropriate
- Identify training and educational needs regarding species-specific and procedural practices based upon auditing activities
- Draft enhancements to the institution's program of training as observations indicate necessary
- Implement and enhance programs of training for animal care and use personnel
- Coordinate the implementation of other global institutional training programs (e.g., orientation to the institution, regulations, rules, protocol preparation) for the animal care and use community
- Coordinate with the animal care unit to assure an integrated and mutually supportive training program for all animal handlers and users on campus

- Serve as a reference resource
- Review technical publications, articles, and abstracts to stay abreast of current regulations, trends, and "official guidance"
- Draft program enhancements in response to the reviewed literature
- Monitor incorporation of actions into workplace environment
- Advise the IACUC on trends within the field
- Collaborate (as necessary) with the Office of Compliance or legal office in matters that require their attention
- Serve as liaison to other institutional monitoring and audit groups

30:15 Is it advisable for a Compliance Liaison to serve as a voting member of the IACUC?

Reg. Neither the PHS Policy nor the AWAR address the specific role of IACUC members in monitoring activities after IACUC approval, however, the institution must always be on-guard for conflicts of interest which are prohibited under the AWAR (§2.31,d,2). The AWAR (§2.31,d,1,xi,3) and PHS Policy (IV,B,3, footnote 8) state that consultants may be used to review proposed activities but that these consultants may not vote with the IACUC unless they are also members of the IACUC. The *Guide* (pp. 33–34) notes that continuing IACUC oversight of animal activities is required by federal laws, regulations, and policies, and includes a review of the value of the PAM process, but leaves absent discussion of the official IACUC relationship with respect to a Compliance Liaison.

Opin. While it is appropriate for a Compliance Liaison to attend IACUC meetings, it is not necessarily in the best interest of the IACUC to have Compliance Liaisons serve as voting members, though there is a wide array of opinion on this matter. The author believes that the Compliance Liaison should be considered an extension of the IACUC (e.g., a consultant). If the Compliance Liaison is regarded as serving in a consultant role to the IACUC, then it is not appropriate (according to the AWAR) for a Compliance Liaison to be a voting member of the IACUC. While the Compliance Liaison is only one member of the Committee, and as with all members has only one vote, his or her role in the process has a real risk to be inflated. It is true that the committee is expected to decide on noncompliance, the IACUC is not obliged to accept any specific recommendation on corrective actions, and the committee may modify any PAM-derived proposal to the committee. Nevertheless, the Compliance Liaison remains the individual who typically has the confidence of the committee, provides guidance to the committee concerning PAM outcomes, and is the individual who will relay the committee's decision to the research staff. All of these activities occur in the context of the Compliance Liaisons' personal bias and individual interpretation of the vote in which he or she participated. These views are based on the author's observation of certain IACUCs where the Compliance Liaison was a voting member. This is neither appropriate nor encouraged and may actually develop into a perceived or actual conflict of interest (i.e., reporting observations with a preferred outcome of the vote). In the author's opinion complete separation of the decision-making options of the Compliance Liaison, as it regards IACUC-determined corrective measures, is preferred and best serves the needs of the IACUC, the institution, and the researcher.

Of those institutions having some form of process oversight, this author is aware of only a few that have Compliance Liaisons as voting IACUC members; most simply have Compliance Liaisons attend the IACUC meetings and provide insight and counsel to the IACUC without voting. Some institutions have reported concerns with the Compliance Liaison having a vote, since Compliance Liaisons often have special or unique knowledge of studies and investigators and that knowledge could introduce bias to a vote.

Serving as an IACUC member may also complicate the job of the Compliance Liaison. It is far more productive for Compliance Liaisons to report the decisions of the IACUC than to defend the decisions of the IACUC as a member who participated in that decision. The purpose of the Compliance Liaison is to serve as the eyes and ears of the IACUC. The IACUC should guard against the development of the Compliance Liaison as the committee's gatekeeper. Institutions should be very cautious of engendering more authority in the Compliance Liaison than in the IACUC.

30:16 Is it advisable for the Compliance Liaison to work directly for the AV?

Reg. There are no specific regulatory or other guidelines to define the work relationship between the AV and the Compliance Liaison.

Opin. While possible, it is nevertheless best if the Compliance Liaison does not report directly to the AV. The AV is responsible for certain programs (e.g., animal husbandry) that should be under the global oversight of the IACUC. For purposes of the institution's compliance process review, the AV (and the husbandry units) should be considered as "another PI" working with the institution's management of animals via SOPs, and can be viewed as another activity under the oversight of the IACUC. The process of SOP development is usually different from protocol development and the approval process is frequently different (e.g., AVs approve husbandry SOPs; IACUCs approve protocols), but at the end of the day, there is no simple method to distance animal care from research. Care affects research and the provision of animal care via SOPs must be taken as seriously as experimentation via an approved protocol is taken. Compliance Liaisons must be free of actual and perceived conflict (e.g., reporting husbandry failures to their supervisor).

If the Compliance Liaison reported to the AV, a significant conflict of interest would exist. The Compliance Liaison would be in the difficult position of reporting process deviations to the individual who was potentially culpable for the deviation, responsible for personnel administrative matters concerning the Compliance Liaison, and obligated to ensure animal welfare and well-being. In other words, the Compliance Liaison would be reporting concerns directly to an individual who could create adverse administrative conditions for the Compliance Liaison, or discourage accurate reporting to the IACUC. This could lead to a biased and unproductive PAM process. Programmatic integrity requires auditing processes which are stable, consistent, and applicable to all animal handlers and users without regard to academic or other credentials. The PI community will become more supportive of a PAM program which is equitably applied to all animal user groups, including the veterinary and animal care staffs.

30:17 Should the Compliance Liaison work directly for the IACUC?

Reg. There are no specific regulatory or other guidelines to define the work relationship between the Compliance Liaison and the IACUC. However, according to the AWAR (§2.31,c) and PHS Policy (IV,B,1-IV,B,5) with respect to activities involving animals, the IACUC, as an agent of the research facility, shall

1. Review, at least once every 6 months, the research facility's program for humane care and use of animals
2. Inspect, at least once every 6 months, all of the research facility's animal facilities, including animal study areas
3. Prepare reports of its evaluations and submit the reports to the IO
4. Review, and, if warranted, investigate concerns involving the care and use of animals at the research facility resulting from public complaints received and from reports of noncompliance received from laboratory or research facility personnel or employees
5. Make recommendations to the IO regarding any aspect of the research facility's animal program, facilities, or personnel training

Opin. If the Compliance Liaison is to serve as the eyes and ears of the committee, then it seems appropriate that the Compliance Liaison report to the IACUC. It is the IACUC that has responsibility and authority for oversight, but the IACUC is composed of a membership that has other duties, unrelated to the IACUC, which consume a great deal of time. Thus, by directly reporting to the IACUC, the Compliance Liaison facilitates fulfillment of the IACUC's responsibility for animal care and use oversight.

In some cases, the Compliance Liaison reports to an institutional compliance official. There are certain synergies which such an arrangement may offer, but it is critical that the institutions' compliance officer recognizes that the IACUC is obligated to fulfill its role as investigator, deliberator, and decision-maker in matters regarding animal care or animal use.

Some institutions choose to have the Compliance Liaison report directly to the IACUC chairperson. This may present problems if the IACUC chairperson is an animal user, serves only part-time as the Chair, or is not always available. It may also present challenges if the Chair is a rotated position, as occurs at some institutions.

Regardless of the administrative relationship of the Compliance Liaison, there must be a clear understanding that all concerns are within the scope of the IACUC to deliberate and determine outcomes. In all cases, the Compliance Liaison must function in support of the IACUC.

30:18 Which type of issues would typically be reported by the Compliance Liaison to the IACUC?

Reg. Notice NOT-OD-05-034[2] provides guidance to assist IACUCs and IOs in determining what, when, and how situations should be reported to NIH/OLAW under the PHS Policy (IV,F,3) and to promote greater uniformity in reporting. PHS Policy (IV,F,3) requires that "the IACUC, through the Institutional Official, shall promptly provide OLAW with a full explanation of the circumstances and actions taken with respect to:

a. any serious or continuing noncompliance with this Policy;
b. any serious deviation from the provisions of the *Guide*; or
c. any suspension of an activity by the IACUC.

Opin. A comprehensive list of definitive examples of reportable situations is impractical in this review. Therefore, the following examples do not cover all instances but demonstrate the threshold at which NIH/OLAW expects to receive a report; thus, it is reasonable that these same types of situations are those that the Compliance Liaison should report to the IACUC. The NIH/OLAW insists that institutions use rational judgment in determining what situations meet the provisions of PHS Policy IV,F,3 and fall within the scope of the examples that follow.[2] NIH/OLAW has also determined that situations that meet the provisions of IV,F,3 and are identified by external entities such as APHIS/AC or AAALAC, or by individuals outside the IACUC or outside the institution, are not exempt from reporting under IV,F,3.[2] This list, while not exhaustive, should be used as a starting point for what should be reported to the IACUC for review and consideration:

- Conditions that jeopardize the health or well-being of animals, including natural disasters, accidents, and mechanical failures, resulting in actual or potential harm or death to animals
- Conduct of animal-related activities without appropriate IACUC review and approval
- Failure to adhere to IACUC-approved protocols
- Implementation of any significant change to IACUC-approved protocols without prior IACUC approval
- Conduct of animal-related activities beyond the protocol expiration date established by the IACUC
- Conduct of official IACUC business requiring a quorum in the absence of a quorum
- Conduct of official IACUC business during a period of time that the committee is improperly constituted
- Failure to correct deficiencies identified during the semiannual evaluation in a timely manner
- Chronic failure to provide space for animals in accordance with recommendations of the *Guide* unless the IACUC has approved a protocol-specific deviation from the *Guide* based on written scientific justification
- Participation in animal-related activities by individuals who have not been determined by the IACUC to be appropriately qualified and trained
- Failure to monitor animals post-procedurally as necessary to ensure well-being (e.g., during recovery from anesthesia or during recuperation from invasive or debilitating procedures)

- Failure to maintain appropriate animal-related records (e.g., identification, medical, husbandry)
- Failure to ensure death of animals after euthanasia procedures (e.g., failed euthanasia with carbon dioxide)
- Failure of animal care and use personnel to carry out veterinary orders (e.g., treatments)
- IACUC suspension or other institutional intervention that results in the temporary or permanent interruption of an activity due to noncompliance with the PHS Policy, AWAR, the *Guide*, or the institution's Animal Welfare Assurance

Note: Of the above list, only those items involving PHS-funded activities or having an impact upon PHS-funded activities require reporting to NIH/OLAW. However, NIH/OLAW will accept reports on noncompliance in activities that are not funded by the PHS to assist the institution in developing policies or procedures for correcting the noncompliant situation and preventing a reoccurrence. If informed, NIH/OLAW is able to respond to inquiries by Congress, the media, and the public regarding the situation. Non-PHS funded activities which are non-compliant and require a corrective action may be reportable to AAALAC (if the institution is accredited), to the funding agency (if so stipulated in the grants agreement), or to the external collaborators (if a Memorandum of Understanding between the institutions necessitates such a report).

Examples of PAM observations that might be reported to the IACUC, but would not result in reporting to NIH/OLAW, include the following:

- Deaths of animals that have reached the end of their natural life spans
- Deaths or failures of neonates to thrive when husbandry and veterinary medical oversight of dams and litters was appropriate
- Animal death or illness from spontaneous disease when appropriate quarantine, preventive medical, surveillance, diagnostic, and therapeutic procedures were in place and followed
- Animal deaths or injuries related to manipulations that fall within parameters described in the IACUC-approved protocol
- Infrequent incidents of drowning or near-drowning of rodents in cages when it is determined that it was caused by the animals jamming the water valves with bedding.[2]

(See 29:52.)

30:19 What process should be used for monitoring and assessing activities assessed as part of PAM?

Reg. There are no specific regulatory or other guidelines to direct an institution with respect to the process used for PAM.

Opin. Over the last few decades, a clear and observable shift has occurred in the philosophy of animal use oversight, from one referred to in the *Guide* (pp. 6–7) as the engineering standard to performance and practice standards. As stated in the *Guide*,

"Ideally, engineering and performance standards are balanced, setting a target for optimal practices, management, and operations while encouraging flexibility and judgment, if appropriate, based on individual situations…" In an operational sense, this means that one size does not fit all, and that programs must be built and managed to achieve an appropriate outcome, rather than have duplicative or nonfunctional institutional offices. Recognizing that the desire is for an appropriate outcome, a single methodology for PAM is likely to be insufficient for the breadth of activities that occur in most programs. Compliance Liaisons typically benefit from using a variety of methods to complete the task of accurate assessment for the IACUC. These include the following:

- Comparative review: A relatively formal process, comparative review is intended to be nonthreatening and collegial in nature. Of the various types of PAM review, comparative review is the closest to an IACUC semiannual review, though it is usually performed with greater intensity than an IACUC semiannual review. The various activities that typically occur as part of comparative review include the following:
  - Review of the approved protocol(s) in advance, to understand fully the focus of the laboratory and activity, and to allow a fair and accurate review of the procedures
  - Comparison of animal usage (purchase or breeding) numbers to date, with the approved numbers in the protocol (to assess laboratory activity from a program perspective)
  - Conduct of an entrance interview with researchers (or senior lab managers) to explain the PAM process and the potential outcomes of the process
  - Inquiry about any observed, unexpected deaths or other adverse events, and the outcome for such observed animals
  - Review of procedure records, looking for consistency of procedures with the approved protocol
  - Review of animal use and study areas to assess proper animal care
  - Review of laboratory training records to determine what training may be needed for the procedures being performed (and subsequent scheduling of that training)
  - Confirming that all members of the laboratory having risk for animal-related injury or illness have participated in the institution's medical surveillance program (Note: the Compliance Liaison usually does not have access to personal health care information; instead, the question is whether the laboratory individuals are enrolled in an appropriate medical surveillance program or not)
- Procedure observation: The procedure observation is a critical aspect of the process of effective PAM. It may be integrated into, or used as a direct follow-on to the comparative review or in some instances as a substitute for the comparative review. The procedure observation provides unique and special information to the IACUC and should be used to the greatest extent possible. It is important to observe activities being performed on animals and ensure that they are consistent with those described in the approved protocol. The

process of procedure observation may be randomly conducted and exercised on all protocols over a given period; or the process may be selective, based on the risk to animal well-being and specific institutional concerns. The procedure observation must be transparent with respect to routine laboratory and research procedures. Injecting too many questions during a procedure or asking for a demonstration of an activity introduces artificial variables into the observed procedure that may alter the normal practice and may create the appearance of a noncompliant activity. A procedure observation can be applied to both IACUC-approved experimental protocols and animal care SOPs (such as those describing routine husbandry and medical care). Procedure observation might review the following aspects of animal care and use:

- Presence and condition of equipment and supplies required to assure care of the research animals
- Husbandry and care practices for animals before, during, and after the procedure
- Animal manipulations: observing specific procedures and confirming they are being performed as approved in the protocol
- Transport of animals to and from housing to the experimental areas to determine whether the transport is being conducted consistently with institutional practices
- Personnel qualifications with respect to the procedures being performed to establish that the individuals working with the animals exhibit the behaviors and expertise expected and consistent with those stated on their personnel qualifications form
- Assessment of findings, potentially sharing any concerns with the IACUC, the AV, and the compliance officer for the institution

- Surgical facility (and use) assessment: The Compliance Liaison should also monitor the use of animal surgical facilities. While the IACUC performs a similar function as part of the IACUC semiannual assessment, rarely will the IACUC see a surgery in progress. The Compliance Liaison should schedule a time to observe ongoing activities and provide the institution with the confidence that the space, equipment use, and the activities that occur in that space and with that equipment are appropriate and are being conducted as approved by the IACUC. The general concerns that would typically be considered as part of this type of review include the following:
  - Physical plant acceptability, to ensure that the operative area has the surfaces and equipment necessary to perform aseptic surgical procedures
  - Housekeeping and sanitation procedures, to assure that the operative areas have been prepared appropriately prior to use
  - Drug inventory and utilization (including controlled substances management)
  - Sterilization method and packs
  - Personnel skills and training
  - Emergency protocols for adverse procedural events

- Anesthetic equipment and scavenging
- Anesthetic and analgesic regimen
- Post-procedural monitoring, nursing, and supportive care
- Methods for peri-procedural (pre-, intra-, and post-procedural) care appropriate for the species in use
- Overall conduct of major operative surgeries
- Surgical records

- Laboratory assessment: The Compliance Liaison may also visit laboratories to ensure the approved practices are being performed as anticipated by the IACUC. These visits may be random or selective (based upon a risk assessment of institutional concerns) and may include both procedure and study areas. The general items that could be considered include the following:
  - Physical plant acceptability (e.g., for rodent surgery performed in the laboratory, verification that the operative area has the surfaces and equipment necessary for aseptic surgery)
  - Housekeeping and sanitation procedures, determination that animal use areas are orderly and prepared appropriately prior to use
  - Drug inventory and utilization
  - Sterilization method and surgical packs
  - Personnel skills and training
  - Emergency protocols for adverse procedural events
  - Anesthetic equipment and scavenging
  - Anesthetic and analgesic regimen
  - Assurance that post-procedural monitoring, nursing, and supportive care are provided
  - Assurance that pre-, intra-, and post-procedural methods are appropriate for the species in use

Other activities such as controlled substance auditing, while not a specific obligation of PAM, provide additional insight into the function of the laboratory, serve the research community by assuring compliance with state and federal regulations, and may prevent institutional fines and citations. Compliance Liaisons may provide a similar service for the institutional safety unit, observing potentially unsafe conditions and sharing their concerns with experts from the safety group. A key attribute of an effective compliance program is integration and communication across lines of responsibly with the focus of building a strong, secure, and functional institutional process.

30:20 How frequently should PAM be performed?

Reg. There are no regulatory directives or other guidelines that suggest that visits by either Compliance Liaison or the IACUC should be scheduled versus unscheduled. The AWAR §2.31,d,5 (continuous annual review) and §2.31,c,3 (semiannual inspection) charges the institution with review of concerns and review and approval of ongoing activities. The PHS Policy (IV,B,7; IV,C,1; IV,C,5) suggests

IACUC should have a process of continuing oversight. The *Guide* (p. 25) identifies an expectation that the institution have a process of ongoing assessment of animal care and use.

Opin. While any cyclic approach could be chosen, the ends of the spectrum are likely the only wrong answers. For example, one every three-year lab visitation would be ineffectual whereas too frequent lab visitations (e.g., quarterly) would be considered harassment and interference. The institution (generally the IACUC) should consider the entire program, the risks associated with various aspects of the program, and measures to identify risk and mitigate risks as the most minimal levels. Activities selected for auditing may be identified by one or more of the following methodologies, each having a different preferred frequency:

- Hazard analysis (e.g., protocol agents, procedures)
- Risk assessment (e.g., prior laboratory history)
- IACUC concerns (e.g., missing documentation/sloppy laboratory procedures observed)
- Anonymous complaints
- Attending Veterinarian requests, and/or
- Directed audits ("For Cause Audits") identified by the IACUC, AV, or Compliance Office

For example, a researcher with many research associates or a high turn-over of associates may have a greater risk factor for protocol departure. A researcher having prior noncompliance issues may be viewed as requiring a more frequent audit cycle than a researcher with several years of stellar performance.

Because protocols have a maximal life span of three years, and PIs may have several protocols, a good "middle-of-the-road" approach to frequency may be reviewing one of the PI's protocols each year. The synergies which often occur with such an approach results in the Compliance Liaison having a visit to each laboratory on campus at least once a year and more frequently in higher risk labs. Since many researchers perform very similar procedures over all of their protocols (e.g., euthanasia is commonly identical, surgeries may be different but the personnel and preparatory or care procedures are often the same) choosing one protocol each year gives a good insight into the culture and behavior in the laboratory—which may serve as an indicator that more frequent assessments would be beneficial.

One approach that has found particular merit is to track all compliance audits over time and trend those kinds of procedures, laboratories, or buildings that bubble to the top of the list of monthly compliance reports. Just as with the use of compliance trends as the basis for education (see 30:31) the same compliance report can be used as a tool to identify those activities, labs, or areas that require an enhanced frequency of monitoring. The goal of this process is to maximize the available Compliance Liaison resources by visiting every PI approximately annually and those with an elevated risk more frequently. This approach is predicated upon the believe that the role of compliance auditing *is not* to catch people doing wrong, but rather to identify problems, engage effective mitigation and corrective measures, and assure a continued allegiance to approved practices.

30:21 Should a PI's self-corrective actions be considered during the compliance review process?

Reg. There are no specific regulatory or other guidelines to define the actions taken by a PI in the overall assessment of a compliance review.

Opin. Were the tasks of compliance monitoring a policing action, then the focus should remain within the PAM office irrespective of any measures taken by a laboratory to correct a non-compliant behavior. There have been occasions where a regulatory agency has cited or fined institutions for actions that were non-compliant but also which were self-identified and self-corrected. It is important to note, however, that the more effective institutional programs of compliance monitoring seek to partner with the researcher, identifying and correcting concerns as they occur. The researcher is obligated to self-report unanticipated outcomes of animal use and subsequently modify a process or procedure through an amendment. The preferred mechanism could involve a PI's self-report to the IACUC of a sub-optimal event which was corrected immediately upon disclosure. The Compliance Liaison could provide additional background and validate the identification and appropriateness of the corrective measure. In this role, the Compliance Liaison is viewed as partnering with the researcher to assure compliant performance of animal activities while also assuring transparency with regard to information available to the IACUC. The IACUC would consider the event, circumstances, and self-directed mitigation and corrective plans. If the approaches are consistent with institutional policies and procedures, it is plausible that the need for IACUC mandated corrective measures could be reduced. It is also probable that the IACUC would view this researcher as one who is focused on compliant research, committed to animal well-being, and determined to be an active partner in regulatory oversight. Yes, PI self-corrective measures should be considered as part of the overall report and evaluation of the event. Note: Events that are NIH/OLAW reportable will still be reported, but with notation of the researcher's proactive performance.

30:22 Should PAM visits be scheduled or unscheduled? (See 23:27.)

Reg. There are no regulatory directives or other guidelines that suggest that visits by either a Compliance Liaison or the IACUC should be scheduled versus unscheduled.

Opin. There are advantages and disadvantages with either option. Scheduling of visits encourages a sense of collegiality and institutional support for the research enterprise by meeting the research team on their grounds at a convenient time. On the other hand, scheduled visits may not provide an unvarnished assessment of the animal care and use procedures. Anonymous concerns reported to the IACUC may require unscheduled visits. These must be handled graciously, presuming innocence of any noncompliance until the IACUC has reviewed a report, deliberated the information and determined an outcome. The Compliance Liaison must be careful to not make any decision or guide a process toward guilt or innocence. Instead, his or her role is to gather information and present that information to the IACUC in a clear and impartial manner. Sudden and frequent appearance by the Compliance Liaison in any laboratory may be viewed as harassment, so the unscheduled option should be employed as a focused event, with a defined purpose.

30:23 Once observations have been made by the Compliance Liaison, how should that information be handled?

Reg. There are no regulatory directives or other guidelines that suggest how Compliance Liaison observations should be handled.

Opin. All Compliance Liaison activities in research or vivarium spaces should be performed in a standard manner, ensuring consistency, equality, fairness, and completeness:

- An initial assessment of the observations: In some instances, the Compliance Liaison may provide a quick exit briefing to the laboratory manager or PI prior to departing the laboratory. In other instances, there may be need for discussion of the observations with other members of the program staff (e.g., IACUC Chair, Compliance Director, AV) prior to making any assessment.
- Consultation with a veterinary member of the institution for clarification on issues related to animal welfare and well-being: If there are observations that are not clear or require background clarification, it is advisable to discuss those observations with members of the institution's veterinary staff.
- Consultation with the PI (or laboratory manager) for accuracy of observations: This should be a written (e.g., e-mail) exchange with the laboratory representative to ensure the observations are accurate. This is not an opportunity to negotiate the corrective actions—that is the responsibility of the IACUC. The goal of this exchange is to establish agreement with respect to the observations made.
- Consultation with the IACUC's leadership (e.g., Chair, Vice-Chair, IACUC manager) if regulatory noncompliance issues have been observed: It is always advisable to keep the IACUC's leadership apprised of any questionable situations. This will provide the IACUC leadership with the opportunity to review relevant background or regulatory information prior to IACUC discussion of the observations.
- Reporting to the IACUC: The Compliance Liaison report should occur as part of a convened IACUC meeting (on occasion, the seriousness of an event may justify a special-called meeting of the committee). The IACUC can then determine whether specified corrective measures are required, the duration or degree of corrective measures needed, or whether the observations confirm the need for focused training or other types of program enhancement tools. The Compliance Liaison (or the veterinary staff) may provide background for the IACUC decision.
- Report to the PI and laboratory manager: This may be a formal report, when dealing with serious issues or informal report for routine or less serious concerns.

Once the steps have been completed, the Compliance Liaison should assist with preparation of the findings, IACUC decision, and the required corrective measure(s), if applicable. The report, along with needed corrective action(s), should be signed by the IACUC or the IACUC chairperson, not the Compliance Liaison, as the final decisions on matters of noncompliance are made by the IACUC. However, the Compliance Liaison can play an integral role in facilitating a laboratory's full completion of the corrective actions.

30:24 Should IACUCs develop a method for the timely self-reporting by PIs of adverse events (e.g., a large number of postoperative complications, excessive mortality rate)?

Opin. Yes. Institutions that have developed a collegial and mutually supportive PAM program have reported that PIs often self-report as a natural extension of the process, especially in institutions with a strong culture of compliance. As with most issues, simple is best. Any list of best practices would be wanting, but ideas for best practices regarding advertising expectations of reporting include:

- A "program business card" that can be distributed to all laboratories at the time of the PAM visit. This tells the research team how and to whom to report concerns and explains what is expected in the report, including a definition of "in a timely manner."
- A special hotline or Web address to collect self-reports as well as anonymous allegations of animal research misconduct.

Regardless of the mechanism developed, all IACUCs should work diligently to encourage communication from the PI to the committee. Active participation in the oversight process is an indicator of a culture of compliance that supports a high quality animal care and use program. (See 29:6.)

30:25 Should the Compliance Liaison audit and evaluate the IACUC?

Reg. The AWAR (§2.31,c,1) and PHS Policy (IV,B,1) require that the IACUC review, at least once every 6 months, the research facility's program for humane care and use of animals.

Opin. Maybe. While all aspects of an institution's program for animal care and use must be evaluated, the IACUC usually performs this task as part of the semiannual program review. The IACUC might consider itself a poor reviewer of its own processes and could utilize the Compliance Liaison, as a consultant, to perform a review of the IACUC process. The committee may also use a subcommittee (e.g., Compliance Liaison plus members of the research community) to perform the program review and provide a report to the IACUC.

Internal audits by non-IACUC members can serve as an effective bench-marking tool: Does the IACUC meet the needs of researchers while assuring regulatory compliance? The IACUC remains responsible for the semiannual program review, but adjunctive assessment tools may enhance the committee's review by identifying shortcomings and recommending reparative measures to enhance the process.

If your institution is near other institutions that also have animal research activities and qualified Compliance Liaisons, one good idea is to ask another institution's Compliance Liaison to serve as a consultant to your IACUC's internal program review. It is amazing what an individual unfamiliar with your process will point out to you. Their observations may or may not be significant but it is usually worthwhile knowing how a new set of trained eyes sees your program. Institutions which are AAALAC accredited receive this sort of IACUC program assessment every three years, so the concept is not new, but the more frequent application of external assessment (perhaps annually or every 18 months) may be novel.

30:26 Should the animal care procedures and facilities be included in PAM?

Reg. The AWAR (§2.31,c,2) and PHS Policy (IV,B,2) state that it is a responsibility of the IACUC to inspect, at least once every 6 months, all of the research facility's animal facilities, including animal study areas. The PHS Policy (IV,C,1,d) also notes that the review should include an assessment to assure that the living conditions of animals will be appropriate for their species and contribute to their health and comfort. The *Guide* (p. 25) states there should be a program review and facility inspection at least annually. The *Guide* (p. 33) further notes that "Postapproval monitoring (PAM) is considered here in the broadest sense, consisting of all types of protocol monitoring after the IACUC's initial protocol approval. PAM helps ensure the well-being of the animals and may also provide opportunities to refine research procedures."

Opin. The better question is, Why should the animal care facilities not be included in a PAM process? The animal program is an institution-wide program for "animal care and animal use," therefore no part of the institution's animal activities should be exempted from monitoring. An effective and appropriately constructed institutional PAM program includes all animal care spaces and processes, including animal care facilities. The author believes that the *Guide* (p. 33) unintentionally suggests that PAM is a strictly a process for "... protocol monitoring after the IACUC's initial protocol approval." The *Guide* (p. 34) appears to offer some clarification by stating that, "As part of a formal PAM program some institutions combine inspection of animal study sites with concurrent review of animal protocols." To the extent that animal care facilities might be regarded as animal study sites, it stands to reason that such facilities would be included in PAM. Why would an institution focus solely on researcher use of animals when the majority of animal lifespan and human interaction is in the animal care facility? It is also worth noting that federal regulations never specify an oversight document of a "protocol" as the authorizing instrument for animal use. Federal regulations refer to the IACUC responsibility for "activities" involving animals (AWAR § 2.31,d; PHS Policy IV,C,5–IV,C,6). Most animals will be enrolled in a protocol and all will be participants in the animal care program of the institution; therefore, all animals come under the oversight of the IACUC. What is also critical (and sometimes sensitive) when reviewing animal care practices and facilities, is that the PAM review for animal care and husbandry focus on processes and procedures and not on professional veterinary judgment of clinical care decisions. The PAM review of animal care facilities must be consistent with the review of experimental procedures. For example, review might include

- Spaces being used and the appropriateness of those spaces for such use
- Adherence to protocols (or SOPs) for care
- Training or qualifications of the personnel performing animal care activities

30:27 How can institutional, researcher, or IACUC resistance to PAM be handled?

Opin. Regardless of the real or perceived value of any process, change is not easy for people. Appealing to the scientific mind and using a clear discussion of the value of the process is usually the most productive method. It is helpful to focus on the advantages of PAM, and how this will protect and defend the rights and privileges of continued animal use while extending the ability of the IACUC to achieve its regulatory obligations.

It is further helpful to discourage the tendency to look at PAM as "another regulation" or "institutional invasion" into the laboratory. The truth is that PAM is not a new regulation; nor is it an invasion. Rather, PAM is evidence of the institution's desire to guarantee the continued use of animals for research through validation of procedural activity which results in clean reports following inspections. It should be the hope of the IACUC that the Compliance Liaison reports far many more examples of "evidence of good performance" than noncompliant behavior; and this author believes that it is the desire of the public that animals receive good care and that the care is confirmed by a program such as PAM.

There are three common questions that arise when introducing PAM into an institution:

- What can PAM do? The PAM process can provide the IO and the IACUC with a level of assurance of overall program performance that cannot be obtained in any other manner. It can identify deficiencies and facilitate corrective measures in an internal and collegial manner. It is a method by which the institution, through the IACUC and the Compliance Liaison, can partner with the researcher, helping them stay fully compliant with federal, state, and local requirements. PAM can allow repair of a deficiency while the issue is small and before it becomes reportable to regulatory or granting agencies; it is generally less costly (e.g., resources and finances) to repair small problems than to address significant deficiencies when they occur.

  The PAM process can facilitate the IACUC program assessment. Many IACUCs struggle with the federal requirement to perform a semiannual program review. The use of the PAM reports generated during the previous 6 months provides a markedly enhanced level of program assessment. In other words, PAM provides the IACUC with information which is not available from other sources. PAM provides the IACUC with the opportunity to direct training toward trended deficiencies in the program.
- Why is it important? The PAM process is critical to the long-term protection and integrity of the animal care and use program. In a world where we hear on occasion of poor judgment or faulty process in animal care and use, PAM can be the least expensive manner to keep the program on track; PAM is institutional insurance which provides assurance that, indeed, things are working or things need repair.

  Having a PAM program makes a strong statement that the institution is dedicated to quality animal care and quality research outcomes, not just to staying out of the news. PAM is an external reflection of internal commitment to assuring that the care and use of animals are valued components of the institution's research program. While some institutions have committed in their Assurance with NIH/OLAW a statement of program integrity through adherence to the *Guide*, without any clear and definable evidence for that assurance, others have claimed that the absence of a programmatic disaster is evidence of a strong and reliable program. Such logic is true, in part, but the absence of an institutional disaster may be just as attributable to pure and simple luck. PAM provides clear and unmistakable evidence, gathered through direct observation, of IACUC-approved activities or SOP practices.

PAM is a daily snapshot, an on-the-ground or in-the-trenches assurance that processes approved by the IACUC or defined in SOPs represent the actual practice.

- What may happen if we do not have a PAM program? Maybe nothing; after all, many institutions have never experienced a program disaster. But the more critical issue to consider is whether it is worth risking the reputation of the institution's research enterprise, especially if the unraveling deficiency had been preventable. While the cost of such a situation is not easily documented, it is clear that the outcome could be catastrophic. Potentially, the institution's animal research program could be suspended, interrupting the research efforts of not only the guilty but also the innocent and well-intended researchers. Possibly, there would be a disruption in core fund support for the institution's research program, due to question of integrity. Potentially, the institution's alumni would be discouraged and benefactor support would suffer. Without question, the prestige of the institution would decay from a preventable program disaster.

  Finally, most importantly, and most critically, are issues related to animal misuse, neglect, or abuse that the institution failed to identify. Such circumstances suggest overall program failure of the most grievous nature. Failure to ensure that principles of humane care and compassionate use are practiced is ethical failure as well. An institution engaged in PAM and focused on looking for "evidence of good performance" will fully ensure a necessary level of engagement in the program. Institutions should establish and monitor benchmarks, without which there are no reliable methods for improving compliance.

30:28 How does PAM encourage compliance with regulatory requirements? How can PAM assist in preventing noncompliance?

Opin. Individuals who dedicate their professional lives to knowing the regulatory requirements are typically not research staff. Conversely, research personnel are not typically well-versed in regulatory requirements. Both communities have special knowledge which provides the opportunity for synergy. Research involving animals can flourish in a symbiotic environment created by varied professionals having a common goal. Even regulations assist research outcomes. Following the standards of care, treatment, and handling leads to healthier animals. Healthier animals reduce variables which lead to reproducible data. Reproducible data is reliable and credible. The Compliance Liaison is a partner of the research team.

Training is a critical component of a successful animal care and use. Training is required by the AWAR (§2.31,d,1,viii), PHS Policy (IV,C,1,f), and the *Guide* (pp. 15–17). The challenge comes when one asks what training is necessary and under what circumstances. At many institutions, the IACUC requires the PI to confirm on the protocol application that the PI and other research personnel are properly qualified to perform the tasks being proposed. However, unless there is a recognized laboratory failure, the qualifications are never questioned. Compliance Liaisons can assist with assuring the integrity of the institutional program by observing on-going activities and providing training, sharing new learning opportunities, and updating the laboratory members on modified or enhanced review or reporting requirements. Compliance Liaisons identify concerns and problems at an early level and provide the institution an opportunity to correct the minor issue before it becomes a significant or costly dysfunction. PAM provides clear and documented evidence that the institution is not simply striving to meet the minimal requirements for animal care and use, but is working toward a reasonable and practical level of care and oversight that also assures integrity of research data. If the assumption is true (and the author believes it is) that research staff do not willfully intend to violate accepted standards of animal care and use, then PAM provides a unique and special bridge over which focused information can be provided to the point where it is in greatest need. When an institution reaches the point where a research staff member calls the Compliance Liaison and says: "I am not sure this is a problem, but I wanted to ask first," the institution will know it has a culture of compliance.

30:29 What internal controls for PAM should an institution implement?

Reg. The PHS Policy (IV,B) indicates that the IACUC has oversight responsibility for the institution's program for humane care and use of animals. The AWAR (§2.31,a) note that "the Chief Executive Officer [CEO] of the research facility shall appoint an Institutional Animal Care and Use Committee (IACUC), qualified through the experience and expertise of its members to assess the research facility's animal program, facilities, and procedures." The *Guide* (p. 13) states that primary program oversight responsibility lies with the IACUC, the IO, and the AV.

Opin. The AWAR, the PHS Policy and the *Guide* agree that oversight responsibility lies not with the PAM staff, but with the AV, the IACUC and the CEO (or IO). For successful performance, the PAM must not be placed in a situation of bias or conflict—either perceived or real. This can best be achieved by the Compliance Liaison's:

- Having sufficient authority to enter and observe all animal care practices at the institution, regardless of location or persons involved
- Performing their task unobtrusively
- Protecting any and all observations with the greatest confidentiality, even if necessary from the PIs in the laboratory area
- Not working directly for an animal user (especially if the animal user is the IACUC chairperson or the AV)
- Not having any authority to determine outcomes and corrective actions. The Compliance Liaison should recommend a corrective action, since they may know the specific laboratory culture and environment better than the IACUC, but any determinations regarding reportable or corrective measures are IACUC functions.

30:30 How is PAM of animal health different from the AV's routine "well-animal" monitoring?

Reg. Both the PHS Policy (IV,C,1,d; IV,C,1,e) and the AWAR (§2.31,d,1,vi; §2,31,d,1,vii) state that the living conditions of animals will be appropriate for their species and contribute to their health and comfort. The housing, feeding, and nonmedical care of the animals will be directed by a veterinarian or other scientist trained and experienced in the proper care, handling, and use of the species being maintained or studied. Medical care for animals will be available and provided as necessary by a qualified veterinarian. The *Guide* (pp. 14,106) notes that the veterinary care program is the responsibility of the AV.

Opin. There is no clear requirement for the AV to perform any monitoring that is separate from the IACUC. The regulations prescribe for the AV the task of implementing a program for animal care and health but do not indicate a requirement to monitor animal care or use. However, a component of any effective program of veterinary care would likely include medical surveillance—a process in which colony health is defined and tracked. Medical surveillance may be used to enhance animal health, improve the conditions for animals, or strengthen the environment of the well-animal. Since the Compliance Liaison typically does not directly participate in clinical animal care, the AV could engage the Compliance Liaison to provide objective oversight of the animal care program. This is valuable in part because the orientation of PAM assessment is different from that of the clinical veterinary staff or the AV. The clinical veterinary staff and AV tend to focus on animal disease prevention, husbandry support, and care facility management. In contrast, PAM tend to focus on adherence to institutional policy or process as defined by the IACUC and institutional/AV SOPs. The process of PAM for clinical care activities also provides benefits to the larger research community. When the research community believes that all program participants are judged by the same standards, then it becomes a community of equals, rather than researchers versus veterinary staff. Confidence and support for PAM activities are critically dependent upon community acceptance of the process and are best accomplished if the AV is considered as another PI whose activities are those of animal care program coordination.

The Compliance Liaison must not consider clinical decisions of the AV as being either right or wrong; rather, the emphasis should be on applying a consistent assessment of practices as compared to approved SOPs and protocols. For example, PAM might determine that animal physical examinations are not being performed as scheduled or in the manner prescribed by veterinary SOPs, or that aseptic technique is not being performed by the veterinary staff in the same rigorous manner required of PIs. Subtle difference in an animal's environment may have a significant impact upon research outcomes; thus, consistency of process—whether involving the PI, AV, or animal care staff—becomes critically important. With respect to routine animal care, in the author's opinion an SOP can be considered reasonably similar to a research protocol for purposes of PAM.

30:31 How do smaller institutions get the funding for Compliance Liaison positions?

Opin. Funding for PAM can be challenging at many institutions, large or small. Every institution has the same basic issue—ensuring that animal care and use (e.g., experimentation, testing, teaching, and exhibition) are conducted as approved by the IACUC. Further, most institutions have the same general regulatory foundation and performance expectations. Though PAM is an effective way to meet these expectations, the ability to resource a program of PAM is often complicated by limited funds.

Some institutions may believe that they do not have a problem and therefore do not need to fund a PAM process. But most institutions which have taken an honest look at their programs have found issues that require attention, in some cases critical and immediate attention. The assertion "We do not have a problem" is a euphemism for "We have not looked or we are unwilling to spend money before a crisis occurs." There is no such thing as a perfect program. Just exactly how much institutional risk are you willing to accept? What is you institution's name worth?

For those institutions that perform federally funded work, the indirect cost recovery (ICR) funds may be used to provide for animal program oversight. But in some instances, ICRs are already over-extended so that institutions may need to consider reassessing the allocation of those resources toward infrastructure, such as PAM, for the overall program needs. For institutions that do not use federal funds, the institutional risk management office may be able to place a dollar value on the cost of significant animal program dysfunction and the relative value of a PAM process. In some cases, reassignment of specified PAM responsibilities to existing staff may be appropriate. Perhaps nearby institutions might consider sharing a common resource and therefore distributing costs of a Compliance Liaison between two or more institutions. Perhaps, if an institution doesn't need a full-time professional Compliance Liaison, it might be possible to "rent" one from a local institution for a specified number of hours per month.

30:32 How can PAM assist in developing programmatic milestones and fulfillment of IACUC obligations for training?

Opin. Probably the greatest stumbling block to programmatic improvements is a lack of program milestones or program direction. While there are many regulations and guidelines, an institution that focuses solely on meeting the regulations and guidelines is setting their standard at the minimum level of satisfactory performance—in other words, they aspire to be an average institution and no better. An institution could define a set of goals toward which to work. For example, if the institution identifies no aseptic technique noncompliances; no instances in which post-procedural pain medication was not provided as described in an approved protocol; or no unapproved personnel performing procedures, a minimum level of compliance has been achieved. In contrast, the IACUC might be encouraged to aspire to higher goals, such as: PIs providing self-reports of adverse events involving animals used in research; a monthly average of, for example, 90% of laboratories found to be fully compliant during PAM visits; no NIH/OLAW-reportable incidents over a defined period of time. Each institution is different; therefore the goals should be locally defined and sufficiently lofty to encourage reaching for the next level. As performance data is collected, the IACUC can compare compliance success over time. Are we doing better this year than last? Do we have more negative PAM reports at a certain season on the year (e.g., associated with new students?) Are there specific procedures or activities that we continue to have problems with? Knowing the institution's shortcomings though collected data provides measurable criteria for the question "Are we improving?"

There are several ways in which the data can be used to foster improvement. For example, many institutions have some form of annual training. The Compliance Liaison might note the "top ten" PAM issues from the previous year and make that information the core of annual training for the current year. One institution that used this approach saw a clear and demonstrable decline in issues from one year to the next, in large part because research faculty became sensitized to the problems occurring elsewhere and wanted to prevent similar occurrence in their own laboratories.

Another way in which the data could be used is as a teaching tool for senior research laboratory staff. Several years ago, the author's institution began a Research Animal Coordinator Certification (RACC) process in which a PI recommends one of his or her lab members to participate in a 10-month long training program, requiring 3–4 hours a month, and focused on PAM-identified deficiencies and other aspects of programmatic process and improvement. The result of this process has been improved quality of protocols submitted by researchers, fewer compliance problems, and lower severity of compliance issues. The goal of this or any program is to identify problems, create training to specifically combat the problem and track the progress, modifying as necessary to achieve the institution's milestone of progress. In this regard, the Compliance Liaison and the PAM activity serve a special and unique role—fostering successful research outcomes with minimal institutional distress while harmonizing the process within regulatory expectations.

References

1. National Institutes of Health, Office of Laboratory Animal Welfare. 2012. Frequently Asked Questions G.6, Is post approval monitoring required? http://grants.nih.gov/grants/olaw/faqs.htm#instresp_6.
2. National Institutes of Health, Office of Laboratory Animal Welfare. 2005. Notice NOT-OD-05-034, Guidance on Prompt Reporting to OLAW under the PHS Policy on Humane Care and Use of Laboratory Animals. http://grants.nih.gov/grants/guide/notice-files/not-od-05-034.html.

31

The European Ethical Review Framework: Collaborative Issues

Javier Guillén

31:1 How is animal research regulated in Europe?

Reg. The 28 countries of the European Union[1] are covered by the European Directive 2010/63/EU on the protection of animals used for scientific purposes.[2] The European Union body responsible for the development and implementation of the Directive is the Directorate-General Environment of the European Commission. As mandated in the Directive (Art. 61,1), Member States had to adopt and publish by Nov. 10, 2012, the laws, regulations and administrative provisions necessary to comply with this Directive, and apply those provisions from January 1, 2013 (this process is called the transposition of the Directive). Although not all Member States complied with that deadline, they currently should have their own piece of legislation transposing the Directive. To help level the playing field across the European Union (which is one of the objectives of the Directive), Member States cannot apply stricter measures than those in the Directive unless they were already in force on November 9, 2010 (Art. 2,1). If they had stricter national measures before that date, the measures were aimed at ensuring more extensive protection of animals falling within the scope of this Directive, and were communicated to the European Commission before January 1, 2013, they could be maintained. In addition to the transposition at national level, there may be countries with a further transposition of the national law at the regional level.

But Europe is larger than the European Union and the remaining European countries have different regulations at the national or regional level. Some of them have strict and well developed regulations (e.g., Norway, Switzerland), while others lack regulations or have very outdated ones.

The Council of Europe (which is not the same as the Council of the European Union referred to above), is an inter-governmental organization (not supranational as is the European Union) covering 47 European countries, including all of the European Union. A very important difference between the Council of Europe and the European Union is that the Council of Europe has no legislative power and seeks voluntary cooperation through recommendations, agreements and conventions. The Council of Europe issued the *European Convention for the Protection of Vertebrate Animals used for Experimental and Other Scientific Purposes*, commonly known as the ETS 123,[3] which includes the *Guidelines for Accommodation and Care of Animals* in Appendix A,[4] which was revised many years later. The content of the revised Appendix A is particularly important because part of it (e.g., cage sizes) has been incorporated as a mandatory requirement in the European Directive (Art. 33,2 and Annex III). The European Commission had already recommended (not mandated) within the European Union the adoption of the full Appendix A in 2007. Members of the Council of Europe are free to sign and ratify the ETS 123, and when a member signs and ratifies the convention it commits itself to implement ETS 123 into its national legislation. However, the actual implementation of the ETS 123, and more specially the revised Appendix A into national legislation is not yet common. The list of members that have signed and ratified the ETS 123 is available online.[5]

ETS 123, except for Appendix A, is much more general than the Directive and does not include detailed engineering standards for the ethical review process. For example, it does not contain reference to IACUC-like bodies. Although compliance with ETS 123 is based on the voluntary agreement by member countries of the Council of Europe (The ETS 123 is not regulatory in nature), it established basic principles that have served as the basis for many of the requirements in both the old and new European Union Directives.

Opin. Europe, and especially the European Union, is a highly regulated environment for animal research, and the new European Directive is trying to level the different regulatory national frameworks. The previous European Directive (86/609/EEC)[6] allowed Member States to go beyond the provisions in the Directive, and this resulted in some Member States just transposing the Directive literally while others implemented stricter measures. However, in spite of the intention of the new Directive (2010/63/EU),[2] and as will be discussed in some of the questions that follow, there may still be differences in the way it is implemented at the national and even the regional level because of the diverse interpretations by applicable competent authorities. Also, one has to take into account that Europe is not only the European Union, and that the rest of European countries may have different requirements. This means that in terms of research collaborations between both sides of the Atlantic, it is essential not only to pay attention to the Directive's requirements, but also to the national (or even regional) transposition and implementation of the Directive (for European Union members), or to the specific legislation of non-European Union members (e.g., Norway, Switzerland, Russian Federation, etc.).

The differences in nature between the Directive 2010/63/EU[2] and ETS 123[3] have led AAALAC International to adopt the ETS 123 and not the Directive as one of the Three Primary Standards for Accreditation.[7] First, the ETS 123 is not regulatory as is the Directive, it is intended more as a recommendation (except in a couple of cases the "should" is used instead of the mandatory "shall" used in the Directive). Second, it is more based on performance standards, especially the very extensive Appendix A containing the guidelines on animal care and accommodation, which focus very much on housing and environmental enrichment. And third, the Directive can apply to European Members only, while ETS 123 can potentially apply to virtually all Europe.

Even though the principles of ETS 123 are the ones present in the Directive and subsequent legislation, it is the Directive that frames, from the legal point of view, the ethical review and the work of the IACUC-like bodies in the European Union. ETS 123, besides not having regulatory nature, does not offer specific recommendations on IACUC-like bodies. Therefore, this chapter will mainly focus on the European Union framework, while acknowledging that this may not apply to other European countries.

31:2 What is the competent authority that oversees animal research in Europe?

Reg. In the European Union, Member States have to designate a so-called "competent authority" (or authorities) to carry out the obligations arising from the Directive[2] (Art. 3,7; Art. 59,1), and ensure that the competent authorities carry out regular inspections of all breeders, suppliers and users, including their establishments, to verify compliance with the Directive requirements (Art. 34). There may be Member States designating a single central national competent authority (e.g., at the Ministry level), and others that designate multiple competent authorities, usually at regional level. In all cases, a national authority must serve as contact point for the purposes of the Directive to the European Commission (Art. 59,2). The same article 59 allows Member States to designate bodies other than public competent authorities for the implementation of specific tasks laid down in the Directive. This can happen only if the designated body has the expertise and infrastructure required to carry out the tasks and is free of any conflict of interests regarding the performance of the tasks. Therefore, there may be Member States that have designated IACUC-like bodies as competent authorities which have the task of project evaluation and even the project authorization, although this last task is generally kept at the public competent authority level. The possibility of this delegation of authority and the potential conflict of interest of an IACUC-like body (e.g., a modified Animal Welfare Body) is discussed further in 31:6. When the delegation is not granted, the IACUC-like bodies (known as Animal Welfare Bodies) have specific tasks (see 31:6) which do not include the project evaluation and authorization. The latter is performed by the public competent authority, or any other competent authority designated by the public competent authority.

Outside of the European Union cases may be diverse. For example, in Switzerland the public authority lies at regional (cantonal) level.[8] Needless to say, any activity related to collaborative research between European and other institutions (e.g., from the US) that involve the use of animals in a European establishment will be subjected to the local competent authority evaluation, authorization, and oversight.

Opin. Due to different national administrative systems, there is wide diversity in the characteristics of European public competent authorities. Only in a few countries do the competent authorities have laboratory animal science specialized full time professionals (as, for example, the Home Office Inspectorate in the United Kingdom (UK).[9] In a majority of cases, competent authorities for animal research issues also have to attend to other non-related areas (e.g., farms, zoos, etc.) and may not have the resources to address all the requirements of the applicable legislation for animal research. A typical scenario may be represented by a competent authority at the regional (local government) level with little time to dedicate to animal research institutions, and moreover, with little chances to get specialized training in laboratory animal science. This may result in a very basic oversight being performed, usually based on engineering rather than performance standards. The heterogeneous character of the competent authorities enhances the importance, in terms of collaborative research, to identify and get a good knowledge of the competent authority in the area to see how the legislation is applied.

31:3 What IACUC-like bodies are required for compliance with regulatory expectations in Europe?

Reg. Until the implementation of the new Directive[2] there was great diversity in the way countries regulated IACUC-like bodies. The old Directive[6] did not mandate the implementation of this type of local body. Therefore, there were countries that had not mandated them by 2012 yet, while others had implemented different oversight bodies or systems, some at local level, others at the government (competent authority) level, and still others having a combination of both. For example, in the UK it took the name of Ethical Review Process (including local review of projects); in Germany it was based on government committees for project review and some specialized personnel at the local level (such as an Animal Welfare Officer). In Spain they were local Ethics Committees. With the new Directive the situation has changed, as it mandates the implementation of an Animal Welfare Body at each breeder, supplier and user establishment (Art. 26). These Animal Welfare Bodies may adopt different names as per the particular transposition in each Member State.

In addition to the Animal Welfare Body, institutions must appoint a person or persons to be responsible for overseeing the welfare and care of the animals in the establishment; ensure that the staff dealing with animals have access to information specific to the species housed in the establishment; and be responsible for ensuring that the staff are adequately educated, competent and continuously trained and that they are supervised until they have demonstrated the requisite competence (Art. 24,1). Also, institutions must appoint a designated veterinarian with expertise in laboratory animal medicine, or a suitably qualified expert where more appropriate, charged with advisory duties in relation to the well-being and treatment of the animals (Art. 25).

In countries outside of the European Union we may find diverse approaches, from countries with no IACUC-like bodies to others with different versions of ethics committees. In some countries with poor or not existing legislation it is possible to find oversight bodies at committed research institutions mirroring the US IACUC composition and function, which in some cases are also called an IACUC.

Opin. The new Directive's mandate to establish institutional Animal Welfare Bodies would seem in theory a way to harmonize the oversight and ethical review process that is currently in a very heterogeneous framework. However, as we will see in the next questions, there is a lot of room for variation in the way their composition and functions are implemented by the Member States of the European Union. It may be possible, depending on national tradition and how the transposition of the Directive has been implemented, to find Animal Welfare Bodies similar to US IACUCs while others are very different in their composition and functions. For example, when the Directive is transposed literally, the Animal Welfare Bodies have a more restricted composition and function than do the US IACUCs.

31:4 What activities may require IACUC-like body and/or competent authority oversight, evaluation and authorization?

Reg. Under the European Union Directive,[2] there is oversight for all activities "where animals are used or intended to be used in procedures, or bred specifically so that their organs or tissues may be used for scientific purposes" (Art. 1,2).

There is a difference with the US framework with regard to the animals covered that has to be taken in consideration: the Directive (Art. 1,3) applies to live non-human vertebrate animals, including independently feeding larval forms and fetal forms of mammals from the last third of their normal development; and also live cephalopods.

The concept of a procedure is crucial, being defined in Article 3,1 as "any use, invasive or non-invasive, of an animal for experimental or other scientific purposes, with known or unknown outcome, or educational purposes, which may cause the animal a level of pain, suffering, distress or lasting harm equivalent to, or higher than, that caused by the introduction of a needle in accordance with good veterinary practice." This includes the creation and maintenance of a genetically modified animal line in any such condition, but the Directive, similar to ETS 123 (see below) and the traditional British system, excludes the killing of animals solely for the use of their organs or tissues. A project can involve one or more procedures, and the project evaluation and authorization is to be performed by the competent authority (Art. 36).

Under Article 5 of the Directive, procedures may be carried out only for the purposes of

a. Basic research
b. Translational or applied research with any of the following aims:
   i. The avoidance, prevention, diagnosis or treatment of disease, ill-health or other abnormality or their effects in human beings, animals or plants
   ii. The assessment, detection, regulation or modification of physiological conditions in human beings, animals or plants
   iii. The welfare of animals and the improvement of the production conditions for animals reared for agricultural purposes
c. For any of the aims in point (b) in the development, manufacture or testing of the quality, effectiveness and safety of drugs, foodstuffs and feed-stuffs and other substances or products
d. Protection of the natural environment in the interests of the health or welfare of human beings or animals
e. Research aimed at preservation of the species
f. Higher education, or training for the acquisition, maintenance or improvement of vocational skills
g. Forensic inquiries

Activities that fall out of the scope of the Directive (Art. 1,5) and therefore are not subjected to oversight, evaluation and authorization are

a. Non-experimental agricultural practices
b. Non-experimental clinical veterinary practices
c. Veterinary clinical trials required for the marketing authorisation of a veterinary medicinal product

d. Practices undertaken for the purposes of recognized animal husbandry
e. Practices undertaken for the primary purpose of identification of an animal
f. Practices not likely to cause pain, suffering, distress or lasting harm equivalent to, or higher than, that caused by the introduction of a needle in accordance with good veterinary practice

The scope of the ETS 123[3] is similar; article 2 restricts the purposes of procedures to

a. i. Avoidance or prevention of disease, ill-health or other abnormality, or their effects, in man, vertebrate or invertebrate animals or plants, including the production and the quality, efficacy and safety testing of drugs, substances or products
   ii. Diagnosis or treatment of disease, ill-health or other abnormality, or their effects, in man, vertebrate or invertebrate animals or plants
b. Detection, assessment, regulation or modification of physiological conditions in man, vertebrate and invertebrate animals or plants
c. Protection of the environment
d. Scientific research
e. Education and training
f. Forensic inquiries

Article 15 of the ETS 123 states that "A procedure for the purposes referred to in Article 2 may be carried out by persons authorised, or under the direct responsibility of a person authorised, or if the experimental or other scientific project concerned is authorised in accordance with the provisions of national legislation."

As for the definition of a procedure, Article 1 of ETS 123 considers them as "any experimental or other scientific use of an animal which may cause it pain, suffering, distress or lasting harm, including any course of action intended to, or liable to, result in the birth of an animal in any such conditions, but excluding the least painful methods accepted in modern practice (that is "humane" methods) of killing or marking an animal."

Opin. First, it is important to notice that in the Directive, not only vertebrate animals but also live cephalopods are included. The inclusion of cephalopods may not have a big impact because they are not used in the majority of European countries, but it raises the issue of the general lack of knowledge about these animals that may affect the quality of the activities by the IACUC-like bodies. The consideration of the independently feeding larval forms may be more important, especially due to the increasing use of aquatic species such as the zebrafish. For example, when is a zebrafish embryo considered to be an independent larval form? This has been discussed in one of the Expert Working Groups established by the European Commission to help harmonizing the implementation of the Directive, and it has been proposed they are considered an independent larval form from 5 days post fertilization under normal laboratory conditions. The implication is huge if we consider that Member States have to collect and make publicly available, on an annual basis, statistical information on the use of animals in procedures (Art. 54,2). Counting such small animals may be impossible and the statistics will have to be approximations. The fetal forms of mammals from the last third of their normal development were already covered in some countries (e.g., UK) but it is now a new requirement in most of Europe and something to consider in collaborative projects. Curiously, they are not to be counted in the statistics. One can argue that there is some inconsistency in this, as they are covered by the Directive but not counted as animals used for research.

Second, the concept of procedure is even more important as it will determine if and how many activities by Animal Welfare Bodies and competent authorities are to be covered. If every single action at the severity level of the introduction of a needle had to be considered a procedure, it would be practically impossible to work from the administrative burden point of view. For this reason, it has been agreed during the discussions at the European Commission level that in practical terms, a procedure be considered as a combination of one or more technical acts carried out on an individual animal for an experimental or other scientific purpose and which may cause that animal pain, suffering, distress or lasting harm. For example, if an experiment involves gavage, injections, laparotomy and sampling, all actions together are to be considered a single procedure.

One of the more important discussions held was with regard to the concept of procedure and the activities that are out of the scope of the Directive, such as practices undertaken for the purposes of recognised animal husbandry. More specifically, the discussion focused on deciding if tissue sampling for genotyping of genetically altered lines is to be considered husbandry (and therefore exempted of all Directive requirements) or not, with all the administrative consequences. Very importantly, the European Commission's interpretation is that tissue sampling for genotyping cannot be considered as a recognized husbandry practice because its primary purpose is to either fulfil scientific needs or confirm scientific suitability. Only the sampling procedures that are below the severity threshold (introduction of a needle) should be exempted. This means that not only the creation of genetically altered lines and the breeding/maintenance of lines with harmful phenotypes are to be considered as procedures, but also that evaluation and authorization of the maintenance of lines with not-harmful phenotypes will be needed, unless the methods used for tissue sampling are non-invasive (e.g., saliva, feces), or are used in conjunction with identification practices, which are explicitly excluded as a procedure. An example is using the ear tissue for both identification and genotyping purposes. The implication of this interpretation is also large, not only for evaluation and authorization issues, but also for the statistics: are all animals subjected to only tissue sampling by invasive methods (e.g., tail snipping, which is still the most common method used) to be counted even they are not used later in any other experimental procedure? Based on the discussions at the European Commission, it may be decided to include only a snapshot of these numbers every 5 years. Again, this may be an example of inconsistency.

A more pragmatic solution might be to consider tissue sampling for genetic characterization as recognized husbandry when the techniques applied are in accordance to international good practice. This way, a lot of administrative burden would be saved and the statistics would not be inflated with hundreds of thousands of animals that in practice have not been used in research projects.

Also for discussion is that the killing of animals solely for the use of their organs or tissues is not considered a procedure and therefore those animals are not to be counted (Art. 3,1). So, apparently no evaluation and authorization by IACUC-like bodies or public competent authorities is required for projects involving only the euthanasia of animals, if this is performed according to any of the established euthanasia methods in Annex IV of the Directive.

But, aren't these animals being used to fulfil scientific needs? Is not killing an animal to be considered as above the threshold of introducing a needle? The same reasons given to include sampling for genotyping as a procedure seem not to apply to killing animals to obtain tissues, and vice versa. Regardless of these practices being considered as procedures or not, the IACUC-like bodies should establish institutional policies to be followed in each case.

31:5 What is the composition of the IACUC-like bodies (Animal Welfare Bodies) in Europe?

Reg. The Directive[2] requires a minimum composition of the Animal Welfare Body. It shall include at least the person or persons responsible for the welfare and care of the animals and, in the case of a user, a scientific member (Art. 26,2). This same article does not require the official participation of the Designated Veterinarian (the equivalent of the US Attending Veterinarian) as a member, though the Animal Welfare Body has to receive input from the Designated Veterinarian. It is important to point out that the position of the Designated Veterinarian in an establishment can be held by a "suitably qualified expert where more appropriate" (Art. 25). This minimum composition does not preclude a Member State to require or open the possibility to have a more complex composition. Previous to the Directive, countries with a requirement to have some kind of local ethics committee or IACUC-like body may not be limited to the minimum composition established in the Directive, and some of them include the obligatory membership of the Designated Veterinarian. Normally, Animal Welfare Bodies may include representation from researchers, persons taking part in the experiments, veterinarians, animal care personnel, lay persons, and others.

Opin. The minimum composition required by the Directive for Animal Welfare Bodies is so short, that in theory breeder establishments could have Animal Welfare Bodies comprising one member only, and only two members in user establishments. This minimum composition differs from the requirements in the US AWAR (§2.31,a; §2.31,b) and the PHS Policy (IV,A,3,a; IV,A,3,b) that require three or five members, respectively. This difference may be explained by the fact that the Directive does not assign the protocol (project) evaluation to the Animal Welfare Bodies, whereas this is an essential function of the IACUC in the US. The Directive assigns that function to the competent authority. Fortunately, some Member States require a more complex composition so they can fulfill correctly the tasks assigned to them (see the next question). There has been a big controversy on why the Directive does not mandate the Designated Veterinarian be a regular member of the Animal Welfare Body, but only requires the Animal Welfare Body to "receive input from the designated veterinarian or the expert referred to in Article 25." There is no indication on what "input" means in practical terms. Other more complex compositions of the Animal Welfare Body that have been implemented by some of the Member States require official veterinary participation. As pointed out above, there is room for establishments not having a veterinarian, but a "suitable qualified expert." Although the intention in the Directive is that this should happen only in establishments using non-conventional species only where other specialists may have a better knowledge of them (e.g., some amphibians), this is not clearly stated and could be potentially challenged from the legal point of view by establishments using conventional species. Therefore, there could be potentially some establishments with no veterinarian and as a consequence, no veterinary input to the Animal Welfare Body. (See 12:28.)

A more complex composition than the minimum required by the Directive should be necessary also to comply with all the tasks assigned to the Animal Welfare Body, and even more when the competent authorities delegate other functions to it. When collaborating with European partners, one must be ready to find different approaches to the IACUC-like bodies and the ethical review process in general (see 31:6).

31:6 Do IACUC-like bodies in Europe have the same functions as US IACUCs, especially with regard to the protocol (project) evaluation?

Reg. Not in theory, especially in those cases where the Directive[2] is followed literally. The functions assigned (as a minimum) to the Animal Welfare Body in Art. 27,1 are

a. Advise the staff dealing with animals on matters related to the welfare of animals, in relation to their acquisition, accommodation, care and use
b. Advise the staff on the application of the requirement of replacement, reduction and refinement, and keep it informed of technical and scientific developments concerning the application of that requirement
c. Establish and review internal operational processes as regards monitoring, reporting and follow-up in relation to the welfare of animals housed or used in the establishment
d. Follow the development and outcome of projects, taking into account the effect on the animals used, and identify and advise as regards elements that further contribute to replacement, reduction and refinement
e. Advise on rehoming schemes, including the appropriate socialization of the animals to be rehomed

One can easily realize that the ethical evaluation of the protocol (project) is not included as one of the basic functions of the Animal Welfare Body. The Directive assigns this task to the competent authority (Art. 36,2). However, Member States may designate bodies other than public competent authorities for the implementation of specific tasks laid down in the Directive only if there is proof that they have the necessary expertise and infrastructure and are free from conflict of interest (Art. 59,1). This means that Member States can potentially designate Animal Welfare Bodies or other types of IACUC-like bodies as competent authorities if they fulfill these conditions, which in practice means that the project evaluation can be delegated to these bodies. And, what has happened is that in the implementation of the Directive some Member States have opened this possibility of delegation and others have retained this task at the level of the government competent authority. The Directive also lists the expertise required for perform a proper project evaluation (Art. 38,3):

a. The areas of scientific use for which animals will be used including replacement, reduction and refinement in the respective areas
b. Experimental design, including statistics where appropriate

c. Veterinary practice in laboratory animal science or wildlife veterinary practice where appropriate
d. Animal husbandry and care, in relation to the species that are intended to be used

Also, the evaluation is to be transparent (Art. 38,4). To undergo an evaluation, the projects must describe the information required in Annex VI. The items to be described include the justification of the species and numbers of animals used; description of procedures; application of the 3Rs in the procedures; anesthesia, analgesia, other pain relief and alleviation of suffering; humane endpoints; experimental and statistic design; re-use (if applicable); classification of severity (see question 31:9); avoidance of duplication of procedures; housing and husbandry practices; euthanasia: and the competence of personnel.

It is very important to understand that a positive project evaluation is not enough to start the experiments because competent authority authorization also is needed (see 31:10). Regardless of the competent authority's nature (i.e., public or IACUC-like body), protocols approved in other countries will have to be evaluated and authorized again in the country where the project is carried out.

Opin. It is clear that the intention of the Directive is to assign to the Animal Welfare Body the development of a culture of care based on the principles of the 3 Rs at the establishment. But the level at which the project evaluation should be performed, either at the government competent authority or delegated competent authority level (i.e., local Animal Welfare Body or other institutional or external bodies) remains controversial. The intention of the Directive of assigning the project evaluation first at the government competent authority level is to ensure the independence of the evaluation, and this is a good intention. However, a large majority of European public competent authorities lack the expertise and human resources to perform an appropriate evaluation of all project proposals. Also, local IACUC-like bodies may have a much better knowledge of the actual situation in the establishment with regard to the applicant, the personnel (training and competence) involved, other resources and facilities, which surely impact the evaluation. Being realistic, the most appropriate approach would be to open the door to the delegation of the project evaluation to those Animal Welfare Bodies/Ethics Committees that have a good level of expertise (with a composition much more complex than the minimum required by the Directive, so they can have the required expertise) and that can ensure transparency and independence of criteria (for example, including members with no connection with the establishment). This is happening in some countries, and the system can be complemented by the oversight by the government competent authority of the performance of the local IACUC-like body.

31:7 In Europe, are written standard operating procedures required for IACUC-like body's review of relevant activities?

Reg. Yes. The tasks assigned to the IACUC-like bodies (Animal Welfare Body) in the Directive,[2] as were described in 31:6, include to "establish and review internal operational processes as regards monitoring, reporting and follow-up in relation to the welfare of animals housed or used in the establishment" (Art. 27,1,c).

Opin. While the assignment of this task seems to be appropriate, the feasibility of correctly performing this task (and the other tasks assigned to the Animal Welfare Body) may depend on the composition of the Animal Welfare Body. Complex institutions that do not go beyond the minimum requirements of the Directive and do not ensure the Animal Welfare Body has the necessary human resources and expertise will never be able to perform appropriately all the tasks assigned. When collaborating with European institutions, the composition and performance of the Animal Welfare Body may serve to evaluate the level of commitment of the institution to animal care and use.

31:8 Are European IACUC-like bodies involved in post-approval monitoring activities?

Reg. Yes, at least partially in the European Union. One of the tasks of the Animal Welfare Bodies is to "follow the development and outcome of projects, taking into account the effect on the animals used, and identify and advise as regards elements that further contribute to replacement, reduction and refinement" (Art. 27,1,d). In addition to this, for all authorized projects there must be "persons responsible for the overall implementation of the project and its compliance with the project authorization" (Art. 40,2,b), and these persons must "ensure that the projects are carried out in accordance with the project authorization" (Art. 24,2,b).

Opin. The IACUC-like bodies, such as the Animal Welfare Body in the European Union, should be clearly involved in post-approval monitoring. The persons referred to in Articles 24 and 40 are meant to be the persons legally responsible for the compliance with the project authorization (i.e., the PI) but not the persons performing an independent post-approval monitoring. However, the authority of the Animal Welfare Body is not clear as its role is mainly to give advice. Therefore, it will remain the institution's discretion to establish an Animal Welfare body with enough resources and authority to perform a complete follow up of the projects and give appropriate advice on the implementation of the 3Rs, as well as to respond positively to that advice. At the very least, records of any advice given by the Animal Welfare Body and decisions taken regarding that advice are to be kept for at least 3 years (Art. 27,2). These records may serve as one of the indicators of the performance of the Animal Welfare Body and of the institutional support of its activities.

The post-approval monitoring activities are going to be very important because of the Directive requirement to provide, in the statistical information, the *actual* severity of the procedures performed (Art. 54,2). This means that for all uses of animals, the severity caused to each animal is to be included and not only the prospective categorized severity (see 31:9). How can the collection of all this information be coordinated? Who is supposed to collect the information? How can the assessment criteria be harmonized? The participation of the researchers seems to be crucial to collect information, but the Animal Welfare Body should play a decisive role too. The European Commission established an Expert Working Group to help harmonize the criteria for this assessment. It could be a very appropriate task for the Animal Welfare Body to coordinate the collection of information, establish the criteria based on international recommendations (such as those from the European Commission) and train all the individuals involved in the process.

This activity may be closely related to the retrospective assessment mandated to all projects using non-human primates and projects involving procedures classified as 'severe' (Art. 39,2). Although this retrospective assessment is, in theory, the responsibility of the competent authority (Art. 39,1), it appears as one of the activities that could be delegated to the local IACUC-like body, as one of its recognized tasks is to follow the development and outcome of projects, taking into account the effect on the animals used (Art. 27,1,d).

31:9 Is there a severity classification of experimental procedures?

Reg. Yes. The new Directive[2] indicates that the procedures included in projects have to be classified as "non-recovery," "mild," "moderate," or "severe" on a case by case basis (Art. 15). To harmonize the classification across Europe, Annex VIII offers examples for each of the categories. Unless a safeguard is applied (Art. 15,2 and 55,3), procedures that involve severe and long-lasting pain, suffering, or distress are prohibited.

Other different severity categories may be found in other European countries (e.g., Switzerland, ranging 0–3).

Opin. The severity classification is a very important issue because it is very difficult to harmonize interpretations of the severity of a procedure performed on an animal and the category assigned to a procedure may have a major influence on other legal requirements in the Directive. For example, all projects involving procedures categorized as severe will have to undergo a retrospective assessment (Art. 39), and animals used in such severe procedures cannot be re-used (Art. 16,1). The main issue here is the potential for different interpretations of severity. This is why the European Commission established an Expert Working Group to produce the examples found in Annex VIII. However, while the annex may help with the prospective severity classification mandated by Article 15,1, it may be considered insufficient to help with the retrospective assessment of projects. The European Commission established another Expert Working Group, after the publication of the Directive, to address the severity assessment, which is particularly important with regard to the elaboration of the statistical information on the use of animals, where the *actual* severity is to be reported. Although in theory only the "severe" projects are to be retrospectively assessed, the actual severity of the procedures is to be reported for all animals (Art. 54), so in practice the severity has not to be categorized only prospectively, but also retrospectively. This makes it even more difficult to harmonize criteria for a retrospective classification. A very good example is related to the creation and maintenance of genetically altered animal lines, not only in the interpretation of harmful phenotypes but also as relates to genotyping sampling methods.

31:10 Is a positive project evaluation by the IACUC-like body or competent authority enough to start an experiment?

Reg. The Directive[2] mandates that Member States shall ensure that projects are not carried out without receipt of prior authorization from the competent authority (Art. 36). The only cases where the Directive opens the possibility to a simplified administrative procedure are "for projects containing procedures classified as 'non-recovery,' 'mild,' or 'moderate' and not using non-human primates, that are necessary to satisfy regulatory requirements, or which use animals for production or diagnostic purposes with established methods" (Art. 42,1).

There is an established period of 40 working days (that can be extended for very complex projects up to 55 working days) for the authorization decision to be communicated to the applicant (Art. 41,1,2). Importantly, this period includes the project evaluation. Although local designated competent authorities (such as IACUC-like body) could be delegated this function in addition to the project evaluation, this has very rarely happen (Belgium is an example where this occurs).

The European Commission interprets that the Directive does not foresee a possibility for a *"tacit agreement,"* neither for a *"tacit refusal,"* when the competent authority fails to make and communicate the authorization decision within the given deadline. In absence of a decision granting or refusing authorization, the applicant cannot start the project and is obliged to wait for the authorization decision.

Opin: The period of 40 working days to obtain the authorization means that in practice any investigator wanting to start a project may have to wait 2 months until it can be started, and almost an extra month in very complex cases. This will apply also to projects that have undergone a previous evaluation in other places such as the US, because only competent authorities of the Member States can grant the authorization. There are some related issues here. First, even when Member States allow the project evaluation to be delegated to other bodies, such as the IACUC-like body, the final authorization will (in most cases) remain with the government competent authority. In those cases, because the 40-day period includes the project evaluation, coordination with the competent authority could be implemented to notify the government competent authority of the starting of the project evaluation at the local level. But what happens if the evaluation lasts for most of the 40 working days? Can the competent authority grant the authorization in the very short time remaining? To avoid these potential scenarios, it is very advisable to get an agreement with the competent authority on the authorization process. The same issue may appear when the public competent authority retains both the project evaluation and the authorization processes. As was discussed in 31:2, many competent authorities in Europe may lack the resources to perform these activities appropriately or very importantly, within the established time period. The interpretation that a failure to produce an authorization communication in time cannot be considered as a tacit approval is a very controversial issue. In many other aspects of life in general, when applicants do not get a response from the authorities in the established period, the request is considered granted. Why not with regard to the project authorization? The responsibility falls on the shoulders of the competent authorities, and it is their obligation to ensure that project applicants get the authorization in time. First, the 40 working days period already can be considered too long in terms of research competitiveness, and second, any further delay can be a disaster for the researchers and therefore their institutions. A potential legal action against the competent authority in case the project authorization is inexcusably delayed will hardly compensate the applicant, whose main aim is to carry out the project before others have the same idea, and when applicable, within the deadlines established by the funding source. It seems essential for research establishments to establish a coordinated and agile communication system with the competent authority to ensure a rapid processing of research project applications. The participation of the IACUC-like bodies in this process may be of utmost importance.

When there is research collaboration with foreign groups (e.g., US), it is very important to consider these timelines and begin communications with the European collaborators much earlier than the anticipated start date of the procedures.

31:11 Are IACUC-like body members required to follow specific training?

Reg. There is no reference in the Directive[2] for specific training of Animal Welfare Body members, though it could be implied in some articles (see opinion below).

Opin. One can argue that there is a legal obligation to implement this training, depending on the interpretation of one of the requirements for personnel of authorized establishments (Art. 24,1,c), that there must be person(s) "responsible for ensuring that the staff are adequately educated, competent and continuously trained and that they are supervised until they have demonstrated the requisite competence." Although this requirement is clearly related to article 24, which states that "the staff shall be adequately educated and trained before they perform any of the following functions: (a) carrying out procedures on animals; (b) designing procedures and projects; (c) taking care of animals; or (d) killing animals," the same principle could be applied to the members of the Animal Welfare Body or other IACUC-like body.

When there is no clear legal indication, the approach is similar to what is done in other areas of the world. That is, institutions may or may not implement training for IACUC-like bodies, and when the training exists it may be in-house, external, or a combination of both.

31:12 To whom does the IACUC-like body report at the typical research institution?

Reg. There is no clear indication about this in the European Directive.[2]

Opin. Authorized establishments have to communicate to the competent authority the person who is responsible for ensuring compliance with the provisions of the Directive (Art. 20,2). That person can be the equivalent to the IO in the US. Having the IACUC-like body report directly to this person would seem to be a reasonable approach, although it is not specified in the Directive. If IACUC-like bodies, such as the Animal Welfare Body, do not report to this person directly, they should report to someone delegated by that individual who has enough authority within the institution to implement actions as a response to the Animal Welfare Body advice.

Until the transposition and implementation of the European Directive January 1, 2013, IACUC-like bodies have been reporting to different individuals/positions, such as Deans or Vice-Principals of research in academic institutions and Site Heads or persons at research management level in private companies and public research institutions.

31:13 Are there special considerations for the use of non-human primates and endangered species?

Reg. Yes. Non-human primates cannot be used for all the aims listed in the Article 5 of the Directive,[2] as is already described in 31:4. Their use is restricted (Art. 8,1) to procedures meeting the following conditions:

a. The procedure has one of the purposes referred to in (i) points (b)(i) or (c) of Article 5 and is undertaken with a view to the avoidance, prevention, diagnosis or treatment of debilitating or potentially life-threatening clinical conditions in human beings; or (ii) points (a) or (e) of Article 5.

b. There is scientific justification to the effect that the purpose of the procedure cannot be achieved by the use of species other than non-human primates. A debilitating clinical condition for the purposes of this Directive means a reduction in a person's normal physical or psychological ability to function.

The definitions in Article 5 include basic research, translational or applied research (related to human disease), safety (regulated) studies, and preservation of the species. Severe restrictions apply to using nonhuman primate endangered species (Art. 8,2) and importantly, the use of great apes is forbidden (Art. 8,3). The latter is subject to a safeguard clause (Art. 55,2) that states that great apes can be used only when a Member State has justifiable grounds for believing that action is essential for the preservation of the species or in relation to an unexpected outbreak of a life-threatening or debilitating clinical condition in human beings. A Member State may adopt a provisional measure allowing the use of great apes in procedures having one of the purposes referred to in Article 5, provided that the purpose of the procedure cannot be achieved by the use of species other than great apes or by the use of alternative methods. Also it is important to remember that projects involving the use of non-human primates are must undergo retrospective assessment (see 31:8).

In addition to all this, non-human primates are not only included in the list of animals that can be used only if they have been bred to be used in procedures (Art. 10,1 and Annex I), but also non-human primates will be allowed to be used in procedures only where they are the offspring of non-human primates which have been bred in captivity or where they are sourced from self-sustaining colonies. This last condition will apply 5 years after a feasibility study to be conducted and published by November 2017 (Art. 10,1). Another study, to be published by November 2022, will evaluate the possibility of sourcing animals only from self-sustaining colonies. A self-sustaining colony is defined as "a colony in which animals are bred only within the colony or sourced from other colonies but not taken from the wild, and where the animals are kept in a way that ensures that they are accustomed to humans" (Art. 10,1). Similar restrictions apply to other endangered species (Art. 7), with the main difference that they cannot be used for basic research and no safeguard clause exists.

Opin. There are three key points to consider in the European Directive about the use of non-human primates. One is that the use of great apes is basically forbidden. This is not a big change, since their use already had been discontinued in practice, at least in the European Union some time ago. They could still be used in other countries where they are not regulated.

The second point refers to the potential different interpretation of the definition of a "debilitating or potentially life-threatening clinical conditions in human beings" included in the Directive as "a reduction in a person's normal physical or psychological ability to function" (Art. 8,1,b). Apparently, this definition covers the large majority of the current non-human primate use, although doubts may arise with regard to some types of research, such as reproductive studies.

The third point, related to the future exclusive use of non-human primates only where they are the offspring of non-human primates which have been bred in captivity or where they are sourced from self-sustaining colonies, is worrying the research community, as it may impact severely the availability of animals for research purposes. Taking into consideration the deadline for the first feasibility study, this requirement would apply in 2022. It will be important to see the outcome of such study. Several European organizations have taken the initiative to submit a proposal to the European Commission on how this study should be performed.

With regard to other endangered species, their use is even more restricted, as no basic research is allowed regardless of its aim, and no safeguard clause is offered. Authorization from competent authorities to use endangered species, even in the cases considered in the Directive, will likely be difficult to obtain.

References

1. European Union. Countries. http://europa.eu/about-eu/countries/index_en.htm. Accessed October 16, 2012.
2. The European Parliament and the Council of the European Union. 2010. Directive 2010/63/EU of the European Parliament and of the Council of 22 September 2010 on the Protection of Animals Used for Scientific Purposes. *Official Journal of the European Union:* L 276:33–79. http://eur-lex.europa.eu/LexUriServ/LexUriServ.do?uri=OJ:L:2010:276:0033:0079:en:PDF.
3. European Convention for the Protection of Vertebrate Animals used for Experimental and other Scientific Purposes. 1986. http://conventions.coe.int/treaty/en/treaties/html/123.htm.
4. Council of Europe. 2006. Appendix A of the European Convention for the Protection of Vertebrate Animals Used for Experimental and other Scientific Purposes (ETS No. 123). Guidelines for accommodation and care of animals (Article 5 of the Convention). Approved By The Multilateral Consultation. Cons 123 (2006) 3. http://conventions.coe.int/Treaty/EN/Treaties/PDF/123-Arev.pdf
5. European Convention for the Protection of Vertebrate Animals Used for Experimental and other Scientific Purposes CETS.:123. Status as of August 1, 2010. http://conventions.coe.int/Treaty/Commun/ChercheSig.asp?NT=123&CM=8&DF=08/01/2010&CL=ENG.
6. Council Directive 86/609/EEC on the approximation of laws, regulations and administrative provisions of the Member States regarding the protection of animals used for experimental and other scientific purposes. 1986. *Official Journal of the European Union* L358:1–29.
7. Association for Assessment and Accreditation of Laboratory Animal Care, International. What You Need to Know about AAALAC's Expectations and Interpretations: AAALAC Now Using Three Primary Standards. http://www.aaalac.org/about/guidelines.cfm. Accessed October 17, 2012.
8. Switzerland. The revised Animal Welfare Act. http://www.google.com/url?sa=t&rct=j&q=&esrc=s&source=web&cd=4&ved=0CDIQFjAD&url=http%3A%2F%2Fwww.bvet.admin.ch%2Fthemen%2Ftierschutz%2Findex.html%3Flang%3Den%26download%3DNHzLpZeg7t%2Clnp6I0NTU042l2Z6ln1ad1IZn4Z2qZpnO2Yuq2Z6gpJCFdHt9e2ym162epYbg2c_JjKbNoKSn6A—&ei=ag9_UJ-rDIW9yQHMsoCgAw&usg=AFQjCNFUe0_l7VEjDyprp7QZyCelO-AGOg. Accessed October 17, 2012.
9. Home Office, UK. Research and testing using animals. http://www.homeoffice.gov.uk/science-research/animal-research/. Accessed October 17, 2012.

Index

E

F

G

H

J

K

L

M

O

Q

R

S

U

V

W

X

Z

Printed in the United States
by Baker & Taylor Publisher Services